Lecture Notes in Physics

Edited by H. Araki, Kyoto, J. Ehlers, München, K. Hepp, Zürich
R. Kippenhahn, München, H. A. Weidenmüller, Heidelberg
and J. Zittartz, Köln

177

Application of High Magnetic Fields in Semiconductor Physics

Proceedings of the International Conference
Held in Grenoble, France, September 13–17, 1982

Edited by G. Landwehr

Springer-Verlag
Berlin Heidelberg New York Tokyo 1983

Editor

G. Landwehr
Max-Planck-Institut für Festkörperforschung
Hochfeld-Magnetlabor
Avenue des Martyrs (CNRS), 166 X, F-38042 Grenoble

ISBN 3-540-11996-5 Springer-Verlag Berlin Heidelberg New York Tokyo
ISBN 0-387-11996-5 Springer-Verlag New York Heidelberg Berlin Tokyo

Printed in Germany

Printing and binding: Beltz Offsetdruck, Hemsbach/Bergstr.
2153/3140-543210

PREFACE

This volume contains the contributions presented at the International Conference on the Application of High Magnetic Fields in Semiconductor Physics which was held in Grenoble from September 13 to 17, 1982. The meeting was organized in conjunction with the 16th International Conference on the Physics of Semiconductors in Montpellier, September 6 to 10, 1982, and was the sixth in a series of meetings held in Würzburg (1972, 1974, 1976), Oxford (1978) and Hakone (1980).

High magnetic fields have continued to be a valuable tool in semiconductor research, particularly in investigations of the properties of two-dimensional electronic systems. One reason for this is that sufficiently strong magnetic fields can fully quantize such systems. The quantized Hall effect which was discovered in silicon MOSFETs about two years ago captured the imagination of both theoreticians and experimentalists. The precision with which h/e^2 can be determined from the Hall resistance in the presence of localization is difficult to understand, and the papers presented on this topic showed that a satisfying explanation of the quantized Hall effect has not yet been achieved.

Magnetic fields have proved indispensable in distinguishing between so-called weak localization in quasi-metallic two-dimensional systems and many-body effects, both of which can cause a logarithmic increase in resistance with decreasing temperature.

Magnetotransport experiments with two-dimensional and three-dimensional systems, and the application of magnetooptical methods have yielded significant new insights in recent years, a circumstance reflected in the conference program. Very detailed information about deep impurities has been obtained by luminescence experiments in high fields, and the interpretation of high excitation spectra has been aided by magnetic fields. Far infrared magnetospectroscopy is increasingly being employed in conjunction with low temperatures, producing interesting new results.

The investigation of narrow-gap semiconductors in high magnetic fields has yielded sophisticated data, and the incorporation of magnetic impurities has also opened up new aspects.

Progress has been made in the generation of continuous and pulsed magnetic fields, as the reports from high field facilities show. The results obtained from the application of megagauss fields to semiconductor problems have been quite impressive.

Conductors for superconducting magnets can now be reliably produced. Such magnets have become readily available and have contributed significantly to the growth of the physics of high magnetic fields.

As in the previous conferences, the emphasis was on invited papers which not only presented new data but also gave a general overview of the respective areas. In addition to the 35 invited papers, 42 contributions were presented in a poster session. The posters were accessible throughout the conference, and consequently the discussions among the participants were quite extensive and fruitful.

While no permanent record had been made of the first conferences, the proceedings of the Hakone meeting edited by S. Chikazumi and N. Miura proved very useful, and it was decided that this practice should be continued.

The meeting was attended by 145 scientists from all over the world. We tried - and I hope successfully - to maintain the informal atmosphere of the previous conferences in the series.

The organizing committee consisted of G.Landwehr (chairman), G.Aubert, D.Bois, Th.Englert, M.von Ortenberg, R.Pauthenet, J.L.Robert, R.Romestain and M.Voos.

Financial support for the conference was provided by:

Centre National de la Recherche Scientifique (C.N.R.S., Paris)
Max-Planck-Institut für Festkörperforschung (Stuttgart and Grenoble)
Direction des Recherches Etudes et Techniques (D.R.E.T., Paris)
European Research Office of the United States Army and European Office of Aerospace Research and Development of the United States Air Force
Laboratoire d'Electronique et de Technologie de l'Information, Centre d'Etudes Nucléaires de Grenoble (L.E.T.I., C.E.N.G.)
Linde AG (München)
Vacuumschmelze GmbH (Hanau)
Bruker Analytische Messtechnik GmbH (Karlsruhe)
Siemens AG (München)
Etablissements SOPRA (Bois-Colombes)
Mairie de Grenoble

On behalf of the participants and the organizing committee I would like to thank the above organizations again for their support.

Grenoble, January 1983 G. Landwehr

CONTENTS

QUANTIZED HALL EFFECT

LOCALIZATION IN 2D SYSTEMS

<u>NARROW GAP MATERIALS</u>

MAGNETO-TRANSPORT

MAGNETIC SEMICONDUCTORS

GENERAL

REPORTS BY HIGH MAGNETIC FIELD FACILITIES, TECHNIQUE

PRECISION DETERMINATION OF h/e^2 AND THE FINE-STRUCTURE CONSTANT FROM MAGNETO-TRANSPORT MEASUREMENTS ON 2D ELECTRONIC SYSTEMS

K. v.Klitzing, B. Tausendfreund, H. Obloh, and T. Herzog

Physik-Department, Technische Universität München,
D-8046 Garching, FRG

Introduction

High magnetic fields are necessary for the observation of the quantized Hall resistance - a quantum phenomenon which leads finally to a new type of resistor R_H with a resistance value depending only on the Planck constant h and the elementary charge e [1]:

$$R_H = h/e^2 i \approx 25812.8\ \Omega/i \qquad i = 1,2,3,... \qquad (1)$$

Such a resistor allows a direct determination of h/e^2 and therefore the fine-structure constant α, because α is given by the equation (SI units)

$$\alpha = (e^2/h)(\mu_o c/2) \qquad (2)$$

with the velocity of light $c = 2.99792458 \times 10^8$ m/s and $\mu_o \equiv 4\pi \times 10^{-7}$ H/m.

A simple experimental arrangement for the determination of α is shown in Fig. 1. The current I is flowing through both the resistor R_H and a standard resistor R_N (resistors calibrated in SI units with an uncertainty of about 10^{-7} are maintained at the national laboratories). From the ratio of the voltage drops across R_H and R_N (the absolute values of the voltages need not be known) a determination of h/e^2 and therefore α is possible. The uncertainty in h/e^2 from such an experiment is at least as large as the uncertainty of about 10^{-7} in the value of the reference resistor R_N. If the experimental uncertainty is smaller than 10^{-7}, then the quantized Hall resistance R_H can be used as a resistance standard. Different national laboratories are investigating such an application. Up to now, experimental problems limit the accuracy of the data, and the main purpose of this article is to discuss some

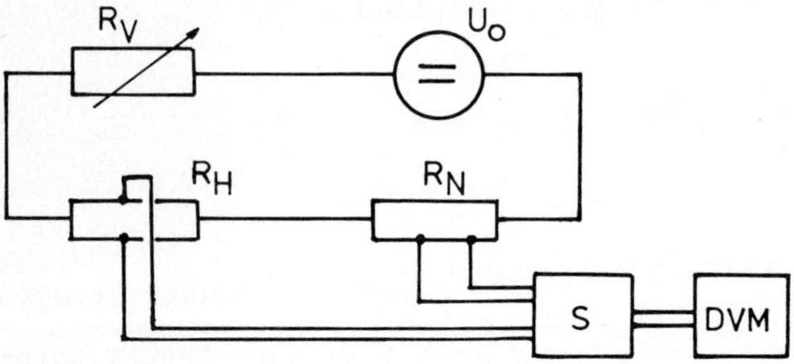

Fig. 1: Experimental set-up for measurements of the fine-structure constant using the quantum Hall effect. R_H is the quantized Hall resistance, R_N the standard resistance, and S is a two-way selector switch.

of the problems related to high-precision measurements of h/e^2 with uncertainties of less than 10^{-6}.

Systems used for h/e^2-measurements

The realization of a resistor with resistance values according to Eq. 1 is based on Hall effect measurements on a two-dimensional electron gas (2DEG) [2]. Up to now, all experimental results are obtained on MOS devices or heterostructures, where a 2DEG can be created at the interface between a semiconductor and an insulator or at the interface between two semiconductors. The two-dimensional character originates from a quantization of the energy for the motion perpendicular to the interface into well separated electric subbands E_o, E_1, ... due to the confinement of the electrons within a narrow channel. Fig. 2 shows cross-sections and top-views of typical devices used in our experiments.

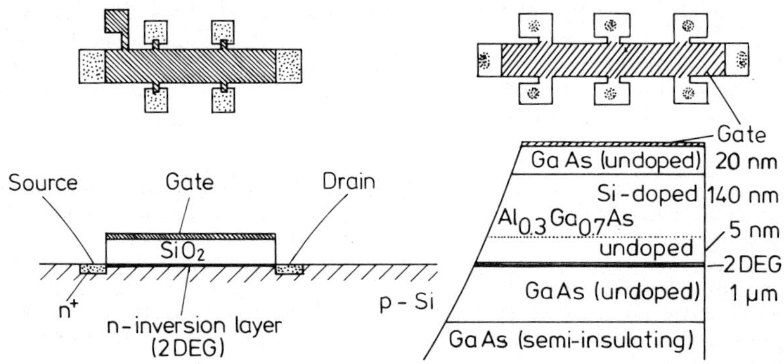

Fig. 2: Top view and cross-section of typical devices used in our experiments. Left side: Silicon MOSFET. The InSb MISFET is built up of lacquer as an insulator on top of p-InSb. Right side: $GaAs/Al_xGa_{1-x}As$ heterostructure with Schottky gate.

The carrier density of the 2DEG can be changed by applying a gate voltage V_g across the metal-insulator (SiO_2 or $Al_xGa_{1-x}As$)-semiconductor (Si, InSb, GaAs) capacitor. The voltage drops between the potential probes for a constant source-drain current are proportional to the components ρ_{xx} and $\rho_{xy} = R_H$ of the resistivity tensor.

A strong magnetic field B perpendicular to the plane of the 2DEG quantizes the energy for the motion in the plane leading schematically to a discrete energy spectrum

$$E_{j,n} = E_j + (n + \tfrac{1}{2})\hbar\omega_c \qquad n,j = 0,1,2,... \tag{3}$$

with the cyclotron energy $\hbar\omega_c = \hbar eB/m$.

The energy splitting between adjacent Landau levels ($\Delta n = 1$) increases with increasing magnetic field, so that well separated energy levels are expected, if this splitting is much larger than the broadening of the energy levels. Since the degeneracy factor N of each Landau level $E_{j,n}$ is equal to [3]

$$N = eB/h \tag{4}$$

a Hall resistance R_H corresponding to

$$R_H = h/e^2 i \tag{5}$$

is expected, if exactly an integer number i of energy levels is occupied (i is identical with g(n + 1), where g is the degeneracy factor due to spin- and valley-splitting and n is the Landau quantum number).

The condition of a fully occupied Landau level is identical with no elastic scattering of an electron from an occupied to an empty state and therefore to a vanishing energy loss. Theoretical calculations based on the self-consistent Born approximation at T = 0 K predict that the resistivity ρ_{xx} is basically proportional to the square of the density of states at the Fermi energy [4], so that the condition $\rho_{xx} = 0$ characterizes the situation that all (mobile) states of a Landau level are occupied, whereas the states of higher energetic levels are empty. Exactly at these carrier densities n_s ($\rho_{xx} = 0$) a quantized Hall resistance $R_H = h/ie^2$ is expected.

In a simple approximation, the Hall resistance R_H should decrease monotonically with increasing carrier density n_s and therefore with increasing gate voltage: $R_H \sim 1/n_s \sim 1/V_g$. This means that the carrier density should be adjusted and stabilized at the value $n_s = iN = ieB/h$ with an accuracy corresponding to the desired accuracy of better than 10^6 for the quantized Hall resistance. This is not possible due to instabilities in the correlation between gate voltage and carrier density (threshold shifts) and a not well defined gate voltage for which the resistivity ρ_{xx} becomes zero.

Fortunately, localized states lead to a stabilization of R_H at the value of $R_H = h/ie^2$ (Hall plateaus). Ando [5] has shown that the tail states in each Landau level are localized. Different authors demonstrated that, if the occupation of localized states is changed, the resistivity components ρ_{xx} and ρ_{xy} remain constant at a value corresponding to fully occupied Landau levels [6,7]. All these theories are based on the assumption of zero temperature, so that the resistivity ρ_{xx} should be exactly zero as long as the Fermi energy lies in the mobility gap between two Landau levels.

Practically, thermal excitations across the energy gap [8] or variable range hopping between "localized" states [9] lead to a finite scattering probability and therefore to a finite resistivity ρ_{xx}. One important question is whether this finite scattering rate leads to a correction in high-precision measurements of the quantized Hall resistance.

Experimental results

Results for the magneto-resistivity and the Hall resistance obtained for different two-dimensional systems (Si, InSb, GaAs) as a function of the gate voltage are summarized in the figures 3-5. The overlap between adjacent energy levels for the InSb MISFET is relatively strong even at B = 12 T, visible as a finite value of the minima in the ρ_{xx}-oscillation. The reason for the unresolved Landau levels is that the quality of the InSb-insulator interface is not very high, which leads to a reduced mobility μ and therefore to an increased broadening Γ of the Landau levels ($\Gamma \sim 1/\sqrt{\mu}$ in the self-consistent Born approximation

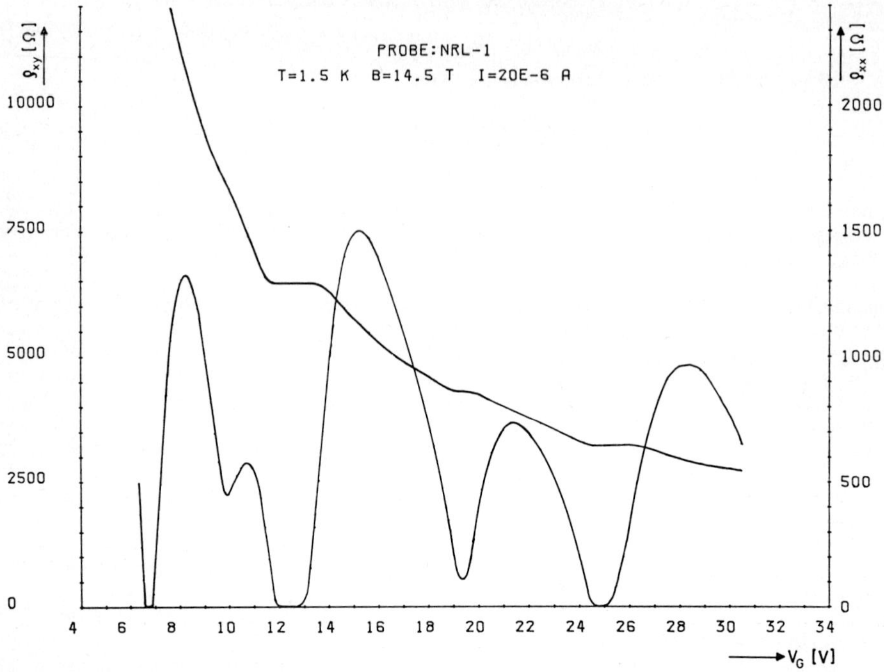

Fig. 3: Gate voltage dependence of the Hall resistance $R_H = \rho_{xy}$ and the resistivity ρ_{xx} at B = 14.5 T for a silicon MOSFET.

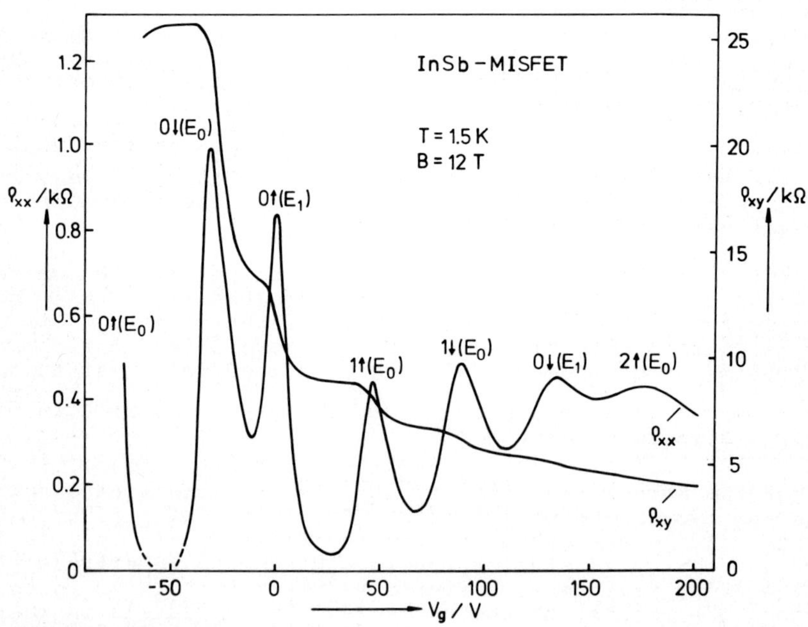

Fig. 4: Gate voltage dependence of the Hall resistance $R_H = \rho_{xy}$ and the resistivity ρ_{xx} at B = 12 T for an InSb-MISFET.

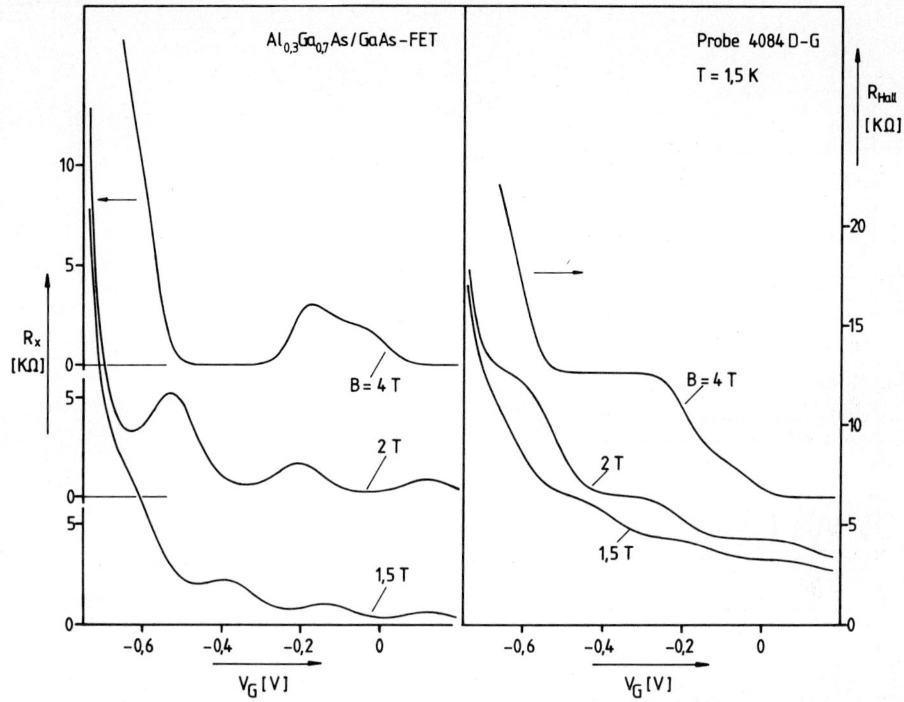

<u>Fig. 5:</u> Hall resistance and resistivity at different magnetic fields as a function of the gate voltage for a GaAs-$Al_{0.3}Ga_{0.7}As$ heterostructure.

[4]). In addition, the occupation of a second electric subband E_1 for this material at a relatively low carrier density leads to a superposition of different Landau levels and probably to an increased level broadening due to intersubband scattering. This means that well separated energy levels are not observed for the InSb system, although the small effective mass for electrons in this material leads to a cyclotron energy larger than oberserved for Si or GaAs.

The most pronounced Hall steps are visible for the GaAs/$Al_xGa_{1-x}As$ heterostructure. Compared to silicon MOSFETs, the mobility of the electrons in this heterostructure is typically one order of magnitude higher and the effective mass a factor of three smaller, so that much better resolved Landau levels are expected. However, the highly doped $Al_xGa_{1-x}As$ layer is not a good insulator, which may lead to leakage currents from the gate to the interface channel and therefore to errors in high-precision measurements of the quantized Hall resistance. Therefore measurements at zero gate voltage are necessary with a fixed carrier density n_s determined mainly by the ionized donors in the $Al_xGa_{1-x}As$ depletion layer. The quantum condition $n_s = ieB/h$ can be obtained by changing the magnetic field. A typical result for such a measurement is shown in Fig. 6. High resolution measurements of the flatness of the Hall steps on a similar sample show that the value of the quantized Hall resistance remains constant within the experimental uncertainty of ± 0.06 ppm even if the magnetic field is changed by 11% (Fig. 7). However, the absolute value of R_H is not known with such an accuracy, because the reference resistors of 6453.2 Ω or 12906.4 Ω in our ac bridge shown in Fig. 8 are not calibrated. Tsui et al. [10]

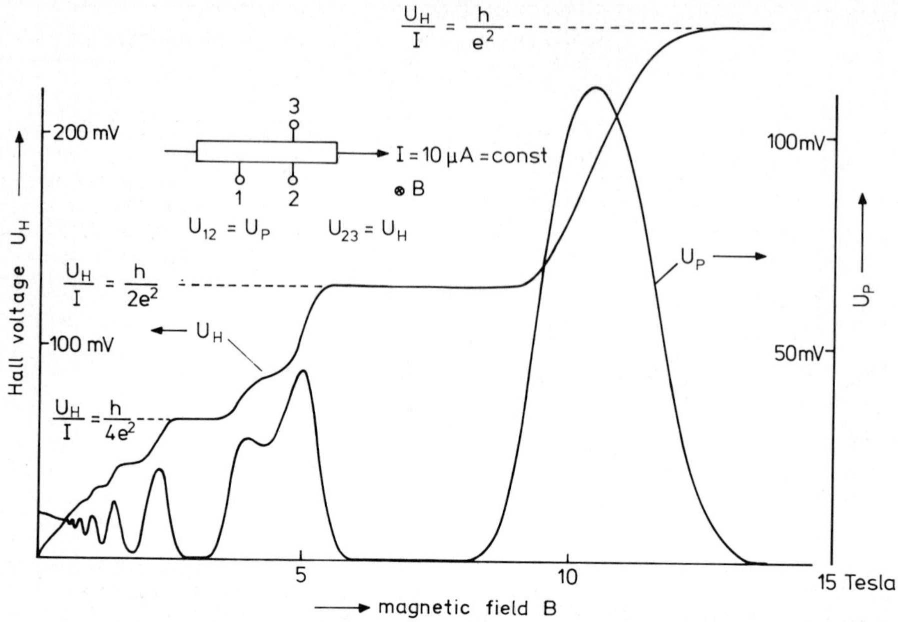

Fig. 6: Hall effect and resistivity measurements on a GaAs-$Al_{0.3}Ga_{0.7}As$ heterostructure as a function of the magnetic field B.

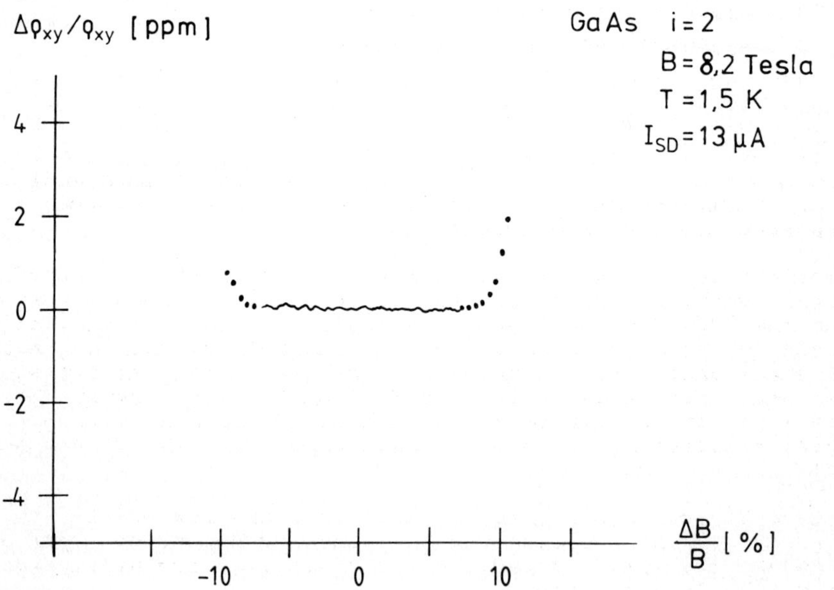

Fig. 7: High-resolution measurements of the flatness of the Hall plateau $R_H = h/2e^2$ for a GaAs-$Al_{0.3}Ga_{0.7}As$ heterostructure (similar sample as shown in Fig. 6).

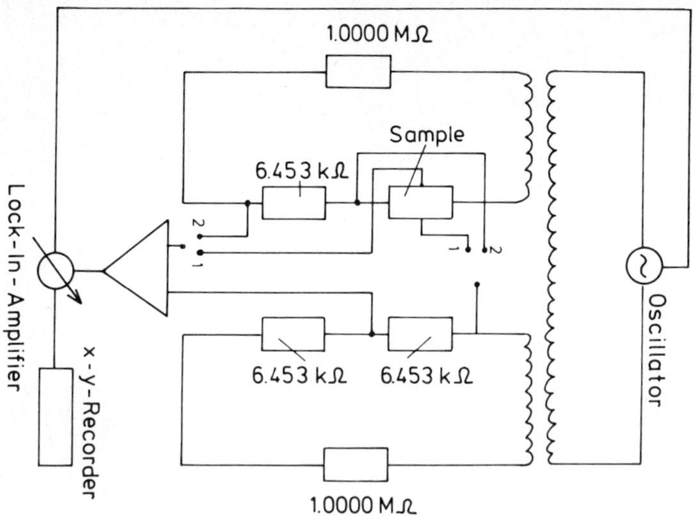

Fig. 8: Experimental set-up for high-resolution measurements of the flatness of the Hall plateau $R_H = h/4e^2$.

reported a value of $h/4e^2 = (6453.2004 \pm 0.0011)\,\Omega$. From this result the following value for the fine-structure constant is obtained, which may be compared with α determined by other methods [11]:

$$
\begin{aligned}
\alpha^{-1} &= 137.035968(23)\ (?) && \text{(quantum Hall effect)}\\
\alpha^{-1} &= 137.035963(15)\ (?) && (\gamma_p\text{-method})\\
\alpha^{-1} &= 137.035993(5)(9) && \text{(anomalous moment)}\\
\alpha^{-1} &= 137.035969(21)(46) && \text{(muonium hfs)}\\
\alpha^{-1} &= 137.036040(110) && \text{(recommended 1973)}
\end{aligned}
\qquad (6)
$$

The first errors listed are experimental and the second theoretical. The question mark in the condensed-matter determinations indicate that the theoretical uncertainties are unknown; they are possibly very small in comparison to the experimental ones.

Systematic errors in high-precision measurements of the quantized Hall resistance using heterostructures may be due to a leakage current through the highly doped $Al_xGa_{1-x}As$ layer. The conductivity of this layer can be increased with infrared illumination, and the result in Fig. 9 shows that after illumination the resistivity minima ρ_{xx} increases and a finite slope for the Hall steps is observed. (Results for the same device without illumination are shown in Fig. 6.) High-precision measurements of h/e^2 on such a device (Fig. 9) are not possible.

Conductive layers parallel to the 2DEG of a silicon MOSFET are practically not present, because the SiO_2 and the depletion layer between the 2DEG and the weakly doped substrate material are nearly ideal insulators at helium temperatures.

Therefore, the silicon MOSFET seems to be the most reliable system for high-precision measurements of h/e^2, but the experimental conditions are more complicated, because magnetic fields higher than used for the $GaAs$-$Al_xGa_{1-x}As$ heterostructure are necessary in order to separate the

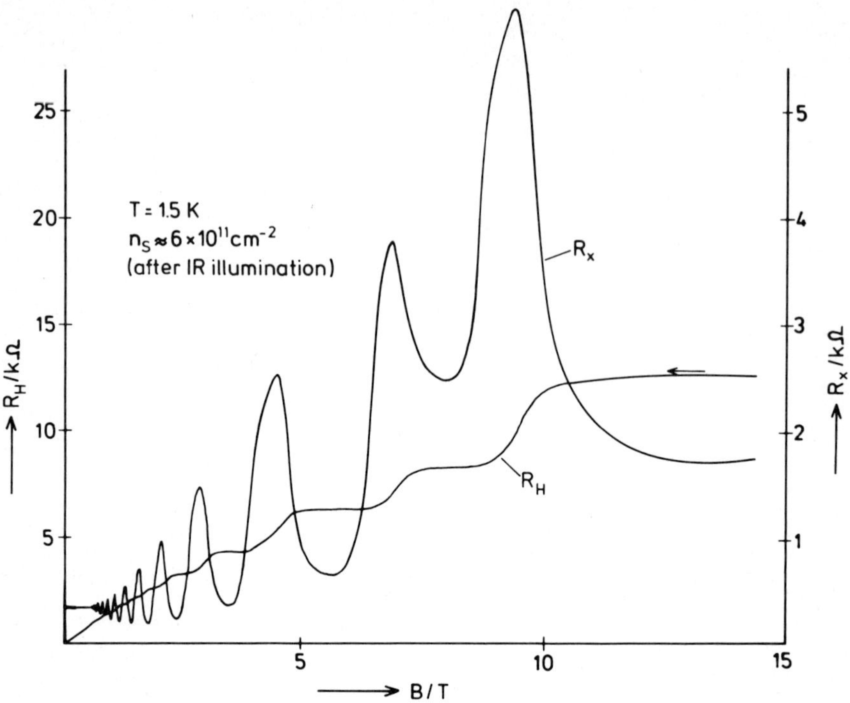

Fig. 9: Hall resistance and magneto-resistivity measurements on a GaAs-$Al_{0.3}Ga_{0.7}As$ heterostructure after illumination with infrared light. The IR illumination leads to an increased carrier density of the 2DEG and to bypass conductivity of the silicon-doped $Al_{0.3}Ga_{0.7}As$ layer. The experimental result without illumination is shown in Fig. 6.

Landau levels. Even at B = 14.5 T the Hall plateau $R_H = h/4e^2$ in Fig. 3 remains constant (within ± 0.2 ppm) only in a relatively small gate voltage region of $V_g = 12.3 \pm 0.2$ V. At lower magnetic fields (B < 14.5 T) or higher temperatures (T > 1.5 K) the gate voltage region where ρ_{xy} stays constant becomes very small and usually a finite slope $d\rho_{xy}/dV_g$ can be resolved at the gate voltage where a fully occupied Landau level is expected. This slope is proportional to the resistivity ρ_{xx} as shown in Fig. 10, which means that for a typical value of ρ_{xx} = 0.01 Ohm and an uncertainty of 1% for the gate voltage a deviation of $\Delta\rho_{xy}/\rho_{xy} \approx 3 \times 10^{-7}$ is expected. However, this uncertainty can be reduced by increasing the magnetic field or reducing the temperature, which leads to smaller ρ_{xx}-values. The thermally activated behaviour $\rho_{xx} \sim \exp(-|\Delta E|/kT)$ observed in the plateau region [8] is also visible in the deviation of the Hall conductivity $\Delta\sigma_{xy}$ relative to the quantized value σ_{qu} (Fig. 11). The activation energy ΔE is the same for both ρ_{xx} and $\Delta\sigma_{xy}$, as long as the activation to <u>one</u> mobility edge dominates (Fermi energy not exactly in the center between the two mobility edges of adjacent Landau levels). The correction to the Hall conductivity at very low temperature, where variable range hopping dominates [12,9], is so small that no quantitative data are available.

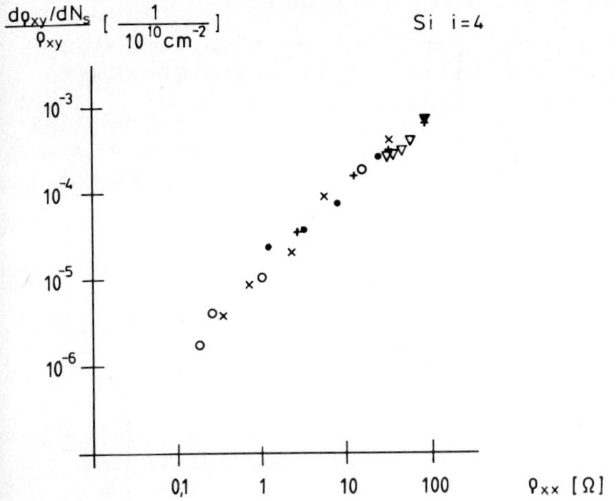

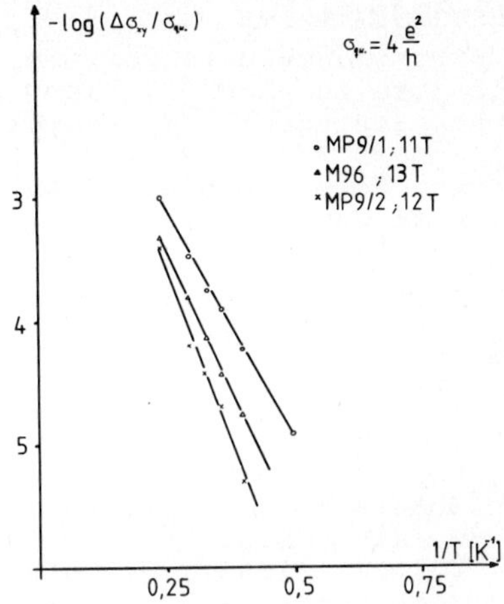

Fig. 10: Slope of the Hall resistance $d\rho_{xy}/dV_g \sim d\rho_{xy}/dN_s$ at the center of the Hall plateaus as a function of the minimal resistivity ρ_{xx}. The magnetic field is varied between $11\ T < B < 13\ T$, the temperature between $1.5\ K < T < 4.2\ K$, and the substrate bias voltage between $-9\ V < V_{SB} < 0\ V$.

Fig. 11: Deviation of the Hall conductivity $\Delta\sigma_{xy}$ relative to the quantized value $\sigma_{xy} = 4e^2/h$ as a function of the inverse temperature $1/T$.

In summary, high-precision measurements of h/e^2 from magneto-transport measurements - this is not only a question of high-resolution measurements of voltages (the problems related to thermoelectric phenomena or high-frequency pick-up have not been discussed in this article), but also a question of a microscopic interpretation of the transport properties of a 2DEG in a strong magnetic field. Corrections to the value of the quantized Hall resistance are always present because the resistivity ρ_{xx} is never exactly zero. Such a finite resistivity leads on the one hand to a geometrical correction in Hall effect measurements (the Hall angle is not 90°, and therefore the finite aspect ratio of the device leads to a reduction in the Hall voltage [13]), and on the other hand to corrections due to the scattering process. A quantitative theory of the Hall effect in the localized regime at finite temperatures is not available. Since all thinkable corrections to the value of the quantized Hall resistance should decrease with increasing magnetic field, measurements at different magnetic fields are the best check of the reliability of the quantum Hall effect for the determination of the fine-structure constant.

At the present level of uncertainty of the order 10^{-6}, magnetic fields of 14 T seem to be high enough for a quantitative determination of h/e^2 from Hall effect measurements on silicon MOSFETs (at $T = 1.5$ K), but still higher magnetic fields are necessary, if a higher precision is desired.

Acknowledgements:

The experiments were performed on a large number of devices, and we would like to thank K. Ploog (Max-Planck-Institut für Festkörperforschung, Stuttgart), M. Pepper (Cavendish Laboratory, Cambridge), R.J. Wagner (Naval Research Laboratory, Washington), and G. Dorda (Siemens Forschungslaboratorien, München) for providing us with samples. This work was supported by the Deutsche Forschungsgemeinschaft via SFB 128.

References:

1. K. v. Klitzing, G. Dorda, and M. Pepper, Phys.Rev.Lett. 45, 494 (1980).
2. For a review see: T. Ando, A.B. Fowler, and F. Stern, Rev. of Modern Physics 54, 437 (1982).
3. R. Kubo, S.J. Miyake, and N. Hashitsume, in Solid State Physics 17, 269 (1965), F. Seitz and D. Turnbull (ed.), Academic Press (New York) 1965.
4. T. Ando, J.Phys.Soc. Japan 37, 1233 (1974).
5. T. Ando, Surface Sci. 113, 182 (1982).
6. R.E. Prange, Phys.Rev. B 23, 4802 (1981).
7. H. Aoki and T. Ando, Solid State Commun. 38, 1079 (1981).
8. Th. Englert and K. v. Klitzing, Surface Sci. 73, 70 (1978).
9. G. Ebert, K. v. Klitzing, C. Probst, and K. Ploog, Solid State Commun. (to be published).
10. D.C. Tsui, A.C. Gossard, B.B. Field, M.E. Cage, and R.F. Dziuba, Phys.Rev.Lett. 48, 3 (1982).
11. G.T. Bodwin, D.R. Yennie, and M.A. Gregorio, Phys.Rev.Lett. 48, 1799 (1982).
12. H.L. Störmer, D.C. Tsui, and A.C. Gossard, Surface Sci. 113, 32 (1982).
13. K. v. Klitzing in: Festkörperprobleme (Advances in Solid State Physics), Vol. XXI, 1, J. Treusch (ed.), Vieweg (Braunschweig) 1981.

Gauge Invariance and the Quantised Hall Effect in Two-Dimensional Systems

Hideo AOKI*

Cavendish Laboratory, University of Cambridge,
Madingley Road, Cambridge CB3 0HE, England

Abstract. The quantised Hall effect in two-dimensional systems in magnetic fields is studied with an emphasis on the importance of gauge invariance in this phenomenon. Theoretical aspects of the problem is reviewed and new numerical results are presented. It is shown that the quantised Hall current is fundamentally related to gauge transformation in the system, and that the quantised plateau in the Hall conduction is a combined effect of localisation and gauge invariance.

1. Introduction

The quantised Hall effect observed recently [1] in two-dimensional(2D) electron systems in strong magnetic fields is, in a sense, as remarkable an effect as the superconductivity. Indeed, it is becoming clearer that the fascinating phenomenon arises from what could be called a quantum mechanical effect on a macroscopic scale: Here the coherence of a wavefunction itself, not a condensate, with a gauge invariance imposed on it, determines macroscopic phenomena. This was first shown by an elegant exposition of the phenomena for a disorder-free case by Laughlin [2]. The quantised Hall effect also offers an interesting ground for examining the localisation of states in this peculiar 2D system in magnetic fields, which is a distinct prototype of disordered system in its own right. In this paper, theoretical aspects of the problem is reviewed, and the gauge-transformation theory for the quantised Hall effect in disordered systems is developed with new numerical results to explore the interplay of randomness and gauge. It is shown that the two points mentioned above, the gauge invariance in the Hall current and the localisation and delocalisation of the states, are in fact closely related, and that the quantised plateau in the Hall conductivity as a function of electron concentration is a combined effect of localisation and gauge invariance. A close analogy of the quantised Hall effect to the superconductivity is also stressed.

2. Electronic Structure

Before discussing the Hall effect, let us first recapitulate the electronic structure of the disordered 2D system in a strong magnetic field. The Hamiltonian of the system is given by

$$H = \frac{1}{2m}\left(\vec{p} + \frac{e}{c}\vec{A}\right)^2 + V(\vec{r}), \tag{1}$$

where $\vec{A}$ is a vector potential, e>0 is the elementary charge and $V(\vec{r})$ is an external random potential. In the Landau gauge, $\vec{A}=(0,Hx)$ with a magnetic field H, we can take the Landau basis function,

$$\psi_{NX}(\vec{r}) = \frac{1}{\sqrt{L}} \exp\left(i\frac{Xy}{\ell^2}\right) \chi_N(x-X), \tag{2}$$

where N is the Landau index, X is the centre coordinate of cyclotron motion, L is the linear dimension of the system, $\ell=(c\hbar/eH)^{1/2}$ is the cyclotron radius and χ_N is the eigenfunction of a linear harmonic oscillator. Note that X, which is related to the y component of wavenumber by $X=\ell^2 k_y$, takes discrete values as $X = \Delta X, 2\Delta X, \ldots, N_L \Delta X$, where $\Delta X=2\pi\ell^2/L$ and $N_L=L^2/2\pi\ell^2$ is the number of states in a Landau subband. If we change the vector potential $\vec{A}$ by $\Delta\vec{A}=\vec{\nabla}f(\vec{r})$, where f is an arbitrary function for the gauge transformation, the wavefunction acquires an additional phase,

$$\psi \longrightarrow \psi \exp\left[i\frac{e}{\hbar c} f(\vec{r})\right] . \tag{3}$$

In the Landau basis, the Hamiltonian is expressed in a second quantised form as

$$H = \sum_{NX} |NX\rangle(N+\tfrac{1}{2})\hbar\omega_c\langle NX| + \sum_{NN'XX'} |NX\rangle\langle NX|V|N'X'\rangle\langle N'X'| . \tag{4}$$

Extensive study of the electronic structure of this system has been done [see a review by Ando, Fowler and Stern, ref.3]. When the energy broadening is less than the cyclotron energy, $\hbar\omega_c=\hbar eH/mc$, the electronic structure comprises well-defined Landau subbands. The localisation of the states due to randomness in this system was first studied by Aoki and Kamimura [4], who discuss the vanishing of d.c. conductivity due to localisation in finite regions of energy spectrum in the Shubnikov-de Haas oscillation. The presence of a strong magnetic field gives rise to unique nature of the system: the dynamics of electrons is governed by the scattering of cyclotron motions. This gives the random potential a complex role, which in turn makes the definition of degree of randomness very different from conventional systems. In the Hamiltonian (4), the randomness occurs in $\langle X|V|X'\rangle$, which is firstly complex and, secondly contains both diagonal(X=X') and off-diagonal(X≠X') randomness. If we regard a random potential as an assembly of impurity potentials, $V(\vec{r})=V_0\sum_i v(\vec{r}-\vec{R}_i)$, it is characterised by a dimensionless concentration of impurities, $c_i=2\pi\ell^2 n_i$, and range, d/ℓ, where n_i is the number of impurities per unit area and d is the range of individual impurity. The electronic structure, and the localisation in particular, of a Landau subband depend on the randomness as well as on the Landau index N. This is confirmed by a numerical calculation by Aoki [5]. Specifically, the localisation depends strongly on d/ℓ as is shown by transport property [4] or by numerical study [6], and the localisation is favoured as the potential range is increased. In Fig. 1(a) we show typical eigenstates in the ground(N=0) Landau subband, one in a localised regime with eigenenergy near the lower band tail and the other in an extended regime near the band centre. We employed periodic boundary conditions in both x and y directions with $N_L=256$, which corresponds to linear dimension of the system $L=(2\pi N_L)^{1/2}\ell=40.1\ell$. The random potential has equal number of attractive and repulsive impurities, which makes the electronic structure symmetric about E=0. We took short-range, delta-function impurity potentials with concentration $c_i=10$, which gives the band width of $4\sqrt{c_i}=12.6$ in the self-consistent Born approximation [7].

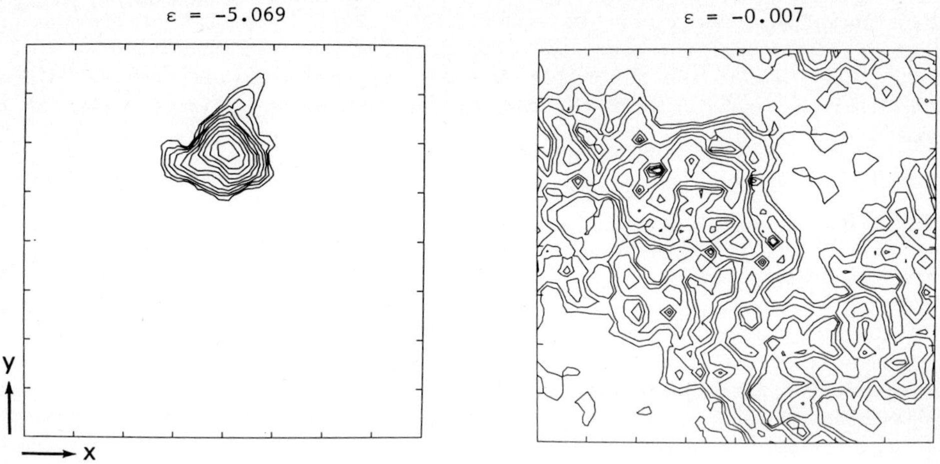

FIG. 1(a) Original eigenstates.

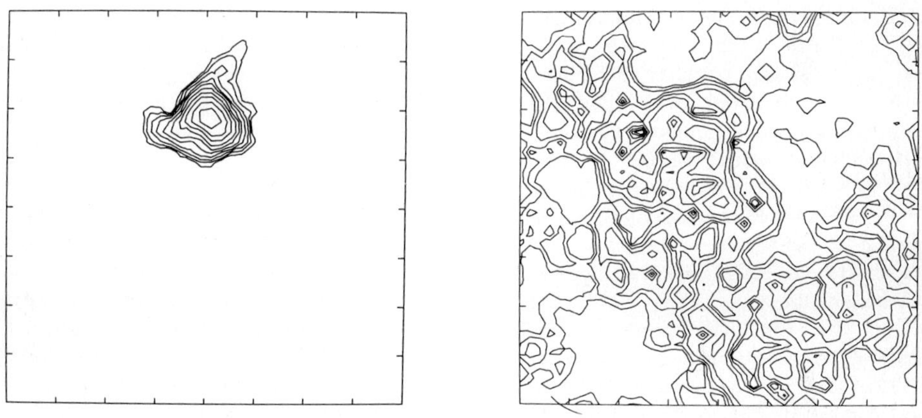

FIG. 1(b) The eigenstates after a gauge transformation with $\Phi=0.4\phi_0$.

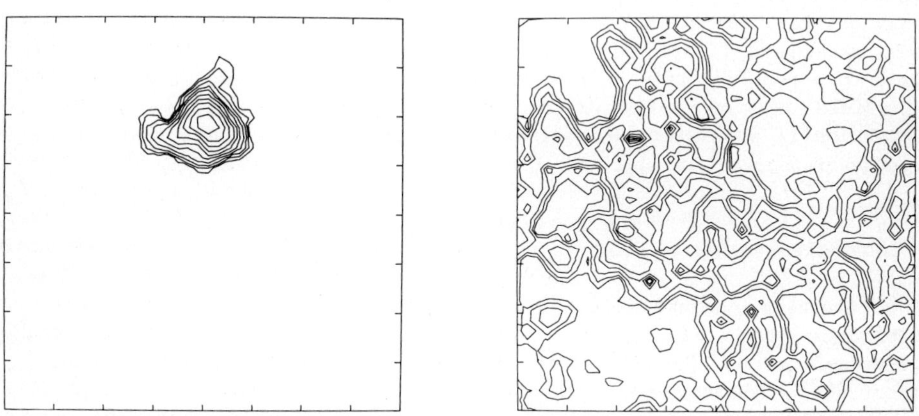

FIG. 1(c) The wavefunctions after an electric field is turned on.

We take hereafter ℓ as a unit of length and $V_0/2\pi\ell^2$ as a unit of energy. The heights of contour for the absolute value of normalised wavefunction are $0.4/2^{n/2}$ (n=0,...,10). One of the noticeable features in the result is that an extended state, although spreading over the sample, can circumvent a finite region of the system.

3. Hall Effect

3.1 Quantised Hall effect

A distinct nature of the present random system manifests itself in the quantised Hall effect [1 and refs in 8] in 2D electrons in Si MOS inversion layers or semiconductor heterojunctions: The Hall conductivity, σ_{xy}, of the system takes the quantised value,

$$\sigma_{xy} = -Ne^2/h \tag{5}$$

with physical constants alone, in finite regions (plateaux) of energy spectrum, which coincide with the mobility gap in σ_{xx} between (N-1)th and N-th Landau subbands. The fact that the formula (5) holds experimentally with a remarkable accuracy of the order of 0.1ppm [9], providing a novel and precise measurement of the fine-structure constant, $\alpha = e^2/\hbar c$, strongly suggests that there must be something fundamental in it.

One of the first theoretical approaches by Aoki and Ando [10] uses the Kubo formula, in which σ_{xy} is given by $-nec/H + \Delta\sigma_{xy}$ with $\Delta\sigma_{xy} = (e^2/2)(\langle\langle\dot{Y}\dot{X}\rangle\rangle - \langle\langle\dot{X}\dot{Y}\rangle\rangle)$. Among their conclusions are that (i) in a localised regime in a finite region of the energy spectrum, the Hall conductivity exactly remains constant to give a plateau, and (ii) the quantised Hall effect is impossible if all the states are localised. The picture is summarised in Fig. 2. Important point here is that both localised and extended states are necessary for the quantised Hall plateau, because the flatness of the plateau comes from the localisation in finite regions of energy spectrum, while the quantised value comes from the existence of extended states as is discussed below. This is a remarkable aspect of the effect, since a randomness usually destroys any sharp transition in a physical system, whereas in the present case a randomness is actually essential for well-defined quantum steps. This is related to a peculiar role of randomness in this system, where scattering exerts a dual effect [4] of producing current and disorder. The important role of localisation in producing the Hall-conductivity plateau is also quite relevant in the gauge-transformation approach discussed below.

3.2 Gauge-transformation theory

Now, Laughlin [2] was first to introduce a gauge invariance approach to the problem in the disorder-free limit. In this section we study a gauge-transformation theory for the disordered systems. Let us first define the geometry of the system, which also makes explicit the Aharonov-Bohm effect [12] at work in the quantised Hall effect. Consider a rectangular sample (Fig. 3(b)) in x-y plane. The centre coordinate of the cyclotron motion, X, takes discrete values, $X_n = n\Delta X$ ($n=1,\ldots,N_L$), with $\Delta X = 2\pi\ell^2/L$. If we vary the vector potential by

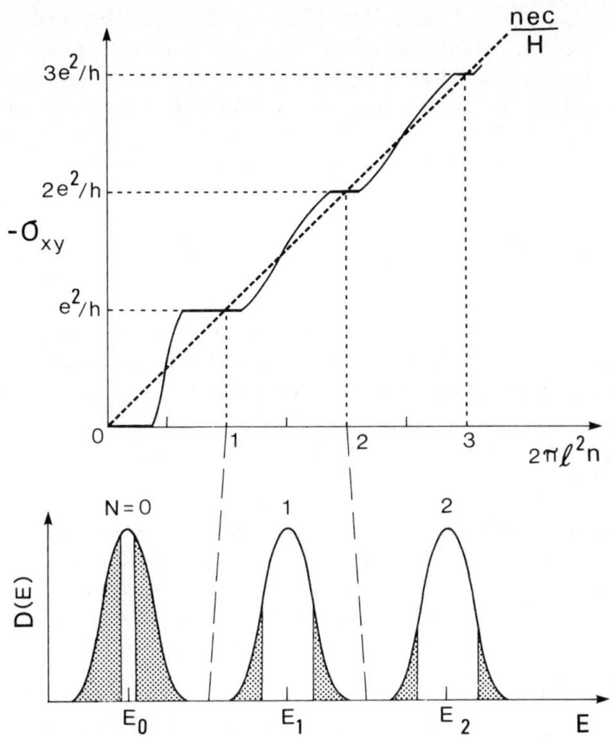

FIG. 2. The Hall conductivity, σ_{xy}, is plotted versus the normalised electron concentration, $2\pi\ell^2 n$ (top). The plateaux come from the electronic structure of Landau subbands (bottom) with both localised regime (shaded) and extended regime. The schematic picture is after ref.10 with the density of states, D(E), by the present numerical result and the position of the mobility edges, E_c, determined by Aoki [16] and Ando [6]. Note that D(E) and E_c are sensitive to the nature of random potential (short-range here), although the plateau behaviour is general. In this figure we employed a symmetric potential, which makes the electronic structure of a Landau subband symmetric. However, a Landau subband is in general asymmetric for a potential with unequal number of attractive and repulsive impurities [see ref.5], and the extended region at which σ_{xy} increases by e^2/h from one plateau to another is not necessarily situated at the half-filled point of a Landau subband.

$\Delta\vec{A}=Ha\hat{y}$, this causes a uniform shift for X as $X_n = n\Delta X + a$. If we employ a periodic boundary condition in y direction, the geometry is topologically equivalent to a looped ribbon with H applied perpendicular to the ribbon, and the change of $\vec{A}$ can be envisaged as a change of another magnetic flux through the opening of the ribbon (Fig. 3(a)). This is a typical geometry for the Aharonov-Bohm effect [12,13], since the flux, which can be changed independent of the magnetic field applied to the ribbon, is entirely confined in an opening of the loop and has no effect classically. The shift of the values for X when Φ is applied is written as

$$X_n(\Phi) = X_n(0) + (\Phi/\phi_0)\Delta X \tag{6}$$

as is depicted in Fig. 3(c). When Φ is changed by an integer multiple of the magnetic flux quantum, $\phi_0 \equiv hc/e$, we have the same set of values for X except at boundaries. The total energy of the system, E_T (i.e. the free energy at T=0), and the total current, I, circulating the ribbon is related [14] by

$$I = c\partial E_T/\partial\Phi \ . \tag{7}$$

Now, if all the states were localised, the right-hand side of this equation would just vanish, since, for a localised state, a change in Φ only amounts to a trivial phase change in the wavefunction, $\psi \rightarrow \psi \exp(i2\pi\Phi / \phi_0)$, and does not affect any physical quantity including eigenenergy. If the state is extended, on the other hand, a change in Φ causes a significant variation of the state: we have to make a periodic continuation of the state around the ring for an extended state, and, unless Φ is varied by an integer multiple of ϕ_0, an extended wavefunction has to adjust itself to the gauge transformation. Figure 1(b) shows the wavefunction after the gauge transformation with $\Phi=0.4\phi_0$ for the two eigenstates shown in Fig.

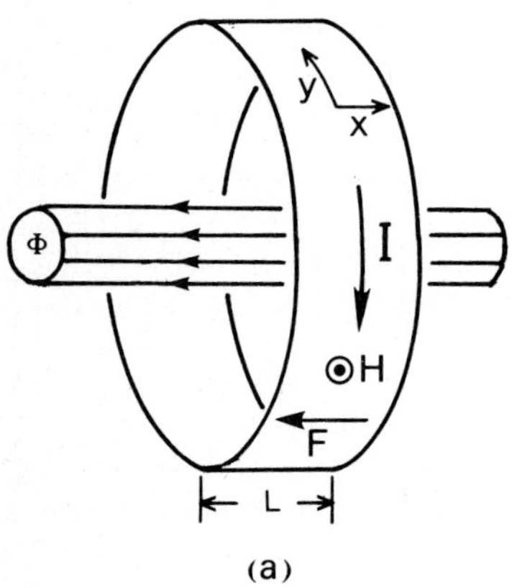

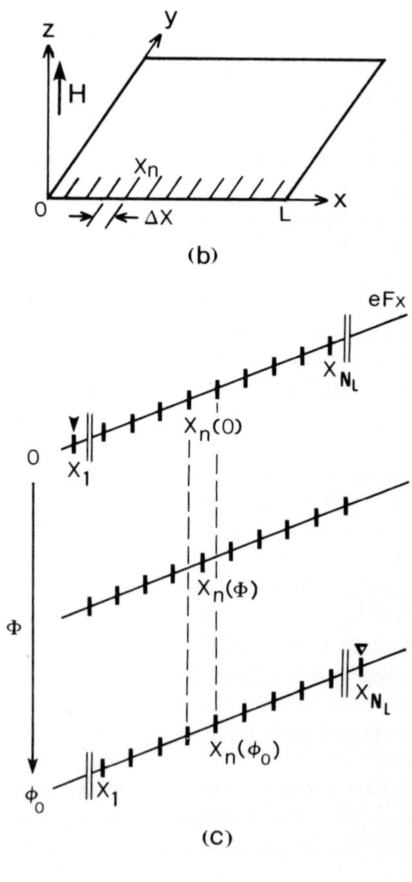

FIG. 3 (a) The Hall current, I, in the Aharonov-Bohm geometry with a magnetic field H, an electric field F and a magnetic flux Φ through the cylinder.
(b) The rectangular geometry with the discrete values allowed for X indicated. If we adopt a periodic boundary condition in y direction, the topology is the same as in (a).
(c) The shift of the set of discrete values for X as a function of the flux, Φ, with a uniform electric field F. The box boundaries are indicated.

1(a). From Eq.(6) it is evident that the state, and consequently the energy level, are periodic in Φ with period ϕ_0. It follows immediately, as was stressed by Imry [15], that E_T cannot be a monotonic function of Φ. Two points arise here: first point is the importance of the radically different behaviour of localised and extended regimes, which gives the Hall conductivity plateau, and the second point is the role of the electrodes as electron reservoir. Let us first assume that the Fermi energy, E_F, is in a gap between (N-1)th and N-th Landau subbands, and that every wavefunction remains rigid against a change in Φ. When Φ is increased from 0 to $\delta\Phi$, every X_n increases by $a=(\delta\Phi/\phi_0)\Delta X$. This will increase the total energy by $NeFLa/\Delta X$, which gives, by Eq.(7), the quantised Hall current, $I=(Ne^2/h)\Delta V$ with $\Delta V=FL$ being the potential drop from one electrode to another. However, this cannot be a correct picture for a disordered system, since (i) an extended wavefunction does change with Φ, and has a more complex Φ-dependence than linear in Φ, and (ii) E_F need not be in a band gap, since we have a quantised Hall current at any point in the plateau region.

To shed light on these points, we performed a numerical study for large but finite samples in the above geometry with a symmetric, short-range potential with $c_i=10$ and $N_L=256$. We actually made a gauge transformation given by Eq.(6), and calculated all the eigenenergies and eigenstates of the system by diagonalising the Hamiltonian matrix, Eq.(4), at each value of Φ, which is varied from 0 to ϕ_0. Figure 4(a) shows the dependence of eigenenergies on Φ. Note that no two levels can cross in a disordered system, so that an energy level should end up with the same eigenenergy after a change of Φ by ϕ_0 unless we consider thermal or inelastic broadening. This periodicity is achieved by a rearrangement of electrons discussed below. But first let us stress that the result clearly shows that the energy level stays essentially constant against Φ for a localised state, while an extended eigenstate has a strong Φ-dependence. The boundary between the two types of behaviour is conspicuous at $|E|\simeq 1.0$ for the randomness and sample size chosen here in agreement with the estimate of mobility edge by other methods [6,19]. The sensitivity of an eigenstate to a gauge transformation can be measured by $\langle D(E)\Delta\varepsilon\rangle$, where $\Delta\varepsilon$ is the shift of energy level, $1/D(E)$ is the mean level spacing and $\langle\ \rangle$ stands for an average over an energy interval. Aoki [16] shows that the sample-size dependence of this quantity may be used as a criterion for localisation as in Licciardello-Thouless study [17], and that $\langle D(E)\Delta\varepsilon\rangle$ is exponentially small for localised states indicating an exponential localisation.

In Fig. 4(b) we plotted the total energy shift, ΔE_T, as a function of Φ, and the Hall current calculated by $I=c\partial\Delta E_T/\partial\Phi$. The result is remarkable since it shows that the total energy behaves as $\Delta E_T=(Ne^2/ch)\delta\Phi\Delta V$ to give the quantised current even in random systems despite (i) the wiggle of the individual energy level, $\varepsilon(\Phi)$, for an extended state and (ii) the virtually constant $\varepsilon(\Phi)$ for the localised states. This behaviour is only possible because of the ability of extended states to compensate the existence of localisation to give the correct E_T for the quantised current. This compensation in the coexistence of localised and extended regimes was first pointed out by Aoki and Ando [10], who emphasise a sum rule by which an extended state carries an enhanced current, and also by Prange [18] for a single-impurity case.

The second point is concerned with the boundaries, at which the set of values for X_n is different for $\Phi=0$ and $\Phi=\phi_0$. To simulate the role of electrode at either edge of the sample, we have employed the following scheme: we take the box boundary condition in x direction which allows electron to move in a region $\Delta X \leq x \leq L$. The values, $X_{min}(0)=0$, ..., $X_{max}(0)=N_L \Delta X$, for $\Phi=0$ will be changed into the set, $X_{min}(\phi_0) = \Delta X$, ..., $X_{max}(\phi_0)=(N_L+1)\Delta X$. Thus $X_{min}(0)$ at the left edge, which is outside the allowed region must be first decoupled from the system and then introduced into it for $\Phi>0$. Likewise $X_{max}(\Phi)$ at the right edge must be decoupled from the system at $\Phi=\phi_0$. The coupling (decoupling) of these basis functions are performed by turning on (off) the matrix elements, $\langle X_{min}|H|X_n\rangle$ ($\langle X_{max}|H|X_n\rangle$) adiabatically with the change of Φ. Thus we have the same set of eigenenergies at $\phi = 0$ and ϕ_0 except for these left-edge level (shown by an black arrow in Figs 3(c)

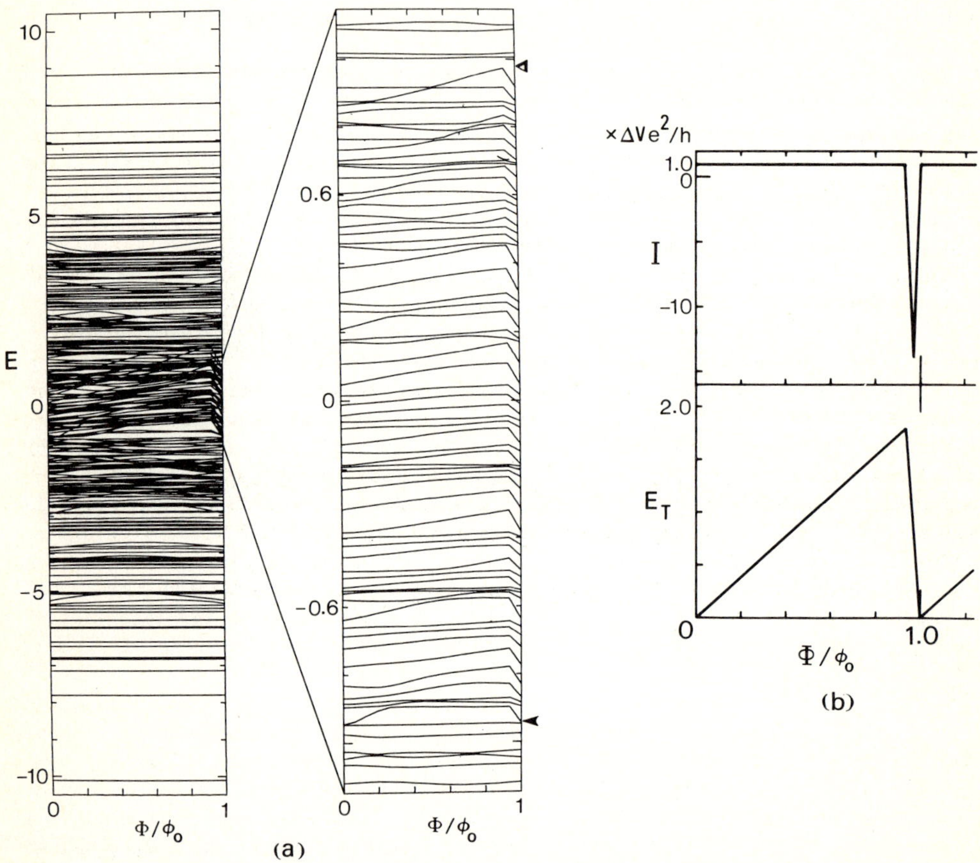

FIG. 4 (a) All the energy levels are plotted versus the flux, Φ, over one period of unit flux change with an electric field, eF=0.05, in the box boundary condition in x direction. To make the wiggle of the lines visible, the energy shift, $\varepsilon(\Phi) - \varepsilon(0)$, is exaggerated by a factor of ten. this makes some of the curves apparently crossing, while no two lines cross in fact, as is shown in the figure on the right, which is an enlarged diagram for $|E| \leq 1.15$ in a real scale. As for a discontinuity of the lines near $\Phi/\phi_0 \simeq 1$, see text.
(b) The total energy, E_T, and the Hall current, I, as a function of the magnetic flux, Φ. The Hall current was calculated by $I=c\partial E_T/\partial\Phi$.

and 4(a)) coupled at ϕ =integer $\times\phi_0$ and right-edge level (white arrow) decoupled at the same time. If this procedure is not done, the eigenenergies will be continuous, but E_T will increase with no periodicity. In fact, this procedure ensures the system being periodic in Φ , and, if we assume that the system for each value of ϕ remains in its ground state, in which no empty state below E_F or occupied state above E_F are allowed at T=0, the rearrangement (or 'pumping' discussed by Halperin [20] or Imry [15]) of electron occupation and, consequently, a discontinuity in E_T and I are inevitable before one cycle of $\Phi \to \phi_0$ is completed (Fig. 4). This makes E_T and I periodic in Φ with period ϕ_0 (resuming the Byers-Yang theorem in ref.14), and gives the quantised Hall current except at the discontinuity. Thus the quantum plateau for σ_{xy} is a combined effect of localisation and gauge transformation. This is just in accord with the picture given in Fig. 2.

In a system with box boundary condition, some edge states may be present. An example of the edge state is shown in Fig. 5. In this figure we show the local current defined by $\vec{j}=(e\hbar/m)|\psi|^2\,[\vec{\nabla}\phi-(e/c\hbar)\vec{A}]$, which is intrinsic to a wavefunction, $\psi=|\psi|\exp(i\phi)$, in a magnetic field [19]. The result shows that the edge state has a current distribution involving the cyclotron motion reflected at the boundary. Halperin [20] argues that these edge states are necessary for the redistribution of electrons discussed above. In real situations, however, we do not actually need edge states, since there is an electrode at either end of the sample, which ensures some current across the two edges with the electrodes acting as electron reservoir.

The different response of localised and extended states to an external electric field can be seen not only for an adiabatic application of the electric field, but also for an electric field applied suddenly. We calculated the time-dependent wavefunction after an electric field of eF=0.05 is turned on at t=0. Figure 1(c) shows the wavefunction at t=2.0h/(LeF) where the two eigenstates shown in Fig. 1(a) are employed, respectively, as an initial state. While an extended state responds to the perturbation, a localised state remains immobile in a finite electric field in a time scale employed in the figure, although the state will eventually make a quantum tunnelling in a sufficiently long time, giving some non-Ohmic effects.

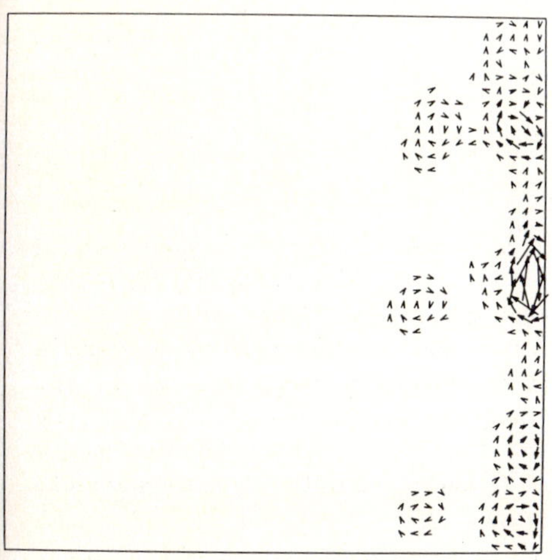

FIG. 5. An example of the edge state in the box boundary condition in x direction. The current density, $\vec{j}(\vec{r})$, is shown for a state with an eigenenergy ε =2.065 for repulsive, short-range impurities with c_i=5 and N_L=256. The vector, $\vec{j}(\vec{r})$, in a continuous space is represented here by arrows in a region where $j(\vec{r})$ has appreciable amplitude ($\geq$0.008eh/m) with the length of the stem of an arrow proportional to $j(\vec{r})$.

It is evident that the existence of extended states is essential in our argument. This poses an interesting question of whether a strong magnetic field can change the universality class for localisation in 2D, which is considered to be the lower critical dimensionality [21] in the absence of magnetic fields. As we mentioned, the extended regime becomes narrower as we increase the range of potential. The delta-function potential employed so far is the short-range limit, where the quantum jump of cyclotron motions, and thus the delocalisation, are most favoured. We can, however, show that even in the opposite limit of an infinitely slowly varying potential, we still have at least one extended state. In this limit, we resume the classical picture, in which the centre of cyclotron motion drifts along an equipotential contour. The percolation theory [22] in 2D continuous space shows that there is always a percolating path (Fig. 6). For a symmetric potential, this path lies at E=0. Percolative nature in this limit was stressed also by other authors [23,24]. Thus the width of extended states may be infinitely narrow in this particular limit, although quantum effects are important in general, where two different processes, localisation and tunnelling, must be considered. In practical experimental situations, however, all we have to have for the quantised Hall effect is the states whose spatial extension exceeds the sample size. The nature of random potential also depends on the system. The GaAs-Al_xGa_{1-x}As heterojunction, for instance, is less disordered than MOS inversion layers, in which charged impurities and interface roughness are considered to be main source of randomness. There is an experimental indication [25] that the extended regime is very narrow in the former system for the lowest few subbands.

FIG. 6. An example of percolating energy contour (heavy line) at E=0 in a continuous random potential in 2D. The local minima and maxima are indicated by different shadings respectively. Here an equal number of attractive and repulsive impurities with range of 3.6ℓ are distributed with concentration c_i=2 in an area of 40$\ell\times$40ℓ in a periodic boundary condition.

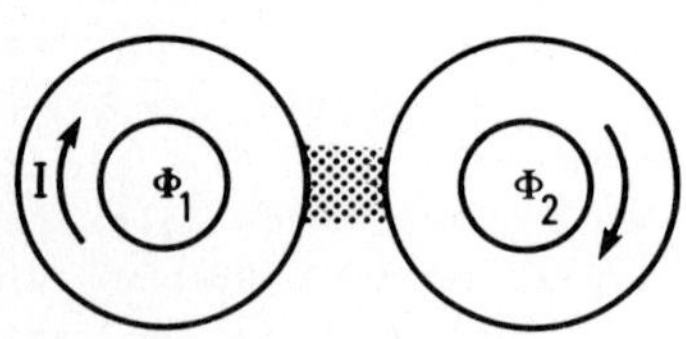

FIG. 7. Two annular samples with a junction (shaded area). When the wavefunctions are connected by the junction, the Hall current should depend on both Φ_1 and Φ_2 penetrating each of the two systems.

4. Discussion

The gauge-invariance theory of quantised Hall effect discussed here has a remarkable analogy with superconductivity. They both come from the coherence of the state, or the single-valued $\psi(\vec{r})$, with ψ being the extended wavefunction in the quantised Hall effect or the order parameter of the condensate. The Corbino geometry used for 2D space-charge layers is a counterpart of a superconducting ring (or a cylinder), which was used to confirm the flux quantisation in superconductors [26]. The current associated with the phase of a wavefunction, which has to be complex in a magnetic field, is similar to a Josephson current. If, for instance, two separated systems in a magnetic field are brought into contact, a definite relative phase between the two systems should be established as is illustrated in Fig. 7 for a branching of a ring just as in a superconducting ring [27]. We could also imagine that, if a magnetic monopole, which may be detected by a superconducting coil [28], goes through the inner region of a Corbino-type sample, the Hall current should respond to it.

Another peculiar aspect, which could be called a devil's staircase in quantised Hall effect, arises if we consider, quite apart from random systems, the gauge invariance in crystalline systems in magnetic fields. As a simplest example, take a system with a periodic potential in one direction in an otherwise perfect 2D continuous space, which was adopted in a study [11] of the Harper broadening. The Hamiltonian is given by $H = (1/2m)[\vec{p}+(e/c)\vec{A}]^2 + U(x)$, where the external potential $U(x)$ has a periodicity of d. The eigenfunction is given by $\psi_{\alpha X}(\vec{r}) \propto \exp(iXy/\ell^2)\,\phi_{\alpha X}(x)$, where $\phi_{\alpha X}$ satisfies a equation for a linear harmonic oscillator in $U(x)$. The eigenenergy is periodic as $\varepsilon_\alpha(X+d) = \varepsilon_\alpha(X)$. We adopt the periodic boundary condition in y direction as before, so that $X=\ell^2 k_y$ takes discrete values with a spacing $\Delta X=2\pi\ell^2/L$. Now, if the ratio, $d/\Delta X$, is irrational, a peculiar phenomenon happens. The eigenenergy $\varepsilon_{\alpha X}$, for instance, has a different set of values when one goes from one unit cell of $U(x)$ to another. This is associated with the eigenstate $\psi_{\alpha X}$ being an anomalous Bloch state in that it never repeat itself. Since $d/\Delta X=\Phi_{uc}/\phi_0$, where Φ_{uc} is the flux within the unit cell of $d\times L$, an irrational $d/\Delta X$ implies an irrational fraction of ϕ_0 contained in the unit cell. If we look at the transformation of a wavefunction under a translation along x axis by nd (n=integer), which can be realised by a gauge transformation, $\vec{A}\to\vec{A}+Hnd\hat{y}$, the wavefunction is transformed as $\psi \longrightarrow \psi\exp[-2\pi i(n\Phi_{uc}y/\phi_0 L)]$. Thus, if Φ_{uc}/ϕ_0 is irrational, there is no way to make the wavefunction after the translation periodic in y direction with periodicity of L.

More recently, Thouless and coworkers [29] investigated the Hall effect in the potential periodic in both x and y directions. They found that the Hall current depends quite sensitively on the commensurability of Φ_{uc}/ϕ_0, whilst the current is still given by one or another integer multiple of the quantised value for any value of E_F.

In summary, we have elucidated the quantised Hall effect in two dimensions in magnetic fields in the presence of localisation using the gauge-invariance property inherent in the system, where a quantum coherence of the wavefunction plays an essential role.

The author wishes to acknowledge Drs J.C. Inkson, J.M.F. Gunn and R. Haydock for valuable discussions. It is also a pleasure to thank Professors Sir N.F. Mott and D.J. Thouless for stimulating conversations. The numerical calculations were done on the CRAY-1 at the SERC Daresbury Laboratory. The work at the Cavendish Laboratory was supported by the Science and Engineering Research Council of U.K.

References

* On leave of absence from Department of Physics, University of Tokyo, Tokyo, Japan.

1. K. von Klitzing, G. Dorda and M. Pepper, Phys. Rev. Lett. 45(1980), 494; K. von Klitzing, Surf. Sci. 113(1982), 1.
2. R.B. Laughlin, Phys. Rev. B 23(1981), 5632; Surf. Sci. 113(1982), 22.
3. T. Ando, A.B. Fowler and F. Stern, Rev. Mod. Phys. 54(1982), 437.
4. H. Aoki and H. Kamimura, Solid State Commun. 21(1977), 45.
5. H. Aoki, J. Phys. C 10(1977), 2583; J. Phys. C 11(1978), 3823.
6. T. Ando in Springer Series in Solid State Sciences Vol.39 (Springer: Berlin, to be published)
7. T. Ando, J. Phys. Soc. Japan 36(1974), 959.
8. F. Stern (ed.), Proc. 4th Intern. Conf. on Electronic Properties of Two-Dimensional Systems, New London, 1981 (North-Holland: Amsterdam, 1982); Surf. Sci. 113(1982).
9. D.C. Tsui, A.C. Gossard, B.F. Field, M.E. Cage and R.F. Dziuba, Phys. Rev. Lett. 48(1982), 3.
10. H. Aoki and T. Ando, Solid State Commun. 38(1981), 1079.
11. H. Aoki and T. Ando, Surf. Sci. 113(1982), 27.
12. Y. Aharonov and D. Bohm, Phys. Rev. 115(1959), 485.
13. M. Peshkin, Phys. Rep. 80(1981), 375.
14. N. Byers and C.N. Yang, Phys. Rev. Lett. 7(1961), 46.
15. Y. Imry, J. Phys. C 15(1982), L221.
16. H. Aoki, to be published.
17. D.C. Licciardello and D.J. Thouless, J. Phys. C 8(1975), 4157.
18. R.E. Prange, Phys. Rev. B 23(1981), 4802.
19. H. Aoki, to be published.
20. B.I. Halperin, Phys. Rev. B 25(1982), 2185.
21. E. Abrahams, P.W. Anderson, D.C. Licciardello and T.V. Ramakrishnan, Phys. Rev. Lett. 42 (1979), 673.
22. R. Zallen and H. Scher, Phys. Rev. B 4(1971), 4471.
23. S.V. Iordansky, Solid State Commun. 43(1982), 1.
24. R.F. Kazarinov and S. Luryi, Phys. Rev. B 25(1982), 7626.
25. M.A. Paalanen, D.C. Tsui and A.C. Gossard, Phys. Rev. B 25(1982), 5566.
26. B.S. Deaver and W.M. Fairbank, Phys. Rev. Lett. 7(1961), 43.
27. F. Block, Phys. Rev. B 2(1970), 109.
28. B. Cabrera, Phys. Rev. Lett. 48(1982), 1378.
29. D.J. Thouless, M. Kohmoto, M.P. Nightingale and M. den Nijs, Phys. Rev. Lett. 49(1982), 405.

SOME REMARKS ON THE PRESENT UNDERSTANDING OF THE QUANTIZED HALL EFFECT IN TWO DIMENSIONS

J. Hajdu

Institut für Theoretische Physik
Universität zu Köln
D-5000 Köln 41, W. Germany

1. Introduction

In this paper I would like to comment - on a very elementary level - on some attempts to explain the observed plateaus in the Hall conductivity of two-dimensional semiconducting systems /1/. For the sake of simplicity, I will disregard spin and multi-valley effects as well as any temperature dependence and consider only the most radical idealization of the experimental observation. This is shown in Fig. 1.

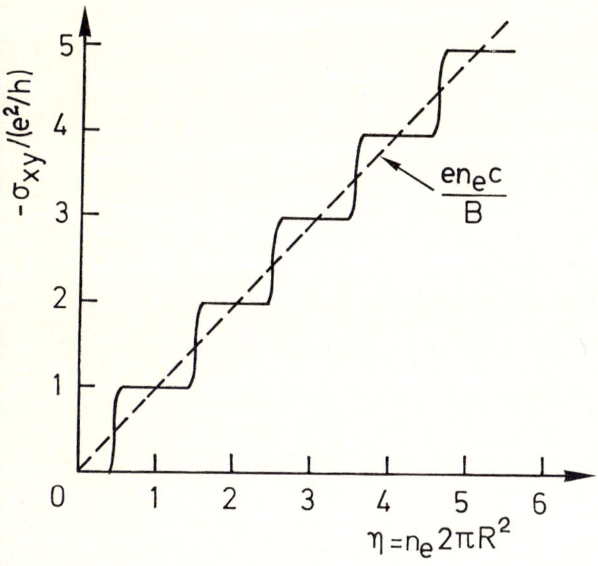

Fig. 1. Idealized picture of the observed Hall conductivity showing equidistant plateaus (quantization in units of e^2/h).

R is the magnetic length, $R^2 = hc/eB$ ($D = 1/2\pi R^2$ is the number of states within a Landau level, see sec. 2). In the plateau regimes the transverse conductivity σ_{xx} vanishes.

It is remarkable that for $\eta = n = 1,2,3,\ldots$, i.e.

$$n_e = \frac{n}{2\pi R^2} \tag{1.1}$$

(the Fermi energy ε_F is between two adjacent Landau levels), the observed values of the Hall conductivity

$$\sigma_{xy} = -n\frac{e^2}{h} \tag{1.2}$$

coincide with the values which follow from the classical formula

$$\sigma_{xy} = -\frac{en_e c}{B} \tag{1.3}$$

For free electrons (and, in the Peierls replacement approximation, even for lattice electrons with parabolic conduction band) this result is well established (see secs. 3 and 4). Ando, Matsumoto and Uemura /2/ have shown, by high order perturbation theory, that under the occupation condition (1.1), eq. (1.2) still holds true in the presence of impurities (cf. also sec. 5 and 6).

The subject set for competition is to explain why σ_{xy} keeps the constant values (1.2) for finite intervals of n_e (or ε_F) corresponding to partially occupied bands (Landau levels broadened by impurity scattering).

2. 2D Electron Gas in a Magnetic Field

Let me very briefly recall the quantum mechanics of a 2D free electron gas in a perpendicular magnetic field $\underline{B} = (0,0,B)$ and a parallel electric field $\underline{E} = (E,0,0)$. The geometry is assumed to be planar and rectangular; the electrons are confined to an area $Ar = L_x L_y$. The Hamiltonian of the system is

$$H = \frac{1}{2m}\left(\underline{p}+\frac{e}{c}\underline{A}\right)^2+e\underline{E}\underline{r} \tag{2.1}$$

Imposing Landau gauge, $\underline{A} = (0,B_x,0)$, $p_y = -X/R^2$ is a constant of motion. Under the normalization condition

$$\int_{-\infty}^{\infty} |\phi|^2 dx=1 \tag{2.2}$$

suitable for $L_x \to \infty$, the eigenfunctions of H are the Landau functions

$$\phi_\alpha(x,y) = e^{iky}\phi_\nu(x-X) \tag{2.3}$$

where $\alpha = (\nu,k)$ with $\nu = 0,1,2,\ldots$; $k = p_y/h$, and $\phi_\nu(x - X)$ denotes oscillator wave functions localized around $x = X$. The eigenvalues are

$$\begin{aligned}\varepsilon_\alpha &= \hbar\omega_c(\nu+\tfrac{1}{2})+eEX \\ &= \varepsilon_\nu+eEX\end{aligned} \tag{2.4}$$

($\omega_c = eB/mc$). Imposing boundary conditions in y-direction

$$\phi_\alpha(x,y+Ly) = \phi_\alpha(x,y) \tag{2.5}$$

the possible values of k are given by

$$k = \frac{2\pi}{L_y}\mu , \quad \mu=0,\pm 1,\pm 2,\ldots \tag{2.6}$$

Counting only states with $0 \leq X \leq L_x$ the density of states for any fixed ν is

$$D = \frac{1}{Ar}\sum_k = \frac{1}{2\pi R^2} \tag{2.7}$$

(for $L_y \to \infty$). The density of states on the energy scale consists of equidistant δ-peaks at each ε_ν. Scattering by impurities leads to level broadening (bands).

3. On Laughlin's Gauge Transformation Approach

In the following summary of Laughlin's approach /3/ I will consider a free electron system confined to a planar rectangular area instead of the less conventional cylindrical belt geometry imposed by him.

The starting point is the formula

$$J_y = -c\,\frac{\delta U}{\delta A_y} \tag{3.1}$$

for the current density, where U is the total energy density. Laughlin replaces the derivative by the ratio $\Delta U/\Delta A_y$ where ΔA_y is the gauge quantum corresponding to the quantization of p_y (cf. (2.6)),

$$\Delta A_y=-B\Delta X=BR^2\Delta k=BR^2 2\pi/L_y=\frac{hc}{e}\frac{1}{L_y} = \phi_o/L_y \tag{3.2}$$

and ΔU is the associated change of U caused by the transfer of n electrons from edge (1) at $x = 0$ to edge (2) at $x = L_x$ ("pumping"),

$$\Delta U = \frac{en}{Ar}EL_x \tag{3.3}$$

Thus,

$$J_y \approx -c\frac{\Delta U}{\Delta A_y} = n\,\frac{e^2}{h}E \tag{3.4}$$

which establishes (1.2).[1)]

---

1) A somewhat different derivation was given by Takemori /4/. Deducing the electric field from a time dependent vector potential, $E = -\dot{A}_x/c$,

$$J_y = \frac{1}{L_x}\frac{dQ}{dt} \approx \frac{1}{L_x}\frac{\Delta Q}{\Delta A_x}\frac{\Delta A_x}{\Delta t}$$

Putting $\Delta A_x = \phi_o/L_x$ and for the corresponding charge transfer $\Delta Q = -en$ we obtain (3.4). On the other hand, $\Delta Q = -e\,(Ar\,n_e)(\Delta X/L_x) = -en_e L_y \Delta X$ and, by gauge invariance, $\Delta A_x = -\Delta A_y = B\Delta X$ (this is equivalent to the macroscopic relation $E = -(1/c)(\underline{v} \times \underline{B})_x$, leading again to (3.8).

In order to see the validity of this replacement let me calculate the derivative explicitely. The total energy density is

$$U = \frac{1}{Ar} \sum_{\alpha} f_\alpha \varepsilon_\alpha \qquad (3.5)$$

where f_α is the Fermi distribution function, normalized as usual,

$$\frac{1}{Ar} \sum_{\alpha} f_\alpha = n_e \qquad (3.6)$$

Under the gauge transformation $\underline{A} \to \underline{A}' = \underline{A} + \delta\underline{A}$, $\delta\underline{A} = -(0, B\delta X, 0)$, ε_α changes by

$$\delta\varepsilon_\alpha = eE\delta X \qquad (3.7)$$

whereas f_α is invariant. Therefore

$$J_y = c\,\frac{eE}{B}\,n_e \qquad (3.8)$$

This confirms (1.3). Since ε_α is a linear function of X, this well established classical result is not affected by the quantization of p_y.

The consistency of the estimated result (3.4) with the exact one (3.8) requires the n level occupation condition (1.1) to be satisfied (i.e. the first n-1 Landau levels have to be completely full and the others have to be empty).

The calculation of the Hall conductivity by gauge transformation, based on (3.1), is, of course, completely equivalent to the traditional procedure of summing up the contributions of the individual states to the current,

$$J_y = \sum_{\alpha} f_\alpha j_{y\alpha} \qquad (3.9)$$

$$j_{y\alpha} = \frac{-e}{Ar} \langle\alpha|v_y|\alpha\rangle = \frac{-e}{Ar} \langle\alpha|\frac{\partial H}{\partial p_y}|\alpha\rangle$$
$$= \frac{-e/\hbar}{Ar}\frac{\partial\varepsilon_\alpha}{\partial k} = \frac{(e/\hbar)R^2}{Ar}\frac{\partial\varepsilon_\alpha}{\partial X} \qquad (3.10)$$

For free electrons, the gauge transformation approach leads - like any other correct argument - to the classical formula (3.8). Laughlin confirms this for the special case characterised by (1.1). The statement that his argument holds exactly true for disordered systems with Fermi energy being anywhere in the range of localized states, as well, is, so far only a hypothesis. In any event, Laughlin's formal derivation of (3.4) is, in such a case, obviously inconclusive; it does not give any indication that the gauge transformation approach is particularly fruitful if localization takes place.

4. On Halperin's Finite System Approach

In reviewing Halperin's treatment /5/ I consider again a free electron gas confined to a rectangular area instead of the angular geometry which he prefers.

The essential requirements are the boundary conditions for a finite system

$$\phi_\alpha(0,y) = \phi_\alpha(L_x,y) = 0 \tag{4.1}$$

The eigenfunctions of H satisfying these conditions are the so-called Weber functions /6/. The energy spectrum (which cannot be given analytically) is shown schematically in Fig. 2 (for zero electric field).

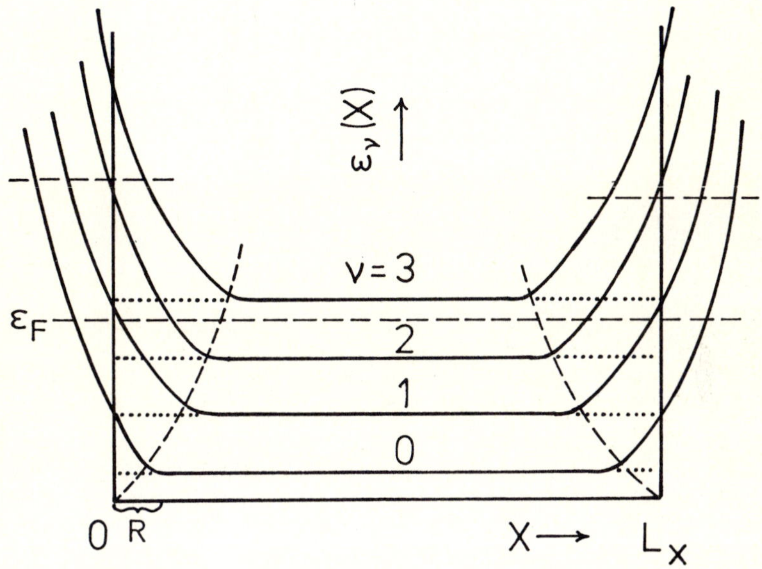

Fig. 2. The energy spectrum of a free electron confined to an infinite strip of width L_x in a strong magnetic field (schematically).

In the interior

$$\varepsilon_\nu(X) \simeq \hbar\omega_c(\nu+\tfrac{1}{2}) + \frac{\hbar\omega_c}{\sqrt{2\pi}}(\tfrac{X}{R})^{2\nu+1}\, e^{-X^2/2R^2} \tag{4.2}$$

Near the boundary (1)

$$\varepsilon_\nu(X) \simeq \hbar\omega_c(2\nu+\tfrac{3}{2}) - \frac{\sqrt{2}}{\pi}h\omega_c\frac{\Gamma(\nu+\frac{1}{2})}{\nu!}(2\nu+1)\frac{X}{R} \tag{4.3}$$

and for $X \simeq -R$

$$\varepsilon_\nu(X) \simeq \hbar\omega_c\,\frac{X^2}{4R^2} + h\omega_c\left[\frac{3\pi}{4}\,(\nu+\tfrac{3}{4})\frac{X}{R}\right]^{2/3} \tag{4.4}$$

/7/. The degeneracy of the energy levels ε_ν is removed. Thus, according to (3.10), a finite current density exists at the boundaries (1) and (2), even in the absence of an external electric field. As pointed out by Teller /8/, this "surface" current is associated with the Landau diamagnetic moment. In a thermally uniform system the integrated current density vanishes. Assuming, however, different values of the Fermi energy at the boundaries (1) and (2), a finite net current results.

If

$$\varepsilon_F = \begin{cases} \varepsilon_F^{(1)} \text{ for } -a<x<b \\ \\ \varepsilon_F^{(2)} \text{ for } L_x-b<x<L_x+a \end{cases} \tag{4.5}$$

where a and b $\sim$ R are suitably chosen distances, the current at zero temperature is

$$J_y = J_y^{(1)} + J_y^{(2)} \tag{4.6}$$

$$J_y^{(1)} = \sum_{\nu}^{n^{(1)}} \frac{L_y}{2\pi R^2} \int_{X_n(\varepsilon_F^{(1)})}^{L_x/2} \frac{e/\hbar}{Ar} R^2 \frac{\partial \varepsilon_\alpha}{\partial X} dX = \sum_{\nu}^{n^{(1)}} \frac{e}{h} \frac{1}{L_x} (\varepsilon_\nu - \varepsilon_F^{(1)}) \tag{4.7}$$

$$J_y^{(2)} = \sum_{\nu}^{n^{(2)}} \frac{e}{h} \frac{1}{L_x} (\varepsilon_F^{(2)} - \varepsilon_\nu) \tag{4.8}$$

Assuming now that, in the interior, both Fermi energies lie between ε_{n-1} and ε_n, i.e. $n^{(1)} = n^{(2)} = n - 1$,

$$J_y = \sum_{\nu}^{n-1} \frac{e}{h} \frac{1}{L_x} (\varepsilon_F^{(2)} - \varepsilon_F^{(1)}) = n \frac{e^2}{h} E \tag{4.9}$$

(since $\varepsilon_F^{(2)} - \varepsilon_F^{(1)} = eEL_x$). Thus, again, (1.2) is established under condition (1.1).

In order to show that for arbitrary values of n_e this derivation as well leads to the classical formula (1.3), we consider the local thermal equilibrium described by the stationary distribution function

$$f_\alpha^L = f(\varepsilon_\alpha - \varepsilon_F(X)) \tag{4.10}$$

with a slowly varying Fermi energy

$$\varepsilon_F(x) = \varepsilon_F + \frac{d\varepsilon_F}{dx} x = \varepsilon_F + eEx \tag{4.11}$$

/9/. With reference to (3.9) and (3.10)

$$J_y = \frac{ec}{B}\{\frac{1}{Ar} \sum_\alpha (-\frac{\partial f}{\partial \varepsilon_\alpha}) \frac{\partial \varepsilon_\alpha}{\partial X} X\} E \tag{4.12}$$

The expression in the bracket is just n_e. Thus we are lead to (1.3)

Our comments on Halperin's approach are similar to those on Laughlin's (sec. 3); we are given no arguments indicating the importance of considering a finite system to explain the quantized Hall conductivity. Nonetheless, the effect of localization on the diamagnetic surface currents seems to be an interesting question which, however, to the best of my knowledge, has not yet been discussed in detail.

5. On the Bound State Argument by Aoki and Ando

A very interesting observation made by these authors /10/ refers to infinite systems. I recall their argument in a slightly different formulation.

Kubo, Hasegawa and Hashitzuma /11/ derived the following general formula for the Hall conductivity

$$\sigma_{xy} = -\frac{en_e c}{B} + \Delta\sigma_{xy} \tag{5.1}$$

$$\Delta\sigma_{xy} = \frac{e^2}{Ar}\frac{1}{i\hbar}\int_0^\infty dt \langle [Y,\dot{X}(t)] \rangle \tag{5.2}$$

X and Y are the center coordinates of the cyclotron orbit and obey the commutation rule

$$[X,Y] = iR^2 \tag{5.3}$$

Both the time development of X and the equilibrium average (denoted by <...>) have to be calculated with the total Hamiltonian including the disorder (impurity) potential. Using the eigen states $|a\rangle$ of the total single electron Hamiltonian

$$\Delta\sigma_{xy} = \frac{e^2}{Ar}\frac{1}{i\hbar}\int_0^\infty dt \sum_a f_a \langle a|[Y,\dot{X}(t)]|a\rangle \tag{5.4}$$

Assuming now that <u>all</u> occupied states are bound states (i.e. the corresponding expectation values of X and Y are bounded) the time integral can be performed to give

$$\begin{aligned}\Delta\sigma_{xy} &= -\frac{e^2}{Ar}\frac{1}{i\hbar}\sum_a f_a \langle a|[Y,X]|a\rangle \\ &= \frac{e^2R^2}{Ar\hbar}\sum_a f_a = \frac{en_e c}{B}\end{aligned} \tag{5.5}$$

Thus, σ_{xy} vanishes. Let us assume now that in the middle of the gap between the $\nu = 0$ and the $\nu = 1$ Landau band $\Delta\sigma_{xy}$ vanishes <u>exactly</u> and, consequently, $\sigma_{xy} = -e^2/h$. If there is now a "mobility edge" in the $\nu = 0$ band such that <u>all</u> states between this and the upper band edge are bound states, then remains constant (e^2/h) as long as the Fermi energy lies in the range of these bound states. This can be proved by repeating the above argument and subtracting the contribution of the bound states one by one. Similarly, if there is a "mobility edge" in the $\nu = 1$ band such that <u>all</u> states between the lower band edge and this "mobility edge" are bound states then, again σ_{xy} remains constant (e^2/h) as long as the Fermi energy is within these states. And so on for higher values of ν. Since σ_{xy} is a single valued function of ε_F there must be a transition within each Landau band connecting adjacent plateaus of σ_{xy}.

The essential assumption on which this argument is based is that

$\Delta\sigma_{xy}$ vanishes exactly in the middle of the gap between adjacent Landau bands. There are several indications /16,17/ and a formal proof by Thouless /12/ for weak impurity potential, supporting this assumption. Nonetheless, it has still to be understood that localization in a disordered system is physically equivalent to bound states in the usual sense. This problem is related to the indispensible task of showing (by transport theory) that the plateaus in σ_{xy} occur simultaneously with vanishing conductivity σ_{xx}.

6. On Percolation Arguments

The effect of localization in the framework of a simple classical percolation model has been investigated by Iordansky /13/. He found that, at the percolation threshold (U_c), there are unbounded trajectories giving rise to a net (Hall) current perpendicular to the applied electric field E. Estimation of σ_{xy}, in the limit of vanishing diffusion and sufficiently weak electric field, yields

$$\sigma_{xy} = -\frac{en_e c}{B}\,(f(U_c)/N_e) \tag{6.1}$$

where $f(U_c)/N_e$ is the relative average occupation number at U_c. Iordansky's observation does not explain the quantized Hall conductivity. But it indicates the possibility of a finite Hall current in spite of localization by some "substrate" potential.

In the quantum limit (n = 1) and arbitrary values of ε_F within the $\nu = 0$ Landau band Ono /14/ obtains for an infinite system at sufficiently low temperatures

$$\sigma_{xy} = -\frac{e^2}{h}\,f(U_c) \tag{6.2}$$

$$= -\frac{e^2}{h}\Theta(\varepsilon_F - U_c) \text{ at } T = 0.$$

The essential assumption leading to (6.2) is that, in the thermodynamic limit, the fluctuating substrate potential at the boundaries (1) and (2) can (exactly) be replaced by the unique percolation energy U_c at which an equipotential line percolates from infinity to infinity. Although intuitive, an exact proof of this assumption is indispensible.

According to (6.2) the transition from the first plateau (zero level) to the second does not necessarily happen at the center of the $\nu = 0$ Landau band.

The importance of the classical percolation aspect in two dimensions has also been pointed out by Tsui and Allen /15/ in an intuitive consideration. It has, however, not yet been demonstrated how the existence of a single percolation threshold can give rise to the multistep structure in σ_{xy}.

7. On Some Other Contributions

(i) Very interesting numerical calculations of σ_{xy} have been performed by Ando /16/. Based on the numerical diagonalization of one and two Landau band Hamiltonians (with finite system boundary conditions) he achieved excellent agreement with experimental observation: the calculated Hall conductivity increases, in the middle of the bands, quite steeply between adjacent plateau values.

A similar investigation of the transverse conductivity appears to be highly desirable.

(ii) Prange /17/ has pointed out that, under the occupation condition (1.1), a bound state due to a δ-potential does not affect the Hall conductivity. This result seems to be covered by an observation of Thouless /12/ according to which σ_{xy} is invariant against a weak local variation of the substrate potential (provided the single electron Green function decays, at the Fermi energy, exponentially for large distances - which seems to be case in strong magnetic fields).

(iii) Thouless et al. /18/ studied the effect of a periodic lattice potential, Streda /19/ that of the density of states on σ_{xy}.

(iv) Imry /20/ lays particular emphasis on the periodicity of the total energy as a function of the flux appearing in the belt geometry considered by Laughlin /3/. He discusses the relation of the quantized Hall conductivity to other macroscopic quantum effects. It should be kept in mind, however, that the cylindrical belt geometry is essentially different from the one realized in experiment.

(v) The importance of strong electron correlation effects in the quantization of the Hall conductivity has been pointed out by Tsui et al. /22/ and by Fukuyama and Platzman /22/. I presume Professor Fukuyama will be reporting on these interesting investigations in his talk.

8. Some Questions - Instead of a Resumé

(i) Is the pumping of electrons from one edge of the sample to the other (parallel to the applied electric field, through localized regions) by gauge transformation real or a sort of linguistic camouflage for something we have not yet understood? Or: Is it possible to learn, only by studying gauge transformation properties, that, with respect to the Hall effect, a real 2D system in a strong magnetic field behaves, when the Fermi energy lies in a range of localized states, exactly in the same way as a 2D free electron gas with Fermi energy situated in one of the gaps between adjacent energy levels?

(ii) Is it essential to consider a finite system with surface currents instead of an infinite system (thermodynamic limit), or is it a matter of taste such as in the theory of conduction electron diamagnetism? If the answer is positive, the promising bound state arguments of Aoki and Ando /10/ and some related arguments of Thouless /12/ have possibly to be reconsidered.

(iii) What are the necessary modifications of the standard quantum transport theory, which has been applied to 2D systems with much success especially by Ando and Uemura, in order to explain the plateaus of σ_{xy} and the disappearance of σ_{xx} in certain ranges of the carrier concentration, in a unique way?[1)]

I thank Patrik Fazekas for many an illuminating discussion.

---

1) Any theory yielding the relation

$$\sigma_{xy} = -\frac{en_e c}{B} + \frac{1}{\omega_c \tau}\sigma_{xx}$$

is obviously misleading since, explaining $\sigma_{xx} = 0$ phenomenologically by $\tau = 0$ or $\tau = \infty$ results in $\sigma_{xy} = 0$ or $\sigma_{xy} = -en_e c/B$, respectively, in contradiction to the observations.

References

1. K. von Klitzing, G. Dorda and M. Pepper, Phys. Rev. Lett. 45, 494 (1980); S. Kawaji and J. Wakabayashi in: S. Chikazumi and N. Miura (Eds.), Physics in High Magnetic Fields, Springer-Verlag, Berlin etc., 1981, p. 284; D.C. Tsui and A.C. Gossard, Appl. Phys. Lett. 38, 550 (1981).
2. T. Ando, Y. Matsumoto and Y. Uemura, J. Phys. Soc. Japan 39, 279 (1975).
3. R.B. Laughlin, Phys. Rev. B23, 5632 (1981); Surf. Sci. 113, 22 (1982).
4. N. Takemori, quoted by T. Ando, ref. 16, as private communication.
5. B.I. Halperin, Phys. Rev. B25, 2185 (1982).
6. cf. e.g. M. Abramowitz and I.A. Stegun, Handbook of Mathematical Functions, Dover, New York 1970.
7. M. Heuser, Diplomarbeit, Köln 1972.
8. E. Teller, Z. Physik 67, 311 (1931).
9. M. Heuser and J. Hajdu, Z. Physik 270, 289 (1974).
10. H. Aoki and T. Ando, Solid State Comm. 38, 1079 (1981).
11. R. Kubo, H. Hasegawa and N. Hashitsume, J. Phys. Soc. Japan 14, 56 (1959).
12. D.J. Touless, J. Phys. C 14, 3475 (1981).
13. S.V. Iordansky, Solid State Comm. 43, 1 (1982).
14. Y. Ono in: Y. Nagaoka and H. Fukuyama (Eds.), Anderson Localization, Springer-Verlag, Berlin etc. 1982, p. 207.
15. D.C. Tsui and S.J. Allen, Jr., Phys. Rev. B 24, 4082 (1981).
16. T. Ando, op. cit. ref. 14, p. 176.
17. R.E. Prange, Phys. Rev. B23, 4802 (1981); R.E. Prange and R. Joynt, preprint, 1981.
18. D.J. Thouless, M. Kohmoto, M.P. Nightingale and M. den Nijs, Phys. Rev. Lett. 49, 405 (1982).
19. P. Streda, J. Phys. C 15, L717 (1982).
20. Y. Imry, J. Phys. C 15, L221 (1982).
21. D.C. Tsui, H.C. Störmer and A.C. Gossard, preprint, 1982.
22. H. Fukuyama and P.M. Platzman, Phys. Rev. B 25, 2934 (1982).

QUANTUM TRANSVERSE AND HALL RESISTIVITIES IN SILICON MOS INVERSION LAYERS AT TEMPERATURES DOWN TO 12 mK

J. Kinoshita, K. Inagaki, C. Yamanouchi and K. Yoshihiro
Electrotechnical Laboratory, Sakuramura, Niiharigun, Ibaraki 305 Japan

and

J. Moriyama and S. Kawaji
Department of Physics, Gakushuin University, Mejiro, Toshimaku, Tokyo 171 Japan

Hall and transverse resistivity measurements for Si-nMOS (100) inversion layers have been made at temperatures between 4.2 K and 12 mK and a magnetic field of 120 kOe. The result indicates that the effect of short range scatterers is still dominant at the lowest temperatures. It has been found from a detailed study of transverse resistivity around the mobility gap region that five quantized Hall resistivity plateaus, corresponding to h/ie^2 for $i = 4$, 6, 8, 12 and 16, are potentially available for the quantum standard resistor under the magnetic field of 120 kOe at ^{3}He - ^{4}He temperatures.

Introduction

The purpose of the present experiment is to study the problems of the electron localization in two dimensional systems under strong magnetic fields and at very low temperatures. The experiment is also aimed at finding how many quantized Hall resistivity plateaus, corresponding to h/ie^2 where i is the integer, are practically available for precision quantum standard resistors at the lowest temperatures.

Theories [1] have shown that the Hall conductivity $-\sigma_{xy}(i)$ in high magnetic fields is exactly ie^2/h whenever the Fermi level lies in the mobility gap region between i th and (i+1) th Landau bands. However they do not suggest the accuracy of the relation, which would break down under real experimental conditions as the precision of the measurement goes up. In the experiment, the accuracy of the absolute value of the Hall conductivity is limited finally by the uncertainty of the unit of 1 Ω; the uncertainty is about 0.5 ppm for the unit of 1 Ω maintained at the ETL, for example. On the other hand, we can examine to what accuracy the relation $\sigma_{xy}(i)/i = \sigma_{xy}(j)/j$ holds regardless of the uncertainty associated with the unit of 1 Ω, by use of two or more well quantized Hall resistance values measured for the same sample. It is thus desirable to find the Hall resistance plateaus of different quantum numbers as many as possible.

The transverse resistivity measurement is quite useful for the present purpose. In contrast to the Hall resistivity, the transverse resistivity is sensitive to low-lying excitations and/or localization of electrons. It is known experimentally [2] that completeness of the quantization is easily examined by measuring the minimum value of the transverse resistivity of the very sample for the Hall resistance measurement, in the gate voltage region aronud the plateau. Thus the completeness of the quantization can be checked even beyond the precision and resolution limits for the present Hall resistance measurements. Under the condition of very good quantization, the most part of the sample would be kept free from the electron heating. This would allow the precision measurement being made at relatively high channel current level, even at temperatures attained with a dilution refrigerator. Here we report a preliminally result.

Experimental Outline

The measurements were made on Si-nMOS devices fabricated on (100) surface.

The maximum mobility of electrons was 14000 $cm^2/V \cdot s$ at 1.4 K under low magnetic fields. The samples are of strip geometry with 600 μm in length, 100 μm in width and 200 μm in distance between the potential probes along the channel current flow. The magnetic field of 120 kOe was applied normal to the inversion layer. The samples were attached by GE7031-like varnish on a copper block, which was screwed into the mixing chamber of a dilution refrigerator: the thermal contact was improved by tightening the screw with another small screw. All the electric leads into the sample were thermally grounded to the mixing chamber. The measuring ac currents of 14 to 23 Hz down to 1 nA were used with a lock-in amplifier to obtain the overall characteristics of ρ_{xx} and ρ_{xy} vs. V_g, and the dc currents up to 10 μA to find various Hall plateaus suitable to the precision measurement. In the lowest temperature region, the transverse resistivity ρ_{xx} and the Hall resistivity ρ_{xy} were independent of the channel current $I_{SD} < 10$ nA at the gate voltages $V_g > 8$ V.

Experimental Results and Discussion

Figure 1 shows ρ_{xx} and ρ_{xy} plotted against V_g, measured at $H = 120$ kOe and $T = 12$ mK and 20 mK respectively. The regions of V_g where $\rho_{xx} \simeq 0$ increase as the temperature is lowered down to 12 mK. The separation between 4 sub-bands of the $N = 1$ Landau level looks almost complete. However, even at the lowest temperatures, the non-zero ρ_{xx} regions are still wide and correspondingly the ρ_{xy} vs. V_g curve looks as a train of plateaus and slopes, in contrast to the result reported on GaAs-$Al_{1-x}Ga_xAs$ hetero structures [3].

For $V_g < 8$ V, a curious irregularly peaked structure is observed which is sample dependent and almost reproducible within the same sample. The amplitude of the structure increases with lowering the temperature, channel current and gate voltage. The amplitude amounts to 30 % of ρ_{xx} at $T \simeq 11$ mK, $I_{SD} \simeq 1$ nA rms, $V_g \simeq 5$ V. These behaviours are similar to the observations at 4.2 K [4,5] and will be primarily resulted from the non-ohmic properties around the electrodes [5].

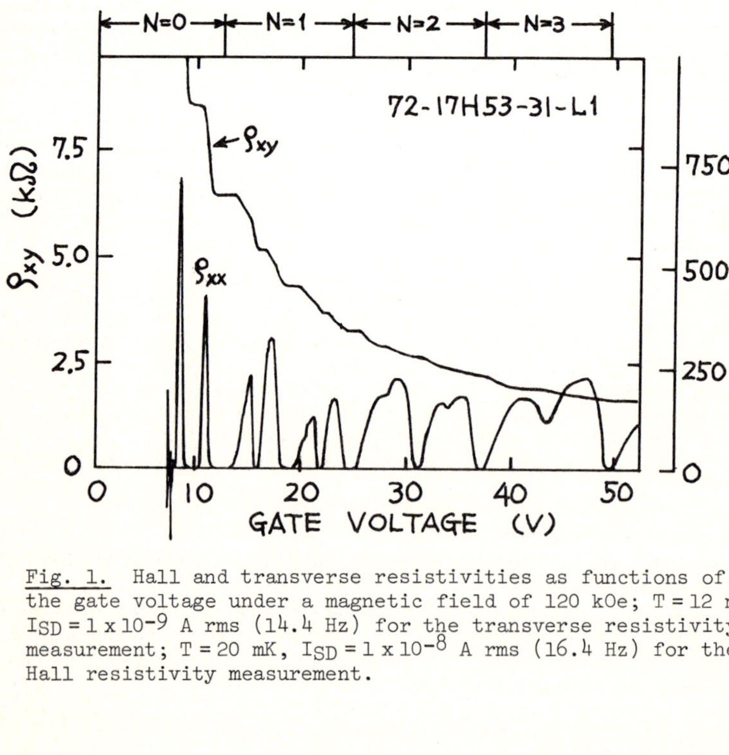

Fig. 1. Hall and transverse resistivities as functions of the gate voltage under a magnetic field of 120 kOe; $T = 12$ mK, $I_{SD} = 1 \times 10^{-9}$ A rms (14.4 Hz) for the transverse resistivity measurement; $T = 20$ mK, $I_{SD} = 1 \times 10^{-8}$ A rms (16.4 Hz) for the Hall resistivity measurement.

Fig. 2. The number of immobile electrons associated **with a** Landau sub-band. The unit in the ordinate is eH/hc, 2.9×10^{11} cm^{-2} for H = 120 kOe. As a mobility gap region we temporalily choose the region of V_g's where $\sigma_{xx} < 5 \times 10^{-7}\ \Omega^{-1}$, the resolution limit of ac ρ_{xx} measurement at $V_g \simeq 50$ V.

△: spin up, valley +;
▲: spin up, valley -;
▽: spin down, valley +;
▼: spin down, valley -;
⅄: spin up, valley +/-;
Y: spin down, valley +/-;
◇: spin up/down, valley +/-;
○: numerical calculation by Ando [6];
—— - —— - —: empirical relation 1/(2N + 1) [7].

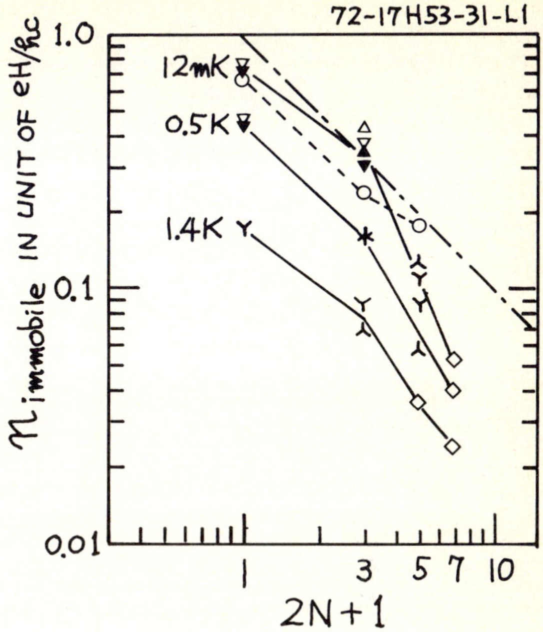

The range of V_g where $\rho_{xx} \simeq 0$ is considered to be a measure of the number of electrons in strong localization. In Fig. 2 the number of localized electrons in each sub-band is plotted against 2N + 1, where N is the Landau level index. The parameters are the temperature and the spin and valley indices. The data for N = 0 and N = 1 at the lowest temperatures are in reasonable agreement with the result of numerical calculation by Ando [6], in which are assumed the short-range delta-potential scatterers and no interactions between different Landau bands. The data also fit the empirical relation obtained by Kawaji and Wakabayashi [7], although the definition of the immobile electron is slightly different. This result suggests that the effect of short range potential is still dominant in the lowest temperature region for the case of electrons in the inversion layer of Si-MOSFET's. It is also shown in Fig. 2 that the number of localized electrons decreases rapidly with increasing N. This trend is in consistent with the theoretical result [6] that the energy range of extended states becomes wider for higher Landau levels in the case of short range scatterers; the steeper slope above N = 2 might be an indication of electron heating.

Figure 3 shows the result of detailed measurement of ρ_{xx} in the mobility gap region. The temperature indicated in the figure is that of the copper mounting block. Since the channel current of 2.5 ∿ 10 μA is applied to get the sufficient resolution of measurements, one would expect considerable heating in both lattice and electrons in the sample. In fact, at V_g's where ρ_{xx} is non-zero, the recorder trace of ρ_{xx} looks the same as that at 1.4 K, indicating the electron heating. However, at V_g's where $\rho_{xx} \simeq 0$, no appreciable electron heating effect, at least in the central area of the channel far from the source/drain electrodes, has been observed.

The minima of ρ_{xx}'s lower than 10^{-2} Ω are realized for i = 4, 6, 8 and 12 at I_{SD} = 10 μA and for i = 16 at I_{SD} = 2.5 μA. This fact indicates that these Hall resistance plateaus are available for the precision measurement of ρ_{xy} and/or as quantum standard resistors of five different values down to 1.6 kΩ on a single device by use of the dilution refrigerator. Some improvement will be needed to cool the device more effectively if the quantum resistor of 1.6 kΩ (i=16) should be used in 10 μA level. For the present the quantum resistor of 1.3 kΩ (i=20) will be hopeless.

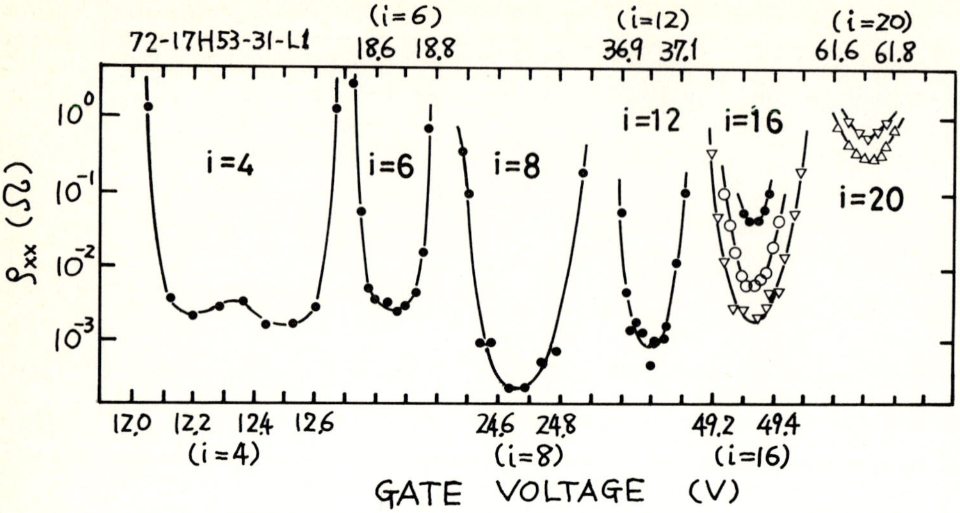

Fig. 3. Transverse resistivity ρ_{xx}'s as functions of the gate voltage in various mobility gap regions between i th and (i+1) th Landau sub-bands, the channel current value being the parameter. •: 9.9 μA; ○: 4.95 μA; ▽: 2.45 μA; △: 1.18 μA. T = ~20 mK.

Acknowledgements

The authors wish to express their sincere thanks to Dr. A. Yagi, Sony Corporation for providing them with samples. They wish to acknowledge the stimulating discussion with Professor T. Ando. They would like to thank Drs. K. Shida, T. Igarashi and S. Iijima for helpful suggestions and encouragements.

References

1. T. Ando, Y. Matsumoto and Y. Uemura: J. Phys. Soc. Jpn. 39 (1975) 279; R. B. Laughlin: Phys. Rev. B23 (1981) 5632.
2. K. Yoshihiro, J. Kinoshita, K. Inagaki, C. Yamanouchi, J. Moriyama and S. Kawaji: in Proc. 16th Int. Conf. Physics of Semiconductors, Montpellir, 1982, to be published.
3. M. A. Paalanen, D. C. Tsui and A. C. Gossard: Phys. Rev. B (1982), to be pubished.
4. R. J. Tidey, R. A. Stradling and M. Pepper: J. Phys. C7 (1974) L353.
5. T. Cole, A. A. Lakhani and P. J. Stiles: Surf. Sci. 58 (1976) 56.
6. T. Ando: Surf. Sci. 113 (1982) 182 and a preprint.
7. S. Kawaji and J. Wakabayashi: Surf. Sci. 58 (1976) 238.

Possible Application of the Quantized Hall Resistance to the Realization of the Electrical Units

E. Braun, V. Kose, F. Melchert and P. Warnecke
Physikalisch-Technische Bundesanstalt
D-3300 Braunschweig, F.R.G.

Abstract

Precision measurements of the quantized Hall resistance in conjunction with those of other fundamental constants allow other, independent realizations of the SI-unit ohm in addition to the classical one based on a calculable capacitance.

Introduction

A new, rather unexpected contribution of special semiconductor devices in high magnetic fields to the basic metrological problem of the realization of the units has become possible by precision measurements of the quantized Hall resistance (QHR) [1]. Since on the one hand a 2-dimensional electron gas with high mobility can be realized in Si-MOSFETs and GaAs-heterostructures at helium temperatures and on the other hand magnetic fields up to about of 15 Tesla can now be generated with superconducting magnets in the persistent mode, high precision measurements of the QHR with uncertainties of 10^{-7} or even less are possible [2]. It is shown that such measurements in combination with accurate determinations of other fundamental constants enable new ways for the realization of the unit of resistance ohm.

Definitions and Realizations in the SI-system

In the Systeme International d' Unites (SI), the units are either defined as base units (e.g. m, s, kg, A) or as derived units (e.g. Ω etc.). In any case the definitions describe ideal conditions with zero uncertainty. For the realization of the units, however, one has to apply appropriate physical laws in order to refer the units to the base ones with the smallest uncertainties possible. These uncertainties will always be finite, but the methods of realization may change with the progress of science.

Classical Realization of the Ohm

The classical realization of the ohm is based on a theorem given by Thompson and Lampard [3], who showed that the capacitance of a special capacitor formed of parallel cylindrical electrodes is given by

$$C_{TL} = \varepsilon_o \frac{\ln 2}{\pi} \ell \tag{1}$$

where ℓ is the length of the cylinders. ε_o is the permittivity of vacuum given by Maxwell's relation

$$\varepsilon_o = 1/c^2\mu_o \tag{2}$$

(c speed of light in vacuum, μ_o permeability of vacuum).
The reactance of this capacitor can be compared - in several steps and with considerable experimental effort - with the resistance of the maintained 1-Ω-standard. The relative uncertainty of 1 Ω derived by this method is of the order of 10^{-7} [4,5].

Realization of the Ohm via Fundamental Constants

Fundamental constants are connected by physical laws, which can be expressed by mathematical equations. The fine-structure constant e.g. is given by the following equations:

$$\alpha^{-1} = \frac{2}{\mu_o c} \cdot h/e^2 \tag{3}$$

$$\alpha^{-2} = \frac{c}{4R_\infty} \frac{\mu'_p}{\mu_B} \cdot \frac{2e/h}{\gamma'_p} \tag{4}$$

In these two equations μ_o is defined, the values of c, of the Rydberg constant R_∞ and of the proton magnetic moment in Bohr-magnetons μ'_p/μ_B are known with very low uncertainties*.

Generally any physical quantity X can be expressed as

$$x = \{x\}\,[x] = \{x\}_{lab}\,[x]_{lab} \tag{5}$$

The {}-brackets mean numerical values based on the SI unit given in []-brackets, whereas the index "lab" indicates a "laboratory system" with a higher reproducibility of its units than that of the SI units.

---

*The prime indicatess that the measurement is based on protons in a spherical sample of H_2O.

A laboratory may have standards with the values "1 Ω_{lab}" and "1 V_{lab}" - their values in SI-units are not known - and any measurement of a resistance R resp. a voltage U will then result in multiples of these "laboratory units":

$$R = \{R\}_{lab} \cdot \Omega_{lab} \quad \text{resp.} \quad U = \{U\}_{lab} \cdot V_{lab} \tag{6}$$

Suppose in this laboratory precision measurements of the QHR are made with the result

$$h/e^2 = \{h/e^2\}_{lab} \cdot \Omega_{lab} \tag{7}$$

of the Josephson-voltage with the result

$$2e/h = \{2e/h\}_{lab} \cdot V_{lab}^{-1}\, s^{-1} \tag{8}$$

and of the gyromagnetic ratio of the proton, yielding

$$\gamma'_p = \{\gamma'_p\}_{lab} \cdot \frac{1}{[\mu_o]} \cdot \frac{m}{s} \cdot \frac{\Omega_{lab}}{V_{lab}} \tag{9}$$

where $[\mu_o]$ means the SI-unit of the defined μ_o.

These experimental data are introduced into the above equations for α. Assuming the validity of Ohm's law for the "laboratory units" one obtains:

$$\alpha^{-1} = \frac{2}{\{\mu_o\}\{c\}} \{h/e^2\}_{lab} \frac{\Omega_{lab}}{\Omega} \tag{10}$$

$$\alpha^{-2} = \frac{\{c\}}{4\{R_\infty\}} \{\frac{\mu'_p}{\mu_B}\} \{2e/h\}_{lab} \{\gamma'_p\}^{-1}_{lab} \frac{\Omega}{\Omega_{lab}} \tag{11}$$

The two equations for α yield for the Ω [6]:

$$\Omega = \sqrt[3]{\frac{16\ \{R_\infty\}}{\{c\}^3\ \{\mu_o\}^2\ \{\mu'_p/\mu_B\}} \frac{\{\gamma'_p\}_{lab}\ \{h/e^2\}^2_{lab}}{\{2e/h\}_{lab}}}\ \Omega_{lab} \tag{12}$$

This equation shows, that the ratio between the arbitrary unit of resistance Ω_{lab} to the SI unit Ω can be derived from precision measurements of the three fundamental constants h/e^2, $2e/h$ and γ'_p

using arbitrary laboratory units Ω_{lab} and V_{lab}. The validity of Ohm's law and the stability during the measurement are the only requirements for the laboratory units. The uncertainty of this method can be estimated to be of the same order as that of the Thompson-Lampard method.

Another equation for α can be taken from ref. [7]:

$$\alpha^{-2} = \frac{c\ \mu'_p/\mu_B}{2R_\infty}\ \frac{m_p}{m_n}\ \frac{1}{\mu'_p/\mu_n}\ \frac{1}{h/m_n} \qquad (13)$$

(μ'_p/μ_n proton magnetic moment in nuclear magnetons, m_n/m_p ratio, neutron mass to proton mass, h/m_n Planck constant divided by the neutron mass).

In combination with equ.(3) an additional expression for Ω is obtained:

$$\Omega = \frac{2^{3/2}}{\{\mu_o\}\{c\}^{3/2}} \sqrt{\frac{\{R_\infty\}}{\{\mu'_p/\mu_B\}}\ \{\frac{m_n}{m_p}\}\ \{\frac{\mu'_p}{\mu_n}\}\ \{\frac{h}{m_n}\}\ \{h/e^2\}_{lab}}\ \Omega_{lab} \qquad (14)$$

Conclusions

In addition to the classical realization of the ohm derived from a calculable capacitance, precision measurements of combinations of fundamental constants including the QHR allow new ways for the realization of the ohm. If the different routes lead to the same value for the ohm within the experimental uncertainties, one can use this result as a strong indication that no correction terms exsist at this level of accuracy in the basic equation for the QHR

$$R_{H,n} = \frac{1}{n} \cdot h/e^2 \qquad (n = 1,\ 2\ \ 3,\ \ldots) \qquad (15)$$

We hope to be able to report soon new experimental data for the ohm derived by the different methods described above.

References

1. K. v. Klitzing, G. Dorda and M. Pepper, Phys. Rev. Lett. 45, 494 (1980)
2. L. Bliek et al., PTB-Mitt.93 (1983), to be published
3. A.M. Thompson and D.G. Lampard, Nature 177, 888 (1958)
4. A.M. Thompson, Metrologia 4, 1 (1968)
5. R.D. Cutkosky, IEEE Trans. Instr. Meas. IM-23, 305 (1974)
6. V. Kose, PTB-Mitt. 92, 249 (1982)
7. W. Wöger, PTB-Bericht E-12, 149 (1979)

TEMPERATURE DEPENDENCE OF THE QUANTUM HALL EFFECT IN $In_{0.53}Ga_{0.47}As$-InP HETEROJUNCTIONS

J.P. Hirtz*, Y. Guldner**, A. Briggs***, J.P. Vieren**, P. Voisin**, M. Voos** and M. Razeghi*

* Laboratoire central de recherches, THOMSON-CSF, 91401 Orsay (France)
** Groupe de Physique des Solides de l'E.N.S., Laboratoire associé au CNRS - 24, rue Lhomond, 75005 PARIS
*** C.R.T.B.T. and S.N.C.I., CNRS, 38042 Grenoble.

Abstract : We have studied the temperature dependence of the quantum Hall effect in modulation-doped InGaAs-InP heterojunctions from 4.2 K to 50 mK. The results are briefly discussed in terms of electron localisation.

The existence of a high-mobility, two-dimensional electron gas (2DEG) has been shown[1] recently to occur at the interface of modulation doped $In_{0.53}Ga_{0.47}As$-InP heterojunction. As previously reported in Si MOSFET[2] and modulation-doped GaAs-AlGaAs[3] structures, we have observed[4] the quantum Hall effect (QHE) in such InGaAs-InP heterojunctions, which manifests, when the magnetic field is swept, in the occurence of quantized plateaus in the Hall resistance ρ_{xy}, while the diagonal resistance ρ_{xx} tends to vanish. We study here the temperature dependence of the QHE from 4.2 K down to 50 mK, and we discuss our data in terms of electron localization.

The InGaAs-InP heterostructures grown by low-pressure metalorganic chemical vapor deposition[5] (MO-CVD) on (100) semi-insulating Fe-doped InP substrates. A sketch of such a structure is shown in Fig. 1 (a).

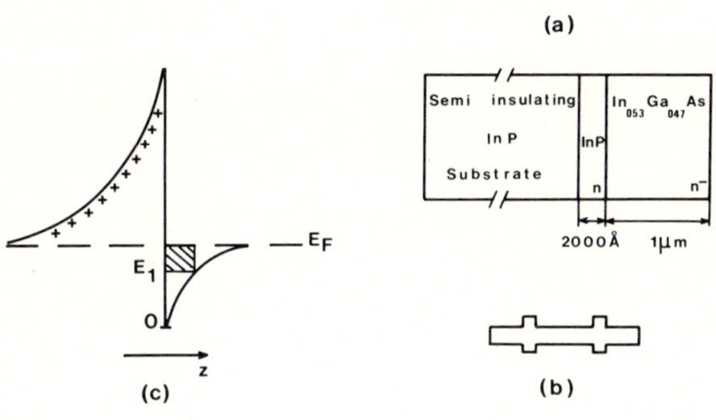

Fig.1 -(a) Sketch of the InGaAs-InP heterojunctions used here. (b) Geometry of the Hall bridge (c) Energy band diagram in a modulation - doped heterojunction, InP and InGaAs being respectively on the left and the right hand side of the interface which is located at $z = 0$. E_1 and E_F are the first electron subband and Fermi energies, respectively. The hatched area corresponds to the two-dimensional electron gas, and the crosses represent ionized donors.

The electron mobility was typically $33000 cm^2V^{-1}s^{-1}$ at 4.2 K, and the InP and InGaAs layers were both n-types with $N_D-N_A = 3x10^{16}$ and $1.5x10^{15}$ cm^{-3}, respectively. Standard Hall bridges (Fig.1(b)) were used to study ρ_{xx} and ρ_{xy}, ρ_{xx} being measured along the length of the sample, and ρ_{xy} perpendicular to it. The sample was cooled in a dilution refrigerator, and the magnetic field, perpendicular to the interface, was provided by a superconducting coil. As in modulation-doped GaAs-AlGaAs heterojunctions[6], electrons from donors in the wide band gap material (InP) are transferred to the narrow gap one (InGaAs), so that the Fermi level is constant throughout the structure under consideration here. As a consequence, a 2DEG is confined in the potentiel well due to the band bending near the interface resulting from the spatial separation of positive (i.e. ionized donors) and negative (i.e. electrons) charges, as schematized in fig.1 (c).

Fig. 2 and 3 show typical data obtained at 1.85 K and 55 mK for ρ_{xy} and ρ_{xx} as a function of the magnetic field B. Using standard procedures, we obtain from the periodicity in 1/B of the ρ_{xx} oscillations an electron density $n_s=4.5\times10^{11}$ cm^{-2}. The spin splitting observed for the n = 1 Landau level results from the enhancement of the effective g-factor due to exchange interactions[7], as previously observed for 2 DEG in Si MOS[8] structures and GaAs-AlGaAs heterojunctions[9]. The estimated g-factor is here ∿ 10, instead of ∿ 3 in bulk InGaAs[10]. The temperature dependence of the QHE is mainly characterized by an increased width of the ρ_{xy} plateaus and by a narrowing of the associated ρ_{xx} peaks when T is decreased. When T is varied from 4.2 K to 50 mK, the $\rho_{xy} = h/4e^2$ plateau width ranges respectively from 30% to 80% of its largest possible value obtained from the midpoints of the adjacent steps. We have also studied the conductivity σ_{xx} at different ρ_{xx} maxima as a function of T and , for quantum numbers n > 3, we have observed between 0.2 and 2 K the logarithmic dependence $\Delta\sigma_{xx} = 0.8\times10^{-5}\ \ell n\ T$ (mho), analogous to what has been obtained in Si MOSFET[11] in the weak localization regime and in GaAs-AlGaAs heterojunctions[12].

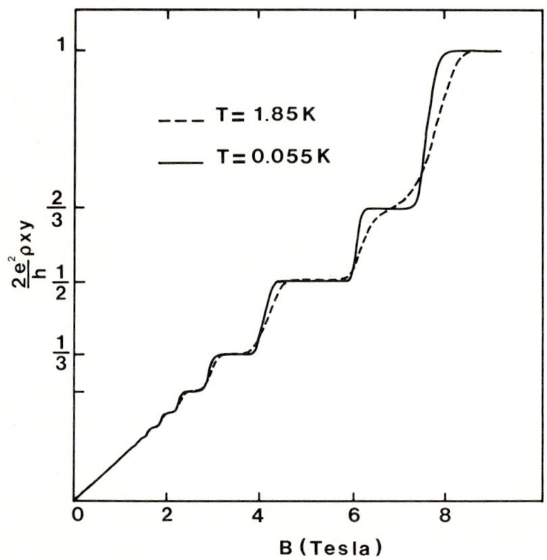

Fig. 2 - Hall resistance ρ_{xy} versus B at different temperatures. The quantized plateau correspond to $\rho_{xy} = h/ie^2$, where i is the number of filled Landau level of the 2 DEG.

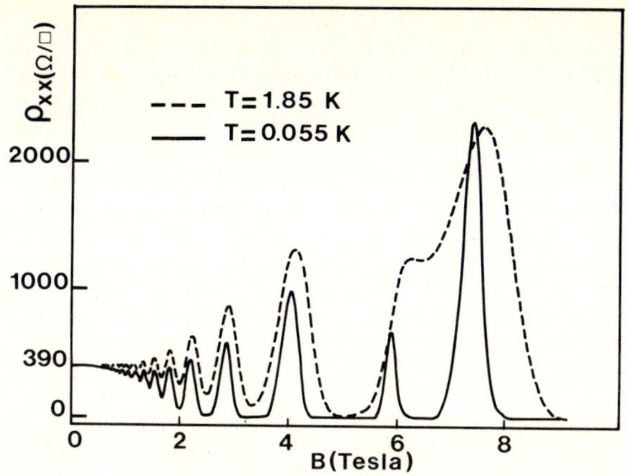

Fig. 3 - ρ_{xx} as a function of B as different temperatures.

The QHE implies that the Fermi level E_F is pinned between Landau Levels over finite ranges of B, instead of jumping from a Landau level to the following one when B is varied. Theoretical models have been proposed to account for the QHE, the pinning of E_F being due to shallow donors in the wide gap material[13] (AlGaAs or InP) or to localized states[14] in the tails of the broadened Landau levels. As noted in Ref.12, the model of Baraff and Tsui[13] is not expected to give a strong temperature dependence of the QHE, and would not explain more than a fraction of the plateau widths observed here. We interpret our plateau widths, as in GaAs-AlGaAs structures[12], as the rate of localized states, and we believe that our data support the idea that the QHE results from electron localization[14].

Finally, we have also studied the QHE at 4.2 and 1.5 K in modulation -doped MBE-grown GaAs-$Al_{0.25}Ga_{0.75}$ and MO-CVD-grown $In_{0.53}Ga_{0.47}As$-InP heterojunctions having comparable electron density ($4x10^{11}$ cm^{-2}) and mobility (50 000 $cm^2V^{-1}sec^{-1}$ at 4.2 K). The results are quite similar in both cases,leading us to believe that the potential fluctuations due to alloy disorder in InGaAs, where the 2 DEG is confined, do not help in localizing electrons.

To conclude, we think that additional experiments in differents systems will be helpful in obtaining interesting information leading to a better understanding of the quantum Hall effect which,in our opinion,is not yet fully explained.We wish also to emphasize that we believe that electron localization is at the present time the most likely interpretation of the origin of this effect

Acknowledgments

We are grateful to J.P. Duchemin and M.A. Poisson for their constant interes in this work. We tank P. Delescluse and Nuyen T. Linh for providing us w th the GaAs-AlGaAs heterojunction which was grown at THOMSON-CSF. We are also grateful to G. Toulouse and B. Souillard for fruitful discussions. The work at THOMSON-CSF was supported by the DGRST and DRET.

References

1 - Y. Guldner, J.P. Vieren, P. Voisin, M. Voos, M. Razeghi and M.A. Poisson, Appl. Phys. Lett. 40, 877 (1982).
2 - K. von Klitzing, G. Dorda and M. Pepper, Phys. Rev. Lett. 45, 494 (1980).
3 - D.C. Tsui and A.C. Gossard, Appl. Phys. Lett. 38, 550 (1981).
4 - Y.Guldner, J.P. Hirtz, J.P. Vieren, P. Voisin, M. Voos and M. Razeghi, J. Phys.-Lettres 43, L613 (1982)
5 - J.P. Hirtz, J.P. Larivain, J.P. Duchemin and T. Pearsall, Electron. Lett. 16, 415 (1980).
6 - H.L. Störmer, R. Dingle, A.C. Gossard, W. Wiegmann and M.D. Sturge, Sol. St. Commun. 29, 705 (1979).
7 - T. Ando and Y. Uemura, J. Phys. Soc. Japan 37, 1044 (1974).
8 - F.F. Fang and P.J. Stiles, Phys. Rev. 174, 823 (1968).
9 - Th. Englert, D.C. Tsui, A.C. Gossard and Ch. Uihlein, Surf. Sci. 113, 295 (1982).
10 - E.H. Perea, E.E. Mendez and C.G. Forstad, Appl. Phys. Lett. 36, 978 (1980).
11 - D.J. Bishop, D.C. Tsui and R.C. Dynes, Phys. Rev. Lett. 44, 1153 (1980).
12 - M.A. Paalanen, D.C. Tsui and A.C. Gossard, Phys. Rev. B 25, 5566 (1982).
13 - G.A. Baraff and D.C. Tsui, Phys. Rev. B 24, 2274 (1981).
14 - H. Aoki and T. Ando, Solid St. Commun. 38, 1079 (1981) ; R.F. Prange, Phys. Rev. B 23, 4802 (1981) ; R.B. Laughlin, ibid., 5632 (1981) ; B.I. Halperin, ibid., B 25, 2185 (1982) ; D.C. Tsui and S.J. Allen, ibid., B 24, 4082 (1981).

THE QUANTIZED HALL RESISTANCE AND THE JOSEPHSON EFFECT

P.E. Lindelof and O.P. Hansen
Physics Laboratory I, H.C. Ørsted Institute
University of Copenhagen
Universitetsparken 5, DK-2100 Copenhagen

In semiconductor inversion layers with an applied strong magnetic field, the resistance can vanish and the Hall resistance take discrete values h/ie^2 ($i = 1,2,\ldots$). We consider the possibility that this might be an occurrence of a macroscopic quantum state.

Inversion layers of silicon MOSFET's constitute at low temperatures a rather ideal two-dimensional electron gas. In high magnetic fields perpendicular to the inversion layer such systems exhibit spectacular and unexpected behaviour in the magnetoresistance and Hall effect.

It is well-known that a magnetic field B changes the rectangular pattern of allowed wavenumbers to rings of Landau levels with energies $(\nu + \frac{1}{2})\,\hbar\,\frac{eB}{m}$, where ν is an integer, $\hbar$ Planck's constant, and e and m the charge and mass. When either the magnetic field or the density of electrons in the two-dimensional sheet is varied, the Fermi energy will sweep through these discrete rings, and this gives rise to a periodic variation of a number of properties, including the resistance of the specimen (the Shubnikov-de Haas effect). Kawaji and Wakabayashi[1] observed that the minima in the resistance were very close to zero for finite ranges of magnetic fields or electron densities. Subsequently it was found by von Klitzing et al.[2] that in the same ranges the Hall resistance pulled into certain constants values $R_H=h/ie^2$.

Since the Hall resistance in high magnetic fields is quite universally given by the expression

$$R_H = \phi/e\,N \quad , \tag{1}$$

ϕ being the flux and N the number of carriers, the finding of von Klitzing et al. means that

$$N = i\,\phi/\phi_0 \tag{2}$$

where ϕ_0 denotes the flux quantum h/e. As ϕ/ϕ_0 equals the number of wavenumber states per Landau level this expression for N indicates that all wavenumber states in i Landau levels are occupied. According to (2), the constant values of the Hall resistance can be interpreted as a locking of the magnetic flux to the number of electrons in such a way that i electrons share one quantum of flux, and the inversion layer can be pictured as a mobile lattice of charge-carrying vortices. This is in contrast to superconductors where the charge connected to each vortex has no discrete character and therefore can flow while the vortices are kept by pinning forces.

If the electronic system of the inversion layer under these conditions can be considered as a macroscopic quantum state with a macroscopically observable phase θ, then the following picture emerges: A current I means that I/ie quanta of flux are crossing a line from one side of the sheet to the other per unit time. And, using the equation of the a.c. Josephson effect, the difference in phase across the sheet slips by 2π for each crossing flux quantum:

$$\frac{d\theta}{dt} = 2\pi\ I/ie \quad . \tag{3}$$

This in turn yields a potential difference V :

$$eV = \hbar \frac{d\theta}{dt} = \frac{h}{ie} I \quad , \tag{4}$$

i.e. the quantized Hall resistance.

The paper by Laughlin[3] also draws on phase coherence of the wavefunctions (i.e. a macroscopic phase), but Laughlin considers a gedanken experiment where a metallic loop is formed in the direction of the current, whereas we consider phase differences caused by flux-flow in the two-dimensional plane of an actual sample. In a discussion in terms of Laughlin's explanation, Imry[4] has recently suggested the existence of an a.c. Josephson effect in normal conductors.

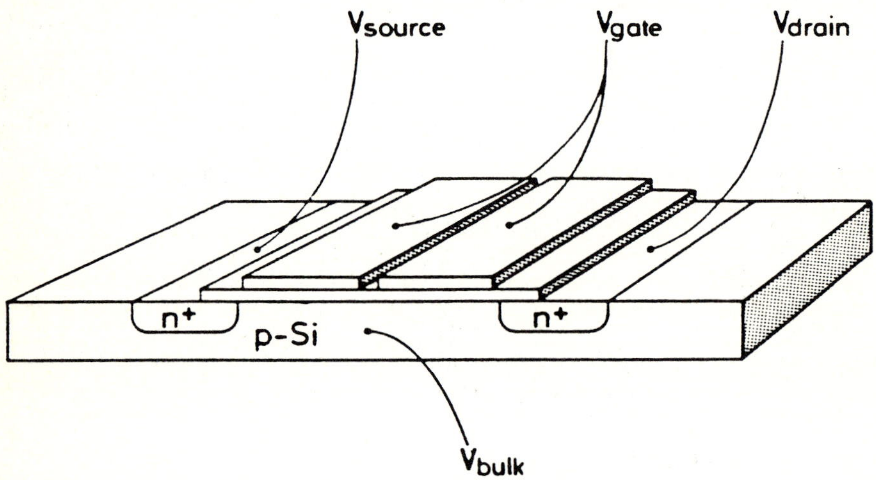

Fig.1. Double-gate geometry for a possible observation of the a.c. Josephson effect in a MOSFET in the quantized Hall regime.

The question naturally comes to mind whether a possible phasecoherent state could be observed experimentally. One way would be to form a Josephson junction and detect the a.c. Josephson effect. This might be done by making a break in the gate electrode as shown in fig.1. The break will bring the underlying semiconductor into a depletion regime over a length determined by the width of the break and this may sustain a phase difference in the direction of the current. A suitable width should then be the diameter of the outer Landau level: $2((2i + 1)\frac{\hbar}{eB})^{\frac{1}{2}}$.

References

1. S. Kawaji and J. Wakabayashi, Surface Science 58, 238 (1976).
2. K. von Klitzing, G. Dorda and M. Pepper, Phys.Rev.Lett. 45,494(1980).
3. R.B. Laughlin, Phys.Rev. B 23, 5632 (1981).
4. Y. Imry, J.Phys.C: Solid State Phys. 15, L221 (1982).

ON THE TWO-DIMENSIONAL WIGNER LOCALIZATION IN HIGH MAGNETIC FIELDS

Hidetoshi Fukuyama
The Institute for Solid State Physics
The University of Tokyo
Roppongi, Minato-ku, Tokyo 106, Japan

Abstract

Based on the recent experiments of the quantized Hall effect in GaAs/GaAlAs, the possible features of the charge density wave state are discussed.

1. Introduction

The possible existence of the charge density wave (CDW) state, or highly correlated state, in two-dimensional systems in high fields, has been a subject of extensive investigation both experimentally [1-6] and theoretically [7-10] . Above all the experiment bv WILSON, TSUI and ALLEN [6] on the frequency dependence of conductivity in Si-MOS indicated the existence of anomalous structure consistent with the impurity pinning of the charge density wave [10]. This possibility became more clearly understood in view of the recent observation of the quantized Hall effect [11-17]. The purpose of this note is to discuss about the possible features of the CDW state based on experiments of this effect in GaAs/$Al_xGa_{1-x}As$. The latest experimental results can be summarized as follows:

(1) The variation of σ_{xy} with respect to the magnetic field in the region $H \lesssim 8T$ is almost step-like at 50mK and σ_{xx} decreases as the temperature is lowered. (PAALANEN, TSUI and GOSSARD [14]). However, whether σ_{xx} actually vanishes as $T \to 0$ or not does not appear to have been answered without ambiguity for the moment [17,18] . It can be said, however, that even if there exists a region of finite σ_{xx} as $T \to 0$, this region will be extremely narrow [17] .

(2) In high fields $H \lesssim 8T$ there exists a plateau of $\sigma_{xy}=3h/e^2$ [15] at $\nu \equiv 2\pi n \ell^2 \sim 1/3$, n and $\ell=\sqrt{c/eH}$ being the electron density and the cyclotron radius, respectively. This plateau is accompanied by a minimum of ρ_{xx}, below $T \lesssim 1K$. This remarkable experimental result by TSUI, STÖRMER and GOSSARD is shown in Fig. 1.

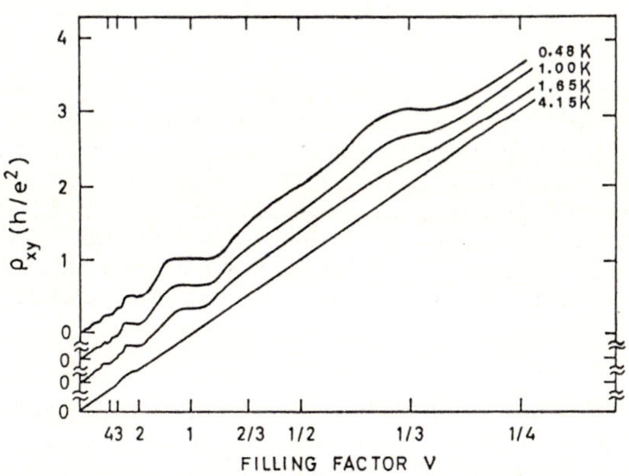

Fig. 1 Experimental results by Tsui, Stormer and Gossard |15|.

In the following we first summarize what theories have predicted about the CDW state and about the localization due to impurities alone.

2. Wigner Transition (CDW State) [19]

Consider interacting two-dimensional electrons in the lowest Landau level represented by

$$\mathcal{H}_c = \frac{1}{2} \sum v(q)\rho(q)\rho(-q) \quad , \qquad (1)$$

where $v(q)=2\pi e^2/\varepsilon_0 q$, $\rho(q)=\int dr\, e^{-iqr}\psi^+(r)\psi(r)$ and $\psi(r)=\sum a_x \phi_x(r)$, $\phi_x(r)$ being the Landau wave functions for the lowest level. This system has a Hartree-Fock instability to the CDW state and the critical temperature, T_c, is found to be [9]

$$T_c = 0.56\ \nu(1-\nu)\ e^2 / \varepsilon_0 \ell \quad . \qquad (2)$$

This T_c has a maximum at $\nu=1/2$, and is appreciabley larger than the melting temperature of classical Wigner solid [20], T_M,

$$T_M = \frac{1}{\Gamma_0} \frac{e^2}{\varepsilon_0 r_0} \quad , \qquad (3)$$

where $r_0=(\pi n)^{-1/2}$ is the average distance between electrons and $\Gamma_0 \sim 137$ is the dimensionless parameter [21,22]. These temperatures are shown in Fig.2 for a choice of $\varepsilon_0=10$ and H=10T

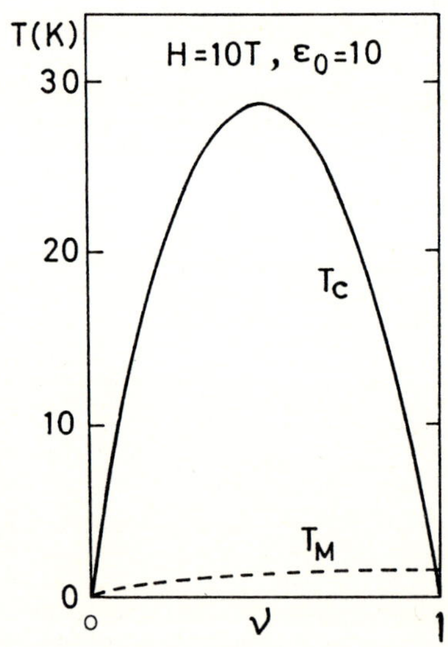

Fig. 2 The critical temperature, T_c, of CDW in Hartree-Fock approximation and the melting temperature, T_M, of classical Wigner solid.

The T_c, eq.(2), in the Hartree-Fock approximation is obviously an overestimation. Moreover its ν-dependence has a characteristic maximum at $\nu=1/2$. This feature is not essentially affected by impurity scattering treated as an average (i.e. within CPA, consequently, without localization effects) and by scattering between different valleys [23].

On the other hand the calculation of the ground state energy of the CDW state [24] resulted in a somewhat different ν-dependence of the cohesive energy, E_{coh}, as shown in Fig.3: $|E_{coh}|$ does not have a maximum at ν=1/2. In this calculation only few fundamental order parameters of CDW are taken into account. (Recently fully detailed calculations of E_{coh} have been performed by YOSHIOKA, LEE and HALPERIN [25].)

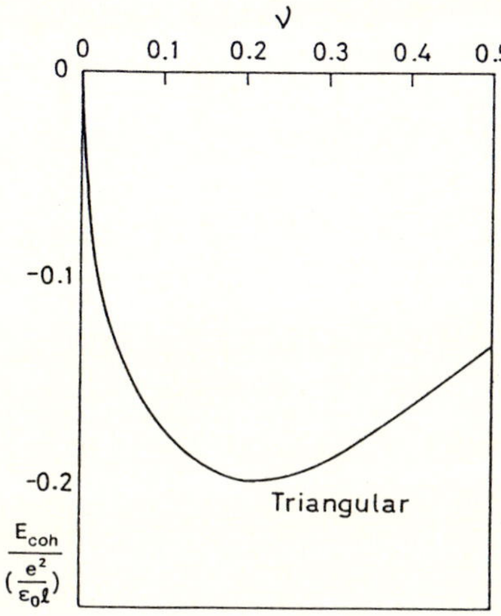

Fig. 3 The ν-depencence of the cohesive energy of the CDW state [24].

In ordinary cases of the phase transition, the condensation energy at T=0 and the critical temperature are almost proportional , which is not the case in our present calculations. Since the Hartree-Fock theory at T≠0 is less reliable than that at T=0 we argued [37] that the actual T_c would have a minimum (not a maximum) at ν=1/2 which is suggested by E_{coh}.

3. Anderson Transition

Next we consider the effect of randomness,

$$\mathcal{H}_R = \sum_{R_\ell} \int dr\ \psi^+(r)\ \psi(r)\ u(r - R_\ell) \quad . \tag{4}$$

Effects of localization caused by $\mathcal{H}_R$ have been discussed by many authors in conjunction to the quantized Hall effect [26-30]. Quite recently localization in the quantum limit, i.e. in the lowest Landau level, has been examined by ONO [31], who constructed a theory based on the memory function formalism. This theory can be considered as a quantum limit version of the VOLLHARDT-WÖLFLE [32] theory. He found that all the states are exponentially localized except at energy E_0 which satisfies $\mathrm{Re}G(E_0)=0$, $G(E)$ being the electron Green function. This result does not depend on details of the approximation employed to determine the density of states function. The numerical results of the localization length, ξ, are shown in Fig.4, which is taken from ref. [31]. In this case $E_0=0$ and ξ is found to depend on the Fermi energy E_F as $\xi \sim \ell \exp(\gamma^2/E_F^2)$ in $|E_F| \sim 0$ where γ is of the order of width of the level.

ONO's results are consistent with those obtained in the weak field region [33], which demonstrated that the localization is not destroyed by magnetic fields. (Similar conclusions for the high field region are drawn by HOUGHTON [34] quite recently).

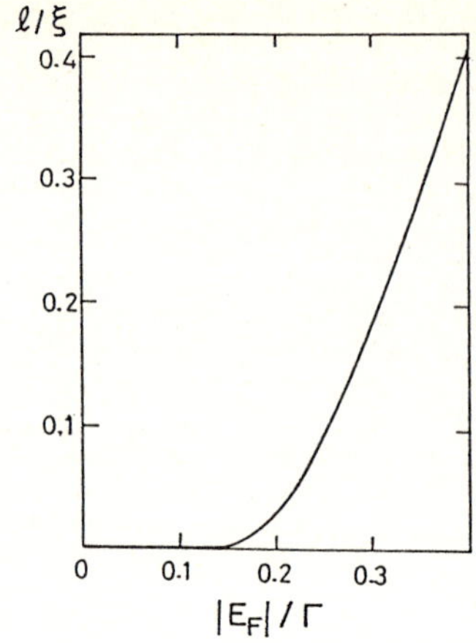

Fig. 4 The energy dependence of the localization length, ξ, by Ono [31].

Considering the effects of Coulomb interaction in the presence of any impurity potential, GIRVIN et al. [35] evaluated the temperature dependence of the conductivity by treating interactions as weak. This will be a proper way to consider higher Landau levels, or $H \lesssim 8T$.

4. Discussion

Since theories of CDW are developed for the lowest Landau level, we will be mainly concerned with the experimental finding of (2) of Sec.2. First the observation of a plateau of ρ_{xy} at $\nu \sim 1/3$ in $T \lesssim 1K$ is remarkable. It is to be noted that ρ_{xy} at around $\nu \sim 1/2$ varies rather smoothly even in this temperature region. Although there does not exist yet any satisfactory explanation of this plateau at $\nu \sim 1/3$, it may be possible that the CDW is especially favored at $\nu \sim 1/3$ as TSUI et al. [15] indicated. For the moment the region of ν where ρ_{xy} keeps a plateau value of $\rho_{xy} = 3h/e^2$ have not been determined experimentally. This region will be a measure of the stability of the CDW state with the commensurability, M=3, relative to the magnetic flux; i.e. the lowest Landau level is split into three sub-bands as shown schematically in Fig.5 with a possible small number of holes (or discommensurations [36]) in the lowest

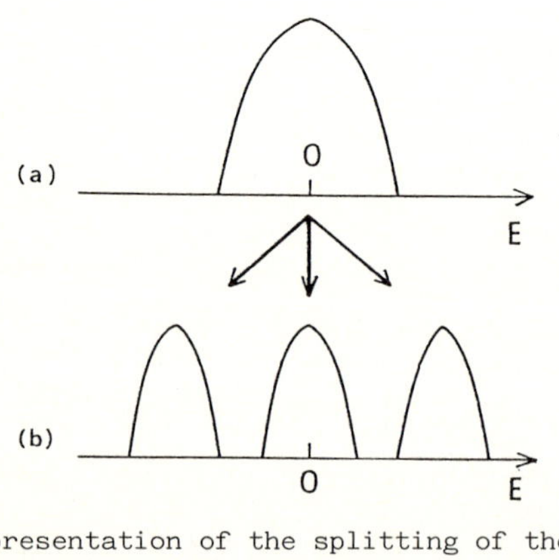

Fig. 5 Schematic representation of the splitting of the density of states into three sub-bands of CDW with the commensurability, M=3.

sub-band or electrons in the second sub-band. Due to the electron-hole symmetry in the lowest Landau level T_c will be appreciable also around $\nu\sim 2/3$. However T_c at $\nu\sim 2/3$ will be much lower than that at $\nu\sim 1/3$ due to more screening.

The situation in the limit of $T\to 0$ around $\nu\sim 1/2$ is intriguing. In the absence of Coulomb interactions ONO's results [31] will imply that σ_{xy} has a step-wise change with one quantum, e^2/h. The simple electron-hole symmetry argument [37] of a CDW state will also result in a similar behavior in σ_{xy}. However, in view of the presence of a plateau with 1/3-quantum, the change of σ_{xy} at $\nu\sim 1/2$ will be either type A or B in Fig. 6. If there actually exists a plateau at $\nu\sim 2/3$, type A is expected. In this case, the CDW with M=3 commensurability is considered to be very stable for any degree of filling, ν, and three sub-bands behave like rigid bands. This CDW state is very distinct from the one which is usually examined under the assumption that all electrons participate to form a CDW. The type B behavior will be expected when the three sub-bands in Fig. 5 merge already at $\nu\sim 1/2$.

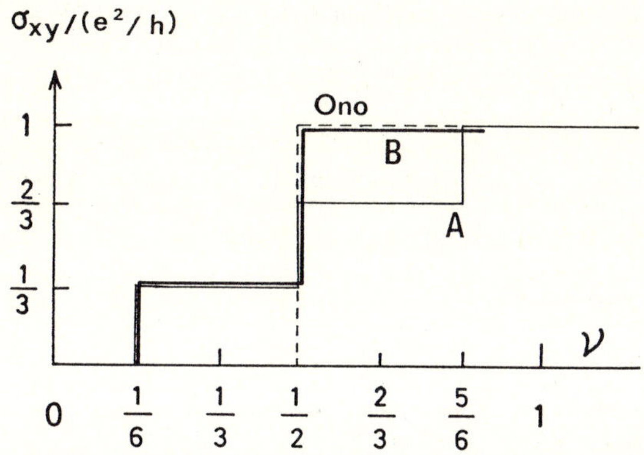

Fig. 6 Possible variation of σ_{xy} in the limit of low temperatures; (A) CDW with M=3 is rigid, (B) three sub-bands merge at around $\nu=1/2$, (C) in the absence of Coulomb interaction (Ono).

At present theoretical investigations are far behind the experimental findings. High field experiments ($H\gtrsim 10T$) at very low temperatures will give decisive clue to sort out possible ground state of this novel system.

References

[1] S. Kawaji and J. Wakabayashi, Solid State Commun. 22, 87 (1977).
[2] D.C. Tsui, Solid State Commun. 21, 675 (1977).
[3] T.A. Kennedy, R.J. Wagner, B.C. McCombe and D.C. Tsui, Solid State Commun. 22, 439 (1977).
[4] R.J. Nicholas, R.A. Stradling, J.C. Portal, P. Perrier and S. Askenazy, Solid State Commun. 31, 437 (1979).
[5] M. Pepper, Phil. Mag. B37, 83 (1978).
[6] B.A. Wilson, S.J. Allen and D.C. Tsui, Phys. Rev. Lett. 44, 479 (1980).
[7] Y.E. Lozovik and V.I. Yudson, JETP Lett. 22, 11 (1975).
[8] H. Fukuyama, Solid State Commun. 19, 551 (1976).
[9] H. Fukuyama, P.M. Platzman and P.W. Anderson, Phys. Rev. B19, 5211 (1979).
[10] H. Fukuyama and P.A. Lee, Phys. Rev. B18, 6245 (1978).
[11] K. von Klitzing, G. Dorda and M. Pepper, Phys. Rev. Lett. 45, 494 (1980).
[12] D.C. Tsui and A.C. Gossard, Appl. Phys. Lett. 37, 550 (1981).
[13] D.C. Tsui, H.L. Stormer and A.C. Gossard, Phys. Rev. B25, 1405 (1982).
[14] M.A. Paalanen, D.C. Tsui and A.C. Gossard, Phys. Rev. B25, 5566 (1982).
[15] D.C. Tsui, H.L. Stormer and A.C. Gossard, Phys. Rev. Lett. 48, 1559 (1982).
[16] K. Yoshihiro, J. Kinoshita, K. Inagaki, C. Yamanouchi, J. Moriyama and S. Kawaji, Surface Sci. 113, 23 (1982).
[17] G. Ebert, K.v. Klitzing, C. Probst and K. Ploog, preprint
[18] G. Kote and N.P. Ong, preprint.
[19] For review, H. Fukuyama, Proc. Kyoto Summer Institute (Kyoto Univ., 1979) p.144.
[20] E.P. Wigner, Phys. Rev. 46, 1002 (1934).
[21] C.C. Grimes and G. Adams, Phys. Rev. Lett. 42, 795 (1979).
[22] R.H. Morf, Phys. Rev. Lett. 43, 931 (1981).
[23] D. Yoshioka and H. Fukuyama, J. Phys. Soc. Jpn. 50, 1560 (1981).
[24] D. Yoshioka and H. Fukuayma, J. Phys. Soc. Jpn. 47, 394 (1979).
[25] D. Yoshioka, P.A. Lee and B.I. Halperin, preprint.
[26] R.B. Laughlin, Phys. Rev. B23, 5632 (1981).
[27] R.E. Prange, Phys. Rev. B23, 4802 (1981).
[28] B.I. Halperin, Phys. Rev. B25, 2185 (1982).
[29] D.J. Thouless, J. Phys C 14, 3475 (1981).
[30] H. Aoki and T. Ando, Solid State Commun. 38, 1079 (1981).
[31] T. Ono, J. Phys. Soc. Jpn. 51, 2055 (1982); ibid, to be published.
[32] D. Vollhardt and P. Wölfle, Phys. Rev. Lett. 45, 842 (1980); Phys. Rev. B22 4666 (1980).
[33] D. Yoshioka, Y. Ono and H. Fukuyama, J. Phys. Soc. Jpn. 50, 3419 (1981).
[34] A. Houghton, private communications.
[35] S.M. Girvin, M. Jonson and P.A. Lee, to appear in Phys. Rev. B.
[36] W.L. McMillan, Phys. Rev. B14, 1496 (1976).
[37] H. Fukuyama and P.M. Platzman, Phys. Rev. B25, 2934 (1982).

Negative Magnetoresistance and Anderson Localization in Si-MOSFETs and Other 2D Systems in Semiconductor Interfaces

S. Kawaji

Department of Physics, Gakushuin University, Mejiro, Toshima-ku
Tokyo, Japan
and

Y. Kawaguchi

Gakushuin Women's Junior College, Toyama, Shinjuku-ku, Tokyo, Japan

1. Introduction

Before 1979, several observations of negative magnetoresistance in two-dimensional (2D) electron systems in semiconductor inversion and accumulation layers had been reported [1∿8] . The negative magnetoresistance observed in tangential magnetic fields by Tansal, Fowler and Cotellessa [2] and Sakaki and Sugano [3] has been understood to be the normal effect due to the Lorentz force described by Uemura and Matsumoto's theory [9]. Other observations had been considered anomalous. The negative magnetoresistance in InAs inversion layers by Kawaji and Kawaguchi [1] and that in ZnO accumulation layers by Goldstein, Grinshpan and Many [8] had shown very weak anisotropy on the field direction which is favourable for a spin scattering model. However, Kawaguchi, Kitahara and Kawaji's observations [6,7] which clearly show that the negative magnetoresistance depends only on the field component perpendicular to the surface had been difficult to explain by a model based on the localized spin scattering which most authors had invoked.

On the basis of the recent scaling theory of the Anderson localization by Abrahams, Anderson Licciardello and Ramakrishnan [10] and Anderson, Abrahams and Ramakrishnan [11], a theory of negative magnetoresistance in the weakly localized regime (WLR) ($\varepsilon_F\tau/h \gg 1$) of 2D systems has been developed by Hikami, Larin and Nagaoka [12] and Altshuler, Kel'nizkii, Larkin and Lee [13]. The theory has successfully explained recent experiments in Si-MOSFETs by Kawaguchi and Kawaji [14,15,16], Uren, Davies, Kaveh and Pepper [17], Wheeler [18] and Dynes [19].

In WLR the quantum interference between states with e^{ikr} and e^{-ikr} causes spatially localized states which produce a correction term σ' to the classical Drude conductivity σ_0. The quantum interference is sensitive to perturbations destroying time reversal symmetry. Therefore, an external magnetic field applied perpendicularly to the 2D systems suppresses the quantum correction to the conductivity due to the weak localization and produces negative magnetoresistance. The quantum correction has a logarithmic dependence on temperature, electric field as well as on magnetic fields. However, of these dependences, the negative magnetoresistance is probably most useful to investigate microscopic parameters in the theory such as the inelastic scattering time.

In addition to the localization theory, the important role of the mutual Coulomb interactions of electrons in 2D systems have been pointed out by Altshuler, Aronov and Lee [20] and by Fukuyama [21]. The effect of the mutual interactions appears as a change in the prefactor of the logarithmic dependences. Therefore, detailed analysis of the negative magnetoresistance provides useful information on electron localization and interactions in 2D electron systems [22,23].

In the present paper we will briefly summarize theoretical results on the negative magnetoresistance in the Anderson localization regime and effects of interactions of electrons in WLR of 2D systems. Then our results of experiments on Si-MOSFETs will be presented and discussed with the theoretical results. A brief description of recent results in 2D systems of compound semiconductor interfaces will be also presented.

2. Theoretical Background

2.1 Magnetoconductivity

Following Hikami, Larkin and Nagaoka [12], the change in conductivity with magnetic field B applied perpendicularly to the surface is written as

$$\Delta\sigma_B = \sigma(B) - \sigma(0)$$
$$= -\frac{n_v \alpha e^2}{2\pi^2\hbar}\left[\psi\left(\frac{1}{2}+\frac{1}{a\tau}\right) - \psi\left(\frac{1}{2}+\frac{1}{a\tau_\varepsilon}\right) - \ln\frac{\tau_\varepsilon}{\tau}\right] \qquad (1)$$

where n_v is the valley degeneracy, α is a constant prefactor, τ is the relaxation time due to normal impurity scattering, τ_ε is the inelastic scattering time and $a = 4DeB/\hbar$ where D is the diffusion coefficient. This formula is applicable for $a\tau < 1$ because $(a\tau)^{-1} \sim N_0$ is the cut-off parameter in the summation of $\Sigma\,\{1/[(N + 1/2)a + 1/\tau_\varepsilon]\}$. Therefore, in the analysis of experiments, the magnetic field has to satisfy the following condition:

$$a\tau = \frac{4\pi\hbar\sigma_0\omega_c\tau}{n_v e^2} = \frac{\sigma_0(\text{mho})\mu B}{1.93\times10^{-5} n_v} < 1 \qquad (2)$$

where μ is the electron mobility. Equation (1) gives also the correct result for the saturation value of the magnetoconductivity in the high field limit,

$$\Delta\sigma_B(\text{sat}) = \frac{n_v\alpha e^2}{2\pi^2\hbar}\ \ln\frac{\tau_\varepsilon}{\tau}\ . \qquad (3)$$

2.2 Effect of Electric Field

An important factor in negative magnetoresistance experiments is the strength of the electric field E. As is discussed by Kaveh et al [24] and Kaveh and Mott [25], the quantum correction σ' to the conductivity depends on the electric field. Detailed theories on the electric field effect on localization in the absence and the presence of magnetic fields have been developed by Tsuzuki [26,27]. Following his results, the conductivity lies approximately in the Ohmic region when $(E/E_0)<(\tau/\tau_\varepsilon)^{7/4}$ is satisfied. Here, the characteristic field E_0 is given by $E_0=\sqrt{6}\hbar/e\tau\ell$. The numerical value of E_0 is written as

$$E_0 = \left(\frac{3n_v}{\pi}\right)^{1/2}\cdot\frac{e}{m^*\mu^2 N_s^{1/2}} = 1.27\times10^4\left(\frac{0.19m_0}{m^*}\right)\left(\frac{10^4\text{cm}^2/\text{v}\cdot\text{s}}{\mu}\right)^2\left(\frac{10^{12}\text{cm}^{-2}}{N_s}\right)^{1/2}(\text{V/m}). \qquad (4)$$

Another mechanism of non-linear effects of applied electric fieldsis the Ohmic heating mechanism as is proposed by Anderson et al [11]. In this mechanism, the electron temperature T_e in stationary state of the systems is given by

$$T_e(E) - T_\ell = \sigma E\ \tau_\varepsilon(e\ell\text{-ph})/C \qquad (5)$$

where T_ℓ is lattice temperature, C is the specific heat of electrons and $\tau_\varepsilon(e\ell\text{-ph})$ the inelastic scattering time determined by the electron-phonon scattering.

2.3 Constant Prefactor α

The constant prefactor α is an empirical parameter which depends on the spin-orbit interaction [12,28], the magnetic scattering [12], the intervalley scattering [29], and the Coulomb interaction [30]. The interaction effect in conjunction with the prefactor has been discussed in detail by Fukuyama [23].

In single valley systems, α is unity when the Coulomb interactions are ignored in the absence of the spin-orbit and the magnetic scattering [12]. Althshuler et al [20] and Fukuyama [21] have examined the effect of mutual Coulomb interactions. Following Fukuyama [23], in single valley systems α is given by

$$\alpha = 1 - \frac{F}{2} \tag{6}$$

where [13]

$$F = \frac{1}{2\pi}\int_0^{2\pi} \frac{d\theta}{1+(2k_F/q_s)\sin\theta/2}, \tag{6'}$$

Here, q_s = inverse screening length and $2k_F/q_s = 0.190(\bar{k}/7.7)(2/n_v)^{3/2}(0.19mo/m) \times (N_s/10^{12}cm^{-2})^{1/2}$. It is easy to see that $F \to 1$ as $2k_F/q_s \to 0$ and $F \to 0$ as $2k_F/q_s \to \infty$.

In many valley systems, we have [23,24]

$$\alpha = [1 + F(\frac{1}{2} - n_v)/n_v] \qquad (\tau' \gg \tau_\varepsilon) \tag{7}$$

$$\alpha = (1 - F/2n_v)/n_v \qquad (\tau' \ll \tau_\varepsilon) \tag{7'}$$

where τ' is the intervalley scattering time. We note here that these results are obtained in the lowest order approximation.

Different from the orbital effect, the Zeeman splitting affects the Coulomb interactions and causes changes in the conductance. Kawabata[31,32] has calculated that a positive magnetoresistance appears in magnetic fields parallel to the surface. The magnitude of the negative magnetoconductivity is twice the positive magnetoconductivity due to the Zeeman splitting in perpendicular magnetic fields.

2.4 Inelastic Scattering Time

In the localization theory of WLR, the quantum correction to the conductivity depends on the characteristic scale length. At finite temperature, Thouless [33] has introduced the inelastic diffusion length $\sqrt{D\tau_\varepsilon}$, the distance an electron diffuses in an inelastic scattering time, as a characteristic scale length. Negative magnetoresistance experiments in 2D systems can deduce the inelastic scattering time τ_ε, the lifetime of the one-electron eigenstates by use of (1). This makes the negative magnetoresistance experiment a powerful tool in the study of the Anderson localization. However, physical processes which account for τ_ε in experiments are still not clear now.

Following Landau's argument, τ_ε due to electron-electron process in the pure metal is written as [33,34]

$$\tau_\varepsilon(\text{el-el}) \simeq \hbar\frac{\varepsilon_F}{(k_BT)^2} \simeq \frac{5.58\cdot 10^{-10}}{T^2}\left[\frac{2}{n_v}\right]\left[\frac{0.19mo}{m^*}\right]\left[\frac{N_s}{10^{12}cm^{-2}}\right](S) \tag{8}$$

Recently Abrahams, Anderson, Lee and Ramakrishnan [35] have calculated electron-electron scattering in the dirty 2D system. Their result at a finite temperature is written as

$$\frac{\tau_\varepsilon\ (\text{el-el})}{\tau} = -\frac{E_F}{k_BT\ln(T/T_1)} \tag{9}$$

where $k_BT_1 = \hbar D\varepsilon^2/e^4$ and $\varepsilon = \hbar Dq_s^{\ 2}$. The numerical value of T_1 is given as

$$T_1 = \frac{2n_v \bar{R}_y^*}{k_B \bar{k}^2} \left(\frac{2\pi\hbar\sigma}{e^2}\right)^3 \qquad (10)$$

$$= 34.1 \left(\frac{n_v}{2}\right) \left(\frac{\overline{R_y^*}}{43.6\text{meV}}\right) \left(\frac{7.7}{\bar{k}}\right)^2 \left(\frac{\sigma(\text{mho})}{3.87\cdot 10^{-5}}\right)^3 \ (\text{K})$$

Calculation of the inelastic scattering time due to the electron-phonon process which is applicable to 2D systems in semiconductor inversion or accumulation layers has not yet been performed. Shinba et al. [36] have calculated the energy relaxation time of the electron system interacting with the phonon system in Si (100) and (111) n-channel inversion layers. Their results show that

$$\tau_{relax}(\text{el-ph}) \propto \frac{(E_F)^{1/2}\, T_e}{T_\ell^3} \quad \text{for } T_e \sim T_\ell \qquad (11a)$$

$$\propto \frac{(E_F)^{1/2}}{T_e^3} \quad \text{for } T_e \gg T_\ell \qquad (11b)$$

at temperatures below 7 K where Te and T_ℓ is the electron temperature and the lattice temperature, respectively. The numerical value of τ_{relax} is 9×10^{-10} s at $T_e = 4.2$ K, $T_e - T_\ell = 0.1$ K for $N_s = 5.3 \times 10^{12}$ cm^{-2} in Si(100) n-channel. The inelastic scattering time due to electron phonon process τ_ε(el-ph) has probably the same dependences on E_F and T to those in τ_{relax}(el-ph).

3. Experimental Results and Discussions

3.1 Silicon Inversion Layers (Fundamental Properties)

Angular dependence of transverse negative magnetoresistance $\Delta\rho/\rho_0$ in a Si MOSFETs in Fig. 1 shows clearly that the negative magnetoresistance depends only on the normal component of the magnetic field [6]. This fact suggests that the origin of the negative magnetoresistance is not a spin mechanism but an effect of obrital motion of electrons. However, we were unable to account for the phenomenon until 1979 when the thory of negative magnetoresistance in 2D systems [12] based on the scaling theory of Anderson localization [10,11] appeared.

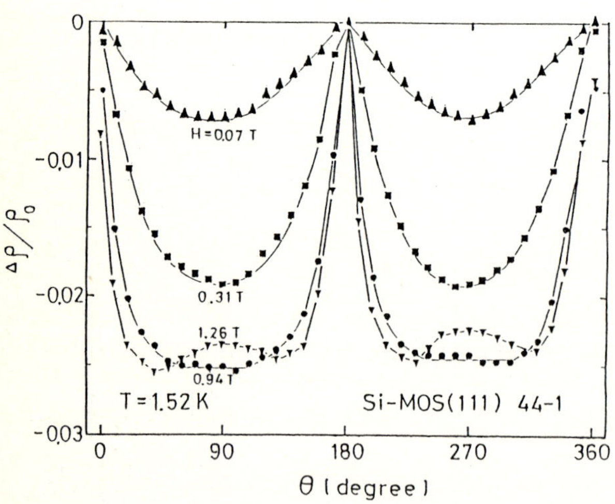

Fig. 1 Angular dependence of negative magnetoresistance in an n-channel (111) Si MOSFET at Ns = 5×10^{12}cm^{-2} at T = 1.5 K. Here, θ = 0 at the field paralell to the surface. [6]

We mention here that Sasaki [37] had observed first an anisotropic negative magnetoresistance which showed evidence of the orbital mechanism of the effect which had been observed in the impurity conduction of Ge in 1965. Kawabata's theory [38] based on the Anderson localization in three dimensions has explained Sasaki's observations.

Changes in conductance with magnetic field observed in a Si(100) MOSFET can well be fitted to the expression of magnetoconductance of Hikami et al. [12] as shown in Fig.2 [14]. In this calculation, an approximation of $\Psi\{1 + 1/a\tau\} \sim \ell n(1/a\tau)$ and $n_v = 2$ are used. Experimental points are well reproduced by the theory in the low field region.

In the following we present the temperature dependence and electron concentration dependence of τ_ε and α which are obtained by fitting experiments to the expression (1). In our analysis both τ_ε and α are adjusted in the fitting of experimental data to the expression (1). Wheeler [18] has employed $n_v\alpha = 1$ in (1) to fit his

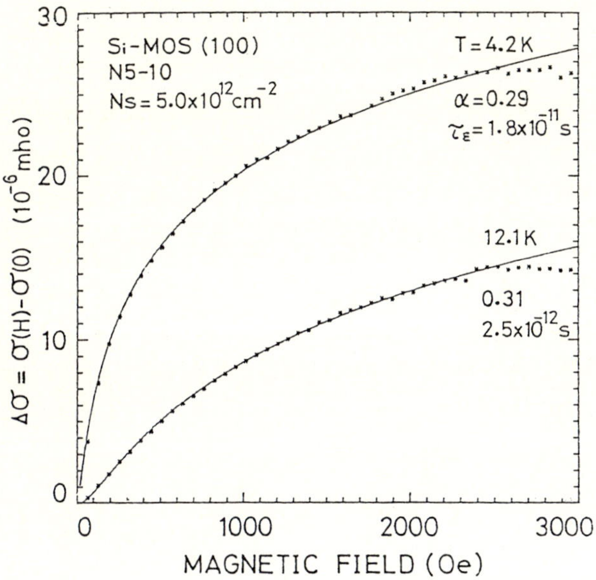

Fig. 2 Change in the conductivity $\Delta\sigma$ with magnetic field in an n-channel (100) Si MOSFET with peak mobility of 13000 $cm^2/V\cdot s$. The experimental points have been fitted to

$$\Delta\sigma = -\frac{n_v \alpha e^2}{2\pi^2 h}\left[\ell n\left(\frac{1}{\tau_\varepsilon a}\right) - \psi\left(\frac{1}{2}\right) + \left(\frac{1}{\tau_\varepsilon a}\right)\right]$$

which is derived by use of $\psi\left(\frac{1}{2} + \frac{1}{\tau a}\right) \approx \ell n(1/\tau a)$ in (1) [14].

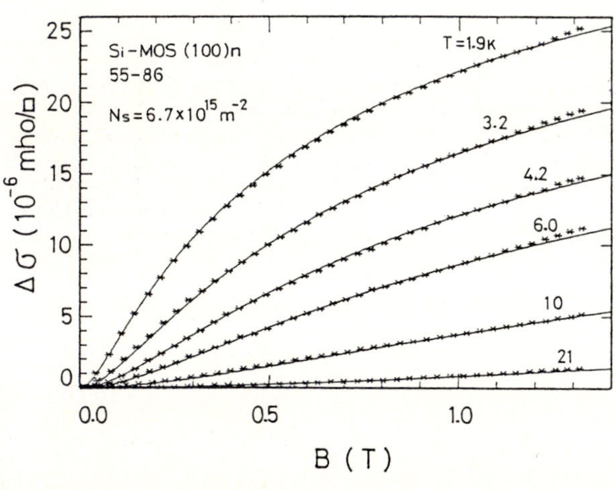

Fig. 3 Change in the conductivity with magnetic field in an n-channel (100) Si MOSFET with low electron mobility. The experimental points have been fitted to the expression (1).

experiments to the theory. However, in our experiments we have obtained better results in the fitting by adjusting τ_ε and $n_v\alpha$. Examples of our fitting of (1) to experiments for a low mobility sample are shown in Fig. 3. In calculations of (1) in Fig. 3, $a\tau$ is kept less than 0.6. It is interesting to note that experiments can be well fitted to (1) for $a\tau > 1$. Variation of the fitting parameters τ_ε and α is less than 10 % for $a\tau \le 4$.

In the following we will present experimental results mainly obtained in low mobility samples having a peak mobility at 4.2 K of 3500 $cm^2/V{\cdot}s$ [16].

Temperature and electron concentration dependences of the inelastic scattering time are shown in Fig. 4 and Fig. 5, respectively. The inelastic

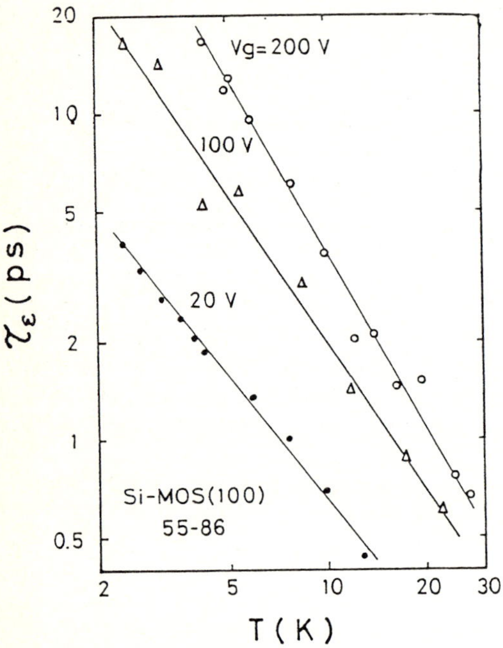

Fig. 4 Temperature (T) dependence of inelastic scattering time τ_ε for $E_F\tau/\hbar$ = 0.4 (V_G = 20 V), 9.5 (100V) and 14.5 (200V), respectively [16].

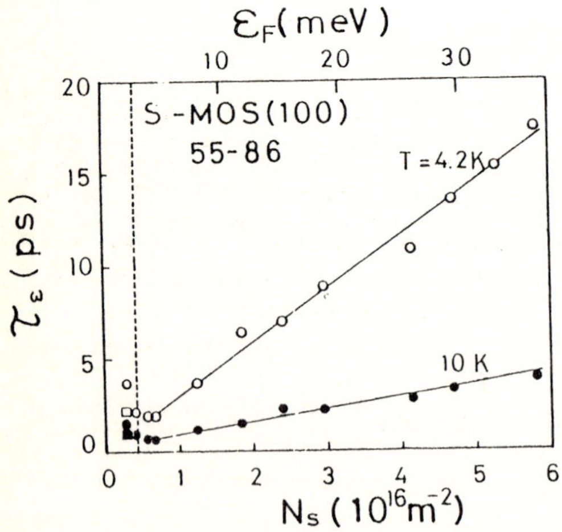

Fig. 5 Electron concentration (Ns) dependence of inelastic scattering time τ_ε at 4.2 K and 10 K. The vertical broken line indicates the electron concentration Ns for $E_F\tau/\hbar$ = 0.5 or $k_F\ell$ = 1 [16]. The open and solid squares are deduced from magnetoconductivity by use of (1) for n_v = 1, open and solid circles for n_v = 2.

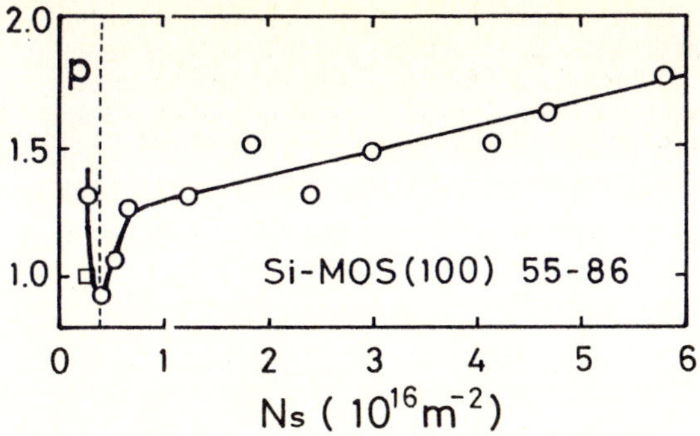

Fig. 6 Electron concentration (Ns) dependence of expotent p in the temperature dependence of inelastic scattering time $\tau_\varepsilon \propto T^{-p}$. The vertical broken line indicates N_s for $E_F\tau/\hbar = 0.5$ or $k_F\ell = 1$ [16]. The open and solid squares are deduced from magnetoconductivity by use of (1) for $n_v = 1$, and the open and solid circles for $n_v = 2$.

scattering time data presented in Fig. 4 are limited to the region where the non-linear electric field effect predicted by Tsuzuki [26,27] is excluded. The Ns-dependence in Fig. 5 is in good agreement with the results of electron-electron interaction mechanism summarized in (8) and (9). For $\tau_\varepsilon \propto T^{-p}$, p = 1.75 at V_G = 200 V (Ns = 5.8 x 10^{12} cm^{-2}) and p = 1.3 at V_G = 20 V (Ns = 6.8 x 10^{11}). The exponent of temperature dependence of τ_ε is shown in Fig. 6. Even at high electron concentration of V_G = 200 V, p is less than 2 which is expected for the pure metal limit. However, τ_ε calculated by use of (8) is not so far from the experimental data; $\tau_\varepsilon(\text{theor})/\tau_\varepsilon(\text{exp}) \simeq 6 \sim 15$ in the temperature range in Fig. 4. Some other scattering processes exist to reduce the magnitude of τ_ε and p. The temperature dependence of τ_ε at 20 V is close to the results for dirty 2D systems calculated by Abrahams et al [35]. However, the τ_ε (theor) is one order of magnitude smaller than τ_ε(exp). Further theoretical investigations of τ_ε which cover the entire region of potential fluctuations from the pure metal limit to the dirty limit are necessary.

The constant prefactor α shows an interesting behaviour as a function of the electron concentration Ns as shown in Fig. 7.

Fukuyama [22] has accounted for the behaviour in the higher Ns region by the

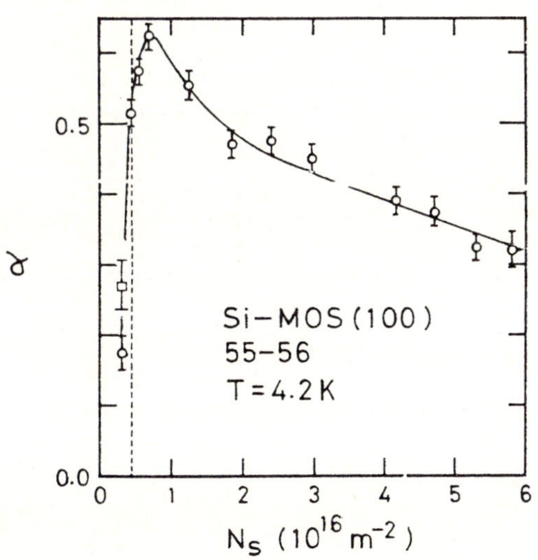

Fig. 7 Electron concentration (Ns) dependence of the prefactor α at 4.2 K. The vertical broken line indicates Ns for $E_F\tau/\hbar = 0.5$ or $k_F\ell = 1$. [16]. The open square is deduced from magnetoconductivity by use of (1) for $n_v = 1$ and the open circles for $n_v = 2$.

effect of intervalley impurity scattering on the prefactor of the logarithmic correction to the conductivity due to mutual interactions. In many-valley systems with no intervalley scattering, the prefactor of the quantum correction in the magnetoconducvity due to the mutual interactions is given by (7). Following Fukuyama's calculation of F for Si(100) inversion layers, $F \simeq 0.8$ at $N_S \simeq 6 \times 10^{12}$ cm^{-2} [30], then we have from (7) $\alpha = 1 - 3F/4 \simeq 0.4$. This is close to our experimental results in Fig. 6. However, the N_S dependence of experiments is opposite to that of the theoretical prediction because the function F is decreasing with N_S as is shown in (6).

Fukuyama [29] has taken into account τ', the life time due to intervalley impurity scattering, and calculated α as a function of τ_ε/τ'. For independent valleys, i.e. $\tau' \gg \tau_\varepsilon$, we have $\alpha = 1$. Once $\tau \leq \tau_\varepsilon$, α is appreciably reduced and in the limit of $\tau'/\tau_\varepsilon \to 0$, α tends to $1/n_v$, i.e. $n_v\alpha = 1$. The result of Fukuyama's numerical calculation [22] is shown in Fig. 8. As shown in Fig. 5, τ_ε decreases with decreasing N_S, and α increases gradually as shown in Fig. 8.

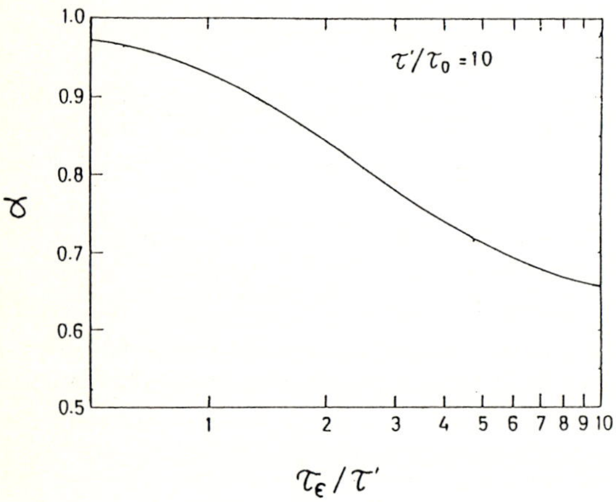

Fig. 8 Effect of intervalley impurity scattering on the prefactor α calculated by Fukuyama [22]. Here, τ, τ', τ_ε are life times due to intravally impurity, intervalley impurity, and inelastic scatterings, respectively.

The effect of intervalley impurity scattering appears also in the temperature dependence of $n_v\alpha$ as shown in Fig.9. As shown in Fig. 4, τ_ε decreases with increasing temperature, and the cross-over of $n_v\alpha$ in Fig. 8 causes increase in $n_v\alpha$ in Fig. 9. At lower gate voltage of $V_G = 20$ V, τ_ε/τ' reduces much faster than at $V_G = 200$ V with increasing temperature. At temperatures where τ_ε -0.8ps at V_G=20 V and 200 V in Fig. 4, the rapid increase of $n_v\alpha$ with increasing temperature occurs as shown by two arrows in Fig. 9. Thus we justify the choice of $\tau' \simeq 0.8$ ps. This is consistent with τ' in the calculation of the structure of conductivity associated with the minigap by Ando [39] .

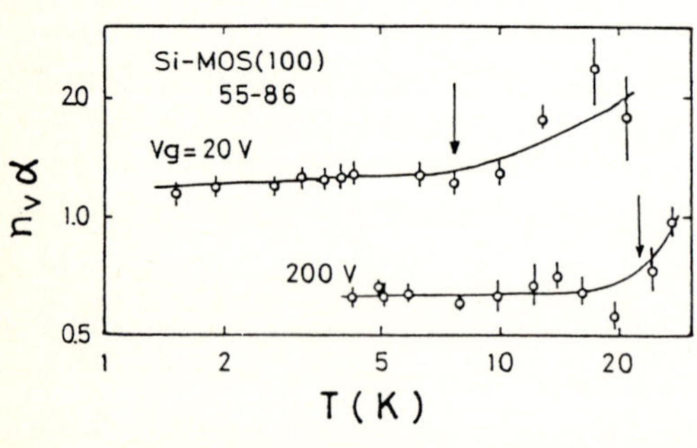

Fig. 9 Temperature dependence of $n_v\alpha$ [16]. At temperatures indicated by two arrows the cross-over in α for τ_ε/τ' in Fig. 8 occurs.

Fukuyama [22] has pointed out that the sharp drop of α with decreasing N_s at low N_s region ($N_s < 5 \times 10^{11}$ cm^{-2}) in Fig. 7 is consistent with the removal of the valley degeneracy, i.e. $n_v = 2 \rightarrow 1$ predicted by Bloss, Sham and Vinter [40] . If $n_v = 1$ in this region - roughly speaking - we have to double the magnitude of α in Fig. 7. We have re-examined the magnetoconductivity data by assuming $n_v = 1$. The results are shown by the squares in Figs. 5,o and 7. They appear more reasonable than the previous data obtained for $n_v = 2$.

The vertical broken lines in Fig. 5,6 and 7 indicate that the electron concentration for $E_F\tau/\hbar = 0.5$ or $k_F\ell = 1$ where $E_F\tau/\hbar \gg 1$ is not satisfied. However, all experimental points near and in the left side of the broken line have been obtained by fitting experimental magnetoconductivity to Hikami et al's expression (1). The correct expansion parameter in this problem appears to be $\lambda = \hbar/2\pi E_F\tau$ rather than $\hbar/E_F\tau$ from the results by Vollhardt and Wölfle [41]. If this is the case, all experimental points lie in the region of $\lambda < 0.3$ and our deduction of τ_ε and $n_v\alpha$ by use of (1) is allowed.

3.2 Silicon Inversion Layers (Application)

As is discussed in the former section, the inelastic scattering time τ_ε is determined by electron-electron interaction. Therefore, we can measure the temperature of 2D electron systems which are not in thermal equilibrium with the lattice by τ_ε deduced from the negative magnetoresistance. The inelastic scattering time depends strongly on temperature. Therefore, measurement of τ_ε itself is probably the best method to obtain the electron temperature. Kawaguchi and Kawaji [42,43] have applied negative magnetoresistance data for measurements of temperatures of electrons in Si MOSFETs at high electric fields to evaluate the deformation potential constant.

Magnetoconductivity at various source-drain electric fields, E_{SD}, are shown in Fig. 10. The inelastic scattering time, τ_ε, is deduced from the magnetoconductivity

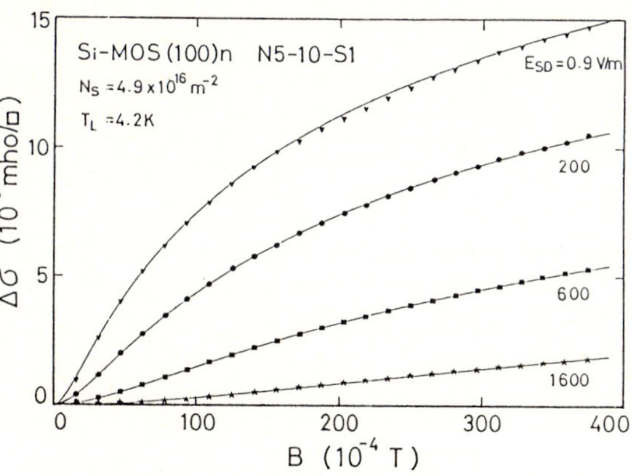

Fig. 10 Magnetoconductivity of an n-channel (100) Si MOSFET at various source-drain fields [42].

by use of (1). The changes in τ_ε with E_{SD} are shown in Fig. 11. At the electron concentration experiments were done, the electron mobility is 5000 cm^2/V·s. Experimental data in Fig. 11 show that the electric field lies within the Ohmic region of quantum correction σ' for E_{SD} in view of Tsuzuki's results [26,27] summarized in section 2. At 4.2 K , τ_ε at E_{SD} = 1000 V/m decreases to one tenth of τ_ε at E_{SD} = 1 V/m. This change is drastic as compared with that the change in the conductivity is only 2 % for the same change in E_{SD}.

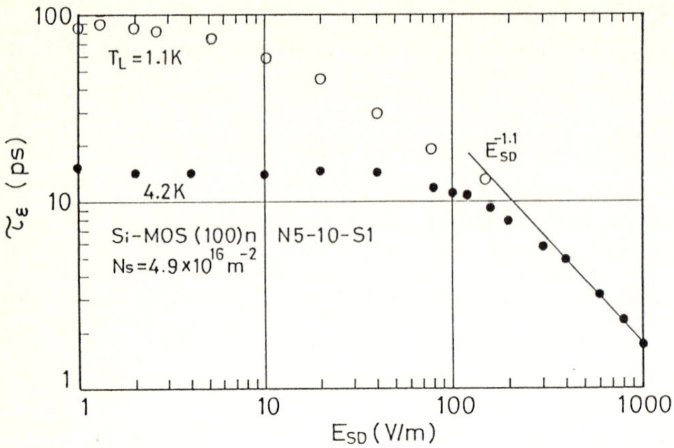

Figl 11 Inelastic scattering time (τ_ε) versus source-drain fields (E_{SD}) at lattice temperatures of 4.2 K and 1.1 K [42].

The energy loss per electron, $-\langle d\varepsilon/dt\rangle$, from the electron system to the lattice by electron-surfon scattering has been calculated by Shinba et al [36,44]. In the steady state, $-\langle d\varepsilon/dt\rangle$ is equal to the energy gain per electron through the field, $\sigma E_{SD}^2/Ns$. The calculated value of $-\langle d\varepsilon/dt\rangle$ and $\sigma E_{SD}^2/Ns$ are plotted against the temperature difference between the electrons and the lattice as shown in Fig. 12. In the calculation, the deformation potential constants of silicon bulk, Ξ = 12 eV, D = $\Xi d/\Xi u$ = -0.67, are used. Good agreement between the theoretical results and experiments in Fig. 12 makes it clear that the deformation potential constants in the inversion layer are same as in the bulk.

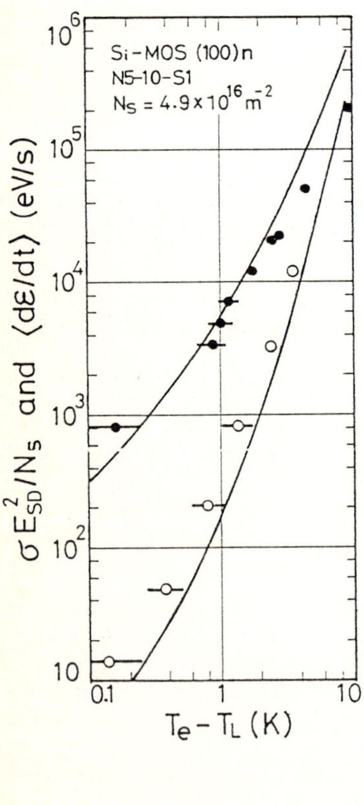

Fig. 12 The energy loss per electron from the electron system to the lattice ($\langle d\varepsilon/dt\rangle$) calculated by Shinba et al [36,44] (lines) and the energy gain by the electric field per electron $\sigma E_{SD}/N_S$ (circles) versus Te (electron temperature)-T_L(lattice temperature). Solid circles and the nearest lines are experiments and theoretical results, respectively, at T_L = 4.2 K. Open circles and the nearest lines are those obtained at 1.1 K [42].

3.3 InAs Inversion Layers and GaAs-GsAlAs Heterojunctions

Interaction effects in the negative magnetoresistance regime should show up more clearly in single-valley systems in comparison with Si inversion layers where inter-valley scattering effects complicate the problem. In order to avoid the non-linear effect predicted by Tsuzuki [26,27], however, we have to use samples having smaller elastic scattering time and smaller electron effective mass.

Yamada and Kawaji [45] have measured a negative magnetoresistance in InAs-SiO_2 interfaces. Results of magnetoconductivity experiments in samples having $\mu \simeq 4000\ cm^2/v{\cdot}s$ have been fitted to Hikami et al's expression (1) as shown in Fig. 13. Fitting parameters are $\alpha = 0.62$ and $\tau_\varepsilon/\tau = 19.6$ at 4.2 K. In view of the prefactor calculated by (5) of 0.84, these parameters look reasonable. However, temperature dependence of τ_ε is very weak; i.e. $p \simeq 0.4$. In this system, the observed negative magnetoresistance in parallel magnetic fields is smaller than that in perpendicular fields in the low field region. Such a weak anisotropy has been observed in InAs surfaces cleaved in atmospheric conditions [1]. Similar weak anisotropy has been observed also in ZnO accumulation layers by Goldstein et al [8]. We note here that their observations in perpendicular fields can be accounted for by Hikami et al's results in weak field limit [12] by assuming $\tau_\varepsilon/\tau \simeq 50 \sim 100$ and $p = 1$.

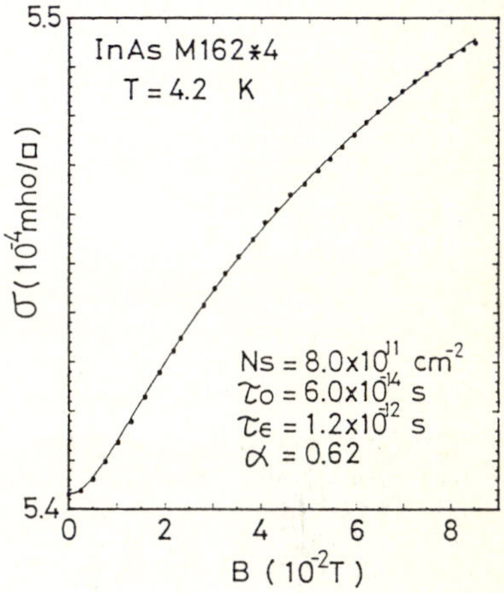

Fig. 13 Change in conductivity vs magnetic field in an InAs surface [45].

Nanbu et al [46] have measured negative magnetoresistance in GaAs-GaAlAs heterojunctions. Their preliminary results in a sample having $\mu \simeq 50000\ cm^2/v{\cdot}s$ show that $\alpha \simeq 0.9$ and $\tau_\varepsilon/\tau \simeq 16$ at 1.1 K and $p = (1 \sim 0.7)$. They have observed almost no magnetorsistance in parallel magnetic fields.

4. Summary and Disscussions

We have described mainly our results of negative magnetoresistance experiments in WLR of Si MOSFETs and some results in compound semiconductor interfaces, too. The results in Si MOSFETs are well understood by localization theory and interaction theory except the inelastic scattering time, τ_ε. Thus our understanding of negative magnetoresistance in WLR of 2D systems is still incomplete.

One problem that remains is the positive magnetoresistance in parallel magnetic fields predicted by Kawabata [31,32]. A positive magnetoresistance in parallel fields has been observed in cesiated Si inversion layers [7] and in low conductance regions ($\sigma < 4 \times 10^{-5}$ mho) of Si MOSFETs. In the low conductance region - including the strongly localized regime of Si MOSFETs -, isotropic positive magnetoresistance seems to coexist with a negative magnetoresistance which depends on the perpendicular component of the field [47]. These magnetoresistance effects, negative and positive, may allow us an approach to the understanding of the strongly localized regime.

The negative magnetoresistance in parallel magnetic fields observed in InAs and ZnO surfaces is another problem that remains to be solved.

Acknowledgements

The authors thank H. Fukuyama, A. Kawabata for helpful discussions and K. Nakamura and Y. Shinba for providing us with results of calculations before publication. This work is supported by the Grants-in-Aids for the Scientific Research from the Ministry of Education.

References

1. S. Kawaji and Y. Kawaguchi, Proc. 9th Int. Conf. Phys. Semiconductors, Moscow, (Nauka, Leningrad), p. 730 , 1968
2. S. Tansal, A. B. Fowler and A. B. Cotellessa, Phys. Rev. 178, 1326 (1969)
3. H. Sakaki and T. Sugano, J. Japan Soc. Appl. Phys. 41 suppl., 141 (1972)
4. I. Eiseleand G. Dorda, Phys. Rev. Lett. 32, 1360 (1974)
5. S. Pollitt, M. Pepper, C. J. Adkins, Surface Sci. 58, 79 (1976)
6. Y. Kawaguchi, H. Kitahara and S. Kawaji, Solid State Commun. 26, 701 (1978)
7. Y. Kawaguchi, H. Kitahara and S. Kawaji, Surf. Sci. 73, 520 (1978)
8. Y. Goldstein, Y. Grinshpan and A. Many, Phys. Rev. B19, 2256 (1979)
9. Y. Uemura and Y. Matsumoto, J. Japan Soc. Appl. Phys. 40 Suppl., 205 (1970)
10. E. Abrahams, P. W. Anderson, D. C. Licciardello and T. V. Ramakrishnan, Phys Rev. Letters 42, 673 (1979)
11. P. W. Anderson, E. Abraham and T. V. Ramakrishnan, Phys. Rev. Letters 43, 1718 (1979)
12. S. Hikami, A. I. Larkin and Y. Nagaoka, Prg. Theoret. Phys. 63, 707 (1980)
13. B. L. Altshuler, D. E. Kmel'nizkii, A. I. Larkin and P. A. Lee, Phys. Rev. B22, 5142 (1980)
14. Y. Kawaguchi and S. Kawaji, J. Phys. Soc. Jpn. 48, 699 (1980)
15. Y. Kawaguchi and S. Kawaji, J. Phys. Soc. Jpn. 49 Suppl. A, 983 (1980)
16. Y. Kawaguchi and S. Kawaji, Surf. Sci. 113, 505 (1982)
17. M. J. Uren, R. A. Davis, M. Kaveh and M. Pepper, J. Phys. C 14, L 395 (1981)
18. R. G. Wheeler, Phys. Rev. B24, 4645 (1981)
19. R. C. Dynes, Surf. Sci. 113, 510 (1982)
20. B. L. Altshuler, A. G. Aronov and P. A. Lee, Phys. Rev. Letters 44, 288 (1980)
21. H. Fukuyama, J. Phys. Soc. Jpn. 48, 2169 (1980)
22. H. Fukuyama, Surf. Sci. 113, 489 (1982)
23. H. Fukuyama, "Anderson Localization" Proc. 4th Taniguchi Int. Symp. 1981 (Springer-Verlag), (1982),
24. M. Kaveh, J. J. Uren, R. A. Davis and M. Pepper, J. Phys. C14, L 413 (1981)
25. M. Kaveh and N. F. Mott, J. Phys. C14, L 177 (1981)
26. T. Tsuzuki, Solid State Commun. 38, 915 (1981)
27. T. Tsuzuki, Solid State Commun. 40, 339 (1981)
28. S. Maekawa and H. Fukuyama, J. Phys. Soc. Jpn. 50, 2516 (1981)
29. H. Fukuyama, Progr. Theor. Phys. Suppl. 69, 220 (1980)
30. H. Fukuyama, J. Phys. Soc. Jpn. 50, 3562 (1981)
31. A. Kawabata, J. Phys. Soc. Jpn. 50, 2461 (1981)
32. A. Kawabata, Surf. Sci. 113, 527 (1982)
33. D. J. Thouless, Phys. Rev. Lett. 39, 1167 (1977)
34. A. A. Abrikosov, "Lecture on the Theory of Normal Metals" (Hidustan Publishing, India, 1968, p.24)
35. E. Abrahams, P. W. Anderson, P. A. Lee, I. V. Ramakrishnan, Phys. Rev. B 24 6783 (1981)
36. Y. Shinba, K. Nakamura, M. Fukuchi and M. Sakata, J. Phys. Soc. Jpn. 51, 157 (1982)
37. W. Sasaki, J. Phys. Soc. Jpn. 20, 825 (1965)
38. A. Kawabata, J. Phys. Soc. Jpn. 49, 628 (1980)
39. T. Ando, Surf. Sci. 98, 327 (1980)
40. W. L. Bloss, L. J. Sham and B. Vinter, Surf. Sci. 98, 250 (1979)
41. D. Vollhardt and P. Wölfle, Phys. Rev. Lett. 45, 842 (1980)
42. Y. Kawaguchi and S. Kawaji, Proc. 16th Int. Conf. on Phys. of Semiconductors, Montpellier, 1982, to be published
43. Y. Kawaguchi and S. Kawaji, submitted to JJAP Letters
44. Y. Shinba and K. Nakamura, private communication
45. Y. Yamada and S. Kawaji, unpublished
46. T. Nanbu, Y. Kawaguchi, S. Kawaji, J. Yoshino and H. Sakaki, unpublished
47. N. Hoshi and S. Kawaji, "Work Book in 4th Int. Conf. on Electron. Properties 2DS, New London, 1981", p. 160 (unpublished)
48. J. Moriyama and S. Kawaji, submitted to Solid State Commun.

ELECTRON-ELECTRON INTERACTION EFFECTS IN P-TYPE GERMANIUM INVERSION LAYERS

S. Uchida [(a)] and G. Landwehr

Max-Planck-Institut für Festkörperforschung, Hochfeld-Magnetlabor
166X, F-38042 Grenoble Cédex, France

The galvanomagnetic properties of p-type inversion layers in germanium bicrystals were studied at helium temperatures T in high magnetic fields B up to 20 Tesla. A logarithmic correction of conductivity and Hall-coefficient together with a positive magneto-resistance $\propto \ln B$ in the parallel and transverse configuration suggests the existence of electron-electron interaction effects. The scaling of the magneto-resistance with B/T can be explained with Zeeman splitting of the energy levels.

In recent years much attention has been paid to the low temperature properties of two-dimensional electronic systems with disorder. A small logarithmic decrease of conductivity with decreasing temperature in thin metal films and semiconductor inversion layers has been explained by a localization model proposed by Abrahams et al.[1]. The logarithmic increase in resistance is caused by quantum interference effects. Somewhat later the so-called interaction model was advanced by Altshuler et al.[2] which can explain a logarithmic divergence of the resistivity at low temperatures by electron-electron interaction effects in conjunction with impurity scattering. There is a fundamental difference between the two mechanisms: Whereas in the model by Abrahams et al.[1] the conductivity change is caused by a variation of the mobility, in the interaction model[2] the density of states is affected resulting in a change of the Hall constant. In order to judge the applicability of the two theories experiments in magnetic fields have been employed[3]. Not only the Hall constant and its dependence on temperature can assist in distinguishing between the two models, but also the magneto-resistance. Because the symmetry of quantum interference effects is broken by a magnetic field B perpendicular to the plane of a two-dimensional electron gas, a negative magneto-resistance[4] can result. On the other hand the balance between Hartree- and exchange effects is influenced by a magnetic field[3] in a very subtle way, and a positive magneto-resistance is expected in the interaction model even in the case when the field is in the plane of the conducting layer. Recently, the theoretical treatment of disordered two-dimensional systems has been advanced by Kawabata[5], Fukuyama[6] and also by Lee and Ramakrishnan[7] who have included the influence of the Zeeman splitting of the electronic energy levels on the magneto-transport. Due to various experimental problems it has not been possible to check the existing theories with sufficient rigour. Well-defined thin metal films are difficult to prepare. For silicon inversion layers a negative magneto-resistance at low magnetic fields has been explained by the localization model[8]. An observed positive logarithmic magneto-resistance contribution at higher B-values has been attributed [9,10] to interaction effects although there is the quantitative difficulty that the interaction theory should not apply[2]. This is expected because the ratio of Fermi wave vector k_F to inverse screening length κ in silicon inversion layers is < 1. Therefore it seems appropriate to look for other systems to test the proposed models. We will show that inversion layers existing in germanium bicrystals can serve this purpose.

A bicrystal consists of two single-crystalline halves which are tilted with respect to each other. In the resulting grain boundary a regular array of edge-dislocations exists. Because of the acceptor character of the dislocations in germanium a p-type inversion layer arises in n-type bulk material. The magneto-transport properties of p-type inversion layers in germanium bicrystals were studied about 20 years ago by

Landwehr und Handler[11] and it turned out that the magneto-resistance was anomalous. At that time no theory was available which allowed the interpretation of the data in a satisfying fashion. Recently, quasi-metallic behaviour at helium temperatures was also demonstrated by Vul and Zavaritskaya[12] for Ge-bicrystals with tilt angles >8°. Our own experiments on germanium bicrystals - with tilt angles of 10° and 15° - showed that they look very promising to obtain information about disordered two-dimensional systems.

Our work was performed on bicrystals grown by Mr.A.Köhler of the Max-Planck-Institut für Festkörperforschung Stuttgart. Samples of about 12mm length with side dimensions of 1.6mm contained a grain boundary in the center. By alloying indium potential- and current leads to the specimens ohmic contacts were made to the p-type inversion layer and rectifying ones to the n-type bulk, which had an antimony concentration of $6\cdot10^{15}cm^{-3}$. This guaranteed that no current was flowing through the bulk.

Low temperature magneto-resistance measurements on 15° samples with a hole concentration of $1\cdot10^{13}cm^{-2}$ in magnetic fields up to B=23 T revealed[13] pronounced quantum oscillations for B>5 T. Two electric subbands were identified. The carriers in the lower subbands showed no quantum oscillations, however. The analysis of data for a sample with two subbands with different mobility complicates the analysis considerably. Therefore we prefer to confine ourselves in this paper to the presentation of the data for specimens with a tilt angle of 10°. The hole concentration for such samples as derived from the low field Hall coefficient is $5\cdot10^{12}cm^{-2}$ and the Hall mobility at 4.2K is about 200 cm^2/Vs. The temperature dependence of the resistance and the low field Hall coefficient are shown in figure 1. It can be recognized that the resistance shows logarithmic behaviour over almost one temperature decade and that the plotted Hall constant has a slope (within an experimental accuracy of better than 5%) twice as large as the resistance. This suggests that we are dealing with interaction effects as proposed by Altshuler et al.[2]. P-type germanium inversion layers should be more suitable to observe interaction effects than the silicon counterpart[3] because $2k_F/\kappa \geq 1$. The product of Fermi wave vector k_F and elastic mean free path l is around 4 in our samples. The Hall-constant was independent of B up to 3T and decreased almost logarithmically at high fields by an amount of 45% per decade.

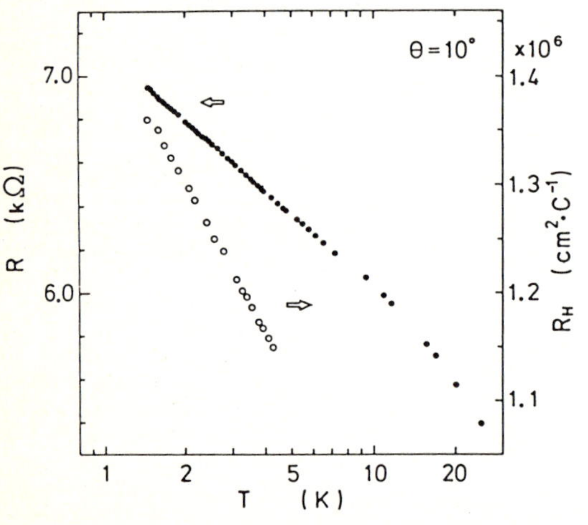

Fig.1: Resistance R per unit area and low field Hall-coefficient R_H for a germanium bicrystal with Θ=10° tilt angle as a function of temperature.

Another indication that we are dealing mainly with interaction effects is the absence of a negative magneto-resistance. This can be visualized from the inset of figure 2 where the magneto-conductance $\Delta\sigma=\sigma(B)-\sigma(0)$ (calculated as usual from the components of the resistivity tensor) has been plotted for transverse magnetic fields parallel and perpendicular to the grain boundary for temperatures of 4.2K and 1.6K. It can be recognized that a considerable temperature dependence exists and that the magneto-conductance in the parallel configuration is by no means negligible. The upper curves which have been drawn with a logarithmic B scale reveal logarithmic behaviour over a wide range. The deviation in the perpendicular configuration at the highest B-values is probably caused by cyclotron motion of the carriers. The slopes of the drawn straight lines are different for transverse- and parallel magnetic fields and they change somewhat with temperature.

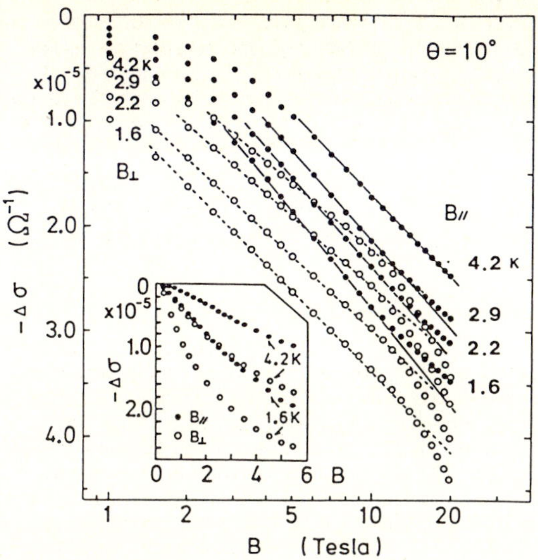

Fig. 2: Change in conductivity $\Delta\sigma=\sigma(B)-\sigma(0)$ as a function of transverse magnetic field B for a germanium bicrystal with the temperature as parameter. B is oriented parallel(||)or perpendicular(⊥) to the grain boundary. The insert at the lower left side shows the magneto-conductance at low magnetic fields on a linear scale.

In figure 3 the conductivity has been plotted for the temperature range 1.6K-4.2K with magnetic fields between 0 and 8 Tesla as a parameter, again for the perpendicular and parallel configuration. It can be seen that the slope in the logarithmic plot increases initially with increasing magnetic field in order to saturate at about 5 Tesla. The shown set of data allows us to compare them with theory.

The temperature dependence of the Hall coefficient and the absence of a negative magneto-resistance even at very small magnetic fields have encouraged us to neglect localization effects[3]. Analysis of the high field magneto-resistance data indicated that they scale with B/T within an accuracy of better than 10%, as shown in Fig.4. The deviation seems to stem partly from the slight T-dependence of the logarithmic slope which can be seen in figure 2.

This behaviour suggests that Zeeman effects are important. Therefore we have analysed our data with the theory derived by Kawabata[5]. In this model, the anomalous positive magneto-resistance in a parallel magnetic field has its origin in the Hartree terms associated with both particle hole (PH) and particle particle (PP) diffusion processes which are sensitive to the Zeeman splitting of the energy levels to be described by the ratio $g^{*}\mu_B B/k_B T$ where μ_B and g^{*} are the Bohr magneton and an effective g-factor, respectively. For large parallel magnetic fields $\Delta\sigma$ can be written as

$$\Delta\sigma=-(g_3+g_4)(e^2/2\Pi^2\hbar)\ \ln B \qquad (1)$$

using the coupling constants g_i (i=1, 2, 3, 4) introduced by Fukuyama[14]. This expression is different from that proposed by Lee and Ramakrishnan[7] who neglected the g_4 contribution. For the perpendicular field configuration all the authors proposed the following expression:

$$\Delta\sigma=-g_3(e^2/2\Pi^2\hbar)\ \ln B, \qquad (2)$$

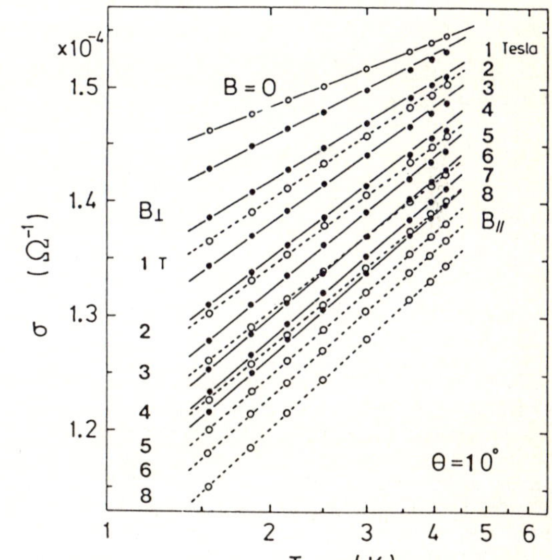

Fig.3: Conductivity σ as a function of temperature T with a transverse magnetic field oriented parallel (||) or perpendicular (⊥) to the grain boundary as a parameter.

when the field is strong enough to suppress the PP processes g_2 and g_4. Actually, a significant difference is observed in the logarithmic slope for the two orientations in the present experiment. It turns out, however, that the experimentally obtained difference in the logarithmic slopes (see fig.2) is fairly small compared with the prediction of a factor 2 based on the first order theory of the screened Coulomb interaction in which $g_3=g_4$. Furthermore, the slight deviation from the logarithmic dependence observable at very high fields as well as the slight T-dependence of the logarithmic slope is not predicted by the theory. According to Fukuyama[14] at the zero field limit the temperature dependence of the change in conductivity can be expressed as

$$\Delta\sigma = (g_1+g_2-2g_3-2g_4)\ (e^2/2\Pi^2\hbar)\ \ln T \qquad (3)$$

and in the limit of a parallel high field $\Delta\sigma$ can be written as

$$\Delta\sigma = (g_1+g_2-g_3-g_4)\ (e^2/2\Pi^2\hbar)\ \ln T \qquad (4)$$

In case of a perpendicular field all the (PP) corrections and the antiparallel spin (PH) corrections are suppressed and only the parallel spin (PH) contributions survive at high B-values. This corresponds to

$$\Delta\sigma = (g_1-g_3)\ (e^2/2\Pi^2\hbar)\ \ln T \qquad (5)$$

All these relations (1)-(5) allow the complete determination of the interaction parameters g_1-g_4 from our data. The obtained values are listed in table 1 :

Table 1 :

g_1	g_2	g_3	g_4
2.2	0.14	0.62	0.26

It has been argued before that for the first order screened Coulomb interaction model the following values should hold : $g_1=1$, $g_2=g_3=g_4=F/2$. In our present case F should be around 0.5. It is obvious that there is a difference between our results and the theoretical estimates. Nevertheless, it should be emphasized that a qualitative agreement does exist.

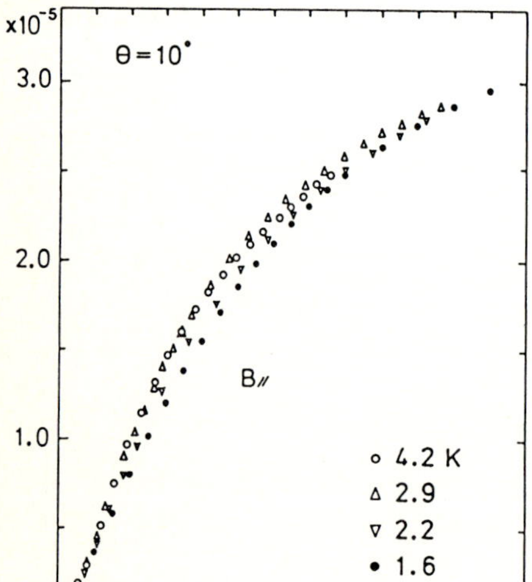

Fig.4: Change in conductivity $\Delta\sigma$ as a function of B/T for a parallel magnetic field.

If one tries to trace possible origins of the deviations one could suspect that the neglect of localization might be the reason. However, this is not probable because in germanium the spin-orbit interaction is quite substantial, resulting in the depression of the third valence band by 290 meV compared with only 44 meV in silicon. It should also be noted that in bicrystals with 20° tilt angle a negative magnetoresistance has never been observed[11] at He-temperature. Therefore we believe that the neglect of localization effects is justified.

Hikami[4] et al. argued that in the case of strong spin-orbit scattering the prefactors of both the log T- and log B-corrections have an opposite sign compared to the case without spin-orbit scattering. It has been noted, however, that in order for the process to become effective some kind of three-dimensional behaviour is required[3]. Due to the strong electric field at the grain boundary of our samples the inversion layer must be considered as a strictly two-dimensional system. Recently subband calculations for p-type germanium inversion layers have been made by E. Bangert[15]. They suggest that inter-subband scattering is unimportant at the temperatures of our experiment. Moreover, the effects caused by intersubband scattering would be opposite to those which we actually observed.

In conclusion we believe that we have demonstrated the existence of interaction effects in p-type germanium inversion layers. Our results are qualitatively quite consistent with recent theoretical models. There exists, however, some quantitative discrepancy in the magnitude of the interaction parameters. Two possibilities have been pointed out by Fukuyama[16] recently. The interaction theory so far proposed takes only the first order terms in $(k_F l)^{-1}$ and F into account, which is valid in case $k_F l \gg 1$ and $k_F \gg \kappa$. The value of $k_F l$ is about 4 and $k_F \lesssim \kappa$ in the present samples, so that the theory may not be quite adequate to explain the experiments. Another reason for the disagreement might be due to the anisotropy of the mass in the p-type inversion layer. The recent calculations by Bangert[15] have indicated that the constant energy contours of the heavy holes in a [100] medium angle grain boundary are substantially warped, resembling those in (100)p-type silicon inversion layers. It has been pointed out recently, that for a multivalley model with elliptical constant energy lines the prefactor of the logarithmic terms depends on the mass anisotropy [17] In order to check the above mentioned possibilities, further theoretical and experimental efforts seem necessary.

Acknowledgement

The efforts of Mr. A. Köhler and Dr. H. Schönherr, Max-Planck-Institut für Festkörperforschung, Stuttgart, in the preparation of the germanium-bicrystals are appreciated. The technical assistance of Mr. A. Klaschka was helpful. The authors enjoyed valuable discussions with Prof. H. Fukuyama.

References

[a] Permanent address: Dept.of Applied Physics, University of Tokyo, Japan

[1] E.Abrahams, P.W. Anderson, D.C. Licciardello and T.V. Ramakrishnan, Phys.Rev. Lett. 42 , 672 (1979)

[2] B.L. Altshuler, A.G. Aronov and P.A. Lee, Phys.Rev.Lett. 44, 1288 (1980)

[3] For a recent review see, e.g., H. Fukuyama, in: Electronic Properties of Two-Dimensional Systems, Surface Science 113, Special Issue, F. Stern Ed. 1982, p. 489; and M.J. Uren, R.A. Davies, M. Kaveh and M. Pepper, J.Phys. C Solid State Phys. 14, 5737 (1981)

[4] S. Hikami, A.I. Larkin and Y. Nagaoka, Progr. Theor. Phys. 63, 707 (1980)

[5] A. Kawabata, Surface Science 113, 355 (1982)

[6] H. Fukuyama, J. Phys. Soc. Japan 50, 3562 (1981)

[7] P.A. Lee and T.V. Ramakrishnan, Phys. Rev. B 26, 4009 (1982)

[8] Y. Kawaguchi and S. Kawaji, J. Phys. Soc.Japan 48, 699 (1980)

[9] R.C. Dynes, Survace Science 113, 510 (1982)

[10] R.A. Davies, M.J. Uren and M. Pepper, J. Phys. C 14, L531 (1981)

[11] G. Landwehr and P. Handler, J. Phys. Chem. Solids 23, 891 (1962)

[12] B.M. Vul and E.I. Zavaritskaya, Phys. of Semiconductors, 1978, Conf. Ser. 43, Inst. Phys. London, 1978, p. 421

[13] S. Uchida and G. Landwehr, to be published

[14] H. Fukuyama, J. Phys. Soc. Japan 48, 2169 (1980)

[15] E. Bangert, to be published

[16] H. Fukuyama, private communication

[17] R.N.Bhatt and T.V.Ramakrishnan, Physica 109 & 110B 2078 (1982).

ANHARMONIC EFFECTS IN TWO-DIMENSIONAL ELECTRON SOLIDS IN MAGNETIC FIELDS

Günther Meissner
Theoretische Physik, Universität des Saarlandes,
Saarbrücken, Federal Republic of Germany* and
Brookhaven National Laboratory, Upton, New York 11973

Making extensive use of sum-rule techniques when time-reversal symmetry is broken we were able to investigate collective excitations of two-dimensional electron solids in the presence of a quantizing magnetic field not resting upon the harmonic approximation.

There has been considerable interest recently in the effect of an intense magnetic field B applied perpendicular to a two-dimensional (2D) system of electrons[1] which is expected to form a Coulomb solid[2] at low temperatures T. Large values of lattice spacing a >100Å are realizable experimentally in semiconductor inversion layers (Si-MOSFET) as well as in layered semiconductor compounds (GaAs-AlGaAs heterojunctions) due to relatively low areal densities $n_s(\approx 10^{11}\ cm^{-2})$ of electrons. Quantum effects and nonlinearities are therefore expected to play an important role for magnetic fields B of the order of 100 kG, when the number of magnetic flux quanta per unit cell area of a triangular lattice[3], $\alpha=(\sqrt{3}\ a^2B/2)/(hc/e)$, becomes larger than one.

Unlike in the absence of magnetic fields, time-reversal invariance is broken in such systems. By extensive use of sum-rule techniques when time-reversal symmetry is broken we were able to investigate collective excitations of such 2D electron solids in the presence of a quantizing magnetic field not resting upon the harmonic approximation (HA). Besides generalizing and proving rigorously results obtained previously in HA[4], the inclusion of anharmonicities is inevitable in order to investigate (1) the temperature dependence of measurable quantities as for instance the shear modulus; (2) contributions to the linewidths due to damping of collective excitations, etc. Exact expressions for such quantities are derived. Their explicit evaluation up to one-loop approximation as well as resulting physical implications are presented.

Let us start with the usual simple, but rigorous way of studying the excitation energies of the electron lattice in a uniform charge compensating background by considering the response to a suitable external probe. Applying sum-rule techniques in the presence of the time-reversal symmetry breaking magnetic field B we have been able to obtain the following exact spectral representation for the inverse of the displacement response function

$$\chi^{-1}_{\alpha\alpha'}(\underset{\sim}{k},z=\omega+i\delta)=m\left\{z^2\delta_{\alpha\alpha'}+iz\Omega_c\varepsilon_{\alpha\alpha'}-M_{\alpha\alpha'}(\underset{\sim}{k},0)+z\int_{-\infty}^{+\infty}\frac{d\omega}{\pi}\,\frac{\gamma_{\alpha\alpha'}(\underset{\sim}{k},\omega)}{\omega-z}\right\} \qquad (1)$$

where m denotes the electron mass, α and α' (=1,2) are Cartesian indices, and the 2x2 matrices

$$\delta_{\alpha\alpha'}=\begin{pmatrix}1&0\\0&1\end{pmatrix} \text{ and } \varepsilon_{\alpha\alpha'}=\begin{pmatrix}0&1\\-1&0\end{pmatrix}.$$

The spectral width function $\gamma_{\alpha\alpha'}(\underset{\sim}{k},\omega)$ in (1) has the same symmetry properties as the spectral function $\tau_{\alpha\alpha'}(\underset{\sim}{k},\omega)$ which is the Fourier transform of the commutator of displacement operators. In the second term on the r.h.s. of (1), $\Omega_c\equiv eB/mc$ denotes the cyclotron frequency of electrons of charge e due to the magnetic field B with c being the velocity of light. Physically, this term therefore represents the Lorentzian force; formally it results from the second frequency moment of $\tau_{\alpha\alpha'}$ which would vanish identically in a time-reversal invariant system. For the static (z=0) limit of the dynamical matrix in $\chi^{-1}_{\alpha\alpha'}$ of (1) we obtained rigorously again by sum-rule techniques

$$M_{\alpha\alpha'}(\underset{\sim}{k},o) = M^{(\infty)}_{\alpha\alpha'}(\underset{\sim}{k}) - \int_{-\infty}^{+\infty} \frac{d\omega}{\pi} \gamma_{\alpha\alpha'}(\underset{\sim}{k},\omega) \tag{2}$$

with the high-frequency dynamical matrix

$$M^{(\infty)}_{\alpha\alpha'}(\underset{\sim}{k}) = \frac{1}{m}\sum_{\ell\neq o}\left[1-e^{-i\underset{\sim}{k}\cdot\underset{\sim}{R}(\ell)}\right]\int d^2r \nabla_\alpha \nabla_{\alpha'} \frac{e^2}{|\underset{\sim}{r}|} \langle\delta(\underset{\sim}{r}-\underset{\sim}{x}(\ell)+\underset{\sim}{x}(o))\rangle . \tag{3}$$

In (3), $x(\ell)$ and $\underset{\sim}{x}(o)$ denote position operators of electrons and the lattice vector $\underset{\sim}{R}(\ell)$ of the ℓth electron is given by its mean equilibrium position $\langle\underset{\sim}{x}(\ell)\rangle$. Zeros of the determinant of the real part of the inverse of the displacement response function of (1), i.e.

$$|\mathrm{Re}\ \chi^{-1}_{\alpha\alpha'}(\underset{\sim}{k},\tilde{\omega}_j(\underset{\sim}{k}))| \equiv o, \tag{4}$$

finally determine the relation between frequencies $\tilde{\omega}_j$ and wave vectors $\underset{\sim}{k}$ of the associated mode $j(=1,2)$. The damping of the jth mode is determined by

$$\alpha_j(\underset{\sim}{k}) = \gamma_j(\underset{\sim}{k},\tilde{\omega}_j(\underset{\sim}{k}))/(\partial\tilde{\omega}_j(\underset{\sim}{k})/\partial|\underset{\sim}{k}|) \tag{5}$$

and the density of states reads

$$g(\omega) = \frac{2\omega}{N}\sum_{\underset{\sim}{k},j=1,2} \mathrm{Im}\chi_j(\underset{\sim}{k},\omega^2-i\delta) \tag{6}$$

where the quantities γ_j and χ_j are related to $\gamma_{\alpha\alpha'}$ and $\chi_{\alpha\alpha'}$, respectively, by introducing left and right polarization vectors which formally diagonalize $\chi^{-1}_{\alpha\alpha'}$. In Eq. (6), N is the number of electrons.

All of the foregoing results are exact. The method of their application to the study of anharmonic effects in 2D electron solids will now be illustrated by using perturbation theory to evaluate various quantities up to one-loop approximation.

For the dynamical matrix in the static limit we may write this perturbation expansion as

$$M_{\alpha\alpha'}(\underset{\sim}{k},o) = M^{(o)}_{\alpha\alpha'}(\underset{\sim}{k},o) + M^{(1)}_{\alpha\alpha'}(\underset{\sim}{k},o) + M^{(2)}_{\alpha\alpha'}(\underset{\sim}{k},o) + \ldots \tag{7}$$

From (2) we find in an iterative procedure

$$M^{(o)}_{\alpha\alpha'}(\underset{\sim}{k},o) \simeq \frac{1}{m}\sum_{\ell\neq o}\left[1-e^{-i\underset{\sim}{k}\cdot\underset{\sim}{R}(\ell)}\right] \frac{\partial^2}{\partial R_\alpha(\ell)\,\partial R_{\alpha'}(\ell)} \frac{e^2}{|\underset{\sim}{R}(\ell)|} \equiv \phi^{(o)}_{\alpha\alpha'}(\underset{\sim}{k},-\underset{\sim}{k}) \tag{8}$$

if we replace operators $\underset{\sim}{x}$ in $\langle\delta(\underset{\sim}{r}-\underset{\sim}{x}(\ell)+\underset{\sim}{x}(o))\rangle$ of (3) by their mean values $\langle\underset{\sim}{x}\rangle\equiv\underset{\sim}{R}$ and put the damping function $\gamma_{\alpha\alpha'}\equiv o$. The excitation frequencies of polarization $j=\pm$ to be obtained in that approximation from (1) according to[4]

$$\tilde{\omega}^{o\,2}_{\pm}(\underset{\sim}{k}) = \frac{1}{2}\left\{\omega^{o\,2}_{+} + \omega^{o\,2}_{-} + \Omega^2_c \pm \left[(\omega^{o\,2}_{+} + \omega^{o\,2}_{-} + \Omega^2_c)^2 - 4\omega^{o2}_{+}\omega^{o2}_{-}\right]^{\frac{1}{2}}\right\} \tag{9}$$

together with their polarization vectors $\underset{\sim}{e}^j(\underset{\sim}{k})$ can then be used to determine in one-loop approximation third-order anharmonicities which have to be included in the damping function $\gamma_{\alpha\alpha'}(\underset{\sim}{k},\omega)$ simultaneously with fourth-order anharmonicities in the instantaneous dynamical matrix $M^{(\infty)}_{\alpha\alpha'}(\underset{\sim}{k})$. We thus find explicitly due to the instantaneous anharmonicities

$$M^{(1)}_{\alpha\alpha'}(\underset{\sim}{k},o) = \frac{\hbar}{Nm}\sum_{\underset{\sim}{q}j} \phi^{(o)}_{\alpha\alpha'\beta\beta'}(\underset{\sim}{k},-\underset{\sim}{k},\underset{\sim}{q},-\underset{\sim}{q}) \frac{e^{*\,j}_{\beta}(\underset{\sim}{q})\,e^{j}_{\beta'}(\underset{\sim}{q})}{2m\tilde{\omega}^o_j(\underset{\sim}{q})} \coth(\beta\hbar\tilde{\omega}^o_j(\underset{\sim}{q})/2) \tag{10a}$$

and due to the dispersive anharmonicities

$$M^{(2)}_{\alpha\alpha'}(\underset{\sim}{k},o) = \frac{\hbar}{2}\sum_{\underset{\sim}{q}j_1j_2} -\phi^{(o)}_{\alpha\beta\gamma}(\underset{\sim}{k},\underset{\sim}{q},-\underset{\sim}{k}-\underset{\sim}{q})\,\phi^{(o)}_{\alpha'\beta'\gamma'}(-\underset{\sim}{k},-\underset{\sim}{q},\underset{\sim}{k}+\underset{\sim}{q}) \frac{e^{*\,j_1}_{\beta}(\underset{\sim}{q})\,e^{*\,j_2}_{\gamma}(\underset{\sim}{k}+\underset{\sim}{q})\,e^{j_1}_{\beta'}(\underset{\sim}{q})\,e^{j_2}_{\gamma'}(\underset{\sim}{k}+\underset{\sim}{q})}{4m^2\tilde{\omega}^o_{j_1}(\underset{\sim}{q})\,\tilde{\omega}^o_{j_2}(\underset{\sim}{k}+\underset{\sim}{q})}$$

$$\times \left\{ \frac{\left[1+n(\tilde{\omega}^{o}_{j_1}(q))+n(\tilde{\omega}^{o}_{j_2}(k+q))\right]}{\tilde{\omega}^{o}_{j_1}(q)+\tilde{\omega}^{o}_{j_2}(k+q)} + 2\,\frac{\left[n(\tilde{\omega}^{o}_{j_1}(q))-n(\tilde{\omega}^{o}_{j_2}(k+q))\right]}{\tilde{\omega}^{o}_{j_2}(k+q)-\tilde{\omega}^{o}_{j_1}(q)} \right\} \qquad (10b)$$

In (10a) and (10b) we have introduced the Fourier transforms of the microscopic expressions for the bare fourth-order coupling constant

$$\phi^{(o)}_{\alpha\alpha'\beta\beta'}(k,-k,q,-q)=\sum_{\ell\neq o}\left[1-e^{ik\cdot R(\ell)}\right]\left[1-\cos q\cdot R(\ell)\right]\frac{\partial^4}{\partial R_\alpha\,\partial R_{\alpha'}\,\partial R_\beta\,\partial R_{\beta'}}\;\frac{2e^2}{|R(\ell)|} \qquad (11a)$$

and for the bare third-order coupling constant:

$$\phi^{(o)}_{\alpha\beta\gamma}(k,q,-k-q) = \sum_{\ell\neq o}\left[1-e^{-ik\cdot R(\ell)}\right]\left[1-e^{-iq\cdot R(\ell)}\right]\frac{\partial^3}{\partial R_\alpha\,\partial R_\beta\,\partial R_\gamma}\;\frac{e^2}{|R(\ell)|}\;, \qquad (11b)$$

respectively. Because of the factor coth $(\beta\hbar\omega/2)$ in (10a) and due to the Bose distribution functions, $n(\omega)=\left[\exp(\beta\hbar\omega)-1\right]^{-1}$, in (10b) where $\beta\equiv 1/k_BT$, expression (7) for $M_{\alpha\alpha'}$ is valid both in the quantum limit and in the classical limit, as well as in the intermediate regime. The scale of splitting the degeneracy in $\tilde{\omega}^{o}_{\pm}(k)$ as compared to the frequencies $\omega^{o}_{\pm}$ of (9) in the absence of the magnetic field is given by the cyclotron frequency Ω_c. From the structure of the rigorous expression (1) for $\chi^{-1}_{\alpha\alpha'}$ we can infer that this behavior holds quite generally, i.e., in the presence of anharmonicities. Therefore, as a first step we have calculated these frequencies $\tilde{\omega}^{o}_{\pm}$ in the main symmetry directions of the Brillouin zone (Fig. 1b). The representation in Fig. 1a shows that this splitting of degeneracy is reflected in the density of states $g(\omega)$, too. There, the magnetic field increases the separation between the van Hove singularities of the transverse (T) excitation branch (j= -) and of the longitudinal (L) branch (j= +). Simultaneously, the intensities of the corresponding peaks are increased. Since the density of states is accessible experimentally

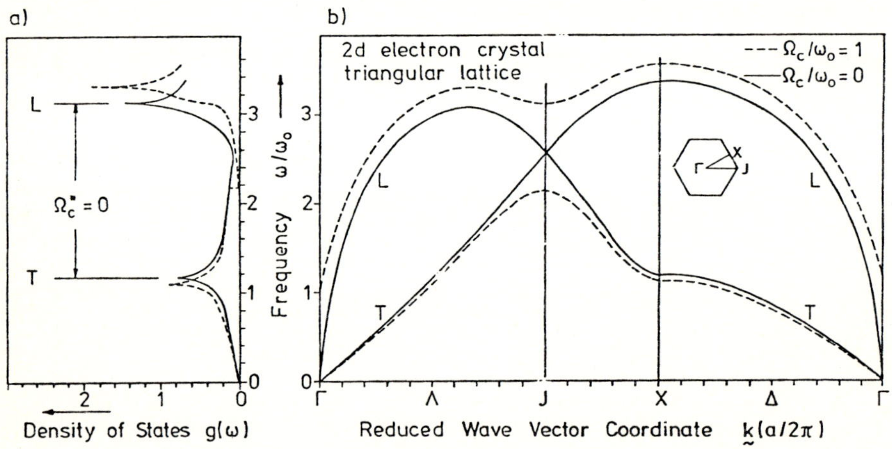

Fig. 1. (a) Density of states of a triangular electron lattice for $\Omega_c^* = 0,1$ of the reduced cyclotron frequency $\Omega_c^* = \Omega_c/\omega_o$ with the characteristic frequency $\omega_o^2 = e^2/ma^3$.

(b) Dispersion curves for a triangular electron lattice at zero magnetic field (solid line) and at finite magnetic field (broken line).

this signature of the cyclotron frequency might be of importance for observing a 2D electron solid in semiconductors.

From a long wave-length expansion of (7), i.e.

$$M_{\alpha\alpha'}(\underset{\sim}{k},o) = \frac{2\pi e^2 n_s}{m} \frac{k_\alpha k_{\alpha'}}{k} + Z_{\alpha\alpha'\beta\beta'} k_\beta k_{\beta'} + 0_{\alpha\alpha'}(k^4) \qquad (12)$$

we have also been able to investigate microscopically anharmonic effects on the temperature dependence of the shear modulus

$$\mu(T) = mn_s \left[Z^{(o)}_{12} + Z^{(1)}_{12} + Z^{(2)}_{12} + \ldots \right] \qquad (13)$$

in the classical limit for vanishing magnetic field[5]. Evaluating carefully lattice sums via Ewald transformation we found a factor 26.7 ± 2 in the linear temperature decrease of $\mu(T) = \mu_o[1-30.7k_BT/e^2(\pi n_s)^{\frac{1}{2}}]$ induced by the instantaneous and dispersive anharmonicities $Z^{(1)}$ and $Z^{(2)}$, respectively, correcting for the temperature independent term $\mu_o = mn_sZ^{(o)}_{12}$. This factor is in good agreement with molecular dynamics calculations of Morf 6). In the present analysis, however, $Z^{(2)}_{12}$ contains no logarithmic divergence, in contrast to investigations reported by Chang and Maki[7]. The renormalization of the shear modulus in our anharmonic approach can therefore be considered to be consistent with dislocation-mediated melting[8].

In conclusion, we would like to remark that the effect of a quantizing magnetic field on the spectral width function $\gamma_{\alpha\alpha'}(\underset{\sim}{k},\omega)$ may further be studied by employing a moment approximation. Frequency moments of $\tau_{\alpha\alpha'}$ and $\gamma_{\alpha\alpha'}$ can be connected by a high-z expansion of (1). Sum-rule techniques can again be used to obtain higher frequency moments of $\tau_{\alpha\alpha'}(\underset{\sim}{k},\omega)$ in terms of higher static electron-correlation functions. Rapidly varying parts which may appear in $\gamma_{\alpha\alpha'}(\underset{\sim}{k},\omega)$ must be extracted. Details of such an analysis, however, have to be deferred to an extended publication.

I would like to thank Dr. V. J. Emery for the hospitality afforded me at Brookhaven National Laboratory.

* Permanent Address.

References

1) D. C. Tsui, H. L. Störmer, and A. C. Gossard, Phys. Rev. Lett. 48, 1559 (1982).
2) G. Meissner, Surface Science 73, 411 (1978).
3) G. Meissner, H. Namaizawa, and M. Voss, Phys. Rev. B13, 1370 (1976).
4) G. Meissner, Z. Physik B23, 173 (1976).
5) G. Meissner and G. Schöner, to be published; the dependence of the actual numerical coefficient in the linear temperature decrease of the shear modulus on the precise form of the dispersion of the phonon frequencies and of the polarization vectors in HA is being discussed in this paper.
6) R. Morf, Phys. Rev. Lett. 43, 931 (1979).
7) M. C. Chang and K. Maki, Surface Science 113, 432 (1982).
8) J. M. Kosterlitz and D. J. Thouless, J. Phys. C6, 1181 (1973).
R. Nelson and B. Halperin, Phys. Rev. B19, 2457 (1979).

LOCALISATION AND SCALING OF RESISTANCE OF DISORDERED SYSTEMS INCLUDING MAGNETIC FIELD EFFECTS

A. MacKinnon
Universität Heidelberg
Institut für Theoretische Physik
Philosophenweg 19
D-6900 Heidelberg, F.R.G.

B. Kramer
Physikalisch-Technische Bundesanstalt
Bundesallee 100
D-3300 Braunschweig, F. R. G.

The concept of localisation is briefly reviewed with special emphasis on the scaling properties of the conductance. The influence of a magnetic field is discussed. Numerical results for the localisation length in two- and three-dimensional systems are reported. It is found that it obeys a one-parameter scaling law, which depends on the dimensionality, and which allows conclusions on the temperature dependence of the resistance. The influence of a magnetic field on localisation is studied with the same numerical procedure as in the case of zero magnetic field. The localisation length is found to obey a two parameter scaling law. A negative magnetoresistance is obtained. Evidence is presented that even for strong magnetic field all quantum states remain localised. An interpretation of recent high field magnetotransport data obtained from inversion layers is attempted.

1. Introduction

Localisation of quantum mechanical particles in random potentials has been an extensively studied problem, since Anderson [1] introduced the concept from the viewpoint of diffusion. Theoretical investigations concentrated on basic concepts [2,3] as well as detailed analytical [4] and numerical [5] studies of various models. Particularly important are one dimensional disordered systems [6] since here all states are localised and many rigorous conclusions are possible.

The relation to the scaling properties of the conductance, although introduced previously [7], was formulated conceptually only in 1979 [8], and gave rise for the study of the connection between localisation and the transmission properties of randomly arranged potential barriers [9]. This idea appeares to be particularly useful for numerical studies [10] since it may be generalised to dimensions greater than one [11,12]. There is now *consensus* that for dimensions $d \leq 2$ all single particle states remain localised for arbitrarily small disorder, whereas there is an Anderson transition in three dimensions at some finite disorder, say W_c, such that only for disorder $W > W_c$ all states are localised. Thus, d=2 is the critical dimension for localisation.

At present, the main emphasis of studies of localisation is on how the picture may be changed when external magnetic fields and interactions are taken into account [13-15]. There is analytical evidence for a negative magnetoresistance, when localisation is important. There are claims that this is equivalent to an increase of the localisation length upon increasing the magnetic field [15]. There is also evidence that interactions in disordered systems lead to similar effects as localisation as far as the temperature dependence of the resistance is concerned. It is therefore desirable to have a tool, which allows for the distinction between the two. This may be the magnetic field [16]. Of particular interest are two dimensions since here, without magnetic field, the localisation length is finite.

Experimentally, localisation in two dimensions can be studied via transport measurements on thin metallic films and inversion layers at low temperatures such that interaction effects are reduced [17-20]. The most spectacular result is a *logarithmic increase of the resistance with decreasing temperature.* Also a *negative magnetoresistance* was observed at low magnetic field with a *dependence on the field strength,* which may be interpreted as *logarithmically* [21-23], consistent with predictions from localisation theory [14,15]. At higher magnetic fields there is evidence that interaction effects become more and more important such that the magnetoresistance becomes positive [24]. For extremely high fields the *quantum Hall effect* is observed [25,26] in inversion layers.

For a better understanding of these effects a thorough investigation of the *interplay between localisation and a magnetic field* is certainly only a first step. With the numerical data, which will be presented in this paper, we want to contribute to this problem.

2. Localisation and the scaling of the conductance

From the point of view of diffusion localisation can be defined by the statement that the dc-conductivity, given by the Kubo-Greenwood formula, of an infinite disordered system tends to zero if the temperature approaches absolute zero. Unfortunately, this is impossible to realise experimentally since none of the limits involved can be achieved in reality. One has to ask, whether one can find a definition of localisation using finite systems [7]. This is indeed possible by considering the *scaling properties* of the conductance with the size of the system [27]. The definition found in this way is not only advantageous from the experimental point of view, since it allows direct predictions about e. g. the temperature dependence of the resistance in the localised regime, but also distinguishes between different regimes of localisation, which give rise for a variety of observable effects [16,28].

In the "classical" limit by which we mean the limit of vanishing disorder, such that localisation is not important, the conductance g of, say, a hypercube of volume L^d, where d is the dimensionality, scales with the length L according to

$$g = L^{d-2} \sigma. \qquad (1)$$

The dc-conductivity σ is here a material constant, which can be calculated, in principle, from the Kubo formula. It is defined via a thermodynamic limit and thus independent of L.

In the limit of strong disorder one can expect that localisation will dominate such that σ vanishes. Eq. (1) no longer describes the scaling properties of the conductance. If the states near the Fermi energy are exponentially localised, we may expect exponential scaling

$$g = g_0 \exp(-L/\lambda_\infty) \qquad (2)$$

where λ_∞ is a localisation length, which depends on the disorder.

In the intermediate regime perturbation theory together with a number of simple assumptions [8], as well as detailed diagrammatic techniques [29] and numerical algorithms [12] may be used in order to obtain information about the scaling properties of the conductance. It is found that g depends on L only as

$$g = g(L/\lambda_\infty). \qquad (3)$$

The localisation length λ_∞ is a material constant and depends only on the disorder. It is defined in the thermodynamic limit. As a consequence there is a universal scaling function $\beta = d \ln g/d \ln L \equiv \beta(g)$.

In the asymptotic regimes of vanishing and strong disorder which we may now characterize adequately by $\lambda_\infty >> L$ ("weak localisation") and $\lambda_\infty << L$ ("strong localisation"), respectively, the scaling function is

$$\beta(g) = \begin{cases} d - 2 - \frac{1}{\pi^2}\,\frac{1}{g} + \ldots & \lambda_\infty >> L \\ \ln(g/g_0)\,[1 + \gamma\, g + \ldots] & \lambda_\infty << L \end{cases} \tag{4}$$

The first terms in these expressions may be derived from eqs. (1) and (2). The corrections are obtained by perturbation expansion [8].

If the β-function is known the dependence of the conductance on the size of the system can be determined by integration. In two dimensions, for instance, one obtaines for $\lambda_\infty >> L$ the celebrated logarithmic length dependence, which may be converted by a number of ingenious arguements [30] identifying L with the length travelled by a particle between two inelastic scattering events into a logarithmic temperature dependence.

In the presence of a magnetic field the conductivity becomes a tensor. There are two components to consider: σ_{xx}, which corresponds to the current parallel to the electric field, and the Hall conductivity σ_{xy}. They can be calculated from the Kubo formula, in principle [31].

Again, for localised states σ_{xx} is zero at absolute zero of the temperature. On the other hand σ_{xy} may be finite if there are some extended states below Fermi energy. This is because the Hall conductivity essentially counts the total number of mobile carriers, whereas σ_{xx} is determined only by the mobile carriers with an energy near Fermi level. If a system contains only localised states σ_{xy} is also zero.

Again, as in the case of zero magnetic field, one may ask for the conductance of a finite system, in particular if localisation comes into play. One may ask whether the definition of a localisation length is still possible. Whether such a localisation length would have to depend on the magnetic field, or, alternatively, only on the disorder, the magnetic field giving rise for a second characteristic length, connected with the cyclotron radius, for instance, is presently not clear. This is closely related the question of the existence of a universal scaling function, which may be used to derive the dependence of the conductance on the size of the system and the magnetic field[32].

3. Localisation and transmission through random potentials

In this chapter we define localisation in terms of the transmission properties of random potentials.

Consider a system of length L in the x-direction and lateral dimensions M, N in the y- and the z-direction, respectively, represented by some random Hamiltonian H, which may contain a magnetic field. We ask for the probability that a particle incident on the plane x = o will be transmitted in x-direction through the system. This is given by the transmission coefficient [12].

$$T = \mathrm{Tr}\, |\langle x = o|\; G(E)\; |x = L\rangle|^2 \tag{5}$$

where

$$\langle x = o|\; G(E)\; |x = L\rangle = \langle x = o|\; (E - H)^{-1}\; |x = L\rangle \tag{6}$$

is the matrix element of the resolvent of H between points at the planes x = o and x = L. The trace in eq. (5) is over the surfaces at x = o and x = L with the areas M·N. For L >> M,N the system is effectively one dimensional.

The transmission probability will then exponentially decrease with L [9,33] and we may define a decay length λ as

$$\lambda^{-1} = -\lim_{L\to\infty}(2L)^{-1} \ln T \tag{7}$$

Via the resolvent, λ depends on the energy of the incident particle, E, as well as the other parameters in the Hamiltonian, i.e. the disorder W, and the magnetic field B. The dimensionality d enters via the lateral dimensions of the system. For M or N equal zero we have d = 2. For both M = N = o d = 1.

The localisation length in the infinite two dimensional system may now be defined as the limit

$$\lambda_\infty(E,W,B) = \lim_{M\to\infty} \lambda(E,W,B,M) \tag{8}$$

This definition is readily generalised to three dimensions. The numerical study of this limit gave rise for the discovery of a scaling law for λ, which depends on the dimensionality [12].

In one dimension the transmission probability may be related to the conductance at absolute zero of the temperature [9,34,35] (in units of e^2/h)

$$g = \frac{T}{1-T} \tag{9}$$

In the limit of large L the conductance is given directly by T and scales therefore exponentially with L.

For $d \neq 1$ (L, M $\neq$ o) there is no such relation [36,37] except in the case of independent "channels" [37,38]. Thus there is no simple recipe how the exponential decay length of T, which can be determined very accurately, can be used to calculate the conductance, even in the limit L >> M,N. However, assuming that the dominant contribution to the current is in the channel with the largest decay length one may use eq. (9) as a working definition [12]. This relates the scaling properties of the conductance, i.e. the β-function, to the scaling properties of λ, which turned out to be very useful in actual calculations.

4. Model Hamiltonians

Before describing our results it is useful to summarize briefly some of the Hamiltonians used in the study of disordered systems, particularly with respect to the inclusion of a magnetic field.

The most simple model is that of a particle moving in a random potential $V(\vec{r})$.

$$H = \frac{\vec{p}^2}{2m} + V(\vec{r}) \tag{10}$$

The potential has to be specified by some probability distribution

$$P = P\,[V] \tag{11}$$

For practical purposes it is often more convenient to work on a lattice with a complete set $\sum_j |j\rangle\langle j| = 1$ associated to the lattice sites. Then

$$H = V \sum_{\langle j\, j'\rangle} |j\rangle\langle j'| + \sum_j |j\rangle\langle j|\, \varepsilon_j \tag{12}$$

is a suitable model for the study of localisation [1]. The first sum is only over pairs of nearest neighbours. For sufficiently localised site states $|j\rangle$ eq. (12) may be derived from eq. (10) [39]. The first term corresponds to the kinetic energy. The second term is the random potential. For convenience one may take ε_j independently at random, i.e.

$$p(\varepsilon_1,\varepsilon_2,\varepsilon_3 \ldots) = \prod_j P(\varepsilon_j) \tag{13}$$

The disorder parameter may then be defined as the width of the site energy distribution $P(\varepsilon)$, W, divided by V, the hopping energy.

A magnetic field may be introduced into eq. (10) by the replacement $\vec{p} \to \vec{p} - e\vec{A}$, where $\vec{A}$ is the vector potential.

$$H = \frac{1}{2m} (\vec{p} - e\vec{A})^2 + V(\vec{r}) \tag{14}$$

This may be converted into a lattice model in analogy to eq. (12) if the magnetic field is sufficiently constant.

$$H = \sum_{<j\,j'>} V_{j\,j'}(\vec{A}) \,|j><j'| + \sum_j |j><j|\, \varepsilon_j \tag{15}$$

Here, the nearest neighbour hopping matrix elements $V_{j\,j'}$ are complex and depend on the vector potential, that is the magnetic field.

$$V_{j\,j'} = V\, e^{-i\, e\vec{A}\cdot(\vec{j}-\vec{j}')/\hbar} \tag{16}$$

In the limit of *very strong magnetic field* it may be more convenient to use the Landau states, the eigenstates of $H_0 = (\vec{p} - e\vec{A})^2/2m$, as a basis set. This has been done in recent numerical work [40,41]. The resulting Hamiltonian has a very complicated structure, particularly concerning the matrix elements of the random potential, and seems only advantageous if the coupling between the Landau levels may be neglected In general cases, especially for numerical purposes, the model of eqs. (15) and (16) is certainly more flexible.

5. Localisation without magnetic field

Using the tight binding Hamiltonian of eq. (12) the resolvent eq. (6) may be calculated iteratively from the recursion relation

$$A_l = (E - H_l)\, A_{l-1} - A_{l-2} \tag{17}$$

where H_l is the part of H which corresponds to an isolated slice l of the system. A_l is the inverse of $<x = o|\; G\; |x = l>$. The matrix elements of A_l are exponentially rising. The numerical implications of this were discussed earlier [12].

We calculated the decay lengths of the transmission probability at E = o, $\lambda(W,M)$ and $\lambda(W,M,M)$, corresponding to two and three dimensions, respectively, for various values of M and W. The statistical accuracy was set to be 1 %. About 10^5 iterations were necessary for M = 32 to achieve this in two dimensions, for instance.

We first discuss the two dimensional case.

In order to investigate the behaviour for large M we introduce a scale transformation $M \to M/\lambda_\infty(W)$, where the *scaling parameter* λ_∞ depends only on the disorder. Minimizing the variance of the values of $\ln(\lambda_\infty/M)$, which correspond to a given value of $\lambda(W,M)/M$ we found the scaling law [12]

$$\lambda(W,M)/M = f_2(\lambda_\infty(W)/M) \tag{18}$$

shown in Fig. 1. No assumption about the form of $f_2(x)$ was made. The influence of the boundary conditions accross the system (periodic or antiperiodic) turned out to be of the order of the statistical accuracy, i.e. 3-4%, when only data for M > 4 where taken

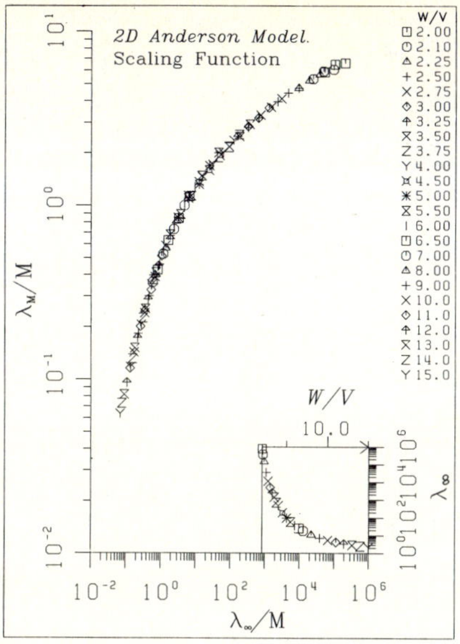

Figure 1

Scaling function $\lambda_M/M \equiv f_2(x)$ versus $x = \lambda_\infty/M$. $\lambda_M = \lambda(W,M)$ is the decay length of the transmission probability of a strip of width M described by the Anderson Hamiltonian. λ_∞ is the scaling parameter, i.e. the localisation length in the infinite two dimensional Anderson model. The insert shows λ_∞ as a function of the disorder parameter W/V.

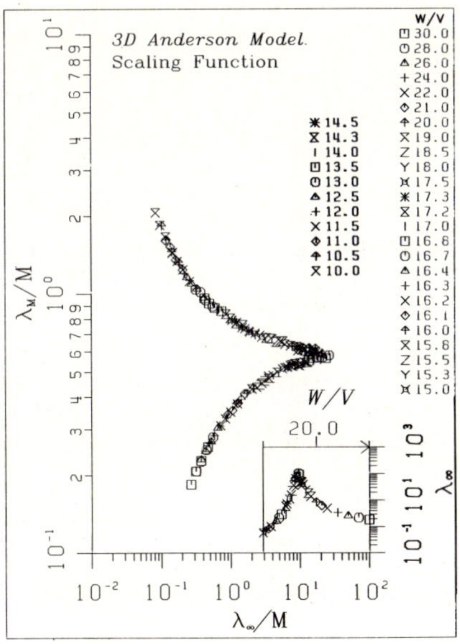

Figure 2

Scaling function $\lambda_M/M \equiv f_3(x)$ versus $x = \lambda_\infty/M$. $\lambda_M = \lambda(W,M,M)$ is the decay of length of the transmission probability of a bar with lateral dimensions M x M described by the Anderson Hamiltonian. λ_∞ is the scaling parameter. For the lower branch of $f_3(x)$ it is the localisation length in the infinite three dimensional Anderson model. In the upper branch it corresponds to the reciprocal dc-conductivity. The insert shows λ_∞ versus the disorder W/V. The singularity is at $W_c = 16.5 \pm 0.5$ V.

into account. This was checked down to W = 5.0 [42]. We note that $f_2(x)$ is monotonically increasing with x and $f_2(x) = x$ for $x \to o$. Thus

$$\lambda_\infty(W) = \lim_{M\to\infty} \lambda(W,M) \tag{19}$$

is the localisation length in the infinite two dimensional system. It is finite for values of the disorder parameter down to W = 2.0, although it increases rapidly with decreasing W, resulting in spectacularly large values as, for instance, $\lambda_\infty \approx 10^6$ at W = 2.0. This increase appears to be consistent with the essential singularity predicted by approximate analytical theory [29] in d = 2. Thus we conclude that no Anderson transition at a finite disorder occurs, where λ_∞ *certainly would have to diverge*. We found this behaviour in three dimensions, Fig. 2, where the scaling function shows a singularity at some critical disorder of $W_c = 16.5\pm0.5$ V. There is a lower branch for $W > W_c$, which corresponds to the localised regime, and an upper branch for $W < W_c$ representing the conducting regime. In the latter λ_∞^{-1} may be identified with the dc-conductivity [12,43].

Using the Landauer definition of the conductance via the transmission probability, eq. (9), we can calculate the β-function from $f_2(x)$ [12]. It is fully consistent with earlier work, and implies in particular that the length dependence of the conductance is logarithmically in the regime of weak localisation $\lambda_\infty >> M$. This is closely related to the asymptotic behaviour of $f_2(x)$ for large x, namely

$$\lim_{x\to\infty} f_2(x) = \frac{\pi}{2} \ln x. \tag{20}$$

Thus, the technique allows the numerical study of large disordered systems with a controllable accuracy, and the distinction between localised and extended states. Also conclusions about the scaling of the conductance with the size of the system can be drawn.

6. Localisation in a magnetic field

To apply the same method to a disordered system in a magnetic field we start from the Hamiltonian of eqs.(15) and (16). This allows the investigation of the limit of weak magnetic field as well as the limit of strong magnetic field, as long as the cyclotron radius is larger than the lattice distance.

First we write the Hamiltonian explicitely for a two dimensional system in a constant magnetic field B parallel to the z-axis. In the Landau gauge we have then the vector potential

$$\vec{A} = (o, -B\,x, o). \tag{21}$$

On a square lattice we may further specify x = aj and y = ak, where a is the lattice constant, and j = 1,L k = 1,M. Then eq. (16) reads

$$\begin{aligned} V_{jk,jk\pm1} &= V\, e^{\pm i\, B\frac{e}{\hbar} a^2 j} \\ V_{jk,j\pm1k} &= V \end{aligned} \tag{22}$$

Conveniently, we measure the magnetic field strength by the period of the exponential L_B, which is defined by

$$\frac{2\pi}{L_B a} = \frac{eBa}{\hbar} \tag{23}$$

Alternatively, one can use the cyclotron radius $l_B = \sqrt{\hbar/(eB)}$, which is connected with L_B via

$$l_B = a\sqrt{\frac{L_B}{2\pi}} \tag{24}$$

Again, from eq. (17) the resolvent, and, via the definition eq. (7), the decay length of the transmission probability can be calculated. In addition to the variation of the disorder we now have to consider variations in the magnetic field. The statistical accuracy of the data to be presented is 2%. Throughout this chapter E is in the center of the band, i.e. E = o.

Fig. 3 shows some examples, which are selected to illustrate the dependence of the decay length on the width of the strip M. We note that the data are monotonically, although not linearly, increasing with M. This behaviour is independent of the magnetic field and similar to that found for B = o [43].

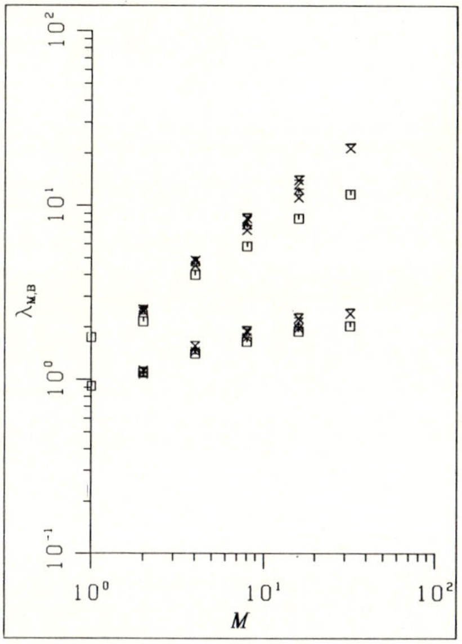

Figure 3

Decay length of the transmission probability, $\lambda_{M,B}$, of a strip of width M with a magnetic field B perpendicular to it versus M for two values of the disorder parameter W/V = 7 (upper data) and W/V = 15 (lower data). The magnetic field strength corresponds to L_B = ∞ (□), 32 (X), 16 (▲), 8 (Z) and 4 (X̄)

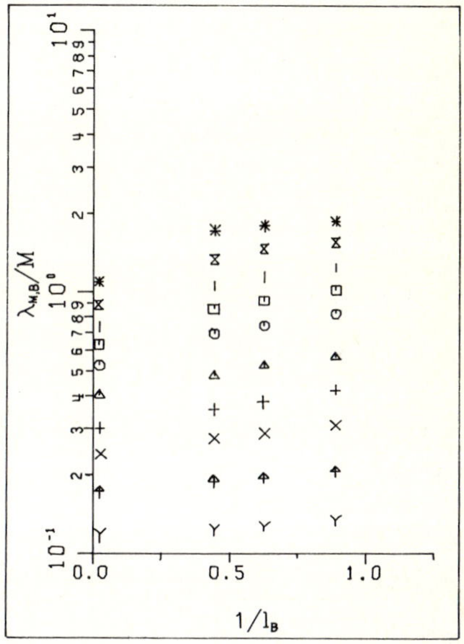

Figure 4

Decay length of the transmission probability, $\lambda_{M,B}$, of a strip of width M = 16 with a magnetic field B perpendicular to it versus the inverse of the cyclotron radius l_B. The disorder parameters are W/V = 15 (Y), 12 (↑), 10 (X), 9 (+), 8 (▲), 7 (⊙), 6.5 (□), 6 (ı), 5.5 (X̄), 5 (*).

In Fig. 4 we have plotted the decay length as a function of the magnetic field, here represented by the cyclotron radius. For field strengths such that $l_B > a$ the data are again monotonically increasing with B. Using again Landauer's definition for the conductance this can be interpreted as a negative magnetoresistance. No quantitative interpretation of the dependence of this magnetoresistance has been attempted yet, since this would need more data in particular in the regime of small field strength.

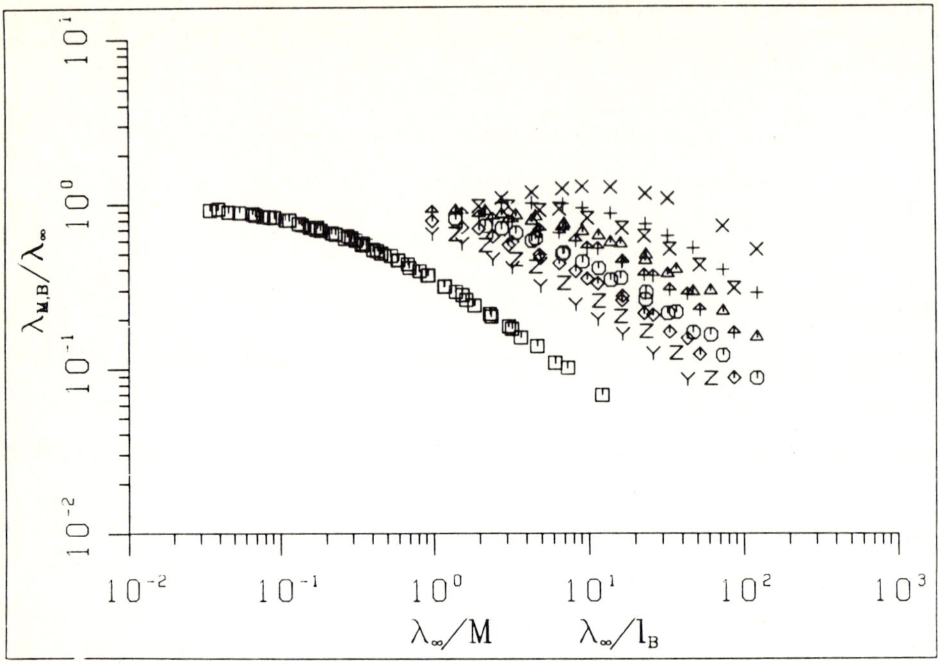

Fig. 5 Scaling functions λ_M/λ_∞ versus λ_∞/M for zero magnetic field B = o (⊡), and $\lambda_{M,B}/\lambda_\infty$ versus λ_∞/l_B for parameters M/l_B = 0.5 $\sqrt{4\pi}$ (Y), $\sqrt{2\pi}$ (Z), $\sqrt{4\pi}$ (◈), 2 $\sqrt{2\pi}$ (⏀), 2 $\sqrt{4\pi}$ (⍏), 4 $\sqrt{2\pi}$ (▲), 4 $\sqrt{4\pi}$ (X̄), 8 $\sqrt{2\pi}$ (+), and 16 $\sqrt{2\pi}$ (X). M and L_B values taken into account in this plot are M = 4, 8, 16 with L_B = 4, 8, 16. For M = 32 only data for L_B = 4 were calculated. Disorder is W/V = 15, 12, 10, 9, 8, 7, 6.5, 6, 5.5, 5.

Fig. 5 shows an attempt to scale the raw data. The assumptions involved are the following: First of all we assumed that the size dependence is described by the same scaling law as with B = o, i.e. the 1/M axis is scaled with the same scaling parameter $\lambda_\infty(W)$ as before. This is illustrated by the data points for zero magnetic field, which tend to unity in our plot of $\lambda(W,M)/\lambda_\infty(W)$ versus $\lambda_\infty(W)/M$ in the limit $M\to\infty$. Secondly, in order to investigate scaling with the magnetic field $\lambda(W,B,M)/\lambda_\infty(W)$ is plotted against $\lambda_\infty(W)/l_B$ for *fixed* M/l_B. This is equivalent to a plot against λ_∞/M.

It can be seen that within the computational accuracy the data lie on smooth curves. This justifies our initial scaling assumption *a posteriori*. In particular, we have for fixed l_B/M

$$\lim_{M\to\infty} \lambda(W,B,M) = \lambda_\infty(W) \tag{25}$$

No quantitative analysis of the data shown in Figs. 3 to 5 was tryed yet. We did not yet succeed in extracting any localisation length for fixed magnetic field, although from Fig. 3 we tend to conclude that the localisation properties are not changed drastically as compared to the case B = o in that the localisation length remains finite. This conclusion is supported by the behaviour shown in Fig. 5. We do not yet have any explanation of the nonmonotonic behaviour of the data for very large B. This may be due to the fact that these data correspond to $l_B < a$, which is possibly beyond the validity of our model.

Preliminary results for other energies, especially for those near the Landau levels are similar, although $\lambda(E,W,B,M)$ seems to increase more rapidly with M.

Thus, as the most important result of our investigations we consider the existence of a two parameter scaling law

$$\frac{\lambda(W,B,M)}{M} = f\left(\frac{\lambda_\infty}{M}, \frac{l_B}{M}\right) . \qquad (26)$$

The scaling properties of the decay length (Fig. 5, eq. 26) can be used to determine the dependence of the conductance on the system size, and the magnetic field strength. However, more data have to be collected before we can make comparison with the analytical results for weak magnetic field, where two parameter scaling for the conductance has been found recently [15].

7. Discussion with respect to magnetotransport in inversion layers at high magnetic field

Inversion layers in MOSFETS and Heterostructures can be considered as a realisation of a two dimensional electron gas subject to some disorder potential due to inhomogeneities at the surface (or interface) of a crystal (between two crystals) [44]. It is difficult to estimate a disorder parameter for such systems. For MOSFETS estimates of the concentration of interface inhomogeneities are 10^{11}-10^{12} cm^{-2}. If W is much larger than the cyclotron energy there should be considerable mixing of Landau levels. Then the assumption of independent Landau bands, broadened by the disorder is certainly no longer valid. Presumably, the quantum Hall effect would not be observable in this case. The conditions under which this happens are still to be determined experimentally.

Theoretically a careful study of the localisation length as a function of the energy near the center of the lowest Landau bands is desirable. For a first estimate we may use the asymptotic behaviour of the scaling function for the decay length found in chapter 5, namely eq. (20) which is also consistent with the data for $B \neq o$ presented in the preceeding chapter.

In order to be able to apply our numerical results we first have to make some comparison between the parameters of the electron gas and our lattice model. This is done in table 1.

Table 1

	free electron	lattice
dispersion law near band edge E(k)	$\frac{\hbar^2}{2m} k^2$	$-4V + V a^2 k^2$
cyclotron energy E_B	$\frac{e\hbar B}{m}$	$\frac{4\pi V}{L_B}$
clyclotron radius l_B	$\sqrt{\frac{\hbar}{eB}}$	$a\sqrt{\frac{L_B}{2\pi}}$

Without disorder the n = o Landau level would be situated at $E = -4V + 1/2\, E_B$, the n = 1 Landau level at $-4V + 3/2\, E_B$, and so on. For small disorder, say especially $W = E_B$, the band edge will be shifted downwards by about W/2. Then the n = o and n = 1 Landau levels are boradened by the disorder by an amount, which will be proportional to W.

We consider the energy $E = -4V + E_B$ and take $a = 0.392$ nm, which corresponds to the lattice constant of the Si (100) surface. Some numbers for the relevant parameters of the system, and the localisation length at this energy are given in table 2 for various strength of the magnetic field. λ_∞ was estimated via eq. (20) from data for the decay length obtained for M = 8 and $W = E_B$.

B/T	L_B	l_B/nm	l_B/a	E_B/meV	E_B/V	λ_∞/a
1	27 000	25.7	65.5	0.116	$4.65\cdot10^{-4}$	$10^{3\cdot10^5}$
20	1 350	5.74	16.6	2.32	$9.31\cdot10^{-3}$	$10^{5\cdot10^3}$
421	64	1.25	3.19	48.7	0.196	10^{14}

Table 2

Some numbers for the parameters describing a two dimensional disordered system in a magnetic field. B (in T) is the magnetic field, L_B the corresponding period in the lattice model. l_B is the cyclotron radius for the free electron model (in nm), and for the lattice model (in units of the lattice constant a). E_B is the corresponding cyclotron energy for the free electron model (in meV) and for the lattice model (in units of V). The last column contains the localisation length for the infinite system for $W = E_B$ at energy $E = -4V + E_B$ estimated from the scaling properties of the decay length. It should be emphasized that λ_∞ is the localisation length of a system with B = o!

It is seen that the localisation length is extraordinarily large and decreases rapidly with increasing magnetic field, which by construction, means increasing disorder. This reveals the essential singularity of $\lambda_\infty(W)$ at W = o described earlier [12,43], i.e. $\lambda_\infty(W) \sim \exp(\text{const}/W^2)$.

If values for W of this order of magnitude were realistic for a Si-MOSFET, for instance, the localisation length at energies near the lowest Landau levels would be astronomically large, even assuming a disorder energy of $W \approx 0.01V$, which is of the order of the cyclotron energy corresponding to the largest accessible magnetic field of about B = 20T. The numbers for λ_∞ of table 2 are certainly only lower limits for the localisation length with $B \neq o$, since they are estimated from the scaling function for B = o. Applying the scaling function for $B \neq o$ would presumably increase the localisation length, although not delocalising the states completely, as explained in section 6 (cf. Fig. 3). In view of actual sample sizes this were the ideal case of "weak localisation". For the explanation of the quantum Hall effect [45-50] it would be a good approximation to neglect the effects of localisation since in any case of practical interest $l_B \ll \lambda_\infty$. Deviations from the ideal behaviour would have to be assigned to interaction effects [51].

On the other hand if the disorder energy in an actual sample is considerably larger than the cyclotron energy the quantum Hall resistance remains essentially unexplained. It would be interesting to have detailed experimental investigations at magnetic field strength, where the steps in the Hall voltage just begin to appear, in particular concerning the dependence on the temperature [51], in order to be able to distinguish between the various mechanisms such as Coulomb correlations [48] disorder [46], edge states [50] etc.

8. Conclusion

We have presented numerical evidence that in a two dimensional disordered system described by a lattice model the localisation length in the center of the band remains finite, even in the presence of a very high magnetic field. It appears,that the decay length, which is via the Landauer relation, a rough measure of the conductance, obeys a two parameter scaling law. System size dependence of the conductance and magnetoresistance may be in principle determined from this scaling law. While the conductance remains logarithmically decreasing with the length of the system in the regime where the localisation length is of the order of the system diameter, the magnetoresistance is negative.

For practical purposes we have estimated the localisation length on the first Landau levels in an inversion layer at a Si-surface. It turns out to be astronomically large for disorder energies of the order of the cyclotron energy, such that in any experiment it would appear as infinite. This gives support to the current theories of the quantized Hall effect.

References

[1] P.W. Anderson, Phys. Rev. 109, 1492 (1958)
[2] D.J. Thouless, Physics Reports C 13, 93 (1974)
[3] N.F. Mott, Philos. Mag. 26, 1015 (1972)
[4] A recent review was given by W. Götze in "Modern Problems in Solid State Physics", Vol. 1, Yu.E. Lozovik and A.A. Maradudin, eds. (North Holland, to appear)
[5] A collection of references may be found in B. Kramer, A.MacKinnon and D. Weaire, Phys. Rev. B 23, 6357 (1981)
[6] K. Ishii, Suppl. Progr. Theor. Phys. 53, 77 (1973)
P. Erdös and R.C. Herndon, Adv. in Physics 31, 65 (1982)
[7] F.J. Wegner, Z. Phys. B 25, 327 (1976)
Compare also the review of D.J. Thouless, in Ill-Condensed Matter, R. Balian, R. Maynard and G. Toulouse, eds. (North Holland, 1979)
[8] E. Abrahams, P.W. Anderson, D.C. Licciardello and T.V. Ramakrishnan, Phys. Rev. Letters 42, 673 (1979)
[9] P.W. Anderson, D.J. Thouless, E. Abrahams and D.S. Fisher, Phys. Rev. B 22, 3519 (1980)
[10] A. MacKinnon, J. Phys. C 13, L 1031 (1980)
[11] J.L. Pichard, G. Sarma, J. Phys. C 14, L 617 (1981)
[12] A. MacKinnon and B. Kramer, Phys. Rev. Letters 47, 1546 (1981)
[13] B.L. Altshuler, A.G. Aronov and P.A. Lee, Phys. Rev. Letters 44, 1288 (1980)
[14] S. Hikami, A.I. Larkin and Y. Nagaoka, Progr. Theor. Phys. 63, 707 (1980)
B.L. Altshuler, D.Khmel'nitzki, A.I. Larkin, P.A. Lee, Phys. Rev. B 22, 5142 (1980)
[15] D. Yoshioka, Y. Ono and H. Fukuyama, J. Phys. Soc. Japan 50, 3419 (1981)
See also the review by H. Fukuyama, Surface Sci. 113, 489 (1982)
[16] M.J. Uren, R.A. Davies, M. Kaveh and M. Pepper, J. Phys. C 14, 5737 (1981)
[17] G.J. Dolan and D.D. Osheroff, Phys. Rev. Letters 43, 721 (1979)
[18] D.J. Bishop, D.C. Tsui and R.C. Dynes, Phys. Rev. Letters 44, 1153 (1980)
[19] L. van den Dries, C. van Hasendonck, Y. Buynseraede and G. Deutscher, Phys. Rev. Letters 46, 565 (1981)
[20] R.S. Markiewicz and L.A. Harris, Phys. Rev. Letters 46, 1149 (1981)
[21] Y. Kawaguchi and S. Kawaji, J. Phys. Soc. Japan 48, 699 (1980)
[22] I. Eisele and G. Dorda, Phys. Rev. Letters 32, 1360 (1974)
[23] Zvi Ovadyahu, Yoseph Imry, Phys. Rev. B 24, 7439 (1981)
[24] R.A. Davies, M.J. Uren and M. Pepper, J. Phys. C 14, L 531 (1981)
[25] K. von Klitzing, G. Dorda and M. Pepper, Phys. Rev. Letters 45, 494 (1980)
[26] D.C. Tsui and A.C. Gossard, Appl. Physics Letters 38, 550 (1981)
M.A. Paalanen, D.C. Tsui and A.C. Gossard, Phys. Rev. B 25, 5566 (1982)
[27] Elihu Abrahams, Physics Reports 67, 9 (1980)

[28] P.W. Anderson, E. Abrahams and T.V. Ramakrishnan, Phys. Rev. Letters 43, 718 (1979)
[29] D. Vollhardt and P. Wölfle, Phys. Rev. Letters 48, 699 (1982)
[30] D.J. Thouless, Phys. Rev. Letters 39, 1167 (1977); D.J. Thouless, Sol. St. Commun. 34, 683 (1980); see also D.J. Thouless, J. Non Cryst. Solids 35-36, 3 (1980)
[31] R. Kubo, S.J. Miyake and N. Hashitsume, in Solid State Physics, Vol. 17, 269, F. Seitz and D. Turnbull eds. (Academic Press, London 1965)
[32] B. Shapiro and E. Abrahams, Phys. Rev. B 24, 4025 (1981)
[33] P.A. Lee and D.S. Fisher, Phys. Rev. Letters 47, 882 (1981) A collection of references can be found in J. Sak and B. Kramer, Phys. Rev. B 24, 1761 (1981)
[34] R. Landauer, Philos. Mag. 21, 863 (1970)
[35] E.N. Economou and C.M. Soukoulis, Phys. Rev. Letters 46, 618 (1981)
[36] D.S. Fisher and P.A. Lee, Phys. Rev. B 23, 6851 (1981)
[37] D.C. Langreth and E. Abrahams, Phys. Rev. B 24, 2978 (1981)
[38] P.W. Anderson, Phys. Rev. B 23, 4828 (1981)
[39] See, for instance, J.M. Luttinger, Phys. Rev. 84, 814 (1951)
[40] H. Aoki, J. Phys. C 11, 3823 (1978)
[41] T. Ando in Anderson Localisation, Vol. 39, 176, Y. Nagaoka and H. Fukuyama eds., Springer Series in Solid-State Sciences (Springer, Heidelberg 1981)
[42] A. MacKinnon and B. Kramer, unpublished
[43] A. MacKinnon, in Anderson Localisation, Y. Nagaoka and H. Fukuyama eds., Springer Series in Solid-State Sciences (Springer, Heidelberg, 1981)
[44] G. Landwehr in Les Editions de Physique, p. 1, M. Savelli and P. Nougier eds. (Orsay 1979)
[45] R.B. Laughlin, Phys. Rev. B 23, 5632 (1981)
[46] D.J. Thouless, J. Phys. C 14, 3475 (1981)
[47] H. Aoki, T. Ando, Sol. St. Commun. 38, 1079 (1981)
[48] H. Fukuyama, P.M. Platzman, Phys. Rev. B 25, 2934 (1982)
[49] R.E. Prange, R. Joynt, Phys. Rev. B 25, 2943 (1982)
[50] B.I. Halperin, Phys. Rev. B 25, 2185 (1982)
[51] D.C. Tsui, H.L. Stormer, A.C. Gossard, Phys. Rev. Letters 48, 1559 (1982)

HIGH FIELD MAGNETOTRANSPORT IN GaAs/AlGaAs HETEROJUNCTIONS AND Si MOSFETs

Thomas Englert

Max-Planck-Institut für Festkörperforschung, Hochfeld-Magnetlabor, 166X, F-38042 Grenoble Cédex, France

A survey of recent experimental results on the transport properties of AlGaAs/GaAs heterojunctions and Si inversion layers is presented.

I. INTRODUCTION

In recent years quasi two-dimensional (2D) electronic systems at semiconductor interfaces have been investigated intensively, partly because of their technological relevance and partly because they provide systems with potentially new physics. In such an electronic system, carriers are confined to a potential well close to the interface due to the presence of a strong electric field, which can either be applied externally, as is the case in a field effect device, or be the built-in field due to ionized impurities in selectively doped semiconductor heterostructures. The electronic motion perpendicular to the 2D layer is quantized leading to 2D electric subbands E_i. Among a number of different experimental techniques to elucidate the electronic structure of such 2D systems magnetic fields play a predominant rule. A strong magnetic field, applied perpendicular to the 2D layer, quantizes the orbital motion of the electrons and thus leads to a fully quantized system with a discrete energy spectrum.

In the present article recent experimental work on 2D electronic systems performed at the Hochfeld-Magnetlabor is presented with emphasis on dc transport properties of GaAs/AlGaAs single interface heterojunctions and Si inversion layers of (100) surface orientation.

II. GaAs/AlGaAs HETEROSTRUCTURES

The samples investigated are heterojunctions of GaAs and $Al_xGa_{1-x}As$ ($x \cong .25$) grown by molecular-beam-epitaxy. In this system, the existence of a space charge layer arises from the discontinuity of the conduction band at the interface. The mobile electrons in the 2D layer at the GaAs side stem from Si donors in the AlGaAs. In so called modulation doped structure [1], the donor ions are placed at a selected distance away from the interface in order to reduce ionized impurity scattering at low temperatures. This leads to high electron mobilities of the order 100,000 cm^2/Vs at low temperatures.

II.1. INTERSUBBAND SCATTERING

When the number of 2D carriers is sufficiently low, so that only the first electric subband E_o is occupied, impurity intrasubband scattering is in general the limiting scattering mechanism for the mobility in this system. At carrier densities of the order 1×10^{12} cm^{-2}, however, depending on the device parameters, the second subband E_1 starts being populated, which opens a new scattering channel, the intersubband scattering. It has been shown theoretically, that this should lead to a strong and discontinuous decrease of the mobility in the system [2]. Recently Stormer, Gossard and Wiegmann [3] have found from an analysis of Hall and Shubnikov-de Haas (SdH) measurements, that intersubband scattering reduces the mobility by about 30 %, when the second subband is occupied. Here we use a different method to investigate the influence of intersubband scattering. We have chosen samples with a carrier density such that the Fermi energy is close to the second subband. The application of a strong magnetic field parallel to the interface allows us to increase the subband separation E_1-E_o due to the different spread of the wave function in the direction perpendicular to the interface [4]. This method has already been used previously in intersubband spectroscopy on Si [5]. Starting from a situation where the second subband is slightly occupied, we increase the subband separation and thus depopulate the excited subband. When the subband separation is sufficiently large so that all the carriers can be accommodated in the lowest subband and the Fermi energy is more than kT below E_1, intersubband scattering should be suppressed. Fig. 1 shows a plot of the sample resistance as a function of the magnetic field

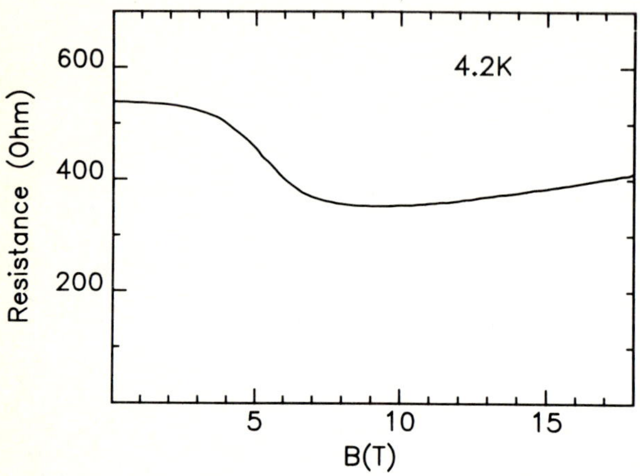

Fig. 1
Sample resistance vs. magnetic field parallel to the interface.

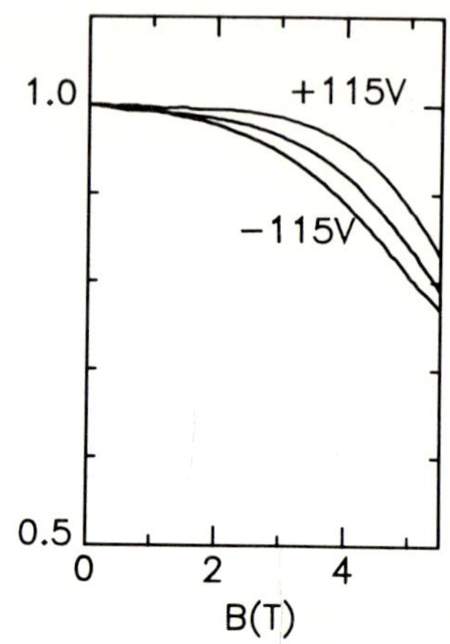

Fig. 2
Magnetoresistance for different gate voltages, -115 V, OV, +115V.

parallel to the interface at 4.2 K. The resistance decreases strongly between about 3 T and 8 T. We attribute this strong negative magnetoresistance, which was not observed in samples of much lower carrier density, to the suppression of intersubband scattering, when the subband separation is increased by the parallel magnetic field. The 2D carrier density in the GaAs layer obtained from the period of SdH oscillations in a magnetic field perpendicular to the interface is $n_s=7.3\times10^{11}$ cm^{-2}. Only one period of oscillations is observed. The density n_H, derived directly from the slope of the Hall voltage is slightly higher, $n_H=7.69\times10^{11}$ cm^{-1}, indicating that a few percent of the carriers are already in the second subband. In order to demonstrate that the negative magnetoresistance is related to the occupation of the second subband, we have changed the carrier density in the system by applying a gate voltage between a backside contact and the 2D layer [6]. A gate voltage of about +115 V (-155 V) leads to a decrease (increase) of the sample resistance at zero magnetic field. The slope of the Hall voltage varies by $\pm$ 4.2 %, indicating that the carrier density in the second subband is varied by a few times 10^{10} cm^{-2}. Fig. 2 shows the influence of the gate voltage on the magnetoresistance in a parallel field. The onset of the decrease in resistance shifts to higher magnetic fields when the gate voltage is increased from -115 V to +115 V. The correlation between the occupation of the second subband and the behaviour of the negative magnetoresistance clearly demonstrates that the decrease in resistance is due to a suppression of intersubband scattering. The mobility determined from the Hall carrier density and the sample resistance at the magnetic field where the minimum occurs is, $\mu=73,400$ cm^2/Vs. The Hall mobility at zero (parallel) magnetic field, where intersubband scattering is present, is 45,300 cm^2/Vs. Mori and Ando [2] have calculated the mobilities in the lowest and first excited subband due to impurity intra- and intersubband scattering for a modulation doped GaAs/AlGaAs superlattice. Since this situation is not very different from a single heterostructure, we compare their theoretical result with out experiment. Fig. 3 shows the calculated average Hall mobility and the two data points obtained in the present study under conditions where intersubband scattering is present (B=0), and where the intersubband scattering is suppressed. The agreement is reasonable, but we do not find a discontinuous change in mobility. A rather

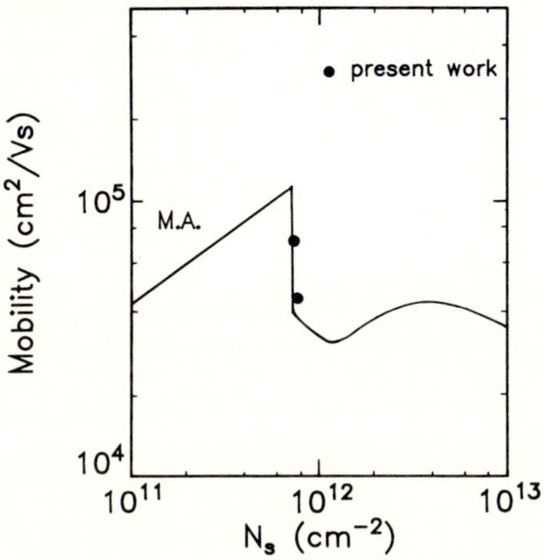

Fig. 3
Theoretical Hall mobility calculated by Mori and Ando [3] and exerperimental data points.

surprising result is obtained when the sample is slightly tilted away from the parallel configuration (ψ=90°) in a strong magnetic field. As can be seen in Fig. 4, only a small component of a perpendicular magnetic field is necessary to reduce the negative magnetoresistance to a large extent. For a total magnetic field of 18.95 T (Fig. 4) the negative magnetoresistance is limited to a range of angles 90°±1.5°. At first glance this result seems to be difficult to reconcile with the interpretation of the negative magnetoresistance as being due to the diamagnetic-like shift of the subbands. In a simple perturbational treatment one would expect that such a diamagnetic term shows an anisotropy as $\sin^2\psi$ [7]. However, it is known that a strong magnetic field of arbitrary orientation mixes electric and magnetic subbands and its influence on the energy levels cannot be inferred in a simple way [8,9]. Such calculations as well as a detailed theory of scattering in a parallel magnetic field are left for further theoretical and experimental studies.

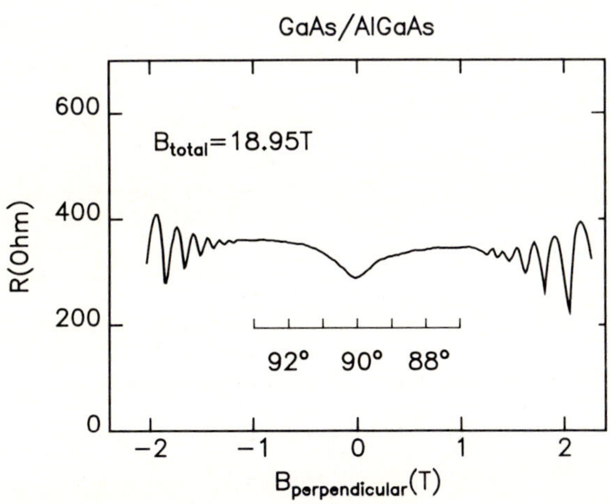

Fig. 4
Sample resistance as a function of a perpendicular magnetic field at a constant total magnetic field of 18.95 T. The additional scale at the bottom shows the corresponding tilt angles.

II.2. NEGATIVE MAGNETORESISTANCE IN LOW MAGNETIC FIELDS - WEAK LOCALIZATION EFFECTS -

In two dimensions the scaling theory [10] and later theoretical work [11] have indicated that metals are not truly metallic. The conductivity of a 2D system with only normal impurity scattering should tend to zero as the temperature approaches absolute zero, no matter how weak the scattering is. As a precursor effect to this complete localization at absolute zero there exists a region, where the conductivity behaves as lnT and this is the so-called weak localization regime in 2D. An external magnetic field is expected to have important effects on the quantum correction to the conductivity, which is responsible for the localization. It tends to suppress this localization leading to a negative magnetoresistance. All these effects have been observed in Si MOSFETs and the theory of weak localization can explain most of the experimental results in this system in detail [12]. We have observed a negative magnetoresistance at low magnetic fields in GaAs/AlGaAs heterostructures, in samples of relatively low carrier density. Fig. 5 shows a plot of the sample resistance as a function of a magnetic field perpendicular to the 2D layer at T=1.9K for a sample of $n_s=2.2\times10^{11}$ cm^{-2} and a mobility of $\mu=20{,}000$ cm^2/Vs. The dashed curve in Fig. 5 represents a model calculation of the resistance according to the theory of weak localization.

$$\sigma = \sigma(T,B=0) + \Delta\sigma$$

where

$$\Delta\sigma = \alpha e^2/2\pi^2\hbar \cdot (\psi(a+1/2) - \ln a)$$

with $a=\hbar/\mu\cdot4\cdot\tau_i\cdot E_F\cdot B$. $\psi(a+1/2)$ is the digamma function and τ_i is the inelastic scattering time. In the calculation the following parameter values have been used: $\alpha=1$, $\tau_i=1\times10^{-11}$ sec. The agreement between the fit and the experimental curve, although not perfect, indicates that the results are compatible with an explanation based on the theory weak localization. For liquid He temperatures such a negative magnetoresistance has not been observed in samples of higher carrier densities.

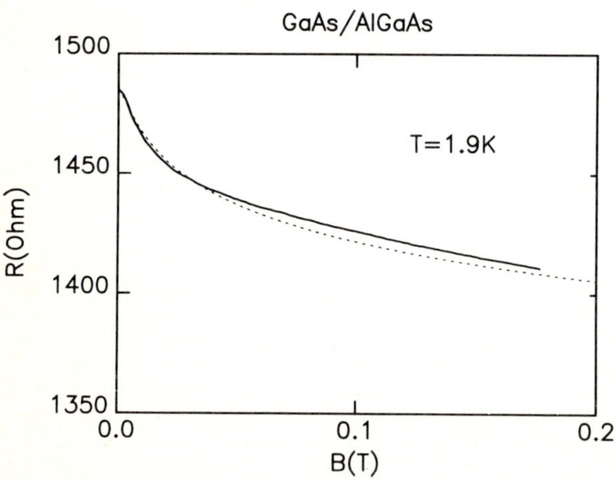

Fig. 5
Sample resistance vs. magnetic field.

II.3. TEMPERATURE DEPENDENCE OF SdH OSCILLATIONS IN LOW MAGNETIC FIELDS

In general a derivation of the cyclotron mass from the temperature dependence of SdH oscillations is more problematic in a 2D system than in 3D and can easily lead to misinterpretation because of the discrete nature of the density-of-states. However, in the weak field limit, where $\omega_c\tau$ is not much larger than 1, and oscillations correspond to high quantum numbers, the SdH effect becomes sinusoidal and the amplitude follows a x/sinhx behaviour as a function of T, where $x=2\pi^2KT/\hbar\omega_c$ [13]. In Fig. 6 the SdH amplitude measured at two different magnetic fields is plotted as a function of temperature for a sample with a nominal mobility of 70,000 cm^2/Vs. From the fit shown as solid lines in Fig. 6, a mass value of $m_c=0.069+0.002\ m_o$ is derived, which is in good agreement with cyclotron resonance measurements [14]. This demonstrates that even in a high mobility 2D system the cyclotron mass can be derived from transport measurements in the weak magnetic field limit.

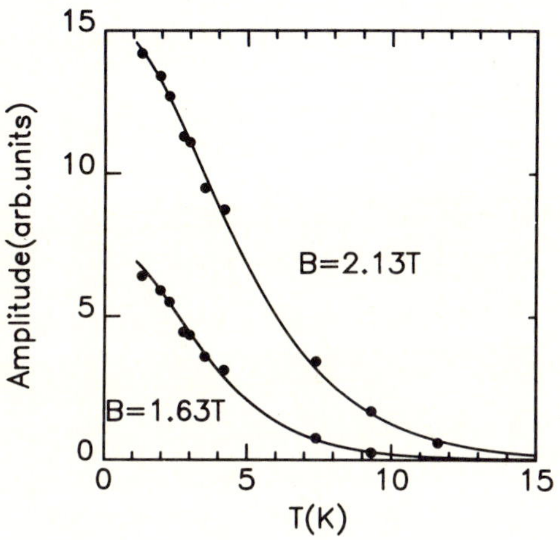

Fig. 6
Temperature dependence of SdH oscillations, experimental points and fit.

II.4. QUANTUM OSCILLATION IN STRONG MAGNETIC FIELDS

It has been known that, at sufficiently high magnetic fields, ρ_{xx} is vanishingly small and ρ_{xy} shows plateaus in finite regions of B, when the Fermi energy lies between two separated Landau levels [15]. The value of the Hall resistance at the plateaus is quantized to $\rho_{xy}=h/e^2 \cdot i$, with i=1,2... Since the discovery that this relation is fulfilled to a very high precision [16,17], a number of theories have been developed in order to explain this quantum Hall effect [18]. Most of the models invoke localization of Landau electrons. In either case, the Hall current carried by the mobile electrons is equal to that carried by the total number of states in the filled Landau levels, in the absence of immobile states, to account for the observed quanta of ρ_{xy}. Fig. 7 shows an example of SdH oscillations in strong magnetic fields at different temperatures obtained on a sample with a carrier density of $7.8\times10^{11}cm^{-2}$. The plateaus in ρ_{xy} and the vanishing of ρ_{xx} extend to larger ranges in B, when the temperature is lowered. Recent experiments at much lower temperatures, in the mK region, have shown, that the Hall effect degenerates into a step-like function indicating that almost all the carriers are localized in the extremely low temperature limit [19]. The additional splitting of the SdH oscillations at high magnetic fields in Fig. 7 is due to the spin. Since the free electron g-factor in GaAs is only of the order 0.5, an experimental resolution of spin splitting was not expected in view of the estimated linewidth and the temperature. Therefore enhancement of the spin splitting has to be invoked. The notion of enhanced spin splitting was first applied to Si inversion layers and has been explained theoretically in terms of a self-energy correction due to exchange [20]. The established method for determining the energy splitting in a 2D system from transport measurements is to produce an accidental degeneracy between spin split levels of different quantum numbers. However, in GaAs this method is not applicable, because the ratio between spin and Landau splitting is unfavourable. In order to obtain g^* from the experiment we used the temperature dependence of the splitting and compared it to a model calculation. The model used is essentially the one outlined by Ohkawa et al. for Si [21]. In order to account for the plateaus it is assumed that immobile carriers exist in the tails of Landau

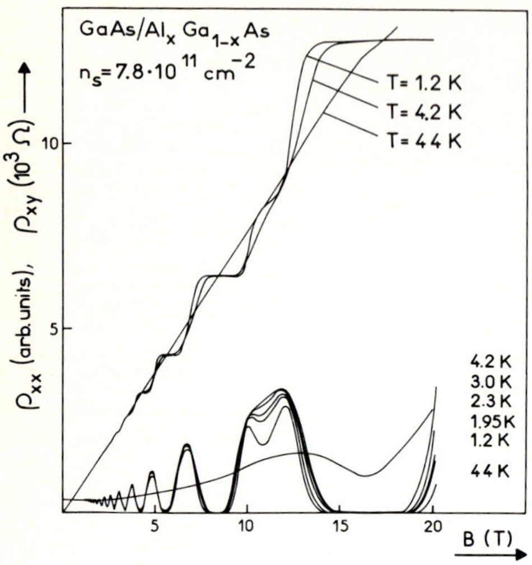

Fig. 7
Plots of ρ_{xx} and ρ_{xy} versus magnetic field at different temperatures.

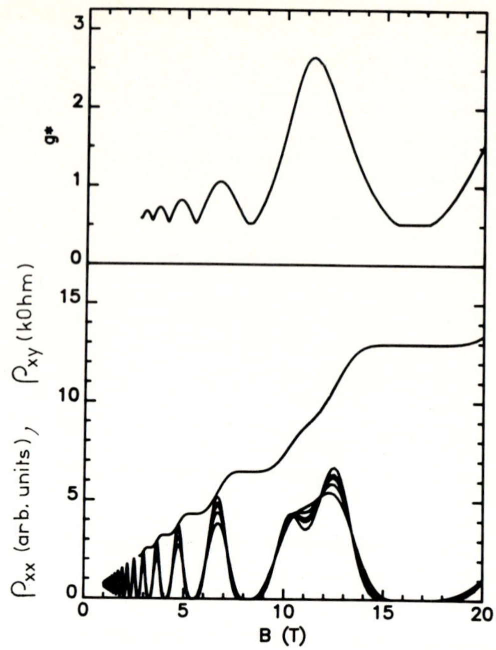

Fig. 8
Calculated ρ_{xx} and ρ_{xy} versus magnetic field. The upper panel shows the effective g-factor.

levels. In the model this is done by cutting off the tails of the density-of-states function at a characteristic energy. The mobile electron density in each level is normalized to the total number of carriers in one Landau level to give the correct behaviour of the Hall effect. Details of the calculation are given elsewhere [22]. Fig. 8 shows the results of the model calculation for the resistance and the Hall-effect calculated for the same temperatures as in Fig. 7. The agreement between the model and the measurement is satisfactory both for the spin splitting as well as for the localization region. The upper part of Fig. 8 represents the effective g-factor as it comes out of the selfconsistent calculation. It oscillates between the asymptotic bulk value and a maximum value, when the Fermi energy is between spin split levels. The maxima are strongly temperature dependent through the occupation dependence of the enhancement mechanism.

II.5. MAGNETOPHONON EFFECT

Because of the spatial separation between mobile carriers and ionized impurities, Coulomb scattering is reduced in modulation doped heterostructures. On the other hand it has been shown that the electron-phonon scattering is enhanced in 2D [23]. Qualitatively, this is due to the fact that, in a scattering process in 2D, only the component of the wavevector parallel to the layer has to be conserved, its perpendicular component can be arbitrary. In principle, electrons can couple to 2D phonons, localized at the interface, as well as to bulk (LO) phonons, and both intrasubband and intersubband scattering have to be considered. In a strong magnetic field, under conditions where $N\cdot\hbar\omega_c = \hbar\omega_o$, resonant coupling of Landau electrons and phonons occurs, which in principle can lead to a splitting of Landau levels. It has been shown theoretically, that this effect is more important in a 2D system, because firstly the effective electron-phonon coupling is enhanced and secondly the missing k_z-term in the kinetic energy makes the Landau levels sharper [24]. This effect should be most pronounced for the fundamental resonance, where N=1. Resonant phonon scattering leads to oscillations in the magnetoresistance, known as the magnetophonon effect. Fig. 9 shows results obtained on a sample with a carrier density of $n_s=3.4\times10^{11}$ cm^{-2} and a nominal mobility of 300,000 cm^2/Vs at T=220 K. The data above 20 T were taken in pulsed magnetic fields at the Laboratoire de Physique des Solides, Toulouse [25]. The maxima in the resistance correspond to the resonances N=1, N=2, and N=3. The peak position of the fundamental (N=1) resonance is observed at B=(22.5±0.5) T. No splitting of the fundamental resonance is observed. Similar oscillations, though weaker in amplitude, have been obtained in multiple-interface heterostructure samples (superlattices). From the period of the magnetophonon oscillations and the known mode frequency of LO phonons in bulk GaAs, a 'magnetophonon mass' of $m_p=0.071$ m_o is determined. This value is in good agreement with the bulk GaAs results. It shows that the resonant scattering is indeed due to the LO phonon mode of bulk GaAs and that corrections due to carrier confinement are not significant. A comparison with mass values obtained in cyclotron resonance measurements indicates that the polaron correction is not significantly larger than in the bulk. The theoretically predicted splitting of the N=1 resonance due to resonant coupling must be smaller than the width of Landau levels, which is roughly given by $\hbar/\tau=1.7$ meV at 200 K. Recently magnetophonon oscillations have also been observed in another weakly polar 2D system, the GaInAs/InP superlattice [26].

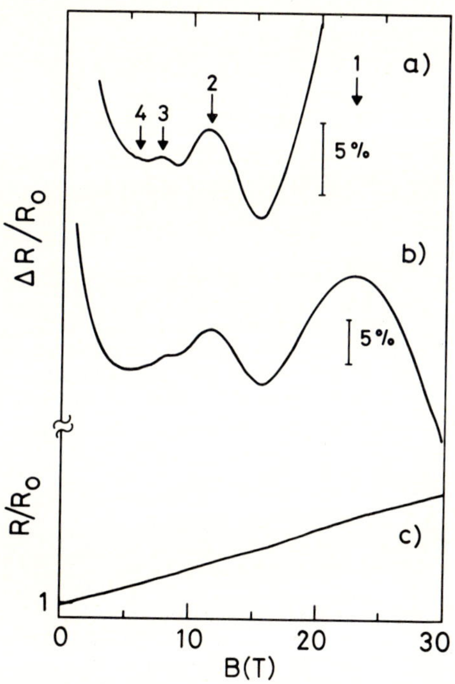

Fig. 9
Magnetophonon resonances in the resistance of a single heterojunction sample. In traces a) and b) the linear increase of the resistance was compensated. a) was taken in a Bitter magnet, b) and c) were measured in a pulsed field.

III. SILICON INVERSION LAYERS

III.1. MAGNETOTRANSPORT AT He-3 TEMPERATURES

It has already been known from measurement on GaAs/AlGaAs heterostructures in the mK region that the plateaus in the Hall resistance and the regions of vanishing becomes considerably wider with decreasing temperature [19]. Moreover it has been found that the peak value for the lowest quantum numbers decreases strongly and may even disappear completely at sufficiently low temperature [19]. We have performed measurements on silicon inversion layers using a He-3 cryostat in combination with a Bitter magnet. The samples investigated are Hall-bar shaped Si MOS transistors of (100) surface orientation. The maximum effective mobility of the samples is of the order 16.000 cm^2/Vs. Fig. 10 and Fig. 11 show experimental data for the sample resistance and the Hall effect plotted as a function of magnetic field for two different temperatures (T=1.5 K and T=0.4 K). The gate voltage was chosen so that the plateau at the highest magnetic field corresponds to a situation where the Fermi level is between spin and valley split levels of the N=0 Landau level. In agreement with the results obtained on GaAs/AlGaAs it is found that the oscillation peaks narrow considerably with decreasing temperature and the Hall resistance approaches a step-like be-

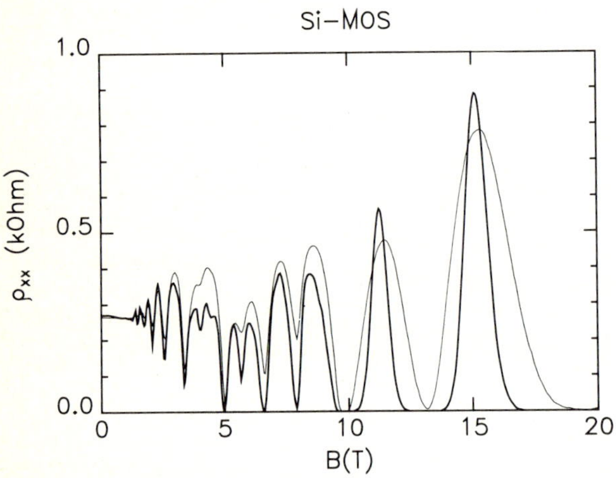

Fig. 10
ρ_{xx} versus B for T=1.5 K (weak curve) and T=0.4 K. The carrier density is 9.65×10^{11} cm^{-2}.

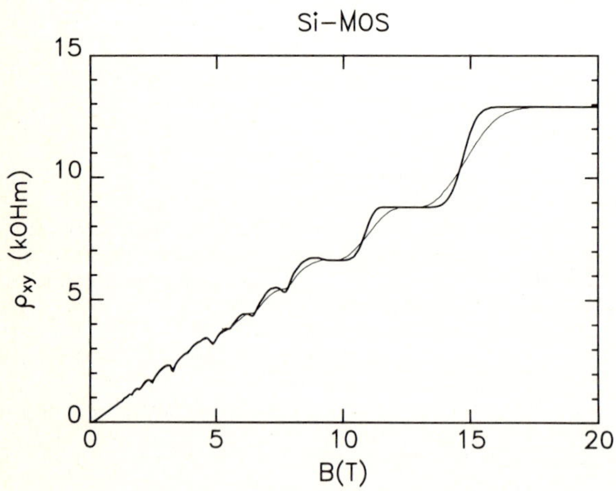

Fig. 11
Hall resistance vs. B for T=1.5 K (weak curve) and T=0.4 K.

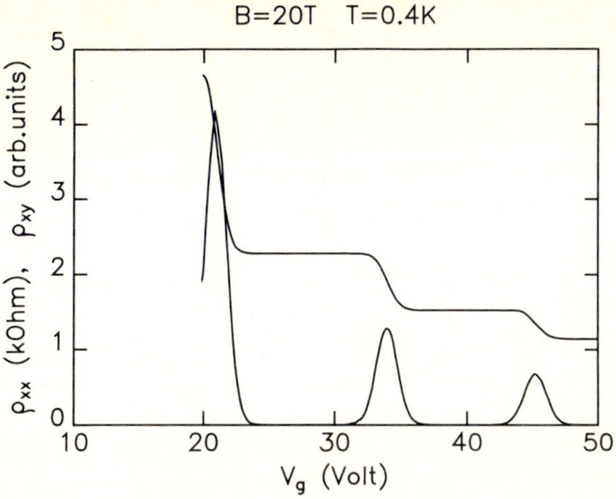

Fig. 12
Resistance and Hall effect vs. gate voltage for the N=0 Landau level.

haviour. The combination of very low temperatures and very high fields allows us to observe plateaus not only between Landau and spin split peaks but also between valley split levels of one Landau level. Contrary to the results obtained in GaAs/AlGaAs, the amplitude of the resistance oscillations does not decrease at lower temperatures. In the low magnetic field part of Fig. 10 the variation of the amplitude with temperature is due to a better resolution of spin and valley splitting at lower T. In Fig. 12 the resistance and the Hall effect are plotted at constant magnetic field as a function of the gate voltage for the N=0 Landau level. Three of the four levels are observed, the lowest level is absent due to contact problems in the low carrier density regime in this sample. The plateaus extend over a wide range of gate voltages and we would like to point out that the first two steps seem to be especially suited for high precision measurements of the fine structure constant. In Fig. 13 the components of the conductivity tensor, evaluated from the measured ρ_{xx} and ρ_{xy} data, are plotted for T=4.2 K and T=0.4 K. When the temperature is lowered the effect of carrier localization is not symmetric with respect to the center of the peak at high temperatures. Indication for such an asymmetric behaviour in Si has already been observed previously by different authors [27], it is, however, not observed in GaAs. The asymmetry is not an artifact due to a shift of the threshold voltage, which can be inferred from SdH oscillations at higher gate voltages, where localization effects are not important. We have also investigated the influence of a substrate voltage on the plateau width and found that the SdH peaks at low temperatures are narrowed further by about 15 %, when a negative substrate bias is applied.

III.2. MODEL CALCULATION

Since a generally accepted theory of these localization effects is still lacking at present, we have made a semi-empirical calculation in order to understand our results. For the energy levels and the conductivity we used the same model, which has been found to describe the quantum oscillations in Si inversion layers at temperatures and magnetic fields, where localization effect are not important [28]. In order to account for the localization effects we assume that potential fluctuations lead to localization of carriers under conditions where a Landau level is almost empty (electron-like) or almost filled (hole-like). Completely filled levels cannot be localized and show the response of all the electrons in one Landau level. This is accounted for by introducing a 'response' function $\alpha(E)$ in the calculation of σ_{xx} and σ_{xy}, which is narrower in energy than the density-of-states function D(E).

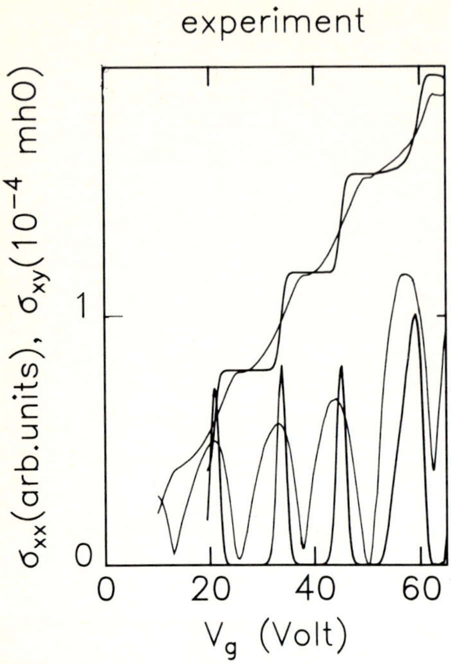

Fig. 13
Measured σ_{xx} and σ_{xy} vs. gate voltage for T=4.2 K and T=0.4 K

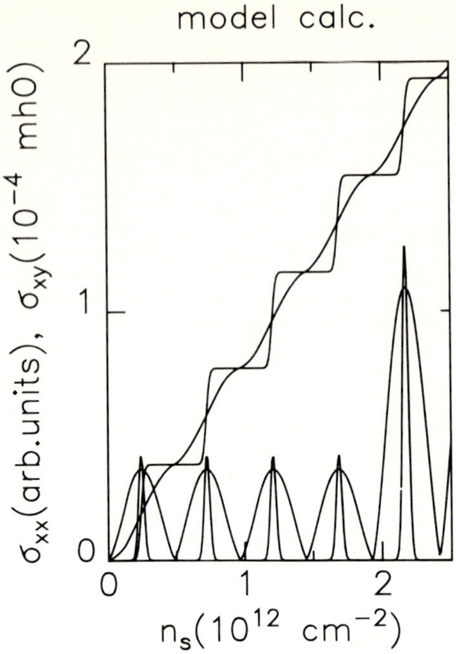

Fig. 14
Calculated σ_{xx} and σ_{xy} for T=4.2 K and T=0.4 K

$$\sigma_{xx} = \frac{e^2}{\pi^2 \hbar} \int dE \left(- \frac{\partial f}{\partial E}\right) \Sigma \ (N+1/2)\ D^2_{N,\sigma,\gamma}(E) \cdot \alpha(E)$$

$$\sigma_{xy} = - \frac{e}{B} \int dE \ \Sigma \ D_{N,\sigma,\gamma} \cdot \alpha(E) \cdot f(E)$$

$$\int dE \, D(E) \cdot \alpha(E) = 1$$

It was found that such a model can describe the observed phenomena at temperatures above about 1.5 K in a reasonable way [22]. The model is, however, not sufficient to explain the results at lower T. Here, one has to assume that the localization itself is temperature dependent in order to account for the strong variation of the plateau width with T. In our model this means that the width of $\alpha(E)$ becomes smaller with decreasing temperature. In order to avoid additional parameters we have tentatively used the following function for $\alpha(E)$, where $f(E)$ is the Fermi function and $\delta(E)$ the δ-function.

$$\alpha(E) \propto \int dE' \ f(E-E') \cdot (1-f(E-E'))\, \delta(E-E')$$

One can imagine that some sort of thermal activation within one Landau level plays a role. In the ansatz made above we assume that in the limit $T \to 0$ only the carriers at the center of one Landau level are mobile. At higher temperatures carriers within an energy range of kT become delocalized. Details of the model will be published elsewhere [29]. Fig. 14 shows the result of the calculations. The parameter values for the linewidth, valley splitting and exchange enhancement are taken from ref.[28]. The agreement between the model and the experimental data shown in Fig. 13 is satisfactory in view of the fact that no adjustable parameter is used to describe the plateau width. The asymmetry can in principle also be produced by employing a different function $\alpha(E)$. The model gives too strong an increase of the peak values of σ_{xx} with decreasing temperature, which is due to the use of the δ-function. However, we would like to point out that experimentally it is very difficult to establish the temperature dependence of the peak values, because the results are very sensitive to small inhomogeneities in the sample once the SdH peaks become very narrow. In summary our model reproduces all the essential experimental features.

ACKNOWLEDGEMENTS

It is a pleasure to acknowledge close collaboration with Prof. D.C. Tsui, Princeton. The high quality GaAs/AlGaAs heterojunction crystals have been grown by A.C. Gossard, Bell Labs.. The Si MOS transistors have been fabricated by G. Kaminsky, Bell Labs.. Finally I would like to thank Prof. G. Landwehr and my colleagues J.C. Maan and Ch. Uihlein for valuable discussions and advice.

REFERENCES

[1] R. Dingle, H.L. Störmer, A.C. Gossard, W. Wiegmann, Appl.Phys.Lett. 33, 665(1978)
[2] S. Mori and T. Ando, J.Phys.Soc.Japan, 48, 865 (1980)
[3] H.L. Stormer, A.C. Gossard, W. Wiegmann, Solid State Commun., 41, 707 (1982)
[4] F. Stern and W.E. Howard, Phys. Rev., 163, 816 (1967)
[5] W. Beinvogl, A. Kamgar, J.F. Koch, Phys.Rev.B, 14, 4274 (1976)
[6] H.L. Stormer, A.C. Gossard, W. Wiegmann, Appl.Phys.Lett., 39, 493 (1981)
[7] Y. Uemura, Proc. 12th Int.Conf.Physics of Semiconductors, ed.: M.H. Pilkuhn, Teubner, Stuttgart (1974)
[8] T. Ando, J.Phys.Soc.Japan, 44, 475 (1978)
[9] J.C. Maan, Ch. Uihlein, L.L. Chang, L. Esaki, in press
[10] P.W. Anderson, E. Abrahams, T.V. Ramakrishnan, Phys.Rev.Lett., 43, 1718 (1979)
[11] see e.g. H. Fukyama, Surf.Sci., 113, 489 (1982), and references therein
[12] R.C. Dynes, Surf.Sci., 113, 510 (1982)
[13] T. Ando, J.Phys.Soc.Japan, 37, 1233 (1974)
[14] Th. Englert, J.C. Maan, Ch. Uihlein, D.C. Tsui, A.C. Gossard, 16th Int.Conf. on the Physics of Semiconductors, Montpellier (1982)
[15] D.C. Tsui, A.C. Gossard, Appl.Phys.Lett., 38, 550 (1981)
[16] K. von Klitzing, G. Dorda, M. Pepper, Phys.Rev.Lett., 45, 494 (1980)
[17] D.C. Tsui, A.C. Gossard, Appl.Phys.Lett., 38, 550 (1980)
[18] see e.g. K. von Klitzing, this issue
[19] M.A. Paalanen, D.C. Tsui, A.C. Gossard, in press
[20] T. Ando, Y. Uemura, J.Phys.Soc.Japan, 37, 1044 (1974)
[21] F.J. Ohkawa, Y. Uemura, J.Phys.Soc.Japan, 43, 925 (1979)
[22] Th. Englert, D.C. Tsui, A.C. Gossard, Ch. Uihlein, Surf.Sci., 133, 295 (1982)
[23] K. Hess, Appl.Phys.Lett., 35, 485 (1980)
[24] S. DasSarma, A. Madhukar, Phys.Rev.B22, 2823 (1980)
[25] Th. Englert, D.C. Tsui, J.C. Portal, J. Beerens, A.C. Gossard, Solid State Commun., in press
[26] M.A. Brummel, R.J. Nicholas, J.C. Portal, M. Razeghi, M.A. Poisson, 16th Int. Conf. on the Physics of Semiconductors, Montpellier (1982)
[27] S. Kawaji, J. Wakabayashi, in: Physics in High Magnetic Fields, eds., S. Chikazumi, N. Miura, Springer-Verlag (1981), p. 284
[28] Th. Englert, Solid State Commun., 40, 893 (1981)
[29] Th. Englert, Ch. Uihlein, to be published

MAGNETOTRANSPORT IN GaInAs-InP HETEROJUNCTIONS AND SUPERLATTICES

R.J. Nicholas+, M.A. Brummell+, J.C. Portal*,
M. Razeghi° and M.A. Poisson°

+ Clarendon Laboratory, Parks Rd., Oxford OX1 3PU, England.
* I.N.S.A. 31077 TOULOUSE, S.N.C.I.-C.N.R.S. 166 X 38042 GRENOBLE
o L.C.R., Thomson-C.S.F., 91401 Orsay, France.

Shubnikov-deHaas oscillations, cyclotron and magnetophonon resonance have been studied in heterojunctions and superlattices grown with $Ga_xIn_{1-x}As$ and InP at the lattice match with x=0.47. A two dimensional electron gas is formed which shows pronounced quantum Hall steps. Cyclotron resonance is observed in the heterojunctions giving $m^*=0.049m_o$, with a linewidth which is determined by Gaussian scattering centres. Weakly enhanced g-factors of up to $g^*=4.6$ are observed. The superlattices show an increase in effective mass with decreasing layer thickness, and are shown to be two dimensional up to 200 K. Evidence for long range Optical Phonon scattering is presented.

1. INTRODUCTION.

Selectively doped semiconductor heterojunctions and superlattices offer the possibility of observing many interesting effects associated with the formation of a two dimensional electron gas (2DEG) due to binding in the quantum wells at the heterojunction interface(1,2) or within the layers of the superlattice(3,4). The studies reported up till now have concentrated very heavily on the GaAs-GaAlAs and InAs-GaSb systems. This paper reports recent work performed at the high magnetic field laboratories in Oxford and Grenoble, on a different system $Ga_xIn_{1-x}As$-InP, where there is a lattice match to InP at x=0.47. A 2DEG has been observed in both heterojunctions(5) and superlattices(6), which is found to show strong Shubnikov-deHaas oscillations, pronounced quantum Hall steps enhanced electron g-factors, cyclotron and magnetophonon resonances.

The energy band gaps of InP and $Ga_{.47}In_{.53}As$ are 1.42 eV and 0.81 e respectively at 4.2 K(7). Measurements of electron affinity for GaAs, InAs(8) and InP(9) indicate that the conduction band discontinuity shoul be of order 0.2 eV with, however, a considerable degree of uncertainty. Recent measurements of photoemission from a number of Ge - III-V compound heterojunctions(10) show that a valence band discontinuity of 0.3 eV would be expected, in agreement with the simple tight binding theory of Harriso (11). We may thus assume that there will be similar discontinuities of ∿0. eV at both the conduction and valence bands, in contrast to GaAs-GaAlAs heterojunctions, where the valence band discontinuity is less than 0.1 eV(12). For an n-type on n-type heterojunction electron transfer will take place from the more heavily doped InP to the GaInAs in corder to maintain a constan Fermi level. This leads to considerable band bending and the formation of a 2DEG in the GaInAs as shown in Fig.1

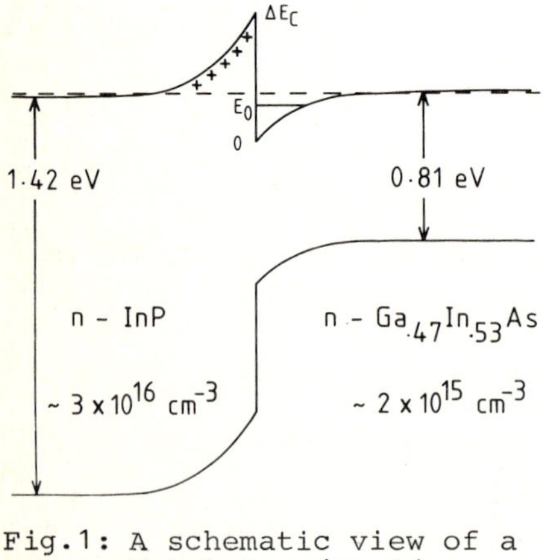

Fig.1: A schematic view of a GaInAs-InP heterojunction.

The electron mobility is enhanced due to the spatial separation of the electrons from the shallow donor impurities. Measurements of cyclotron and magnetophonon resonance in bulk GaInAs(13) give a band edge effective mass $m^*=0.041\ m_o$ while a g-factor of 3.5 has been deduced from $\underline{k}.\underline{p}$ theory and measurements of electroreflectance(14). For the heterojunctions electron concentrations of $3 - 5 \times 10^{11}\ cm^{-2}$ are found leading to the occupation of a single electric subband, which is in contrast to recently reported results on $Ga_{.47}In_{.53}As\text{-}Al_{.48}In_{.52}As$ heterojunctions(15,16) where by $7 \times 10^{11}\ cm^{-2}$ the second subband is occupied.

The samples were grown by low pressure metal organic chemical vapour deposition (LP-MOCVD), at the Laboratoire Central de Recherche, Thomson-C.S.F. The undoped GaInAs is n-type and has been shown to have a background doping level of $\sim 2 \times 10^{15}\ cm^{-3}$, with mobilities up to 12,000 $cm^2V^{-1}s^{-1}$ at 77 K(17). For the heterojunctions a layer of GaInAs with a thickness of order 1 μm was grown on top of a layer of nominally undoped InP, $\sim$ 1500 Å thick, with an electron concentration of $\sim 3 \times 10^{16}\ cm^{-3}$, on a substrate of semi-insulating InP:Fe. In the superlattices twenty alternate layers of InP and GaInAs were grown, with the InP layers 300 Å thick, while the GaInAs was 80 Å, 100 Å and 150 Å thick in the three samples studied.

2. HETEROJUNCTIONS.

The sample characteristics are shown in Table I. Magnetotransport measurements were performed at temperatures between 1.5 K and 30 K on samples cut into both Van de Pauw geometry and long bars with current and potential contacts. Typical experimental recordings of the magnetoresistance component ρ_{xx} at 4.2 K are shown in Fig.2, for magnetic fields applied perpendicular to the heterojunction interface. These show very strong and well resolved Shubnikov-deHaas oscillations down to 0.4 Tesla. Rotation of the samples through an angle θ shows that the maxima and minima occur at constant values of $B\cos\theta$, the component of field perpendicular to the interface, as is characteristic of two dimensional systems. The angular dependence is shown in Fig.3. The periodicities of the oscillations, $(\Delta 1/B)^{-1} = nh/2e$, all lie between 6.2T and 10.2T, giving two-dimensional electron concentrations between $3.0 \times 10^{11}\ cm^{-2}$ and $5.0 \times 10^{11}\ cm^{-2}$, as shown in Table I. Rather higher electron concentrations are deduced from Hall effect measurements at 300 K and 77 K, and in some cases 4.2 K. This indicates some bulk conduction in the GaInAs layer at higher temperatures, and in some samples at 4.2 K also. Further evidence of this at 120 K is provided by the observation of magnetophonon resonances from the bulk GaInAs layer, which do not change when the sample is rotated, hence being shown to have a three dimensional character.

TABLE I. HETEROJUNCTION CHARACTERISTICS

Sample	μ ($cm^2V^{-1}s^{-1}$)			n (cm^{-2}), 4.2 K	
	300 K	77 K	4.2 K	Shubnikov-deHaas	Hall effect
9	9020	21000	8900	4.9×10^{11}	8.0×10^{11}
13	8700	41000	50000	3.0×10^{11}	
55	4700	2500	2900	4.1×10^{11}	8.4×10^{11}
58				3.7×10^{11}	
68	9732	32150	32000	5.0×10^{11}	5.2×10^{11}
102	3800	18000		4.3×10^{11}	

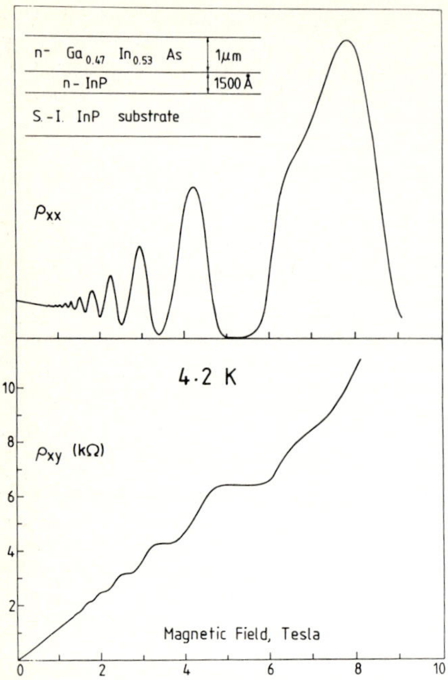

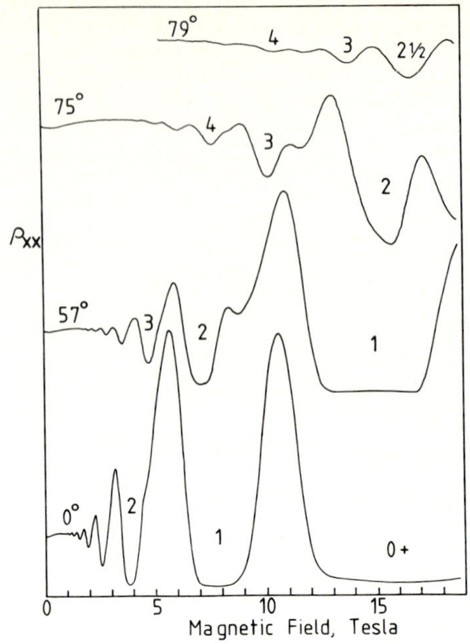

Fig.2: Recordings of the Shubnikov-deHaas oscillations in ρ_{xx} and Hall resistivity ρ_{xy} at 4.2 K for sample 68. The weak splitting observed at 7T becomes much more pronounced at 1.6 K. The inset to the ρ_{xx} curve shows a schematic view of the sample configuration.

Fig.3: The magnetoresistance ρ_{xx} as a function of angle for sample 58 at 1.6 K. The numbers indicate the quantum number of the last completely filled Landau level, with the 0+ level being filled below 15.4T. Further spin splittings are shown as half integers.

The two dimensional electron concentrations are in good agreement with the known doping level of the InP and the estimated barrier hight ΔE_c of the GaInAs-InP interface, of order 0.3 eV. This leads to a depletion of the InP by N_{dep} carriers, where

$$N_{dep} = \left(\frac{2\varepsilon\varepsilon_o N_D \Delta E_c}{e} \right)^{1/2} = 3.5 \times 10^{11}\ cm^{-2}. \qquad (1)$$

This gives a depletion depth of $\sim$ 1200 Å, indicating almost complete depletion of the n-InP layer.

The Shubnikov-deHaas oscillations appear as strong minima in the resistivity at integer sub-multiples of the fundamental field $(\Delta 1/B)^{-1}$. This is due to the minima in the conductivity σ_{xx} which occur when a Landau level is just completely filled. The resistivity is given by

$$\rho_{xx} = \frac{\sigma_{xx}}{\sigma_{xx}^2 + \sigma_{xy}^2} \qquad (2)$$

and in high magnetic fields $\sigma_{xy} >> \sigma_{xx}$, so that $\rho_{xx} \propto \sigma_{xx}$. The accurate appearance of the minima at integer sub-multiples of the fundamental field, and the fact that no further series of oscillations are observed indicate clearly that only one quantum level is occupied at electron concentrations up to 5.0×10^{11} cm^{-2}.

Fig.2 also shows recordings of the Hall resistivity ρ_{xy}. This shows that ρ_{xy} becomes constant over a range of field values around the region where ρ_{xx} approaches zero. In this region it was first shown by vonKlitzing, Dorda and Pepper(18) that the Hall resistivity is given accurately by the relation

$$\rho_{xy} = \frac{h}{e^2 i} = \frac{\alpha^{-1}\mu_o c}{2i} \qquad (3)$$

where i is the number of completely filled Landau levels. This has enabled very accurate measurements of the fine structure constant α to be made(18,19,20). The quantum Hall steps observed in the GaInAs-InP system are very pronounced and extend over a wide region of magnetic field, comparable in extent to GaAs-GaAlAs heterojunctions of considerably higher mobility. This is probably due to the presence of a higher degree of potential fluctuations, which lead to the localisation in the Landau level tails(21), thought to be the origin of the Hall steps(22-5). In addition, the lower electron effective mass in GaInAs leads to larger Landau level separations, thus rendering the quantum effects more significant even in the presence of increased disorder. Measurement of the Hall resistance at the steps corresponding to complete filling of the second, third and fourth Landau levels are in agreement with the expected values, within the limits of the measuring equipment used ($\sim$0.1%).

The electron effective mass was determined for two of the heterojunctions by a study of the temperature dependence of the amplitudes of the low field oscillations given by the relation(26)

$$A(T) \propto X/\sinh X, \qquad \text{where } X = \frac{2\pi^2 kT}{\hbar\omega_c} \qquad (4)$$

and ω_c is the cyclotron frequency. The values deduced are $m^*=(0.049\pm 0.002)m_o$ and $m^*=(0.052\pm0.002)m_o$ for samples 102 and 68 respectively. The band edge mass in $Ga_{0.47}In_{0.53}As$ is $0.041m_o$ (13), and the difference is caused by the strong non-parabolicity of the conduction band. The energy of the lowest quantum level E_o is given by the variational calculations of Stern(26) as 58 meV above the band edge for sample 102 with $n=4.3 \times 10^{11}$ cm^{-2} and 64 meV above the band edge for sample 68 with $n= 5.0 \times 10^{11}$ cm^{-2}. The Fermi levels for these electron concentrations are 21 meV and 24 meV respectively, giving $(E_o + E_F)$ as 79 meV for sample 102 and 88 meV for sample 68. The cyclotron effective mass in three band $\underline{k}.\underline{p}$ theory(27) is given by the expression,

$$\frac{1}{m^*} = \frac{1}{m_o^*}\left[1 + \frac{2K_2}{E_g}\left[(N+1)\hbar\omega_c + E_{||}\right]\right] \qquad (5)$$

where E_g is the band gap, N is the quantum number of the initial Landau level, $E_{||}$ is the energy due to motion parallel to the field and K_2 is a factor dependent upon the band gap and spin-orbit splitting, which is $K_2 = -0.85$ for GaInAs. For low fields we may put $(N + 1)\hbar\omega_c = E_F$, and identifying $E_{||}$ with E_o gives an effective mass for electrons at the Fermi level of

$$\frac{1}{m^*} = \frac{1}{m_o^*}\left[1 - \frac{1.7(E_F + E_o)}{E_g}\right], \qquad (6)$$

so that $m^* = 0.049m_o$ for sample 102 and $m^* = 0.050m_o$ for sample 68, in reasonable agreement with the experimental values.

Cyclotron resonance has been observed using far infra-red lasers operating between 337 μm and 79 μm. Typical results are shown in Fig.4, for sample 13, with $n=3 \times 10^{11}$ cm^{-2}, and a substrate which was wedged in order to avoid interference effects. The measured effective mass values rise from $0.0484m_o$ at 337 μm to $0.0496m_o$ at 78.6 μm, in quite good agreement with equations (4) and (5). There is thus no evidence for a

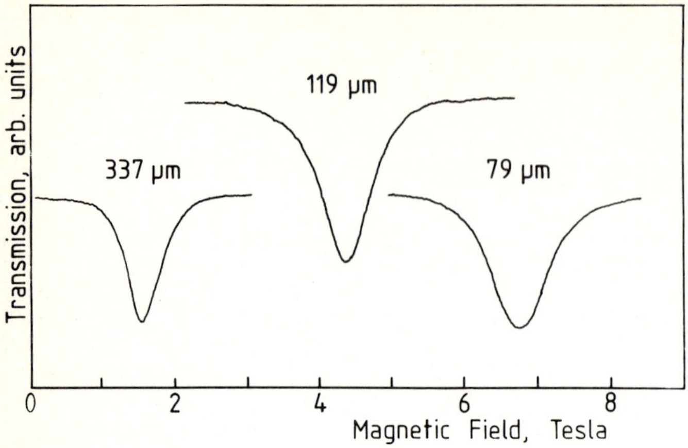

Fig.4: Cyclotron resonance absorption at 15 K for sample 13, with $n = 3 \times 10^{11}\ cm^{-2}$. Typical values of the absorption at resonance are around 30% of the cyclotron active component, as measured by reversing the magnetic field direction.

difference between the Shubnikov-deHaas and cyclotron effective masses as has been observed in silicon, and attributed to many-body effects(26)

Both the cyclotron mass and the resonance linewidth, Γ, are almost independent of temperature up to 60 K, but the linewidth clearly increases with increasing magnetic field. This is illustrated in Fig.5 where Γ is plotted as a function of $\sqrt{B}$. For short range scattering centres it has been shown both theoretically and experimentally for the case of silicon inversion layers(28), and GaAs heterojunctions at low fields(29) that the linewidth, Γ_s, is given by $\Delta B_{1/2} \sim 0.6\sqrt{B/\mu}$, where μ is the low field mobility. This relation holds reasonably well for the lowest field resonance, however the linewidth does not increase as rapidly as predicted with magnetic field. This behaviour may be accounted for in terms of a finite range for the scattering centres which, as predicted by Prasad and Fujita(30) and by Sarkar and Nicholas(31), reduces the linewidth by a factor such that:

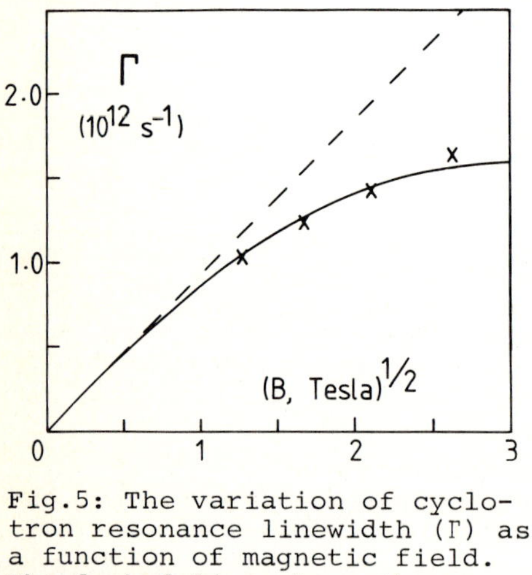

Fig.5: The variation of cyclotron resonance linewidth (Γ) as a function of magnetic field. The dashed line shows the expected dependence for short range scatterers, while the solid line comes from Eq. 7, with d=40 Å, the range of the Gaussian scatterers.

$$\Gamma = \frac{\Gamma_s}{(1 + \beta^2)^{3/2}}, \qquad (7)$$

where $\beta^2 = 2d^2/l^2 = 2ed^2B/\hbar$, d is the range of the Gaussian scattering potential, and l is the Larmor radius. Using a range of 40 Å in equation (7) predicts very well the observed magnetic field dependence of the linewidth, as seen in Fig.5. Such a range is not unreasonable, particularly in view of the known interface spreading of 30 - 40 Å for samples grown by LP-MOCVD (32). One further example of such finite range scattering are electrons bound to defects in InSe layer crystals(33). Scattering ranges of order 100 - 200 Å have been observed, which cause the cyclotron resonance linewidth to decrease with increasing magnetic field.

3. SPIN EFFECTS.

A final feature of considerable interest in the heterojunctions is the appearance of a well resolved spin splitting, which may be seen as a splitting of the last peak in Fig.2, and a very strong splitting of the last (N=0) Landau level in Fig.3, indicated as 0+. Since the spin splitting has its origin as a band structure effect it is determined by the total magnetic field, B, while the Landau splitting depends only upon the perpendicular component of field, $B_\perp$(26). The spin splitting thus becomes relatively more pronounced as the sample is rotated relative to the magnetic field, as is clearly demonstrated in Fig.3. This fact may be used to measure the effective g-factor, g^*, by observing the relative amplitudes of the oscillations and their angular dependence, as first shown by Fang and Stiles(34). In particular as the magnitude of the spin splitting $g^*\mu_B B$ approaches one half of the Landau splitting $\hbar\omega_c$, then the spin and Landau splittings in the magnetoresistance curves will become equally well resolved, as this condition will result in a ladder of equally spaced levels. When the levels are not sufficiently well resolved then the amplitude of the Shubnikov-deHaas oscillations will pass through a minimum at this condition, and exhibit a 180° phase change, as the largest level seperation passes from that between two adjacent Landau levels, to that between the spin split levels of a single Landau level.

Examples of both types of behaviour are shown in Figs.3, 6, 7 and 8. For the trace at 79° in Fig.3 the spin split minimum labelled $2\frac{1}{2}$ has achieved an amplitude intermediate between those of the N = 2 and 3 minima, giving $g^* = 3.8$. For the oscillations at lower field the spin splitting is much less well resolved, as is illustrated in more detail in Fig.6, where it is clear that the characteristic behaviour of a minimum in the oscillatory amplitude and an inversion of phase is observed at

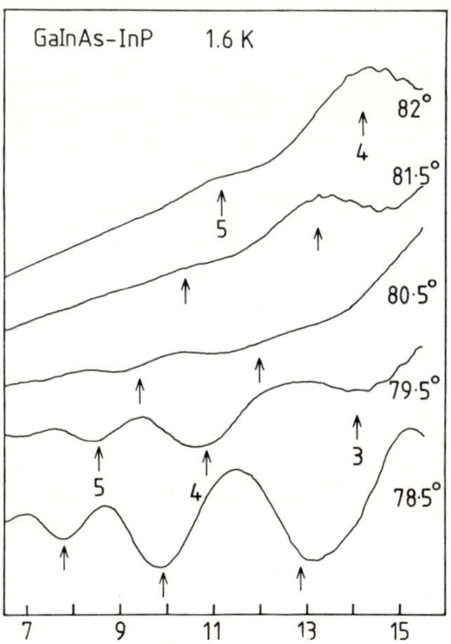

Fig.6: Shubnikov-deHaas oscillations at large angles in sample 58. The arrows show the values of magnetic field corresponding to complete filling of 3, 4 and 5 Landau levels.

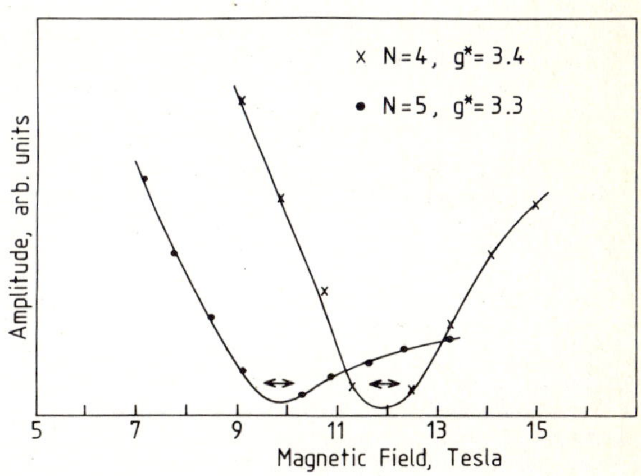

Fig.7: Plots of the amplitude of the Shubnikov-deHaas oscillations for the fields corresponding to completely filled Landau levels (and hence fixed values of $B_\perp$), as a function of the total field, (and hence spin splitting). Some of the experimental traces used are shown in Fig.6, where it can be seen that around the minimum amplitude the phase of the oscillations reverses.

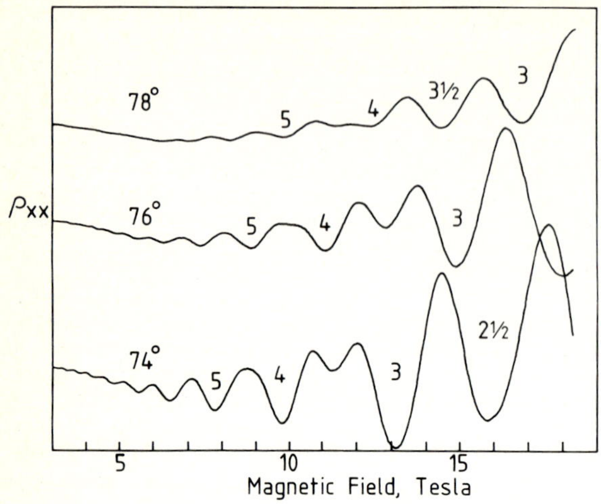

Fig.8: Shubnikov-deHaas oscillations in tilted magnetic fields for sample 68 at 1.6 K, in the region where the spin splitting is resolved, and is close to one half of the cyclotron energy. The half integer minima correspond to the points where the Fermi level lies between the spin up and spin down states of one Landau level.

$\sim 80.5^o$. A plot of the amplitude of a particular oscillation as a function of the total field, and hence spin splitting, enables g-factors of 3.3 and 3.4 to be deduced, as shown in Fig.7. A more detailed picture for the case where the spin splitting remains well resolved is shown in Fig.8, for sample 68. In this case the condition of equal resolution gives $g^* = 4.6$ for $N = 2\frac{1}{2}$ at 77° and $g^* = 4.2$ for $N = 3\frac{1}{2}$ at 78°. These results may be summarised by saying that the effective g-factor becomes larger as the spin splitting becomes more resolved.

The bulk band edge g-factor for GaInAs has been deduced to be 3.5 (14), which will be reduced to a value of 3.0 at the Fermi level by nonparabolicity, using the same $\underline{k}.\underline{p}$ theory (27) as was found to predict accurately the effective mass. There is thus clear evidence that the g-factor has been enhanced above its bulk value, as has been reported by a number of authors for the case of silicon inversion layers(34-7) and more recently for GaAs-GaAlAs heterojunctions(38-40). The theoretical origin of this effect was first explained by Janak(41) and by Ando and Uemura(42), as due to a many-body exchange enhancement, with both a constant weak increase of the same order of magnitude as the effective mass enhancement, and an oscillatory enhancement which could become very large for a large difference in spin population. The silicon results indicate an almost constant enhancement, while for GaAs curve fitting(38) suggests a very strong oscillatory effect. The present results indicate that the oscillatory enhancement clearly occurs in the 2DEG in GaInAs, with the enhancement being strongly dependent on the spin population difference. This is supported by the abscence of any many-body enhancement of the Shubnikov-deHaas mass, and by recent results(also reported at this conference, 43) on GaInAs-AlInAs heterojunctions, where g-factors up to 9.3 have been observed in samples with somewhat higher mobilities and hence much more well resolved spin splittings.

One consequence of the oscillatory variation of g-factor is that the deduction of g^* gives a relation between the g-factor (g_S^*) when the Fermi level lies midway between the (N,+) and (N,-) levels and the g-factor (g_L^*) when it lies between the (N,-) and (N+1,+) levels(26):

$$g_S^* \mu_B B = \frac{\hbar e B}{m^*} \cos\theta - g_L^* \mu_B B \qquad (8)$$

so that in practice a mean value $(g_S^* + g_L^*)/2$ is measured. Assuming that g_L^* is the unenhanced bulk value of 3 this will then give values of g_S^* of up to 6.2 for the GaInAs-InP structures, and of up to 15.6 for the GaInAs-AlInAs structures.

4. SUPERLATTICES.

Three GaInAs-InP superlattices were studied each with ten active layers of nominally undoped GaInAs, 80 Å, 100 Å and 150 Å thick respectively, seperated by 300 Å of InP with $n \sim 3 \times 10^{16}$ cm^{-3}. The system is thus 'modulation doped' in the manner first described by Störmer et al(1), with the InP layer thickness and doping level being expected to give an electron concentration of $\sim 1 \times 10^{11}$ cm^{-2}. The carriers will fall into the potential wells formed by the GaInAs layers where they will be bound for the purposes of motion perpendicular to the interfaces. The considerable thickness of the InP layers means that the potential wells may be considered as independent, and the carriers will behave as a 2DEG(44). The Shubnikov-deHaas measurements shown in Fig.9, demonstrate that the carriers are in fact two dimensional, with electron concentrations of 1.7, 2.2, and 1.1×10^{11} cm^{-2} for the 80, 100 and 150 Å layer thicknesses respectively. Hall effect measurements indicate that all ten active layers are contributing to conduction, and the electron mobilities lie in the range 6000 - 18000 for the three samples.

Fig.9 also shows the influence of electron spin, which causes the very pronounced splitting of the lowest Landau level indicated by 0+. Further splitting of the N = 1 Landau level is observed in tilted magnetic fields, however the oscillations are not sufficiently strong in order to enable an accurate measurement of the g-factor to be made by the method of coincidences, as described in the previous section. An approximate measurement was made, indicating a value in the range 3 - 3.5, for the low fild region. This does not, however seem to be the case for the N = 0 level. The strong spin splittings occur around a field of $\sim$ 6T, where a g^* of 3 would give a total spin splitting of 1.0 meV. The Landau level broadening parameter Γ = 2 meV, as determined both from cyclotron resonance and mobility measurements, and the use of a simple Rayleigh criterion for the resolution of a minimum (39) would thus imply a total spin splitting of order 3 meV or greater, and thus g-factors of 10 or more. This deduction is thought to be in agreement with the ideas of many-body enhancement discussed in the previous section(35-42). Such an effect should be very pronounced for the 0+ minimum, since at this point the system has reached the quantum limit, with only the 0- level remaining occupied, the 2DEG

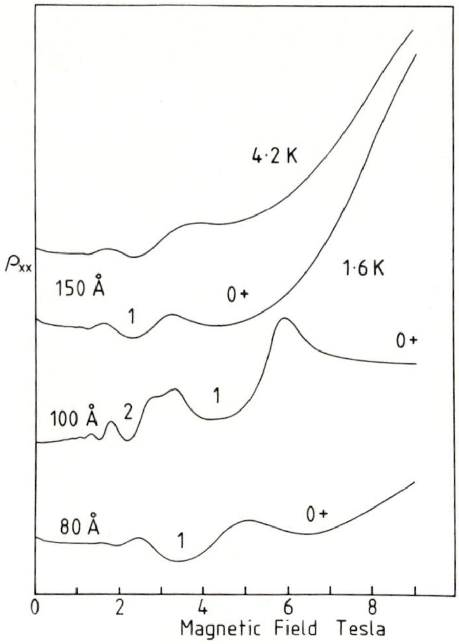

Fig.9: Shubnikov-deHaas oscillations in the superlattices at 1.6 K and 4.2 K.

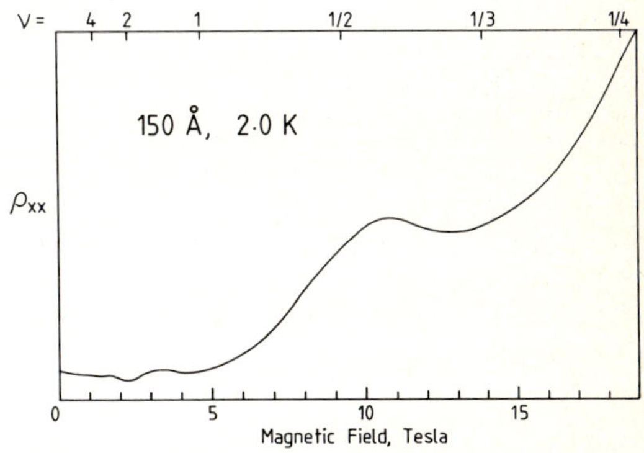

Fig.10: Magnetoresistance of the 150 Å superlattice in the high field regime.

being thus completely spin polarised. The magnetoresistance in the region where the Landau level degeneracy, ν (=nh/eB) is less than 1 is shown in Fig.10 for the 150 Å sample where it is possible to reach values of $\nu = 0.24$. The magnetoresistance rises rapidly showing a peak around $\nu = 1/2$ and a minimum around $\nu = 1/3$. Similar behaviour has been reported recently by Tsui et al (45), who observed a sharp minimum at $\nu = 1/3$ in GaAs-GaAlAs heterojunctions, which they attributed to the possible formation of a Wigner solid or Charge Density Wave, with triangular symmetry. It is not however impossible in the present case to explain the curves in terms of Eq.2, with the peak around $\nu = 1/2$ being due to the peak in σ_{xx} for a half filled level, and the rising background coming from the rapid fall in σ_{xy}^2, $\sim (ne/B)^2$.

Cyclotron resonance experiments were also performed over the temperature range 15 K - 150 K. Typical experimental recordings are shown in Fig.11. Despite wedging of the substrates some weak interference effects are still visible. For a given sample some weak non-parabolicity was observed, with effective masses rising by approximately 3% over the range of laser energies used, and an increase of 1 - 2% on raising the temperature to 150 K. These changes are approximately what would be expected from simple three band k.p theory(27), but are hardly above the level of experimental error (1-2%), due to the relatively broad linewidths. Values of $\omega_c\tau$ for 119 μm are of order 4 - 6. The linewidth was almost independent of magnetic field, a feature which is characteristic of ionised impurity scattering(31), which is rather surprising in view of the very short screening lengths in a 2DEG(26). The values of effective mass are however a strong function of layer thickness, as shown in Table II; the mass rising from 0.040 m_o to 0.054 m_o between the 150 Å and 80 Å samples. Since the band edge mass of bulk $Ga_{.47}In_{.53}As$, the composition at which the perfect lattice match to InP occurs, is 0.041 m_o, the value of 0.040 m_o implies a small enhancement of the In content of these layers, caused by the need to switch rapidly the growth conditions. The effective mass is expected to rise for thin layers due to the increased energy of the lowest bound state in the potential well, and the resulting non-parabolicity. The first energy level for square wells of 150, 100 and 80 Å, is at 40, 85 and 120 meV respectively. Using Eq.5 for the non-parabolicity, the measured effective masses lead to values of $E_{||}$, which we assume to be the energy of the first quantum level, of 0, 50 and 120 meV. These values are in reasonable agreement with the square well values, particularly when the possible grading and compositional fluctuations of the interfaces are considered. Recent measurements on a double heterostructure grown in the same equipment(32), indicate an interface spreading of ± 30 - 40 Å, which would lead to a potential well approaching a triangular or even Lorenzian form.

Magnetophonon resonance was also observed in the same samples. This occurs at high lattice temperatures due to the resonant absorption of a Longitudinal Optic (LO) phonon, which scatters an electron between one or more Landau levels. The resulting series of maxima in the transverse

TABLE II.

Layer th. Å	n $\times 10^{11} cm^{-2}$	m^*/m_o (C.R.) 119μm 125 K	m^*/m_o (M.P.R.) Estimated	NB_N T	$\hbar\omega_{LO}$ cm^{-1}
80	1.7	0.0543	0.0560	15.7	262
100	2.2	0.0458	0.0465	13.3	267
150	1.1	0.0407	0.0422	12.4	274
			0.0428	16.0	349

magnetoresistance occur at magnetic fields given by the equation,

$$\frac{NeB_N}{m^*} = \omega_{LO} \qquad (9)$$

where ω_{LO} is the LO phonon frequency(46).

Recordings of the second derivative of the magnetoresistance for the 150 Å sample are shown in Fig.12, where a characteristic series of oscillations is observed. The oscillations are two dimensional in character at lattice temperatures up to 200 K and beyond, as has been observed in GaAs-GaAlAs heterojunctions(47), with the resonant fields thus determined by the normal component of field $B\cos\theta$. The main series of oscillations arises due to scattering from the 'GaAs-like' LO phonon mode of the GaInAs, which has been shown by reflectivity(48,49), Raman(50) and magnetophonon(13) measurements in bulk material, to be dominant for the alloy composition with x = 0.47. The oscillations are also two dimensional for the thinner layers, and move to higher fields due to the increasing effective masses. The values of the fundamental fields (NB_N) are also given in Table II, along with the appropriate magnetophonon effective masses, deduced from the measured high temperature cyclotron masses, taking account of the additional non-parabolicity at the phonon

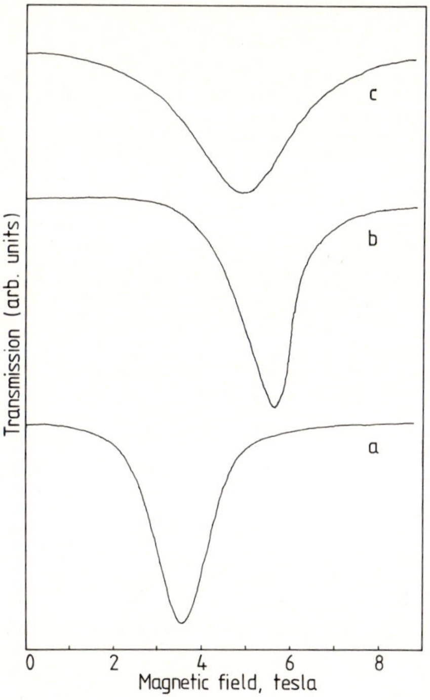

Fig.11: Typical cyclotron resonance results. (a) and (b) show transmission of the 150 Å superlattice at 15 K for 119 μm and for 79 μm radiation respectively, while (c) shows the 119 μm resonance for the 80 Å superlattice at 122 K. The greater linewidth in this sample reflects the lower mobility.

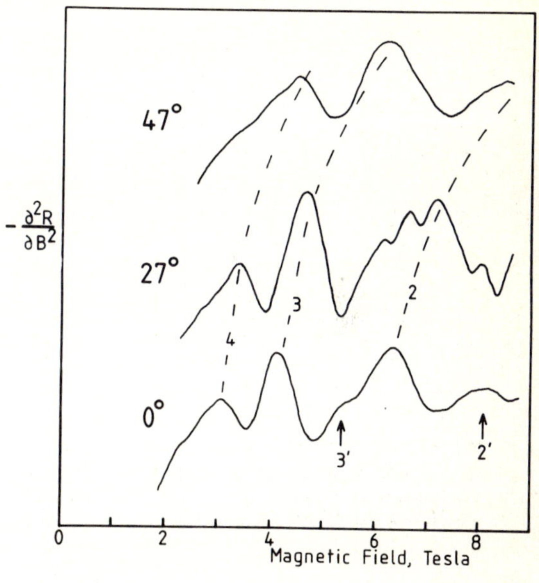

Fig.12: Magnetophonon oscillations in the 150 Å superlattice at 150 K. Sample rotation shows that the periodicity is constant in $B\cos\theta$. The main series, 2,3,4, is due to the 'GaAs-like' LO mode in GaInAs, and the two weaker peaks, 2' and 3', are caused by penetration of the InP LO phonon field into the GaInAs.

energy by using Eq.5. The resonant polaron contribution to the effective mass is very small since in alloys the polaron coupling constant α (46) relates only to the particular optic mode involved in scattering(51) and for the 'GaAs-like' mode $\alpha = 0.01$, giving an increase of only 0.25% through the factor $(1 + \alpha/4)$, (46). Using these masses the LO phonon energies shown in Table II were deduced. It may be seen that these decrease slightly for the thinner layers, possibly due to the admixture of more Phosphorous in the finite layer thickness, which will lead to a decrease in the LO phonon energy (49,50).

A further interesting feature of the curves in Fig.2 for the 150 Å superlattice is the appearance of two weaker features labelled 2' and 3'. These correspond to scattering by an optical phonon of frequency 349 cm^{-1} which is characteristic of the LO phonon in InP(49). This implies a long range phonon interaction between the LO phonons in the InP layers and the carriers bound in the GaInAs potential wells. Such a long range phonon interaction has been proposed recently by Hess and Vogl(52), as a mechanism for polar scattering in MOS devices. The exact suggestion of Hess and Vogl was that interface phonons should be involved, however in the present case the graded interfaces will not lead to any well defined interface phonons. The finite screening of the InP optical phonons will lead to some penetration into the GaInAs layers, and at the same time there will be some penetration of the carrier wavefunction into the InP layer. Both factors will enable the scattering to occur. To our knowledge this is the first direct experimental verification of this effect.

In conclusion we may say that the 2DEG in GaInAs shows a wide variety of interesting effects and offers considerable scope for further study. Spin effects are of particular interest due to the relatively large free electron g-factor of bulk GaInAs.

ACKNOWLEDGEMENTS: We wish to thank the C.N.R.S. and S.N.C.I. Grenoble for the use of the magnet facilities, Professors J.L. Robert and J.C. Thuillier for the loan of experimental equipment, and Dr. J.P. Duchemin for useful discussions. M.A.B. was supported by a S.E.R.C. CASE Studentship in collaboration with the G.E.C. Hirst Research Centre.

REFERENCES.

1) R. Dingle, H.L. Störmer, A.C. Gossard and W. Weigmann, Appl. Phys. Lett. 33 665 (1978).
2) H.L. Störmer, R. Dingle, A.C. Gossard, W. Weigmann and M.D. Sturge, Solid State Commun. 29 705 (1979).
3) L.L. Chang and L. Esaki, Surf. Sci. 98 70 (1980).
4) G.A. Sai-Halasz, R. Tsu and L.Esaki, Appl. Phys. Lett. 39 651 (1977).
5) R.J. Nicholas, M.A. Brummell, J.C. Portal, M. Razeghi and M.A. Poisson, Solid State Commun. in press (1982).
6) M.A. Brummell, R.J. Nicholas, J.C. Portal, M. Razeghi and M.A. Poisson, Physica B, Montpelier,1982, in press (1982).
7) R.E. Nahory, M.A. Pollack, W.D. Johnston Jr. and R.L. Barns, Appl. Phys. Lett. 33 659 (1978).
8) G.W. Gobeli and F.G. Allen, Phys. Rev. 137 A245 (1965).
9) T.E. Fischer, Phys. Rev. 142 519 (1966)
10) G. Margaritondo, A.D. Katnani, N.G. Stoffel, R.R. Daniels, and T-X. Zhao, Solid State Commun. 43 163 (1982).
11) W.A. Harrison, J. Vac. Sci. Technol. 14 1016 (1977).
12) R. Dingle in Festkörperprobleme Vol XV p21, Ed. H.J. Queisser, Pergamon Vieweg, Braunschweig (1975).
13) R.J. Nicholas, S.J. Sessions and J.C. Portal, Appl. Phys. Lett. 37 178 (1980).
14) E.H. Perea, E.E. Mendez and C.G. Fonstad, Appl. Phys. Lett. 36 978 (1980).
15) A. Kastalsky, R. Dingle, K.Y. Cheng and A.Y. Cho, to be published.

16) J.C. Portal, R.J. Nicholas, M.A. Brummell, A.Y. Cho, K.Y. Cheng and T.P. Pearsall, Solid State Commun. in press (1982).
17) M. Razeghi, M.A. Poisson, B. deCremoux and J.P. Duchemain, Electron. Lett. in press (1982).
18) K. vonKlitzing, G. Dorda and M. Pepper, Phys. Rev. Lett. 45 494 (1980).
19) D.C. Tsui and A.C. Gossard, Appl.Phys. Lett. 38 550 (1981).
20) D.C. Tsui, A.C. Gossard, B.F. Field, M.E. Cage and R.F. Dziuba, Phys. Rev. Lett. 48 3 (1982).
21) R.J. Nicholas, R.A. Stradling and R.J. Tidey, Solid State Commun. 23 55 (1977).
22) T. Ando, Y. Matsumoto and Y. Uemura, J. Phys. Soc. Japan 39 279 (1975).
23) H. Aoki and T.Ando, Solid State Commun. 38 1079 (1981).
24) R.E. Prange, Phys. Rev. B23 4802 (1981).
25) R.B. Laughlin, Phys. Rev. B23 5632 (1981).
26) T. Ando, A.B. Fowler and F. Stern, Rev. Mod. Phys. 54 437 (1982).
27) E.D. Palik, G.S. Picus, S. Teitler and R.F. Wallis, Phys. Rev. 122 475 (1961).
28) G. Arbstreiter, J.P. Kotthaus, J.F. Koch and G. Dorda, Phys. Rev. B14 2480 (1976).
29) P. Voisin, Y. Guldner, J.P. Vieren, M. Voos, P. Delescluse and N.T. Linh, Appl. Phys. Lett. 39 982 (1981)
30) M. Prasad and S. Fujita, Physica 91A 1 (1978).
31) C.K. Sarkar and R.J. Nicholas, Surf. Sci. 113 326 (1982).
32) R. Bisaro, G. Laurencin, A. Freidrich and M. Razeghi, Appl. Phys. Lett. 40 978 (1980).
33) E. Kress-Rogers, R.J. Nicholas and A. Chevy, to be published.
34) F.F. Fang and P.J. Stiles, Phys. Rev. 174 823 (1968).
35) M. Kobayashi and K.F. Komatsubara, Solid State Commun. 13 293 (1973)
36) G. Landwehr, E.Bangert and K. vonKlitzing, in Physique sous Champs Magnetiques Intenses, Grenoble, 1974 (Colloques Internationaux No. 242, Editions du C.N.R.S., Paris) p. 177 (1975).
37) Th. Englert K. vonKlitzing, R.J. Nicholas, G. Landwehr, G.Dorda and M.Pepper, Phys. Stat. Solidi B99 237 (1980).
38) Th. Englert, D.C. Tsui, A.C. Gossard and Ch. Uihlein, Surf Sci. 113 295 (1982).
39) R.J. Nicholas, M.A. Brummell and J.C. Portal, Surf. Sci 113 290 (1982).
40) S. Narita, S. Takeyama, W.B. Luo, S. Hiyamizu, K. Nanbu and H. Hashimoto, Surf. Sci. 113 301 (1982).
41) J.F. Janak, Phys. Rev. 178 1416 (1969).
42) T. Ando and Y. Uemura, J. Phys. Soc. Japan 37 1044 (1974).
43) R.J. Nicholas, M.A. Brummell, J.C. Portal, A.Y. Cho, K.Y. Cheng and T.P. Pearsall, this conference, Grenoble, 1982, and submitted to Solid State Commun. (1982).
44) M. Razeghi, M.A. Poisson, J.P. Larivain, B. deCremoux, J.P. Duchemain and M. Voos, Electron Lett. 18 339 (1982).
45) D.C. Tsui, H.L. Störmer and A.C. Gossard, Phys. Rev. Lett. 48 1559 (1982).
46) P.G. Harper, J.W. Hodby and R.A. Stradling, Rep. Prog. Phys. 36 74 (1973).
47) D.C. Tsui, Th. Englert, A.Y. Cho and A.C. Gossard, Phys. Rev. Lett. 44 341 (1980).
48) M. Brodsky and G. Lucovsky, Phys. Rev. Lett. 21 990 (1968).
49) C. Pickering, J. Electron. Mater. 10 901 (1981).
50) A. Pinczuk, J.M. Worlock, R.E. Nahory and M.A. Pollack, Appl. Phys. Lett. 33 461 (1978).
51) R.J. Nicholas, R.A. Stradling, J.C. Portal and S. Askenazy, J. Phys. C: Solid State Phys. 12 1653 (1979).
52) K. Hess and P. Vogl, Solid State Commun. 30 807 (1979).

QUANTUM TRANSPORT IN GaInAs-AlInAs HETEROJUNCTIONS

R.J. Nicholas+, M.A. Brummell+, J.C. Portal*, A.Y. Cho°, K.Y. Cheng° and T.P. Pearsall°.

+ Clarendon Laboratory, Parks Rd., Oxford OX1 3PU, England.
* I.N.S.A. 31077 TOULOUSE, S.N.C.I.-C.N.R.S. 166 X 38042 GRENOBLE
o Bell Laboratories, Murray Hill, New Jersey 07974, U.S.A.

Shubnikov-deHaas and cyclotron resonance results are presented for GaInAs-AlInAs heterojunctions in both perpendicular and tilted magnetic fields. Tilted fields cause magnetic shifts and mixing of the electric subbands. The g-factor is measured by the method of coincidences and shown to be exchange enhanced to values of order 9.

Strongly enhanced electron mobilities have been reported recently for $Ga_{.47}In_{.53}As$-$Al_{.48}In_{.52}As$ heterojunctions grown by MBE(1). Low field Shubnikov-deHaas measurements(2,3) have shown that the electrons are accumulated on the GaInAs side of the interface and behave as a two dimensional electron gas (2DEG) at 4.2 K. In the samples studied with electron concentrations of order 7×10^{11} cm^{-2} two electric subbands were found to be occupied(2,3), while for GaInAs-InP structures with $n < 5\times10^{11}$ cm^{-2} only one electric subband has been found to be occupied(4,5). We report here recent measurements on this system made at the high magnetic field laboratories in Oxford and Grenoble.

The formation of a 2DEG at a semiconductor heterojunction interface was first reported by Störmer et al(6) for GaAs-GaAlAs, who showed that electron confinement takes place in the smaller gap material (GaAs). In the present case $Ga_xIn_{1-x}As$-$Al_yIn_{1-y}As$ samples were grown lattice matched to InP with x=0.47, y=0.48. The sample characteristics have been described in detail by Cheng et al(1). At 4 K the band gaps of GaInAs and AlInAs will be 0.81 eV and 1.7 eV respectively. There will be only a small discontinuity in the valence bands of order 100 meV, since the effect of passing from ∿ 50% GaAs to 50% AlAs will introduce a step of half of the difference of valence band energies of GaAs and AlAs which is known to b of order 200 meV(7). The conduction band discontinuity will thus be of order 0.8 eV. The 2DEG concentration is then determined by the depletio of the AlInAs which in the present samples gives values of order 8×10^{11} cm^{-2} with electron mobilities of 93,000 $cm^2V^{-1}s^{-1}$ at 4.2 K (1).

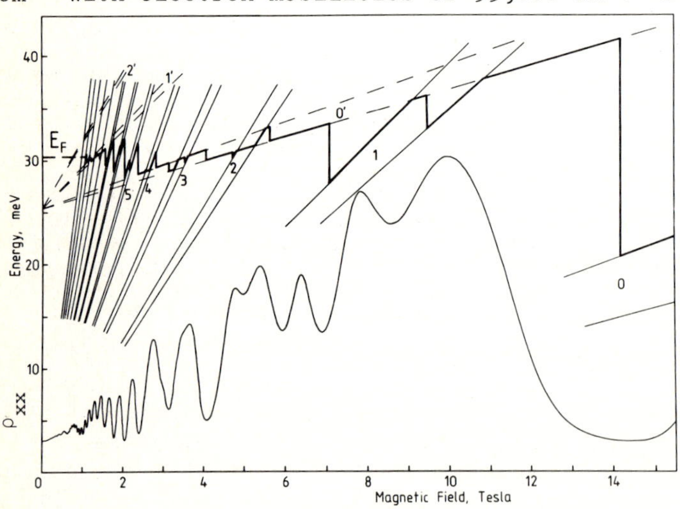

Fig.1: A Plot of the relative positions of the Landau levels arising from the two occupied electronic subban E_0 and E_1 with an applied magnetic field perpendicular to the surface. The position of the Fermi level is also shown as a function of the field, assuming infinitely sharp Landau levels. The lower half shows an experimental recording of the magnetoresistan at 1.6 K.

Magnetoresistance measurements were made on long bar shaped samples at temperatures down to 1.6 K. Very pronounced Shubnikov-deHaas oscillations appear directly in the magnetoresistance as is shown in Fig.1. The oscillations clearly reveal the occupation of the two electric subbands (E_o and E_1) which produce the two obviously very different periodicities(2). The low frequency oscillations visible at low field give a periodicity of $(1.9\ T)^{-1}$ which gives a population of the higher subband = $0.92 \times 10^{11}\ cm^{-2}$. The periodicity of the high frequency (E_o) series oscillates, with the periodicity of the second. When the Fermi level (E_F) is in the region between two Landau levels of E_1 then its contribution to the density of states is no different from that of a filled level of E_o, so that the periodicity measures the total number of electrons = $6.9 \times 10^{11}\ cm^{-2}$. When E_F lies within a Landau level of E_1 then both sets of Landau levels are filled simultaneously and the periodicity of the E_o series falls. We may thus deduce the low magnetic field populations of E_o and E_1 to be 0.92 $\times 10^{11}\ cm^{-2}$ and $6.0 \times 10^{11}\ cm^{-2}$ respectively. In high fields however the well resolved nature of the Landau levels leads to oscillating populations in the two electric subbands. This is illustrated in Fig.1 which shows the Landau levels of both subbands, together with the resulting movement of the Fermi energy, calculated assuming infinitely sharp levels.

Using tilted magnetic fields introduces a parallel magnetic field component $B_\parallel$. This has been shown(8,9) to shift the relative positions of the two subbands by an amount proportional to the square of the magnetic field and the spread of the wavefunction. At large angles between the surface normal and the applied magnetic field, the parallel component is sufficient to totally depopulate the E_1 subband for magnetic fields of order 6 T and above. This effect is seen directly in the magnetoresistance at 90° where a fall in resistance of approximately 25% occurs around 6 T, due to the suppression of the inter subband scattering. A similar fall has been seen for GaAs-GaAlAs(10), as a function of electron concentration. The magnetic field induced shift in the E_1-E_o seperation may be deduced from the changes in population(3) as being given by:

$$(E_1 - E_o) = 0.26\ meV/T^2.$$

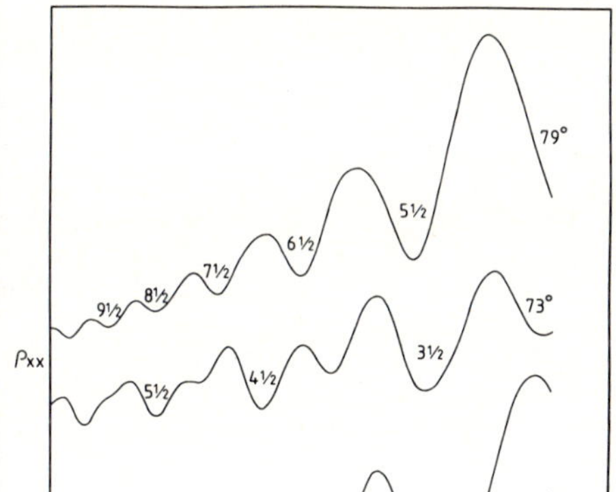

Fig.2: Shubnikov-deHaas oscillations in tilted fields at 1.6K. The half integer minima occur when the Fermi level lies between spin split levels with the same quantum number.

It is also necessary to invoke the presence of a very large spin splitting of the Landau levels in order to explain the higher field peaks in Fig.1. This spin splitting becomes more pronounced in tilted magnetic fields, as may be seen in Fig.2, since the Landau level separation is determined only by the perpendicular component of magnetic field $B_\perp$, while the spin splitting comes from the total field. By observing the relative amplitudes of the oscillations and their angular dependence(11,12), the relative values of the spin and Landau splittings may be deduced at one or more of the conditions:

$$r = \frac{g^* \mu_B B}{h w_c} = \frac{g^* m^*}{2 m_o} \cdot \frac{B}{B_\perp} = 1/2,\ 1,\ 3/2,\ 2 \ldots$$

where for unresolved spin splittings the half integer conditions correspond

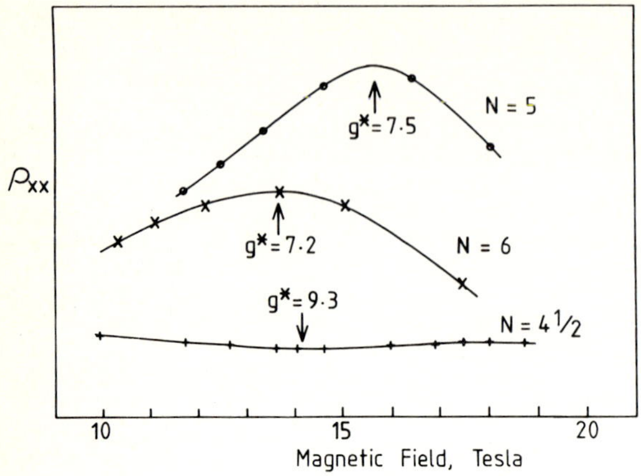

Fig.3: Plots of the values of given resistivity maxima (N=5, 6) and minima (N=4 1/2), as a function of the total magnetic field, which are obtained as the sample is rotated relative to the field direction.

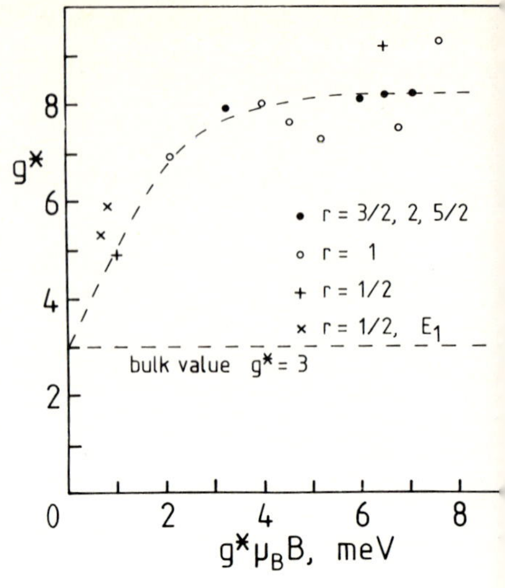

Fig.4: The measured g-factors, g^*, as a function of the spin splitting at which they were measured, $g^*\mu_B.B$.

to a minimum in the oscillation amplitudes, and the integers give a maximum. For large spin splittings or high mobility samples the half integ condition will lead to equally resolved spin and Landau structures in th conductivity, as may be seen in Fig.2 at 66°. Using $m^*=0.048m_o$ as measured by cyclotron resonance, we find $g^*=9.2$ for this case. On further rotation the minima corresponding to complete filling of one or more Landau levels become weaker and then dissappear, as the spin up and spir down states from adjacent Landau levels approach each other. A plot of the amplitude of a given maximum shows a peak (Fig.3) where the two spir states (N,+) and (N+1,-) coincide, corresponding to the condition r=1, and giving $g^*=7.5$. If the spin split minima are considered then a very weak minimum in the 'minimum value' is observed (Fig.3) giving $g^*=9.3\pm0$. Further minima and maxima of the oscillation amplitudes are observed at the conditions r=3/2, 2, 5/2 at angles up to 85.5°, all three conditions giving $g^*=8.2\pm0.2$. At lower total magnetic fields the oscillations are more complex due to the occupation of both electric subbands, however it was possible to deduce g-factors which lay in the range 5 - 7.

The measured values of g^* are summarised in Fig.4, where they are plotte as a function of the total spin splitting $g^*\mu_B.B$. They are clearly mucl larger than the bulk value of 3 (13) and increase with the spin splittir and hence the spin population difference. Evidence for g^* enhancement has been given for the 2DEG in both silicon(11,12) and GaAs(14-16), and is thought to be due to many body effects(17), arising through differen in spin populations(18). This idea is supported by the present results with the two differing values of g^* (7.5 & 9.3) deduced at r=1 for differing positions of the Fermi level, being particularly noticeable. Mucl lower g-factors down to 3.3 are observed in GaInAs-InP heterojunctions, where the spin splitting is no longer fully resolved due to the somewhat lower mobilities (19).

We have also observed cyclotron resonance absorption in the same sample using 337, 195, 119 and 79 μm radiation. The cyclotron effective mass found to be $0.048m_o \pm 0.001m_o$ and typical experimental recordings are sh in Fig.5. In tilted fields the parallel magnetic field component also

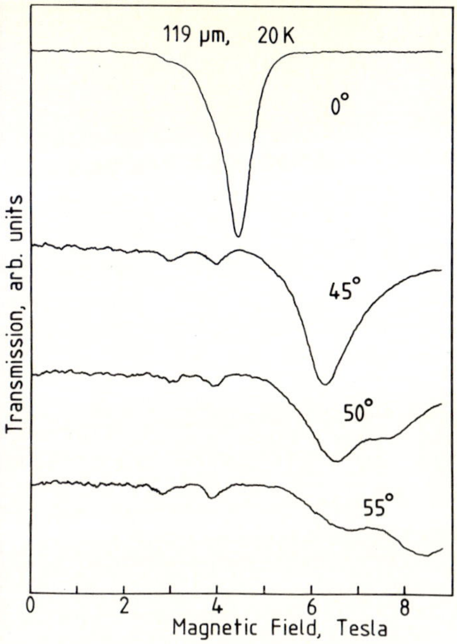

Fig.5: Recordings of the cyclotron resonance absorption using tilted fields. The two weak absorptions at 3.0 and 3.8 T, are due to the bulk GaInAs.

a significant effect on the cyclotron resonance. By 50°, for 119 μm, the parallel field component has shifted the N'=0 level of the E_1 subband up, until it is coincident with the N=3 level of E_o. The Fermi level lies around the N'=0 level. There is then a strong resonant increase in both inter-subband scattering and mixing as these two Landau levels overlap, and the resonance shows a typical coupled mode behaviour (Fig.5). Two absorptions occur with the intensity gradually being transferred from one to the other. This effect is probably due to mixing of the subbands by the parallel magnetic field component, which may also influence the measurements of g-factor.

ACKNOWLEDGEMENTS: To C.N.R.S. and S.N.C.I. Grenoble for the use of the magnets, to Professors J.L. Robert and J.C. Thullier for the use of equipment, and Dr. A. Kastalsky for discussions. To S.E.R.C. and G.E.C. Hirst Research Centre for providing a CASE Studentship for M.A.B.

REFERENCES:

1) K.Y. Cheng, A.Y. Cho, T.J. Drummond and H. Morkoc, Appl. Phys. Lett. 40 147 (1982).
2) A. Kastalsky, R. Dingle, K.Y. Cheng and A.Y. Cho, submitted to Appl. Phys. Lett. (1982).
3) J.C. Portal, R.J. Nicholas, M.A. Brummell, A.Y. Cho, K.Y. Cheng and T.P. Pearsall, Solid State Commun. in press (1982).
4) Y. Guldner, J.P. Vieren, P. Voisin, M. Voos, M. Razeghi and M.A. Poisson, submitted to Appl. Phys. Lett. (1982).
5) R.J. Nicholas, M.A. Brummell, J.C. Portal, M.Razeghi and M.A. Poisson, Solid State Commun. in press (1982).
6) H.L. Störmer, R. Dingle, A.C. Gossard, W. Wiegmann and M.D. Sturge, Solid State Commun. 29 705 (1979).
7) R. Dingle in Festkörperprobleme, XV, Pergamon Vieweg (1975).
8) F. Stern and W.E. Howard, Phys. Rev. 163 816 (1967).
9) W. Beinvogl, A. Kamgar and J.F. Koch, Phys. Rev. B14 4274 (1976).
10) H.L. Störmer, A.C. Gossard and W. Wiegmann, Solid State Commun. 41 707 (1982).
11) F.F. Fang and P.J. Stiles, Phys. Rev. 174 823 (1968).
12) T. Ando, A.B. Fowler and F. Stern, Rev. Mod. Phys. 54 437 (1982).
13) E.H. Perea, E.E. Mendez and L.G. Fonstad, Appl. Phys. Lett. 36 978 (1980).
14) T. Englert, D.C. Tsui, A.C. Gossard and Ch. Uihlein, Surf. Sci. 113 295 (1982).
15) R.J. Nicholas, M.A. Brummell and J.C. Portal, Surf. Sci. 113 290 (1982).
16) S. Narita, S. Takeyama, W.B. Luo, S. Hiyamizu, K. Nanbu and H. Hashimoto, Surf. Sci. 113 301 (1982).
17) J.F. Janak, Phys. Rev. 178 1416 (1969).
18) T. Ando and Y. Uemura, J. Phys. Soc. Japan 37 1044 (1974).
19) R.J. Nicholas, M.A. Brummell, J.C. Portal, M. Razeghi and M.A. Poisson, this conference, Grenoble, (1982).

BIAS VOLTAGE DEPENDENCE OF THE TEMPERATURE OF HOT n-Si INVERSION LAYER ELECTRONS IN HIGH MAGNETIC FIELDS

E. Vass
Institut für Experimentalphysik
University of Innsbruck, AUSTRIA

ABSTRACT

The bias voltage (U_B-) dependence of the hot electron temperature T_c in quantized (100) n-Si inversion layers is calculated from the power balance equation. It is shown that the experimental observed increase of T_c /1,2,3/ in the low U_B range $0 \leq -U_B \leq 4V$ can be interpreted in terms of two dimensional (2D) acoustic phonon scattering and the U_B dependence of the effective electron mass. The decrease of T_c in the high U_B range $-U_B > 4V$ is due to the enhanced surface roughness and surfon scattering.

1. INTRODUCTION

Under the influence of a strong electric surface field the motion of the electrons in the n-inversion channel of a Si-MOSFET is quantized perpendicular to the channel. As a consequence of this the electrons behave like a quasi-two dimensional (2D) gas. Such 2D electronic systems have been the subject of extensive studies in the past decades /4/ and represent a useful tool to investigate the electron phonon interaction as well as many body effects. Recently /1,2,3/ Shubnikov-de Haas measurements have been performed to determine the influence of the substrate bias voltage U_B on the temperature T_c of hot electrons in the n-inversion channel of Si-MOSFETs on (100) p-Si substrates. It was observed that T_c increases when the electrons are pushed closer to the interface by applying an increasing negative bias voltage U_B. As it was pointed out in ref. /2/, this behavior cannot be explained by the theoretical model of ref. /5/, because the matrixelements used in that model increase with increasing U_B. In order to interpret their data the authors of ref. /2/ attributed the observed increase of T_c to the anomalous thermal and dielectric properties of the amorphous SiO_2-layer adjacent to the inversion layer.

In the presented paper it is shown that the discrepancy between the theoretical predictions of the model in ref. /5/ and the experimental results of ref. /2,3/ can be resolved if the unphysical U_B dependence of the 2D phonon matrixelement is eliminated and the U_B dependence of the effective electron mass m is additionally taken into account.
The paper is organized as follows: In section 1 the calculation of the electron temperature, the modification of the 2D phonon matrixelement and the calculation of the effective mass are described. The theoretical results are compared with the experimental data taken from ref./3/ in the last section.

2. CALCULATION OF THE ELECTRON TEMPERATURE T_c

The electron temperature was calculated from the power balance equation

$$\frac{e^2}{m} <<\tau>> F^2 = (<E> - E_o)\tau_\varepsilon^{-1} \tag{1}$$

where e is the electron charge, m is the effective electron mass and F is the electric field applied parallel to the n-inversion layer. The total momentum relaxation time τ was obtained from

$$\tau^{-1} = \tau_1^{-1} + \tau_2^{-1} + \tau_3^{-1} \tag{2}$$

where τ_1^{-1}, τ_2^{-1}, τ_3^{-1} are the scattering rates for 2D phonon, surfon and surface roughness scattering.
The energy relaxation time τ_ε is defined by

$$(<E> - E_o)\tau_\varepsilon^{-1} = < \frac{dE}{dt} > \tag{3}$$

where E is the electron energy and dE/dt is the total energy loss rate to the phonon system. E_o is the thermal equilibrium energy of an electron. The brackets << >> and < > denote appropriate 2D energy averages over the Fermi-Dirac distribution $f(E) = 1/(\exp((E-E_F)/k_BT)+1)$ where E_F is the Fermi level and k_B is the Boltzmann constant.
From equ. (1) and equ. (3) the electron heating temperature T_c-T is calculated to be

$$T_c - T = \frac{e^2}{m} \frac{<<\tau>>\tau_\varepsilon}{k_B} F^2 \tag{4}$$

where T is the lattice temperature.
Because of the complicated energy dependence of τ and dE/dt, T_c-T cannot be evaluated analytically in general. Therefore equ. (1) or equ. (4) have to be solved numerically for given F, T and substrate bias voltage U_B.
In this paper the total energy loss rate $\frac{dE}{dt}$ was calculated with the phonon matrixelements given in ref. /5/. In order to eliminate the unphysical U_B-dependence of the 2D-phonon-matrixelement, which is proportional to $(\omega z_o)^{-1/2}$ (ω: angular frequency of a phonon with the wavevector q, z_o: average channel thickness in the triangular approximation), the quantity ωz_o was replaced by the sound velocity c_s. This modification of the 2D matrixelement is based on the assumption that at low temperatures the electrons in the inversion channel prefer to interact with phonons with the wavevector $q = 1/z_o$ and that the dispersion relation is given by $\omega = c_s q$.
In ref. /5/ the unphysical z_o-dependence of the 2D phonon matrixelement was eliminated by introducing an area mass density $\varrho_{2D} = \varrho \cdot z_o$ where z_o represents here an appropriate constant channel thickness. ϱ is the volume mass density. The use of the quantity ϱ_{2D} however is problematic because the Si-lattice atoms in the inversion channel cannot be considered as "two-dimensional". Therefore this quantity has no real physical meaning and was not used in this paper.

In addition to the modification of the 2D phonon matrixelement outlined above, the effective electron mass m was considered to be U_B dependent. The U_B dependence was calculated from the quasiparticle energy in the random phase approximation by using an appropriate repulsive potential in order to describe the interaction between the inversion channel and depletion charge distribution. It was found that the correlation energy between the electrons is enhanced due to the applied negative bias voltage U_B and therefore the effective mass m decreases with increasing U_B as observed in ref. /6/. A crude estimation shows that the relative decrease of m is in the order of $2\delta z_o/z_o$, where δz_o is the shift of the inversion layer charge distribution at the bias voltage $-U_B$. The U_B dependence of m is essential for the heating of the inversion layer electron gas at low U_B values as it is shown in the next section.

3. RESULTS AND DISCUSSION

Fig. 1 shows the theoretical and experimental electron heating temperature T_c-T versus $-U_B$ for two different electric fields F, where N is the electron density in the inversion channel and T is the lattice temperature. As it is seen, T_c-T increases with increasing F. Furthermore T_c-T increases in the low bias voltage range $0 \leq -U_B \leq 4V$ and decreases again in the high U_B range $-U_B > 4V$. The theoretical (full and dashed) curves were calculated by taking into account 2D phonon, surfon and surface roughness scattering. The fitting parameters used in the calculation are the deformation potential constants $D_{2D} = 6eV$, $D_u = 9eV$, $D_d = -6eV$ for 2D phonons and surfons, the surface roughness parameters $\Delta = 4Å$ and L = 15Å as well as the effective mass m at $U_B = 0$. The values of these parameters are in reasonable agreement with those found in literature and were chosen so to obtain the best fit to the experimental data points taken from ref. /3/. All other physical quantities are the same as in ref. /5/.
The full curves were calculated with the U_B-dependent 2D phonon matrixelement given in ref. /5/ by neglecting the U_B-dependence of the effective electron mass m. In this case the theoretical T_c-T curve decreases in the whole U_B range. Therefore we can conclude that the 2D phonon matrixelement usually used in literature cannot be correct and that the U_B dependence of m has to be taken into account in the calculation of the electron temperature T_c.
The dashed curves in fig. 1 were calculated with the modified U_B-independent 2D phonon matrixelement which was described in the previous section. The effective mass used in the calculation of the energy relaxation time τ_ε is shown in fig. 2 and decreases with increasing $-U_B$. This behavior is due to the fact that the correlation energy of the inversion layer electrons increases if a negative bias voltage is applied to the MOSFET sample. As a consequence of this τ_ε as well as T_c-T (given by equ. (4)) increase in the low U_B range $0 \leq -U_B \leq 4V$. The presented analysis shows that in this range 2D phonon scattering is the predominant energy loss mechanism. With increasing $-U_B$ surfon scattering becomes more and more important since the surfon amplitudes decay exponential into the bulk. In the range $-U_B > 4V$ this scattering process is the dominant energy loss mechanism and causes the decrease of τ_ε as shown in fig. 2.
It was found that the experimental observed U_B-dependence of T_c-T cannot be explained by 2D phonon and surfon scattering alone. Since the experimental T_c-T values shown in fig. 1 were obtained at high electron densities ($N = 2{,}5 \times 10^{12} cm^{-2}$) the total momentum relaxation time τ is mainly determined by surface roughness scattering. If $-U_B$ increases, the electrons in the inversion layer are pushed closer to the Si-SiO_2 interface irregularities. Therefore $\langle\langle\tau\rangle\rangle$ as well as the right side of equ. (4) must decrease. The detailed numerical analysis has shown that in the range $0 \leq -U_B \leq 4V$ the decrease of $\langle\langle\tau\rangle\rangle$ due to the U_B-dependence of the surface roughness matrixelement is smaller than the increase of τ_ε/m due to the U_B dependence of the effective electron mass.
Therefore we can conclude that the bias voltage at which the maximum of the T_c-T curves occur is completely determined by the U_B-dependence of m and the matrixelements for surfon and surface roughness scattering.

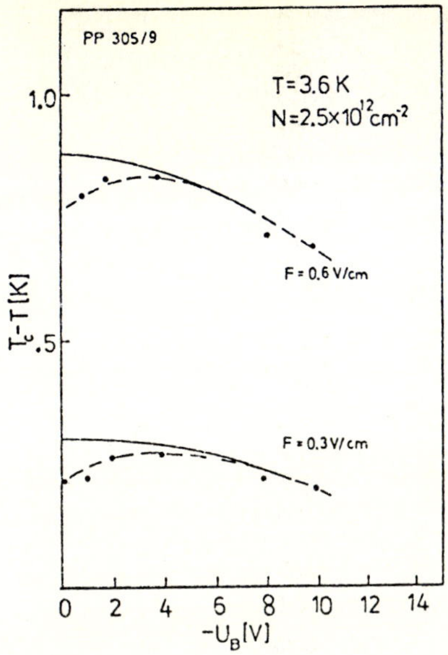

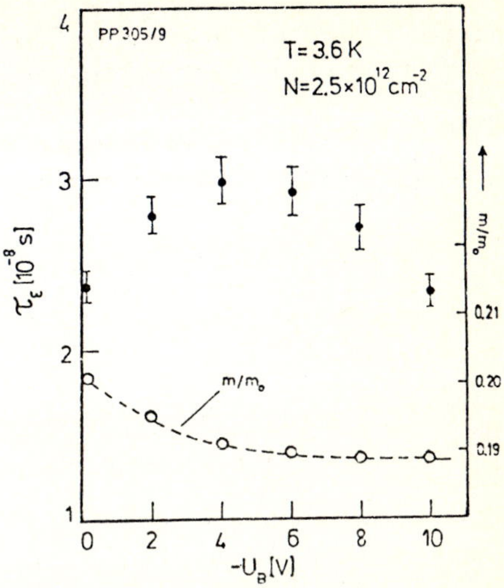

Fig.1: Electron heating temperature T_c-T versus substrate bias voltage U_B at constant electron density N

• : Experimental data points taken from ref. /3/

—— : Theoretical curve calculated with the 2D matrixelement taken from ref. /5/

---- : Theoretical curve calculated with the modified U_B independent 2D matrixelement and the U_B dependence of the effective mass taken into account

Fig.2: Calculated energy relaxation time and effective mass versus substrate bias voltage at T = 3.6K. Both curves were used in the calculation of the T_c-T curves shown in fig.1.

N : Electron density in the inversion channel

T : Lattice temperature

m_o : Free electron mass

REFERENCES

/1/ Cham K.M. and Wheeler R.G., Surf. Sci. 98 210 (1980)

/2/ Hönlein W. and Landwehr G., Proc. of the 4th Int. Conf. on Electronic Properties of Two Dimensional Systems, New London, New Hampshire, USA, 24-28 August 1981, p 62

/3/ Hönlein W., Thesis, University of Würzburg 1981, FRG

/4/ Ando T., Fowler A.B. and Stern F., Rev. of Modern Phys. 54 437 (1982)

/5/ Hess K., Englert Th., Neugebauer T., Landwehr G. and Dorda G., Phys. Rev. B16 3652 (1977)

/6/ Fang F.F., Fowler A.B., Hartstein A., Phys. Rev. B16 4446-4454 (1977)

Acknowledgements: The author would like to thank Prof. Dr.E. Gornik and Prof. Dr. G. Landwehr for valuable discussions and Dr. W. Hönlein for providing his thesis and stimulating comments.

MAGNETORESISTANCE IN A TWO-DIMENSIONAL IMPURITY BAND

A. Hartstein and A. B. Fowler

IBM Thomas J. Watson Research Center
Yorktown Heights, New York 10598, USA

and

K. C. Woo*

Department of Physics, Brown University†
Providence, Rhode Island 02912, USA

The effect of a magnetic field perpendicular to the surface on the conductance of the two-dimensional (2D) Na^+ induced impurity band in silicon inversion layers has been studied. The temperature dependence of the conductance at low fields (H < 5T) is $\sigma = \sigma_o' \exp\text{-}(T_o'/T)^{1/3}$, which is expected for variable range hopping in 2-D when the impurity wave functions are hydrogenic. At higher fields (H > 10T), $\sigma = \sigma_o \exp\text{-}(T_o/T)^{1/2}$, which should result for Landau (Gaussian) impurity wave functions. In this high field regime we observe $T_o = aH$, where a depends only on the density of states at the Fermi level and fundamental constants. The density of states derived in this way is about an order of magnitude larger than the density of states inferred from earlier measurements and is even approaching the magnitude of the free electron density of states.

The conductance in the 2-D Na^+-induced impurity band in silicon inversion layers has been studied in some detail.[1-4] Previous work has included the temperature dependence of the conductance, the effect of different Na^+ concentrations, and the effect of different band fillings. In this paper we present the results of a magnetic field perpendicular to the surface on the conductance. The related problem of the magnetoresistance in a 3-D impurity band (shallow impurities in doped semiconductors) has been studied.[5] However, those results are qualitatively different in that in 3-D electrons are not quantized along the magnetic field direction. In the 2-D case the electrons are fully quantized, resulting in quite different conductance properties.

The samples used in this study have been described elsewhere.[6] The samples were MOSFET devices with an oxide thickness of 100 nm. They were circle samples, with a width to length ratio of 50, made on <100> p-type silicon. The conduction path then averaged over all of the crystallographic directions in the <100> surface. NaCl was evaporated onto the gate oxide prior to gate metallization. Na^+ ions were subsequently drifted to the $Si\text{-}SiO_2$ interface at elevated temperatures under the influence of an applied electric field. In this study the Na^+ concentration was set to be 5×10^{11} cm^{-2} so that the impurity band conductance could be easily measured and would overlap the band tail as little as possible.

Following the Na^+ drift, the samples were cooled to 77K and the trans- conductance was measured. As discussed elsewhere[1] the Na^+ concentration was determined from the observed threshold at 77K. The sample was then further cooled to 4.2K and the conductance was measured using an a.c. technique. These measurements were made at 500 Hz with the source-drain voltage held to 1 mV rms. In the impurity band electron heating is not a problem at these voltages.

With the temperature held at 4.2K, a magnetic field was applied perpendicular to the surface, and the conductance was measured in the range 5T $\leq$ H $\leq$ 27T. The conductance as a function of gate voltage is shown in Fig. 1 for several values of the magnetic field. It is clear that the shape of the impurity band peak does not change with magnetic field, except perhaps at 23T, but that the magnitude of the conductance does.

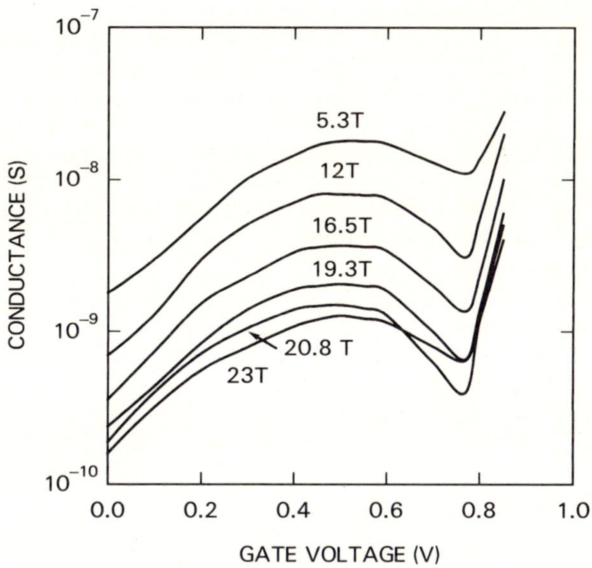

Fig. 1 - The conductance of the 2-D impurity band at 4.2K for various values of the magnetic field.

The magnetic field dependence of this peak is shown in Fig. 2 for two values of the the applied substrate bias. The data show two ranges of magnetic field dependence. The high field range fits quite nicely to a $\ln\sigma \propto -H^{1/2}$ dependence. As we shall see this range can be understood quite readily. The lower field region is a transition region and will not be analysed in any detail.

A subsequent set of data was obtained in the temperature range 1.6K $\leq$ T $\leq$ 4.2K and the magnetic field range 0T $\leq$ H $\leq$ 15T. These data were obtained by pumping on the liquid helium bath in a different magnet. These data are shown in Figs. 3 - 6. The number of Na^+ ions at the $Si\text{-}SiO_2$ interface was slightly different (< 2%) from the number in the first data set, and therefore, direct quantitative comparisons are not possible between the two data sets.

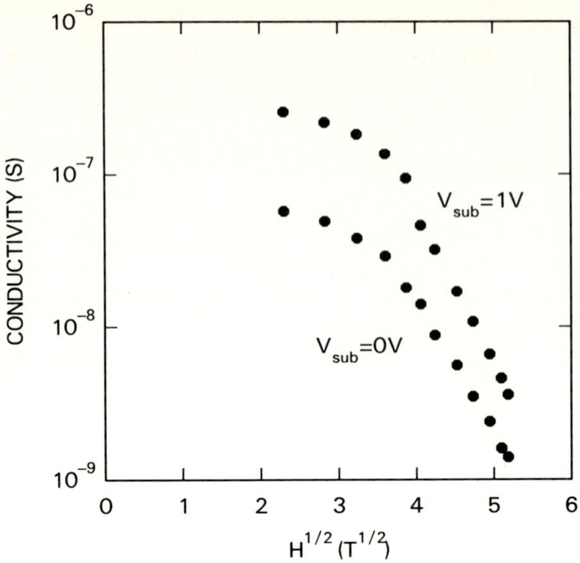

Fig. 2 - The magnetic field dependence of the impurity band conductivity peak for two values of the substrate bias. The data were taken at 4.2K.

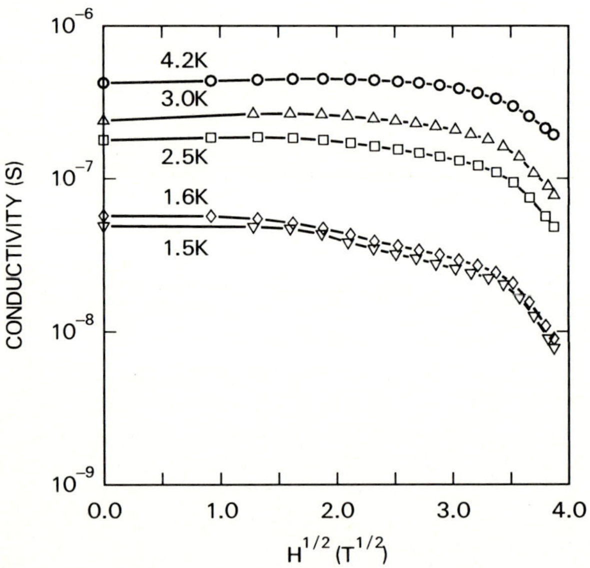

Fig. 3 - The magnetic field dependence of the impurity band conductivity peak at five different temperatures.

Fig. 3 shows the magnetic field dependence for the second data set. Five different temperatures are shown on this figure. The high magnetic field region of these curves shows the beginning of the $H^{1/2}$ dependence shown in Fig. 2. In the low field region, particularly at the higher temperatures, a weak negative magnetoresistance can be observed. This is more evident in Fig. 4 where we show the data for T = 4.2 K on a linear plot.

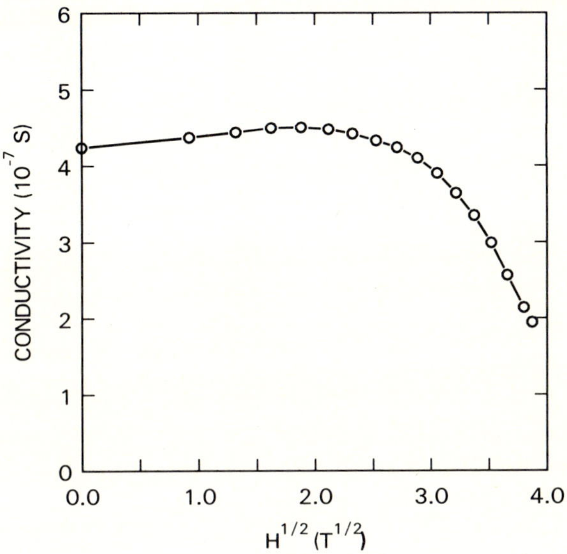

Fig. 4 - A plot of the 4.2 K data of Fig. 3 on a linear scale. This emphasizes the negative magnetoresistance effect.

At the lowest temperatures additional structure, which is perhaps oscillatory, is evident in the curves. Fig. 5 shows the 1.5 K data on a linear plot to demonstrate the magnitude of the effect. We have not obtained extensive data on the dependence of the effect on magnetic field, the average distance between Na^+ ions, or temperature. This will have to be done before any serious attempt can be made to ascertain the origin of the observed structure.

Fig. 6 shows the temperature dependence of the conductivity for various values of the applied magnetic field. The data below 2.0K show a systematic downturn. We believe that this is due to a systematic error in the measurements and we have omitted these data from the data analysis that follows. Nevertheless, a least squares fit to the data above 2.0K indicates that the temperature dependence observed is $\sigma = \sigma_0 \exp\text{-}(T_0/T)^n$ with $n = 1/3$ for $H < 5T$ and $n=1/2$ for $H > 10T$. The temperature range used is large enough to distinguish between an exponent of 1/2 or 1/3, but an extended temperature range should be used to refine the data further, as well as to clear up the question of the origin of the observed downturn in the low temperature region. It is well known that the $n = 1/3$ dependence arises from variable range hopping in two dimensions.[4]

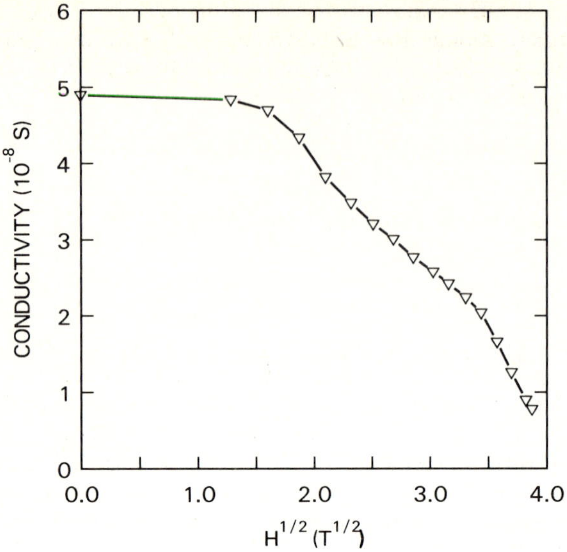

Fig. 5 - A plot of the 1.5 K data of Fig. 3 on a linear scale. This emphasizes the oscillatory character of the data.

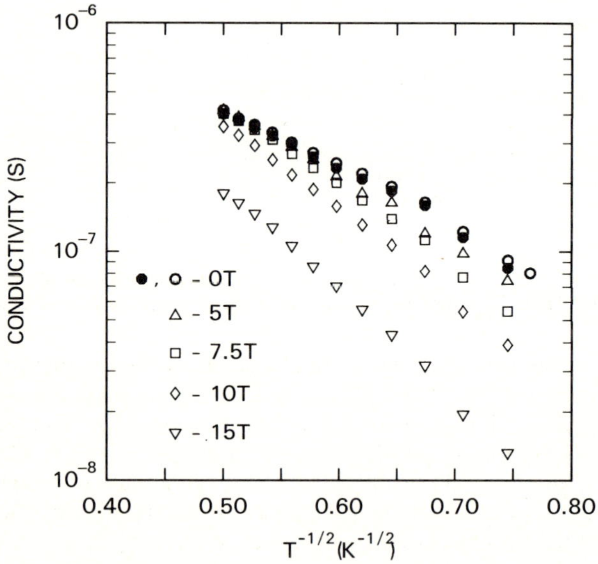

Fig. 6 - The temperature dependence of the impurity band conductivity peak for five magnetic field values. Data for H=0 were taken at the beginning and end of the run to check consistency.

The theory for the conductance in a strong magnetic field in the variable range hopping regime is a fairly straightforward extension of Mott's original derivation of variable range hopping.[7] The only difference is that the impurity wave function changes from the normal hydrogenic wave function $\Psi \propto \exp(-\alpha r)$ to

$$\Psi \propto \exp\,(-r^2/2r_c^{\,2}), \qquad (1)$$

where $r_c=(\hbar c/eH)^{1/2}$ is the cyclotron radius. The conductivity is then given by

$$\sigma = \sigma_o \exp(-r^2/2r_c^{\,2} - W/kT), \qquad (2)$$

where r is the electron hopping distance, W is the energy separation between hopping sites and σ_o is a constant. Eq. 2 essentially says that the conductivity is given by the overlap of wave functions between two sites and the energy separation between the same sites. The actual conductivity is obtained by determining the hopping distance r which maximizes Eq. 2 with a subsidiary condition which relates the density of states, $N(E_F)$, to W: $\pi r^2 N(E_F) W=1$ in 2-D. This condition counts the number of available hopping sites within a radius, r, and energy, W, from the initial electron site. This number must be 1 for a hop to occur.

The result of this analysis in a strong magnetic field is

$$\sigma = \sigma_o \exp\text{-}(T_o/T)^{1/2} \qquad (3)$$

$$T_o = \frac{2eH}{\pi N(E_F) k\hbar c}\,. \qquad (4)$$

We note that one obtains an exponential $T^{-1/2}$ dependence in the strong magnetic field case rather than the usual $T^{-1/3}$ dependence found in zero field. We also note that the dependence on magnetic field is $\exp(-H^{1/2})$. Both of these predictions are seen experimentally in Figs. 2, 3 and 6. A similar theoretical result has been obtained by Ono.[8] In addition, he finds that $\sigma_o \propto 1/T$. Our temperature data (Fig. 6) would fit this result equally well. The above considerations do not properly take into account the percolation problem,[9] and therefore T_o must be scaled by a numerical factor of order unity.

Eq. 4 shows that in the strong magnetic field case T_o depends only on the magnetic field, the density of states at the Fermi level, and fundamental constants. There are several ways to analyse the data to take advantage of this fact. We have chosen to determine the slopes of the high magnetic field region in Fig. 3 and to plot these slopes against $T^{-1/2}$. This analysis is independent of the temperature dependence of the prefactor in Eq. 3. The result of this analysis is shown in Fig. 7. The slope, b, plotted in Fig. 7 can be evaluated using Eqs. 3 and 4 and is given by $b^2 = 2e/\pi N(E_F)k\hbar cT$. The constant a used in the abstract is related to b by $a = b^2T$. As predicted b depends linearly on $T^{-1/2}$. The slope of the curve shown in Fig. 7 gives the density of states at the Fermi level, in this case for a half filled impurity band. The value obtained is $N(E_F) = 1.2 \times 10^{14}\ eV^{-1}cm^{-2}$. This is a factor of 5 larger than our previously determined values from other experiments,[3,4] and is even approaching the value of the free electron density of states. When percolation effects are taken into account (they have not yet been calculated for the magnetic field case) the effect will be to increase this discrepancy. This discrepancy indicates a fundamental problem with the theory.

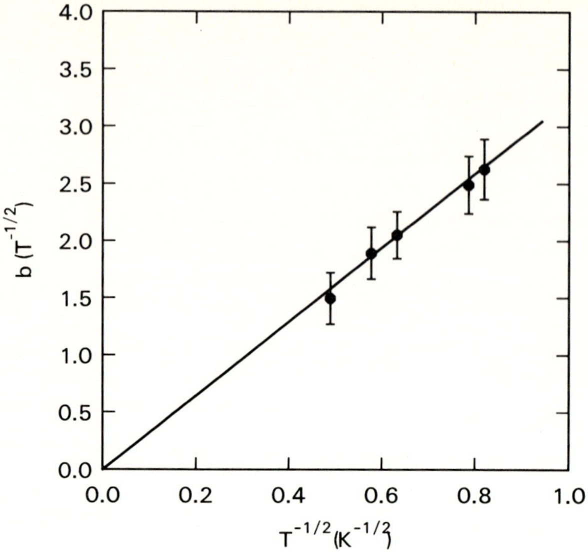

Fig. 7 - The slopes of the high magnetic field data shown in Fig. 3 plotted against $T^{-1/2}$.

The extrapolations of the high magnetic field data in Fig. 3 contain information on the preexponential factor, σ_0, as well as the exponent. The preexponents were extracted from the fits, and their temperature dependence is shown in Fig. 8. The error bars are necessarily large because of the long extrapolations needed and the small magnetic field range available for the extrapolations. However, the data show a trend consistent with Ono's prediction[8] of a 1/T dependence. This dependence is not derived in Ono's paper and it is not clear how he arrived at the result. A straightforward use of Mott's approach would predict that σ_0 varies as $T,^{-1/2}$ which is also consistent with our data. Further, it would predict that σ_0 is proportional to $H,^{-1/2}$ so that the experimental extraction of the dependence of σ_0 on H and T would be more complicated that the procedure described above.

The weak negative magnetoresistance observed in Figs. 3 and 4 is somewhat surprising in that the data were taken in the variable range hopping regime. No adequate theory exists for the magnetic field dependence in this regime. We have observed that the effect is extremely sample dependent, perhaps due to slight variations of either the amount or location of Na^+ ions at the interface. The effect is also angle dependent. The magnetoresistance becomes positive as the magnetic field is tilted parallel to the surface. This would seem to indicate that the negative magnetoresistance is not a spin effect.

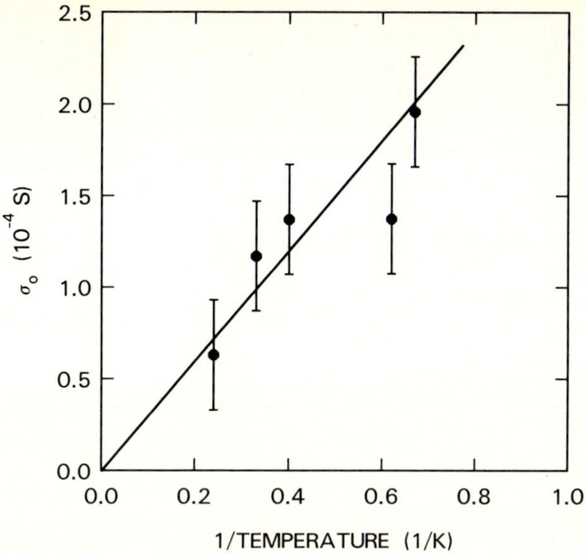

Fig. 8 - The preexponential conductivity, σ_0, as a function of 1/T. The error bars are estimated from the fitting procedure.

Movaghar and Schweitzer[10] and more recently Kamimura, Kurobe and Takemori[11] have proposed models to explain both positive and negative values for the magnetoresistance in a hopping system. Both models depend on the intrasite Coulomb interaction and the spin of the hopping electrons. Since our results indicate that we are not observing a spin effect, these models appear to be incapable of explaining this effect in the 2-D impurity band system.

In conclusion we have measured the magnetoresistance of a 2-D impurity band. In the high magnetic field regime we observe both temperature and magnetic field dependences which are consistent with variable range hopping theory. In the low field region we observe a weak negative magnetoresistance which remains unexplained. At low temperatures the onset of an unexplained oscillatory behavior is observed.

The authors would like to thank Frank Stern for many helpful discussions. We would also like to thank J. Tornello for valuable technical assistance. Part of this work was performed at the Francis Bitter Magnetic Laboratory at MIT which is supported in part by the National Science Foundation.

REFERENCES

* Permanent Address: Moore School of Electrical Engineering, University of Pennsylvania, Philadelphia, Pa. 19104.

† Work supported in part by the National Science Foundation under grant No. DMR 79-13787.

1. A. Hartstein and A. B. Fowler, Phys. Rev. Lett. **34**, 1435 (1975).

2. A. Hartstein and A. B. Fowler, Proc. of 13th Int. Conf. on Phys. of Semicond., Rome, p.741 (1976).
3. A. Hartstein and A. B. Fowler, Proc. of 14th Int. Conf. on Phys. of Semicond., Edinburgh, p.1001 (1978).
4. A. B. Fowler and A. Hartstein, Phil. Mag. B **42**, 949 (1980).
5. See for example, J. A. Chroboczek, Phil. Mag. B **42**, 933 (1980).
6. A. Hartstein and A. B. Fowler, J. Phys. C **8**, L249 (1975).
7. N. F. Mott, J. Non-cryst. Solids **1**, 1 (1968).
8. Y. Ono, J. Phys. Soc. Japan **51**, 237 (1982).
9. K. J. Hayden and P. N. Butcher, Phil. Mag. B **38**, 603 (1979).
10. B. Movaghar and L. Schweitzer, J. Phys. C**11**, 125 (1978).
11. H. Kamimura, A. Kurobe and T. Takemori, to be published in the Proc. of the 16th Int. Conf. on the Phys. of Semiconductors, 1982.

MAGNETOCONDUCTIVITY AND CYCLOTRON RESONANCE STUDIES ON Na^{+}-CONTAMINATED $Si\text{-}SiO_2$-INTERFACES

H.R. Chang *) and F. Koch

Physik-Department, Technische Universität München
D-8046 Garching, FRG

Na^{+}-ions at the $Si\text{-}SiO_2$-interface are a readily controlled impurity. We investigate the electronic properties of these Na^{+}-states using the techniques and methods of electron subband studies in a magnetic field. Data on the influence of interface Na^{+} on the cyclotron mass m_c^* and the electron scattering rate are presented. The strong damping of the resonance at Na^{+}-concentration below $10^{11}cm^{-2}$ is correlated with negative magnetoresistance and unusually strong scattering in the dc-magnetoconductivity. We suggest that the effects are caused by a Na^{+}-induced defect complex. For Na^{+}-concentration above $10^{11}cm^{-2}$ the scattering rate measured here confirms that derived from transport studies in the absence of a magnetic field.

I. INTRODUCTION

The surface conductivity on (100) Si has been studied extensively in an effort to identify the various different contributions to electron scattering in the $Si\text{-}SiO_2$ interfacial region. Such experiments have been complemented by cyclotron resonance to separately establish the dynamical mass m_c^* of the carriers.

A particular perturbation of the interface, introduced and studied in some detail by Fowler, Hartstein and coworkers [1-5], is the controlled doping with Na^{+}-ions. The Na^{+}-impurity scattering contribution has been identified and its rate measured in surface mobility experiments [3]. At low surface carrier density N_s, evidence for binding to the impurity potential and the creation of an impurity band has been presented [4,5]. Ref. [2] points to difficulties in the determination of m_c^* from Shubnikov-de Haas (SdH) data. The scattering rate from this magnetotransport work differs by about a factor of four from Ref. [3]. It is difficult to relate the many observations on m_c^* in that paper. At the risk of oversimplifying a bit, however, one may conclude from Ref. [2] that the SdH m_c^* decreases with Na^{+} contamination at fixed N_s in the range $1 - 4 \times 10^{12}cm^{-2}$. Some theoretical work has been done on this problem [6,7].

This possible lowering of m_c^* in SdH experiment is all the more surprising, when contrasted with observations on surface cyclotron resonance (CR) by the NRL group [8]. For the same electron densities N_s and a similar range of the scattering parameter $\omega_c\tau$ as in Ref. [2], the cyclotron work [8] finds an increase of m_c^*. Typical numbers are $\sim 20\%$ increase, when $\omega_c\tau$ reduces from 5 to 1. The magnetotransport work at $N_s = 2 \times 10^{12}cm^{-2}$, for example, has m_c^* decreasing from $0.225\ m_o$ to 0.20 with Na^{+} concentration of $7 \times 10^{11}cm^{-2}$. The latter corresponds roughly to the $\omega_c\tau$ range above. The situation is further confused by other CR work that fails to confirm the $\omega_c\tau$ dependence [9] and has m_c^*

constant. For N_s below $1 \times 10^{12} cm^{-2}$, there is generally observed a downward shift of the resonance field at fixed frequency (equivalent to a decrease of m_c^*). The effect is thought to result from the interaction of surface carriers with a random distribution of attractive potentials [10,11].

Our investigation of Na^+-doped interfaces was initiated with an awareness of the conflicting m_c^* evidence. It sought to elucidate the role of this particular interface perturbation on m_c^* and collision damping in CR. In the course of our work an unexpectedly strong damping was noted to occur at low Na^+ coverage [12]. In order to better characterize this effect and the nature of the interface defect that was responsible, related magnetoconductivity work on the same specimens was carried out. Results are given in this report together with additional CR data that are intended to complete the account of the resonance in the presence of interface Na^+.

Section II below gives some remarks on the experimental techniques. The experimental results follow in Sec. III. The paper concludes with a discussion and summary in Sec. IV.

II. SAMPLE PREPARATION AND MEASUREMENT TECHNIQUES

Results on five different n-type, (100) Si, MOS capacitor samples are reported. The choice of n-type material ($\rho \sim 10\ \Omega$ cm) for the experiments was made, because it is relatively easier to achieve surface electron modulation through the substrate contact of n-type capacitor samples. Samples N 1, N 2, N 3 are from a single wafer, oxidized to our specifications (dry oxide, $d_{ox} \sim 2000$ Å) in a university-associated institute. Samples N 4, N 5 are from a second wafer processed similarly in another laboratory. Typical sample dimensions are $6 \times 8\ mm^2$, with the gate area a centered $4 \times 4\ mm^2$ square. Each of the samples was ground into the form of a wedge to avoid interference effects.

Na^+-ions are introduced according to the procedure followed in Refs. [1-5]. The sample edges were masked during the NaCl evaporation. A density of only $\sim 10^{12}$ ions was introduced under the gate. This number proved sufficient for the small range of concentrations intended in this work. The drift involved heating to $\sim 60^\circ$ C and applying + 30 V to the gate for the length of time necessary to achieve the desired concentration. The samples were immediately cooled and held at least at 77 K in the course of the experiments. The capacitance C(V) was measured at both 77 and 4.2 K, with the sample flooded by 300 K thermal radiation. Typical frequencies are 1-10 Hz.

Cyclotron resonance is measured in the apparatus employed in previous work [12]. The measurement involves an on-off chopping ($\sim$ 3 Hz) of the carriers. The sample charging is monitored to assure full modulation for each of the N_s values.

The magnetoconductivity is obtained in the microwave scheme used for the earlier Ge work [13]. It is a measurement of $\sigma_{xx}(H)$, because the rf $\vec{E}$-field is defined by the wave-guide geometry. rf fields of order 0.1 - 1 V/cm are necessary to achieve sufficient sensitivity in this apparatus.

III. EXPERIMENTAL RESULTS AND DISCUSSION

1. Capacitance C(V), flat-band voltage V_{FB}, and the Na^+-count

A primary concern of the experiments is the quantitative evaluation of the Na^+-contamination. In most previous work [1-5] the shift in the on-set voltage for conduction at T ~ 77 K, where carriers are activated, has been used to measure the Na^+ at the interface. It has been found that this measure relates well to the shift of the C-V curve. The lack of surface contacts on our samples has forced us to work with the capacitance criterion to identify V_{FB} shifts in the CR apparatus. The Na^+ concentration is $C \Delta V_{FB}/e$.

There are three factors related to the Na^+ count that have an influence on the outcome and physical interpretation of the experiments. In the order of importance for the present experiments they are - inhomogeneity of the Na^+ concentration over the sample area, the possible formation of surface donor states N_{SS}, and the fact that the C-V voltage displacement measures only the shift of the centroid of charge. We choose to consider these in some detail, because differences in the results between researchers could relate to the above factors. Fig. 1 is intended to illustrate the possible, experimentally realized Na^+ configurations.

CR measurements have a signal strength proportional to sample area and require adequately sized samples. This makes Na^+ inhomogeneities, introduced during the initial contamination or subsequent drift procedure, a real concern. Little is known of the scale of possible inhomogeneities. With the average Na^+ spacing of the order of 100 Å and the fact that each ion has drifted through the oxide thickness ~ 2000 Å, one may expect clustering to occur. The cluster scale could be determined by local strains, by interface growth defects, etc. That our samples have some degree of Na^+ inhomogeneity over the 4 x 4 mm^2 area is suggested by the sampling of C-V curves in Fig. 2. With each successive drift above ~ 1×10^{11} Na^+ cm^{-2} the curves for sample N 1 get broader. V_{FB} cannot be marked with certainty. The distortion is much less for the capacitance curves of sample N 4 in the lower part of Fig. 2. Generally the width has been found to vary from sample to sample, and for a given sample in successive drift cycles. Fowler and Hartstein [14] have found

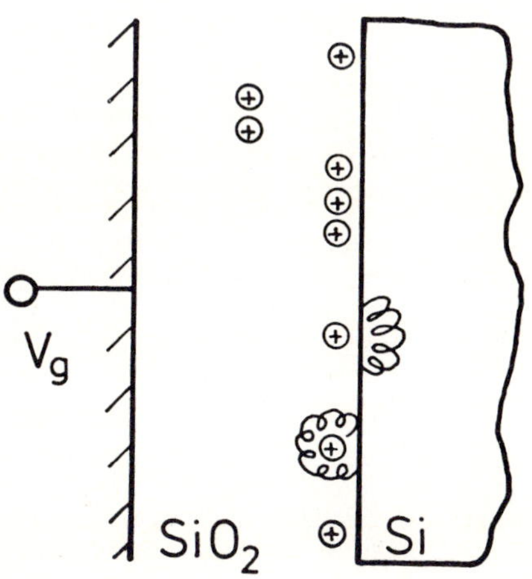

Fig. 1: Na^+ in MOS samples. Ion positioning along the interface surface can be inhomogeneous. Some Na^+ will bind an electron in a localized orbital in the oxide (N_{SS}). Ions deep inside the insulator will be registered by the capacitance without binding an interface electron.

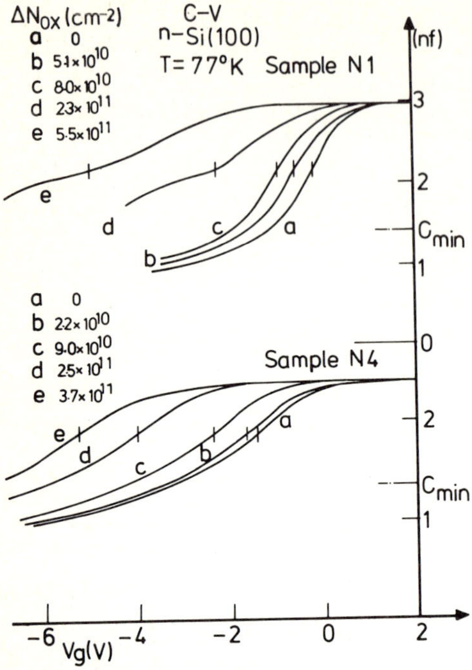

Fig. 2: Typical capacitance-voltage curves and identification of oxide charge ΔN_{ox} from the midpoint $1/2(C_{max} + C_{min})$. C_{min} is the equilibrium value determined at a higher temperature.

that the distinct impurity band signal discussed in Refs. [4,5] is not resolved and identified clearly in big samples. Other circumstantial evidence for inhomogeneity comes in the conductivity-onset curves shown below. The rounding suggests inhomogeneity. Work with Na^+-doped p-type samples, where the 4.2 K capacitance on the accumulation side continues to rise after the hole conductivity threshold has been reached, also hints at inhomogeneity. We recognize that the distortions and broadening of C(V) can result from both the creation of interface states N_{ss} and inhomogeneity, but believe the latter to dominate. Consistent with this supposition we mark the midpoint $1/2\ (C_{min} + C_{max})$ to measure the Na^+ contamination. Because of the absence of minority carriers the curves in Fig. 2 dip below C_{min} and indicate extended depletion. The real C_{min} is determined at higher temperature.

It is conceivable that Na^+ also creates fast interface donor states N_{ss} which exchange electrons with the semiconductor. The consequence, if these states lie in the gap, would be a broadening to that in Fig. 2. N_{ss} cannot really be distinguished from inhomogeneities by our measurements alone. If these states exist and are filled and emptied at 4.2 K, their effect on electron scattering would be minor because for n-accumulation they are occupied and neutral. The Na^+ count via the capacitance average, as above, would be an overestimate.

A serious concern for all measurements involving Na^+ interface doping is the positioning of the ions relative to the interface surface. The scattering rate and binding energy of electrons depends crucially on the exact location of the ions [15]. The C-V curve measures the shift of the centroid of the Na^+ charge and obviously cannot distinguish an ion at the interface (z = 0) from one a few lattice constants away. Work by the IBM researchers suggests all the Na^+ is within ~ 50 Å of the surface [16] and probably closer. Nevertheless, as a matter of principle, the voltage shift generated by the shift of 2 ions from the gate to the middle of the oxide is equivalent to one at z = 0, a situation indicated in Fig. 1. The scattering rate, however, is negligible for Na^+ which is that far away from the interface.

Because the Na^+ positioning plays such a decisive role, it is necessary to check out limits on the fraction of Na^+ not at the immediate interface. The results in Fig. 3 on the comparison of the C-V shift with the conductivity onset $\sigma(V_g)$, provide a measure of the Na^+ positioning. To the extent that each interface Na^+ ion binds a single electron (for

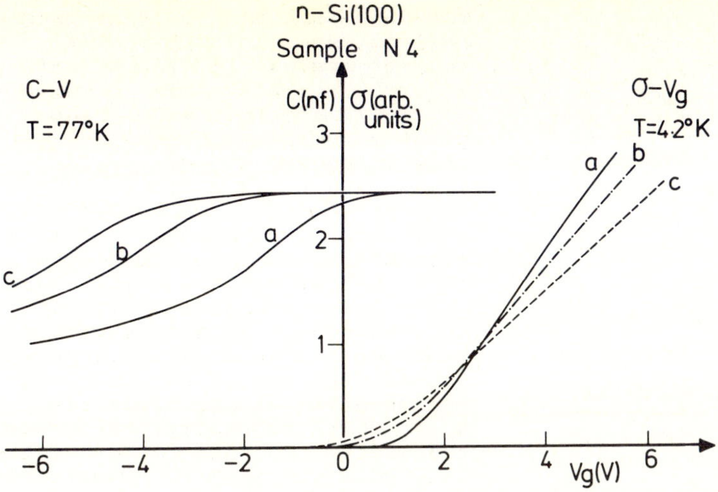

Fig. 3: Comparison of C-V and $\sigma(V_g)$. The conductivity onset voltage remains nearly constant, whereas the capacitance has shifted by more than 4 V.

the low Na^+ coverage considered here the electron binding length is less than the average Na^+ spacing), one expects the conductivity to start at a fixed point on the V_g scale in Fig. 3. Each ion that is at some distance from the interface will contribute to the capacitance-shift but fail to bind an electron. From data such as Fig. 3 we conclude that more than 80% of the Na^+ is of the interface variety. The rounding of the foot of $\sigma(V_g)$ can be taken as evidence for inhomogeneity. The contactless microwave measurement registers inhomogeneities more sensitively than a dc measurement, because it sums the absorption contributions of small conducting patches. Measurements with leads require a continuous source-to-drain current path, before a signal is registered.

2. Cyclotron Resonance

A first report on CR at Na^+-doped interfaces is contained in Ref. [17]. That work was concerned primarily with the Na^+-induced collision-damping. The most remarkable result was the disappearance of the resonance for quite small ΔN_{ox} (below $10^{11} cm^{-2}$). With subsequent larger ΔN_{ox} drifts, the resonance was found to reappear and broaden gradually. The gradual collision-broadening followed the ΔN_{ox} dependence expected from the scattering rate in Ref. [3]. A tendency for the resonance mass m_c^* to decrease with rising ΔN_{ox} at fixed N_s was noted.

The damping and reappearance of the resonance is a perfectly reproducible phenomenon for the entire batch of the N samples. Shown in Fig. 4 is a set of curves, which illustrates the effect for N 1 at a carrier density $N_s = 3 \times 10^{12} cm^{-2}$. The initial drift with $\Delta N_{ox} = 0.3 \times 10^{11} cm^{-2}$ completely removes the resonance. Fit according to the usual lineshape theory, the damping parameter $\omega\tau$ would be less than 0.7. The relative amplitude of the signal, as it was used in Ref. [17], is not a good criterion for $\omega\tau$. We comment on this below. With the additional drift to $\Delta N_{ox} = 2.7 \times 10^{11} cm^{-2}$ the resonance is recovered with $\omega\tau = 1.35$. More Na^+ gradually causes increased broadening and finally the disappearance of the signal.

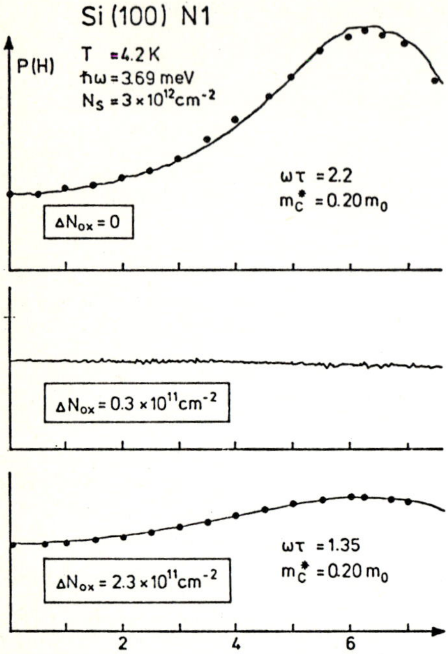

Fig. 4: Damping and recovery of the cyclotron resonance with successive, increasing Na^+ drifts.

From data such as Fig. 4 it is possible to derive an approximate value of the Na^+ scattering rate for high N_{ox} values ($\Delta N_{ox} \gtrsim 10^{11} cm^{-2}$). The relation of the increased scattering rate $\Delta \Gamma$ and ΔN_{ox} is nearly linear. We write $\Delta \Gamma = B \ \Delta N_{ox}$ and evaluate $\Delta \Gamma$ from the changes in $\omega\tau$. The major difficulty in doing so is that the resonance value of $\omega\tau$ and the equivalent dc scattering lifetime τ_o are related in a complicated way. The effect is referred to as $\sqrt{H}$ -broadening. We follow the empirical relation in Ref. [12] to translate the $\omega\tau = 2.2$ in Fig. 4 as $\omega\tau_o = 2.8$. The scattering rate is $\Gamma = 2.0 \times 10^{12} sec^{-1}$. With $\Delta N_{ox} = 2.7 \times 10^{11} cm^{-2}$, as in the lower curve of Fig. 4, $\omega\tau = 1.35$. This value, being less than 2, requires no correction and gives $\Gamma = 4.1 \times 10^{12} sec^{-1}$. It follows that $B \sim 8 \ cm^2/sec$. The procedure obviously involves too many uncertainties to allow a critical comparison with the work in Refs. [2] and [3]. The following section will derive B from the magnetoconductivity.

A further point about the CR data concerns amplitudes. For a wedge-type sample with a high resistance gate, theory gives a unique signal amplitude as a percentage change of transmitted power for a given $\omega\tau$ and N_s [8,12]. The signal at H = 0 in Fig. 4 as calculated should be ~ 10%. That is more than twice the measured value of ~ 4%. A check for possible preferential circular polarization proved negative. The leakage signal was not significant. However, the fact that the relative signal amplitude could be varied with the adjustment of the 45°-mirror at the top of the light-pipe, suggests to us standing wave behaviour and reflections in the sample holder assembly. The problem was not resolved satisfactorily in this work.

For the high Na^+ concentrations we have measured the CR in several different samples in order to examine the influence of Na^+ on the m_c^* value. With increasing N_{ox} the collision lifetime is lowered and we automatically scan the range of the $\omega\tau$ parameter, where in Ref. [8] had been found a notable rise of m_c^* (typically $\omega\tau$ varies from 4 to 1). With the addition of oxide charge, by and large we find a lowering of m_c^*. The complete set of data for sample N 1 for three representative oxide charges is shown in Fig. 5. The undoped sample follows an m_c^* (N_s) which is quite similar to that reported earlier in Refs. [10,12]. With the addition of N_{ox} there is a significant reduction of m_c^* for all N_s values. The effect is strongest at low N_s. Other samples showed qualitatively similar behaviour. The error bars marked in the figure represent reasonable limits of the fit m_c^* parameter for the given set of data curves, taking account of the signal to noise and the sensitivity of the fit of m_c^* for differing $\omega\tau$. m_c^* in Fig. 5 is a parameter that

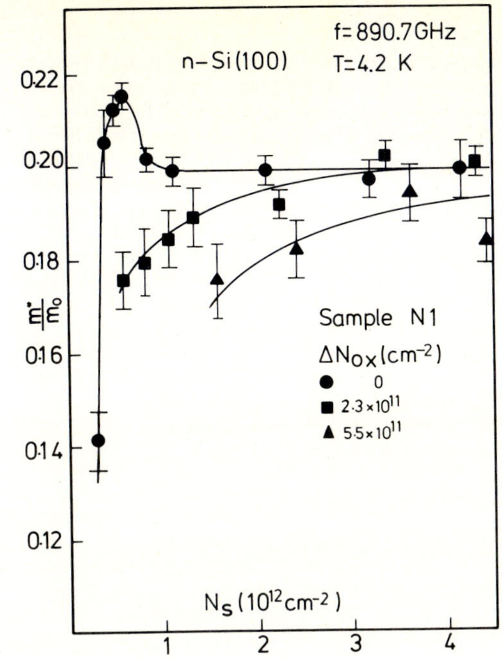

Fig. 5: Variation of the mass parameter $m_c^*(N_s)$ in a fit to the classical lineshape theory for three different Na^+-doping levels.

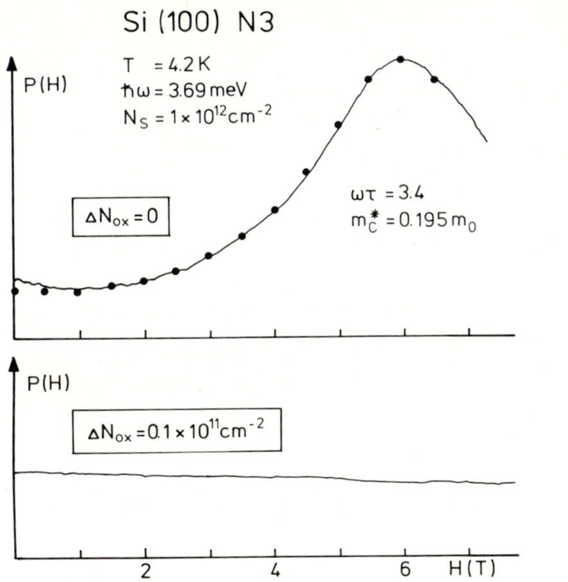

Fig. 6: Cyclotron resonance before and after the drift of only $\sim 10^{10}$ Na^+-ions to the interface for sample N 3.

describes the position of the resonance. It is not necessarily the cyclotron mass.

The magnetoconductivity results of the section below involve mostly sample N 3. In order to relate the measurements to CR, it is good to demonstrate the low N_{ox} damping effect for CR in sample N 3. Not only does Fig. 6 demonstrate that the disappearance of the resonance is a quite general effect in our samples, but it illustrates how large the change in $\omega\tau$ with ΔN_{ox} can be. Prior to the drift the value $\omega\tau = 3.4$ applies in Fig. 6. With ΔN_{ox} of only $\sim 1 \times 10^{10} cm^{-2}$, which corresponds to only a 0.1 V shift of the C-V curve, the resonance is completely damped out. With increased drift the signal is recovered in this sample also. In several forward and reverse drift cycles, we always found the strong damping condition for N 3, even though it would occur at quite different ΔN_{ox}. Obviously the damping condition is created with good homogeneity over the entire active area. If as little as 10% of the sample area had remained with $\omega\tau = 3.4$, a small rise near 6.2 T would have been observed superposed on the straight line in Fig. 6. Hysteresis in the up-down cycles shows that ΔN_{ox} is not the proper parameter to describe the configuration of interfacial scatterers.

For the relative magnitude of the scattering effect in Fig.6, we note that prior to the ΔN_{ox} drift, the $\omega\tau$ value is 3.4. Scaled according to Ref. [12] into dc scattering rate, this is equivalent to $\omega\tau_o \sim 5$. When the damping occurs, $\omega\tau$ is less than 0.7, probably substantially less. We conclude that the relative change of the scattering rate, measured at this high cyclotron frequency of $\sim$ 890 GHz, is at least a factor of 7.

3. Magnetoconductivity $\sigma_{xx}(H)$

The surface conductivity in a perpendicular magnetic field provides a convenient measure of the mobility of our MOS-capacitor samples. We make use of such measurements in order to provide a reliable determination of the Na^+ scattering rate in our samples for the high ΔN_{ox} regime. In addition, the observations are intended to correlate with the peculiar strong damping of the CR at low ΔN_{ox}.

The first part of this work concerns the rate parameter B in the relation $\Delta\Gamma = B\ \Delta N_{ox}$ for $10^{11} \leq N_{ox} \leq 10^{12}cm^{-2}$. Typically 5-6 values of ΔN_{ox} are used and the conductivity $\sigma_{xx}(H)/\sigma_{xx}(0)$ is recorded for fixed N_s values between 0.5 and 4 x $10^{12}cm^{-2}$. Fig. 7 is a sampling of such data for specimen N 3. The curves are fit to the relation $\sigma_{xx}(0)/(1+\mu^2H^2)$ to give an experimental value of the mobility μ. The scattering rate is evaluated as $\Gamma = e/(m^*\mu)$ with m^* as a constant 0.2 m_o. The scatter of the data points and the relatively few values of N_{ox} preclude anything but a straight-line fit for the relation between Γ and N_{ox}. It has been argued [3] that for concentrations of Na^+ for which the average spacing between ions is less than the Fermi-wavelength λ_F, multiple scattering will occur. In Ref. [3] both a linear and quadratic term has been employed to fit the relation between Γ and N_{ox}. For the average N_s of 2 x $10^{12}cm^{-2}$, and average N_{ox} of 2 x $10^{11}cm^{-2}$, as used in this work, λ_F equals the average impurity spacing both being of the order of 250 Å. Nevertheless, an approximately linear relation provides a reasonable description for the dependence of Γ on N_{ox}.

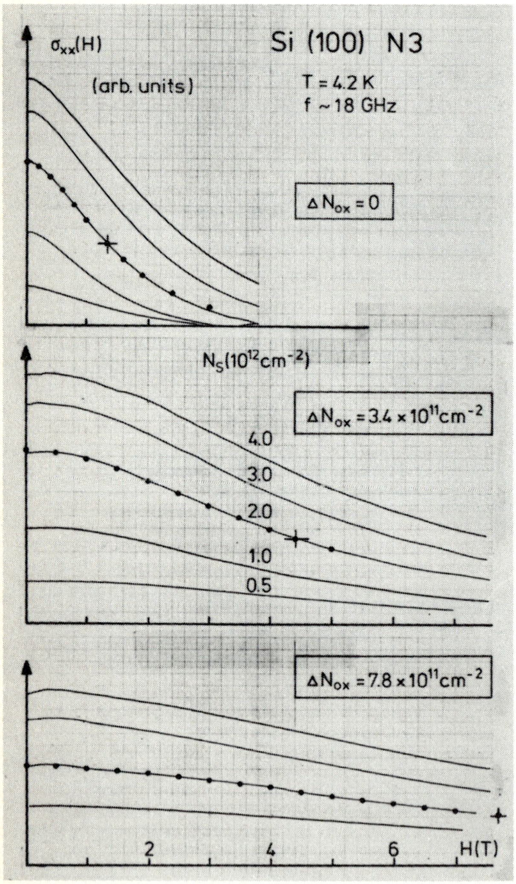

Fig. 7: Magnetoconductivity curves $\sigma_{xx}(H)$ for three different Na^+ concentrations.

Magnetoconductance data has been obtained on samples N 2, N 3, as well as from those from the different wafer, N 4 and N 5. The points entered in Fig. 8 for $B(N_s)$ are values that have been averaged for all the four sets of data. The error bar in that figure covers the scatter from different runs and samples. It includes the estimated uncertainty of the N_{ox} determination. The open circles are points taken from Ref. [3]. They represent the magnitude of the linear term in that work. The deviation and in particular the rise of B at low N_s must be expected, because the contribution of the N_{ox}^2 term is simulated by an increased B.

Drifting the Na^+ in the range below $10^{11}cm^{-2}$ we recover in the magnetoconductance the same sort of anomalously strong scattering

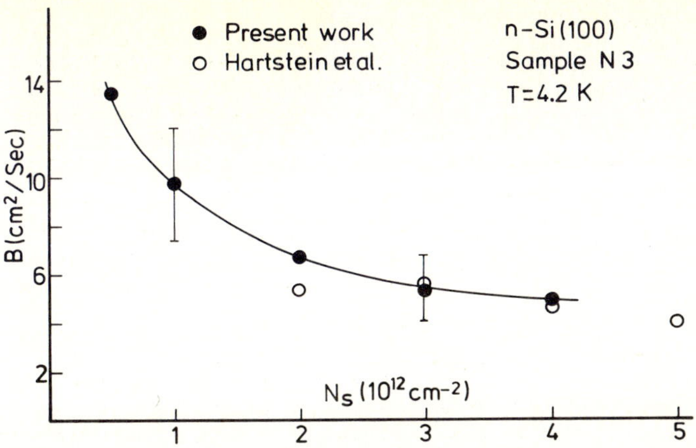

Fig. 8: The linear Na^+-scattering coefficient in the relation $\Gamma = BN_{ox}$ vs. electron density N_s. For most of the data, N_{ox} has been varied between 0 and $5 \times 10^{11} cm^{-2}$.

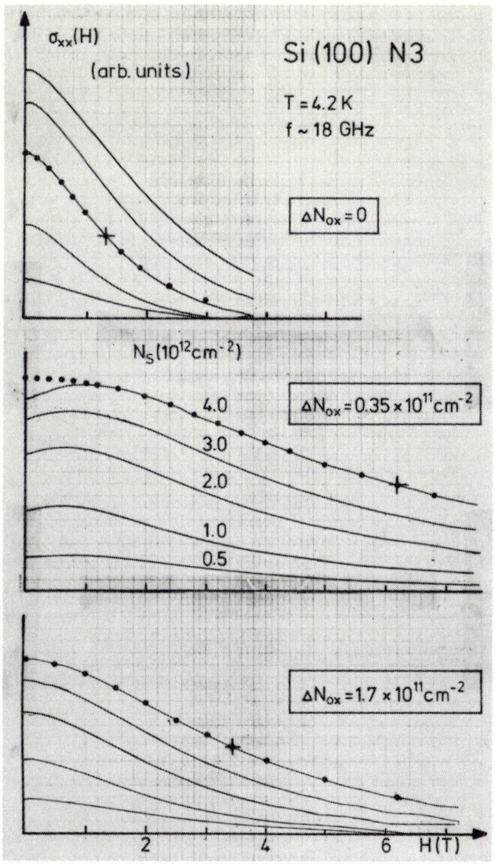

Fig. 9: Magnetoconductivity curves for low Na^+ coverage. Note the rising conductivity at low fields for the middle set of data.

that had characterized the CR work. Fig. 9 shows how the $\sigma_{xx}(H)$ of sample N 3 responds to a Na^+-drift of 0.35 x $10^{11} cm^{-2}$. The middle set of curves in the figure are for a mobility reduced by a factor of ~ 3 at all densities N_s. With the addition of more Na^+ in a second drift, the mobility is once again increased. The scattering rate has returned to a value only about 50% higher than at the outset.

The qualitative behaviour of the magnetoconductivity $\sigma_{xx}(H)$ in the "low" Na^+ regime is quite similar to that noted for the CR. However, the magnitude of the scattering rate increase is considerably less. Whereas in CR the estimated reduction of $\omega\tau$ for sample N 3 is more than a factor of 7, here the mobility decrease amounts to at best a factor of ~ 3 in successive runs. There is the same kind of hysteresis in the effect. The exact value of ΔN_{ox} for the strong scattering condition depends on the drift cycle history. For several different up-down drift cycles the strong scattering effect was observed for sample N 3, without ever exceeding the above quoted factor 3. It seems that the magnetoconductivity Γ, assumed as frequency independent, cannot account for the vanishing CR.

The experimental problem in this work is that the CR and $\sigma_{xx}(H)$ are never measured concurrently. The different spectrometer systems require warming to room temperature and remounting the specimen in a new apparatus. The condition of strong damping usually would disappear, but could be recovered in the drift procedure. Future work will aim to develop measuring procedures that allow simultaneous recording of $\sigma_{xx}(H)$ and CR in order to resolve the frequency-dependence question.

Invariably, and Fig. 9 is a good example of this, the anomalous damping is accompanied by a sizeable negative magnetoresistance. That is $\sigma_{xx}(H)$ first rises with the application of H and then falls according to the expected free-electron behaviour. The fact that, so-to-say, there is a missing conductivity contribution $\Delta\sigma$ at H = 0 follows from a fit of the overall curve $\sigma_{xx}(H)$, in which the high field part is taken to obey the relation $\sigma(0)/(1+\mu^2H^2)$. As in Fig. 9, this results in an experimental curve for σ_{xx} which falls substantially below $\sigma(0)$. We conclude from data such as this that there is a conductivity reduction at H = 0. Its magnitude is quite large. Typical values are $\Delta\sigma/\sigma \sim -10\%$, but up to 20% have been observed at 4.2 K. The effect exists over most of the relevant N_s range, with a tendency to increase with N_s. In this respect, the effect is wholly different from the one-percent type, small logarithmic correction linked to weak localization [18]. The latter has been found to exist mostly for low N_s and at the very lowest T.

Our concern, that the strong damping effect is caused by a chance contamination of a single wafer, led us to prepare the additional samples N 4, N 5 from another wafer oxidized by a different laboratory. We use the same Na^+ contamination and drift procedures. The four curves in Fig. 10 give $\sigma_{xx}(H)$ for sample N 4 at $N_s = 2 \times 10^{12} cm^{-2}$ with different Na^+-doping. As was the case for sample N 3, the initial broadening is followed by a mobility increase with rising Na^+. For the $\Delta N_{ox} = 1.5 \times 10^{11} cm^{-2}$ the virgin mobility is recovered in N 4. The strong scattering is accompanied by negative magnetoresistance. $\Delta\sigma/\sigma$ is 17% in this case. These observations leave us to conclude that the phenomeno of the nonmonotonic damping is one with a certain degree of generality.

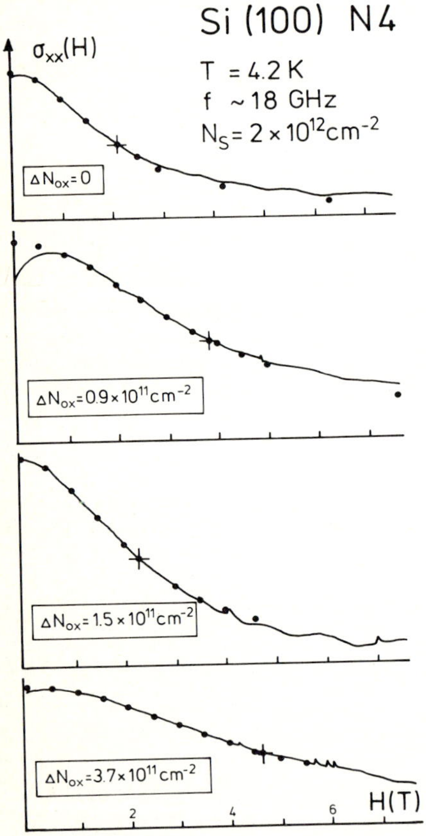

Fig. 10: The nonmonotonic scattering rate increase and negative magnetoresistance observed in sample N 4.

IV. DISCUSSION AND CONCLUSIONS

We recapitulate what has been learned in the course of the experiments and what may be concluded from them.

First the m_c^* considerations. The introduction of Na^+ ions at the interface causes mainly collision damping and has no really large effect on the cyclotron mass value. At the really metallic densities of N_s = $2 \times 10^{12} cm^{-2}$ or greater, the $\omega\tau$ is reduced from typically 3 to about one, without changing m_c^* in any significant way. At lower N_s, the relative $\omega\tau$ change will be even greater. The result on m_c^* is a major decrease, which in a great many ways resembles the down-shift found in earlier CR work [10]. We conclude that the shift is caused by the interaction of the electron with the binding potential of the ions.

The down-shift obtained in CR should not be compared directly with the SdH results [2], although magnitude and sign of the change of m_c^* are quite similar. The m_c^* in these two types of experiments are fundamentally different quantities, so that an immediate comparison is not warranted. The CR value of m_c^*, if the hypothesis of binding is correct, would be frequency-dependent. It would rise with increasing excitation energy $\hbar\omega$. This effect has been found in experiments on InP [19].

The scattering rate for the large Na^+ concentrations (typically 1 - 5×10^{11} Na^+ cm^{-2}) as determined from CR or magnetoconductance measurements, agrees basically with the value derived from mobility measurements in the absence of a field H [3]. For CR it is difficult to make an exact comparison. The magnetoconductance, however, gives a reliable measure that disagrees with the value quoted from SdH work [2], but closely equals the H = 0 mobility value. We have chosen to express the rate as a scattering per Na^+ ion by making a straight line fit for our Na^+ range (about 1 - $5 \times 10^{11} cm^{-2}$). The collision rate increases at low N_s. The error bars on our measurements of B should be considered ± 25%, mostly because of uncertainty with the Na^+ count. The value 6 ± 1.5 $cm^2 sec^{-1}$ for B agrees well with that recently calculated by Vinter [15]. The reasons for the increase at low N_s need to be explored further.

The experiments here have provided additional evidence for the unusually strong and nonmonotonic damping of CR at low Na^+ coverage in this type of sample. The $\omega\tau$ reduction in CR by a factor ~ 10 is correlated with a mobility decrease in the magnetoconductivity. The latter is observed to be at most a factor of 3. We speculate that the scattering effect is frequency-dependent. A distinct "negative magnetoresistance", of the order of magnitude 10 - 20% in $\Delta\sigma/\sigma$, accompanies the strong scattering.

We suppose that in these initial drifts of Na^+ a scattering complex, that has a much greater effect on the surface electrons than a single Na^+ ion, is formed at the interface. In order to account for the frequency-dependence, it would have to have some internal degree of freedom that could be excited in the ~ meV range of frequencies. The scattering condition is achieved with amazing uniformity over the interface. As little as 5 - 10% of the area left in the virgin state after the drift could have been observed. With the addition of more Na^+ ions it seems that the complex is stabilized and then the scattering per Na^+ ion is that observed also in dc mobility measurements. One should keep in mind the hysteresis effect in the up-down Na^+-drift cycles.

More experimental work is required in order to completely identify the scattering complex and the conditions at the interface. The work here serves as an example of how the experimental machinery and theoretical know-how of subband physics can be applied to an examination of specific defects at the Si-SiO_2 interface.

ACKNOWLEDGEMENTS

We thank Alan B. Fowler from IBM Research Laboratories, Yorktown Heights, for many fruitful discussions. The financial support by Alexander von Humboldt Stiftung and Deutsche Forschungsgemeinschaft (SFB 128) is gratefully acknowledged.

REFERENCES

*) A.v.Humboldt-Fellow on leave of absence from the Institute of Metallurgy, Shanghai, PR China.

1. F.F. Fang, A.B. Fowler, and A. Hartstein, Phys.Rev. B 16, 4446 (1977).
2. A. Hartstein and F.F. Fang, Phys.Rev. B 18, 5502 (1978).
3. A. Hartstein, A.B. Fowler, and M. Albert, Surface Sci. 98, 181 (1980).
4. A. Hartstein and A.B. Fowler, Phys.Rev. Lett. 34, 1435 (1975).
5. A.B. Fowler and A. Hartstein, Philos.Mag. B 42, 949 (1980).
6. A.W. Overhauser, Phys.Rev. B 18, 2884 (1980).
7. B.E. Sernelius and K.F. Berggren, Surface Sci. 98, 191 (1980).
8. R.J. Wagner, T.A. Kennedy, B.D. McCombe, and D.C. Tsui, Phys.Rev. B 22, 945 (1980).
9. G. Abstreiter, F. Koch, P. Goy, and Y. Couder, Phys.Rev. B 14, 2494 (1976).
10. J.P. Kotthaus, G. Abstreiter, F. Koch, and R.Ranvaud, Phys.Rev. Lett. 34, 151 (1975).
11. H.J. Mikeska and H. Schmidt, Z. Physik 20, 43 (1975).
12. G. Abstreiter, J.P. Kotthaus, F. Koch, and G. Dorda, Phys.Rev. B 14, 2480 (1976).
13. J. Binder, K. Germanova, A. Huber, and F. Koch, Phys.Rev. B 20, 2382 (1979).
14. private communication.
15. B. Vinter, Phys.Rev. B (in press).
16. See for example: D.J. Di Maria, J.Appl.Phys. 52, 7251 (1981).
17. H.R. Chang and F. Koch, Solid State Commun. 38, 1189 (1981).
18. See for example: R.C. Dynes, Surface Sci. 113, 510 (1982).
19. H.C. Cheng and F. Koch, Solid State Commun. 37, 911 (1981).

ON THE SCREENING OF IMPURITIES BY A TWO DIMENSIONAL ELECTRON GAS IN HIGH MAGNETIC FIELDS

K. Heift and J. Hajdu

Institut für Theoretische Physik
Universität zu Köln, Köln, Germany

We consider the effective potential of an impurity charge placed into a two dimensional non-interacting electron gas at zero temperature and in the presence of a perpendicular, quantum limit magnetic field. Restricting ourselves to a one band model and describing the electronic self-energy due to the impurity scattering in the generalized Born approximation (GBA), it is possible to derive a selfconsistency equation for the RPA-dielectric function.

It is shown that the selfconsistent effective potential can approximately be described by the dimensionless screening parameter $s_0\ell$ (ℓ is the magnetic length). Numerical results are presented for $s_0\ell$ as a function of the filling factor ν at fixed values of the electron concentration and the magnetic field, respectively. The effective potential in real space is depicted for several values of $s_0\ell$, representative for the situation met in experiments performed at fixed fields.

1. Selfconsistency equation for the dielectric function in RPA

At zero temperature the polarization function in RPA is given by

$$\Pi_{RPA}(\underline{q}) = \frac{1}{\pi A} \operatorname{Im} \sum_{\alpha\alpha'} \rho^{*}_{\alpha\alpha'}(\underline{q}) \int_{-\infty}^{\mu} d\varepsilon\, F^{\underline{q}}_{\alpha\alpha'}(\varepsilon+io), \tag{1}$$

$\alpha := (n, k_y)$ denotes the Landau quantum numbers, μ the chemical potential, A the area of the system, and $\rho_{\alpha\alpha'}(\underline{q}) := \langle\alpha|\, e^{i\underline{q}\hat{\underline{r}}}|\alpha'\rangle$. The vertex function $F^{\underline{q}}_{\alpha\alpha'} := \langle\hat{G}(\varepsilon+io)\, e^{i\underline{q}\hat{\underline{r}}}\, \hat{G}(\varepsilon+io)\rangle_I$ is determined by the Bethe-Salpeter-equation:

$$F^{\underline{q}}_{\alpha\alpha'} = G_\alpha \rho_{\alpha\alpha'}(\underline{q}) G_{\alpha'} + G_\alpha \Big[N_i \sum_{\beta,\beta'} \frac{1}{A} \sum_{\underline{p}} |u_{\underline{p}}|^2 \rho_{\alpha\beta}(\underline{p}) \rho^{*}_{\alpha'\beta'}(\underline{p}) F^{\underline{q}}_{\beta\beta'} \Big] G_{\alpha'} \tag{2}$$

where we use the interaction part consistent with GBA (N_i is the density of impurities, and $u_{\underline{p}}$ the effective impurity potential). The impurity - averaged one particle Green function $G := \langle\hat{G}(\varepsilon+io)\rangle_I$ is diagonal in the Landau representation and does not depend on k_y.

Since the effective impurity potential has rotational symmetry around the axis of the magnetic field, the dielectric function depends only on $q := |\underline{q}|$. Thus it is sufficient to calculate $\Pi_{RPA}(q_x, q_y=0)$ and to replace q_x by q in the final result. Eq. (1) then reads:

$$\Pi_{RPA}(q_x, q_y=0) = \frac{1}{\pi A} \operatorname{Im} \sum_{k_y} \rho^{*}_{ok_y, ok_y}(q_x) \int_{-\infty}^{\mu} d\varepsilon\, F^{q_x}_{ok_y, ok_y}(\varepsilon+io) \quad , \tag{1a}$$

where we have used $\rho_{nk_y, n'k_y'}(\underline{q}) \propto \delta_{q_y, k_y-k_y'}$. (Here and in the following we take into account only terms which stem from the lowest Landau level). For our needs, eq. (2) reduces to

$$(F^{q_x}(k_y) := F^{q_x}_{ok_y,ok_y}(\varepsilon+io)) \quad : \tag{2a}$$

$$F^{q_x}(k_y) = G_o^2\,\rho_{ok_y,ok_y}(q_x) + N_i\,G_o^2\,\frac{1}{A}\sum_{\underline{p}} u_p^2\,|\rho_{ok_y,ok_y-py}(\underline{p})|^2 F^{q_x}(k_y-p_y),$$

which can be solved immediately for the combination $\sum_{k_y}\rho^{*}_{ok_y,ok_y}(q_x)\,F^{q_x}(k_y)$, that is needed in (1a):

$$\Pi_{RPA}(q) = e^{-\ell^2 q^2/2}\,\tilde{s}_{q\ell}\;,\;\tilde{s}_{q\ell} := \frac{1}{2\pi^2\ell^2}\,\mathrm{Im}\int_{-\infty}^{\mu} d\varepsilon\,\frac{G_o^2(\varepsilon+io)}{1-c_qG_o(\varepsilon+io)}\;,$$

where
$$c_q := \frac{N_i}{2\pi}\int_0^\infty dp\;pu_p^2\,e^{-\ell^2p^2/2}\,J_o(\ell^2pq) \quad .$$

The energy integral can be evaluated with the help of the GBA-equation for the electronic self-energy ,

$$\Sigma_o(\varepsilon+io) = \frac{c}{\varepsilon-E_o-\Sigma_o(\varepsilon+io)}\;, \qquad c = c_{q=0}$$

(E_o is the energy of the lowest Landau level). Inserting into the result the implicit expression for the effective potential of an impurity charge Q_i in a medium with background dielectric constant κ_o,

$$u_q = \frac{2\pi(-e)\,Q_i/\kappa_o}{q+e^{-\ell^2q^2/2}s_{q\ell}}\;, \qquad s_{q\ell} := \frac{2\pi e^2}{\kappa_o}\,\tilde{s}_{q\ell} \tag{3}$$

we get the following selfconsistency equation for the dimensionless quantity $s'_{q'} := \ell\cdot s_{q\ell=:q'}$:

$$s'_{q'} = \alpha\,\frac{\sqrt{I_o}}{I_{q'}}\left[\beta - \gamma_{q'}\,\mathrm{arctg}\,\frac{\beta}{\gamma_{q'}}\right] \tag{4}$$

where

$$I_{q'} := \int_0^\infty dp\,\frac{pe^{-p^2/2}\,J_o(pq')}{(p+s'_p\,e^{-p^2/2})^2}\;, \qquad \gamma_{q'} := \frac{1}{2}\sqrt{\frac{I_o}{I_{q'}}}\left(1-\frac{I_{q'}}{I_o}\right),$$

$$\alpha := (\sqrt{2}\,\pi^{3/2}\ell\,|\frac{Q_i}{e}|\sqrt{N_i})^{-1}, \qquad \beta := \sqrt{1-(\frac{\mu-E_o}{2\sqrt{c}})^2}$$

and $\sqrt{I_o/I_{q'}}$ denotes the principal root. For q'=0, eq. (4) represents the compressibility sum rule $\Pi(q=0) = \rho(\mu)$, (ρ is the density of states).

The sample parameters enter the sc.-eq. via $\alpha \propto \sqrt{(B/N_i)}/|Q_i/e|$ and β. $x:=(\mu-E_o)/(2\sqrt{c})$ follows from the equation, which determines the chemical potential in GBA:

$$\pi\nu = x\sqrt{1-x^2} + \arcsin x + \frac{\pi}{2} \quad .$$

So β is a function of the filling factor $\nu := 2\pi\ell^2 N \propto (N/B)$ (N is the electron concentration, and $0 \le \nu \le 1$), which is even with respect to $\nu = 1/2$.

2. Approximate solution

Our aim is the determination of the effective potential u_q. As the knowledge of

$s_{q\ell} \lesssim \sqrt{2}$ is sufficient for an approximate description of u_q, we have to consider eq. (4) für $q' \lesssim \sqrt{2}$. The main contribution in $I_{q'}$ stems from $p \lesssim \sqrt{2}$. Thus $J_0(pq')$ may be replaced by unity, since its smallest value $J_0((pq')_{max}) \geq J_0(pq'=2) = 0.22$ and the accuracy of $s_{q\ell}$ becomes less important with increasing q. The rhs of the sc.-eq. is then independend of q', and thus $s'_{q'} \approx s'_0$ for $q' \lesssim \sqrt{2}$. In conclusion, the effective potential can approximately be described by the screening parameter $S_0\ell$ $(=s'_0)$.

Considering now the sc.-eq. for s'_0, in I_0 we may put $s'_p = s_0$ for all p. So s'_0 is approximately determined by the equation

$$s'_0 = \alpha\beta\left(\int_0^\infty dp \frac{pe^{-p^2/2}}{(p+s'_0\, e^{-p^2/2})^2}\right)^{-1/2} \quad . \tag{5}$$

3. Numerical results for the screening parameter and the effective potential in real space

Figs. 1, 2 show the solution of the sc.-eq. (5). The screening parameter $s_0\ell$ is plotted as a function of the filling factor ν at fixed values of the magnetic field and the electron concentration, respectively. (Explicitely, in terms of the sample parameters,

$$\nu = 4.14\,\frac{N}{B}\left[\frac{10^{10}\ \mathrm{cm}^{-2}}{\mathrm{kG}}\right] \;).$$

In both cases the screening becomes less effective, if we fix the filling factor and increase the impurity density. The vanishing of screening at the band edges reflects a general property of our selfconsistency problem for the dielectric function: $s_{q\ell} = 0$ for all q (bare coulomb potential), if ν equals 0 or 1.

At fixed carrier concentration $s_0\ell$ decreases monotonically with increasing ν, there is no wiggle as found by Das Sarma (Solid State Commun. 36, 357 (1980)) in a treatment in which the vertex corrections are not taken into account. This shows the importance of vertex corrections in a selfconsistent description of the electron-impurity system. For $q \to 0$ it means that the compressibility sum rule must be fulfilled.

Fig. 3 shows the effective impurity potential in real space as given by the approximation discussed in section 2:

$$\varphi(r) = \frac{1}{(2\pi)^2}\int d^2q\, \varphi_q\, e^{i\underline{q}\underline{r}} \;, \qquad \varphi_q := \frac{u_q}{-e} = \frac{2\pi Q_i/\kappa_0}{q+s_0 e^{-\ell^2q^2/2}} \quad .$$

The values of the screening parameter are chosen such that one can see the variation of the effective potential in an experiment in which the carrier concentration is changed at constant field. Roughly speaking, with increasing $s_0\ell$ the oscillations of the potential get stronger and are damped more rapidly.

Fig. 1: Dimensionless screening parameter as a function of the filling factor at constant magnetic field.

Fig. 2: Dimensionless screening parameter as a function of the filling factor at constant electron concentration.

Fig. 3: Dimensionless effective impurity potential as a function of the dimensionless distance.

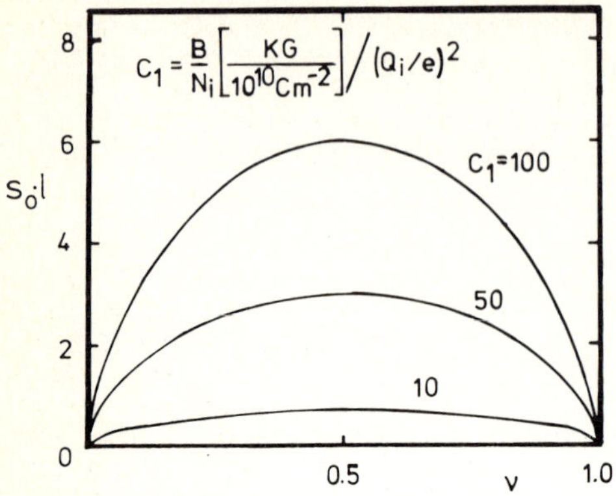

Fig. 1

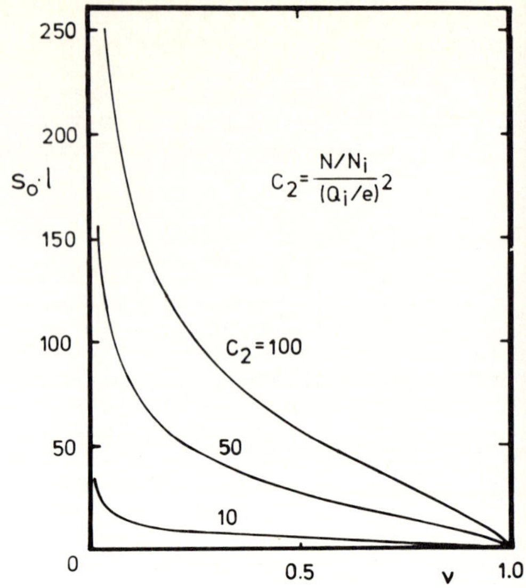

Fig. 2

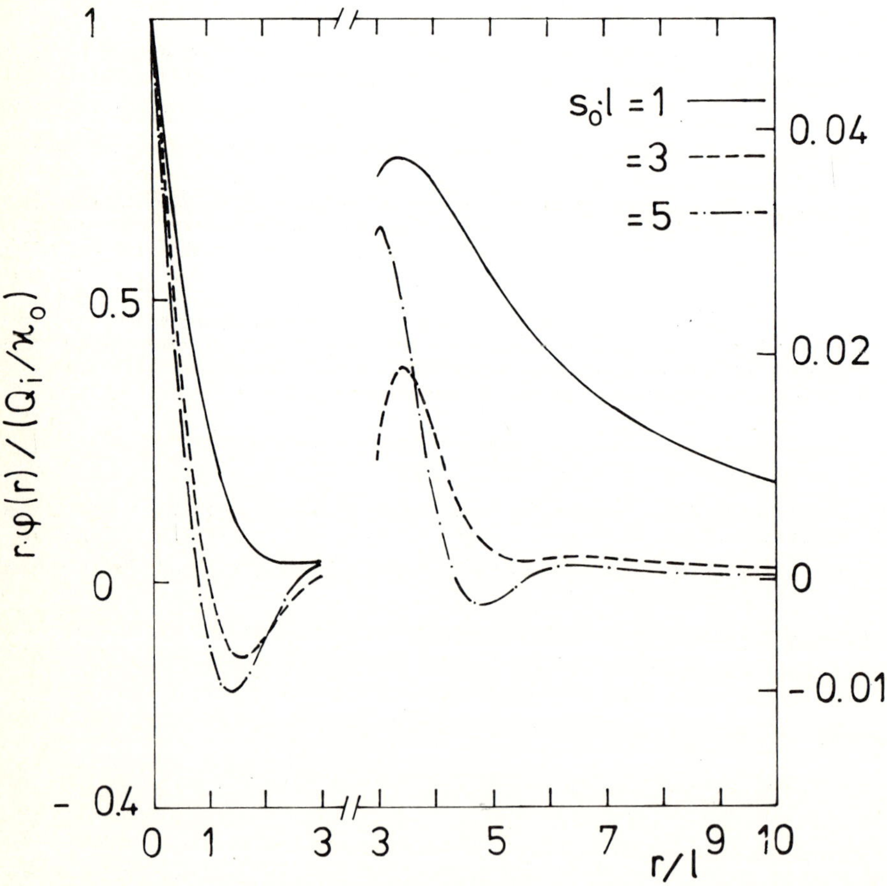

Fig. 3

VALLEY PHASE TRANSITION OF Si INVERSION LAYERS IN HIGH MAGNETIC FIELDS

Ute Gummich[+] and L.J. Sham
Department of Physics, University of California
San Diego, La Jolla, California 92093, USA

1. Introduction

In a silicon n-channel inversion layer - depending on the surface orientation - a number of conduction band valleys are equally populated. In the absence of magnetic fields Bloss, Sham and Vinter /1/ have shown that the valley degeneracy may be lifted at low densities due to the intravalley exchange and correlation energy, which dominate over the kinetic energy. From qualitative considerations Bloss et al. predicted the appearance of additional Shubnikov-de Haas oscillations at low densities. A confirmation for this are the measurements by Englert, Tsui and Landwehr /2/, showing an additional splitting of the lowest spin-split Landau level in an n-channel inversion layer on Si(111) surface. This additional splitting is interpreted as valley splitting and was not only observed in samples which have a six-fold valley degeneracy but also in samples with a two-fold degeneracy. There are many explanations /3/ for the occurance of the two-fold degeneracy at normal densitities, but the origin of the reduction from a six-fold to a two-fold degeneracy does not concern us here. In this paper the ground state energy of a valley degenerate system is calculated including exchange and correlation in the presence of a high magnetic field and the resulting conductivity is compared with the experiment.

2. Model

We study the effects of intravalley exchange and correlation in the presence of a high magnetic field perpendicular to the interface. We restrict our attention to the limit of such a high magnetic field and low densities that only the lowest spin-split Landau level is occupied and that the excitation to higher levels may be neglected. In the density range of interest in Si only the lowest electrical subband is occupied. For simplicity, the model system is limited to having two conduction valleys. The Landau level has a finite width Γ due to electron impurity scattering, included in a selfconsistent Born approximation.

3. Outline of the Calculation and Results

The phase transition is determined by the competition of the kinetic energy, characterized by the Landau level with Γ and the exchange and correlation energy characterized by the size of the Coulomb interaction α. The total energy:

$$E = (L^2/2\pi R^2)\ \sum_v [(\varepsilon_o - \tfrac{1}{2}\alpha n_v)\, n_v - (2\Gamma/3\pi)\,(1-\tilde{\varepsilon}_v^2)^{3/2}]$$

$$n_v = [\tilde{\varepsilon}_v (1-\tilde{\varepsilon}_v^2)^{1/2} + \frac{\pi}{2} + \sin^{-1}\tilde{\varepsilon}_v]/\pi \qquad (1)$$

$$\tilde{\varepsilon}_v = (\varepsilon_F - \varepsilon_o + \alpha n_v)/\Gamma$$

(ε_F being the Fermienergy and $\varepsilon_o = \dfrac{\hbar\omega_c}{2}$)

is minimized allowing unequal valley occupation numbers n_v. There are only two types of solutions. Depending on the values of the energy ratio $\gamma = \Gamma/\alpha$ and the filling factor $\nu = N/(eB/hc)$ the electrons are either confined to one valley or distributed equally in both valleys. The phase diagram (solid curve Fig. 1) shows that for small γ the one valley phase

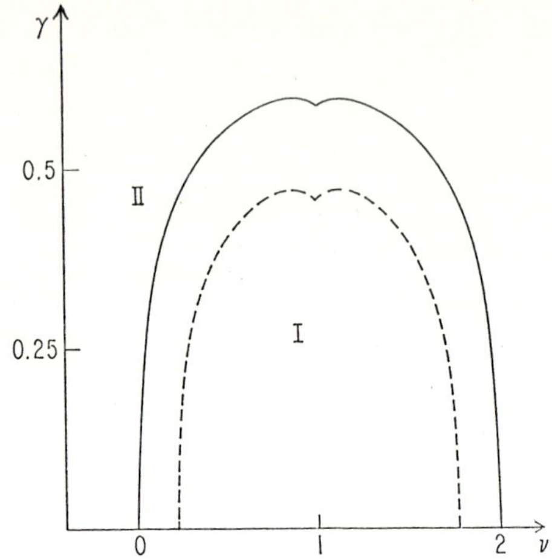

Fig. 1. Phase diagram in Γ/α versus plane at T = O K. I denotes the single valley phase and II the double valley phase. Solid line is the phase boundary in the Hartree-Fock approximation. Dashed line is when correlation is added.

occurs almost in the whole density range $0 \leq \nu \leq 2$. When Γ and α are comparable, the low density regime for electrons (ν small) or for holes (ν close to 2) favors the kinetic energy term in (1), hence the double valley phase. When the density for the electrons (or holes) is increase - according to (1) - the exchange energy increases more rapidly than th kinetic energy and the transition to a single valley state occurs. The small dip of the phase boundary at $\nu = 1$ and large γ appears not to be the result of the rapidly varying square root density of states (SBA) since a Gausian density of states distribution leads to the same featur

Based on this phase diagram the static magneto-conductivity is calculated at T = O using the formula /4/

$$\sigma_{xx} = (e^2/2\hbar\pi^2)\sum_v \int dE\; f(E)\{\mathrm{Im}\, G_v(E)\}^2/[\{\mathrm{Im}\, G_v(E)\}^2+\{\mathrm{Re}\, G_v(E)\}^2] \qquad 2)$$

Fig. 2 shows σ_{xx} versus ν at two values of γ. The sharp change of σ_{xx}, when the system undergoes a transition from the double valley phase to the single valley phase or vice versa is noticable only when γ is large The dip of the conductivity at $\nu = 1$ due to the single-valley phase effectively gives the conductivity a two-peak appearance, which occurs at all values of γ. With this result we are able to explain qualitatively the observation by Englert et al. /2/, where in the lowest Landau level for the two valley or six valley degeneracy two peaks are observed. For the samples measured, the width of the Landau level, Γ, is about 3-4 meV. Thus γ ($=\Gamma/\alpha$) is about .2. The double to single valley transition occurs close to $\nu = 0$ and $\nu = 2$. The discontinuity in σ_{xx} at the phase transition is too small to be observable and only the two peak structur is seen.

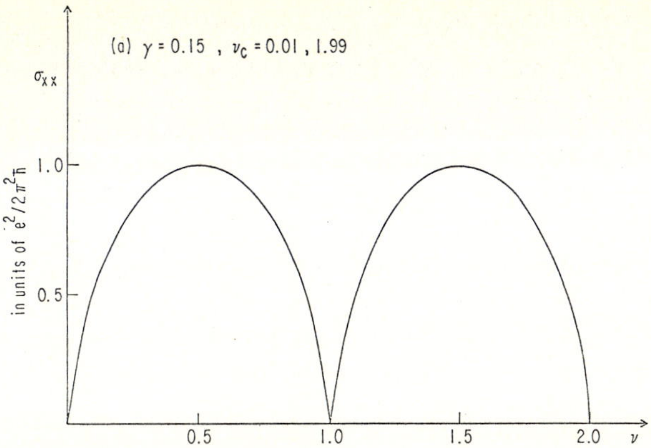

Fig. 2a. Magnetoconductivity σ_{xx} (in units of $e^2/2\pi^2\hbar$) versus Landau level occupation ν at two values of γ, the ratio of impurity to electron interaction. ν_c is where the phase transition occurs.

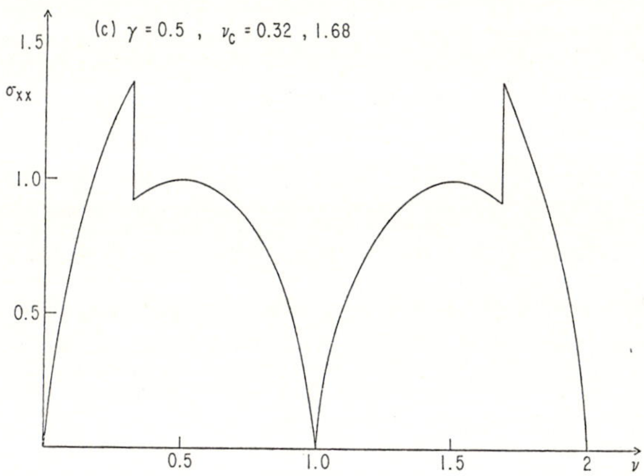

Fig. 2b. As above

Since the correlation energy usually tends to reduce the effect of the exchange energy, we have additionally calculated the correlation correction in the RPA plasmon pole approximation. The dashed curve in Fig. 1 shows the phase boundary, when correlation is included. The phase boundary is lowered in γ for a fixed occupation ν. There is however no qualitative change and the explanation of the observation given previously stands. The investigation on CDW formation in MOS systems by Yoshioka and Fukuyama /5/ also show the occurance of valley splitting.

References

+ Permanent Address: Institut für Theor. Physik, Universität zu Köln, D-5000 Köln 41, W. Germany.

/1/ W.L. Bloss, L.J. Sham and B. Vinter, Phys. Rev. Lett. 43, 1529, (1979).
/2/ Th. Englert, D.C. Tsui and G. Landwehr, Solid State Commun. 33, 1167 (1980).
/3/ M.J. Kelly and L.M. Falicov, Phys. Rev. Lett. 37, 1021 (1976); D.C. Tsui and G. Kaminsky, Solid State Commun. 20, 93 (1976); D.C. Tsui and G. Kaminsky, Phys. Rev. Lett. 42, 595 (1979); B. Vinter and A. Overhauser, Phys. Rev. Lett. 44, 47 (1980); W.L. Bloss, S.C. Ying and J.J. Quinn, Phys. Rev. B23, 1839 (1981).
/4/ W. Götze and J. Hajdu, Solid State Commun. 29, 89 (1978). T. Ando and Y. Uemura, J. Phys. Soc. Japan 36,959 (1974).
/5/ D. Yoshioka, H. Fukuyama, J. Phys. Soc. Japan 50, 1560 (1981).

ELECTRICAL PROPERTIES OF LAYERED MATERIALS IN HIGH MAGNETIC FIELDS

Sei-ichi Tanuma

Institute for Solid State Physics, University of Tokyo
Roppongi, Minatoku, Tokyo, 106 Japan

The electrical properties, especially the magnetoresistance (MR), of layer-type crystals were measured in high magnetic fields employing stationary and pulsed field magnets. Three topics are presented in this report; (1) anomalous stepwise increase of MR of graphite in increasing fields of several ten teslas at liquid helium temperature, (2) large positive and negative MR in the Anderson localized state in the commensurate CDW phase of 1T-TaS_2 and (3) field-linear MR in several kinds of layer-type crystals such as C_6Li, $C_{36}K$, $C_{48}K$, $C_{24}SbCl_5$ and 1T-Li_xTaS_2 are described and discussed.

1. Introduction

In the preceding two conferences on physics in high magnetic fields, we had a chance to present findings on some electrical properties of layered materials, i.e., we reported on electronic properties of graphite intercalation compounds, C_8K, $C_{36}K$, $C_{48}K$ and C_xBr, and on 1T-TaS_2 in the Oxford Conference on the Application of High Magnetic Fields in Semiconductor Physics held in 1978 [1]. In the Hakone Conference on the Physics in High Magnetic Fields in 1980, we presented data on the newly discovered anomalous stepwise increase of magnetoresistance (abbreviated as MR hereafter) of single crystal graphite in increasing magnetic fields perpendicular to the layer plane as well as on topics similar to the ones covered at the Oxford conference [2].

In this paper, we present the subsequent experimental developments and theoretical explanations of some of the above mentioned problems. In the section 2, we report new experimental data on the anomalous MR of graphite being achieved in our institute and in the Francis Bitter National Magnet Laboratory at MIT. In the section 3, we describe extended measurements of MR of 1T-TaS_2 at lower temperatures and give a systematic explanation of the shift of MR from small negative to large positive and again to large negative values as the temperature decreases. Finally in section 4, we note the linear MR of several graphite intercalation compounds and lithium-intercalated TaS_2.

2. Anomalous magnetoresistance of graphite

We reported at the Hakone Conference on the Physics in High Magnetic Fields the discovery of anomalous sudden changes of the MR of single crystal graphite at field intensities of several ten teslas in the liquid helium temperature range [1]. H. Fukuyama pointed out that the phenomenon could be explained by a Peierls instability. However, the authors were not fully confident of the intrinsic nature of that anomaly at that time, because the number of specimens was small. After further experiments with the same pulse field method which had been used before, it became clear that the anomaly was not an artifact. Thus a subsequent report was presented at the 15th Biennial Conference on Carbon held in Philadelphia in 1981 by Tanuma et al. [3]. On the other hand, Yoshioka and Fukuyama [4] developed a theory capable to explain the anomaly that graphite becomes a quasi-one dimensional conductor rather than a two dimensional one after the application of magnetic fields higher than the quasi-quantum limit. It is assumed that the spanning $2k_F$ vector makes an effective electron nesting for electron-phonon coupling causing a CDW-type instability. By the formation of a CDW gap at a certain field intensity graphite becomes a poorer conductor and gives rise to a sudden increase of resistivity. Since then, the authors wanted to confirm the effect by measurements in stationary magnetic fields.

Fortunately, Y.Iye could perform the MR experiment in continuous fields up to 28.5T at MIT in collaboration with M.S.Dresselhaus and her group. The work was the first research use of the newly set up (superconducting and normal) hybrid-magnet in the Francis Bitter National Magnet Laboratory, which generates a maximum field of 30 T.

In this experiment by Iye et al.[5], three kinds of graphite crystals were investigated: single crystal graphite (SCG), segregated from molten iron (so called Kish graphite or KG), Madagascar natural SCG (MG) and highly oriented pyrolytic graphite (HOPG). Seven specimens were measured. The residual resistance ratio RRR ranged from 38 to 1.5, and the specimens showed an anomalous MR except two specimens of HOPG (RRR=1.5) and MG (RRR=18). An example of the relationship between MR and magnetic field intensity is shown as curve KG21 in Fig.1 for a broad field range and in Fig. 2 for the field range from 18 to 28.5T. In Fig. 1, the curves denoted as SCG514 are presented for comparison, which were obtained by the pulsed field method of K. Nakamura et al. in Miura's laboratory in the ISSP [6]. In their experiment, the maximum field was extended to 42T compared to the value of 27T [2] reported before. As a general tendency of the MR shown in Fig. 1, it increases approximately linearly with the field up to about 15 T and then tends to saturate at higher fields. The wavy structures below 7.3T are Shubnikov-de Haas oscillations. Calculation of Landau levels of graphite show, that the dip at 7.3T occurs at such a field strength that the level (n=0, σ=-) meets the Fermi level. No more oscillations can be expected in higher fields up to about 70T, when the level (n=-1, σ=-) leaves the Fermi level; the relative heights of these levels are shown in figure 4.

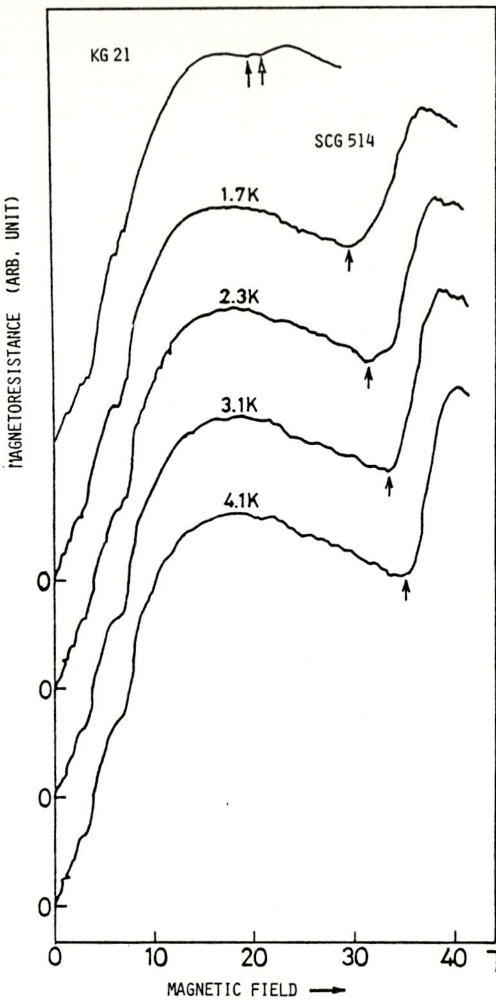

Fig. 1 Magnetoresistance of graphite single crystals.
KG21: static field measurements
SCG514: Pulse field measurements

As can clearly be seen in Figs.1 and 2, the anomaly occurs in both static and pulsed magnetic fields and the critical field value does not show any hysteresis for increasing and decreasing H. These facts are evidence for the intrinsic nature of the phenomenon. Furthermore, the reversibility of the anomaly suggests the transition to be of second order. By the more precise measurements in static fields it was found that (i) the MR causes a two-step anomalous change in some of specimens as seen in Fig.2 by the solid and empty arrows. The other findings are: (ii) the decrease of MR with increasing field after passing through a broad maximum, and (iii) the temperature dependence of the height of maximum characterized by a decrease of height with temperature. The features (i) to (iii) have not yet been explained in terms of a CDW transition.

The onset of the anomaly, indicated by the solid arrows in Figs.1 and 2 as well as in previous data have been plotted against temperature in Fig.3. The universality of the dependence of the onset point on temperature and magnetic field is clearly seen in the curve (A) in the figure. Curve (B), representing the pulsed field data, is shifted relative to curve (A). This could be due to a possible error in the temperature calibration. The temperature was measured by the vapor pressure of liquid helium, but the sample immersed in the liquid was probably heated by the eddy currents during the field pulse, so that the temperature was raised due to the finite heat conductance of the system. If the specimen temperature would be about 0.5K higher than the temperature of the liquid, (A) and (B) would form a single curve.

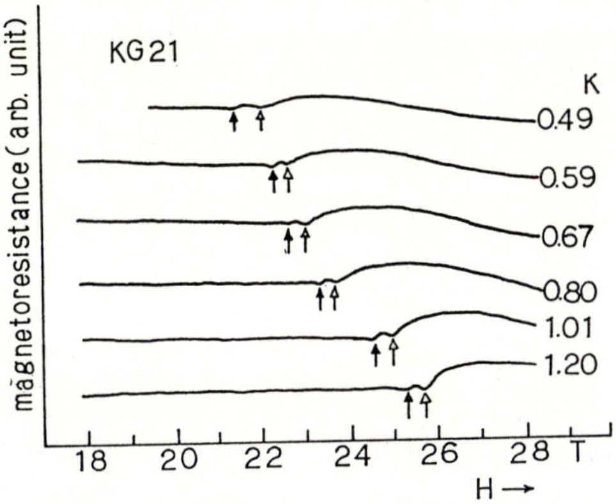

Fig.2 Magnetoresistance anomaly. Two-step changes are recognized.

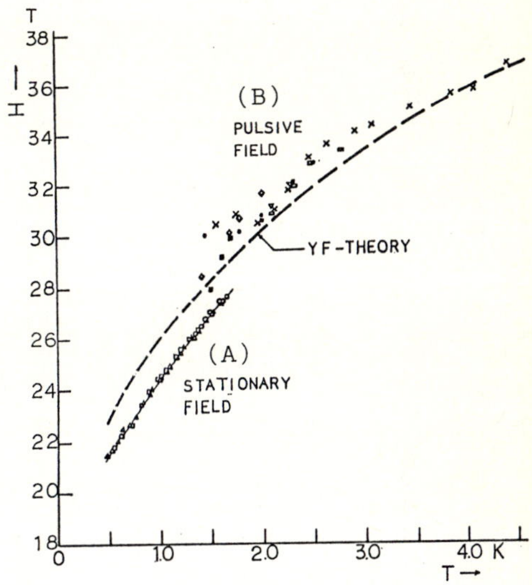

Fig. 3 Critical field of anomaly versus temperature relation.

It is believed that this MR anomaly is hard to understand in the one electron approximation scheme, because of the strong temperature dependence of the critical field. The data do not fit the de Haas-Shubnikov effect. In such high fields as shown in Fig.3, the Landau levels are almost in the quantum limit, as be seen in Fig.4. This figure was calculated using the Slonczewski-Weiss-McClure band model of graphite, taking the most appropriate band parameters [7]. The four levels in Fig.4, n=-2, σ=-; n=-1, σ=+; n=-1, σ=-; and n=0, σ=+ are very insensitive to the field. The higher Landau levels n=1, σ=+ and n=0, σ=- pass through the Fermi level at 6.6 and 7.3T, respectively; above the field of 7.3T the system is in the quasi-quantum limit. The extreme quantum limit will be attained at about 70T when the upper two levels move above the remaining two. In the relevant high fields along the c-axis, the electronic motion within the c-plane becomes localized and the electron system becomes quasi-one dimensional. Yoshioka and Fukuyama [4] investigated what interaction should give the lowest transition temperature: charge density wave, spin density wave or excitonic interaction, in the band scheme of Fig. 4. They concluded that the Coulomb-interaction-derived CDW instability corresponding to the wave vector designated as ξ_{0+} in Fig. 4 is the most probable, giving a critical temperature as follows:

$$T_c = \frac{8\gamma}{\pi} E_F(H) \frac{\cos^2(\pi\xi_{0+})}{\cos^2(2\pi\xi_{0+})} \exp\left[\left\{ \frac{2\pi U(\varepsilon)}{c_o(2\pi\ell)^2} \frac{d\xi_{0+}}{dE_{o,+}} \right\}^{-1} \right]; \tag{1}$$

where $U(\varepsilon)$ is the electron-electron interaction as a function of the dielectric constant ε, E_F is the Fermi level as a function of H through ξ_{0+} and ℓ is the Larmor radius given by $\ell^2=c/eH$. The constants γ and c_o are 1.78 and 6.74Å, respectively. Assuming $\varepsilon=10$, YF evaluated Eq.1 numerically. The results are shown by the dashed curve in Fig.3. Eq. 1 can be rewritten in the BCS type form as follows:

$$T_c = \alpha(H)\exp[-1/\lambda(H)], \tag{2}$$

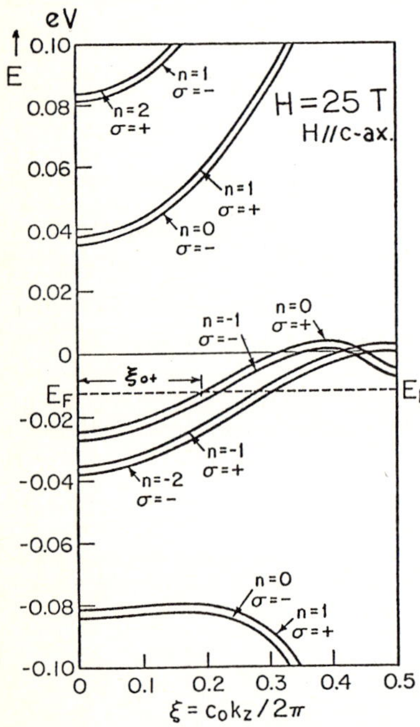

Fig. 4 Calculated Landau levels.

where λ is given by the product of the density of states and the effective interaction; the density of states could be a linear function of H, i.e., $\lambda=\zeta H$. The curve (A) was obtained from Eq. 2 by choosing the parameters as α=94.9K and ζ= $8.93\times10^{-3}T^{-1}$. Thus, comparing Eqs. 1 and 2, the value of ε which was obtained from 5.9. This value is not far from other reported experimental values ranging 5.8 to 3.4.

So far, the anomaly seems to be explained by the CDW instability. The quasi-quantum limit band scheme represented in Fig.4 suggests another possible successive transition with longer spanning vector associated with n=-1,σ =+,- and n=-2, σ=- at higher magnetic fields.

3. Magnetoresistance of 1T-TaS_2 in Anderson localization regime

In the preceding conferences on the physics in high magnetic fields, the authors reported the MR of 1T-TaS_2 at low temperatures [1],[2]. The material is in the commensurate CDW state below 200K and enters the Anderson localized state at temperatures below the liquid helium range. The authors found a very large positive MR between 4 and 0.5K. After that, Kobayashi et al.[8] found an even larger negative MR in the same material but at a lower temperature range from 0.4 to 0.05K. The general behavior was reported in a preceding report[1].

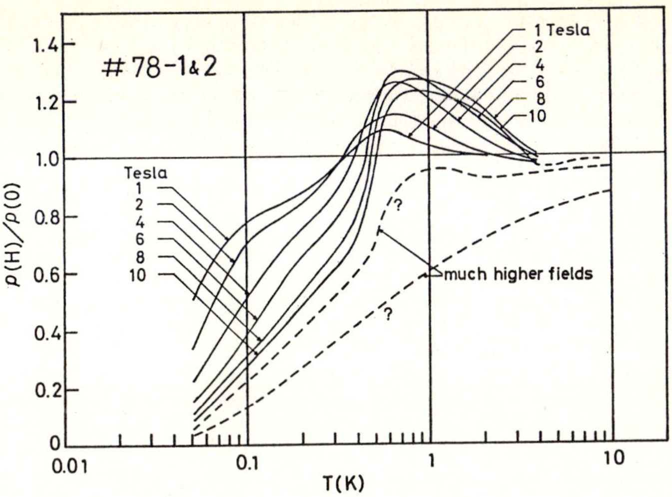

Fig. 5 Magnetoresistance versus temperature relation for 1T-TaS_2.
Specimens #78-1 and # 78-2 combined

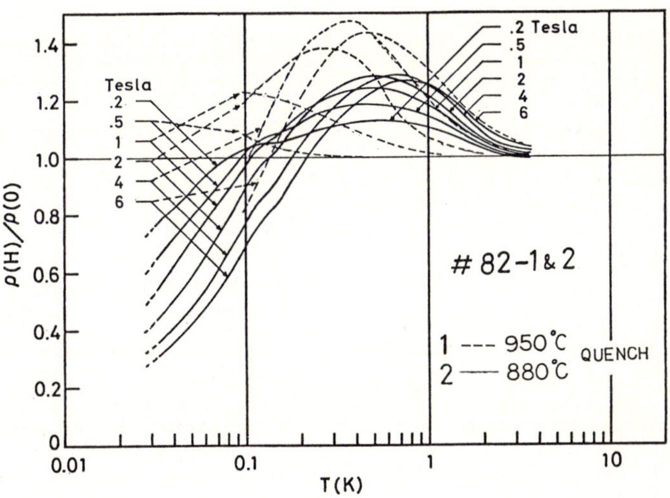

Fig. 6 Magnetoresistance versus temperature relation for 1T-TaS_2.
Specimens #82-1 and #82-2 which were grown in the same ampule.

As the two specimens described in the report had almost the same quality, a new plot was prepared, using the positive and negative MRs of these two specimens as the relation between ρ_H/ρ_O versus temperature with a fixed magnetic field strength as the parameter. This plot is shown in Fig.5. The figure seems to show that the negative MR is increasing in its magnitude with decrease of temperature and the

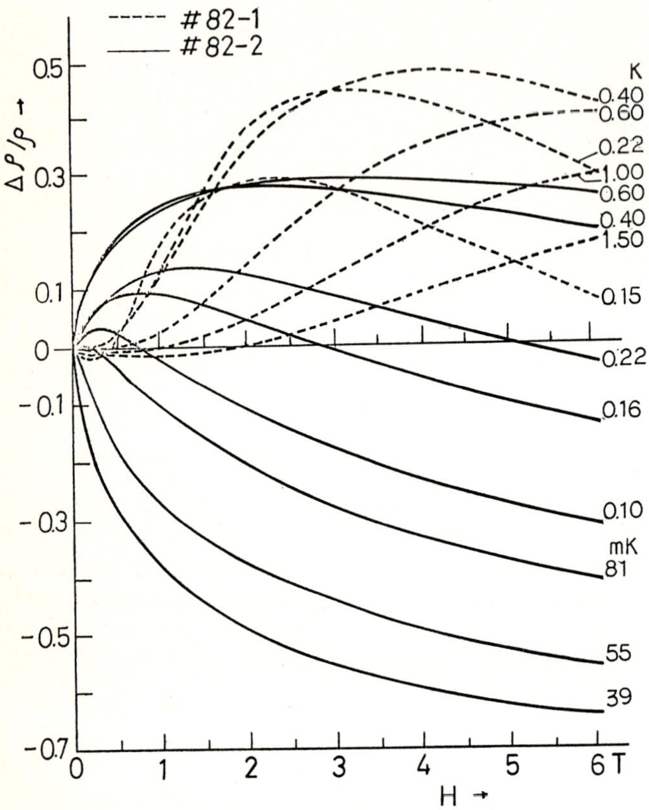

Fig. 7 Magnetoresistance versus magnetic field relation. Specimens #82-1 and #82-2 which were grown in the same ampule.

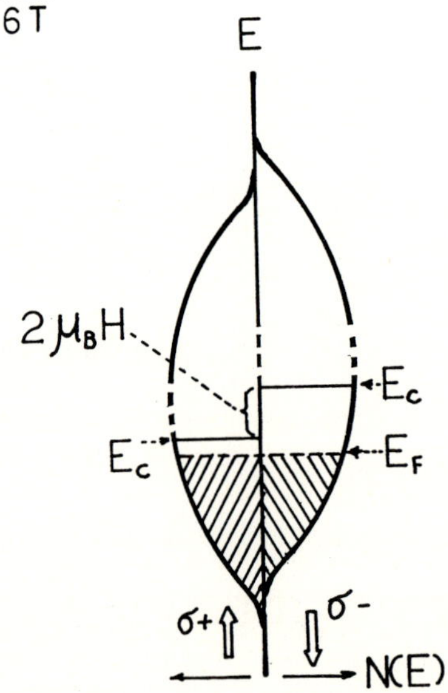

Fig. 8 Energy scheme of a band containing Anderson localized states under a magnetic field. Zeeman shifts of parallel and anti-parallel spins are illustrated.

positive MR is superposed on the negative one in the temperature range of 4 to 0.4K. Furthermore, other specimens are investigated in ISSP down to 25mK, using a dilution refrigerator and a 6T superconducting magnet [9]. The results of these measurements are given in Figs. 6 and 7. It is a remarkable fact that both the positive and negative MRs, $\rho_H-\rho_O$, are so large in absolute magnitude that they amount to several tens per cent of the resistivity ρ_O at zero field.

Comparing Figs. 5 and 6, the overall behavior is the same, but there exist quantitative differences; # 78-1&2 has a larger negative MR at temperatures lower than 0.4K. For a closer comparison, the plots in Fig.6 contain the data of two samples #82-1 and #82-2. These were grown in the same ampule but quenched at different temperatures, i.e., #82-1 was quenched at 950°C and #82-2 at 880°K. Therefore the former specimen could have more defects and more randomness in the crystal potential. As a difference in the MR which characterises these two specimens, Fig.6 shows the following: #82-1 has larger values of positive MR the peak of which occurs around 0.3K, while the smaller positive peak for #82-2 arises around 0.6K. In other words, the crystal containing more randomness has a larger positive MR which decays more quickly with rising temperature.

(a) Fukuyama and Yosida gave a pertinent explanation based on the Zeeman effect for the negative MR [10]. With the localization length α^{-1} in the Anderson localized state, the envelope function for the energy E state is written as $\Psi_E(R)\sim\exp(-\alpha R)$, and $\alpha(E)$ is proportional to $(E_c-E)^\beta$, where E_c is the mobility edge and β is about unity. The conductivity in the absence of a magnetic field is given as,

$$\sigma_O = A\exp[-(T_O/T)^{1/n}] \tag{3}$$

where T_O is proportional to $\alpha(E_F^O)^d$ of which d is the dimension of the system and E_F^O is the Fermi energy. The exponent n is 3 for the two-dimensional system. This formula holds well except that n is 2 rather than 3 in the experimental data of 1T-TaS_2 at the temperature lower than about 4K. The application of a magnetic field H introduces Zeeman shifts depending on the spin directions, and causes the repopulation among the localized states. This energy scheme is shown in Fig.8. Thus a magnetic field affects the envelope function through α, and also the conductivity $\sigma(H)$ through T_O which is proportional to α^d. Then $\sigma(H)$ is written as follows:

$$\sigma(H) = \frac{A}{2}\sum_{\sigma=\pm1}\exp[-(T_\sigma/T)^{1/n}], \tag{4}$$

$$T_\sigma \propto [E_c - (E_F + \frac{1}{2}g\mu_B H\sigma)]^{\beta d}, \tag{5}$$

where g, μ_B and σ are the effective g-factor, the Bohr magneton and the sign of up or down spin, respectively. Although E_F and E_c are field dependent, the leading change is linear in the Zeeman shift $1/2\cdot g\mu_B H\sigma$. Due to the exponential dependence of conductivity on $E_c - E_F^+ \cdot g\mu_B H\sigma/2$, Eq.4 always gives a negative MR. For the approximation of $|1/2 \cdot g\mu_B H\sigma| \ll E_c - E_F$, the expression of MR is as follows,

$$\frac{\rho_H}{\rho_O} = \mathrm{sech}[\frac{\beta d g\mu_B H}{2n(E_c-E_F)}\;\frac{T_O}{T}\;1/n] \tag{6}$$

with this formula derived by Fukuyama and Yosida, the magnitudes of large negative MRs in Figs. 5-7, i.e., ρ_H/ρ_O=0.1 to 0.3 at H=10 to 5T, are matched, when the experimental fit of Eq.3 is done with $T_O\sim$1K.

(b) For the understanding of the positive part of the MR, the necessity of taking into account the correlation energy of the Anderson localized state was pointed out by Kamimura and his coworkers. Kamimura, Takemori and Kurobe developed a theory of correlation effects on variable range hopping conduction with and without magnetic field [11],[12],[13]. They claim that the intra- and inter-site interactions play an important role. There are three types of electronic states in the presence of intra-site interaction: unoccupied state (UO), singly occupied state (SO) and doubly occupied state (DO). Corresponding to these three electronic states, there appear the following four kinds of hopping processes as shown in Fig.9: (i) hopping from SO state to UO state, (ii) from SO to SO state, (iii) from DO to UO state and (iv) from DO to SO state. These four hopping processes are all contributing to the conductivity in the absence of a magnetic field, but the circumstances changed in a magnetic field. The problem of magnetoresistance will be reduced to the calculation of conduction in a random network of resistors in the field and the percolation theorem is applied for the network. The total transition rate from (i) to (j) centers of localization in a magnetic field, with neglect both of spin flip processes and a change of localization length under the field, can be formulated as follows:

$$\Gamma_{ij} = \sum_{\sigma=\pm 1} 1/(Z_i Z_j) \{\exp[-\beta(E_i+\sigma\mu_B H)]\gamma_{ij}(i)+\exp[-\beta(E_i+E_j)]\gamma_{ij}(ii)\}$$

$$+ \exp[-\beta(2E_i+U)]\gamma_{ij}(iii)+\exp[-\beta(2E_i+E_j+U-\sigma\mu_B H)]\gamma_{ij}(iv)\}, \qquad (7)$$

where $\beta=1/(k_B T)$, $\gamma_{ij}(\ell)$ represents an intrinsic hopping rate for the four kinds of above processes (ℓ=1,2,3 and 4), and Z_i the grand partition function in a magnetic field H which is given by

$$Z_i = 1 + 2\cosh(\beta\mu_B H)\exp(-\beta E_i) + \exp[-\beta(2E_i+U)]. \qquad (8)$$

When a magnetic field is applied, the spin of SO state becomes parallel to the field. Therefore the probability to find a pair of SO states whose spins are antiparallel becomes smaller with increase of the field. The effective hopping processes between SO states should occur when the spins of both states are antiparallel, as far as the spin flip processes are neglected, so that these hoppings are liable to be suppressed by a magnetic field. The DO to UO hoppings are also suppressed because this process is connected to the SO to SO process by the detailed balance relation. The suppression of these hopping processes by the field application give rise to the positive MR. The MR saturates for a field value larger than a certain value at which the processes (i) and (iv) remain only.

(c) Another interaction, the inter-site correlation, should be considered. The above mentioned intra-site correlation causes the singly occupied states which are distributed randomly. Thus two SO states having a close distance may interact with each other to form a spin-pair. A major number of SO states causes spin-pairs at such

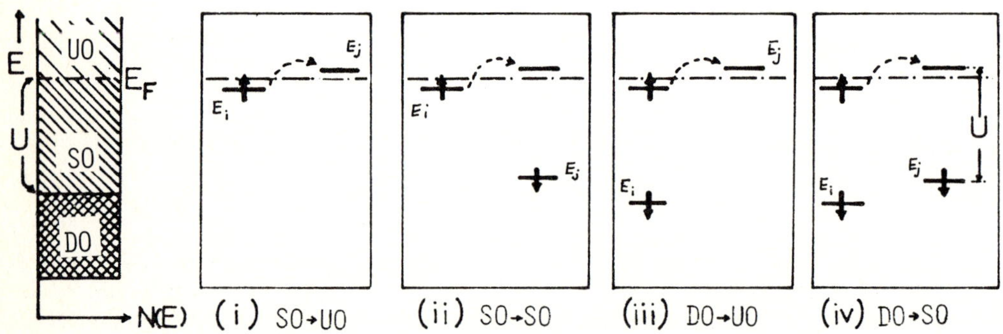

Fig. 9 Energy scheme of the four kinds of transitions between two sites, when intra-site correlation is taken into account.

low temperatures that $k_B T$ is smaller than the average interaction energy. Both kinds of interactions, i.e., ferromagnetic and antiferromagnetic, depending on each spin-pair, are needed for the explanation of the low temperature specific heat anomaly of the Schottky-type. The spin-pair formation produces spin locking, therefore in the antiferromagnetic spin-pairs, the SO→SO and DO→UO hoppings are not suppressed unlike the case (b). Thus the positive MR disappears in such a temperature range,

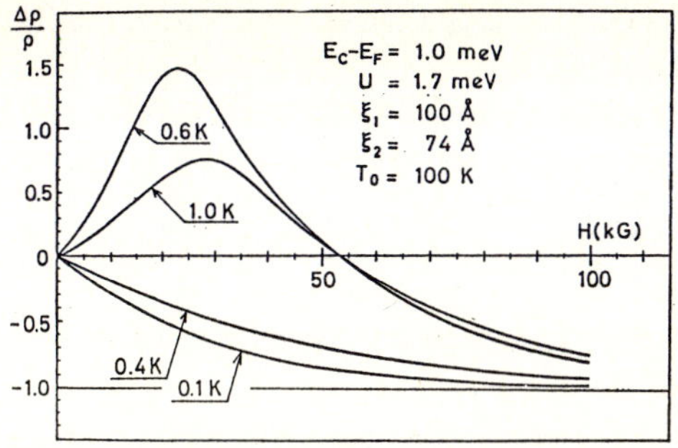

Fig. 10 Calculated negative and positive magnetoresistances of a simulated $1T\text{-}TaS_2$ under the Anderson localization

and again the Zeeman effect plays a role for the negative MR.

The measured MR behavior should be the combination of the above described two mechanisms, thus the MR can be written conventionally in the following form:

$$\Delta\rho/\rho_0 = N\ (H) + P\ (H),$$

where N(H) and P(H) represent the negative and positive part of MR, respectively.

The numerical calculation of an overall behavior of MR is exemplified in Fig.10. In this example, the spin locking between singly occupied states by the inter-site interaction is assumed to occur below 0.5K. The numerical values for various parameters are inserted in the figure, where ξ_1 and ξ_2 are the localization lengths of SO and DO states, respectively. When compared with Fig.7, Fig.10 seems to give reasonable agreement. As for Figs. 5 and 6, the positive MR in the intermediate

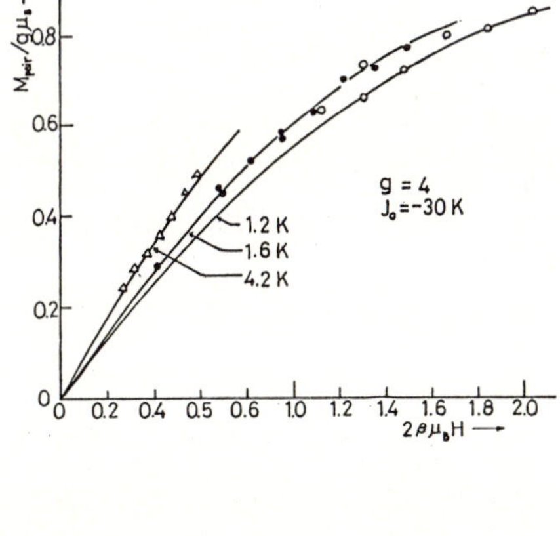

Fig.11 Magnetization versus magnetic field relation on $1T\text{-}TaS_2$. Curves are calculated by the antiferromagnetic spin pair model to get a fit to the measured points.

temperature range corresponds to the intra-site interaction mechanism treated in subsection (b), and the negative trend in the higher temperature range is caused by the vanishing of such a mechanism by the thermal excitation. The large negative MR at lower temperature range is due to the suppression of the positive MR effect by the spin locking.

The antiferromagnetic inter-site interaction turns out to be predominant in the theoretical fitting of the magnetization measurement. Fig. 11 shows an example of the fitting. The solid line is calculated by the following formula for the antiferromagnetic spin-pairing model:

$$\frac{M}{g\mu_B} = \sum_{\text{pair}} \frac{2\sinh(\beta H')}{1+2\cosh(\beta H')+\exp(\beta J)}\,, \tag{10}$$

where $H' \equiv g\mu_B H$, $J = J_0 \exp^{-2\alpha R}$ and α^{-1} is the localization length. The three curves should be unified if the simple paramagnetic Brillouin function is employed; this is not the case. Thus the fitting shows clearly the predominance of antiferromagnetic interaction ($J_0<0$).

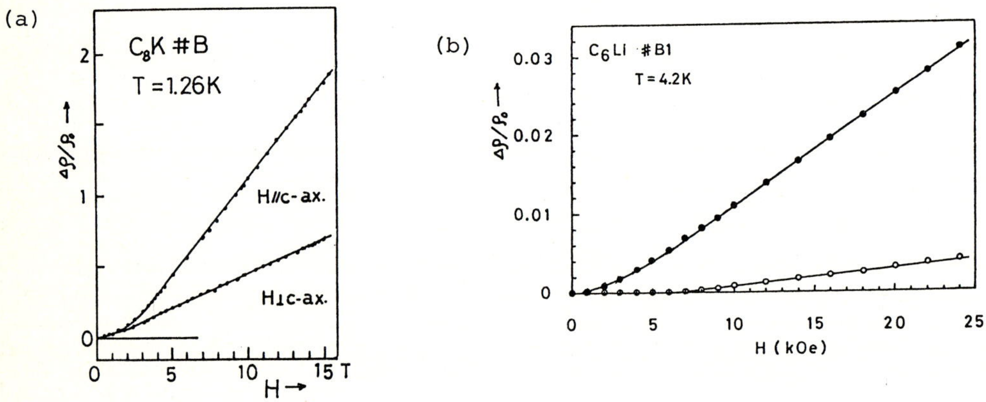

Fig.12 H-linear relationship of magnetoresistance in alkali metal graphite intercalation compounds.
a) C_8K
b) C_6Li

4. H-linear magnetoresistance appearing in layer-type crystals

In the MFSC-78 Conference, the authors reported that graphite intercalation compounds C_8K, $C_{36}K$ and $C_{48}K$ showed the H-linear transverse MR at low temperatures. Especially the first stage compound C_8K showed the linearity from about 4T to the maximum measured field 15T in both field contigurations of H// c-axis and H⊥ c-axis. In the present report, a few intercalation compounds are added which show similar properties. The transverse MR of C_6Li, which was studied by Furukawa in the author's laboratory [14], is shown in Fig. 12(a), while the already reported MR of C_8K is cited in Fig. 12(b). Both compounds are the first stage alkali metal intercalation compounds, and the H-linearity enters at lower fields for C_6Li, although the magnitude of MR is larger for C_8K. So far, the graphite intercalation compounds yielding the H-linear MR are the donor type compounds; acceptor type compounds have been examined by Inada and Tanuma [15] and the second stage $C_{24}SbCl_5$ has been proved to yield the H-linearity up to 10T, as shown in Fig. 13(a),(b). As the figure (b) shows, the linearity starts in such a small field as 0.2T or less.

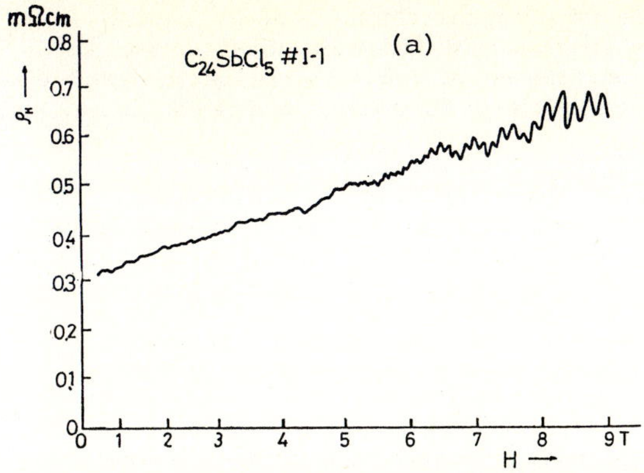

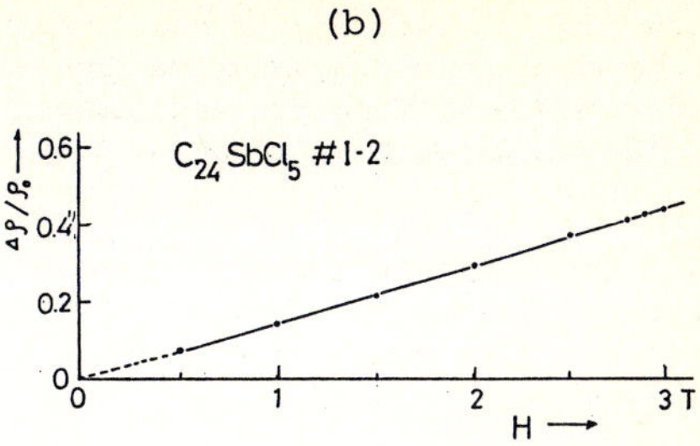

Fig. 13 H-linear relationship of magnetoresistance in the second stage $SbCl_5$ graphite intercalation compound.
(a) #I=1, 4.2K
(b) #I-2, ca. 10K

$1T$-TaS_2 forms an intercalation compound with lithium, as $1T$-Li_xTaS_2 of which x value is not determined exactly but estimated as 0.1, by dipping the host specimen into hexane solution of n-butyllithium. The temperature dependence of resistivities of the materials before and after intercalation are shown in Fig. 14(a)

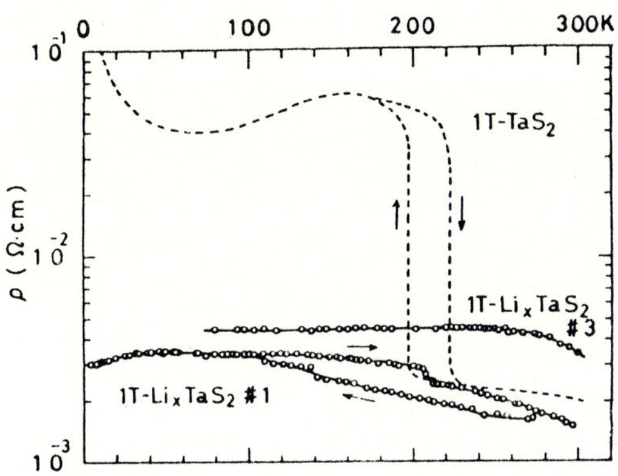

Fig.14 (a)
Resistivity versus temperature relations of $1T$-TaS_2 and lithium intercalated $1T$-TaS_2. The resistivity hysteresis in $1T$-TaS_2 (unintercalated) corresponds to the nearly commensurate to commensurate phase transition of CDW.

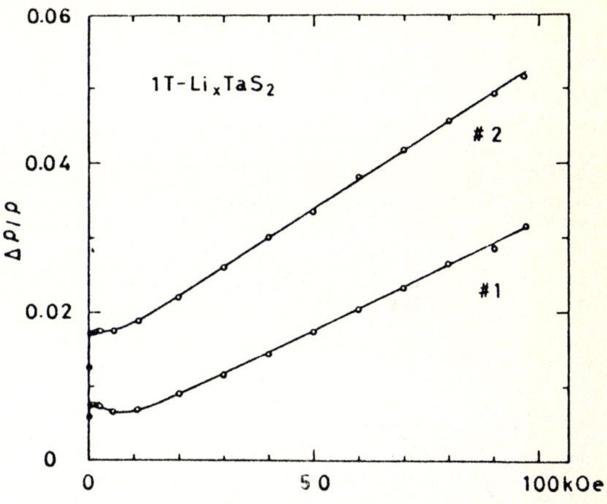

Fig.14 (b) H-linear relationship of $1T$-Li_xTaS_2 at 4.2K.

The large resistivity gap at the transition temperature, 200K, between nearly commensurate and in commensurate CDW is suppressed. The Andersón localization is also suppressed by the lithium intercalation. The MR of the lithium intercalated $1T-TaS_2$ is shown in Fig. 14(b). The MR shows H-linear dependence above ca. 1T.

It is thus proved that a fair number of layered materials have a feature of H-linear MR, although the mechanism for this phenomenon is not yet clarified.

Acknowledgement

The author is indebted to his collaborators, Drs. Y.Iye, R. Inada and Y.Onuki, to Prof. H. Nagano and his collaborators Dr. Y. Oda, Mr. A. Furukawa and to Prof. N.Miura and his collaborators Dr. G. Kido, Messrs. K.Nakamura and T.Osada for every contribution to the works presented here. He is grateful to Profs. M.S.Dresselhaus and G. Dresselhaus for giving a favor of joint work in MIT, and to Drs. L. Rubin B.Brandt and Y. Iwasa of the Francis Bitter National Magnet Laboratory for their advice and support for the use of the hybrid magnet. He also thanks Prof. H.Suematsu for supplying Kish graphite samples and Dr. A.W. Moore of Union Carbide Co. for supplying HOPG samples. His thanks is also due to Profs. H. Kamimura and H. Fukuyama and to Mr. A.Kurobe for their helpful discussions. The work is supported by Grant in Aid of Ministry of Education and Culture.

References

1. S. Tanuma, H. Suematsu, K. Higuchi, R. Inada and Y. Ōnuki: Booklet of the Application of High Magnetic Fields in Semiconductor Physics, Oxford, ed. J.F. Ryan (1978) 85.
S.Tanuma, R. Inada, A. Furukawa, O. Takahashi and Y. Iye: Splinger Series in Solid-State Sciences 24, Physics in High Magnetic Fields, ed. S. Chikazumi and N. Miura, Proceedings of the Oji International Seminor in Hakone (1980) 316.
3. S. Tanuma, A. Furukawa and Y. Iye: Extended Abstract of the 15th Biennial Conference on Carbon, Philadelphia, (1981) 16.
4. D. Yoshioka and H. Fukuyama: J. Phys. Soc; Jpn. 50 (1981) 725.
5. Y. Iye, P.M. Tedrow, G. Timp, M. Shayegan, M.S. Dresslhaus, G. Dresselhaus, A. Furukawa and S. Tanuma: Phys. Rev. B25 (1982) 5478.
6. K. Nakamura, T. Osada, G.Kido and N. Miura: to be reported in this Conference.
7. E. Mendez, A. Misu and M.S. Dresselhaus: Phys. Rev. B21 (1980) 827.
8. N. Kobayashi and Y. Muto: Solid State Commun. 29 (1979) 67.
9. A. Furukawa, Y. Oda, H. Nagano, R. Inada, Y. Onuki and S. Tanuma: to be published in J. Phys. Soc. Jpn.
10. H. Fukuyama and K. Yosida: J. Phys. Soc. Jpn. 46 (1979) 1522.
11. H. Kamimura, T. Takemori and A. Kurobe: Anderson Localization, Y. Nagaoka and H. Fukuyama Eds., Springer Verlag, 1982, p 156
12. A. Kurobe and H. Kamimura: J.Phys. Soc. Jpn. 51 (1982) 1904.
13. A. Kurobe: private communication.
14. A. Furukawa: Master Thesis of Tokyo University (1981)
15. R. Inada and S. Tanuma: to be published in J. Phys. Soc. Jpn.

QUANTUM OSCILLATORY EFFECT IN GRAPHITE INTERCALATION COMPOUNDS

I. ROSENMAN, F. BATALLAN, Ch. SIMON
Groupe de Physique des Solides de l'Ecole Normale Superieure*
Universite Paris 7, T23, 2, place Jussieu, 75251 Paris cedex 05, France
G. FURDIN
Laboratoire de Chimie Minerale Appliquee (A. Herold)*
Universite Nancy 1, CO140, 54037 Nancy cedex, France

* Laboratoire associe au CNRS

We present a study of the bromine second stage graphite intercalation compound by the de Haas-van Alphen effect and its relation with some recent structural work. These two combined studies suggest the possibility of charge density waves which are discussed.

-1- INTRODUCTION

The graphite intercalation compounds are layered materials characterized by a charge transfer process between the intercalate and the graphite layers. Intercalated molecules are classified in two families: electron donors and acceptors. Their electrical properties are quite different and we are interested here in the acceptor compounds (GAC) which are two-dimensionnal (2-D) metals. The experimental method that we have used is the de Haas-van Alphen (dHvA) effect which gives information about the carriers at the Fermi level. Such dHvA studies have been previously realized in various GACs and in particular in the second stage (the stage is the number of carbon planes between two nearest intercalate layers) nitric acid GAC (1). In that study we have proposed an electronic structure model which is in good agreement with our experimental results. Some details of the interpretation of the second stage nitric acid GAC results depended on the crystallographic structure of the GAC which was then unknown. However the overall good agreement had led us to undertake more detailed studies on such 2-D metals. Indeed nowadays the crystallographic structure of the bromine GACs is quite well understood and it seems interesting to combine dHvA results with the structural studies.
In this paper we present the study of the second stage bromine GAC $C_{14}Br$ by dHvA effect and its relation with the structural studies.

-2- STRUCTURAL STUDIES ON BROMINE GACs

Many problems about the bromine GACs structure have been elucidated in the last years. Thermogravimetric studies (2) have confirmed that the richest compound of this family is C_8Br and corresponds to the stage two. The desorption of this compound gives a family of compounds of chemical formula $C_{7s}Br$ (in which s is the stage) starting from s=2.

We have made neutron diffraction studies (3) at room and liquid helium temperatures on the C_7Br family. The in-plane unit cell that we have found is the same for all the family and is shown in the figure 1. The results

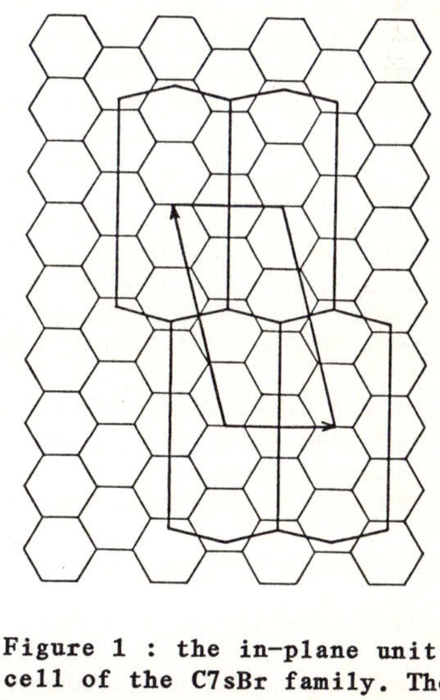

Figure 1 : the in-plane unit cell of the $C_{7s}Br$ family. The hexagonal array is the graphite lattice.

obtained by other authors (4,5) correspond to this cell. However a more detailed analysis shows the existence of incommensurate satellites and diffuse streaks in the diffraction pattern (5,6). A phase transition between room and liquid helium temperatures, visible on these satellites, has been observed (6), whose origin is not very clear (periodic lattice displacement, mass density waves, orientationnal order...). If the origin of this modulation is related to a Fermi surface effect (i.e. charge density waves) it will be interesting to observe the structure of this Fermi surface.

-3- THE DE HAAS-VAN ALPHEN EFFECT

1 Experimental aspects

The dHvA effect allows to study the extremal cross-sections of the Fermi surface (related to the dHvA frequencies) and the band structure at the Fermi level by the study of the effective masses.

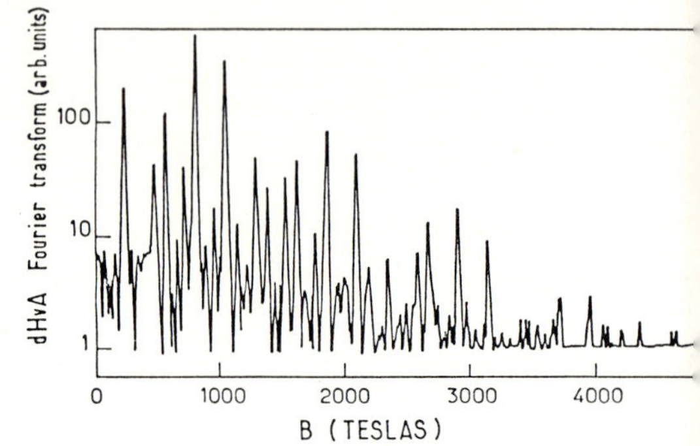

Figure 2 : Typical spectrum of dHvA effect on the second stage bromine intercalation compound.

The sample that we have used is prepared by starting from HOPG (highly oriented pyrolitic graphite) which is a set of single crystals (of about 1μm) randomly oriented with the same c axis (the axis perpendicular to the layers). In addition crystallographic studies show, even in samples prepared from single crystalline graphite, the presence of many domains in the intercalate layers. Despite the poor quality of the samples, which is transposed to the electronic properties by a very small (around 10) residual resistivity ratio, the dHvA peaks are very well defined and easy to observe. This can be explained by the presence of a term of anisot in the dHvA formula (1) which considerably enhances the amplitude of the effect in a 2-D material. Therefore the dHvA effect is a very good technique for studying the Fermi surface of such 2-D metals.

The experimental setup was previously described (1). The samples were held in sealed pyrex containers in order to realize in-situ experiments. We have studied two samples of the compound with an excellent reproducibility.

2 Experimental results

We have studied the second-stage bromine GAC. The stage of the samples was characterized by 001 X-rays diffraction. We could not determine directly the in-plane structure because the study of HOPG samples held in a pyrex container is quite difficult. However the study of other samples, prepared in the same conditions, allows us to attribute the chemical formula $C_{14}Br$ to them.

Figure 2 shows a typical spectrum of $C_{14}Br$ obtained by dHvA effect. This spectrum is very rich (more than 20 peaks) and many peaks are distant of 240 Teslas from each othe We classify then the frequencies in five families. Table 1 summarizes these frequencie with the corresponding effective masses.

3 The coupling of the dHvA orbits

We have observed here as in many other GACs many combinations of frequencies which we have attributed to orbits coupling. This leads us to study the possible mechanisms of orbits coupling in a magnetic field. Two mechanisms which can be used here have been proposed: the magnetic breakdown among crossing semiclassical orbits (7) and the

dHvA families				
I	II	III	IV	V
240(0.05)	580(0.14)	1380(0.20)	2280	3720
475	810(0.19)	1620(0.03)	2500	3960
710(0.14)	1050(0.22)	1850(<0.01)	2750(0.03)	4290
940	1280	2100	2980(0.13)	
	1520	2320	3220	
		2560		

Table 1 : dHvA frequencies (in Teslas) and in brackets the corresponding effective mass (in free electron mass units) of second stage bromine intercalation compound

non-linear coupling between two related orbits in a magnetic field (8). These two mechanisms give different kinds of combination which mainly differ by the corresponding effective masses. Unfortunately, it was not possible to determine all the effective masses because the peaks are very close to each other. For this reason, For this reason, the effective masses given in table 1 are uncertain and should be taken with caution.

- 4 - A MODEL FOR THE ELECTRONIC STRUCTURE

In order to interpret these experimental results we have elaborated an electronic structure model which is an improvement of the one we have previously proposed for the second-stage nitric acid GAC (1).

1 The 2-D unit zone

The charge transfer which occurs between graphite layers and the intercalated molecules results in charged intercalate layers in which the carriers are localized on the intercalate molecules. We assume that this layer of ions prevents the free carriers left in the graphite zones from moving across the layer. The graphite zones which lie on the two sides of this layer are therefore uncoupled. The consequence is the existence of the 2-D unit zone, made of s adjacent carbon layers, with which one can describe all the electronic properties of the GAC.

2 The band structure

Blinowski et al. (9) have computed the band structure of the GACs by using the same assumption. In a stage two compound, the Fermi surface obtained is made of two concentric circles centered at the U points of the graphite Brillouin zone (BZ). It is possible to add to this calculation, which only takes into account the resonance integrals γ^0 and γ^1 (between the nearest carbon atoms lying in and out of the layers), the γ^3 and γ^4 parameters (between the second neighbour carbon atoms lying out of the layer plane) which introduce the trigonal warping. In this case, the Fermi circles are warped, which is quite important in the interpretation that we propose subsequently.

3 The effect of the in-plane order

The intercalate layer is considered in this calculation as a homogeneous distribution of electrons, the only requirement is the electrical neutrality of the GAC. Actually, the molecules of the intercalate layers are ordered as we have previously seen. This results in a new periodic potential which is seen by the graphite holes. We take this into account in a pertubation method applied to the Schroedinger equation of the free carriers. New energy gaps appear which modify the Fermi surface. In the case of C14Br which is commensurable with the graphite cell, we can fold the the graphite BZ into that of the GAC which is smaller.

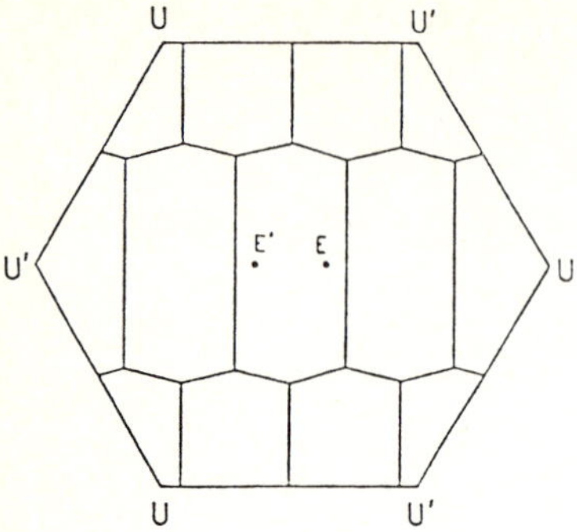

Figure 3 : The folding of the graphite Brillouin zone into the GAC Brillouin zone.

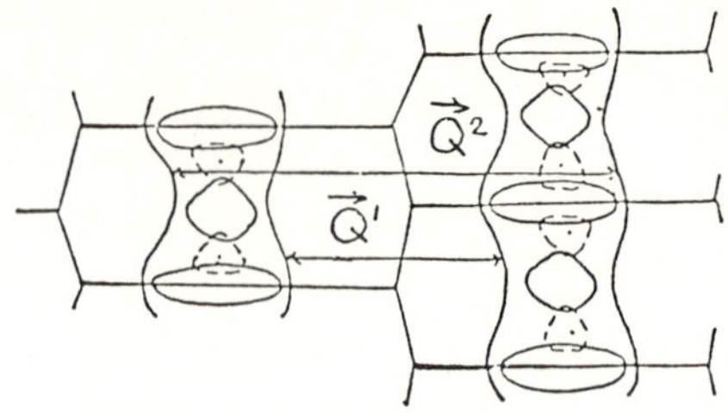

Figure 4 : The Fermi surface of C14Br. Q1 and Q2 are the possible nesting vectors of the charge density waves.

Figure 3 shows this folding. The U points of graphite BZ are folded into two points E and E' of the new BZ which are not high symmetry points. Figure 4 shows the schematic Fermi surface of the compound. One can see in this figure the two warped circles centered on the E and E' points of the BZ. In the approximation of the model, the two kinds of circles are uncoupled by the folding (respectively symmetric and antisymmetric wave functions (9)). The coupling of the orbits is possible across the gaps which appear between two crossing orbits.

4 Charge density waves

In such a 2-D Fermi surface, many mechanisms can be foreseen in order to explain the existence of charge density waves, as for example the nesting (10) and the coupling between two saddle points (11).

The trigonal warping of the circles forbids the existence of a nesting between two parts of the Fermi surface distant of EE', which is the most obvious coupling.The only two vectors which can couple dHvA orbits are drawn on figure 4 (Q1 and Q2). These vectors just correspond to those of the structural modulations observed in the X-rays studies (satellites and diffuse streaks).

The coupling of these two experiments is not a definite proof of the existence of charge density waves in this compound, although many features suggest it, even if some problems remain. Though we understand the overall topology of the Fermi surface, the quantitative explanation of the frequencies remains to be elaborated. The explanation of the observed diffraction patterns is also still uncomplete.

1 Ch. Simon, F. Batallan, I. Rosenman, H. Fuzellier, Phys. Rev. B23,2836 (1981)
2 G. Furdin, private comm.
3 F. Batallan, I. Rosenman, Ch. Simon, H. Lauter, G. Furdin, presented at the 16th international conference on semiconductors physics, Montpellier (France), to be published by North Holland (1982)
4 J. Moret, private comm.
5 W.T. Eeles, J.A. Turnbull, Proc. Roy. Soc. A283, 179 (1965)
6 D.D.L. Chung, M.S. and G. Dresselhaus, Mater. Sci. Engng. 31, 107 (1977)
7 L.M. Falicov, M.J. Zukermann, Phys. Rev. B160, 372 (1967)
8 P. Schlottmann, L.M. Falicov, Phys. Rev. Lett. 38, 855 (1977)
9 J. Blinowski, N. H. Hau, C. Rigaux, J.P. Vieren, R. Letoulec, G. Furdin, A. Herold, J. Melin, J. de Phys. (Paris) 41, 47 (1980)
10 J.A. Wilson, A.D. Joffe, Adv. Phys. 18, 193 (1969)
11 T.M. Rice, G.K. Scott, Phys. Rev. Lett. 35, 120 (1975)

MAGNETO-OPTICAL EXPERIMENTS ON A THIN InAs LAYER CONFINED BETWEEN GaSb IN A PARALLEL AND PERPENDICULAR MAGNETIC FIELD

J.C. Maan

Max-Planck-Institut für Festkörperforschung, Hochfeld-Magnetlabor, 166X, F-38042 Grenoble Cédex, France

ABSTRACT

Far infrared magneto-optical experiments on a sample consisting of a thin (1000 Å) InAs layer confined between GaSb show for the magnetic field perpendicular to the plane of the layers transitions between hole subbands localized in the GaSb and electron subbands localized in the InAs. This localization is induced by potentials wells created at the interface through the band bending due to the transfer of electrons from the GaSb to the InAs. In addition a double cyclotron resonance of the electrons and a single for the holes is observed. These observations are explained with a simple model for the subband structure at the InAs-GaSb interface.
A magnetic field parallel to the interface leads to a change in the subband structure and hybrid electric and magnetic subbands are formed. Transitions between these subbands are observed experimentally and explained theoretically. In this context the influence of a parallel magnetic field on the subband structure in two dimensional systems in general is discussed.

I. INTRODUCTION

In two dimensional electron systems, like those at semiconductor surfaces and interfaces or on liquid helium, the electrons are mobile in a plane and confined in the perpendicular direction. This confinement leads to the formation of two dimensional subbands in the direction of the confining potential. A magnetic field perpendicular to the plane splits each subband into Landau levels and quantum effects such as Cyclotron Resonance (CR) or the Shubnikov-de Haas effect (SdH) can be observed. From this type of experiments information about the subband structure can be obtained. As the magnetic field is tilted away from the perpendicular direction, the presence of a magnetic field component parallel to the plane mixes the electric and magnetic subbands and a, in general complicated, subband structure results. The importance of a parallel magnetic field component on the subband structure depends on the distance between subbands compared to the Landau level splitting, or otherwise expressed, on the cyclotron orbit radius compared to the extent of the wavefunction perpendicular to the plane.

In this paper we will discuss the results of magnetooptical experiments on a thin InAs layer (1000 Å) confined between GaSb. From measurements with the magnetic field B perpendicular to the plane of the layers the subband structure has been determined and explained with a simplified model. It is found that the spatial extend of the wavefunction perpendicular to the layer together with the InAs-CB non-parabolicity gives rise to different masses of different subbands. Measurements with B tilted away from the perpendicular configuration show the formation of hybrid magnetic and electric subbands. Transitions between these subbands have been observed experimentally and explained qualitatively. Both these results show the importance of the finite width of the potential because deviations from a strict two dimensional behaviour are observed. It turns out that the system studied here is very interesting in this respect.

In the InAs-GaSb system the top of the bulk GaSb Valence Band (VB) is 0.15 eV above the bottom of the InAs Conduction Band (CB). Electrons transfer from GaSb to InAs, the number of transferred electrons of course being equal to the number of holes left behind. This process causes band bending which leads to the formation of quantum wells at each GaSb-InAs interface and quantum subbands in these wells [1,2].

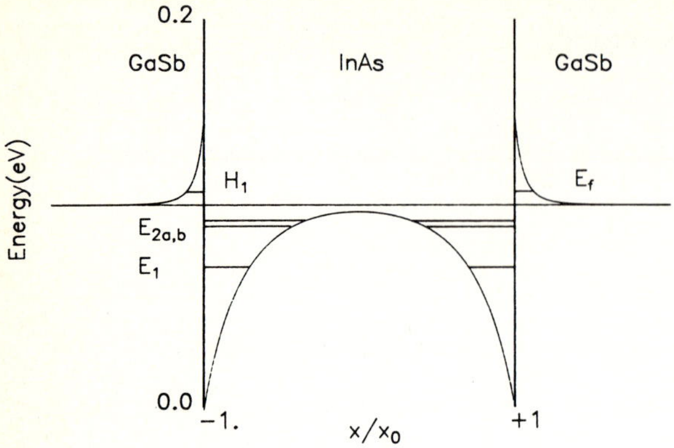

Fig. 1.

Energy level structure of the double GaSb/InAs/GaSb heterostructure. x_o is the characteristic length of the InAs band bending and is half the layer thickness (∽500 A).

The resulting energy diagram is illustrated in fig. 1. and is obtained from calculations described later. Throughout this paper the interface is parallel to the yz plane and therefore the potential is in the x direction. The magnetic field is in the xz plane, therefore B parallel is B//z and B perpendicular is B//x. As is clear from fig. 1. several electron subbands are involved (E_1, $E_{2a,b}$) and one hole subband (H_1) at each interface. Therefore two dimensional gases of both type of carriers are formed at the interface.

II. EXPERIMENT WITH B PERPENDICULAR TO THE PLANE

Magnetooptical experiments have been performed on a sample consisting of 20 alternatly grown layers of InAs and GaSb, each 1000 Å thick. (2) The transmission of far infrared radiation (FIR) at fixed frequency is observed as a function of the magnetic field perpendicular to the plane of the layers (B//x). Radiation energies between 1 meV and 21 meV have been used with magnetic fields up to 20 Tesla. The spectra (fig. 2) show several minima for each photon energy, which are, of course, absorption maxima. The position of these minima are plotted vs. photon energy and magnetic field in fig. 3. These results have been analyzed and interpreted as being typical for a single InAs-GaSb heterojunction. The sets of transitions which extrapolate to zero energy for zero magnetic field are interpreted as CR of the electrons (the two curves with the steep slope) and of the holes (the flatter curve). The split electron CR is due to the different masses of different subbands, which is well known for space charge layers in non-parabolic materials (3). The other transitions are attributed to transitions between Landau levels from the heavy hole subband and those originating from the two electron subbands. Since the electron subbands are below the hole subbands at zero magnetic field these latter set of transitions extrapolate to a negative energy.

A simplified model has been used to calculate the subband structure of the InAs-GaSb heterojunction and to analyze these experimental results. The calculation is performed in several steps.

(i) The band bending due to the transferred charge is calculated using the Thomas Fermi approximation:

$$E_b = E_b^o \ / \ (1 + \frac{x}{x_o})^4 \tag{1}$$

where E_b is the energy, measured from the InAs cq. GaSb band edge, and E_b^o is the band bending at x = 0, at the interface. x_o is a parameter and is given by:

$$x_o = \left(\frac{(4\pi)^2\, 225\hbar^6\, \pi^2\, \varepsilon^2}{8m^{*3}\, e^4\, E_b^o} \right)^{1/4} \qquad (2)$$

where m^* is the effective mass, and ε the dielectric constant.
(ii) The subbands are calculated in a similar way as originally done by

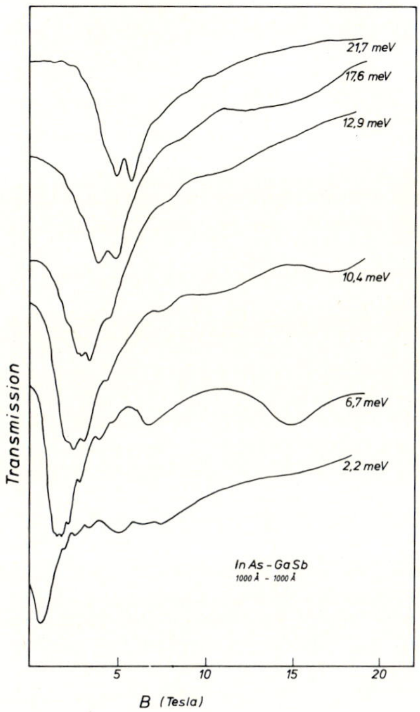

Fig. 2.

Measured far infrared radiation transmission as a function of the magnetic field for different FIR energies. The Maximum absorption observed in typically 30 % of the zero field transmission (after ref. 2).

Fig. 3.

Position of the observed resonant transitions (fig. 2.). The drawn lines represent the theoretical calculations as described in the text (after ref. 2).

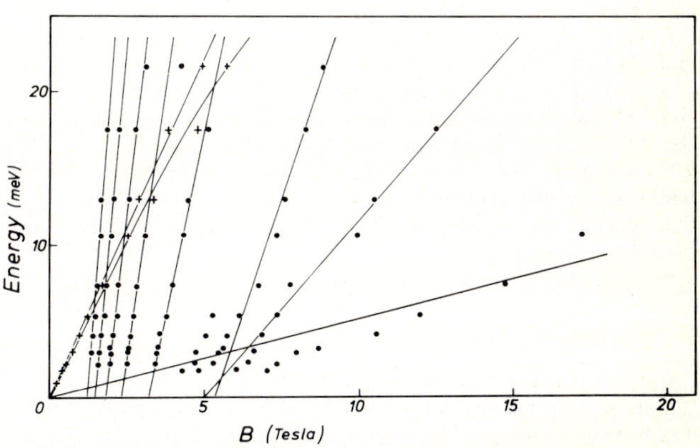

Antcliffe et al. (4) using the Bohr-Sommerfeld semiclassical quantization rule:

$$\int_{o}^{x_t} k_x dx = \pi \ (n - 1/4) \qquad (3)$$

Here n (1,2,3..) is an integer to designate the different subbands, k_x is the wavevector perpendicular to the interface which is obtained from the dispersion relations of the InAs CB and GaSb-VB respectively and x_t is the classical turning point. For InAs the non-parabolic dispersion relation,

$$\frac{\hbar^2}{2m^*} (k_x^2 + k_{/\!/}^2) = E \ (1 + \frac{E}{E_G}) \qquad (4)$$

is used, where $k_{/\!/}$ is the wavevector in the plane. The advantages of this calculational procedure lies in the fact that the InAs-CB non-parabolicity can be incorporated in a simple manner.

(iii) The number of electrons and holes is obtained by integrating over the density of states. Since the number of electrons transferred from the GaSb is of course equal to the number of holes left behind the Fermi energy is fixed by this condition. Therefore the calculation is repeated for several trial values of E_b^o for the InAs and 0.15 eV-E_b^o for the GaSb until charge neutrality is obtained. For $k_{/\!/} = 0$ and using $m_e^* = 0.023\ m_o$, $m_h^* = 0.33\ m_o$; $\varepsilon = 14.5$ and 15.7 $E_G = 410$ meV and 810 meV for InAs and GaSb respectively and the overlap between the GaSb-VB and the InAs-CB 150 meV, these calculations yield three electron subbands $E_1 = 72$ meV, $E_2 = 95$ meV, $E_3 = 102.5$ meV, a single hole band $H_1 = 107.5$ meV and the Fermi energy $E_F = 104$ meV for the subbands at the single InAs-GaSb heterojunction. The characteristic length x_o is calculated to be $\backsim$500 Å for InAs and $\backsim$100 Å for GaSb. These values justify the assumption that the sample can be considered as a 1000 Å thick InAs layer confined between GaSb, with the subbands at the interface essentially the same as those of the single heterojunctions.

This method to calculate the subband structure using the Bohr-Sommerfeld quantization can easily be used to calculate the subband and the Landau levels for B perpendicular. In this case k_x is calculated using the dispersion relation (4) but replacing $\frac{\hbar^2 k_{/\!/}^2}{2m^*}$ by $\frac{\hbar eB}{m^*}$ (N + 1/2) where N is the Landau level number. The thus calculated Landau levels for the electron and the hole subbands can be used to compare the experiments with the theory and the result is shown as the drawn lines in fig. 3. As can be seen a good agreement is obtained for the observed intersubband transitions. Also the split CR at higher radiation energies is adequately described by the model. From this agreement the mass at the subband edge is determined as $m_{E_1}^* = 0.029\ m_o$ and $m_{E_2}^* = 0.026\ m_o$ for E_1 and E_2 respectively. That electron subbands at higher energy have a lower mass, contrary to what one would expect from non-parabolicity, is a direct consequence of the band bending. The mass observed for a subband will depend on the average position of the electron in the well as is determined by its wavefunction. The lower subbands are on the average closer to the interface and therefore deeper inside the InAs band and have a higher mass. The Bohr-Sommerfeld calculation incorporates this effect through the integration over k_x. Therefore the observation of the split CR peak is a direct consequence of the finite width of the potential.

Two experimental observations have not yet been discussed; the broad but clearly observable transition at low energies and high magnetic fields (fig. 2.) and the weak transitions around 5 T at low radiation energies. The former transition has been attributed tentatively to CR of the heavy hole subband. The mass deduced from the line drawn in fig. 3. corresponds to a mass of 0.26 m which is quite close to that expected for the heavy holes in (100) GaSb. On the other hand it is known that the valence band subband structure is in general very complicated (5) and that one cannot in a simple manner decouple the k dependence of the heavy and the light holes. Furthermore the band structure is complicated by the fact that hole and electron subbands cross in the plane and an anticrossing is expected to occur (6,7).

III. TILTED FIELD AND B PARALLEL EXPERIMENTS

The influence of a parallel magnetic field on the subbands in two dimensional electronic systems has been considered by several authors. (For a review of several magnetic field effects see for instance F. Koch (8)).

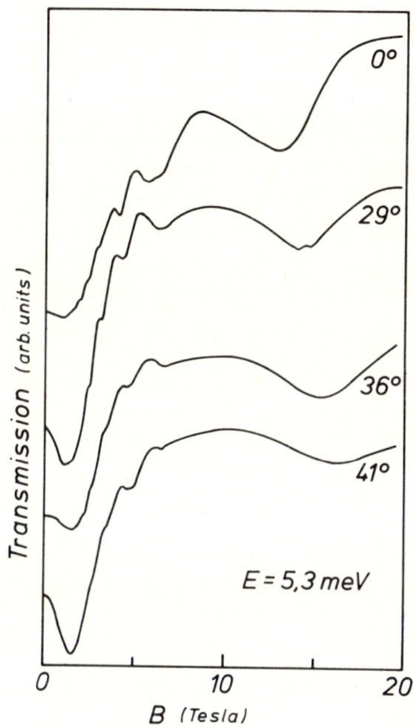

Fig. 4.

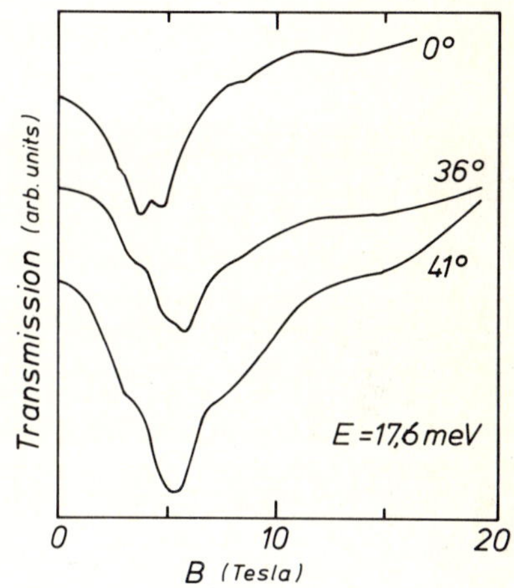

Fig. 5 .

Fig. 4.
Measured transmission at hν=5.3 meV as a function of the magnetic field for different angles between the field and the normal of the layers.

Fig. 5.
The same as fig. 4 but at hν=17.6 meV.

Fig. 6.
The same as fig. 4 but at hν=21.7 meV.

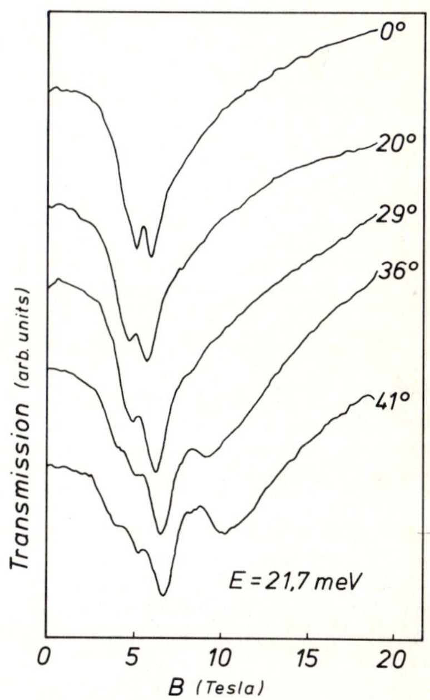

Fig. 6.

For n-Si inversion layers it has been shown that the SdH effect and CR follow a strictly 2D behaviour, where only the perpendicular component of the magnetic field is important. On the other hand a broadening and a shift and the appearance of satellite peaks at $\pm\hbar\omega_c$ have been observed in the intersubband resonance (10,11). Furthermore a 3D behaviour of the CR in PbTe inversion layers (12) and quantum oscillations for B parallel in InAs accumulation layers have (13,14) been observed. Here we will discuss the effect of a parallel magnetic field on the subbands in the GaSb-InAs-GaSb double heterostructure and we will show how mixed electric and magnetic subbands are formed which, at high magnetic field, evolve into a Landau level like band structure. Preliminary results of this work are published elsewhere (15)

In figs.4,5 and 6 the evolution of the spectra with the angle between the plane of the layers and the magnetic field as B is tilted away from the perpendicular configuration for three different frequencies are shown and in fig. 7. the experimental results for B parallel. The angular dependence at $h\nu$=21.7 meV shows that the CR doublet becomes less well defined; one peak shifts slightly to higher magnetic fields as the angle increases while the other becomes weaker and hardly moves with the angle (fig. 6.). At about 20° a new transition appears at 10 T and the intersubband transitions broaden and become more pronounced. For B parallel (fig.7.) at the same radiation energy a single sharp resonance is observed preceded by some weak structure and with a shoulder on the high field side. A similar behaviour is observed for

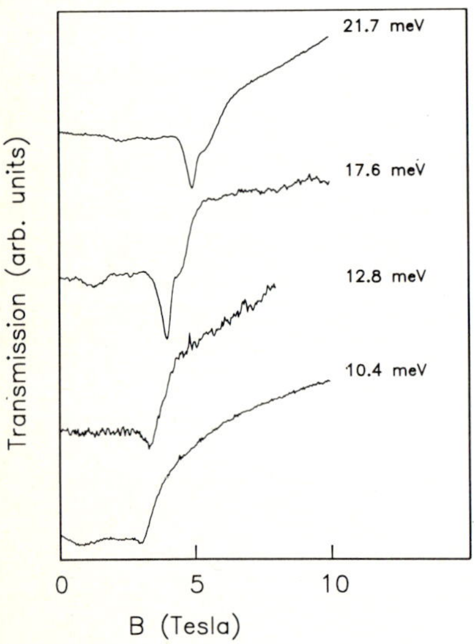

Fig. 7.
Measured transmission for different radiation energies as a function of the magnetic field parallel to the layers.

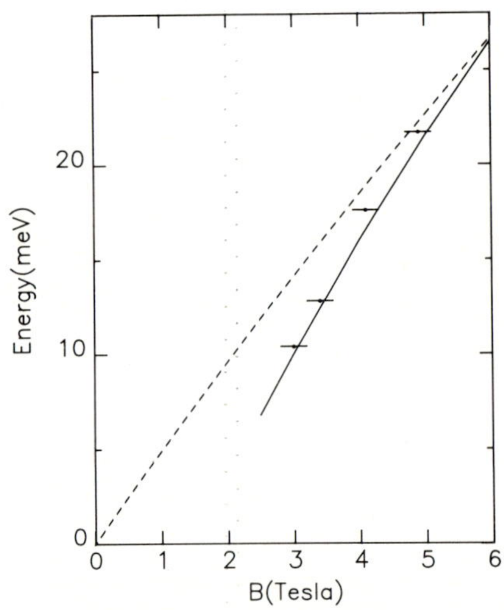

Fig. 8.
Position of the resonance (Fig. 7.) (data points), calculated CR position for bulk InAs (dashed line) and calculated hybrid subband transitions.

$h\nu$=17.6 meV, and 5.3 meV but no new transition appears (fig. 4,5). The only transition which follows a simple angular dependence in the sense that only the perpendicular component of the magnetic field determines the resonance position is the broad transition at high magnetic field (fig. 4.) which has been attributed to CR of the holes. For B//z a sharp resonance is observed which shifts to lower magnetic fields, weakens and broadens for lower radiation energies. For radiation energies below 7 meV the transmission becomes monotonically increasing with the magnetic field. A plot of the position of the sharp resonance as a function of the magnetic field (fig. 8.) shows that the observed transition does not extrapolate to zero magnetic field for zero energy, and tends to approach the calculated CR energy in InAs bulk (the dotted lined fig. 8.) for higher radiation energies. These results clearly show the influence of a parallel magnetic field on the subband structure of the system under investigation.

To discuss this effect we write the Hamiltonian for the system with a magnetic field in the xz plane, and adopt a gauge $\underline{A}=(0,xB_z-zB_x,0)$:

$$H = \Big[-\frac{\hbar^2}{2m^*}\Big(\frac{\partial^2}{\partial x^2}+\frac{\partial^2}{\partial z^2}\Big) + \frac{\hbar^2 k_y^2}{2m^*} + e(xB_z-zB_x)\frac{\hbar}{m^*}k_y + \frac{e^2(xB_z-zB_x)^2}{2m^*} \qquad (5)$$

$$+ V(x)\Big] e^{ik_y y}\,\psi(x,z) = E e^{ik_y y}\,\psi(x,z)$$

for B//x (perpendicular configuration) H=H(z)+H(x); H(z) leading to Landau levels and H(x) leading to electric subbands. Neglecting at present non-parabolicity the zero magnetic field case gives simply $E = \frac{\hbar^2(k_y^2+k_z^2)}{2m^*} + E_n$ with E_n the bottom of the subband n. A magnetic field induces the usual degeneracy for k_y and leads to states $E_N=E_n+(N+1/2)\frac{\hbar eB}{m^*}$. For arbitrary angles of the magnetic field (B_z and $B_x\neq 0$) the Hamiltonian (5) cannot be separated in two parts which depend on one coordinate only and therefore in principle magnetic and electric subbands cannot be distinguished anymore. For B//z (5) reduces to:

$$\Big[-\frac{\hbar^2}{2m^*}\frac{\partial^2}{\partial x'^2} + \frac{e^2B^2x'^2}{2m^*} + V(x)\Big] e^{ik_y y} e^{ik_z z}\psi(x) = \Big(E + \frac{\hbar^2 k_z^2}{2m^*}\Big) e^{ik_y y} e^{ik_z z}\psi(x)$$

$$\text{with } x' = x + \frac{\hbar k_y}{eB} \qquad (6)$$

In this case both the magnetic and electric term depend on x. Therefore the eigenvalues of (6) will depend on the magnetic field and the potential together and a set of hybrid magnetic-electric subbands are formed. In general the eigenvalues of (6) depend on k_y. The physical reason for this effect is the destruction of the translational symmetry due to the potential term V(x) in eq. 6. In this context it is good to remember that $\hbar k_y/eB$ defines the cyclotron orbit center in the x direction, which is the direction of the potential. For small magnetic fields that is when $l=eB/\hbar$, the cyclotron orbit radius is large compared to the width of the potential well, or similarly, as long as $\hbar\omega_c$ is small compared to the subband spacing, a solution of eq. 6 has been given in a perturbation approach (9,16) as:

$$E = E_n + \frac{\hbar^2 k_z^2}{2m^*} + \frac{1}{2m^*}[\hbar k_y + eB\langle x\rangle_n]^2 + \frac{1}{2m^*}e^2B^2[\langle x^2\rangle_n - \langle x\rangle_n^2] \qquad (7)$$

The terms $\langle x\rangle_n$ and $\langle x^2\rangle_n$ are the expectation values of the respective operators of the subband n and essentially parameterize the extent in x of the wavefunction. Eq. 7 leads to a parabolic dispersion relation in k_y shifted an amount $-\frac{eB}{\hbar}\langle x\rangle_n$ from $k_y=0$ and associated with a rise in the minimum energy by a diamagnetic shift (10):

$$\frac{1}{2m^*}e^2B^2(\langle x^2\rangle_n - \langle x\rangle_n^2) \qquad (8)$$

In the present case this perturbation result cannot be used but instead eq. (6) has to be solved exactly. In this calculation non-parabolicity is neglected, the potential V(x) is described in eq. (1), and E_F is used from the selfconsistent calculation as described before. However, now the potential at both interfaces is taken into account corresponding to the double heterostructure in the 1000 Å InAs layer. The result for B=0 is shown in fig. 1. Electron subbands are found at E_1=72 meV and $E_{2a,b}$ at 93,96 meV. These values are very close to those found for the single heterojunction using the Bohr-Sommerfeld method. The first subband is degenerate in the sense that the coupling of the two wells through the barrier of the band bending is not sufficiently strong to observe a splitting. The second electron subband is split due to the interaction between the the wells

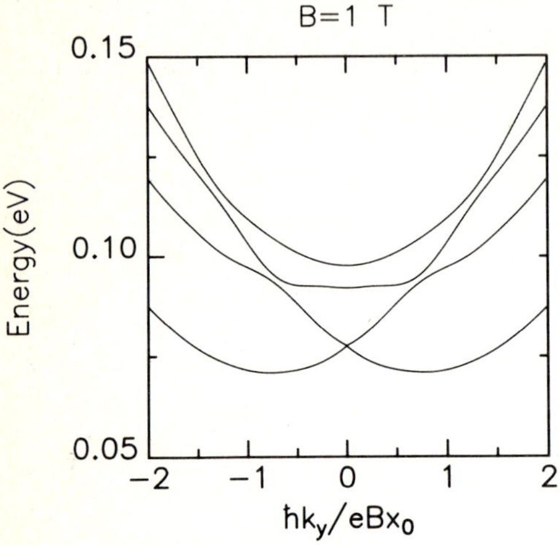

Fig. 9.

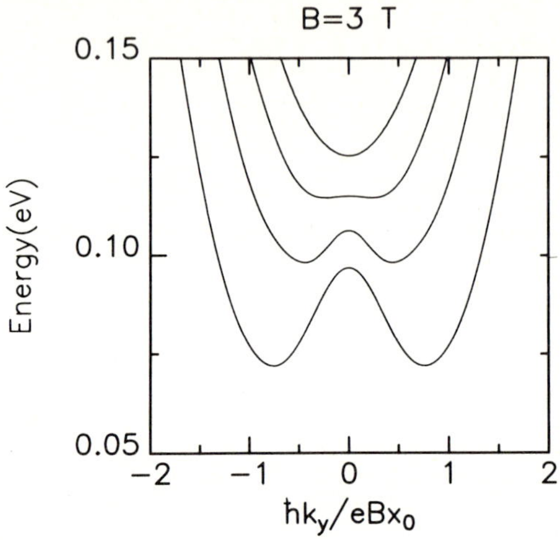

Fig. 10.

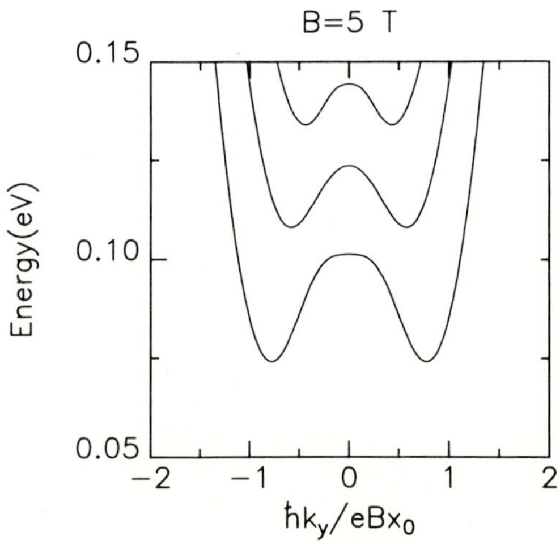

Fig. 11.

Fig. 9.
The dispersion of the electron subbands in a parallel magnetic field at 1 T. $\hbar k_y/eBx_o$ is the CR orbit center coordinate normalized to the characteristic length x_o, taken to be zero in the center of the InAs layer.

Fig. 10.
The same as fig. 9 but at B=3 T.

Fig. 11.
The same as fig. 9 but at B=5 T.

(tunneling) and no third subband is found. The result of the calculation for B//z is shown in fig. 9,10,11 for 1 T, 3 T and 5 T respectively. The horizontal axis in the figures $\hbar k_y/eBx_0$ represents the position of the cyclotron resonance orbit center normalized to the characteristic length of the well x_o which is half the InAs layer thickness. As with the perturbation calculation one observes that the k_y degeneracy is lifted and that a parabolic dispersion relation results for each electric subband. The minimum energy is shifted to different k_y values for different subbands and experiences a small diamagnetic energy shift. However, since we deal with two potential wells for the two sides of the InAs layer, there are two sets of parabolas formed which show an anticrossing behaviour. This effect leads to a very complex level structure at low magnetic fields. At higher magnetic fields the degeneracy for the lowest subband at $k_y=0$ is lifted and a more regular level structure is observed. This degeneracy is lifted because the magnetic field breaks the symmetry for the motion along the interface. Electrons moving in the same direction in the two wells are bent towards the interface in one well and towards the center of the layer in

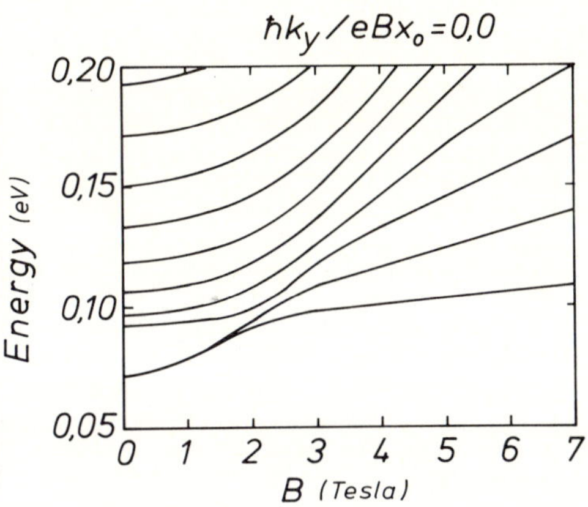

Fig. 12.
Evolution of the hybrid subbands with the magnetic field parallel to the layers at $\hbar k_y/eBx_o=0$; CR orbit center in the middle of the InAs layer. For B>4T a linear dependence is obtained with a difference of $\hbar\omega_c$ between the levels.

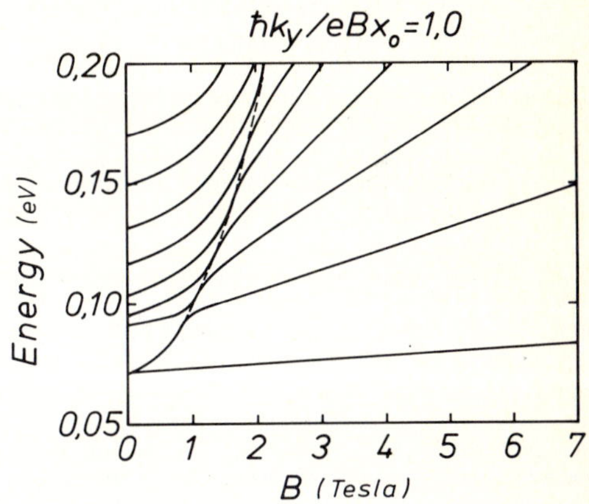

Fig. 13.
Idem as Fig. 12. with $\hbar k_y/eBx_o= 1$; CR orbit center coordinate at the interface. For B>4T a linear magnetic field dependence is observed with a difference of $2\hbar\omega_c$ between the levels.

the other well. It is interesting to follow the magnetic field dependence of the subbands at $\hbar k_y/eBx_o=0$ (cyclotron orbit in the center of the layer) and at $\hbar k_y/eBx_o=1$. (orbit center at the interface) This dependence is shown in fig. 12. and 13. It can be seen that at $\hbar k_y/eBx_o=1$ the degeneracy of the lowest subband is immediately lifted by the magnetic field; at $\hbar k_y/eBx_o=0$ this takes place at about 3 T. This behaviour can be understood in terms of the electronic motion along the interface as discussed before. One further sees that at high magnetic field the magnetic field dependence becomes linear and that the distance between subbands approaches $\hbar\omega_c$ for $k_y=0$ (fig.12.) and $2\hbar\omega_c$ for $\hbar k_y/eBx_o=1$. (fig. 13.) That is, at high magnetic fields bulk properties dominate in the center and magnetic surface states are formed at the interface. Magnetic surface states are localized states at the surface through the magnetic potential. These states for B parallel to a metal or semimetal surface do not depend on an additional electric potential inside the material but only on the existence of a high potential step at the surface which reflects the electrons pushed to the surface by the magnetic field back into the material, thus creating a potential well at the surface caused by the magnetic field only. (17,18) In the calculation the potential of the GaSb is considered to be very large for the electrons in the InAs, therefore at high magnetic fields these magnetic surface states will appear. It can easily be understood that $2\hbar\omega_c$ is found for the level spacing at the interface. Setting V(x) to zero inside the material, eq.(6) reduces to an asymmetric harmonic oscillator (for $x>0$; $V(x)=0$ and $x<0$, $V(x)=\infty$). Only uneven solutions of the harmonic oscillator, which have wavefunctions that are zero for x=0 are found, resulting in $2\hbar\omega_c$ for the level spacing.

Returning to the experimental data, the results for B//z can be compared with the calculations. Assuming a selection rule $\Delta k_y=0$, sharp transitions can be expected between hybrid subbands (figs.10,11) adjacent to the Fermi energy which are parallel. Therefore the position of the sharp transitions in fig. 7. is compared with the calculated energy difference at $k_y=0$, and the result is shown as the line drawn in fig. 8. The agreement between the calculated and measured results is quite satisfying and indicates that clearly hybrid subband transitions and no simple CR is observed. It is interesting to note that for high magnetic fields parallel to the interface no difference exists anymore between transitions between electric subbands or Landau levels. The reason that for lower radiation energies and magnetic fields the transition broadens and that below 7 meV no transitions are observed anymore can also be understood. It is clear from fig. 9,10,11 that at low magnetic fields there are no parallel bands in the neighbourhood of E_F and therefore no sharp transitions can be expected.

It should be noted in this context that the observed transition is in fact sharper than would be expected, since the bands are not at all equally spaced for $k_y\neq 0$. However, the observed linewidth will be the result of all possible transitions weighted with their transition probabilities which need not to be equal for different k_y values. The weak structure at fields below the sharp peak in fig. 7. may very well be due to transitions for $k_y\neq 0$. These transitions, as expected should be broad, have a stronger magnetic field dependence, and occur at lower fields. The broad shoulder at the high field side cannot be explained on this basis and it is speculated that it may arise from a combined hybrid intersubband spin transition.

It is interesting to note the differences between optical absorption measurements as presented here and magneto-transport measurements. In the former case the transition intensity is determined by the density of states of different subbands at the same k_y value, due to the $\Delta k_y=0$ selection rule. Therefore the k_z dependence of the eigenvalues (eq. 6) is unimportant. For magneto-quantum oscillations the density of states at the Fermi energy is determining and in addition a different scattering rate for different subbands, or a change in the carrier density has to be invoked to explain the oscillations (13,19). This can explain why SdH measurements on the same sample, for low magnetic fields and small angles show that essentially B is important, contrary to the optical result (19).

The effect of a parallel magnetic field on the hole subbands in the GaSb may be expected to be neglegibly small since the well width is five times smaller (100 Å) and the effective mass is fourteen times heavier. Therefore the diamagnetic shift (eq. 8) is expected to be almost two orders of magnitude smaller. Bearing this fact in mind the angular dependence of the spectra can qualitatively be understood. The intersubband transitions are broadened as the tilt angle increases and are not at all observed for B//z, as would be expected from the strong dispersion in k_y. Other features of the angular dependence can also be explained. The weakening of the CR doublet due to CR of different subbands and the complete disappearance of this doublet for B//z is explained by the fact that for B//z no distinction can be made between different electric subbands, but that instead only transitions between two hybrid subbands can be observed. The new transition observed at larger angles for $h\nu=21.7$ meV and not observed for lower energies, has its origin associated probably with an E_1-$E_{2a,b}$ transition. Their energies differ by 21-24 meV at B=0 and vary differently with B to account for the angular dependence. As is clear this transition can only be seen for $h\nu>21$ meV.

Finally it may be noted that the angular dependence adds further evidence to the interpretation of the hole CR for $B_\perp$ in fact this transition is the only transition which follows a strict cos ψ dependence which would be expected due to the small effect of $B_{/\!/}$ on the hole subbands.

IV. CONCLUSIONS

Magnetooptical measurements for B perpendicular to a layer of 1000 Å InAs confined between GaSb, show the coexistence of a hole gas and an electron gas at the GaSb-InAs interface. In addition,the importance of the spatial extend of the potential in the InAs is shown from the observation of a different mass of different subbands. These observations are quantitatively explained with a simple model for the subband structure of a InAs-GaSb heterojunction. Measurements in a tilted field configuration and for B parallel to the interface, clearly show the formation of mixed electric and magnetic subbands. This subband structure is calculated numerically for B parallel and a quantitative agreement between experiments and theory is found.

ACKNOWLEDGEMENTS

It is my pleasure to thank L.L. Chang and L. Esaki of IBM, T.J. Watson Research Center and Y. Guldner, J.P. Vieren, P. Voisin and M. Voos from the Ecole Normale Supérieure for the collaboration in this work. In addition I would like to thank Prof. G. Landwehr for his interest in this work and also to acknowledge pleasant and numerous discussions with T. Englert and Ch. Uihlein at the HML in Grenoble. Finally I would like to thank W. Schmid for his excellent technical assistance and fine art work.

REFERENCES

1. L.L. Chang, N.J. Kawai, E.E. Mendez, C.A. Chang, L. Esaki, Appl. Phys. Lett. 38, 30,(1981)
2. Y. Guldner, J.P. Vieren, P. Voisin, M. Voos, J.C. Maan, L.L. Chang, L. Esaki, Solid State Commun. 41, 755,(1982)
3. A. Daerr, J.P. Kotthaus, J.F. Koch, Solid State Commun. 17, 455 (1975)
4. G.A. Antcliffe, R.T. Bate, R.A. Reynolds, Proc. Conf. On the Physics of Semimetals and Narrow Gap Semiconductors, Dallas 1970, Ed. D.L. Carter and R.T. Bate, p. 499, Pergamon, New York (1971)
5. E. Bangert, K. von Klitzing, G. Landwehr, Proc. 12th Int. Conf. on the Physics of Semiconductors, Stuttgart, 1974, ed. B.G. Teubner, Stuttgart (1974)
6. G.H. Dohler, Surface Sci. 98, 108,(1980)
7. M. Alterelli, this conference
8. F. Koch in 'Physics in high magnetic fields', Solid State Sciences 24, ed. S. Chikazumi, N. Miura, Springer, Berlin, (1981)
9. F. Stern, W.E. Howard, Phys. Rev. 163, 816, (1967)
10. W. Beinvogl, A. Kamgar, J.F. Koch, Phys. Rev. B14, 1610, (1976)
11. W. Beinvogl, J.F. Koch, Phys. Rev. Lett. 40, 1736, (1978)
12. H. Schaber, R.E. Doezema, Phys. Rev. B20, 5257, (1979)
13. R.E. Doezema, M. Nealon, S. Whitmore, Phys. Rev. Lett. 45, 1593, (1980)
14. M. Nealon, S. Whitmore, R.R. Bourassa, R.E. Doezema, Surface Sci. 113, 282, (1982).
15. J.C. Maan, Ch. Uihlein, L.L. Chang, L. Esaki, Solid State Commun., to be published
16. T. Ando, J. Phys. Soc. Japn. 39, 411, (1975)
17. M. Wanner, R.E. Doezema, U. Strom, Phys. Rev. B12, 2883, (1975)
18. R.E. Prange, T.W. Nee, Phys. Rev. 168, 779, (1968)
19. L.L. Chang, E.E. Mendez, N.J. Kawai, L. Esaki, Surface Sci. 113, 306, (1982)

ELECTRONIC STRUCTURE OF SEMICONDUCTOR SUPERLATTICES

M. Altarelli

Max-Planck-Institut für Festkörperforschung,
7000 Stuttgart 80, Fed. Rep. of Germany

A new method for the calculation of the electronic structure of semiconductor superlattices in the envelope-function approximation is presented. The method is particularly suited to a description of type-II superlattices (e.g. InAs - GaSb). The band-edges of the constituent materials are realistically described, and appropriate boundary conditions are derived from the envelope functions. To determine the energy levels, a variational principle is obtained, which automatically produces correctly matched solutions and reduces the numerical work to a standard matrix diagonalization. Results are presented for InAs-GaSb superlattices. They reveal novel aspects of the band-crossing transition which occurs with increasing superlattice period and which produces a zero-gap or very-narrow-gap semiconductor state, rather than a true semimetal.

Introduction

There has recently been much interest in the electronic properties of artificial semiconductor superlattices,[1] and in particular of the so-called type-II systems such as InAs-GaSb[1]. Type-I (e.g. GaAs-GaAlAs) and type-II superlattices differ in the relative band line-up of the constituent systems. In the latter case, the valence band top of one material (GaSb) lies higher in energy than the conduction band bottom of the other one (InAs). It is then to be expected that there will be transfer of charge across the interfaces, and that novel electronic properties will result. In particular, experiments seem to indicate a semiconductor-semimetal transition[2] occurring as the superlattice period is increased, in agreement with theoretical arguments based on the expected crossing of the lowest conduction-like subband with the highest heavy-hole-like subband[3]. The semimetallic state, in an intrinsic sample, would be characterized by interfacing two-dimensional electron and hole gases, a novel and stimulating situation. Another interesting result of magneto-transport[4] and magneto-optical[5] experiments is a relationship between effective mass and energy for carriers in the conduction-like subbands which differs appreciably from the non-parabolic law of the InAs conduction band.

For all these reasons, it is interesting to investigate the electronic structure of these systems theoretically. There are some stringent requirements on a band-structure calculation for these systems. In fact, (a) the narrow-gap nature of the constituent materials, and the strong mixing expected at the interfaces between the valence bands of one material and the conduction band of the other, make a many-band treatment necessary; (b) the investigation of the transverse dynamics ($\vec{k}$ parallel to the superlattice layers) requires a realistic description of the constituent band-edges, and in particular of the degenerate valence band; (c) the occurrence of charge transfer across the interfaces requires a method suitable to be extended to self-consistent calculations. On the other hand, the large size of the superlattice unit cell makes a full microscopic calculation intractable, and the only attempt in this direction[6] was restricted to very small periods, far from the interesting region of the band crossing (periods $\gtrsim$ 160Å in InAs-GaSb)[7]. Among the semiempirical methods, tight-binding[3,7] is especially simple but is unsuitable for the requirement (c) (as well as for (b), if oversimplified schemes are used). The envelope-function approximation appears to be the most convenient framework for this problem as pointed out by White and Sham[8] and by Bastard[9].

In the envelope-function approximation the wave-function in each layer is written as the product of a slowly varying envelope function and of the $\vec{k}$ = 0 Bloch function of the corresponding semiconductor. The envelope functions satisfy effective-mass-like equations and are joined at each interface with appropriate boundary conditions, which must be derived from the smoothness conditions for the total wave-function. In the method presented here, the boundary conditions are embodied in a variational principle, which automatically produces correctly matched wave-functions. The numerical work is thus greatly simplified and reduces to the usual diagonalization of matrices. It is then possible to handle the problem with a realistic description of the constituent bands, including fully coupled heavy-hole, light-hole and conduction bands, and to calculate the subband dispersion in the direction parallel to the layers. Band coupling in this direction affects the nature of the semiconductor-semimetal transition, which, for intrinsic samples, reduces actually to a transition to a zero-gap (or very narrow-gap) semiconductor.

Equations for the Envelope Functions

Consider an A-B superlattice composed of layers of thickness a and b of semiconductor A and B, respectively, i.e. with period d = a + b. Let x denote the direction of the superlattice axis, and y,z two arbitrary directions in the x = 0 plane, parallel to the layers. We describe each constituent with a three-band model, including the s-like conduction band, and the j = 3/2 valence band manifold, split into a heavy-hole and a light-hole band at $\vec{k} \neq 0$. The split-off band j = 1/2 is thus neglected, a reasonable approximation in compounds like InAs or GaSb, in which the spin-orbit splitting is as large as the fundamental gap. Following Marques and Sham[10], we choose z as the direction of quantization of angular momenta, and consider $\vec{k}$-vectors in the (x,y) plane. With this choice our six (including spin) basis states decouple into two 3 x 3 sets ("spin up" and "spin down") given by

$$u_1 = |s\uparrow\rangle$$

$$u_2 = |j=3/2,\ j_z=3/2\rangle = (x + iy)\uparrow/\sqrt{2}$$

$$u_3 = |3/2,\ -1/2\rangle = -(x-iy)\uparrow/\sqrt{6} - \sqrt{2/3}\,z\downarrow$$

and

$u_4 = |s\downarrow>$

$u_5 = |3/2,\ 1/2> = -\sqrt{2/3}z\uparrow + (x + iy)\downarrow/\sqrt{6}$

$u_6 = |3/2,\ -3/2> = -(x - iy)\downarrow/\sqrt{2}$

respectively. For k in the x,y plane, the band dispersion for "spin up" is given by the eigenvalues of (atomic units are used throughout the paper):

$$H^o_{jj'} = \begin{bmatrix} E_c + \frac{1}{2}k^2 & ik_+P/\sqrt{2} & -ik_-P/\sqrt{6} \\ -ik_-P/\sqrt{2} & E_v - \frac{\gamma_1+\bar{\gamma}}{2}k^2 & \sqrt{3}\bar{\gamma}k_-^2/2 \\ ik_+P/\sqrt{6} & \sqrt{3}\bar{\gamma}k_+^2/2 & E_v - \frac{\gamma_1-\bar{\gamma}}{2}k^2 \end{bmatrix} \qquad (1)$$

where $E_c(E_v)$ is the conduction (valence) band edge energy, $k_\pm = k_x \pm ik_y$, P is Kane's matrix element[11] $iP = <s|p_x|x>$, and γ_1 and $\bar{\gamma}$ are Luttinger-type parameters[12] describing the effect of all other bands not included in the j = 1, ..., 6 set on the valence band dispersion, in the spherical ($\gamma_2 = \gamma_3 = \bar{\gamma}$) approximation. A similar matrix can be written for the "spin-down" states. The parameters P, γ_1, $\bar{\gamma}$ are determined, for each material, by magneto-optical experiments[13,1

Following the standard effective-mass theory for multiple bands, as used in magneto-optics[15] or in the theory of acceptor impurities[16], we now write the wave-function of a "spin up" electron in a given layer in terms of envelope functions $F_j(x)$, as:

$$\psi(\vec{r}) = \sum_{j=1,3} e^{ik_y y} F_j(x)u_j(\vec{r}) + \sum_{j=1,3} \sum_{i>6} \frac{i\vec{k}\cdot\vec{P}_{ij}}{E_j - E_i} e^{ik_y y} F_j(x)u_i(\vec{r}) \qquad (2$$

where the second term represents the first-order $\vec{k}\cdot\vec{p}$ correction due to all higher bands not included in the j = 1, ..., 6 set, and can, for most purposes, be neglected. In Eq.(2) and in the following, $\vec{k}$ is an operator acting on envelope functions, given by $\vec{k} = (-i\,\partial/\partial x, k_y, 0)$. k_y is of course a good quantum number in the problem.

To write the effective-mass equation for the envelope functions F_j in Eq. (2), we first rewrite the $H^o_{jj'}$ in Eq. (1) in a self-explanatory compact notation:

$$H^o_{jj'} = \sum_{\alpha,\beta=x,y} D^{\alpha\beta}_{jj'}k_\alpha k_\beta + \sum_{\alpha=x,y} P^\alpha_{jj'}k_\alpha + E_{j'}\,\delta_{jj'} \qquad (3)$$

$(j=1,2,3)$

and follow the standard effective-mass procedure to obtain:

$$\sum_{j'=1,3} H_{jj'}\,F_{j'} = E\,F_j \qquad (j=1,2,3) \qquad (4)$$

where

$$H_{jj'} = \sum_{\alpha,\beta=x,y} D^{\alpha\beta}_{jj'} k_\alpha k_\beta + \sum_{\alpha=x,y} P^{\alpha}_{jj'} k_\alpha + (E_{j'} + V(x))\delta_{jj'} \qquad (5)$$

In Eq. (4-5) V(x) is the potential arising from charge redistribution in the system. In the "flat-band" approximation V(x) is neglected. Eq. (5) can be written in each layer, and the boundary conditions at the interfaces must be written. Given the structural similarity between III - V compounds, it is reasonable to assume[8] that the u_j basis functions (j = 1, ...,6) are the same in both materials. It is also easy to see that the u_j functions are linearly independent in the unit cell of the interface plane, so that the continuity condition $\psi_A(x=0^-) = \psi_B(x=0^+)$ for the total wave-functions Eq. (2) implies continuity of the envelopes

$$F_j(0^-) = F_j(0^+) \qquad (j=1,2,3) \qquad (6)$$

The other three boundary conditions are obtained by observing that, in a stationary state, the probability current in the x direction, J_x, must be the same across all planes perpendicular to x, in particular across the interfaces. In ordinary quantum mechanics this requirement yields the continuity of the derivative of the wave-function. If we require $<\psi|J_x(x)|\psi>$ = constant, where

$$<\psi|J_x(x)|\psi> = \frac{1}{2i}\left(\psi^* \frac{\partial}{\partial x}\psi - \psi\frac{\partial}{\partial x}\psi^*\right) \qquad (7)$$

and $\psi(\vec{r})$ is the total wave-function Eq. (2), it follows that also the integral of Eq. (7) across a unit cell of the constituent semiconductors must be constant. Taking this integral we find after some lengthy but straightforward algebra:

$$<\psi|\bar{J}_x|\psi> = \sum_{j,j'} F^*_j \left[(D^{xy}_{jj'} + D^{yx}_{jj'})k_y + P^x_{jj'}\right] F_{j'}$$
$$-i \sum_{j,j'} \left[F^*_j D^{xx}_{jj'} \frac{\partial}{\partial x} F_{j'} - F_{j'} D^{xx}_{jj'} \frac{\partial}{\partial x} F^*_j\right] \qquad (8)$$

where the bar over the current indicates this unit cell averaging. Notice that the sums are over j,j' = 1,2,3 only and that all sums on i > 6, coming from the second term in Eq. (2), result in the presence of the $D_{jj'}$ matrices. If we now write that the value of Eq. (8) must be the same on both sides of an interface and exploit Eq. (6), the condition that $P_{jj'}$ is the same in both materials (because the u_j's are the same), and the hermiticity of D and P, we finally get

$$\sum_j F^*_j \left[\sum_{j'}(D^{xy}_{jj'} + D^{yx}_{jj'})k_y F_{j'} - 2i\, D^{xx}_{jj'}\, \partial/\partial x\, F_{j'}\right] \text{ continuous} \qquad (9)$$

To obtain 3 independent boundary conditions, we make the reasonable assumption, in analogy with usual quantum mechanics, that each of the coefficients of the continuous F^*_j (see Eq. (6)) is separately continuous. We thus obtain the set of matching conditions for j = 1,2,3

$$F_j \quad \text{continuous} \qquad (10a)$$

$$\sum_{j'} \left[(D^{xy}_{jj'} + D^{yx}_{jj'}) k_y - 2i\, D^{xx}_{jj'}\, \partial/\partial x \right] F_{j'} \quad \text{continuous} \tag{10b}$$

Eq. (10) reduce, in particular cases, to the boundary conditions previously derived by other authors. In fact in the absence of k^2-terms in the band Hamiltonian (vanishing D matrices) one obtains the simple continuity of the envelope functions (as obtained in Ref. 8 for a 2 x 2 Kane model); and the coefficient of $\partial/\partial x$ reflects the jump in effective-mass that must be included in the derivative continuity condition as pointed out in Ref. 9.

Entirely similar results are of course found for the spin-down states with j=4,5,6

Variational Principle for the Envelope Functions

We must now find an efficient scheme to solve Eq. (4), with boundary conditions (10) at the interfaces, and the Bloch condition in the x-direction. Straightforward numerical approaches, involving a solution of Eq. (4) in A and B layers, and a search of the energies at which Eq.(10) can be satisfied are lengthy and impractical. Especially in view of the need to generate eigenfunctions efficiently and in a manageable form for a self-consistency loop calculation, it would be much better to express the problem in a secular form, i.e. to calculate all eigenvalues and eigenvectors simultaneously by a matrix diagonalization in an energy independent set of basis functions. This goal can be achieved by reformulating the problem in a variational form, by extension of a procedure developed in the band theory of metals[17,18].

The superlattice cell consists of one A and one B layer. Let us write a wave-function in this cell as a two-component object:

$$\Psi = (\psi_A, \psi_B) \tag{11}$$

When A or B band parameters are plugged into Eq. (5), Hamiltonians H_A acting on ψ_A, and H_B, acting on ψ_B, are obtained. Thus the Hamiltonian is a diagonal operator in this 2-component basis. It is possible to introduce an "interface operator" F, which is not diagonal and has the following property:

If $\Psi = (\psi_A, \psi_B)$ is an eigenstate of H, $H\Psi = E\Psi$, and satisfies the proper boundary conditions at the A - B interface, then Ψ is a stationary function for the functional $\langle\Psi|H + F|\Psi\rangle$, against arbitrary variations $\delta\Psi$, and furthermore $\langle\Psi|H + F|\Psi\rangle = E\langle\Psi|\Psi\rangle$.

It is important to stress that the stationarity condition is for arbitrary variations $\delta\Psi$, that is also for trial functions $\Psi + \delta\Psi$ that do not satisfy the boundary condition. This is in contrast to the usual variational principle of quantum mechanics, which is valid only for trial functions $\Psi + \delta\Psi$ that satisfy the correct boundary conditions (H is otherwise non-hermitian and hermiticity, in the form $\langle\delta\Psi|H|\Psi\rangle = \langle\Psi|H|\delta\Psi\rangle^*$ is crucial in the proof of the variational theorem).

It is then clear that if one knows the interface operator F, one can take two suitable sets of N basis functions in A and B, into which to expand ψ_A and ψ_B, respectively, and then proceed to build and diagonalize the 2N x 2N matrix representing H + F at a particular $\vec{k}$-point of the superlattice Brillouin zone.

To determine the operator F, or more precisely the functional $\langle\Psi|F|\Psi\rangle$, we generalize the procedure of Schlosser and Marcus[17] for the multicomponent nature of the wavefunctions, and for the boundary conditions of our problem. This task turns out to be straightforward if one first rewrites Schlosser and Marcus' functional in terms of matrix elements of current operators. Then one has simply to replace the ordinary current (Eq.(7)) with the $\bar{J}_x$ operator of Eq.(8). One then obtains immediately

$$<\Psi|F_{SM}|\Psi> = \frac{i}{2}\,(<\psi_A|\bar{J}_x(0)|\psi_A> - <\psi_B|\bar{J}_x(0)|\psi_B>$$

$$+ <\psi_B|\bar{J}_x(0)|\psi_A> - <\psi_A|\bar{J}_x(0)|\psi_B>) \tag{13}$$

It is on the other hand obvious that the functional F is not unique. This can be exploited to select one which gives better convergence and stability in the computation. We choose therefore a functional obtained by adding one term to the Schlosser-Marcus one:

$$<\Psi|F|\Psi> = <\Psi|F_{SM}|\Psi> + \lambda\,|\psi_A(0^-) - \psi_B(0^+)|^2 \tag{14}$$

where λ is an arbitrary real number. The additional term has the effect of making the eigenvalues and eigenvectors converge much faster with respect to the number of basis functions.

Results for InAs - GaSb Superlattices

In this section results for InAs-GaSb superlattices obtained with the method described are reported. For the expansion of each component of the wave-function in each constituent a set of 12 basis functions (6 sines and 6 cosines) is used. The system is clearly overcomplete, but a compromise must be struck and some overcompleteness must be allowed for in order to guarantee accuracy at the edges of the layers, where boundary conditions are enforced, and where phenomena of the Gibbs type may otherwise occur. Therefore the size of the H + F matrices is 72 x 72. The band parameters used in the calculation are shown in Table I and are derived from the literature. No special effort to optimize the parameters was made; it was only verified that none of the important features of the results are sensitive functions of them.

The calculations presented here are performed in the non self-consistent flat-band approximation ($V(x) = 0$ in Eq. (5)). A more complete account including fully self-consistent results is in preparation. However, the

Table I

Band parameters used in the calculations (from Ref. 8, 13, and 14). The values of γ_1 and $\bar{\gamma}$ need not be equal for the two materials, but they turn out to be within experimental error.

	InAs		GaSb
E_c (eV)	0.		0.985
E_v (eV)	- 0.420		0.175
P (at.u.)		0.719	
γ_1		3.7	
$\bar{\gamma}$		0.6	

main results obtained here on the nature of the semiconductor-semimetal transition are probably not affected by inclusion of the self-consistent potential.

In Fig. 1, the band structure of an a = b = 60Å superlattice is shown. The period (d = 120Å) is small, and the system is therefore a semiconductor. Notice however the strong non-parabolicity of the subbands

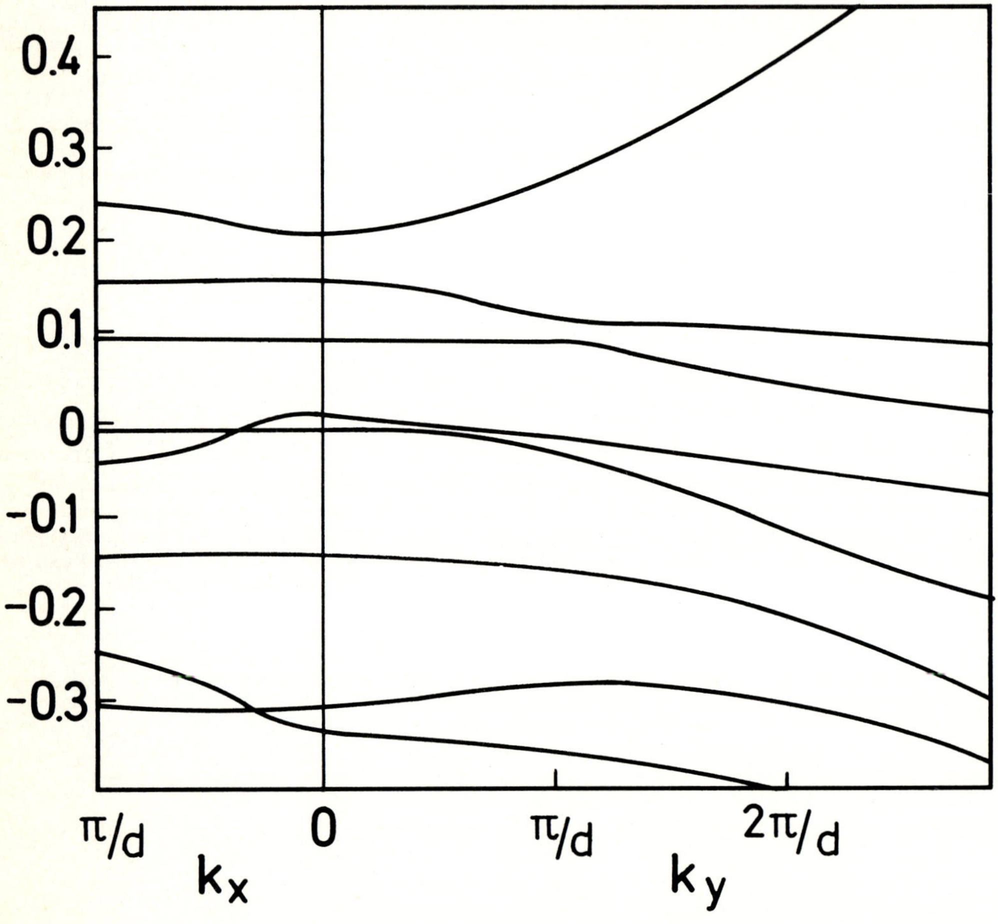

Fig. 1 Band structure of a d = 2 x 60Å InAs-GaSb superlattice. The energy scale is in eV, the zero is the bottom of the InAs conduction band. The x direction is the superlattice axis, y is along the layers.

in the direction parallel to the layers. For $k_y \neq 0$ all three bands of our model are coupled and there is obvious repulsion (non-crossing behavior) of the levels, in particular when heavy- and light-hole-like subbands are very close in energy. On the other hand it is readily seen from Eq. (1) that for $k_y = 0$ an exact decoupling into a heavy hole band and a Kane-coupled set of light-hole + conduction bands is always possible, irrespective of the value of k_x. This implies that light- and heavy-hole bands are distinguishable along the k_x direction, and that crossing may occur.

To explore what happens when the lowest conduction band drops below the highest hole band, we consider a system with a = b = 80Å (fig. 2). Along the k_x direction, the lowest conduction subband actually crosses the highest heavy-hole subband. This crossing is however forbidden as soon as $k_y \neq 0$, and in fact strong hybridization and repulsion are clearly noticeable along the k_y axis. Therefore, the bands really cross in one point only, along k_x, so that the Fermi surface consists only of this point, and the system is, strictly speaking, a zero-gap semiconductor and

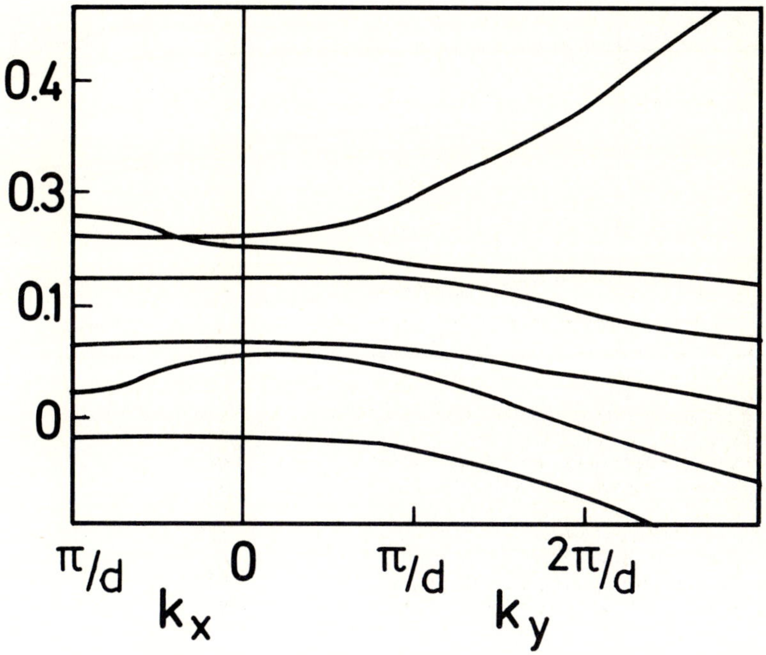

Fig. 2 Band structure of a d = 2 x 80Å superlattice

not a semimetal. When the layer thickness is further increased, the conduction subband continues to drop until it is completely below the heavy-hole one everywhere along k_x. Fig. 3 and 4 show the d = 2 x 90Å and d = 2 x 100Å systems. In this range of thicknesses one has really very narrow-gap semiconductors, with gaps up to ∿20 meV. The Fermi-level is indeed entirely in a gap, as indicated in the figures. A true semimetal is occuring only for much larger thicknesses, via a mechanism incipient for d = 200Å (Fig. 4), the formation of minima at $k_y \neq 0$ in the lowest empty subband and of maxima for the highest occupied one.

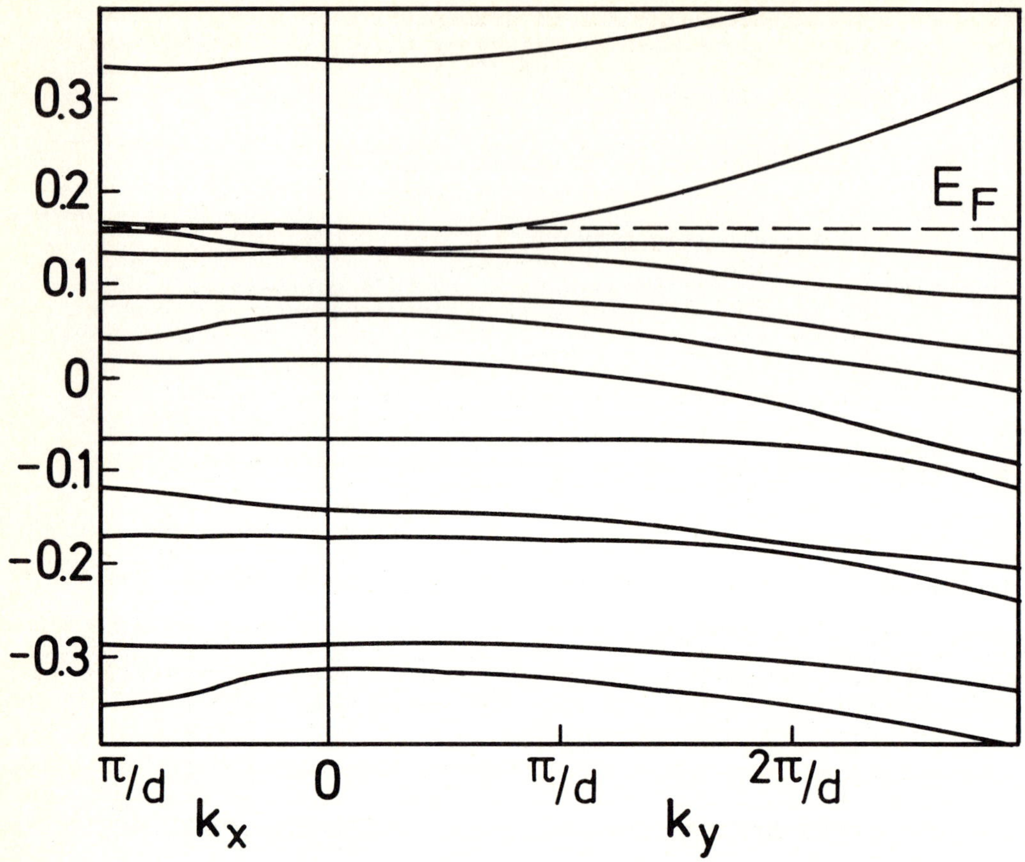

Fig. 3 Band structure of a d = 2 x 90Å superlattice. The dashed line denotes the Fermi energy E_F

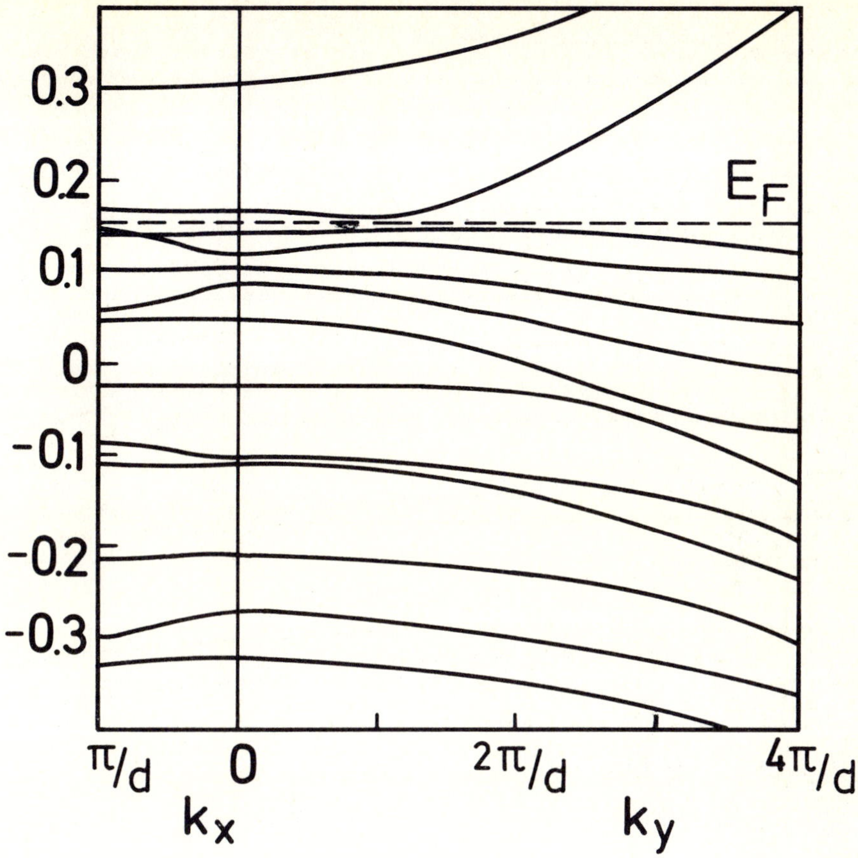

Fig. 4 Band structure of a d = 2 x 100Å superlattice. The dashed line denotes the Fermi energy.

A few comments to the results discussed here are in order. The first is that the accurate description of the k_y dispersion presented here has revealed novel aspects of the band structure. These have been unnoticed so far mostly because of the oversimplifications adapted in other calculations: the neglect of spin-orbit splitting in tight-binding calculations, for instance, or the neglect of the heavy- and light-hole coupling at $k_y \neq 0$. The anticrossing behavior of the subbands could however have been inferred from the results of Döhler[19] and Bastard[9].

The second comment concerns the observability of the effects predicted here. Detailed experimental investigations gave results apparently reconcilable with the traditional model of the semiconductor to semimetal transition. We must however remark that the inversion of the order of the heavy-hole and conduction-like subbands at Γ is present in our results as well, and that the small gaps at $k_y \neq 0$ are quite difficult to observe, particularly when the samples are non-intrinsic and the Fermi level is indeed outside the gap. It appears that the preparation of highly intrinsic samples is a precondition for observation of these effects.

Finally, it is important to realize that the accuracy in the prediction of the exact value of the gap for a given thickness is limited by several factors. The input parameters (band discontinuities, band parameters) are not known with high accuracy; there was no attempt to take into account the effect of charge transfer self-consistently, although the non-metallic nature of the band structure is expected to reduce it considerably. It should however be emphasized that the existence and the order of magnitude of the gap are certainly not affected by these factors, and that in particular charge transfer effects would increase the magnitude of the gap rather than decrease it.

It is a pleasure to acknowledge discussions with G. Döhler, W. Harrison, J.C. Maan, P. Vogl, and in particular with O.K. Andersen.

References:

1. See e.g. L.L. Chang, J.Phys.Soc.Japan 49, Suppl.A, 997 (1980); P. Voisin et al, ib. 1005 (1980)

2. G.A. Sai-Halasz, L.L. Chang, J.M. Welter, C.-A. Chang and L. Esaki, Solid State Commum. 27, 935 (1978)

3. G.A. Sai-Halasz, L. Esaki and W. Harrison, Phys.Rev. B 18, 2812 (1978)

4. H. Sakaki, L.L. Chang, G.A. Sai-Halasz, C.-A. Chang and L. Esaki, Solid State Commun. 26, 589 (1978)

5. J.C. Maan, M. Altarelli, H. Sigg, P. Wyder, L.L. Chang and L. Esaki, Surface Sci. 113, 347 (1982)

6. J. Ihm, P.K. Lam and M.L. Cohen, Phys.Rev. B 20, 4120 (1978)

7. A. Madhukar, N.V. Dandekar and R.N. Nucho, J. Vac. Sci. Tech. 16, 1507 (1979)

8. S.R. White and L.J. Sham, Phys. Rev. Lett. 47, 879 (1981)

9. G. Bastard, Phys. Rev. B 24, 5693 (1981)

10. G.E. Marques and L.J. Sham, Surface Sci. 113, 131 (1982)

11. E.O. Kane, in "Semiconductors and Semimetals", R.K. Willardson and A.C. Beer, eds., (Adademic, New York, 1966) Vol.1, p. 75

12. J.M. Luttinger, Phys. Rev. 102, 1030 (1956)

13. C.R. Pidgeon, D.L. Mitchell and R.N. Brown, Phys. Rev. 154, 737 (1967)

14. K. Suzuki and N. Miura, J. Phys. Soc. Japan 39, 148 (1975)

15. L.M. Roth, B. Lax and S. Zwerdling, Phys. Rev. 114, 90 (1959)

16. See e.g. M. Altarelli and F. Bassani, in "Handbook of Semiconductors",Vol. 1, W. Paul ed., (North-Holland, Amsterdam, 1982) p. 269 and references therein

17. H. Schlosser and P.M. Marcus, Phys. Rev. 131, 2529 (1963), especially Appendix 1

18. T. Loucks, "The Augmented Plane Wave Method", (Benjamin, New York, 1967)

19. G. Döhler, Surface Sci. 98, 108 (1980)

INELASTIC LIGHT SCATTERING BY TWO DIMENSIONAL ELECTRON SYSTEMS IN HIGH MAGNETIC FIELDS

by

J. M. Worlock,[a] A. C. Maciel,[b,c] C. H. Perry,[b,c]

Z. J. Tien,[d] R. L. Aggarwal,[b,e] A. C. Gossard,[d] and W. Wiegmann[d]

Two-dimensional electrons in semiconductors have been popular for some years, not only because of their importance in devices, but also because of their intrinsic interest as systems of low dimensionality.[1] Originally, nearly all attention went to inversion and accumulation layers in silicon and a few compound semiconductors. But as molecular beam epitaxy (MBE) has matured as a growth process,[2] electrons in semiconductor heterostructures have been more widely studied.

Light scattering spectroscopy has emerged as one of the more sensitive and fruitful techniques for the study of two dimensional electrons, at least in the systems involving quantum wells in GaAs-AlGaAs heterostructures.[3] A fortuitous correspondence between the spin-orbit split-off band gap of GaAs and the output of standard argon-laser-pumped dye lasers has simplified the observation of resonant inelastic light scattering involving very small numbers of electrons. Additionally, the polarization selection rules inherent in the light scattering process have allowed the observation separately of single particle and collective transitions. Pinczuk and his coworkers have been particularly vigorous in using this technique for the study of many interesting phenomena.[4]

More recently, we have begun a systematic study of these same electrons in GaAs-AlGaAs superlattices, using the technique of inelastic light scattering spectroscopy, but adding the new component of large magnetic fields.

This paper will be divided into three parts. The first part will review work done over the past two years which already appears in accessible publications.[5,6] In this section we establish the observation of two kinds of electronic excitation: one is a single particle transition between Landau levels which can be called cyclotron resonance, and the other is a collective mode in which cyclotron motion is coupled with a plasma wave, and which is called the magnetoplasma (MP) mode. We are interested in these modes at least in part in the hope that they will give hints of electron ordering or crystallization. In the second part, we deal with the magnetoplasma mode, and show how its behavior is related to the dispersion of the plasma frequency of a layered two-dimensional electron system. The

a Bell Telephone Laboratories, Holmdel, NJ 07733
b National Magnet Laboratories, MIT, Cambridge, MA 02139
c Physics Department, Northeastern University, Boston, MA 02115
d Bell Telephone Laboratories, Murray Hill, NJ 07974
e Physics Department, MIT, Camridge, MA 02139

third part is devoted to a discussion of some recent measurements of the resonance of the scattering cross section, and its change with magnetic field. These show directly the nature of the scattering process: in spite of this, we still do not know what are the matrix elements for this forbidden process. It is our guess that these experiments are trying to teach us something about the nonlocal properties of plasmas in two dimensions.

The experiments we discuss here have all been done on multiple quantum well or superlattice samples of GaAs-AlGaAs, grown by MBE, and modulation doped with Si to give electron areal densities in the range of 2 to 7 $\times$ 10^{11} cm^{-2}. The separations between layers are such that there are no effects of electron tunneling, and the interactions between layers are taken to be electrostatic. The thicknesses of the confining quantum wells are on the order of two hundred angstroms, which for GaAs gives intersubband spacings of several tens of meV. When a magnetic field is imposed parallel to the superlattice axis, the in-plane motion is quantized into Landau levels, and the resulting energy level structure for electrons is:

$$E_n = E_n + (\ell+1)\hbar\omega_c \text{ with } \omega_c = eH/m^*c$$

where E_n is the energy of the bottom of the nth subband, largely unaffected by the magnetic field H, ℓ is the index of the Landau level, and ω_c is the cyclotron frequency. For GaAs the cyclotron frequency corresponds to an energy of 1.7 meV per Tesla. The Landau radius, R_ℓ, a measure of the in-plane extent of an electron's wavefunction, is given by

$$R_\ell^2 = (\ell+1)\hbar c/eH \ ,$$

and is typically on the order of 100A. In the discussion to follow, we will discuss further the change of Landau radius with field.

One more important quantity is the degeneracy parameter. We cannot pack as many electrons as we wish into a given Landau level, but as the field increases, because of the shrinkage of Landau radius, we can crowd the electrons closer and closer. The areal density of a full Landau level (with no spin degeneracy) is

$$N_s(full) = eH/hc, \text{ or } (2\pi R_o^2)^{-1}.$$

so the degeneracy ν, defined as the number of filled Landau and spin levels, is given by

$$\nu = N_s/N_s(full) \ .$$

The quantum limit is defined by the absence of degeneracy, $\nu < 1$, and this occurs at 10 T for an areal density of 2.4 $\times$ 10^{11} cm^{-2}. ν is the parameter to watch in searching for electron ordering or crystallization,[7] and is of fundamental importance also in the phenomena of quantized Hall effect and zero resistance anomalies.[8]

Figure 1 shows a "typical" inelastic light scattering spectrum of a modulation doped superlattice, taken at low temperature. In addition to the peak labeled $\hbar\omega_c$, we see the intersubband transition E_{10}. This latter transition has been extensively studied by Pinczuk and his coworkers[4] as well as by Abstreiter and his colleagues.[9] The spectra we will discuss were all taken in the nearly backscattering configuration shown in the inset.

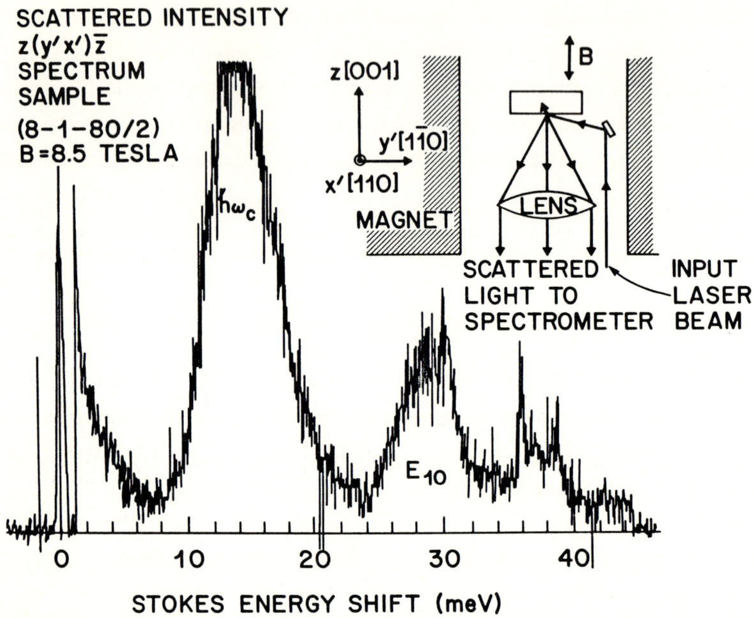

Figure 1 Inelastic light scattering spectrum of typical modulation-doped GaAs-AlGaAs superlattice in a magnetic field of 8.5 tesla. Peak labeled $\hbar\omega_c$ is cyclotron resonance; that labeled E_{10} is intersubband scattering. Inset shows scattering geometry.

With crossed incident and scattered polarizations, $(y'x')$, we have come to expect single particle excitations, and so we take the peak labeled $\hbar\omega_c$ to be single particle $\Delta\ell = 1$ Landau level Raman scattering or cyclotron resonance. This hypothesis is strengthened by the plot shown in Figure 2, indicating the linear dependence of the energy on the magnetic field: the slope gives an electron effective mass $m^*/m_e = 0.068 \pm 0.002$, which agrees well with accepted effective mass for GaAs. We should note here that the intensities of these features are dependent on the frequency of the incident laser light as well as on the magnetic field strength: these are strongly resonant processes. Additionally, although the Landau transition $\Delta\ell = 1$ is formally forbidden in light scattering, it gives spectra of strength comparable to the intersubband transition.

Spectra with parallel optical polarizations are generally interpreted in terms of collective excitations. The peak E_{10}, for example, has two companions, $E-$ and $E+$, collectively shifted and stiffened by resonant screening and interaction with the LO phonon.[10] Similarly, we find that the cyclotron

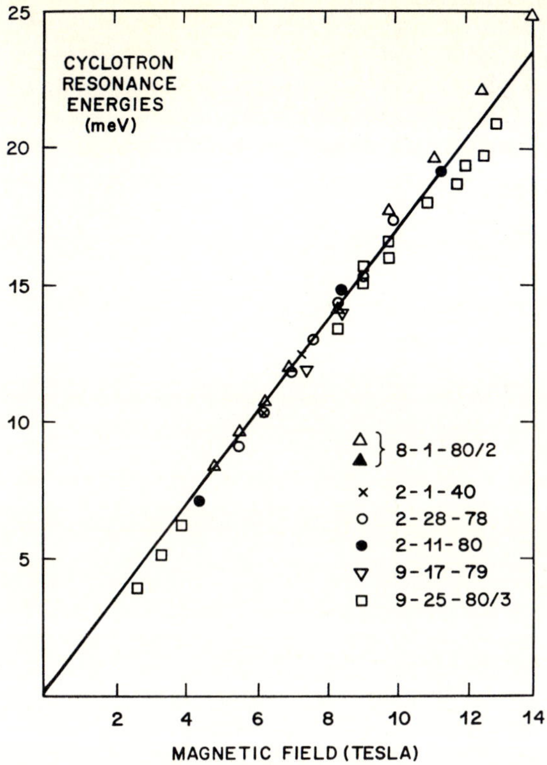

Figure 2 Plot of cyclotron resonance energy shift versus magnetic field, for many samples. The straight line corresponds to $m^*/m_e = 0.068$

resonance has a companion at a somewhat higher frequency, and tuning similarly with magnetic field. Figure 3 shows a plot of peak frequencies for cyclotron resonance and its companion, which we wish to identify as the magnetoplasma mode, as a function of field. Note that for this particular sample, we can follow the MP mode to zero field, where it becomes simply the plasma mode. The plasma modes of layered two dimensional electron systems have been observed also in the light scattering experiments of Olego and Pinczuk,[11] who studied MBE superlattice samples very similar to ours. They were able to observe the dispersion of the plasmon by varying the scattering angle, and to verify the predicted dispersion relation:[12]

$$\omega_p^2(k_{\parallel},k_{\perp}) = \left(\omega_p^{3D}\right)^2 \times \frac{dk_{\parallel}}{2} \; \frac{\sinh dk_{\parallel}}{\cosh dk_{\parallel} - \cos dk_{\perp}}$$

where d is the period of the layers, $k_{\parallel}$ and $k_{\perp}$ are respectively the in-plane and perpendicular wave vectors, and ω_p^{3D} is the equivalent three dimensional plasma frequency, given by:

$$\left(\omega_p^3 D\right)^2 = \frac{4\pi n_s e^2}{\epsilon m^* d}$$

where ϵ is the dielectric constant of the medium in which the layered electrons are embedded. Olego and Pinczuk covered a range of wavevectors: $0.25 < dk_{\parallel} < 0.72$, with $dk_{\perp} \sim 5$, using light nearly resonant with the E_o gap of GaAs, at about 7850A.

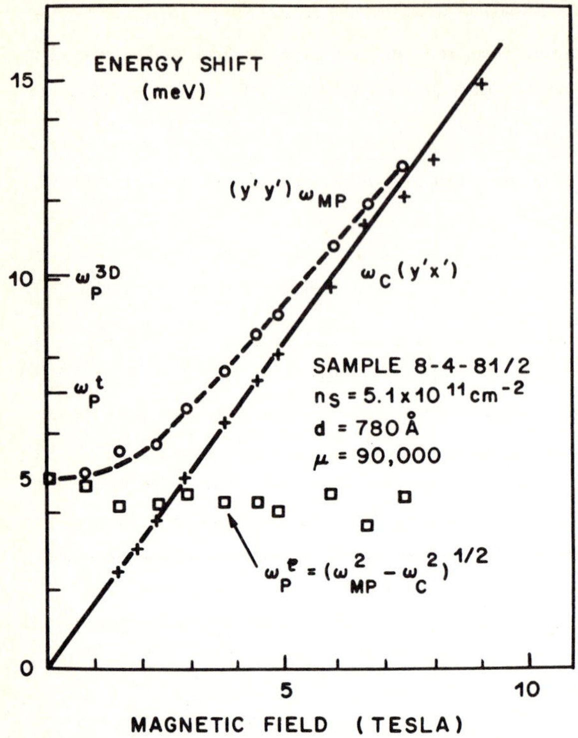

Figure 3 Plot of energy shifts for cyclotron resonance and its companion mode, labeled ω_{MP} Shown also are the theoretical, (ω_p^t), and empirical, (ω_p^e), zero-field plasma frequencies.

In our experiment, using light of wavelength 6500A, incident at an angle of 74 degrees, and collecting light in approximately f/2 optics, we get $dk_{\parallel} = 0.72 \pm 0.18$, and $dk_{\perp} = 5.5$. (The sample has a superlattice period of 780A.) If we evaluate the expected plasma frequency, ω_p, given the parameters quoted above, and a dielectric constant of 13.1, we get a value of 60 cm^{-1}, somewhat higher than the observed zero-field frequency of 40 cm^{-1}. This disagreement is large and disturbing, and puzzling in view of the fact that the behavior of this mode with magnetic field is just what one would expect. By analogy with the three dimensional case,[13] we expect that the magnetoplasma frequency ω_{mp} will be related to the plasma frequency ω_p and the cyclotron frequency ω_c by[14]

$$\omega_{mp}^2 = \omega_p^2 + \omega_c^2 .$$

The figure shows some points derived from the expression

$$\omega_p\,(empirical) = \left(\omega_{mp}^2 - \omega_c^2\right)^{1/2}$$

and it is clear from their constancy that as the field increases, the observed mode frequency does increase as expected for the MP mode.

The principal difference between our experiment and that of Olego and Pinczuk is that our penetration depth into the superlattice is only about 3000A, whereas theirs is probably on the order of

20000A. It is difficult to know the effect of nonuniform excitation of the layered electron system, so it will be useful to have long wavelength scattering measurements in a magnetic field, and to vary the incident and scattering angles in the short wavelength experiments. These will help us to accept or reject the mode in question as a magnetoplasma mode.

We turn now to the third section and the question of resonance. In some recent measurements on a family of samples of excellent mobility, we have been able to observe the cyclotron resonance mode at fields as low as 1.5 tesla, and to follow the resonance as a function of field out to the point where only one Landau level is occupied.

Figure 4 shows a curve of the peak intensity of CR scattering at 2 tesla, as a function of the laser frequency. The maximum occurs at 1.884 eV, which is, within error, equal to the band gap energy $E_o + \Delta_o = 1.861\ eV$, plus the Fermi energy, $E_f = 18\ meV$, plus five or ten meV for the confinement energy of the lowest subband. This is consistent with our original contention[5] that in the scattering mechanism, the incident photon promotes an electron from some state in the V_3 valence band to an unoccupied state in the conduction band (leaving behind a valence hole), and the inelastically scattered photon is generated in the process of recombination of an electron from some occupied state with the same hole. The problem, theoretically,[15] is to find some kind of hole state with matrix elements to both conduction band states, which in the case of Landau level Raman scattering or CR simply differ by $\Delta\ell = 1$.

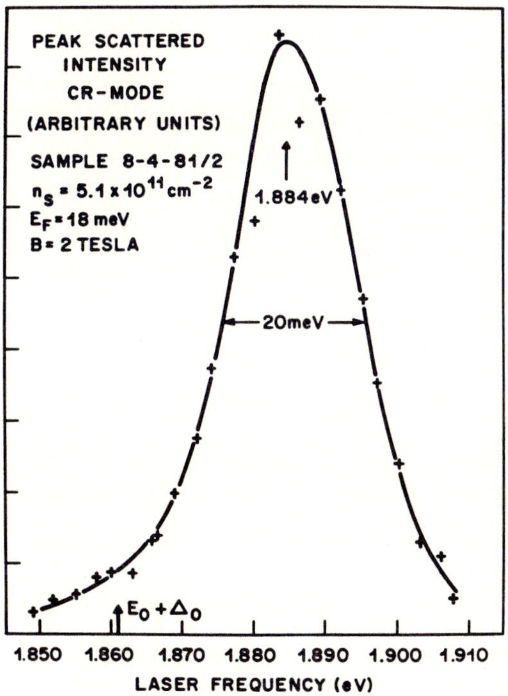

Figure 4 Resonant excitation curve for cyclotron resonance scattering at 2 tesla.

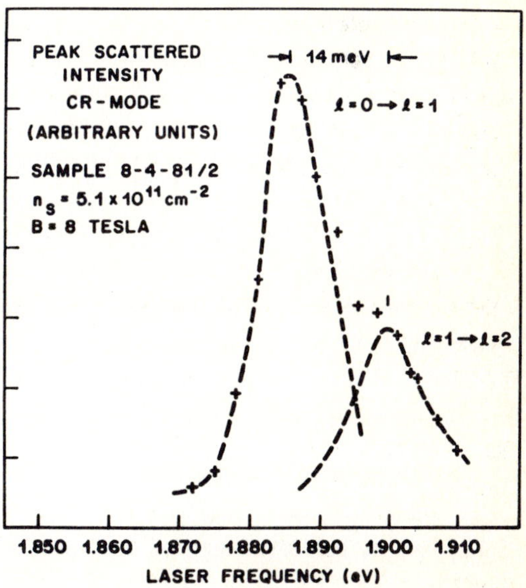

Figure 5 Resonant excitation curve for cyclotron resonance scattering at 8 tesla, when two Landau levels, separated by 14 meV, are occupied and active in the scattering process.

That such a state exists is attested to by the existence of the scattering, and the mechanism is further solidified by the resonance curve in Figure 5. Here we see evidence of two resonances separated by 14 meV, which is just the energy of cyclotron resonance. In this particular sample, at 8 Tesla, the lowest Landau level is completely filled, having its allowed complement of 3.8×10^{11} electrons square cm^{-2}. The $\ell=1$ level contains the remaining 1.3×10^{11} electrons cm^{-2}. There are therefore two possible inelastic transitions with $\Delta\ell = 1$, namely $\ell = 0$ to $\ell = 1$, and $\ell = 1$ to $\ell = 2$, and each has its own resonance peak. They are separated by $\hbar\omega_c$ because that is the separation of the $\ell = 1$ and $\ell = 2$ Landau levels. All this implies, of course, that the two processes share a common valence band state, or at least that the corresponding valence band states are not resolved in energy. They do not, for example, differ by as much as the cyclotron energy in the valence band. At the lower fields, we could also expect to see double resonances, but only if the curves were narrower than their separation.

We do not yet have complete results on the variation of the scattering cross section for CR with magnetic field. Nevertheless there are indications that the process becomes weaker at higher fields. This is a surprising result, since at low fields most of the electrons are hidden and presumably silent: only those just above and just below the Fermi level can participate in the scattering process. Once at or above the quantum limit, of course, all electrons are in the lowest level, and can contribute, but for small fields, the proportion of participatory electrons increases as H^2. We must therefore invoke a process with a very strong inverse field dependence in order to correspond with our observations.

In a speculative mood, we point out that one quantity which is certainly shrinking as the field increases is the size of the Landau orbit: not only is the whole structure of wave functions shrinking, but as Landau levels emerge from the Fermi level, they become inaccessible, and the transitions take place at lower and lower points in the Landau ladder. Thus, the orbit size for the transition decreases as H^{-1}, and we suppose that the larger orbits lead to stronger non-local effects, which are not foreseen in any of the extant theories.

We have presented results of inelastic light scattering experiments on electrons in layered two dimensional structures subjected to large magnetic fields. We believe that we have observed cyclotron resonance, and probably a magnetoplasma mode, but there are several questions and anomalies remaining to be answered and understood. We have not yet identified a mechanism for our scattering, nor have we identified with absolute certainty the mode we call magnetoplasma. It is possible that some of the effects we see can be related to non-local electron behavior, and experiments are planned which will probe different regimes of wave vector transfer, to study this possibility. It is our hope to refine our understanding and observations so as to be sensitive to small anomalies in in-plane motion which might betray electron ordering or crystallization.

We thank P. A. Wolff for encouraging discussions and support; A. Pinczuk, D. Olego, and H. L. Stormer for discussions and for sharing the results of measurements; R. Feigenblatt and M. Smith for help with setting up the experiments; and L. Rubin and the staff of the Francis Bitter

National Magnet Laboratory for cooperation and hospitality. The National Magnet Laboratory is supported by the National Science Foundation. This work was also supported in part by the Office of Naval Research, under contract N-00014-81-K-651.

REFERENCES

[1] T. Ando, A. B. Fowler, and F. Stern, Revs. Mod. Phys. *54* 437 (1982).
Surface Science *113* (1982), Proc. of IV International Conf. on Electronic Properties of Two-Dimensional Systems, New Hampshire (1981), F. Stern, ed.

[2] A. C. Gossard, in *Thin Films:Preparation and Properties*, K. N. Tu and R. Rosenberg eds. Academic Press (1981).

[3] E. Burstein, A. Pinczuk, and S. Buchner, in *Physics of Semiconductors, 1978*, B. L. M. Wilson, ed. Inst. of Physics, Bristol, (1979).

[4] For a recent review, see A. Pinczuk and J. M. Worlock, Surface Science *113*, 69 (1982).

[5] J. M. Worlock, A. Pinczuk, Z. J. Tien, C. H. Perry, H. L. Stormer, R. Dingle, A. C. Gossard, W. Wiegmann, and R. L. Aggarwal, Solid State Commun *40* 867 (1981).

[6] Z. J. Tien, J. M. Worlock, C. H. Perry, A. Pinczuk, R. L. Aggarwal, H. L. Stormer A. C. Gossard, and W. Wiegmann, Surface Science *113* 89 (1982).

[7] Magnetic field stabilization of an ordered state has been predicted by a) Yu. E. Lozovik and V. I. Yudson, JETP Letters *22* 11 (1975) and b) H. Fukuyama, P. M. Platzman, and P. W. Anderson, Phys. Rev. *B19* 5211 (1979), and Surface Science *73* 374 (1978).

[8] K. von Klitzing, G. Dorda and M. Pepper, Phys. Rev. Letters *45* 494 (1980), and D. C. Tsui, H. L. Stormer, and A. C. Gossard, Phys. Rev. Letters *48* 1559 (1982).

[9] G. Abstreiter, Ch. Zeller and K. Ploog, in: Proc. 8th Intern. Symp. on GaAs and Related Compounds, Vienna, 1980, Ed. H. W. Thim, Inst. Phys. Conf. Ser. 56 Inst. Phys., London (1981) p. 741.

[10] A. Pinczuk, J. M. Worlock, H. L. Stormer, R. Dingle, W. Wiegmann and A. C. Gossard, Solid State Commun. *36* 43 (1980).

[11] D. Olego, A. Pinczuk, A. C. Gossard, and W. Wiegmann, Phys. Rev. B *26* 7867 (1982).

[12] M. Apostol, Z Phys. B *22* 13 (1975).

[13] C. K. N. Patel and R. E. Slusher, Phys. Rev. Letters *21* 1563 (1968).

[14] M. Kobayashi, J. Mizano, and I. Yokota, J. Phys. Soc. Japan *39* 18 (1975), and G. Meissner, Z. Physik. B *23* 173 (1976).

[15] Y. Yafet, Phys. Rev. *152* 858 (1966), and G. B. Wright, P. L. Kelley, and S. H. Groves in *Light Scattering Spectra of Solids* G. B. Wright, ed. Springer, New York (1969) p. 335.

ANOMALOUS CYCLOTRON RESONANCE

OF 2D ELECTRONS IN $AL_xGa_{1-x}As/GaAs$ HETEROJUNCTIONS

S. Narita, K. Muro, S. Mori, S. Hiyamizu† and K. Nanbu†

Department of Material Physics, Faculty of Engineering Science,
Osaka University, Toyonaka, Osaka 560, Japan

†Fujitsu Laboratories Ltd., 1015, Kamikodanaka, Nakahara, Kawasaki 211, Japan

Far-infrared cyclotron resonance of a two-dimensional electron gas in $Al_xGa_{1-x}As/GaAs$ heterjunction interfaces has been studied at various electron densities achieved by changing the gate voltage. In the quantum limit, an anomalous shift of mass and line-width change has been observed at low electron densities and low temperatures. The data are discussed in conjunction with previous Si-MOS results in the framework of existing theories.

1. Introduction

Cyclotron resonance (CR) experiments [1-3] performed on Si-MOS inversion layers have shown some anomalous properties of the 2D electron system at low densities especially in the quantum limit. The anomalous behavior has been discussed on the basis of a number of physical models: electron localization by static potential fluctuations [4], Wigner crystalization, a pinned charge density wave [5], localization in Landau levels [6] and others. However, no convincing evidence for any of the theories has been obtained by now.

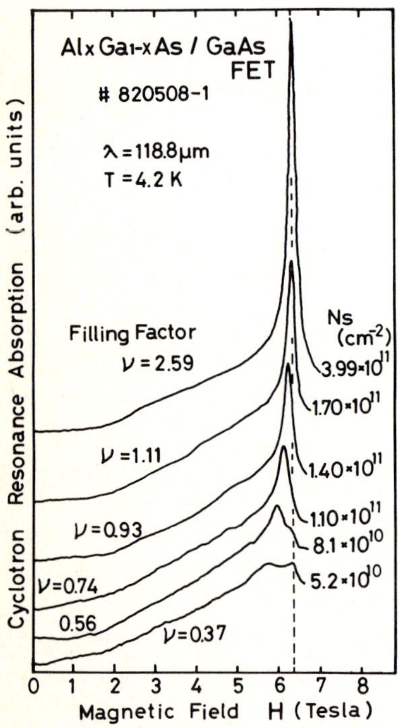

Fig.1 2D electron density or filling factor dependence of the cyclotron resonance spectra

On the other hand, a preliminary experiment on a 2D-electronic system in AlGaAs/GaAs heterojunction-FETs has been performed by some of the present authors (7). In the experiment a gate voltage modulation technique was employed. The data were supposed to contain some errors due to the following origin: 1) a leakage channel 2) a non-linear relationship between 2D-electron density and gate voltage.

Therefore we improved the experimental method; the present measurements were done with fixed gate voltages and a two-detector system over a wide range of the electron density ($0,35-4 \times 10^{11} cm^{-2}$) and temperature 2K-80K .

The anomalous CR data will be compared with those obtained in Si-MOS and dicusssed with respect to proposed physical models. In addition the scattering mechanism of the 2D electrons in the heterojunction FETs will be treated from a viewpoint different from that adopted in Si-MOSFETs.

2. Experimental

The details of the experimental procedures have been already described in a previous report (7). However, in the present work, we used two-detectors placed in front and rear of the sample. The signal ratios are recorded as the relative CR absorptions. The impurity concentration in

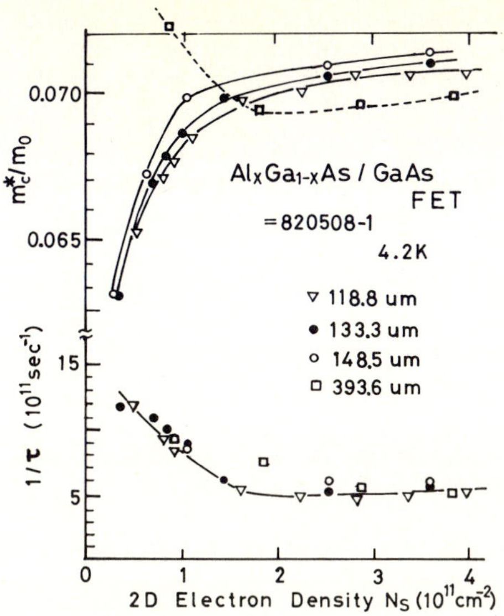

Fig.2 Cyclotron mass ratio and reciprocal relaxation time as functions of 2D electron density for various wavelengths.

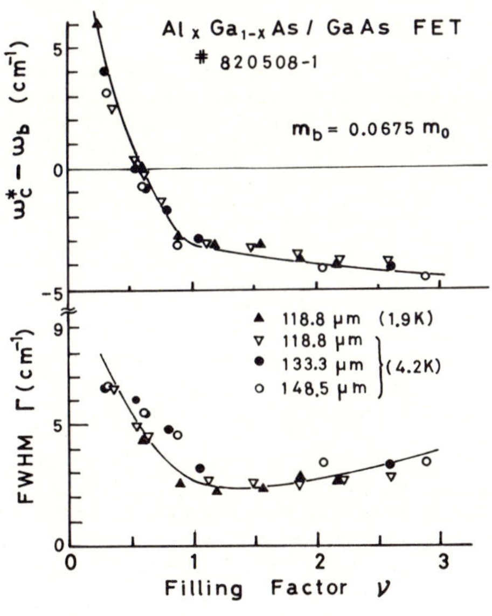

Fig.3 CR frequency shift and linewidth as a function of filling factor. CR frequency shift is obtained from the relation: $\omega_c^*-\omega_b = eH/C \times (1/m_c^* - 1/m_b)$ (H:resonant field, m_c^*:CR mass and m_b: bulk mass).

the AlGaAs layer was about $8\times10^{17}cm^{-3}$. The 2D electron mobility turned out to be about 150,000 cm^2/Vs and the electron density at the interface as determined by Shubnikov-deHaas measurements was $4\times10^{11}cm^{-2}$ at 4,2K, without gate voltage.

The 2D electron density dependence of the CR spectra observed at 118.8 μm laser wavelength at 4,2K is shown Fig.1. The background shift in the spectra is ascribed to the effect of the magnetic field on the detectors.

By decreasing the electron density from 4 to $1.7\times10^{11}cm^{-2}$, the resonant field shifts gradually to lower fields and CR line becomes sharper as shown in Fig.2. With further decrease of the density the CR field shifts toward lower fields and the line width becomes broader. At very low densities, the CR peak seems to split into two peaks, the higher field peak staying at the same B value. The high field peak was ascribed to the existence of a gate-voltage-uncontrolled area at the gate edges, this was confirmed by reducing the FIR radiation transmitting area.

Similar results as shown in Fig.1 were obtained using several other laser lines. The cyclotron resonance masses m_c/m_0 and reciprocal relaxation times $1/\tau$ estimated from the spectra are plotted against the 2D-electron density N_s in Fig.2. The rapid decrease of the CR mass occurs below a certain electron density depending on the laser wavelength.

These experimental results are plotted with the filling factor ν of the Landau levels as a variable (Fig.3.) The frequency shift, $\omega_c^*-\omega_b$ corresponds to the CR mass shift in Fig.2 and shows a critical dependence on ν, though the line width (FWHM) corresponding to $1/\tau$ is not necessarily normalized to ν. For the sake of clarity, $1/\tau$ has been plotted vs Ns in the lower part of Fig.2.

The temperature dependence of the cyclotron resonance spectra in the quantum limit $\nu\simeq0,5$) is shown in Fig.4. In this figure,the line splitting is seen in the low temperature spectra which are attributed to the same origin as discussed in conjunction with Fig.1. The low field peak shifts to higher fields and becomes broader with increasing temperature; above 20K both peaks coincide.

Figure 5 shows the CR mass- and the line-width shifts as a function of temperature obtained from similar spectra as shown Fig.4. At low ν (~0.5,~0.9), with decreasing temperature the cyclotron mass shifts remarkably in the low

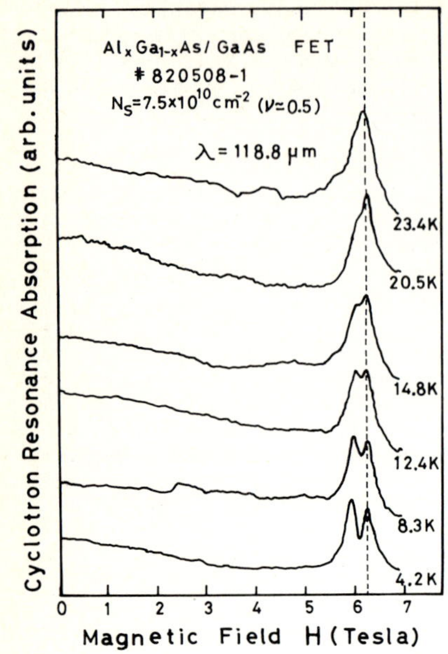

Fig.4 Temperature dependence of cyclotron resonance spectra

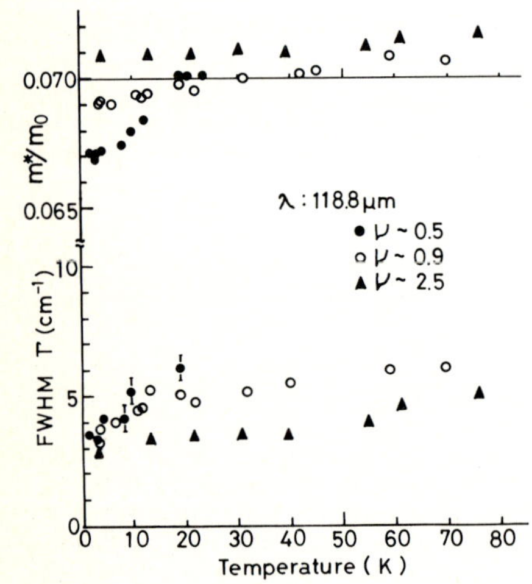

Fig.5 Wide range temperature dependence of cyclotron resonance mass ratio and line width at various filling factors.

temperature region towards smaller values. However, by raising temperature to 20K, the mass becomes temperature-independent (see the range between 20 and 45K). With further increase of the temperature above 45K another gradual increase of mass appears. On the other hand, at relatively high ν (2,5) the mass is hardly changed in the low temperature range, though it exhibits a similar gradual increase above 45K with increasing temperature.

The line-width in the low temperature range for small ν (~0,5,~0,9) shows clear narrowing with decreasing temperature below 20K in spite of the almost unchanged line width for high ν (~2,5) spectra.

The line width (FWHM) for the spectra has also been plotted against magnetic field for small ν (~1 and 1-2) as shown in Fig.6. The experimental data are compared with the theoretical field dependence of the line-width as discussed later.

3. Discussion

The value of the mass and its shift with decreasing N_s at relatively high 2D electron densities (Fig.2) can be ascribed to band non-parabolicity. However, the mass shift in the low density region cannot be explained by this way because the mass becomes appreciably smaller than the bulk mass ($m^*=0.0675_0{}^k$). The mass shift plotted against N_s depends substantially on the laser wavelength, especially the curve for 393.6 μm moves in the opposite direction as the other data when N_s is reduced. At very low N_s the mass seems to be independent of the wavelength.

Figure 3 shows the representation of the mass shift by the CR frequency shift; $\omega_c^*-\omega_b=(eH/c)\cdot(1/m_c^*-1/m_b)$ (H: the CR magnetic field and m_b: the bulk mass), as a function of the filling factor ν of the Landau levels. The figure shows that the normalisation of the data using the filling factor gives a better presentation of the experimental results. For $\nu<1$ the anomalous frequency shift is clearly exposed. Although this result is consistent with the Si-MOS data by Wilson et al. (3), the displacement of $\omega-\omega_c$ towards the negative values approaching $\nu=1$ from high values as observed in Si-MOS is not seen in our case; the frequency shift continues to larger positive values for ν smaller than 0,5 (Fig.3).

On the other hand, the filling factor dependence of the line width in the present data is opposite to that observed in Si-MOSFETs; that is, the FWHM increases as the factor ν decreases from unity. However, this difference may be ascribed to a fundamental difference in the scattering mechanisms between the two systems. As described in a previous paper, the 2D electrons at the interface are scattered by ionized impurities with long range. The scattering becomes more frequent with decreasing N_S because of the screening. This is in contrast to the short range scattering in Si-MOSFETs. In other words: $1/\tau$ or Γ depends on N_s rather than on ν as plotted in Figs. 2 and 3.

As shown in Fig.3, the CR-frequency-shift exceeds the line width and can be well normalized by ν. This fact means that the anomalous mass shift is difficult to explain by the conventional localization model (4) or by Anderson localization in Landau levels, assuming short-range scattering (6). It can better be understood by the pinned charge density wave (CDW) theory (5). Further evidence for that can be deduced from the temperature dependence of mass- and line-width-shifts in Figs. 4 and 5. With increasing temperature at ν = 0,5 for instance, the mass increases gradually and at 20K it coincides with the temperature independent-mass in the range of 20 45K. This temperature change may be explained by a second order phase transition of 2D electrons which is also consistent with the assumption of CDW (5). The critical temperature T_c is calculated according to the theory by Fukuyama et al. (9) which gives T_c of 13K, being not so far from the experimental value of 20K. However it is inconsistent with the theory that the value at $\nu \simeq 0,9$ is close to that at ν = 0,5. In addition, though the present line-width narrowing with decreasing temperature for $\nu<1$ is consistent with the similar result for Si-MOSFETs by Wilson et al. (3), The present filling factor dependence of the line-width for $\nu<1$ is apparently opposite to their data for the Si-MOSFETs. Therefore, from the present results, it cannot be concluded that the CDW formation is w unique explanation of the phenomena.

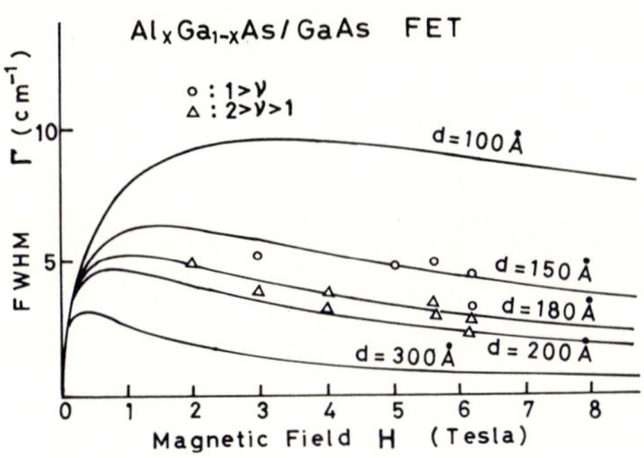

Fig.6 Line width of cyclotron resonance as a function of magnetic field. The solid curves represent the results of the calculation by the theory [10].

Prasad and Fujita (10) have performed a calculation of the CR line shape in 2D electron systems. According to their theory and assuming an Gaussian scattering potential, the line width is given as a function of magnetic field by the formula: $\tau\ [2\hbar^3\omega_o/\pi\tau_o]^{1/2}/\ [1+(d/r_o)^2]^{3/2}$, where r_o is the radius of the cyclotron orbit, d the range of the scattering force, τ_o the relaxation time under zero magnetic field and ω_o=eH/mc (m is the effective mass of 2D electrons at high ν). The results of the calculation are shown by the solid curves in Fig.6 for various d. From the results, it is known that the force has a range of about 150-200A and long-range scattering is dominant. The results do not contradict the model previously mentioned: the long range scattering of 2D electrons by ionized impurities in AlGaAs layers. This mechanism which is different from the case of Si-MOSFETs results in different CR behaviour. We can see in Fig.5 that both mass and the line-width show an additional gradual increase above 45K. These phenomena may be explained by phonon scattering and population of higher energy levels by the 2D electrons, respectively.

References

1. J.P. Kotthaus, G. Abstreiter, J.P. Koch and R. Ranvaud, Phys.Rev.Lett. 34, 151 (1975)
2. T.A. Kennedy, R.J. Wagner, B.D. McCombe and D.C. Tsui, Solid State Commun. 21, 459 (1977)
3. B.A. Wilson, S.J. Allen, Jr. and D.C. Tsui, Surface Sci. 98, 262 (1980); Phys.Rev. 24, 5887 (1981).
H.J. Mikeska and H. Schmidt, Z. Physik B20, 43 (1975)
5. H. Fukuyama and P.A. Lee, Phys. Rev. B18, 6245 (1978)
6. T.Ando, Surface Sci. 113, 182 (1982)
7. K. Muro, S. Narita, S. Hiyamizu, K. Nanbu and H. Hashimoto, Surface Sci. 113, 321 (1982)
8. S. Narita, S. Takeyama, W.B. Luo, S. Hiyamizu, K. Nanbu and H. Hashimoto, Surface Sci. 113, 301 (1982)
9. H. Fukuyama, P.M. Platzman and P.W. Anderson, Phys. Rev. B19, 5211 (1979)
10. M. Prasad and S. Fujita, Physica 91A, 1 (1978)

FARADAY ROTATION AND ELLIPTICITY IN ELECTRON INVERSION LAYERS OF Si MOS STRUCTURES

H. Piller*

Department of Physics and Astronomy
Louisiana State University
Baton Rouge, Louisiana 70803 USA

R. J. Wagner
Naval Research Laboratory
Washington, D.C. 20375 USA

Abstract: We report the first observation of Faraday rotation and ellipticity in the two-dimensional electron gas in the surface of a p-type (100) Si. The surface densities were varied between 1.06×10^{12} and 2.3×10^{12} cm^{-2}. The Faraday rotation changes sign and the ellipticity has a maximum at the cyclotron resonance magnetic field. Both the density dependence of the mass enhancement and the magnetic field dependence of the relaxation time were observed in the Faraday rotation. From the data measured near cyclotron resonance we obtained an effective mass $m^* = 0.19\, m_e$ and a relaxation time of 6×10^{-13} s for a surface density of 2.3×10^{12} cm^{-2}.

I. Introduction

Various magneto-optical techniques have been used to determine band parameters and scattering effects of electrons in inversion layers of Si MOS devices. Many investigations can be made by varying the electric field on a Si MOSFET (metal-oxide-semiconductor-field-effect-transistor). By decreasing the electric field a corresponding decrease in the electron surface concentration is produced. Many-body and localization effects become dominant at very low concentrations and temperatures. At high surface concentrations, high density effects become important. Disagreements are found in the magnetic field- and density dependence of the effective masses and the relaxation times. One of the goals of the Faraday rotation (FR) measurements is to resolve some of the observed discrepancies in these parameters.

II. Experimental

The FR is the rotation angle of the plane of polarization of linearly polarized light. The FR can be determined from the ratio of transmissions for different analyser positions. Multiple internal reflection effects in the observed FR can be prevented by wedging the Si substrate of the MOSFET. The cyclotron resonance absorption of localized charges in the oxide layer and in the metal layer does not exist or is not observable in the far infrared (FIR) wavelength region; therefore, it need not be considered in a theoretical treatment of FR in the inversion layer. Contributions of oxide-Si boundary effects to the FIR FR are also negligibly small when multiple internal reflection effects are negligible as in our experiment. The imperfection of the polarizer-analyser system can be calibrated for zero magnetic field and must be incorporated in the equations describing the experiment. In the magnetic field free case and ideal polarizers one obtains the transmitted intensities $I_{min} = 0$ for crossed polarizers and $I_{max} = 100\%$ for parallel polarizers. The magnetic field causes a shift in the extrema and also increases the transmitted intensity to I_a at the minimum and decreases the intensity to I_b at the maximum. Two intensities are measured, one with the electric field on (space charge, FR) and one with the electric field off (no space charge, no FR). This intensity ratio Y_Φ is given by the following expression:

$$Y_\Phi = \frac{I_a + (I_b - I_a)\cos^2(\Phi + \Theta)}{I_{min} + (I_{max} - I_{min})\cos^2\Phi} \quad (1)$$

Φ is the angle between polarizer and analyser.
Figure 1 shows two measurements of such intensity ratios. The transmission (-ratio) is set equal to 100% for B = 0 (Θ = 0). Curve (a) shows a large increase in the transmission ratio at high magnetic fields because of the increase in the FR Θ.

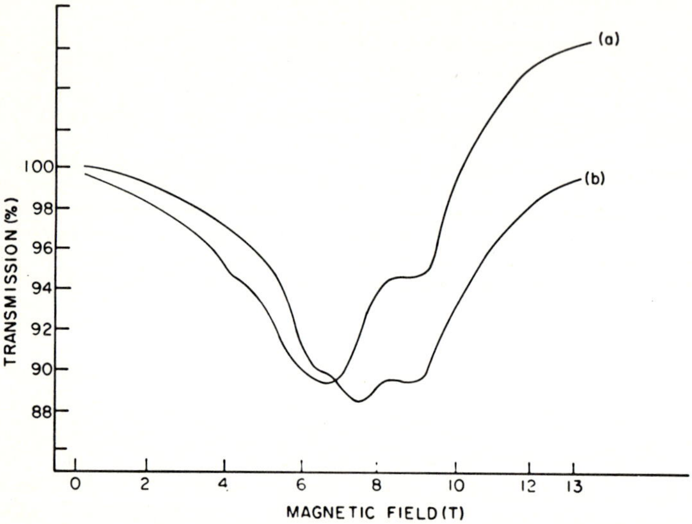

Fig. 1. Cyclotron-resonance transmission in the Faraday rotation experiment at two analyser settings with respect to the polarizer. (a) 45 deg., (b) 0 deg. N = 2.3×10^{12} cm^{-2}, λ = 292.5 μm, T = 4.2 K.

The FR Θ can be calculated from equation (1) using the ratio of intensity (Y_Φ) measurements for different angles Φ. Equation (2) gives the result for Φ and Φ = 0, where Φ is an arbitrary angle. Here $X = I_a/I_b$ and $M = I_{min}/I_{max}$.

$$\cos^2(\Phi+\Theta) = \frac{Y_\Phi/Y_o\{0.5(M+1)\}\{X+(1-X)\cos^2\Theta\}-X}{1-X} \quad (2)$$

For imperfect polarizers the ellipticity is given by

$$\varepsilon^2 = (M-X)/(MX-1) \quad (3)$$

X can be determined from equation (1) and M can be obtained experimentally. The type of samples used in these experiments have been described by Wagner et al.[1] The wedges on our samples are part of the substrate. The older samples had separate wedges attached to them.

III. Theory

The FR per unit length is given by

$$\Theta = (\omega/2c)(n_- - n_+) \quad (4)$$

where ω is the angular frequency and c is the speed of light. n_+ and n_- are the refractive indices for the right and left circularly polarized waves.[3] In weakly absorbing regions the FR can also be written in terms of the real part of the off-diagonal element of the conductivity tensor[2]

$$\Theta = 2\pi\sigma_{xyR}/nc \tag{5}$$

where n is the mean of n_+ and n_-. The assumption of little absorption in the FIR is well justified for free carriers in pure semiconductors[3]. This fact has been verified for inversion layers in p-type Si[4]. Under the same conditions the ellipticity is given by

$$\varepsilon = \tanh\{(\omega/2c)(k_- - k_+)\} \tag{6}$$

where k_+ and k_- are the extinction coefficients for right and left circularly polarized waves.
Ando[5] introduced the following conductivities:

$$\sigma_\pm(\omega) = \underline{\sigma}_\pm(\omega)\{1 - (m/m^*)(m^*/m - 1)\omega/(\omega \mp \omega_c)\}^{-1} \tag{7}$$

m is the bare mass and m* is the dressed mass of the electrons. $\omega_c = eH/m^*c$ is the cyclotron resonance frequency. e is the charge of the electron. $\underline{\sigma}_\pm(\omega) = iNe^2/m^*(\omega \mp \omega_c)$. N is the number of carriers per cm^2.

For this model we obtain

$$\sigma_{xyR} = \frac{(Ne^2\tau^2\omega_c/m^*)\{2\omega^2(1 - m/m^*)-\{1- \tau^2(\omega^2-\omega_c^2)\}(\omega^2- \omega_c^2)\}}{\{\omega^2(m/m^*)^2 - \omega_c^2\}\{4\omega^2\tau^2 + \{1 - \tau^2(\omega^2 - \omega_c^2)\}^2\}} \tag{8}$$

From equation (6) one can determine the ellipticity in terms of the imaginary conductivity tensor element.

$$\varepsilon = \tanh(2\pi\sigma_{xyI}/cn) \tag{9}$$

Ting et al.[6] used the memory function technique to calculate the conductivity of the interacting electron gas:

$$\sigma_\pm(\omega) = iNe^2/m\{\omega \mp \omega_c + M(\omega)\} \tag{10}$$

$$\text{where}\quad M(\omega) = M_1(\omega) + iM_2(\omega) \tag{11}$$

and by transforming equation (10) in to the Drude-Zener form they obtain:

$$\sigma_\pm(\omega) = \{iNe^2/m^*(\omega)\}/\{(\omega \mp eH/m^*(\omega)c) + i\tau^{-1}(\omega)\} \tag{12}$$

where the frequency-dependent mass and relaxation time are given by:

$$m^*(\omega) = 1 + M_1(\omega)/\omega\ , \qquad \tau(\omega) = M_2(\omega)^{-1}\{1 + M_1(\omega)/\omega\}\ .$$

From equations (5) and (12) one obtains the FR

$$\Theta = \{2\pi\omega_c(\omega^2 - \omega_c^2 - \tau^{-2})Ne^2/m^*\}/\{nc(\omega^2 - \omega_c^2 - \tau^{-2})^2 + 4\omega^2\tau^{-2}\} \tag{13}$$

IV. Results and discussion

The FR shown in figure 2 was calculated using equation (2) from the data shown in figure 1. The ellipticity was determined from the same data using equation (9). The maximum value of the ellipticity is 5.5% and occurs at the cyclotron resonance field B_c as indicated in the figure 2. The FR approaches zero at the same magnetic field.

A simple Drude-Zener type model for the conductivity[4] does not explain the magnitude or structure of the observed FR, especially at magnetic fields larger than B_c. The FR data for this large surface density can be explained on the basis of the model by Ting et al.[6], equations (10) to (13). We used the calculated M-values of Ting et al. to estimate the FR. Since the mobility in our sample was 4700 cm^2/Vs, as compared to the mobility of 6500 cm^2/Vs in their sample, we adjusted τ accordingly to $\tau = \tau_{tr/2.5}$. τ is the single particle and τ_{tr} is the transport lifetime.

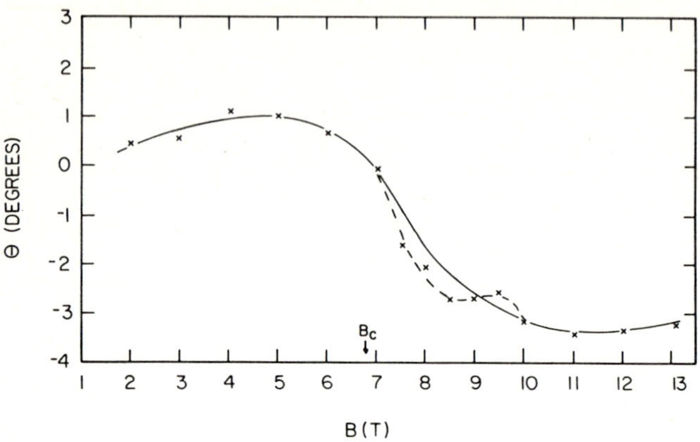

Fig. 2. Faraday rotation in a Si-MOSFET. $N=2.3\times10^{12}$ cm^{-2}, λ=292.5 μm, T=4.2 K, B_c=6.86 T ($m^*=0.19m_e$). The structure in the line shape (dotted line) is caused by oscillations of the Fermi energy as a function of B.

The computed FR agrees well with the data, also in the high field region, where all other models seem to fail. When adjusting the relaxation time, best agreement was found for values larger than 2.5. The sample was also studied at lower surface density ($N=1.06\times10^{12}$ cm^{-2}) and at a different light frequency ($\omega=4.34\times10^{12}$ s^{-1}). The FR approaches zero at 5.4 T and also shows a different shape than the FR at high surface density. The application of equations (6) to (8) shows that the data can be explained by a mass enhancement of 8% and a magnetic field dependence of the relaxation time according to Ando[7]. In summary, the FR is a sensitive tool to measure mass enhancement and the field dependence of the relaxation time. More theoretical and experimental work in magneto-optics is needed, especially in the low density region, in order to resolve the many remaining discrepancies in the data and in the theoretical models.

V. Acknowledgement

We like to acknowledge the support of the staffs of the NRL High Magnet Field and Microelectronics Processing Facilities and the assistance of L. M. Barbier in digital computer work. * Work supported in part by the Office of Naval Research.

References

1. R. J. Wagner, T. A. Kennedy, B. D. McCombe, and D. C. Tsui, Phys.Rev.B 22, 945 (1980).
2. M. Balkanski and E. Amzallag, Phys.Stat.Sol. 30, 407 (1968).
3. E. D. Palik and George B. Wright, Free Carrier Magnetooptical Effects, Ch. 10 in Semiconductors and Semimetals, Vol. 3, edited by R. K. Willardson and A. C. Beer (Academic Press, 1967).
4. H. Piller and L. M. Barbier, to be published; also in: Bull.Am.Phys.Soc. 27, 250 (1982).
5. T. Ando, Phys.Rev.Lett. 36, 1383 (1976).
6. C. S. Ting, S. C. Ying, and J. J. Quinn, Phys.Rev.B 16, 5394 (1977).
7. T. Ando, J.Phys.Soc.Jpn. 38, 989 (1975).

MAGNETOOPTICAL STUDIES OF DEEP IMPURITIES IN III-V COMPOUNDS

Christoph Uihlein

Max-Planck-Institut für Festkörperforschung, Hochfeld-Magnetlabor,
166X, F-38042 Grenoble Cédex, France

Zeeman spectroscopy provides information about the symmetry and the charge state of a defect. The Zeeman splitting of the initial and final state can be described by a most general effective Hamiltonian. A dynamical Jahn-Teller effect in one of the states leads to an extremely weak effective spin-orbit coupling. The resulting complicated energy level scheme can be successfully analyzed in the high magnetic field limit. This is demonstrated in the case of the 0.84-eV no-phonon structure in GaAs:Cr and the 0.786-eV no-phonon line in InP:Co.

I. INTRODUCTION

Zeeman spectroscopy is a well-established and extremely useful technique to study point-defects in semiconductors. This technique is not restricted to a particular type of defect although the theoretical models applicable to specific situations might differ considerably. The only conditions are that the defect manifests in a sharp optical transition and that time-reversal symmetry can be lifted by application of a magnetic field. Among the different types of defects deep impurities, especially transition metals in III-V compounds, have attracted increasing interest in recent years.

One of the most important informations which can be obtained by Zeeman spectroscopy is the symmetry of the defect. The symmetry of a defect may differ from that of the host crystal for several reasons. For instance an impurity may be located on an interstitial position having lower symmetry or the symmetry can be perturbed by the presence of other point defects in the vicinity (complexes). In addition the defect might be subject to a strong Jahn-Teller distortion. The local symmetry of a defect gives rise to a more or less pronounced anisotropy of the Zeeman splitting. Though the Zeeman splitting depends then on the orientation of the field, equivalent orientations should give rise to the same splitting pattern. In other words the eigenvalue spectrum, when regarded as a function of field orientation, should transform according to the identity representation of the point group. In general an anisotropy plot obtained by sweeping the magnetic field through a (110)-plane of the crystal suffices to determine the symmetry of the defect.

Another essential information which can be obtained by Zeeman spectroscopy is the charge state of the defect. It is known that transition metals can exist in a number of different charge states. Which of the charge states are the stable ones depends on the position of the Fermi-energy but generally two of the charge states can coexist at the same time. The charge state can be deduced from the magnitude of the total electron spin S and the g-value associated with it. Sometimes it is difficult to determine the total electron spin but in most cases one is able to distinguish between an integer spin and a half-integer spin situation (extra representations of the double group), when time reversal symmetry is lifted. This provides at least the information whether an even or odd number of electrons is involved in the transition.

Zeeman spectroscopy is often complicated by the fact that an energy splitting is present in the initial and final set of states. Spectroscopic data give only information about the relative energy separation between initial and final states but no information about the absolute energy position. Different combinations between initial and final state Zeeman components are needed to decompose the spectrum into an initial and final state Zeeman splitting. Selection rules are very useful in this context, since not all of the possible combinations are optically allowed. Knowledge of the symmetry of the defect is therefore of extreme importance. A study of the thermalization properties is also very helpful to distinguish between initial and final state Zeeman splitting.

II. THEORETICAL CONCEPT FOR TRANSITION METALS

Among deep impurities in III-V compounds, transition metals occupy an exceptional position not only since a lot of different stable charge states are known but also due to the extremely high degeneracy associated with a partially filled d-shell. It is known that chromium in GaAs for instance, when substitutional on a Ga site, can exist in three possible charge states [1-4]. Since atomic chromium has the electronic configuration $3d^5 4s^1$, the neutral ion substituted on the trivalent Ga-site has the configuration $3d^3$ if it is assumed that three of its electrons participate in the covalent binding with the nearest neighbour As atoms. The neutral state is called Cr^{3+} ($3d^3$) since three of its electrons occupy a delocalized valence band state. The chromium ion can act as an electron trap to give $Cr^{2+}(3d^4)$ or as a hole trap to give $Cr^{4+}(3d^2)$. The high configurational degeneracy of a partially filled d-shell is lifted by electron-electron repulsion giving rise to a splitting into ionic levels which can be classified within the L-S coupling scheme by a total angular momentum L and a total spin S. The degenerate angular momentum states are split in addition by the crystal field leading thus to a partial quenching of angular momentum. Spin-orbit interaction gives rise to a coupling between the effective angular momentum L' associated with the crystal field state and the total electron spin S. Since the spin-orbit interaction can be expected to be strong compared with the magnitude of a typical Zeeman splitting it suffices to confine to a subspace which is spanned by the eigenstates of the effective total angular momentum operator J=L'+S. The (2J+1)-fold degeneracy will generally be lifted by the presence of small cubic or axial terms which reflect the symmetry of the defect and arise from second or higher order spin-orbit interaction. The most general situation can be described by an effective Hamiltonian consisting of all linearly independent (2J+1)x(2J+1)-matrices which transform according to the idendity representation of the point-group and which are invariant with respect to time-reversal. The number of independent terms grows as J is increased or the symmetry is reduced.

The effect of a magnetic field can in most cases be described by a simple isotropic Zeeman term of the type $g\mu_B\vec{J}\cdot\vec{H}$. The anisotropy of the Zeeman splitting is then only a consequence of the non-spherical symmetry of the zero-field Hamiltonian. A diamagnetic shift can be neglected because of the small extension of the wavefunction when the electron is in the d-shell.

Experimentally, in many cases sharp no-phonon transitions are observed between crystal field states of transition metals in III-V compounds. The behaviour of these structures in a magnetic field can be compared with the Zeeman splitting expected for the initial and final states on the basis of most general Hamiltonian model. This will not only test whether the transition has been assigned to the correct crystal field states, but also provide the parameters of the effective Hamiltonians for the initial and final states. Information about the more fundamental crystal field parameters can be deduced in addition. The same procedure can be applied in the case of zero or completely quenched angular momentum. The effective total angular momentum operator J has in such a case to be replaced by the total spin operator S.

It has been found that many transition metals can be described within this concept. However, the situation is much more complex when a dynamical Jahn-Teller distortion is involved. It is worth noting that the degeneracy of a ground state is not really lowered by the presence of a Jahn-Teller distortion. The orbital degeneracy is lifted of cause, but at the expense of an additional degeneracy which is inevitably associated with the existence of symmetry-equivalent Jahn-Teller distortions. Tunneling between equivalent Jahn-Teller configurations gives rise to dynamical Jahn-Teller states which have to be classified by the representations of the undistorted defect. Coupling between such 'electronic-vibronic' states and the total electron spin can lead to a 'fine structure splitting', the magnitude of which may be comparable with a typical Zeeman energy. In the following two examples are discussed in some detail in order to demonstrate that Zeeman spectroscopy is almost indespensable for an understanding of the physics in such cases.

III. THE 0.84-eV LUMINESCENCE STRUCTURE IN GaAs:Cr

As was already mentioned before, Cr can exist in GaAs, when substitutional on a Ga site, in three different charge states. In addition to the simple substitutional form, Cr also occurs as an ion exhibiting a strong trigonal perturbation. This form gives rise to the well-known intense no-phonon luminescence line at 0.84-eV with accompanying phonon sidebands at lower energy [5-16]. In the early work, the 0.84-eV photoluminescence band was attributed to an internal transition of the simple substitutional Cr^{2+} ion. The high resolution electroluminescence and optical absorption measurements of Lightowlers and co-workers [7] revealed considerable fine structure in the 0.84 eV no-phonon line (Figs. 1 and 2). The ground state splitting was found to be conflicting with that of a simple substitutional Cr^{2+} ion deduced from EPR measurements [2]. More recent high resolution Zeeman spectroscopical work has undoubtedly shown that the defect exhibits trigonal (C_{3v}) symmetry characteristic for [111]-axial defects [8,9].

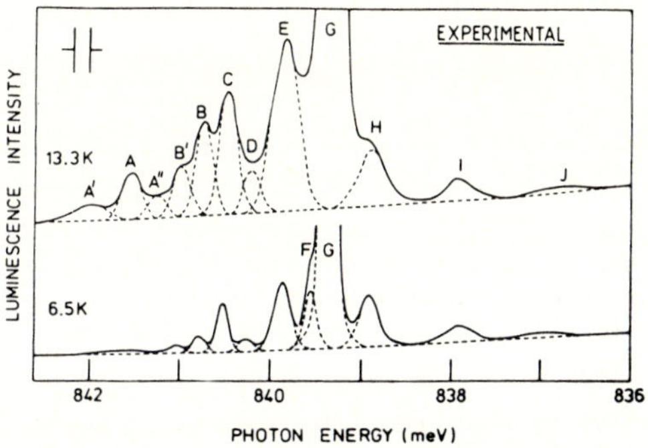

Fig. 1
Cathodoluminescence spectra of GaAs:Cr after Ref. 9

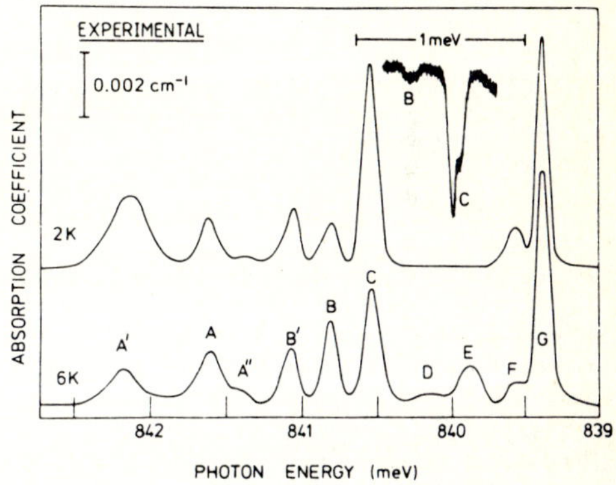

Fig. 2
Absorption coefficient of GaAs:Cr (Cr at 10^{17} cm^{-3})

The experimental observations obtained by Zeeman spectroscopy can be explained by assuming substitutional Cr^{2+} on a Ga site. The site symmetry of this defect is reduced to C_{3v} by the presence of a nearby attractive center, presumably a positively charged donor at the nearest As site ([111]-axial defect) [8,11]. The ionic ground state of $Cr^{2+}(3d^4)$ is a ^5D-state. This state splits as is depicted in Fig. 3 due to the presence of the cubic and trigonal crystal field into two ^5E-states and one $^5A_1$-state. The two-fold degeneracy of the ^5E-ground state is further lifted by Jahn-Teller distortion due to strong $T_2 x \varepsilon$ coupling. Axial symmetry in configurational space - the contour of the electronic energy resembles that of a mexican hat - gives rise to a highly dynamical Jahn-Teller effect. The electronic vibronical ground state, belonging to the E representation of C_{3v}, is again two-fold degenerate. The relatively broad I and J lines observed by Lightowlers et al. can be attributed to the first excited rotational states A_1 and A_2, the degeneracy of these states being lifted by trigonal warping terms [10]. The 0.84-eV no-phonon line can be assigned to the internal transition between the excited ^5E-state and the Jahn-Teller ground state which transforms according to the same representation. It is the interaction between the two-fold degenerate E-states and the total electron spin (S=2) which gives rise to the rich fine structure of the 0.84-eV luminescence line.

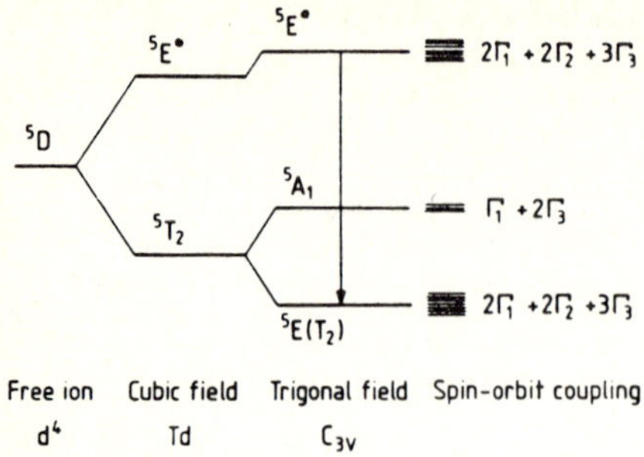

Fig. 3
Schematic energy level scheme for trigonal Cr^{2+} (d^4)

The 'fine structure splitting', though different in the excited and ground state, can be principally described by the same type of effective Hamiltonian. The most general Hamiltonian containing spin-orbit interaction to first and second order has the following form [11]:

$$H_o = D[S_z^2 - \frac{1}{3} S(S+1)] + \lambda' \sigma_z S_z$$

$$+ \mu_1 (\sigma_- \{S_z S_+\} + \sigma_+ \{S_z S_-\})$$

$$+ \mu_2 (\sigma_- S_- S_- + \sigma_+ S_+ S_+)$$

The index z refers to the trigonal symmetry axis. The operators denoted by σ are similar to Pauli matrices but since they are operators for a two-fold degenerate E state they differ from the latter ones with respect to time reversal.

Once the Hamiltonian has been derived by making use of the symmetry of the defect one can in principle try to fit the parameters to the observed fine structure splitting of the 0.84-eV no phonon line. As was already mentioned before, spectroscopic energies give only part of the information which is necessary to decompose a spectrum into a ground and excited state fine structure splitting. Lightowlers and co-workers have studied in detail the thermalization of the different lines both in luminescence and absorption but not all of the ground and excited energy levels could be determined with sufficient accuracy [7]. Here, a different experimental technique is employed: i.e. the zero-field Hamiltonian is tested by application of a strong magnetic field. The effect of a magnetic field on the transition ion can be described by a simple spin Zeeman term of the type $g\mu_B\vec{S}\cdot\vec{H}$. There might also be magnetic moment associated with the two-fold E states but they are expected to be extremely small. The excited state originates from a cubic E state for which a dipole moment is strictly symmetry forbidden and the magnetic moment is in the ground state almost completely quenched due to a strong Jahn-Teller distortion. Anisotropy is usually described by a g-tensor reflecting the symmetry of the defect. Such a description is only possible in the limit of small magnetic fields (up to 2 T). One can then attribute a g-tensor to each doublet state of the fine structure by evaluating the expectation value of the Zeeman term with respect to the eigenstates of H. A much simpler description is possible in the limit of high magnetic fields. The Zeeman term is in this limit the dominant term and H_o can be regarded as a small perturbation. The spin orients parallel to the magnetic field axis and the anisotropy is simply given by the expectation value of H_o with respect to the eigenstates of the magnetic field term. In first order perturbation theory one derives the following expression:

$$H' = g\,\mu_B\,M_S\,H + D/2\,[M_S^2]\,(3\cos^2\theta - 1)$$
$$+ \lambda'\cos\theta\,M_S\,\sigma_z$$
$$+ 3/4\,\mu_1\,[M_S^2]\,\sin 2\theta\,e^{-i\phi}\,\sigma_+ \quad + \text{h.c.}$$
$$+ 3/2\,\mu_2\,[M_S^2]\,\sin^2\theta\,e^{2i\phi}\,\sigma_+ \quad + \text{h.c.}$$

The symbol $[M_S^2]$ stands for $M_S^2 - 1/3\,S\,(S+1)$.

It is seen that every term of H' has its own characteristic angular dependence. This allows to evaluate the parameters of H_o from the anisotropy of the Zeeman splitting in a straightforward way. The eigenvalues of H' give the splitting of the two-fold E-state in the high field limit caused by the effective spin-orbit coupling with the total electron spin. The spin quantization axis is given by the field orientation. The magnetic field is parallel to the trigonal symmetry axis for $\theta=0$ and within one of the reflection planes for $\phi=0$. The degeneracy of the system is completely lifted since a doublet of states exists for every spin state labeled by M_S.

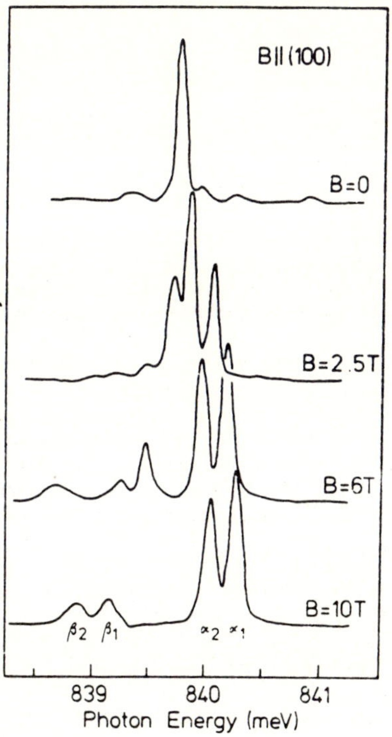

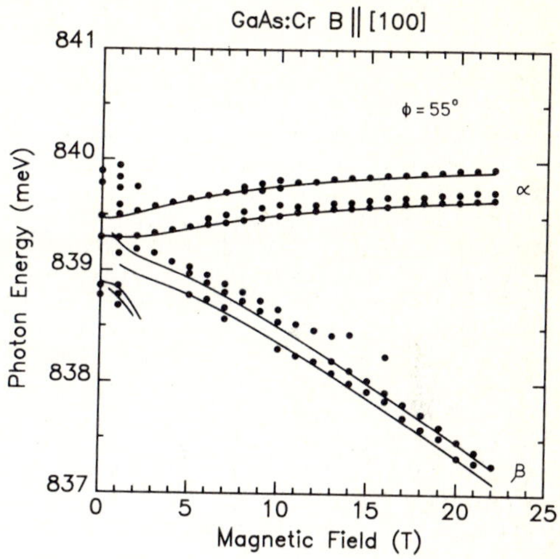

Fig. 5
Zeeman splitting for B || [100]

Fig. 4
Development of the 0.84-eV PL of GaAs:Cr for B || [100]

Fig. 4 shows the evolution of the Zeeman spectra with increasing magnetic field for B parallel to the [100] crystalline axis. The spectra have their simplest form in this configuration since all the trigonal defect orientations are equivalent with respect to B. The magnetic field dependence is depicted for the same field orientation in Fig. 5. One observed a dominant pair of lines denoted by $\alpha_{1,2}$ and another pair of lines $\beta_{1,2}$ the intensity of which becomes weaker with increasing magnetic field. These results can be understood easily. In the excited state thermalization favours the low energy component of the doublet belonging to the lowest spin state ($S=2, M_S=-2$). Because of spin selection rules ($\Delta M_S=0$), the dominant transition to the ground state will be that into the doublet belonging to the same spin ($S=2, M_S=-2$). This explains the dominant $\alpha_{1,2}$-lines. The energy separation between the α_1 and α_2 components results from the splitting of the electronic-vibronic E ground state due to spin-orbit interaction. The $\beta_{1,2}$-lines arise from a transition into the ground state doublet belonging to the $M_S=-1$ spin state. Since $\Delta M_S=1$ for this transition, the intensity becomes weaker the more the high field limit is approached.

Only α-lines have been depicted in the anisotropy plot (Fig. 6). When the magnetic field is rotated in the $(0\bar{1}1)$-plane from [100] ($\theta=0°$) to [110] ($\theta=90°$) the different possible orientations of a trigonal defect become magnetically inequivalent. This explains a splitting into three anisotropy branches when the magnetic field is swept away from the [100]-direction.

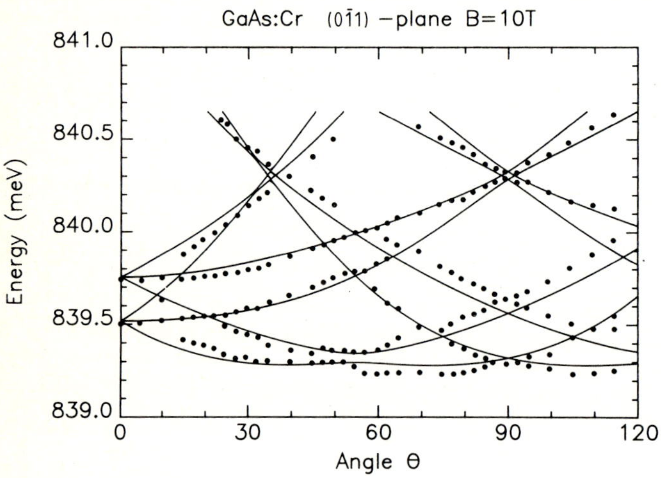

Fig. 6
Observed anisotropy of the Zeeman splitting at 10 T for B rotated in the $(0\bar{1}1)$ plane. Angle $\theta=0°$ corresponds to [100], $\theta=55°$ to [111] and $\theta=90°$ to [011]

For comparison with the experiment the 10x10-matrices for the excited and ground states have been exactly diagonalized. A best fit to the experimental data was obtained with the following set of parameters:

Table I. Parameters for the effective Hamiltonians of the ground and excited states (energies in meV)

	D	λ'	μ_1	μ_2
excited state	-0.12	0.69	0	0
ground state	-0.09	0	0.09	0.007

With this set of parameters one can easily calculate the fine structure splitting at zero field. The results are compared with the experimental observations of Lightowlers et al. in Table II.

Table II. Energy separation of lines in the 0.84-eV no-phonon system relative to line G.

Line	Experiment (meV)	Theory (meV)
A'	2.85	2.92
A	2.31	2.23
A''	2.02	1.99
B'	1.74	1.66
B	1.48	1.42
C	1.20	1.21
D	0.92	0.85
E	0.56	0.64
E'	0.49	0.58
F	0.19	0.16
H	-0.43	-0.41
H'	-0.49	-0.47

A further rigorous test of the parameter set is to calculate the oscillator strengths of the various transitions. Taking into account the thermalization in the initial set of states the luminescence and absorption spectra were calculated by superimposing lines with the experimentally observed linewidths. The result of such a procedure is shown in Fig. 7 and Fig. 8. Only one additional parameter, the ratio of dipole matrix elements for parallel and perpendicular polarization ($d_\parallel/d_\perp=5$), was used to get these results.

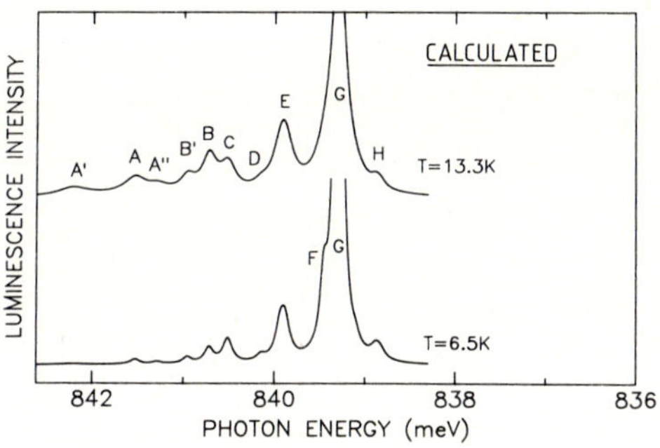

Fig. 7
Calculated luminescence spectra (to be compared with Fig. 1)

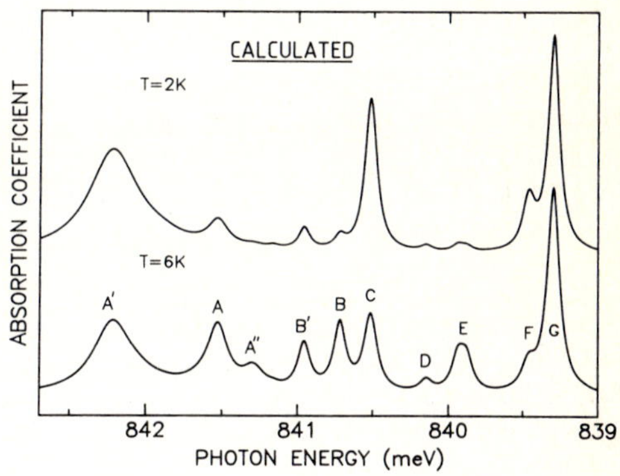

Fig. 8
Calculated absorption spectra (to be compared with Fig. 2)

The good agreement between theory and experiment - the solid lines in Figs. 4-6 result from a calculation with the same parameter set - confirms the correctness of the model.

IV. THE 0.786-eV NO-PHONON LINE IN InP:Co

The 0.786-eV no-phonon line in InP:Co is a further illustrative example showing the importance of Zeeman spectroscopy for the analysis of an optical transition. The 0.786-eV no-phonon line can be observed in optical absorption and in photoexcitation by monitoring the intensity of the 474-meV photoluminescence line. The photoluminescence line at 474 meV is ascribed to the internal transition between the $^4T_2$ excited and the $^4A_2$ ground state and the 786-meV absorption line to the transition between the $^4A_2$ ground state and the $^4T_1$ excited state of Co^{2+} [16]. The three cubic crystal field states arise from the same ionic ^4F ground state of the d^7-electron configuration. The assignment is based on the form of the spectra and the energy positions and supported by the similarities with corresponding crystal field transitions in GaP:Co [17,18] and GaAs:Co [19].

Zeeman experiments on the 0.786-eV no-phonon line were carried out in optical absorption using light from a tungsten-halogen lamp. Fig. 9 shows the evolution of the Zeeman spectra for a magnetic field within the (110)-plane 7° away from the [011] direction. A doublet structure of the no-phonon line is resolved at zero field with an energy separation of 0.1 meV between both components. A study of the thermalization properties has undoubtedly shown that the doublet splitting is in the excited state.

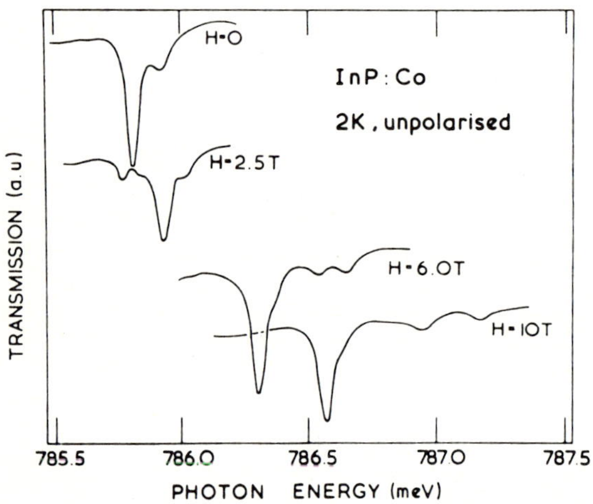

Fig. 9
Magneto-absorption spectra for H at 97° (Fig. 10)

As is the case in GaP:Co or GaAs:Co one may have expected to observe a cubic anisotropy in this system. Instead, the Zeeman splitting shows the typical characteristics of a trigonal defect with symmetry axis parallel to a [111]-direction of the crystal. The essential features of the anisotropy (Fig. 10) and magnetic field dependence (Fig. 11) can be described by an isotropic Zeeman splitting in the ground state and an anisotropic Zeeman splitting in the excited state with $g_\perp > g_\parallel$. The only representation of C_{3v} for which $g_\perp$ is different from zero is Γ_4 corresponding to a quasiangular momentum of 1/2. It follows that the system must have an odd number of electrons.

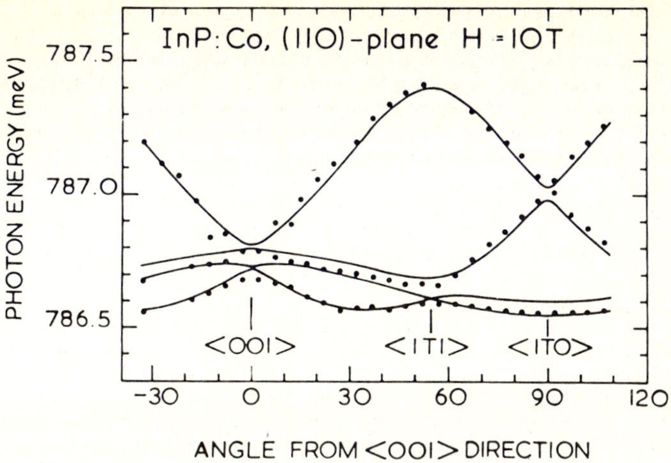

Fig. 10
Anisotropy plot at H=10T in the (100) plane. Circles-experimental points, lines-theoretical fit assuming a dyanmical J-T effect in the excited $^4T_1$ state.

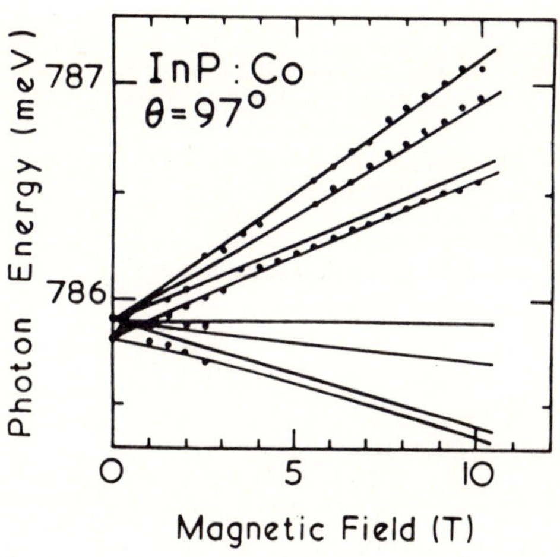

Fig. 11
Magneto-absorption from 0 T to 10 T.
Magnetic field at θ=97°.
Circles and lines as for Fig. 10.

For $H||[001]$ the four possible orientations of a trigonal center should remain equivalent, whereas for $H||[111]$ and $[110]$ three and two center orientations respectively should be equivalent. This is seen not to be obeyed precisely on closer inspection of Fig. 10. Anticrossings of the Zeeman components are found at the main symmetry directions rather than direct crossing points expected on the simple model of non-interacting C_{3v} centres. This behaviour together with the fact that there exists a zero-field splitting of 0.1 meV in the excited state can be explained by the presence of a strong dynamic Jahn-Teller coupling to T_2 (trigonal), C_{3v} distortion modes. The anticrossings then arise from interconfigurational scattering between magnetically equivalent distortions. A schematic energy level diagram is depicted in Fig. 12.

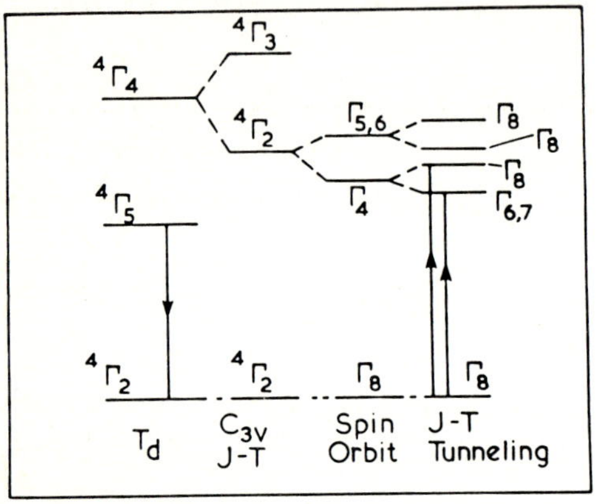

Fig. 12
Schematic energy level diagram for Co^{2+} (d^7). Energies not to scale.

The energy separation between the Γ_4 and $\Gamma_{5,6}$ excited state components is expected to be large compared to the Zeeman energy. One can therefore confine the excited state Hamiltonian to the space which is spanned by the two-fold Γ_4 state ($S=3/2, M=\pm 1/2$) and the four equivalent Jahn-Teller configurations. The mere fact that the spin quantization axis is fixed to the trigonal axis of the Jahn-Teller distortion leads to an effective spin-orbit interaction between the quasispin $S=1/2$ of the Γ_4-state and linear combinations within the JT-configuration space which can be assigned by an quasiangular momentum $L=0$ and $L=1$, $M=0, \pm 1$. The effective Hamiltonian describing the Zeeman splitting in the excited state has following form:

$$H_{exc} = \sum_{\nu} (g_{\parallel} - g_{\perp}) \mu_B (\vec{\sigma}\cdot\vec{X}_{\nu}) (\vec{H}\cdot\vec{X}_{\nu}) |\nu\rangle\langle\nu|$$

$$+ g_{\perp}\mu_B \vec{\sigma}\cdot\vec{H} + \alpha\vec{L}\cdot\vec{\sigma} + \beta(L^2 - 3/2)$$

The first two terms describe the anisotropic Zeeman splitting of the Γ_4-state and the last two terms the tunneling effect between equivalent JT configurations. The index ν denotes the four equivalent JT distortions and X_{ν} is a unit vector pointing along the C_{3v} symmetry axis. The effect of a magnetic field on the spin-only $^4A_2$ ground state can be described by a simple Zeeman term for $S=3/2$ with $g=2.10$. The excited state g-values are found to be $g_{\parallel}=0.96$ and $g_{\perp}=3.76$.

V. CONCLUSION

Both examples, Zeeman spectroscopy on the 0.84 eV no-phonon structure in GaAs:Cr and on the 0.786 eV no-phonon doublet in InP:Co, have demonstrated that the anisotropy of the Zeeman splitting contains much more information than that about the symmetry of a defect and the g-tensor of the effective spin state. It was shown that the parameters of the zero-field Hamiltonian can be directly determined from the Zeeman spectra provided the fine structure of the transition can be studied in the high field limit. This is of special importance in all those cases where a small fine structure is observed as a result of the coupling between an effective angular momentum state and the total electron spin. Such a small effective spin-orbit coupling is often a consequence of a dynamical Jahn-Teller effect in one of the crystal field states.

ACKNOWLEDGEMENT

The author would like to acknowledge close collaboration with Prof.L.Eaves, University of Nottingham, and Dr.M.S.Skolnick, R.S.R.E. Gt. Malvern. Finally I would like to thank H.Krath for expert technical assistance during the experiments.

REFERENCES

[1] J.J.Krebs and G.H.Stauss, Phys.Rev.B15, 17 (1976)
[2] J.J.Krebs and G.H.Stauss, Phys.Rev.B16, 971 (1977)
[3] J.J.Krebs and G.H.Stauss, Phys.Rev.B20, 795 (1979)
[4] U.Kaufmann and J.Schneider, App.Phys.Lett. 36, (9),747 (1980)
[5] W.H.Koschel, S.G.Bishop, and B.D.McCombe, in Proceedings of the 13th Intern. Conference on the Physics of Semiconductors, Rome 1976, ed. by F.G. Fumi, (North-Holland, Amsterdam, 1976), p. 1065
[6] H.J.Stocker and M.Schmidt, J.Appl.Phys. 47, 2450 (1976)
[7] E.C.Lightowlers, M.O.Henry, and C.M.Penchina, in Proceedings of the 14th Intern. Conference on the Physics of Semiconductors, 1978, Edinburgh, ed. by B.L.H.Wilson (IOP, Bristol, 1979), p. 307
[8] L.Eaves, Th.Englert, T.Instone, Ch.Uihlein, and P.J.Williams, in Proceedings of the Semi-Insulating III-V Materials Conference, ed. by G.J.Rees (Orpington, Shiva, 1980), p. 145
[9] N.Killoran, B.C.Cavenett, and W.E.Hagston, Solid State Commun. 35, 333 (1980)
[10] G.Picoli, B.Deveaud, and D.Galland, J.Phys. (Paris) 43, 133 (1981)
[11] L.Eaves, Th.Englert, and Ch.Uihlein, in Proceedings of the Oji Intern.Seminar, Hakone, Japan, ed. by S.Chikazume and N.Miura (Springer, Berlin, 1981), p.130
[12] B.Clerjaud, A.M.Hennel, and G.Martinez, Solid State Commun. 33, 983
[13] P.J.Williams, L.Eaves, P.E.Simmonds, M.O.Henry, E.C.Lightowlers, and Ch.Uihlein, J.Phys.C 15, 1337 (1982)
[14] L.Eaves, P.J.Williams, and Ch.Uihlein, J.Phys.C 14, L693 (1981)
[15] F.Voillot, J.Barrau, M.Brousseau, and J.C.Brabant, J.Phys.C 14, 1855 (1981)
[16] Ch.Uihlein and L.Eaves, Phys.Rev.B (1982) - in press
[17] M.S.Skolnick, P.J.Dean, P.R.Tapster, B.Cockayne and W.R.MacEwan, J.Luminesc. 24/25, 241 (1981)
[18] M.G.Clark and P.J.Dean, Proceedings of the International Conference on Semiconductors, 1978, Edinburgh, edited by B.L.H.Wilson (IOP, Bristol,1979)p.291
[19] J.Weber, H.Ennen, U.Kaufmann, and J.Schneider, Phys.Rev.B21, 2394 (1980)
[20] H.Ennen, U.Kaufmann and J.Schneider, Solid State Commun. 34, 603 (1980)
[21] M.J.Kane, Ch.Uihlein, M.S.Skolnick, P.J.Dean, and W.Hayes, to be published 1983

MAGNETO OPTICS OF II-VI COMPOUNDS UNDER LOW AND HIGH EXCITATION

C. Klingshirn
Physikalisches Institut der Universität, D-6000 Frankfurt a.M.

Abstract: Some recent results of magneto-optical investigations in II-VI semiconductors are reviewed. For the low excitation regime we consider the B-field dependence of free and bound excitons and of their excited states. In the high excitation case we discuss the behaviour of polariton-polariton interaction processes, of biexcitons and of the electron-hole plasma in fields up to 20T.

I Introduction

The optical properties of semiconductors in the spectral region close to the fundamental absorption edge are determined by excitons. Excitons are electron-hole pair states which are either free or localized at a lattice imperfection. The latter ones are called bound excitons. In the II-VI compounds which will be discussed in the following, the radii of the excitons are always large as compared to the lattice constant i.e. we are dealing with Wannier-excitons.

The properties of free and bound excitons and of their behaviour in magnetic fields have been studied over roughly five decades. Some earlier review articles are found e.g. in [1-3]. The fundamental properties of excitons have been investigated and understood at that time. In recent years, new interest arose for the exciton problem. Some of the reasons of this development are:

- in the field of bound excitons excited states of these complexes have been found in a large number of materials,
- concerning free excitons and excitonic polaritons, new experimental techniques of momentum space spectroscopy became possible through the availibility of spectrally narrow, tunable dye lasers. Nonlinear optical phenomena and high excitation effects gave an additional impetus for the investigation of optical properties of semiconductors.
- Last but not least the theoretical description of excitons made significant progresses. E.g. it became possible to overcome the simple two band models and to take into account the rather complex bandstructures of real semiconductors

Some recent review articles describing in detail some of the aspects mentioned above are e.g. ref. [4-8].

The aim of this paper is to review some selected topics of the magneto-optical properties of II-VI semiconductors. Concerning the materials, we selected CdS and ZnO as representatives for the hexagonal (Wurtizite-) type compounds and ZnTe as well as ZnSe for the cubic (Blende-) type compounds. The subjects discussed in the following are

- bound excitons including excited states of these complexes,
- free excitons and excitonic polaritons. Here we put special emphasis on the combination of classical experimental methods (like transmission or reflection measurements) and new linear and nonlinear spectroscopic techniques like resonant Brillouin-scattering or two photon Raman scattering,
- biexcitons and the electron hole plasma.

Due to the lack of space, it is not possible to review all of the contributions made to this field recently nor to cite all relevant literature.

II Bound Excitons

In the field of bound exciton complexes (BEC) we restrict ourselves to rather simple complexes namely to excitons bound to neutral donors (D^oX) and neutral acceptors (A^oX), respectively.

We assume that the radii of these complexes are large as compared to the lattice constants. Therefore electron and hole wave functions have to be classified according to the pointgroup of the lattice rather than according to that of the local symmetry of the center. Furthermore we assume a j-j coupling scheme for the angular momenta of the particles in the complex.

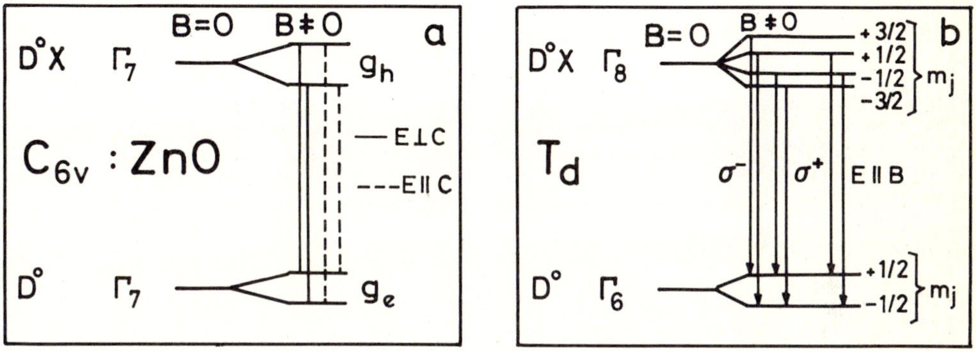

Fig. 1: Level scheme of D^o and D^oX complexes in C_{6v}:ZnO (a) and T_d symmetry (b), indicating dipole allowed transitions.

In Fig. 1a we give the symmetries, dipole allowed transitions and Zeeman splittings of the neutral donor (D^o) and the D^oX complex in ZnO, respectively. The electron g-factor is assumed to be isotropic, the hole g-value is given by

$$g_h = (g_{h\perp}^{\ 2} \sin^2\Theta + g_{h||}^{\ 2} \cos^2\Theta)^{1/2} \qquad (1)$$

where Θ is the angle between the magnetic field and the c-axis. $g_{h\perp}$ is assumed to be rather small, resulting in a significant assymmetry in the Zeeman splitting patterns for $\vec{B}\perp\vec{c}$ and $\vec{B}\,||\,\vec{c}$, respectively (see e.g. [9]). The transitions marked by a solid line are dipole allowed with $\vec{E}\perp\vec{c}$. The transitions marked --- are dipole allowed for $\vec{E}\,||\,\vec{c}$. However, they involve a spinflip and are therefore very weak. This explains why in experiments only a twofold splitting is observed, the lines being polarized with $\vec{E}\perp\vec{c}$. The other II-VI compounds with C_{6v} symmetry (e.g. CdS) are different from ZnO in their electronic structure in so far, as they have an uppermost valence band with Γ_9 symmetry instead of Γ_7. This modifies eq. (1) to

$$g_h = g_{h||} \cos\Theta \qquad (2)$$

since $g_{h\perp} \equiv o$ from symmetry reasons. The selection rules for this case are given e.g. in [1o]. The simplest approach to the Zeeman-splitting of D^o and D^oX complexes in T_d symmetry is shown in Fig. 1b. The transitions which are circularly polarized with σ^+ and σ^-, show up together in $\vec{E}\perp\vec{B}$. The splitting patterns of the A^o and A^oX complexes are similar to the ones shown in Fig. 1. In this case, however, the splitting of the ground state (A^o) is governed by the g-value of the hole whereas the magnetic field behaviour of the A^oX complex reflects the properties of the unpaired electron. The exchange in the roles of electron and hole may result in distinctly different thermalisation behaviour of the correspondi luminescence and absorption lines, provided that the relevant relaxation times are short as compared to the lifetime of the complex. As an exampl of the Zeeman splitting and the diamagnetic shift, we show in Fig. 2a an 2b the splitting patterns of the A^oX and D^oX complexes in ZnTe, respectively, for an orientation of the magnetic field $\vec{B} \parallel [1\bar{1}o]$.

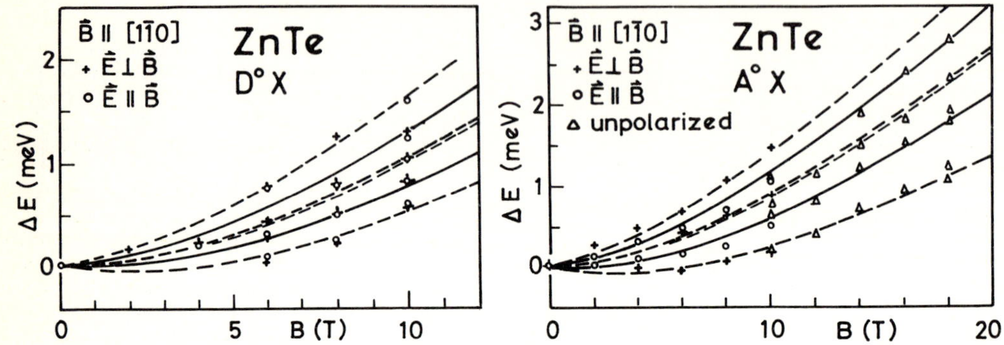

Fig. 2: Shift and splitting of the D^oX and the A^oX emissions in ZnTe situated for B=o at 2.3781 eV and 2.375o eV, respectively. From Ref. [11].

The circles, crosses and triangles are the experimental points for different polarization as indicated in the figures. The full and dashed lines are calculated curves using the parameters given in Table 1.

	g_{el}	$g_h^{1/2}$	$g_h^{3/2}$	$\alpha_{1/2}$ [meV T^{-2}]	$\alpha_{3/2}$ [meV T^{-2}]
A^oX	- o.47	o.54	o.53	o.o1o5	o.o1o
D^oX	- o.47	o.42 $\pm$ o.o5	o.48 $\pm$ o.o5	o.o13	o.o125

Table 1: Values of the g-factors and the diamagnetic shift factors used for the calculated curves in Figs. 2a and b. (ZnTe, B ∥ [1,$\bar{1}$,o])

It should be noted that for B ≳ 7T the regime of the low field approximation is left in ZnTe. Corresponding corrections [12] have been added t the diamagnetic shift terms. More details about these experiments, their evaluation and the comparison with data of other authors are found in [11

There are many other exciting features in the magnetooptic properties of bound exciton complexes (BEC) which are, however, beyond the scope of this paper. The reader is referred to the excellent review article published recently by Dean and Herbert [1o]. We shall concentrate in the se ond part of this chapter on excited states of bound exciton complexes (EBE Such states have been reported e.g. for the A^oX and D^oX complexes in CdS ZnO, ZnSe and ZnTe, e.g. [9,13-17] and the literature cited therein. The generally used experimental technique is photoluminescence excitation spectroscopy (PLE). The luminescence intensity of a certain bound excito line is recorded as a function of the quantum energy of exciting photons $\hbar\omega_{exc}$. Excited states of the bound exciton complexes show up as distinct resonances in the PLE spectrum. Furthermore photoluminescence spectroscopy and Raman experiments gave information about EBEC.

We discuss in the following the example ZnO. One has here detailed information about the magnetic properties of the EBEC [9]. As indicated above, the A^oX and D^oX emission lines split in a magnetic field into two Zeeman components. The diamagnetic shift is negligible for B $\lesssim$ 2oT. The PLE spectra of both components have been measured at low temperature (1.8K) using a tunabe, N_2-laser pumped dye laser as an excitation source. The PLE spectra of both components where found to be different, indicating that the spin relaxation time is larger than the energy relaxation time from the EBEC to the ground state and the ground state live time. In Fig. 3 we give the excitation energies ΔE i.e. the distances of the resonances from the corresponding Zeeman component as a function of $\vec{B}$ for a A^oX and a D^oX complex (I_9 and I_6 lines in Reynolds-notation). The most striking feature is probably the resonance with a $\vec{B}$-independent value for ΔE of 4.5 meV observed both for A^oX and D^oX BEC. Since this value of ΔE is comparable to the $A(\Gamma_7)$ and $B(\Gamma_9)$ valence-band (VB) splitting in ZnO (5.4 meV [19]) and since A- and B-bands are rather similar concerning g-values and effective masses [e.g. 19,2o], the 4.5 meV resonance is attributed to a D^oX or a A^oX complex containing a B-hole instead of an A-hole. Corresponding EBEC are known for CdS, too [14]. In the case of the A^oX complex even two holes from the A-VB can be replaced by holes from the B-band. The appearance of a resonance with ΔE = 9 meV (Fig. 3) can possibly be explained by the following mechanism: The ground state is a A^o complex with an A-hole. The incident photon creates an electron-B-hole pair with some excess energy. This pair relaxes by transferring the A-hole to the B-band and thus effectively a A^oX complex is formed with two B-holes. Data for $\vec{B} \parallel \vec{c}$ are in agreement with this model [9,18]

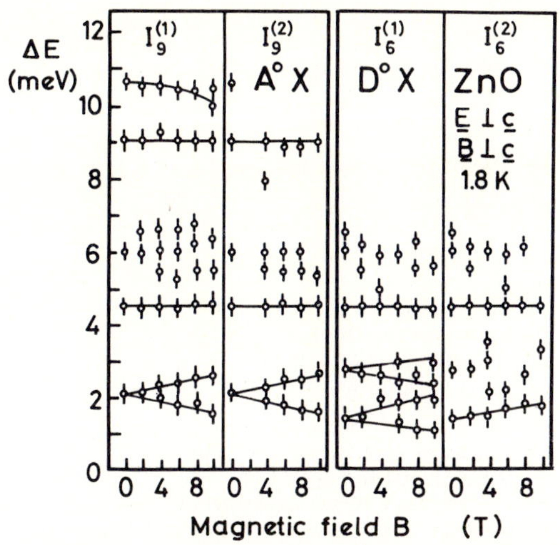

Fig. 3: $\vec{B}$-field dependence of the distances ΔE of resonances in the photoluminescence excitation spectra of a A^oX and a D^oX emission line. The superscripts $^{(1)}$ and $^{(2)}$ mean lower and upper Zeeman-component. For B=o the emission lines are situated at 3.3566 eV and 3.36o6 eV for I_9 (A^oX) and I_6(D^oX), respectively. From ref. [9,18].

The other resonances in Fig. 3 situated for B=o at ΔE values of about 2 meV, 6 meV and 11 meV, respectively, cannot be explained in terms of excitations within the exciton bound to the complex i.e. bound excitons with main quantum number n=2,3...One would expect for ZnO values around 4o meV. More likely, they are due to excitations of the exciton relative to the impurity center [16], to an excitation of the hole in a D^- potential in the case of D^oX [15] or to an excitation in the two hole system in the case of A^oX [16]. Unfortunately, there are to our knowledge until now almost no calculations for such EBEC involving realistic C_{6v} or T_d bandstructures and yielding excitation energies or $\vec{B}$-field splitting parameters which could be compared with experimental data. Maybe the increasing wealth of experimental results gives some stimulus to theoreticians to tackle this obviously not trivial problem.

III Free Excitons and Excitonic Polaritons

In the first part of this chapter, we shall shortly outline the basic co cepts of the theoretical description of excitons and polaritons, startin from the symmetries of the bands. A more detailed discussion is found in [8] and the literature cited therein. In a second part we review the cur rently used, experimental techniques for the observation of (magneto-) op tical properties of excitons and in the third part, we present several se lected results.

1. Theoretical considerations

The conduction bands (CB) of the compounds under consideration here result mainly from s-like electron states. This results in symmetries Γ_7 and Γ_6 for $k \equiv o$ in C_{6v} and T_d symmetries, respectively. The valence bands (VB) are formed from p-states with a more or less pronounced admix ture of lower lying d-levels. Spin orbit coupling splits this level in T symmetry into a fourfold degenerate Γ_8-VB and a twofold degenerate Γ_7-VB The spin orbit coupling is generally so large that only the upper band has to be taken into account in the spectral region close to the fundame tal absorption edge. In the T_d-II-VI compounds the upper band is the one with Γ_8 symmetry. It splits for $k \neq o$ into a light and heavy hole band. C_{6v} symmetry one has already at the Γ-point a splitting into three twofold degenerate VBs with symmetries Γ_9, Γ_7, Γ_7 due to spin-orbit coupling (SOC) and the hexagonal crystal field. The lowest band (Γ_7) can be neglected for our purposes. We are then generally left with an upper Γ_9 and a lower Γ_7-VB. In ZnO the SOC is smaller than the crystal field splittin and negative. So we have an upper valence band with symmetry Γ_7 and a fe meV below the Γ_9 band. The symmetries of the eight excitons formed with an s-like (Γ_1) envelope function are then given by eq. (3) and (4) in C_{6v} and T_d symmetries, respectively.

$$\Gamma_1^{env} \otimes \Gamma_7^{el} \otimes (\Gamma_7^{h} + \Gamma_9^{h}) = \Gamma_6 + \Gamma_5 + \Gamma_1 + \Gamma_2 + \Gamma_5 \tag{3}$$

$$\Gamma_1^{env} \otimes \Gamma_6^{el} \otimes \Gamma_8^{h} = \Gamma_3 + \Gamma_4 + \Gamma_5 \tag{4}$$

For main quantum numbers $n \geq 2$ of the envelope function also p, d... envelopes become possible. Several models have been developed to treat the influence of external perturbations (magnetic, electric or stress fields) including relastic band structures. See e.g. [4,6,7] and the literature cited therein.

We have chosen here the so-called invariant expansion as outlined for exc tons by Cho [21]. This theory regards the exciton states given in e.g. (3) or (4) as degenerate without perturbation. The external and interna perturbations which (partly) lift this degeneracy are treated on an equa footing. Time independent perturbation theory for degenerate states leads to a secular equation for the perturbed eigenenergies given by the equation (5)

$$\det \| H_{ij} - E\delta_{ij} \| = o \tag{5}$$

with $H_{ij} = \langle \theta_i | H | \theta_j \rangle$ where H is the perturbation Hamiltonian and the θ_i are the m unperturbed wavefunctions belonging to the degenerate eigenstate. Here m = 8. For every perturbed eigenenergy E_l deduced from eq. (5) a new eigenstate Φ_l is calculated i.e. a linear combination of the unperturbed wavefunctions.

$$\Phi_l = \sum_{i=1}^{m} a_{li}\, \theta_i \tag{6}$$

In the invariant expansion one starts with functions θ_i transforming according to irreducible representations of the point group of the crystal (here T_d or C_{6v}), so-called invariants. If then a perturbation of definite symmetry is applied to the system, group theoretical arguments allow to know, which of the H_{ij} in eq. (5) will be zero and which of the non zero elements are correlated. The perturbations used in the following are:

- the spin orbit splitting of the VB and in case of C_{6v} the hexagonal crystal field,
- k-linear and k^2 terms including warping terms for T_d symmetry and neglecting the k-linear terms of the CB and the hexagonal warping terms of the VB for C_{6v} symmetry. Also k-induced exchange terms are neglected,
- analytic and nonanalytic exchange terms of the excitons, describing singlet-triplet and transverse-longitudinal splitting, respectively,
- $\vec{B}$ linear (Zeeman-) and $\vec{B}^2$ terms, however, neglecting $\vec{B}$ induced exchange terms and $\vec{k} \cdot \vec{B}$ terms which may occur in C_{6v} symmetry.

Because of lack of space we do not give the complete matrices here. They are found for the general case in [21] and for the applications here in [11,18,2o,22]. Once the Hamiltonian is constructed, the eigenenergies and eigenfunctions (eq. (5) and (6)) are calculated numerically. This gives the dispersion of the various exciton branches in dependence of $\vec{k}$ and $\vec{B}$. From the eigenfunctions the oszillator strength of these exciton branches can be deduced (e.g. [2o]), and with this, the dispersion of the excitonic polaritons is calculated according to the polariton equation for the dielectric function ε.

$$\varepsilon(\omega,\vec{k}) = \varepsilon_b + \sum_l \frac{\varepsilon_b F_l(\vec{k},\vec{B})}{E_l^2(\vec{k},\vec{B}) - (\hbar\omega)^2 - i\hbar\omega\Gamma_l} \qquad (7)$$

$$\varepsilon = \frac{c^2 k^2}{\omega^2}$$

where ε_b is the background dielectric constant, $F_l(\vec{k},\vec{B})$ the oscillator strength of the exciton branch l with eigenenergy $E_l(\vec{k},\vec{B})$. Γ_l is a phenomenological damping term, which is in high quality samples rather small ($\hbar\Gamma \lesssim$ o.1 meV).

Once the dispersions of excitons and excitonic polaritons are calculated, they can be compared to experimental data. From fitting the calculation to the experiment the parameters entering in eq. (5) are determined.

As already mentioned in the introduction, the above outlined method is limited to the low field case, i.e. fields for which the cyclotron energy $\hbar\omega_c$ is small as compared to the exciton binding energy E_{ex}^b. In the high field case ($\hbar\omega_c >> E_{ex}^b$) one is left with series of Landau-levels for the CB and VB, the electron hole pair states being modified by the Coulomb interaction i.e. by the exciton effect. The intermediate field case should describe how the various splitting and shifting exciton states go over in a steady way to the modified e-h pair Landau levels. In contrast to the low field case, the theoretical understanding of the high field limit and especially of the intermediate field case is presently still considerably less developed. See e.g. the discussion in [4]. In contrast e.g. to many III-V compounds [4], the low field limit is left for most II-VI semiconductors only in the region between 1oT and 5oT. At these fields, the splittings and shifts of the excitonic levels are sufficiently well developed to be measured precisely by various experimental techniques. For the evaluation one profits from the highly developed low field theories.

2. Experimental techniques

In the following we shall shortly review the experimental techniques use to investigate the magneto-optical properties of excitons and polaritons starting with classical methods and ending with modern linear and nonlin ar techniques of momentum space spectroscopy:

- transmission measurements are very valuable for the detection of excitonic states with small oscillator strength (e.g. triplet or longitudi nal excitons). Strongly dipole-allowed levels (absorption constant $\gtrsim 10^5 cm^{-1}$) result generally in broad absorption bands which cannot be re solved except for extremely thin samples ($\lesssim 1 \mu m$),
- reflection spectra exhibit pronounced signals for strongly dipole-allowed states. The detailed analysis is, however, not trivial because of exciton free layers, spatial dispersion resulting in additional boundary conditions and a detuning of the exciton eigenenergies close t the surface. [23]. For some recent magnetoreflectance measurements in CdS and CdTe, see e.g. [24],
- luminescence features of free excitons are in II-VI materials generall considerably weaker than those of bound excitons. In spite of this fac and additional problems connected with reabsorption and the transmissi through the sample surface, magnetoluminescence measurements may yield detailed information about excitons (see e.g. [25]),
- two-photon absorption (TPA) measurements e.g. allow to detect states, which are one-photon forbidden. It gives valuable results of the fine-structure of excited excitons (e.g. n = 2 states with p-envelope [26]) and it makes momentum space spectroscopy possible [27],
- Perot-Fabry modes are equidistant in $\vec{k}$-space. The determination of their energetic positions in transmission or reflection thus enables one to reconstruct the $E(\vec{k})$ relation. They may reveal even simultaneous propagation on the inner and outer polariton branch above the longitudinal eigenenergy. However, extremely thin ($< 1 \mu m$) samples are necessary, (see e.g. [28] and the references therein).
- the diffraction of a tunable (laser-) beam through prism-shaped sample gives directly $n(\omega)$ or $E(\vec{k})$. If sufficiently thin prisms are available data can be even obtained in the resonance region [29],
- time of flight measurements of p sec light pulses through samples give the group velocity $d\omega/dk$ in the transparent region. In the strongly ab sorbing and more interesting resonance region the signal velocity may differ from the group velocity. This involves rather complex evaluatio procedures (e.g. [3o]),
- resonant Brillouin scattering (RBS) results from the scattering of polaritons in the bottelneck region under emission or absorption of acoustic phonons. The velocity of sound being known, the energetic Brillouin shift can be translated in a momentum transfer $\Delta\vec{k}$ and, if the involved polariton branches are known, these measurements allow to reconstruct the polariton dispersion in a rather large region of the Brillouin zone ($k \lesssim 0.01 \frac{\pi}{a}$) [5,31].
- two photon Raman scattering (TPRS) results from a nonlinear interactio between polaritons. Two polaritons from the incident (laser-) beam virtually form a biexciton which then decays under energy and momentum co servation in an observable photon-like polariton $\hbar\omega_R$ and another excit or polariton. By measuring $\hbar\omega_{exc}$, $\hbar\omega_R$ and the scattering geometry and applying the laws of refraction and of energy and momentum conservatio the dispersion of excitons and excitonic polaritons may be reconstruct from TPRS measurements in a self-consistent way (see e.g. [8,31] and t references therein). It should be pointed out, that TPRS is a high-exc tation phenomenon involving nonlinear polariton-polariton interaction.

Apart from the investigation of excitons it can be used to detect excitation induced renormalization effects in the polariton dispersion. (e.g. [33,34]).

After this short review of the experimental techniques, we present various selected results first for the T_d materials ZnTe and ZnSe and then for the C_{6v} compounds CdS and ZnO.

3. Results

We start with the dispersion curve of excitons and polaritons for the eightfold exciton state in ZnTe (Fig. 4). The transverse and longitudinal eigenenergies deduced from the analysis of reflection spectra are indicated. TPRS measurements gave a point on the intermediate polariton branch appearing because of the heavy and light hole band. In contrast to C_{6v} compounds, where generally the longitudinal exciton branch is involved in backward scattering experiments (see below), the process with an intermediate polariton in the final state seems to be more likely in T_d symmetry. For ZnSe see [35]. The corresponding part of the discussion in [36] for ZnTe has consequently to be modified. The dispersion of the lower polariton branch has been verified in foreward scattering in the region between 2,375 eV and 2,35o eV , not shown here [11]. For the data used in the calculation and for the comparison with those of other authors see [11].

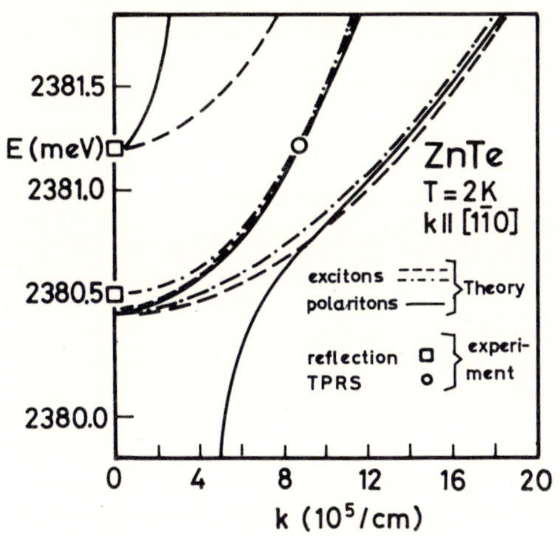

Fig. 4: Exciton and polariton dispersion in ZnTe. Data deduced from reflection- and TPRS experiments are indicated. From Ref. [11]

It is generally stated that the reflection minimum occurs close to the longitudinal eigenenergy (i.e. for $\hbar\omega_l$ given by $\varepsilon(\hbar\omega_l, k=o) = o$). This is only approximately true. Bearing in mind the simple formula

$$R = [(n - 1)/ (n + 1)]^2$$

one expects without damping and spatial dispersion R = o for n = 1. Taking into account the latter two effects, a reflection minimum is expected very close to the polariton energy for which n = 1. For ZnTe one gets n = 1 for k = 1.2 x $1o^5 cm^{-1}$. We make now a cut through Fig. 4. parallel to the ordinate and plot the energies of the corresponding exciton and polariton branches as a function of $\vec{B}$ in Fig. 5 for two different orientations [11]. The experimental points give the positions of reflection minima. They are nicely fitted by calculated polariton curves. For the magnetic field parameters used see [11]. For B > 7T corrections according to [12] are added to the diamagnetic shift.

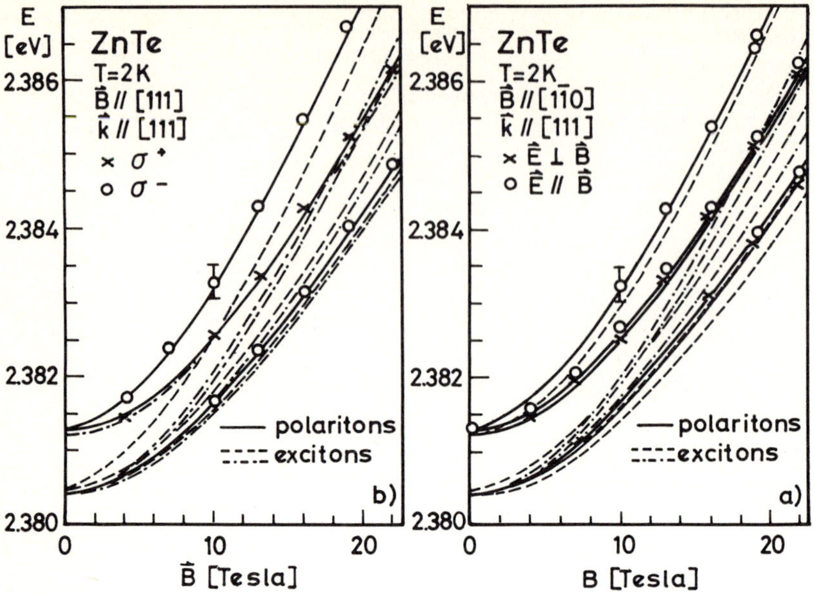

Fig. 5: The calculated $\vec{B}$-field dependence of exciton and polariton states at $k = 1.2 \times 10^5 cm^{-1}$ as compared to experimentally observed reflection minima (x and o). From ref. [11].

TPRS measurements now allow for an independent test of the set of parameters. As seen from Fig. 4, one reaches the intermediate polariton branch in the geometry given in [11] at $k = 8.5 \cdot 10^5 cm^{-1}$. In fig. 6 we plot now the various exciton and polariton branches at this value of k as a function of $\vec{B}$. The experimental points from TPRS nicely confirm the calculated curves. The abrupt transition from exciton to polariton branches and vice versa in fig. 6 is an artefact of the calculation: a threshold for the oscillator strength is used below which the quasi particle is regarded as a pure exciton and above it is subject to the polariton treatment.

In ZnSe magnetoreflectance measurements have been published by different authors ([37] and the literature cited therein). We should like to draw the attention of the reader here on a special aspect, the so-called magnetic circular dichroism (MCD). For $\vec{k} \| \vec{B}$ the radiation field couples differently to the various excitons branches for left and right circularly polarized light (σ^+ and σ^-, respectively). The resulting different polariton dispersions lead to a difference in the absorption and reflection spectra for σ^+ and σ^-, the MCD [37]. For the reflection, the MCD signal is defined as

$$\frac{\Delta R}{R} = \frac{2(R_{\sigma^+} - R_{\sigma^-})}{R_{\sigma^+} + R_{\sigma^-}}$$

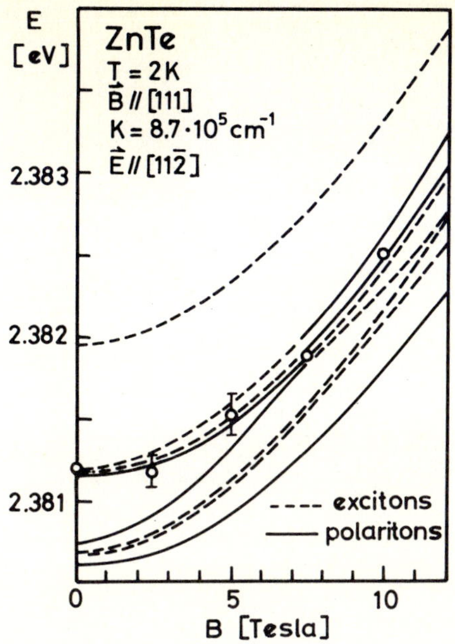

Fig. 6: The calculated $\vec{B}$-field dependence of exciton- and polariton states at $k = 8.5 \times 10^5 cm^{-1}$ as compared to experimental data from two-photon Raman scattering ⚲. From ref. [11].

In Fig. 7 we show MCD spectra for three different fields. The information contained in these spectra goes beyond the one of (wave length)-modulation techniques [37]. The zero- transitions marked by a dashed and a solid arrow give with high precision the k = o energies of the intermediate and the upper polariton branches, respectively.

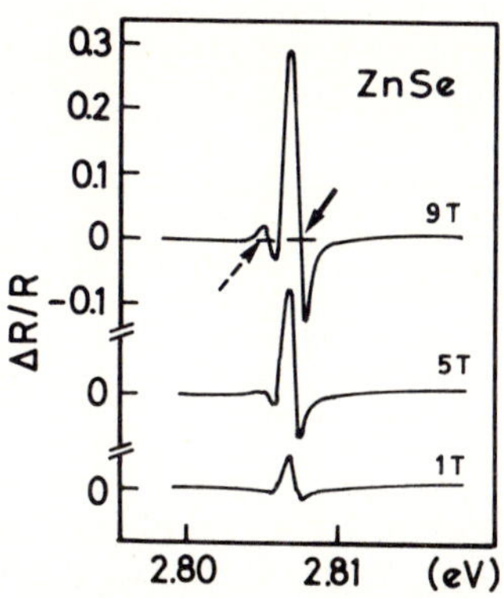

Fig. 7: Spectra of the magnetic circular dichroism for three different fields. The positions indicated by a dashed and a solid arrow give the k = o energies of the intermediate and the upper polariton branches. From Ref. [37].

As an example for the magneto-optic properties of excited free exciton states we present recent results of two-photon absorption measurements made at the n = 2 excitons with p envelope in ZnSe [26]. There are 24 of these states. Apart from the complex band structure, additional features show up due to the non-zero angular momentum and g-value of the envelope function. Fig. 8 shows experimental points for changes of the total magnetic quantum number of o, ± 1 and $\pm$ 2. All three cases are realizable by TPA. The solid lines are the result of a fit. The band parameters deduced from this fit are given in [26].

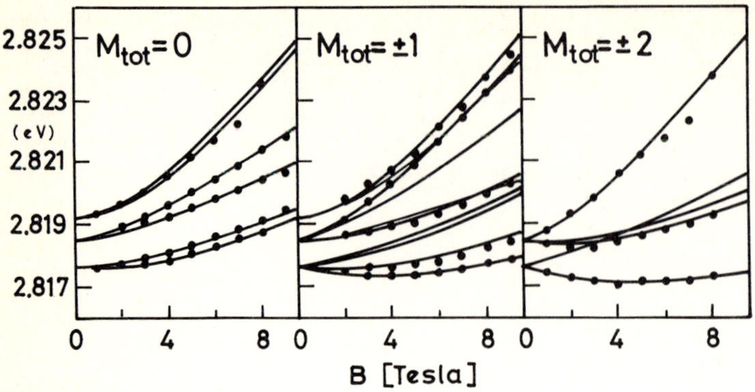

Fig. 8: $\vec{B}$-field dependence of n=2 exciton states with p-envelope. Solid lines: calculation, dots: experiment. From Ref. [26].

There is eventually one open question: 2p states are spread at 9T over a region of about 7 meV. This corresponds to one third of the exciton binding energy. Excitons with s- and p-envelope do not mix via the envelope function. If the internal structure is taken into account s-excitons ($\Gamma_3 + \Gamma_4 + \Gamma_5$) will mix with p-excitons ($\Gamma_5^{env} \boxtimes (\Gamma_3 + \Gamma_4 + \Gamma_5)$) via terms linear or quadratic in the magnetic field $\vec{B}(\Gamma_4)$. It is not clear in the moment, to which extend a mixing of the 2p-states with 1s, 2s and higher exciton states might influence the parameters deduced from the results shown in Fig. 8.

Magneto-Brillouin scattering has been performed for the first time in CdS [31]. In Fig. 9a the dispersion curves of the Γ_5 polariton are plotted for B = 0T. For finite magnetic field Γ_5 and Γ_6 states are mixed. This results among others in the appearance of a new intermediate polariton branch (Fig. 9b). The arrows in Fig. 9a and b give the Stokes and anti-Stokes RBS-transitions, involving LA-phonons for a backward scattering geometry. Fig. 9c and d show the observed Stokes-shifts for various laser energies $\hbar\omega_{exc}$ together with calculated curves for B = 0 and B = 10T. The fit has been used to determine precisely e.g. the exchange splitting between the Γ_6 (triplet) and the Γ_5 (singlet) state to $\Delta_{56}=(0.2 \pm 0.03)$ meV.

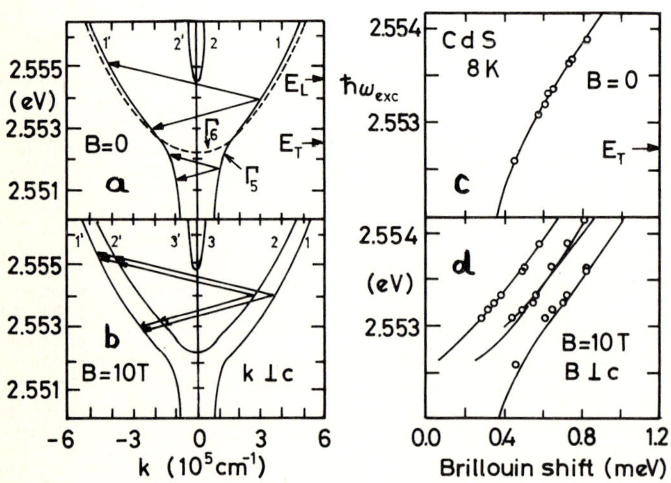

Fig. 9: Polariton dispersion in CdS, for B=0T (a) and B=10T (b). Stokes and anti-Stokes Brillouin scattering transitions with LA-phonons are indicated. Stokes Brillouin shift of 0T (c) and 10T (d) in dependence of $\hbar\omega_{exc}$. Solid lines: theory, dots: experiment. From ref. [31].

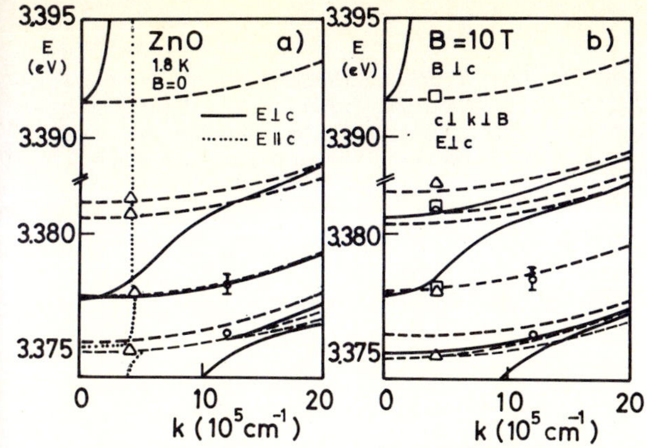

Fig. 1o: Review of magneto-optic exciton data for ZnO. (a): dispersion of excitons and of polaritons for $\vec{E}\perp\vec{c}$ and $\vec{E}\parallel\vec{c}$ and B=o; (b) for B=1oT and $\vec{E}\perp\vec{c}$. (c) and (d): field-dependence of exciton and polariton eigenenergies for k=o and two different orientations. From Ref. [18, 2o]

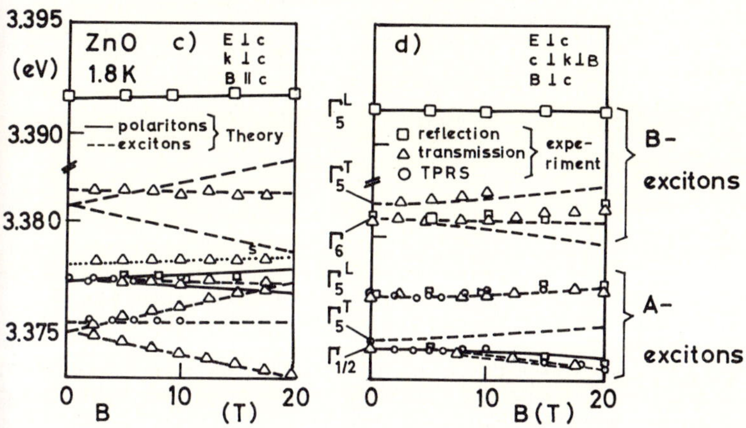

In Fig. 1o we show the combination of magnetooptic results from reflection-, transmission- and TPRS spectroscopy for ZnO [2o]. The combination of these techniques allows to get data for almost each exciton or polariton branch. This wealth of data can generally not be obtained by one technique alone. Fig. 1oa gives the dispersion of excitons and of polaritons for $\vec{E}\perp\vec{c}$ and $\vec{E}\parallel\vec{c}$ for B=o. Fig. 1ob shows the more complex pattern for B=1oT ($\vec{B}\perp\vec{c}$, $\vec{E}\perp\vec{c}$). The B-field induced intermediate polariton branch of the A-exciton in CdS (fig. 9) appears in ZnO for the B-exciton (around 3.381 eV). Fig. 1oc and d finally give the exciton and polariton eigenenergies for two different orientations of $\vec{B}$ and for k=o. Experimental data are extrapolated to k=o. The dotted line in Fig. 1oc corresponds to an isotropic point. Similar data have been obtained for CdS [2o].

We may conclude this chapter with the statement, that theoretical models are sufficiently well developed as to describe magneto-excitons and polaritons in real semiconductors. On the other hand the combination of various experimental techniques gives sufficiently precise data, so that many important material parameters may be deduced from a fit of theory to these magnetooptic data. Starting from this save platform progresses for the intermediate- and high field cases may be expected for the future. For some recent theoretical and experimental results see e.g. [38] and [48].

IV High Excitation Phenomena

The polariton-polariton interaction in the TPRS process, which has been mentioned in the proceeding chapter, is a typical nonlinear, high-excitation phenomenon. Another process in this field is the fusion of two polaritons to form a biexciton. In the following, we discuss magneto-optical properties of the biexciton and of the electron-hole plasma.

The fusion of two polaritons forming a biexciton (or the two photon-absorption to the biexciton in the language of perturbation theory) has a high transition probability (or giant oscillator strength) due to resonance terms and the large radius of Wannier excitons [39].

The first experimental investigation of the magneto-optical properties of biexcitons has been performed in CdS by luminescence spectroscopy [4o] and luminescence-assisted two photon spectroscopy (LATS) [41]. In the luminescence experiments, an emissionline in the spectral region of the P-bands is attributed to the decay of a biexciton into a photon and a 2p-exciton. The binding energy deduced from this model (E^b_{biex} = 2.5 meV) is by a factor two samller than the generally accepted value (4.8 ± o.3) meV [33] and the diamagnetic shift is larger than reported below. So eventually an excited state of the molecule has been observed in [4o]. In LATS experiments one creates by a tunable dye laser a polariton field in the sample with well defined energy ($\hbar\omega_{exc}$) and wavevector. The LATS-process shows up in a dip in the luminescence spectrum shifting in the opposite direction as $\hbar\omega_{exc}$ and fulfilling the relation

$$\hbar\omega_{exc} + \hbar\omega_{dip} = E_{biex} \qquad (1o)$$

By varying the scattering geometry it was even possible to measure the dispersion of the biexciton in CdS [33]. Here we present results where LATS has been used to determine the diamagnetic shift of the Γ_1 biexciton ground state in CdS [41]. In Fig. 11 we give data for $\vec{B}\|\vec{c}$ and $\vec{B}\perp\vec{c}$. A diamagnetic shift of 2.5 x 10^{-6} eVT^{-2} and of 2.o x 10^{-6} eVT^{-2} has been found for these two orientations. Together with exciton data, this results in $\alpha^{biex}/\alpha^{ex}$ = 2.2 in reasonable agreement with theoretical predictions [41]. Due to the strong Zeeman-splitting of the $A\Gamma_6$ exciton for $\vec{B}\|\vec{c}$ the biexciton binding energy decreases with increasing $\vec{B}$ and tends to zero for B ≈ 27T. Since such (steady) fields will be available in the near future (e.g. [42]), a system can be investigated in which E^b_{biex} can be continuously varied without changing the band parameters.

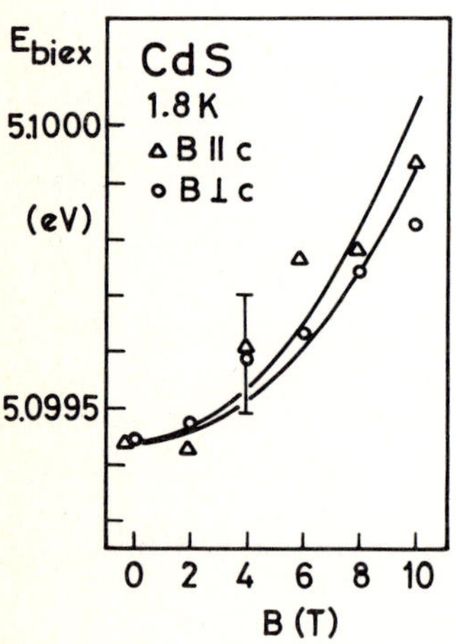

Fig. 11: $\vec{B}$-field-dependence of the biexciton eigenenergy in CdS. From Ref. [41].

If semiconductors are subject to a sufficiently strong pumping, excitons and biexcitons cease to exist and a new collective phase is formed, namely an EHP. The appearance of the EHP is connected with a strong density dependent decrease of the gap $E_g'(n_p)$ and - at least at low temperatures - a degeneracy of the occupation of the bands [8]. Quasithermodynamics predict for plasma temperatures T_p below a critical value T_c a first order phase transition from a low density exciton phase to a high density liquid plasma phase (EHL). The predictions have been verified e.g. in Si or Ge [43]. In magnetic fields the CB and VB split in series of Landau levels. They have been observed in the case of indirect materials e.g. [44]. In CdS one expects a splitting of the Landau levels of about 15 meV in fields of 2oT. This value is considerably larger than the damping, at least around the chemical potential μ [8]. In experiments where the optical gain has been studied, neither a modulation of the spectra due to the Landau levels, nor a oscillation of μ linear in B^{-1} has been observed [45]. The origin of this negative result is seen in the fact, that in the direct gap II-VI compounds the lifetime of the e-h pairs in the plasma is so short, that the phase separation cannot develop. One has strong fluctuations of n_p resulting in a fluctuating $E_g'(n_p)$. This in turn smears out the Landau-structures. For a detailed discussion of the influence of the lifetime on the phase-transition in an EHP see [46].

In closing this chapter, I should like to state, that the investigations of the influence of magnetic fields on high-excitation and nonlinear optical effects is still at its very beginning. Exciting results may be expected in the future e.g. in the fields of spin alligned EHP (for first results in CdS see e.g. [47] or biexcitons.

Acknowledgements: The author wants to thank his corworkers Dr.G. Blattner, Dr. G. Kurtze and Dr. W. Maier for performing many of the above mentioned experiments, and Dr. G. Schmieder for calculating the field influence on excitons and polaritons. Thanks are due to Prof. Landwehr and the Max-Planck-Institut für Festkörperphysik, for giving the possibility to perform the experiments at the high magnetic field facility in Grenoble. Fruitful and stimulating discussions with Dr. B. Hönerlage (Strasbourg), Dr. Ch. Uihlein (Grenoble), Dr. P.J. Dean (Malvern), Prof. Dr. U. Rössler (Regensburg) and many other collegues are gratefully acknowledged. Last but not least, the author wants to express his thanks to the Deutsche Forschungsgemeinschaft for financial support.

Literature references:

[1] R.S. Knox, Theory of Excitons, Solid State Physics,Suppl. 5, eds. F. Seitz and D. Turnbull, Academic Press (1963)
[2] J.O. Dimmock, in Semiconductors and Semimetals Vol III, 259, Academic Press (1967)
[3] S. Nikitine, in Optical Properties of Solids, p. 197, Plenum Press, New York (1969)
[4] D. Bimberg, Advances in Solid State Physics Vol. XVII, 195 (1977)
[5] R.G. Ulbrich and C. Weisbuch, ibid. Vol. XVIII, 217 (1978)
[6] U. Rössler, ibid. Vol. XIX, 77 (1979)
[7] Topics in Current Physics, Vol. 14, ed. K. Cho, Springer (1979)
[8] C. Klingshirn and H. Haug, Physics Reports 7o, 315 (1981)
[9] G. Blattner, C. Klingshirn, R. Helbig and R. Meinl, phys. stat. sol. b 1o7, 1o5 (1981)
[1o] P.J. Dean and D.C. Herbert in ref. [7] p. 55
[11] W. Maier, Ph.D. Thesis, Karlsruhe (1982) to be published
[12] D. Cabib, E. Fabri and G. Fioro, Il Nuovo Cim. B1o, 185 (1972)
[13] H. Malm and R.R. Haering, Can. J. Phys. 49, 2432 (1971)
[14] D. Munz and M.H. Pilkuhn, Solid State Comm. 36, 2o5 (198o)
[15] J. Puls, H. Redlin and J. Voigt, phys. stat. sol. b 1o7, K71 (1981)

[16] P.J. Dean, D.C. Herbert, C.J. Werkhoven, B.J. Fitzpatrick and R.N. Bhargawa, Phys. Rev. B23, 4888 (1981)
[17] C. Klingshirn, W. Maier, G. Blattner, P.J. Dean and G. Kobbe, J. Crystal Growth 59, (1982) in press
[18] G. Blattner, Ph.D. Thesis, Karlsruhe (1982)
[19] R. Helbig, Habilitation Thesis, Erlangen (1976)
[2o] G. Blattner, G. Kurtze, G. Schmieder and C. Klingshirn, Phys. Rev. B 25 , 7413 (1982) and G. Schmieder, Ph.D. Thesis, Karlsruhe (1981)
[21] K. Cho, Phys. Rev. B14, 4463 (1976) and ref. [7] p. 1 and p. 15
[22] G. Kurtze, Ph.D. Thesis, Karlsruhe (1982)
[23] J. Lagois, Phys. Rev. B23, 5511 (1981)
[24] I. Broser and M. Rosenzweig, Phys. Rev. B22, 2ooo (198o); W. Dreybrodt, K. Cho, S. Suga, F. Willmann and Y. Niji, Phys. Rev. B21, 4692 (198o)
[25] K. Hümmer, R. Helbig and M. Baumgärtner, phys. stat. sol. b86, 527 (1978)
[26] H.W. Hölscher, A. Nöthe and Ch. Uihlein, Proc. 16th Intern. Conf. Phys. Semicond., Montpellier (1982)
[27] D. Fröhlich, Advances in Solid State Physics Vol. XXI, 363 (1981)
[28] I.V. Makorenko, I.N. Uraltsev and V.A. Kiselev, phys. stat. sol. b 98, 773 (198o)
[29] I. Broser, R. Broser, E. Beckmann and E. Birkicht, Sol State. Comm. 39, 12o9 (1981)
[3o] T. Itoh, P. Lavallard, J. Reydellet and C. Benoit à la Guillaume, Sol. State. Comm. 37, 925 (1981)
[31] I. Broser and M. Rosenzweig. Sol. State. Comm. 36, 1o27 (198o)
[32] J.B. Grun, J. Phys. Soc. Japan, 49, Suppl. A,563 (198o)
[33] K. Kempf, G. Schmieder, G. Kurtze and C. Klingshirn, phys. stat. sol. b 1o7, 297 (1981) and V.G. Lyssenko, K. Kempf, K. Bohnert, G. Schmieder, C. Klingshirn and S. Schmitt-Rink, Solid State Comm. 42, 4o1 (1982)
[34] R. Baumert and I. Broser, Sol. State Comm. 38, 31 (1981)
[35] Y. Nozue, M. Itoh and K. Cho, J. Phys. Soc. Japan 5o, 889 (1981)
[36] W. Maier, G. Blattner, G. Kurtze and C. Klingshirn, Z. Physik 4o, 15 (198o)
[37] S. Feierabend, Ph.D.-Thesis, Dortmund (1979); S. Feierabend and H.G. Weber, Sol. State Comm. 26, 191 (1978); Ch. Uihlein and S. Feierabend, phys. stat. sol. b 94, 153 and 421 (1979) and the literature cited therein.
[38] N. Miura, G. Kido, H. Katayama and S. Chikamuzi, J. Phys., Soc. Japan 49, Suppl. A, 4o9 (198o)
[39] E. Hanamura and H. Haug, Physics Reports 33C, 2o9 (1977)
[4o] B.S. Razbirin, I.N. Uraltsev and G.V. Mikhailov, Sol. State Comm. 25, 799 (1978)
[41] G. Kurtze, V.G. Lyssenko and C. Klingshirn, phys. stat. sol. b 11o, K1o3 (1982)
[42] G. Landwehr, in Springer series in Solid State Sciences Vol. 24, p. 2 (1981)
[43] T.M. Rice, Solid State Physics, Vol. 32, p. 1 and J.C. Hensel, T.G. Philips and G.A. Thomas ibid. p. 88. Academic Press (1977)
[44] H. Nakata and E. Otsuka, Phys. Rev. B23, 5428 and the literature cited therein.
[45] H. Haug, D.B. Tran Thoai, S. Schmitt-Rink, K. Bohnert and C. Klingshirn, J. Phys. Soc. Japan 49, Suppl. A, 5o3 (198o)
[46] K. Bohnert, M. Anselment, G. Kobbe, C. Klingshirn, H. Haug, S.W. Koch, S. Schmitt-Rink and F.F. Abraham, Z. Physik B 42, 1 (1981)
[47] S.I. Gubarev, V.G. Lysenko, V.I. Revenko and V.B. Timofeev, JETP-Lett, 26, 32o (1977)
[48] N.O. Lipari and M. Altarelli in Physics in High Magnetic Fields, Springer Series in Solid State Sciences, Vol. 24, p 18o (1981)

Magneto-Reflectance of the B-Exciton of CdS

I. Broser and M. Rosenzweig
Institut für Festkörperphysik der Technischen Universität Berlin

A complete lineshape analysis of magneto-reflectance spectra for the B-exciton of CdS showing perfect agreement between theory and experiment is presented for the first time. A strong difference in the Zeeman effect of the Voigt- and the Faraday-configuration is observed for H⊥c. Furthermore the spectra change with the sign of the magnetic field (pole-changing-effect). The g-value of the B-valence band and the exchange interaction energies - parameters unknown up to now - are determined.

1. Introduction

In the lowest order the exciton ground state (n=1) is degenerate corresponding to the product of the degeneracy of the valence and the conduction band. The degeneracy of the exciton states is lifted by different internal perturbations like exchange interaction and k-linear terms and by an external magnetic field. In addition these perturbations mix the excitonic wavefunctions leading to the possibility to observe former forbidden exciton states (1). The determination of the splittings and mixings from experiments becomes difficult because the strong interaction of excitons and photons results in relatively broad structures in the reflectance and transmission spectra, from which the energy parameters can only be deduced by lineshape analysis using the well known model of excitonic polaritons (2,3).

In the special case of the B-exciton of CdS the magneto-reflectance of the n=1 state of normal incidence of light has been observed by Suga et al. (4). The authors discussed the spectra qualitatively only, neglected k-linear terms and exchange interactions and did not distinguish between Voigt- and Faraday-configuration.

We present measured magneto-reflectance spectra of the B-exciton of CdS for H⊥c and different angles between the magnetic field H and the wavevector k. The experiments are analyzed by calculated spectra along the lines of excitonic polaritons considering the exciton dispersion from perturbation theory, a damping constant and a surface layer.

2. The States of the B-Exciton in Magnetic Fields.

The B-exciton of CdS has four states Γ_{5T}, Γ_{5L}, Γ_1 and Γ_2 which are special products of the valence and conduction band states (1). The Γ_2 state is the lowest one, from which the Γ_5 and Γ_1 states are separated by the analytic parts of the exchange interaction Δ_{25} and Δ_{21} respectively. For $\vec{k}=(0,k,0)$ the Γ_{5L} state is shifted to even higher energies by the nonanalytic part of the exchange interaction, the so called longitudinal-transverse splitting Δ_{LT}. The Γ_{5T} is optically allowed for E⊥c, the Γ_1 state for E∥c, while transitions to Γ_2 and to Γ_{5L} are dipole forbidden. The perturbation matrix (5) considering simultaneously k-linear terms and external magnetic fields shows a difference between a field parallel or perpendicular to the optical c-axis and also between the Voigt- and the Faraday-configuration. Here we restrict to the case H⊥c.
In the Voigt-geometry (H⊥k) the external magnetic field creates a splitting and mixing in the same way as the k-linear term:

$$\Gamma_{5T}/\Gamma_2: \quad E_{1,2} = E_o + \tfrac{1}{2}\Delta_{25} \pm \tfrac{1}{2}\sqrt{\Delta_{25}^2+((g_B^{\perp}-g_e^{\perp})\mu_B H_x-\varphi k_y)^2+\varphi^2 k_x^2} \qquad (1)$$

$$\Gamma_{5L}/\Gamma_1: \quad E_{3,4} = E_o + \tfrac{1}{2}(\Delta_{21}+\Delta_{25}+\Delta_{LT})$$
$$\pm \tfrac{1}{2}\sqrt{(\Delta_{25}-\Delta_{21}+\Delta_{LT})^2+((g_B^{\perp}+g_e^{\perp})\mu_B H_x-\varphi k_y)^2+\varphi^2 k_x^2} \qquad (2)$$

with $E_o=E(\Gamma_2)$ for k=0. The Γ_{5T} and the Γ_2 states split for k=0 proportional to the difference of the g-values of the valence and the conduction bands. For $k \neq 0$ the splitting depends on the sign of H_x in comparison to k_y, the field lowers or enlarges the effect of the k-linear term. As a consequence of the perturbation both states (E_1, E_2) have parts of the Γ_{5T}-wavefunction and get oscillator strength f_1, f_2 for $E \perp c$. For large k or H the ratio f_1/f_2 converges to unity. The splitting and mixing of the Γ_{5L}/Γ_1 states is quenched by the large longitudinal transverse splitting compared to H and k.

In the Faraday configuration ($H \parallel k$) the splitting and mixing of exciton states are more complicated since the k-linear term and the magnetic field have different effects. The k-linear term creates the splitting of Γ_{5T}/Γ_2 and Γ_{5L}/Γ_1 while the magnetic field produces a splitting of Γ_{5T}/Γ_1 and Γ_{5L}/Γ_2. As a result of the corresponding mixing of wave function four exciton states exist having oscillator strength for $E \perp c$.

$$E_{1,2} = E_o + \frac{1}{2}(\Delta_{25}+\Delta_{21}) \pm \frac{1}{2}\sqrt{(\Delta_{25}-\Delta_{21})^2+((g_B^{\perp}+g_e^{\perp})\mu_B H_y+\varphi k_x)^2+\varphi^2 k_y^2} \quad (3)$$

$$E_{3,4} = E_o + \frac{1}{2}(\Delta_{25}+\Delta_{LT}) \pm \frac{1}{2}\sqrt{(\Delta_{25}+\Delta_{LT})^2+((g_B^{\perp}-g_e^{\perp})\mu_B H_y+\varphi k_x)^2+\varphi^2 k_y^2} \quad (4)$$

(Note: The equations are only valid for small k-values when the Zeeman-effect is dominant; for higher k-values the perturbation matrix can only be solved by computer iterations.)

For k=0 E_1/E_2 splits proportional to the sum and E_3/E_4 proportional to the difference of g-values of the valence and the conduction band. Again the splitting depends on the direction of the magnetic field compared to the wave vector. But in contrast to the former case now the tangential component of the wave vector has to be considered in the brackets, which only appears for oblique incidence of light. Then it can be transformed into a constant energy shift using the law of refraction

$$\varphi k_x = \varphi k \sin\Theta' = \varphi \frac{\omega}{c} n \sin\Theta' = \varphi \frac{\omega}{c} n_o \sin\Theta = \mathrm{const}(\Theta, \varphi) \quad (5)$$

3. Experimental Results

Magneto-reflectance spectra of CdS were measured in a conventional way. The light of a high pressure Xenon-lamp was monochromized by a double grating spectrometer, reflected at the crystal immersed in liquid helium (T = 1,8 K) and detected by a photomultiplier. The high magnetic field up to H = 10 T was produced by a superconducting coil. Fig. shows the experimental results for different orientations and magnitudes of the magnetic field in full lines. The spectra were analyzed by computer calculations using the well-known model of excitonic polaritons including damping and surface layer (3); the calculated spectra are drawn in dashed lines.

In the case of the Voigt configuration two excitonic oscillators have to be considered in the dielectric function $\varepsilon(\omega,k)$ with eigenfrequencies corresponding to eq.(1). Any k- or H-dependence of the oscillator strength f_i can be neglected, as the large splittings of k and H compared to the zero splitting Δ_{25} creates a ratio of $f_1/f_2 \approx 1.0$. An excellent agreement between experiments and calculations (fig. 1a) can be achieved with the difference of g-values of $g_B^{\perp}-g_e^{\perp}$ = 2,55±0.05 and an exchange interaction of Δ_{25} = (0.1±0.05)meV. The eigenfrequency of the former Γ_{5T}-state (ω_o) shows a diamagnetic shift of δ_{dia}=(2.0±0,3)μeV/T². In the frame of experimental uncertainty the spectra are completely independent of the sign of the magnetic field although eq.(1) includes in principle a pole-changing effect.

In the Faraday configuration four excitonic oscillators exist with eigenfrequencies given in eq.(3) and eq.(4). The complicated dispersions of the exciton energies $\omega_i(H,k)$ resulting from the perturbation matrix have to be considered explicitely in the resonance terms of the dielectric function. In addition a

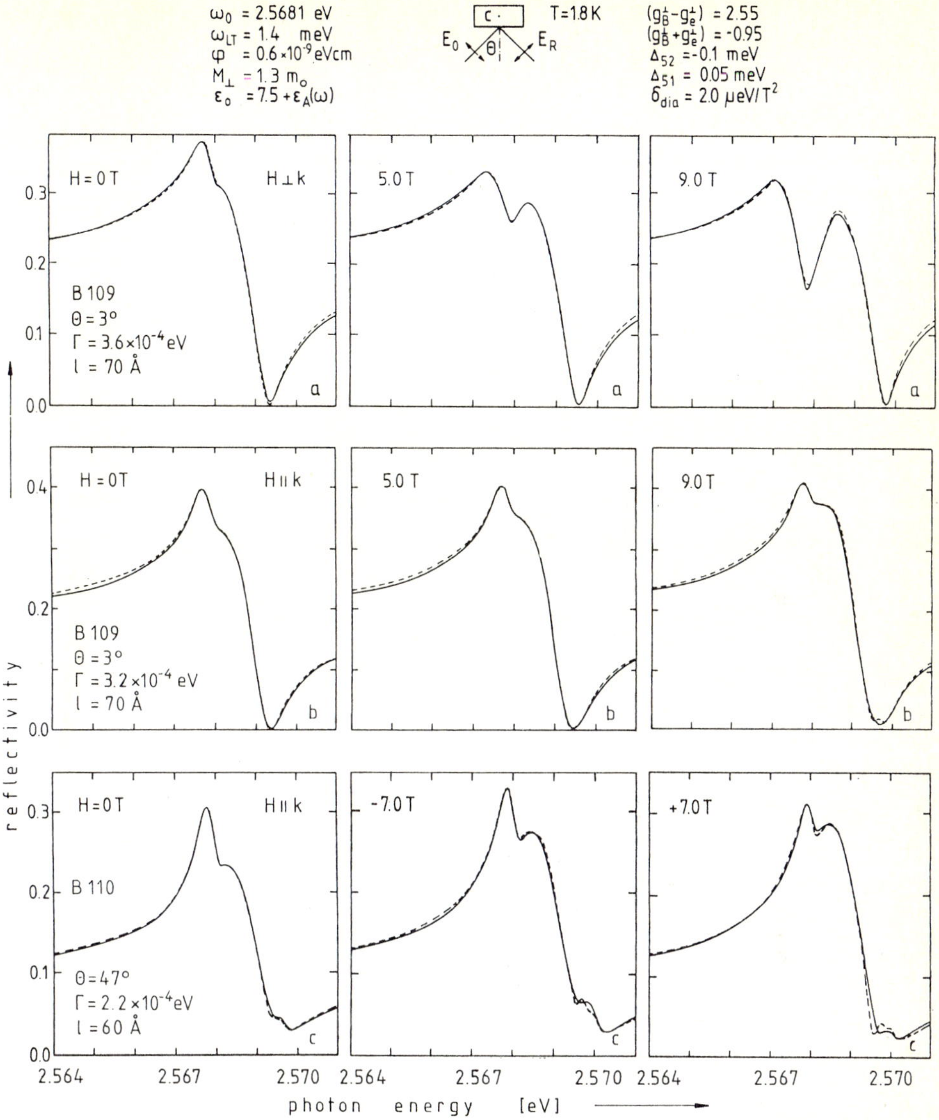

Fig. 1: Lineshape analysis of magneto-reflectance spectra of the B-exciton of CdS for different directions and amounts of the magnetic field, H⊥c.
a) H⊥k, Θ = 3°; b) H∥k, Θ = 3°; c) H∥k, Θ = 47°.
——— experimental results, ----- calculations.

k^2-dependence of the oscillator strength has to be used to describe the spectra. (The method of calculations will be discussed in detail elsewhere (5)). The Zeeman-splitting being much smaller than for the Voigt-geometry can be described by $g_B^\perp+g_e^\perp= -0,95\pm0.1$. The diamagnetic shift is again $\delta_{dia}=(2.0\pm0.3)\mu eV/T^2$. Although the spectra have a quite simple shape they can only be fitted using the calculated exciton dispersions $\omega_i(H,k)$ and considering all four oscillators. The spectra are independent of the direction of the magnetic field which is in accordance to eq. (3) and eq. (4) for normal incidence of light. Fig. 1c shows magneto-reflectance spectra of the Faraday-configuration for oblique incidence of light (Θ=47°) for H=+7.0 T and H=-7.0 T. A distinct change in the spectra is observable. Now the tangential component of the wavevector (eq.(5)) creates a constant energy shift which can be interpreted as an internal magnetic field H_{int}, which lowers or enlarges the external field. The exciton dispersion $\omega_i(H,k)$ and the oscillator strengths depend on the amount of the external field and also on its sign, $|H_{eff}|=|\pm H_{ext}+H_{int}|$. From eq.(5) the pole-changing effect is directly proportional to the k-linear term $\mathcal{S}$, which leads to $\mathcal{S}_H=(0.6\pm0.15)\times10^{-9}$ eVcm. The spectra of fig. 1c are measured with a crystal of smaller damping (Γ=2.2x10^{-4}eV). Thus the maximum of the former longitudinal exciton as the third oscillator can be clearly observed. The fourth oscillator influences only the relative minimum between both maxima but do not produce a separate extremum. The field dependence of the oscillator strengths of these both oscillators and their pole-changing shows an agreement with theory for an exchange interaction of $\Delta_{21}=\Delta_{25}+\Delta_{51}=(0.15\pm0.05)$meV.

4. Discussion

The most interesting problem of the magneto-optic of the B-exciton of CdS is the simultaneous appearance of two vectorial perturbations, the k-vector and the magnetic field, which should lead to pole-changing effects. In the Voigt-configuration a change in the spectra with sign of H could not be observed. This is in agreement with the calculated reflection spectra. The main reason for this result is the fact, that the normal component of the wave-vector, which should be responsible for a possible effect, passes for both field directions through all values of k with increasing photon energy corresponding to the polariton dispersion. In the Faraday-configuration on the other hand a pole-changing-effect has been observed but only for oblique incidence of light. In contrast to the former case now the tangential component of the wavevector appears, which can be transformed into a constant energy shift which is parallel or antiparallel to the external field. Quantitatively one would expect a spectrum which is similar to that of a lower or a higher external field. In eq.(3) and eq.(4) however the signs of $g_B^{\perp}+g_e^{\perp}$ and $g_B^{\perp}-g_e^{\perp}$ are opposite, which means that for $E_{1,2}$ the effective field is higher (smaller) than for $E_{3,4}$.

Concerning the energy parameters the k-linear term deduced from the pole-changing effect $\mathcal{S}_H=(0.6\pm0.15)10^{-9}$ eVcm agrees with the value in the lineshape analysis of the zero-field reflectance spectra $\mathcal{S}_o=0.6\times10^{-9}$ eVcm. The g-value of the conduction band $g_e^{\perp}=-1.75\pm0.1$ is in agreement with the results of the magneto-optic of the A-exciton (6), while the g-value of the B-valence band $g_B^{\perp}=0.8\pm0.1$ is new and has been determined for the first time. The same is valid for the exchange interaction energies $\Delta_{21}=(0.15\pm0.05)$meV and $\Delta_{25}=(0.10\pm0.05)$meV. They can only be compared with theoretical estimations. Δ_{21} and Δ_{25} can be calculated from the exchange interaction energy of the A-exciton (1) $\Delta_{65}=(0.20\pm0.03)$meV well-known from our magneto-Brillouin scattering (7). The values $\Delta_{21}^T=(0.19\pm0.03)$ meV and $\Delta_{25}^T=(0.11\pm0.02)$meV are almost identical with the experimental results.

In the presented paper the results of exciton perturbation theory and exciton-photon-interaction (polariton) are successfully combined leading to a complete understanding of the magneto-optical behaviour of the B-exciton of CdS. From a carefully performed lineshape analysis some new parameters could be determined.

References:

1. K. Cho; Phys. Rev. B 14, 4463 (1976)
2. S.I. Pekar; J. Phys. Chem. Sol. 5, 11 (1958)
3. I. Broser, M. Rosenzweig, R. Broser, M. Richard, E. Birkicht; phys. stat. sol. (b) 90, 77 (1978)
4. S. Suga, M. Bettini, W. Dreybrodt, K. Cho; Colloq. Int. Cent. Nat. Rech. Sci. No 242, 217 (1975)
5. M. Rosenzweig; Dissertation TU-Berlin (1982) D 83
6. I. Broser, M. Rosenzweig; Phys. Rev. B 22, 2000 (1980)
7. I. Broser, M. Rosenzweig; Solid State Commun. 36, 1027 (1980)

MAGNETO-OPTICAL STUDIES OF DONOR IMPURITIES UNDER HIGH PRESSURE

S.Wasilewski†, A.M. Davidson & R.A. Stradling
Physics Department, University of St Andrews, Fife KY16 9SS
†On leave from UNIPRESS, Warsaw, supported by SERC Visiting Fellowship.

S. Porowski
High Pressure Centre, UNIPRESS, Polish Academy of Sciences, Warsaw, Poland.

Magneto-optical studies are presented of the shallow donor states and cyclotron resonance of free carriers in n-InSb at hydrostatic pressures of up to 15 kbar. The deepest of the shallow donors shows an interaction with L-associated resonance states in the conduction band. The admixture of L-states permits intervalley coupling by a single TA-(L) phonon of energy of 35 cm^{-1}. Resonant metastable states of predominantly X-character also exist and, when these are occupied, the L- and Γ-like transitions involving the same donor disappear. Preliminary results for the donors in GaAs suggest that similar anticrossing effects occur between the deepest of the shallow donors and resonant states in the conduction band. Both the cyclotron resonance and impurity lines in InSb show a dramatic narrowing which can amount to almost an order of magnitude on increasing the pressure above 5 kbar and central cell structure can be observed on all three of the 1s-2p transitions observed in a magnetic field.

1. INTRODUCTION

Shallow impurities have binding energies of the order of 10 meV and are generally assumed to be described by effective mass theory with a character given by the nearest band extrema. For the donors in InSb this would seem to be a particularly good approach in view of the extremely small effective mass of the Γ conduction band. However in magnetic fields in excess of about 5T central cell structure becomes apparent on the $1s-2p_-$ line in the purest samples, indicating a significant perturbation by the local potential around the donor impurities and, at the highest fields, the magnitude of the chemical shift can correspond to about 10% of the binding energy. This paper demonstrates that resonance states can exist in the conduction band of InSb which are not of Γ character and shows how the existence of these states can modify the shallow donor spectrum which was previously thought to be described extremely well by simple effective mass theory. The non-Γ admixture in the shallow states can modify the electron-phonon interaction and the associated lattice relaxation effects.

The first indication that donor states might not be associated completely with the Γ conduction band came from Hall and resistance measurements undertaken as a function of hydrostatic pressure and temperature by the group led by Porowski (1,2).

The pressure technique developed by this group is particularly simple and well suited to experiments in the confined space provided by a magneto-transport cryostat. The pressure is generated at room temperature in a Be-Cu cell containing a light hydrocarbon as the pressure transmitting medium. The cell is then sealed and cooled to low temperature. Tests have shown that the resultant pressure in the solidified hydrocarbon is uniform to about 1%. The present experiments employ pressure cells utilising this technique which have been modified for magneto-optical experiments by adding a sapphire window.

The intial transport measurements were with samples so heavily doped with Te, Se and S (3,4) that the Fermi energy was more than a band-gap above the conduction band edge. With this degree of doping it was possible to detect resonance states up to 0.5 eV above the bottom of the conduction band and by applying pressure it was possible to study deionisation of the resonance states associated with higher order minima. By this means it was possible to determine the precise position of the resonance levels as a function of pressure. It was found that the depth of these states with respect to the higher order conduction band minima increases with decreasing atomic number (increasing electronegativity) and that the slopes for Se and S show distinct breaks with the resonance levels at low pressure following quite closely the L-minima and at higher pressures the X-minima. The results for Te which could only be studied with the highest doping levels were too fragmentary to reach a similar conclusion but are likely to have the same dependence as this impurity is more likely to be effective mass-like than the other two. The breaks in slope suggest a crossing of two levels from the same impurity, one having predominantly L-character and the other X-character.

With high purity material (ie donor concentrations below 10^{15} cm^{-3}), deionisation can be found in similar (5) experiments except that the two levels lie much closer to the conduction band edge. These levels arise from a commonly occurring residual donor which, from the position of the levels with respect to those produced by sulphur, selenium and telurium and from a knowledge of the likely contaminants, is believed to be oxygen (6). These oxygen levels again move linearly with respect to the bottom of the conduction band with pressure coefficients of -20 and -10.5 meV/kbar and cross into the gap at about 7 and 8 kbars respectively. Again the coefficients are close to those expected for the X- and L- minima: a surprising result in view of the more than 0.5 eV separation between the localised states and extrema providing their dominant characters. Nevertheless the separations are far too great for the levels to be described by simple effective mass theory and contributions from many points of the zone must be substantial. The transfer of electrons from the Γ band to the X-like impurities may be inhibited by large energy barrier ($\sim$ 0.3 eV) which appears below 120K due to a strong relaxation of the lattice around the impurity site. The time constant for transfer of electrons into or out of this state can become extremely long at low temperatures (eg 270 days

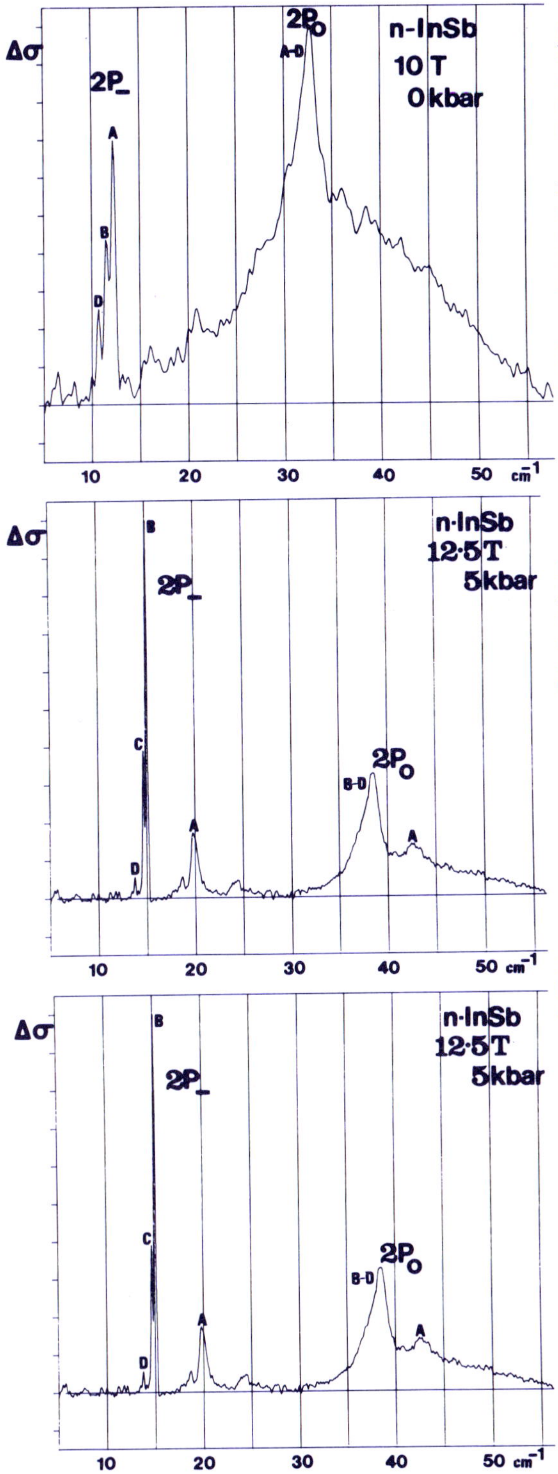

Fig. 1 shows the photoconductive spectrum of the donors in InSb in the region of the $1S \rightarrow 2P_-$ and $1S \rightarrow 2P_o$ lines for three different combinations of field and pressure.

A comparison of (a)-(c) shows a pronounced narrowing of the $1S-2P_-$ lines on increasing pressure and field. In (c) the line widths are instrumentally limited on the $1S-2P_-$ lines for donors B-D.

(a) is taken at zero pressure and 10T and shows a spectrum similar to those in reference (42) and the separation of donors A and B is 0.8 cm^{-1}.

(b) is taken at 4.9 kbar and 9T. The integrated intensity of donor A is about four times that of B and their separation is 2.5 cm^{-1}. As donor A splits off from the remaining donors, it becomes resolved on the broader $1S-2P_o$ line at 34 cm^{-1}.

(c) is taken at 4.9 kbar and 12.5T. Donor A now has less than half of the integrated intensity of B and their separation is 5 cm^{-1}.

at 77K) and this state can be considered to be metastable. A magnetic field may be used in a similar way to pressure to increase the value of the direct gap ($dE/dB \sim 2meV/T$).

If the L-like level was associated solely with the coresponding minima, without any X- or Γ-admixture, then it would be expected to be fourfold degenerate without spin. However valley-orbit interactions will split the states into two sets with the lower level being singly degenerate and of the same symmetry as the hydrogenic states associated with the Γ-minimum. According to the non-crossing rule for states of the same symmetry an interaction is expected as the states approach one another (7). In transport studies of magnetic freeze-out onto the shallow donors an admixture of the L- and Γ-levels was apparent when the two levels approached and anticrossing with an offset of 2 meV between the two branches was deduced. (8) The extension of the electrical measurements by optical measurements described in this paper has now permitted greater detail to be discovered about the nature of the interaction and its associated effects

2. MAGNETO-OPTICAL EVIDENCE FOR AN INTERACTION BETWEEN Γ and L-ASSOCIATED STATES

Magneto-optical experiments have been performed in order to investigate the interaction between the L- and X-associated states and the shallow donors reported in transport experiments (5,6,8). Preliminary results have been reported elsewhere(9,10). The experimental recording shown in figure 1(a) is typical of the magneto-optical spectrum at zero pressure expected for a high-purity sample of InSb without any intentional doping. The group of lines between 11 and 13 cm^{-1} are $1s\text{-}2p_-$ transitions from the four common residual donors (labelled A,B,C and D in this paper). At the magnetic field employed for this recording (10T), B and C are not resolved. The broader line at 33 cm^{-1} is the $1s\text{-}2p_0$ transition and central cell structure is not resolved on this line at zero pressure. The two weaker lines between the $1s\text{-}2p_-$ and $1s\text{-}2p_0$ peaks result from transitions starting and terminating on excited states. Figure 1(b) is taken at a slightly lower field but with a pressure of about 5 kbar applied. The lines have narrowed substantially so that donor C is resolved from donor B and, in addition, line A has more than doubled its separation from line B. The increased resolution which results enables the $1s\text{-}2p_0$ line due to donor A to be observed separately. Figure 1(c) shows the effect of a small increase in magnetic field at the same pressure. A further narrowing of the lines is apparent but the effect on donor A has been dramatic. It has dropped in intensity by almost an order of magnitude with respect to the other donors and a further substnatial increase in separation from B has occurred. In addition another weak line can be observed on the low frequency side of the $1s\text{-}2p_-(A)$ line. The anomalous deepening of the ground state of donor A which can be deduced from increased central cell splitting and the accompanying fall-off in signal intensities can be interpreted as due to the interaction between the Γ- and L-levels and the associated admixture of wave-functions. The fact that the anomalous behaviour is only observed on the deepest of the shallow donors is extremely signficant as the transport experiments suggested that oxygen was responsible for the resonance states which could be introduced into the gap by applying pressure (oxygen is the most electronegative of the possible substitutional donors from group IV or group VI). Furthermore there is a strong correlation between the relative strength of donor A in the central cell spectrum and the amount of freeze out than can be induced onto the L- or X-levels. In samples where A is by far the strongest feature in the central cell spectrum before pressure is applied, sample resistances in excess of 100 MΩ can be obtained at pressures in excess of 10 kbar without any magnetic field being applied.

Figures 2 and 3 show the positions of the $1s\text{-}2p_-$, $2p_-\text{-}3d_{-1}$ and $1s\text{-}2p_0$ lines as a function of pressure at a magnetic field of 11.5T and of field at a pressure 6.1 kbar respectively. The anomalous deepening of the A lines with respect to the B,

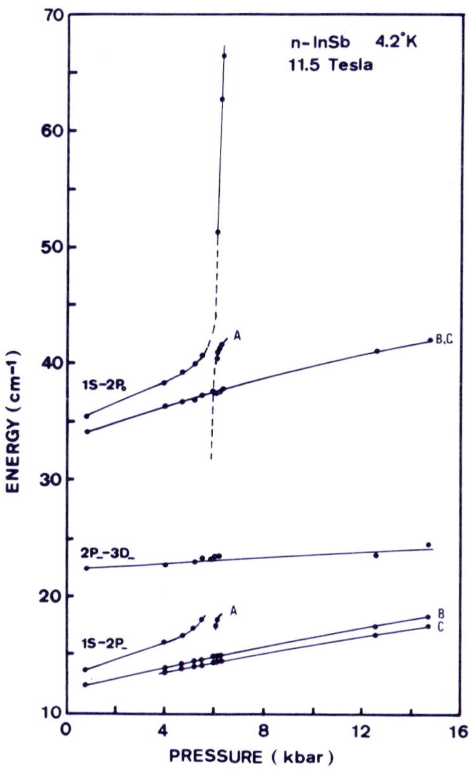

Fig.2. Impurity magneto-absorption peaks in InSb as a function of pressure at 11.5T. At about 6kbar a crossing of the L and Γ associated levels is seen for donor A. The lack of any splitting of the $2p_-\text{-}3d_{-1}$ line shows that hybridisation occurs on the ground state.

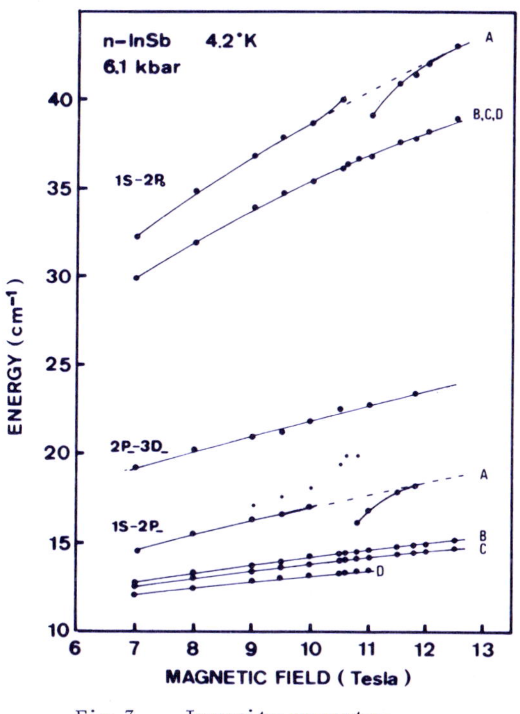

Fig.3. Impurity magneto-absorption peaks in InSb as a function of magnetic field at a pressure of 6.1 kbar. Two branches of the L-Γ interaction are observed on both the $1s\text{-}2p_-$ and $1s\text{-}2p_0$ transitions. On the $1s\text{-}2p_-$ line an extra splitting is observed between 9 and 10T.

C and D components is clearly displayed. In addition two new features can be seen. Firstly two branches of the coupled mode interaction can be observed very clearly on both the 1s-$2p_{-1}$ and 1s-$2p_0$ lines with a minimum separation of approximately 3 cm between the two branches. For the 1s-$2p_0$ transition a rapidly moving component can be followed to about 30 cm^{-1} away from the resonant cross-over as can be seen from Fig.4

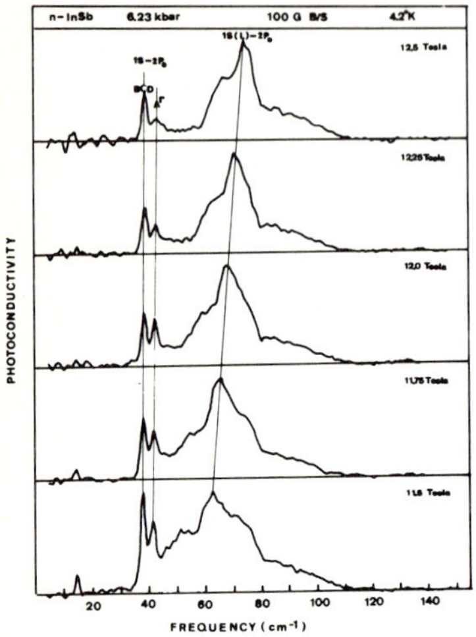

Fig.4 shows the 1s(Γ)-$2p_0$ and 1s(L)-$2p_0$ transitions of donor A at different magnetic fields at a pressure of 6.23 kbar together with the normal 1s-$2p_0$ transitions of donors B-D.

which shows the change in magneto-optical spectrum taken at a pressure of 6.23kbar as the field is changed in steps of 0.25T. The doublet structure at ∿ 40 cm^{-1} represents the B-D and A(Γ) components of the 1s-$2p_0$ transition. The A(Γ) component is weak because population of this level is small as the L-ground state is so much lower at these pressures and fields. The two rapidly moving lines in the region 60-80 cm^{-1} are thought to be the 1s(L)-$2p_0$ line 1s(L) to conduction band transition for donor A. A rapidly moving L-like component can also be seen on the 1s-$2p_+$ transition. Figure 5 shows results taken at a pressure of 6.87 kbar and at magnetic fields where both the 1s-$2p_0$ and 1s-$2p_+$ lines occur in spectral regions which can be studied simultaneously by the Fourier spectrometer. Under the conditions stated the 1s-$2p_+$ lines are well above the Restrahl frequency and consequently are broadened by LO phonon emission No cyclotron resonance is seen because of freeze-out onto the donor sites. Figure 6 shows results taken at higher pressure but much lower field where the cyclotron resonance is clearly seen and the lines are much sharper. Although the main lines are being broadened somewhat by the noise from the sample which has a high resistance under these conditions the $\omega\tau$ values of the cyclotron resonance are superior to any results reported at zero pressure. The true line width can be obtained by using a spectrometer employing a far-infrared laser as source and sweeping the field instead of employing a Fourier spectrometer. Results taken by this means are displayed in figure 7. The 1s-$2p_+$ line is now so sharp that the A(Γ), B and D components are resolved as separate peaks. The improved signal-to-noise figures given by the laser spectrometer enables as-yet-unidentified structure to be detected on the 1s(L)-$2p_+$ line for the A donor including a very sharp line at lower fields (higher energy).

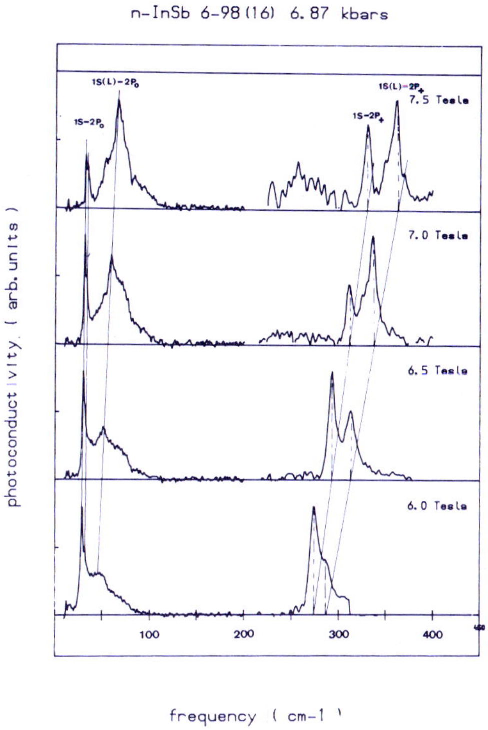

Fig.5 shows the $1s(L)$-$2P_0$ and $1s(L)$-$2p_+$ transitions of donor A simultaneously at a pressure of 6.87kbar and various magnetic fields. The lower frequency component of each doublet contains unresolved components of $1s(\Gamma)$-2P transitions for the B-D donors as well as the $1s(\Gamma)$ contribution from donor A.

Fig.6. Experimental recordings of the cyclotron resonance (CR) and the impurity-shifted cyclotron resonance or 1s-2p+ transition (ICR) at a pressure of 7.63 kbar. The L-component of donor A is seen to move rapidly away on increasing the field.

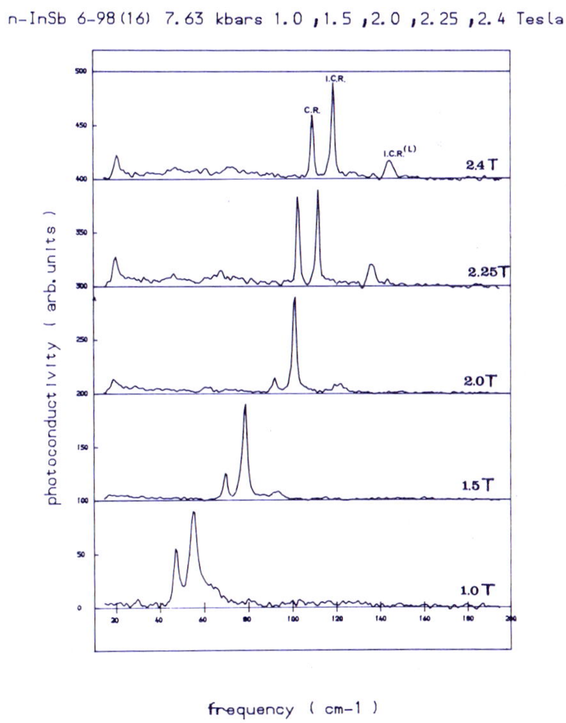

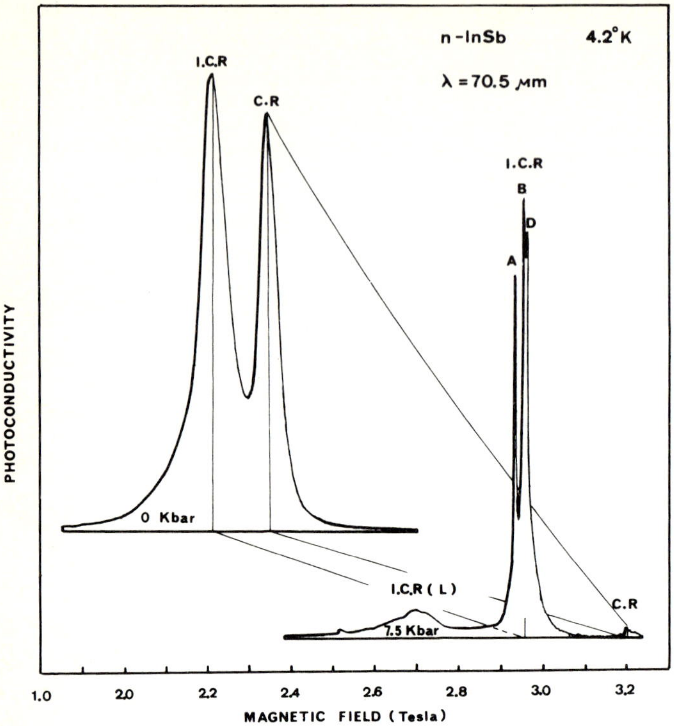

Fig.7 shows cyclotron resonance and impurity shifted cyclotron resonance lines taken at 70.5μm wavelength demonstrating the narrowing of the lines obtained on increasing the presence to 7.5kbar and the appearance of L- and Γ-components of donor A together with the resolution of donors B and D.

Fig.8 shows cyclotron resonance and impurity shifted cyclotron resonance lines taken at 251μm wavelength demonstrating nearly an order of magnitude narrowing on going to a pressure of 7.56 kbar.

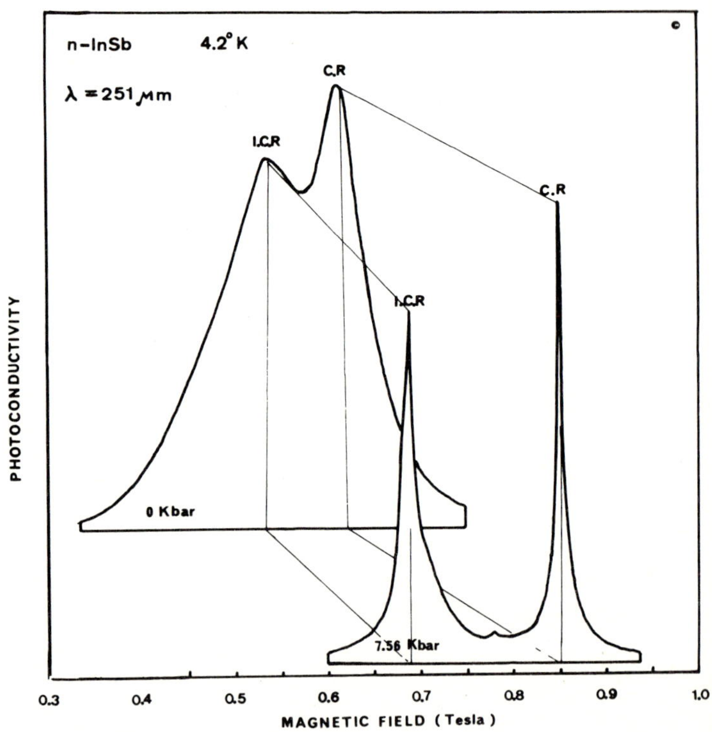

Experiments will be performed with oriented samples to assist the interpretation of this structure. The $\omega\tau$ of the cyclotron resonance line is 650 which compares with a value of 100 at zero pressure which is similar to the best value reported by McCombe (11) et al under similar conditions. An even more dramatic narrowing from $\omega\tau$ values of 25 to 200 is found on increasing the pressure from zero to 7.6 kbar at 251 μm wavelength (figure 8). In order to eliminate the possiblity of over-absorption broadening the lines shown in figures 7 and 8 a sample of thickness of only 30 μm was used. Some of the narrowing observed may result from neutralisation of the donors with increasing pressure but the number of ionised donors only drops by ∿30% between the recordings shown in figures 7(a) and 7(b). Furthermore the change in effective mass with pressure is too small to account for the changes in line width observed. Consequently it appears likely that a redistribution between donor sites located differently with respect to charged acceptors is responsible for the line narrowing. This correlation effect was previously put forward as a mechanism for explaining an unexpected temperature dependence of impurity lines found at very low temperatures (12). The idea is associated with the tendency for the positively charged ionised donors to be found close to the negatively charged acceptors in compensated material. The resultant dipolar fields have a smaller perturbing effect on the impurities or cyclotron motion than the corresponding number of randomly-distributed monopole fields. However, as the temperature is raised, the electrons on the neutral donors can be excited across onto the previously ionised sites with a resultant increase in broadening. The narrowing found on increasing the pressure can therefore be explained if the energy differences between donor sites located in different regions is increased.

An analysis of the pressure dependence of the observed positions of the cyclotron resonance and impurity lines has been given in a previous publication(10).

The experiments described here demonstrate an interaction between Γ- and L-associated states with a minimum offset of 3 cm^{-1} (∿0.4meV) which is much less than deduced from the transport measurements (8). However the upper branch seen in figures 2 and 3 would have complicated the freeze-out analysis. Furthermore it should be noted that in the region of the resonant cross-over the mean separation of the A-component from the B-line is 4 cm^{-1} and that this separation then grows rapidly with increasing pressure or magnetic field. This compares with a separation of 0.8 cm^{-1} at 11T between the same impurities with no pressure applied.(13) This very pronounced deepening of donor with respect to the other donors probably arises from a rapid increase of its chemical shift due to compression of the wavefunction into the strongly attractive local potential for this donor (14) and would certainly contribute to the deepening of the state involved in the transport measurement which goes some way to explain the discrepancy between the thermal and optical results.

3. Polaron Coupling involving single zone-edge phonon modes

In early magneto-optical studies of the donors in InSb Kaplan and Wallis found

resonant modes at energies Δ and 2Δ above the LO phonon energy and Δ below this energy (Δ = 35 cm^{-1}). It can be seen from the markedly different offsets for the two crossings that the interaction strength is considerably greater for the +Δ resonance than for the +2Δ resonance. In the work of Kaplan et al (13) similar coupled mode behaviour was found on the 1s-$2p_0$ line. However, in both sets of experiments the nature of the coupling was unclear as the only excitation which could give a resonant mode near this frequency involved single TA phonons whose frequencies are 34 cm^{-1} at the X-point.(15) Two-phonon interactions with both these modes have been detected in magnetophonon experiments (16) but no other single phonon interaction with zone edge phonons has been reported in either magneto-optics or transport experiments, nor would a coupling be expected for electrons associated solely with the Γ minimum. The admixture of Γ and L states invoked to explain the anomalous behaviour of the A component of the central cell spectrum in section 2 provides a natural mechanism for appearance of resonant mode behaviour associated with single TA phonons at the L-point. Such an admixture also explains the observation in the experiments of Kaplan and Wallis that the coupling to the Δ modes is stronger than to the 2Δ modes and the fact that coupling to single TA modes at the X-point is not observed although the density of states is greater for these modes as evidenced by the strength of 2TA(X) series in hot electron magnetophonon experiments. The 35 cm^{-1} resonances can be thought of as involving intervalley transfer between localised states mediated by these phonons or as an intervalley resonant polaron. As the A-component of the 1s-$2p_0$ line becomes resolved from the B, C and D lines on the application of pressure, an elegant confirmation of this mechanism for coupling becomes possible. On tuning the A component and the unresolved B-D components successively through the 35 cm^{-1} frequency it is apparent that the resonant coupling can only be found on the A-component. This is consistent with the observations detailed in the previous section which show that the anomalous deepening of t
of the ground state is only found for the A donor and that only L-like states are involved in the resonance. Further evidence that the 35 cm^{-1}(Δ) mode is indeed a phonon mode is provided by the observation that the resonant energy is pressure independent up to at least 6kbar. If the resonance mode was electronic in character it would be strongly pressure dependent. However, it should be noted that at zero pressure, under the conditions that the Δ modes were observed on the 1s-$2p_+$ and 1s-$2p_0$ lines, the L-levels were between 40 and 70 meV above the bottom of the Γ-band, suggesting that some L-admixture is retained for some considerable distance from cross-over. This would also be consistent with the ability to follow the 1s-$2p_0$ (L) branch to more than 30 cm^{-1} from cross-over.

4. An unidentified mode-crossing

Whereas the pressure experiments provide an explanation for the mechanism for the appearance of Δ-modes in magneto-optical experiments, they have also thrown up at least one further mode-crossing which remains to be identified. This is concerned only with 1s(A)-2p_(A) transition and can be seen in figure 9 between 9 and 10T. Two branches with a minimum offset of 0.8 cm^{-1} can be observed between these fields.

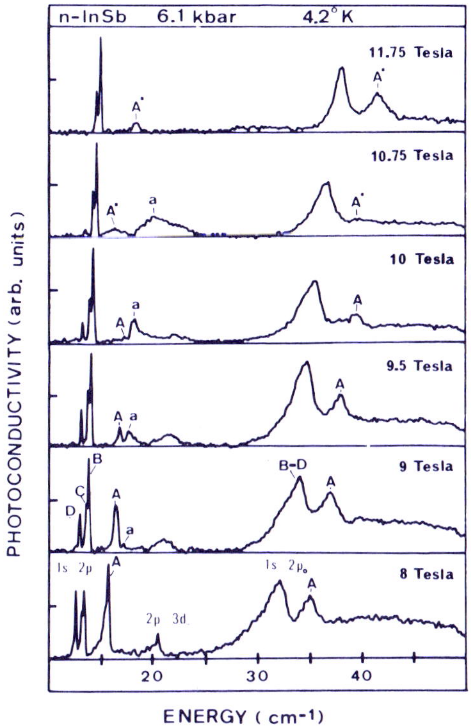

Fig.9. Experimental recordings of the shallow donor spectra in n-InSb at 6.1 kbar taken at narrow intervals of field showing an abrupt splitting into two components (A and a) of the 1s-2p_ line between 9 and 10T.

Typical coupled-mode behaviour is found with the lower energy branch losing intensity and the upper increasing in strength with increasing field. The field at which both branches are of equal intensity varies rapidly with pressure and is strongly correlated with the centre position of the subsequent resonant cross-over of the 1s(Γ) and L states. These features are not inconsistent with the resonance being associated with a crossing of the 2p-(Γ) state with the L-state, although the rather weak dependence of the positions of the two branches on field shown in fig.3 would argue against this interpretation. Another possibility might be a phonon-mediated replica of the 1s-2p_ transition. The very low energy of phonon mode required (0.1meV) would at first sight make this possibility unlikely. However the increasingly local character of the ground state might mix-in a very low energy local mode. Studies of the temperature dependence of the intensities of the components give some credance to the latter possibility. Further experiments with samples having different magnetic field orientations are in progress in an attempt to distinguish between these two possibilities.

5. Changes in the central cell spectrum produced by the occupancy of the X-like states.

It is possible to follow the magneto-optical spectrum through the region where the X-like and L-like states are nearly degenerate provided that sufficient electrons are left in either the L and Γ-like states of donor A or in the other donors (B-D). This can happen with uncompensated material even if the sample is cooled slowly at pressure in a manner where all of X-associated states of donor A are filled. However compensated samples become infinitely resistive if all the X-levels are filled.

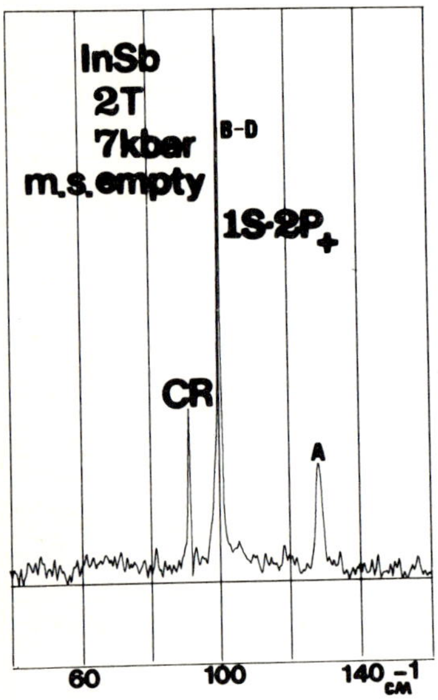

Fig.10(a) Experimental recording of the cyclotron resonance and impurity-shifted cyclotron resonance lines taken when the thermal cycling has been performed to keep the metastable X-state unoccupied.The L-associated component of donor A is clearly visible about 30 cm^{-1} above the main ICR line.

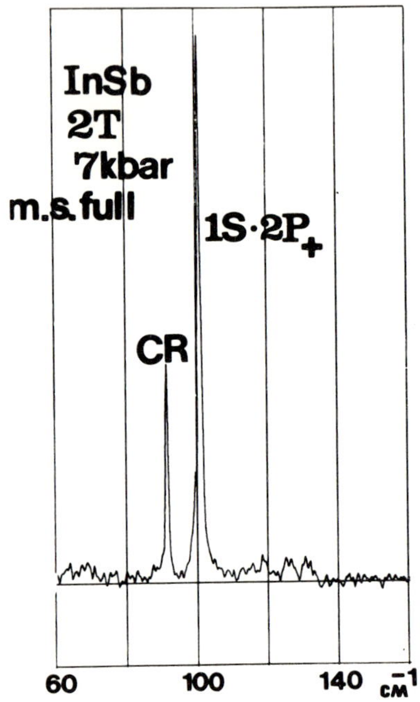

Fig.10(b) The experimental conditions are similar to 10(a) except that the sample has been cooled in such a way that the metastable X-states are almost completely occupied. The L-associated component of donor A has disappeared and the height of the main $1s\text{-}2p_+$ line has decreased with respect to the cyclotron resonance.

Figure 10 shows the $1s\text{-}2p_+$ or impurity-shifted cyclotron resonance (ICR) and the cyclotron resonance (CR) lines in a relatively uncompensated sample where the total concentration of B, C and D donors exceeds that of the acceptors at a pressure of 7.63 kbar where the X-level is actually below the L-level. As the magnetic field is increased from 1.0T to 2.4T the L-component of the ICR line for donor A rapidly separates from the component which is unresolved from the ICR lines from the other donors (B-D). As a consequence of the energy barrier due to the large lattice relaxation around the impourity, the population of the X-state is determined by the thermal history of the sample. Thus, if the state is populated at higher temperatures (∿>100K) by choosing a certain combination of pressure and magnetic field, the population at 4.2K can remain practically constant even if there are subsequent changes in the Fermi level with respect to the bottom of the Γ-band. In a magneto-optical experiment the effective number of A-centres appearing in the shallow donor spectrum

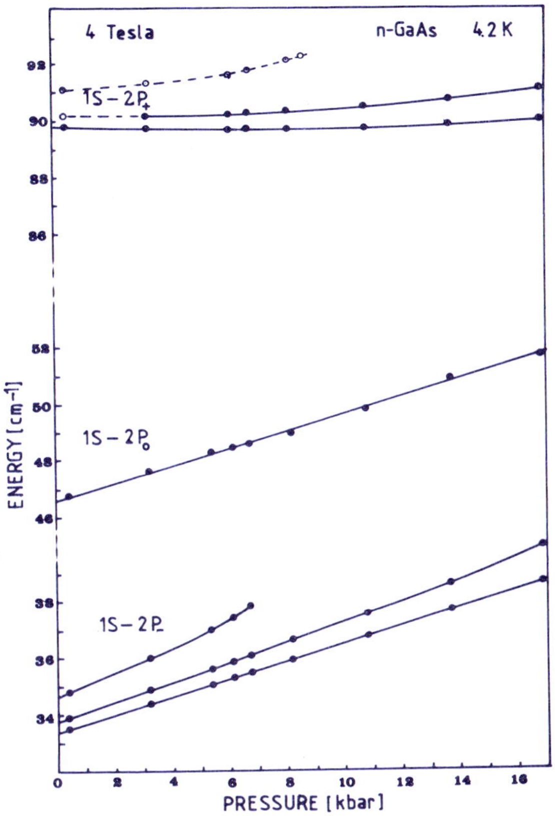

Fig.11 shows the positions of the three 1s-2p transitions for the donors in GaAs as a function of pressure at a magnetic field of 4T. A VPE sample is employed and three donors are resolved on the 1s-2p$_+$ and 1s-2p$_-$ lines. The deepest of the donors (believed to be germanium or carbon) moves more rapidly than the other two donors and disappears at pressures above 8 kbar in a manner reminiscent of the first reports of the Γ-L interaction for the deepest donor (A) in InSb.(9). The pressure dependence of the Zeeman mass is in reasonable agreement with that expected on a four-band model. The GaAs results will be the subject of a future publication.

will be the total number of A centres less those which have electrons occupying the X-states. Figure 10 demonstrates the changes in the $1s\text{-}2p_+$ spectrum which can be brought about by inducing occupancy of the X-states. The pressure and magnetic field have been chosen so that the L-associated states of donor A are about 30 cm^{-1} below the $1s(\Gamma)$ levels giving rise to the impurity shifted cyclotron resonance line ($1s\text{-}2p_+$) from these states, which can be seen at 130 cm^{-1} in the first recording taken with the X-states unoccupied. As can be seen from the second recording the effect of populating the X-states is to bleach out the L-associated lines entirely and there is also a noticeable decrease in intensity of main ICR line due to the loss of $1s(\Gamma)$ component of A.

6. Conclusion

In all the experiments described above we have the remarkable situation in which the shallowest of all donor systems yet detected unambiguously for any semiconductor are being used to extract information with spectroscopic precision about states which are strongly localised and deep with respect to the bands which provide their dominant character. The same type of interaction should exist for semiconductors other than InSb. Indeed transport measurements with GaAs have demonstrated the presence of resonant states in the conduction band which can be brought below the Γ-band edge at experimentally accessible pressures.(17) Preliminary studies of the central cell spectrum in GaAs as a function of pressure have indicated an anomalous deepening and loss of signal intensity for the deepest of the shallow donors similar to the first experimental observations of the Γ-L interaction for donor A in InSb as can be seen from figure 11.

References

1. E Litwin-Staszewska, S Porowski, A S Filipchenko Phys.Stat.Solidi b48 525(1971)
2. M Konczykowski, S Porowski and J Chroboczek Proc.11th Int.Conf. on Phs.Semicond. Warsaw) p1050 1972)
3. E Litwin-Staszewska, A Jedrzejczak, S Porowski and A S Filipchenko Proc.11th Int.Conf.on Phys.Semicond.(Warsaw) p952 (1972).
4. L Konczewicz, E Litwin Staszewska and S Porowski Proc.3rd Int.Conf. on Narrow Gap Semiconductors (Warsaw) p211 (1977).
5. S Porowski, L Konczewicz, M Konczykowski, R Aulombard and J L Robert Proc.15th Int.Conf. on Phys.of Semicond.(Kyoto) p271 (1980).
6. L Dmowski, M Konczewicz, R Piotrzkowski and S Porowski Phys.Stat.Solidi b73 1131 (1976).
7. M Altarelli and G Iadonisi Il Nuovo Cimento 5 21 (1971).
8. S Porowski Proc.4th Int.Conf. on Narrow Gap Semicond.(Linz) p420 (1981).
9. A M Davidson, P Knowles, P Makado, R A Stradling, S Porowski and Z Wasilewski Solid State Sciences 24 84 (Springer-Verlag 1981).

10. Z Wasilewski, A M Davidson, P Knowles, S Porowski and R A Stradling Proc. 4th Int.Conf. on Narrow Gap Semicond.183 (1981).

11. B D McCombe, R Kaplan, R J Wagner, E Gornik and W Muller Phys.Rev.B13 2536 (1976).

12. J Golka, J Trylski, M S Skolnick, R A Stradling and Y Couder Solid State Comm. 22 623 (1977).

13. R Kaplan, R A Cooke and R A Stradling Solid State Chem. Solids 1 137 (1956).

14. W Trzeciakowski and J Krupski to be published.

15. D L Price, J M Rowe and R M Nicklo Phys.Rev. B3 1268 (1971).

16. R A Stradling and R A Wood J.Phys.C3 2425 (1970).

17. G D Pitt and J Lees Phys Rev B2 4144 (1970).

MAGNETICALLY TUNABLE FAR INFRARED EMITTERS AND DETECTORS

E. Gornik
Institut für Experimentalphysik
Universität Innsbruck
6020 Innsbruck, AUSTRIA

ABSTRACT

The present state of the art in magnetically tunable detectors and emitters in the FIR range is reviewed.
On the detector side two materials are widely used: InSb and GaAs. New aspects for tunable narrowband detection provide experiments with hydrostatic pressure and uniaxial stress in InSb.

Landau emission has been investigated in several semiconductors and a useful source for spectroscopy with a resolution below 1 cm^{-1} has been developed.
New promising concepts for tunable sources have been demonstrated recently: The broadband and tunable emission from streaming carriers in crossed electric and magnetic fields and the radiative decay of two-dimensional plasma oscillations as present in Si-MOS structures via a grating antenna.

1. INTRODUCTION

The far infrared (FIR) defined as the wavelength range between 1 mm and 50 µm can be considered today as an experimentally rather settled range. A variety of sources and detectors are available to the experimentalist.
On the detector side the photoconductive response from shallow impurities and free carriers is used in Ge /1/, GaAs /2,3/ and InSb /4/. The detectors show high sensitivity and are easy to operate. In addition bolometers made out of Ge and Si and the well known Golay detector are widely used. The rather unusual feature in the FIR is the existence of narrowband and tunable detectors which allow the detection and the analysis of radiation. In the present paper the performance of magnetically tunable detectors with special emphasis on InSb and GaAs will be summarized.

The dominant sources are optically pumped gas lasers and the Fourier Transform Spectrometer which allow a wide range of basic investigations. However, the search for a simple tunable solid state source has been of great interest since the proposals of B. Lax /5/ and P. Wolff /6/ for a solid state magnetically tunable laser in the early 1960's. The first step in this direction was the observation of spontanous tunable radiation from Landau levels in the early 1970's /7/. In the meantime a useful tunable source has been developed out of this effect /8/, but several attempts to obtain stimulated emission have failed /9/. In the last years new impact was given to the field by the discovery of voltage tunable emission from inversion layers /10/ as present in Si-MOSFETs and GaAs/GaAlAs heterostructures and by the observation of FIR emission from streaming holes in Ge /11/. The effect of streaming carriers in crossed electric and magnetic fields is a rather old idea /12,13/.

In this paper the latest developments in FIR solid state emission with emphasis on magnetically tunable sources will be reviewed: Landau emission from bulk n-GaAs /14/ and Ge /15/ with linewidth below 1 cm^{-1} has been observed. Emission from n-InSb shows promising line narrowing with hydrostatic pressure /16/. FIR radiation was observed from streaming holes in p-Ge and broad-band and narrowband emission has been detected /11/. The radiative decay of 2-dimensional plasmon oscillations has turned out to be a potential candidate for a new tunable FIR source with power levels in the range of 10^{-5}W /17,18/.

2. DETECTORS

The most commonly used photoconductive detectors are InSb, GaAs and Ge (doped with different shallow impurities such as Ga,Sb,In,P). While Ge detectors are highly sensitive and broadband, InSb and GaAs exhibit a narrowband and tunable response in a magnetic field.
The properties of InSb detectors have been investigated in great detail by Putley /19/. The material shows a rather broadband response without magnetic field due to free carrier absorption. The application of a magnetic field splits the conduction band into Landau levels, and transitions between neighbouring levels and associated impurity levels result in a narrow response for a given magnetic field. Since the peak cyclotron resonance absorption is considerably stronger (absorption coefficients α between 100 cm^{-1} and 300 cm^{-1}) than free carrier absorption (α between 10 cm^{-1} and 30 cm^{-1}) the narrowband response can be optimized by detector design: The response will be dominated by narrowband cyclotron transitions when thin platelets (4x4 mm^2) with a thickness between 50 μm and 100 μm are used.

Fig. 1 shows the relative signal of an n-InSb detector as a function of magnetic field for several incident frequencies. The narrowband radiation is generated with a GaAs Landau emitter (linewidth $\sim$ 0.8 cm^{-1}). The best performance of the detector is achieved at a temperature of 2K where the response consists only of one line due to transitions from the impurity ground state to an excited impurity state. This line with a linewidth of about 3 cm^{-1} can be tuned according to the full curve in fig. 1. The sensitivity has its maximum between 1.0 and 1.5T depending on mobility, geometry and carrier concentration of the material used. Best performance is achieved with high purity samples (lowest doping level $\sim 10^{14}cm^{-3}$) and rather high compensation. This yields a low free carrier concentration and results in a high detector impedance. At 4.2K an additional line appears in the spectra due to transitions between the two lowest Landau levels which is split by about 8 cm^{-1} from the impurity line. The best responsivity achieved with alloyed In (3 % Te) contacts is between 1×10^3V/W and 5×10^3V/W. The main advantage of InSb is that the response of the detector itself can be tuned due to the rather small resistance change with increasing magnetic field. It is sometimes an advantage to orient the detector bias parallel to the magnetic field to reduce the magnetoresistance.

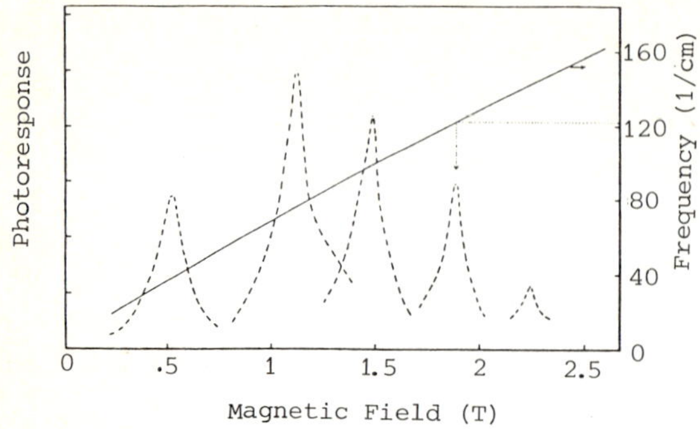

Fig. 1: Photoconductive response of a n-InSb detector as a function of magnetic field at 2K for several frequencies. The source is a GaAs-Landau emitter with a linewidth of 1 cm^{-1}. The full curve gives the tuning characteristic. The resonant frequency is determined as indicated by arrow .

The performance of InSb detectors can be significantly improved by the application of hydrostatic pressure. The group of Stradling /20,21/ has performed extensive studies of the photoconductive response under hydrostatic pressure. A significant narrowing of the impurity and cyclotron resonance lines was observed above 7 kbar together with an increase in sensitivity by an order of magnitude.

A new type of photoconductive response in InSb has been demonstrated by F. Kuchar /22/ with the application of uniaxial stress along the <110> direction. With stress values in the order of a few kbar a response due to spinflip transitions is observed. Its amplitude is comparable with the cyclotron resonance signal. This response is narrowband with a linewidth below 1 cm^{-1} and continuously tunable. As the absorption coefficient is rather small detector thicknesses of few mm have to be used. This effect seems extremely promising for spectroscopy and has the advantage that it can be extend to wavelength up to 1 mm, while the InSb cyclotron resonance detector extends only to 0.5 mm.

High purity n-GaAs has been used as a highly sensitive photoconductive detector since for than a decade /2,3/. At zero magnetic field the response at 4.2K consists of a dominant impurity line (1s-2p) and a broad continum (1s- conduction band transitions). The peak response at 35 cm^{-1} can be higher than 10^6V/W. In a magnetic field the dominant line splits into 3 lines (Zeeman effect), and the continum response vanishes. Photoconductive spectra of a high purity GaAs detector with a total doping of $7x10^{13}cm^{-3}$ and a free electron concentration of $1x10^{13}cm^{-3}$ for OT and 3.OT is shown in fig. 2. Extremely narrow lines with a linewidth of 0.25 cm^{-1} are observed for magnetic fields higher 1.5T. This are the narrowest photoconductive lines yet observed in the FIR. They are due to the fact that only one residual impurity is left in the material (identified as Si; the sample was grown by E. Bauser, MPI Stuttgart).

The dashed curve in fig. 2 shows the tuning rate of the 3 lines with the field. As the spectrum consists of more than one line an application in which the detector is tuned, is not meaningful. The best performance of the GaAs detectors is therefore achieved by setting the detector lines to a constant frequency with the magnetic field according to the dashed curves. In this case the detector acts as a narrowband filter. This allows the analysis of radiation only when the source is tunable. In addition it should be mentioned that the sample resistance and contact noise increase drastically with the magnetic field which makes the tuning of the detector by itself very difficult.

For the analysis of emission experiments in the FIR a combination of InSb and GaAs detectors and access to magnetic fields up to 10T represent a complete range of spectroscopic tools with resolution of 1 cm^{-1}.

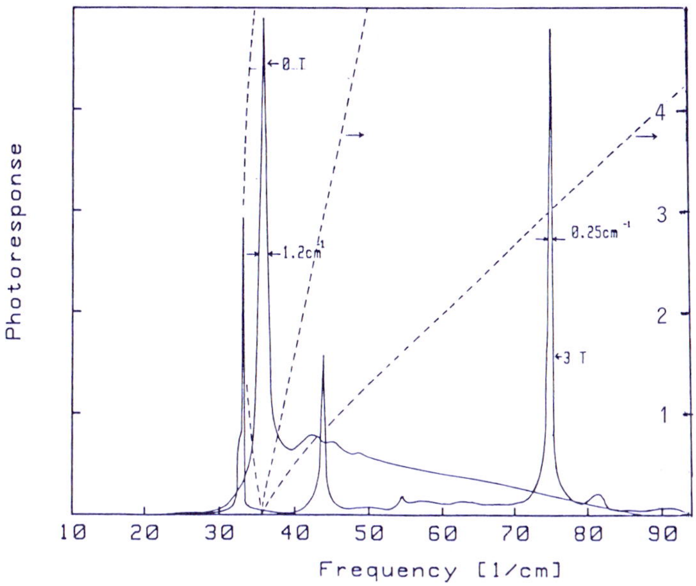

Fig. 2:
Photoresponse of a n-GaAs detector at 4.2K for OT and 3.0 T. The tuning characteristic of the three lines is given by the dashed curves.

3. EMITTERS

Tunable emission requires the existence of energy levels which are variable by external fields. Magnetic quantum levels called Landau levels allow a linear tuning of neighbouring levels with the external magnetic field. In semiconductors with low effective carrier masses the tuning range can be extended over the whole FIR with moderate fields.

Electric quantization appears in systems where high surface fields can be applied. In Si-MOS structures one-dimensional quantum levels have been found which are tunable by an external d.c. gate voltage /23/. However not only an energy level system can lead to tunable emission but also a system oscillating in a collective mode. Two-dimensional plasma oscillations exist in an electrically quantized carrier system as present in inversion layers or thin metallic films /24,25/. Tuning is achieved by varying the plasma frequency with external fields such as the gate voltage or a magnetic field.

Most of the FIR emission from semiconductors has been generated by carrier heating with an electric field. But also concepts with different optical pumping schemes have been applied.

3.1. Landau emission

Tunable emission between Landau levels has been observed in several semiconductors /8/. A detailed study of the emission spectrum excited by electrical current heating has been performed in InSb and GaAs. From a source aspect GaAs turns out to be the more favourable candidate since linewidths below 1 cm^{-1} and intensities comparable with those in InSb in the order of 10^{-7} to 10^{-6}W/cm^2 have been achieved.

A typical emission spectrum with a resolution of 0.25 cm^{-1} from a high purity GaAs sample for different bias electric fields is shown in fig.3. The spectrum is obtained by tuning the emission line through one of the narrow lines of the GaAs detector in a magnetic field. At low electric fields (below 15 V/cm) the spectrum consists only of one narrow line due to transitions between the n = 1 and the n = 0 Landau levels. With increasing field additional lines due to transitions between higher levels (n = 2 to n = 1, n = 3 to n = 2) appear. The splitting of the lines is a consequence of nonparabolicity and polaron effects as has been demonstrated by Lindemann et al. /26/. For source applications only the spectrum with the single line (lowest field) can be used which

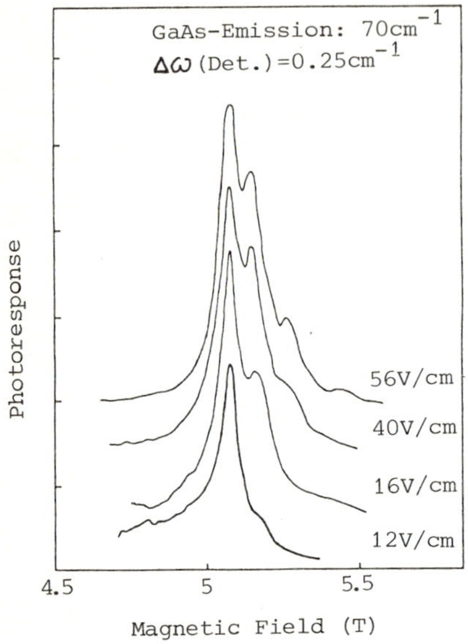

Fig. 3: Emission spectrum of a high purity GaAs emitter as a function of magnetic field for several bias electric fields as observed with a GaAs detector.

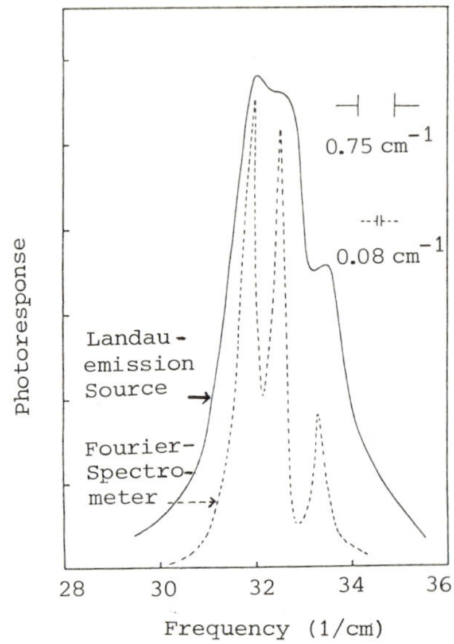

Fig. 4: Photoconductive spectrum of the lowest impurity line (see fig. 2) of a GaAs sample (doping level $N_I \sim 3x10^{14} cm^{-3}$) as obtained with a narrowband GaAs Landau source and a Fourier Spectrometer.

limits the usable intensities to a factor of 5 below the maximum values.

An example of a spectroscopic application with a tunable GaAs-source with a linewidth of 0.75 cm^{-1} is shown in fig. 4. The photoresponse of a GaAs sample at 32 cm^{-1} in a magnetic field is plotted in comparison with the result of an analysis with a high resolution Fourier Transform Spectrometer (FTS). The 3 apparent lines are due to central cell splittings of different shallow impurity species /27/. The lines are well-resolved with the FTS and at the edge of resolution with the GaAs source.

However this example demonstrates that Landau emission sources can be used for spectroscopy with resolution down to 1 cm^{-1}. Especially in the area of detector testing they provide a very practical tool. In addition the fact that the linewidth is proportional to the square root of the total ionized impurity content enables the determination of doping levels by analysing the emission spectrum, which is of significant interest for material growth /14/.

New impact has come to the field by the application of hydrostatic pressure to n-InSb /28/. From conductivity and photoresponse measurements it was found that the active ionized impurity concentration can be reduced above certain pressure levels. As a consequence a line narrowing for Landau emission is expected. An emission experiment performed for different values of hydrostatic pressure is shown in fig. 5 /16/. The photosignal of a GaAs detector at 2.0T is plotted as a function of the InSb emitter magnetic field. The positions of the detector lines in terms of frequencies can be found from fig. 2 by drawing a horizontal line at 2.0T. The emission spectrum consists of two lines due to transitions between the two lowest Landau levels for different spin orientations. The individual linewidth is determined by a curve fitting procedure. The result is shown in the insert of fig. 5. A drastic linenarrowing with increasing pressure by about a factor of two is observed. This demonstrates that the source properties of InSb can be improved. It is interesting to note that the obtained linewidth is lowered to values which cannot be explained with the residual shallow impurities. This means that the dominant impurity scattering process must be changed with pressure.

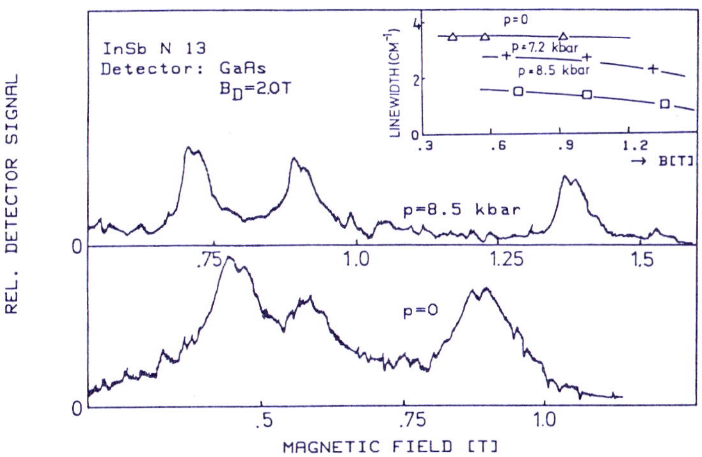

Fig. 5: Landau emission spectra of a n-InSb sample (doping level $N_I \sim 5\times10^{14}cm^{-3}$, $n = 8\times10^{13}cm^{-3}$) as a function of magnetic field detected with a GaAs detector at 2.0T. The applied hydrostatic pressure is indicated. The insert shows the obtained linewidth.

An extremely promising experiment not yet performed is the application of uniaxial stress to InSb: Emission due to spin-flip transitions is expected. This process is enhanced considerably under uniaxial stress according to studies by Kuchar et al. /22/. Spin flip transitions have been observed in absorption in the meantime and reveal extremely narrow lines /29/ which is further encouraging.
An interesting emission experiment has been reported recently by Otsuka et. al /30/: Tunable Landau radiation is generated by illuminating n- and p-type InSb samples with a high intensity Xenon flash lamp. The emission is due to intrinsically generated free carriers high in the band. Intensities comparable with electrical excitation is achieved. However this is quite remarkable since the excitation is restricted to a very thin surface layer, where a high carrier density is generated. The radiation is probably due to carriers diffusing in the bulk.

Landau emission has been and still is a candidate of attention for a narrowband tunable and coherent source. Some concepts including resonant 2 photon pumping across the gap and tunnel injection have not been tried extensively. A mechanism which predicts population inversion by electrical exitation is the streaming motion of carriers in crossed fields. The possibilities in connection with Landau emission are discussed in the next section.

3.2. Emission from streaming carriers

In strong electric fields the carriers can be accelerated to reach the optical phonon energy without suffering any scattering event and will emit immediately a phonon and return to the ground state. This process will be repeated and is referred to as streaming motion /12,13/. The resultant momentum distribution is expected to be highly anisotropic. The application of a transverse magnetic field results in a more remarkable carrier distribution as a consequence of carrier accumulation into a limited area in momentum space.

The free acceleration of a carrier in crossed fields (magnetic field B in z-direction, electric field in x-direction) is a cyclotron oscillation with the center point at $C = (0, -E/B, v_z)$ in velocity space. The curved trajectory is an arc on the plane $v_z = 0$ with the center point at C. The streaming motion is determined by the ratio of the drift velocity in crossed fields $v_y = E/B$ to the velocity v_{op} at the optical phonon energy. The behaviour is therefore best discussed with the parameter $\zeta = v_{op}/v_y$: For $\zeta < 1$ the motion is similar to the motion without B, just curved. For the case $\zeta > 2$ the trajectory for streaming motion is closed before the carrier reaches v_{op} and streaming motion becomes accordingly impossible. The most interesting case is for $1 < \zeta < 2$, which is sketched in fig. 6. The center point C is chosen to allow several Landau orbits inside v_{op}. Streaming motion starts from the ground state along the drawn trajectory. Some carriers will penetrate the optical phonon energy and get scattered into a cyclotron orbit around C. In this orbit the carriers do not reach anymore the optical phonon energy and are accumulated in a limited area of momentum space. The accumulation rate in an individual Landau level is proportional to the orbit area. The resulting distribution will be a population inversion in the Landau level below the optical phonon energy as shown in fig. 6 b). This system would directly lead to stimulated Landau emission. Experimental evidence for carrier accumulation for $\zeta > 1$ has been convincingly demonstrated by Komiyama in AgCl, AgBr /31/ and p-Ge /32/.

Another quite interesting phenomenon has been predicted by Kurosawa /33/ in the system where $1 < \zeta < 2$: The appearance of superradiance is expected when an alternating electromagnetic field $E(\omega)$ is additionally applied to the d.c. field. The net effect of $E(\omega)$ is to make the point C rotate elliptically with ω. Carriers thus make cyclotron oscillations with the cyclotron frequence $\omega_c = eB/m^*$ around C. If $E(\omega)$ is not too large and $\omega = \omega_c/n$ (with n an integer and > 2) the cyclotron oscillation of accumulated carriers is in a partially coherent state which leads directly to superradiance.

Stimulated emission has not been achieved in this system yet. The main drawback is the requirement of extremely high mobilities ($\sim 10^6 cm^2/Vs$ for InSb; $\sim 5x10^5 cm^2/Vs$ for GaAs at 4.2K) which have not been realized yet.

However quite promising far infrared emission has been observed from light holes in p-Ge /11/. The accumulation of holes within a limited range of velocity space has been demonstrated and broadband emission due to transition of light holes to the heavy hole band has been de-

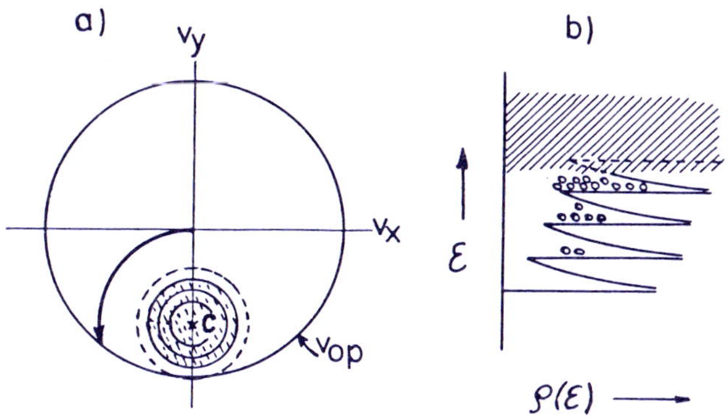

Fig. 6: Schematic description of streaming motion for $1 < \zeta < 2$. Fig. a) shows electron trajectories in velocity space with the center point C at (O,-E/B,O). Streaming starts at $v_{x,y} = 0$ as described by the large arrow. Carriers which are scattered in the shadowed Landau orbits below the optical phonon velocity get accumulated. The resulting carrier density as a function of energy is sketched in fig. b).

tected. The emission results from an increased population in the light hole band due to streaming. In addition a strong Landau emission signal from light holes has been found but only for $\zeta > 2$. In this range streaming is not possible anymore. In the range $1 < \zeta < 2$ the magnetic fields are small (< 1T) so that the Landau level splittings are not larger than the linewidths. The question whether streaming motion can create population inversion within Landau levels in available semiconductors is still open.

3.3. Plasmon emission

A two dimensional electron gas exists in Si-inversion layers /23/ and GaAs/GaAlAs heterostructures. The electrical quantization perpendicular to the plane of free motion is caused by a high surface electric field. Electrical subbands are present which result in voltage tunable emission /10/. However the intensity is extremely weak.

A new type of emission mechanism has been found in this system /17/. A two dimensional carrier gas exhibits the existence of longitudinal plasma waves with a different dispersion relation from 3 dimensional carrier systems. The dispersion relation $\omega_p(k)$ has been verified with high accuracy by FIR absorption /24,25/ and emission experiments /18/ in Si-MOS structures. For strictly two-dimensional plasmons a dispersion relation $\omega_p^2 \propto k$ is obtained. It has been shown previously that plasmon oscillations can be excited with the help of a d.c. current /17,18/. The radiative decay is obtained with the help of a metallic grating in the close vicinity of the 2 dim. electron gas. The grating couples the non-radiative plasma oscillation to the electromagnetic field. The wavevector of the grating defines according to the dispersion relation, the emitted frequency.

Very recently Höpfel et al. /34/ have found that the basic plasmon generation process is a thermal and isotropic excitation. The current passed through the sample heats the carrier gas and creates an excitation proportional to the temperature. The spectral distribution of plasma excitations can be calculated applying Bose-Einstein statistics to the dispersion relation.
The maximum possible emission intensity is given by the time derivative of the spectral function. The emission intensity in a frequency inter-

val of 1 cm^{-1} is plotted for a Si-inversion layer with a given density in fig. 7 for different excitation temperatures. The antenna efficiency is assumed to be 1. The intensities derived are extremely high, when compared with other FIR solid state sources. Antenna efficiencies of about 0.01 have been achieved with gratings in the μm range resulting in observed emission intensities of ∿ $\mu W/cm^2$. In a MOS-system the emission frequency is tunable by varying the channel concentration n_s via the gate voltage due to $\omega_p^2 \propto n_s$.
In a magnetic field the dispersion of a two-dimensional plasma wave is modified to $\omega_{mp}(k) = (\omega_p^2(k) + \omega_c^2)^{1/2}$ /25/. This leads to a shift of the magneto-plasmon frequency to higher values with increasing field.

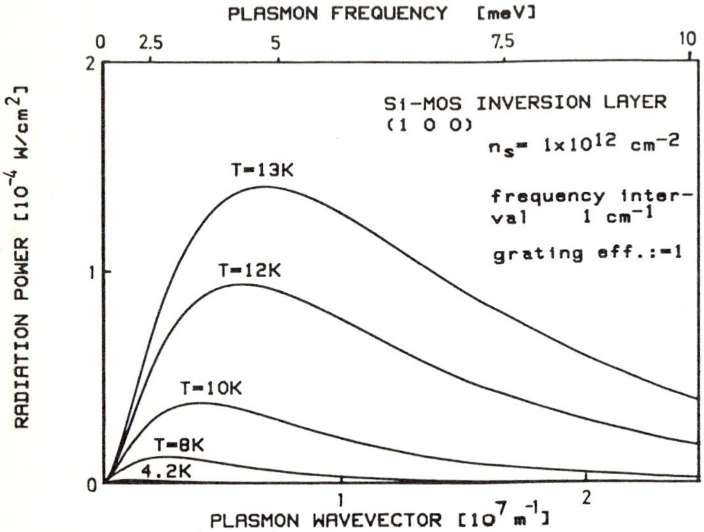

Fig. 7: Calculated plasmon intensity emitted in a frequency interval of 1 cm^{-1} as a function of wavevector and frequency for different excitation temperatures. The antenna efficiency is assumed to be unity.

Fig. 8 shows plasmon emission spectra for several magnetic fields detected with a GaAs detector at B = 0 /35/. The emission frequency increases with B. The resonant detector peak at 35 cm^{-1} shifts to lower n_s values as shown by arrows in fig. 8 in agreement with the prediction.
The application of a magnetic field for tuning the plasma frequency is rather limited since the intensity decreases strongly with increasing field, as appearent in fig. 8 by the decay of the resonant signal. The decrease in intensity is due to Landau damping by single particle excitations. However for a system which does not allow the variation of the channel carrier concentration by a gate voltage tuning via magnetic field within a few cm^{-1} seems meaningful.
The radiative decay of two-dimensional plasma oscillations has to be considered as a potential candidate for a narrowband and tunable FIR source with intensities up to $10^{-4} W/cm^2$ with a linewidth below 1 cm^{-1}.

4. SUMMARY

The field of tunable detection and emission in the FIR has obtained new impact by several recent discoveries:
The growth of GaAs material with a single residual impurity has provide extremely narrowband detection. The application of hydrostatic pressure or uniaxial stress to InSb improved the detectivity significantly. The

observation of high photosignals from spin-flip transitions in InSb extends the narrowband tunable detection into the mm-range.

Landau emission sources excited by electric fields are used now by several laboratories as a spectroscopic tool. Their main drawback is their limited resolution.
Experiments with streaming carriers have provided evidence for carrier accumulation in a limited area of velocity space. As a result FIR emission from these carriers has been observed in p-Ge. Ideas to achieve population inversion within Landau levels resulting in stimulated emission or superradiance (with an additional alternating field) seems possible.

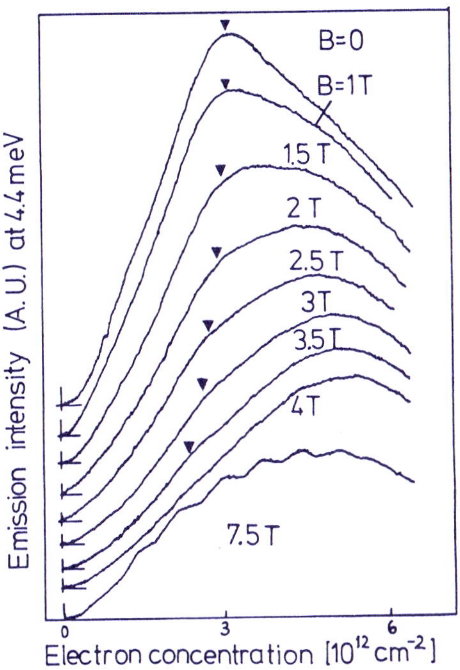

Fig. 8: Spectra of the magneto-plasmon emission as observed with a GaAs detector at OT as a function of carrier concentration in a Si-MOSFET. The triangles give the position of the plasmon resonance. The parameter of the curves is the magnetic field.

The thermal excitation of plasmon oscillations provides a completely new way of generating tunable emission. The potential of this principle is very promising. Intensities up to 10^{-4}W/cm^2 and linewidths below 1 cm^{-1} are expected with operation temperatures well above liquid Helium.

Acknowledgements

The work is supported by the Fonds zur Förderung der wissenschaftlichen Forschung, Austria, (Projekt S22/05), the Jubiläumsfonds der österreichischen Nationalbank and by the European Research Office, London.

REFERENCES

/1/ W.J.Moore, H.Shenker, Infrared Physics 5(1965) 99.
/2/ G.E. Stillman, C.M.Wolfe, J.O.Dimmock, Solid State Comm.7(1969)921.
/3/ G.E. Stillman, C.M.Wolfe, J.O.Dimmock in "Semiconductors and Semimetals" Vol.12 ed. Willardson and Beer, Academic Press 1976.
/4/ B.V.Rollin, Proc.Phys.Soc.(London) 77(1961)1102.
/5/ B.Lax, Quantum Electronics,p.428, Columbia University Press 1960.
/6/ P.Wolff, Physics 1(1964)147.
/7/ E.Gornik, Phys.Rev.Lett. 29(1972)595 and K.L.Kobayashi, K.E. Komatsubara, E.Otsuka, Phys.Rev.Lett. 30(1973)702.
/8/ E.Gornik in "Lecture Notes in Physics" Vol.133 (1980) p. 160-175 ed. W.Zawadski, Springer Verlag, and J.of Magnetism and magnetic materials 11 (1979) 39.
/9/ W.Müller, E.Gornik, T.J.Bridges, T.Y.Chang, Solid State Electron. 21(1978) 1455.
/10/ E.Gornik, D.C.Tsui, Phys.Rev.Lett. 37(1976)1425 and D.C.Tsui, E. Gornik, Appl.Phys.Lett. 32 (1978) 365.
/11/ S.Komiyama, Phys.Rev.Lett. 48 (1982) 271.
/12/ W.Shockley, Bell Syst.Tech.J.30 (1951) 990.
/13/ T.Kurosawa, H.Maeda, J.Phys.Soc.Japan 31 (1971) 668.
/14/ G.Lindemann, E.Gornik, R.Schawarz, D.C.Tsui, Inst.Phys.Conf.Ser. Nr. 56, Chapter 8 (1981) p.631.
/15/ R.A.Stradling, private communication.
/16/ E.Gornik, R.A.Höpfel, M.Baj, S.Porowski, Z.Wasilewski, A.M. Davidson, R.A.Stradling to be published.
/17/ D.C.Tsui, E.Gornik, R.A.Logan, Solid State Comm. 35 (1980) 875.
/18/ R.Höpfel, G.Lindemann, E.Gornik, G.Stangl, A.C.Gossard, W. Wiegmann, Surface Science 113 (1982) 118.
/19/ E.H.Putley, Phys.stat.sol.6 (1964) 571.
/20/ A.M.Davidson, R.A.Stradling, S.Porowski, Sold State Science 24 (1981) p.84, Springer Verlag.
/21/ Z.Wasilewski, A.M. Davidson, P.Knowles, S.Porowski in "Lecture Notes in Physics" 152 (1982) 183 (Springer Verlag).
/22/ F. Kuchar, R.Meisels, M.Kriechbaum in "Lecture Notes in Physics" 152 (1982) p.197 ed. E.Gornik, Springer Verlag.
/23/ T.Ando, A.B.Fowler, F.Stern, Rev. of Modern Physics 54 (1982) 437.
/24/ S.J.Allen, D.C.Tsui, R.A.Logan, Phys.Rev.Lett. 38 (1977) 980.
/25/ T.N.Theis, J.P.Kotthaus, P.J.Stiles, Solid State Comm.26(1978)603.
/26/ G.Lindemann, W.Seidenbusch, R.Lassnig, E.Gornik to appear in Phys.Rev.B.
/27/ P.G.Harper, J.W.Hodby, R.A.Stradling, Reports on Progress in Physics 36 (1973),1.
/28/ S.Porowski in "Lecture Notes in Physics" Vol.152,p.420, Springer Verlag 1982.
/29/ F.Kuchar, see article in this book.
/30/ E.Otsuba, T.Ohyama, F.Fuji, J.de Physique C7, Suppl.10 (1981)393
/31/ S.Komiyama, T.Masumi, K.Kajita, Phys.Rev.B20 (1979) 5192.
/32/ for a review see: S.Komiyama to be published in Advances in Physics 1982.
/33/ T.Kurosawa, J. de Physique C7, Suppl.10 (1981) 377.
/34/ R.A.Höpfel, E.Vass, E.Gornik to be published.
/35/ R.A.Höpfel, E.Gornik, A.C.Gossard, W.Wiegmann, Proc.16th Int.Conf. on the Physics of Semiconductors to be published in Physica B.

MAGNETOOPTICAL STUDIES OF PHASE TRANSITION IN IV-VI COMPOUNDS

G. Bauer

Institut für Physik, Montanuniversität Leoben, A-8700 Leoben, Austria

Investigations on the change of the electronic bandstructure due to the cubic-rhombohedral phase transition in IV-VI compounds are reviewed with special emphasis on the PbGeTe system. Data on magnetooptical intra and interband transitions yield precise information on the phase transition and are analysed with a $\vec{k}.\vec{p}$ model developed by Mitchell and Wallis for the O_h-phase and by Bangert for the C_{3v}-phase.

I. Introduction

The pseudobinary alloys of the PbTe-SnTe, PbTe-GeTe and SnTe-GeTe systems exhibit a structural phase transition from a high temperature cubic NaCl (O_h) to a low temperature rhombohedral (C_{3v}) structure [1-5]. In the binary compounds SnTe and GeTe this transition occurs at about 100 K and 650 K, respectively, PbTe remains cubic even for temperatures $T \to 0$. In the pseudobinary compounds $Pb_{1-x}Ge_xTe$, $Pb_{1-x}Sn_xTe$ and $Sn_{1-x}Ge_xTe$ the critical temperature T_c increases steadily with increasing x.

The phase transition in these systems has been studied by several methods, among them X-ray and elastic neutron scattering investigations of the crystal structure and lattice parameters [5,6] inelastic neutron scattering [7] far infrared reflectivity measurements of the softening of the TO modes [8] investigations of the T-dependence of the static dielectric constant by various methods [9,10] and by Raman scattering [11]. Several reviews by Kawamura [1], Murase [13] and Jantsch [14] were devoted to a discussion of these investigations.

Below T_c, the phase transition temperature, in the rhombohedral structure, these compounds are ferroelectric and with just two atoms per unit cell the simplest conceivable ferroelectrics. Below T_c, the dipole moment in the unit cell arises from a relative displacement of the two (Pb,Ge) and Te sublattices in the cubic [111] direction which is transformed in a rhombohedral c-direction, accompanied by a relatively small decrease of the rhombohedral angle below 90° without indication for a discontinuity. These compounds thus constitute an example for a displacive ferroelectric phase transition of second order.

The change in crystal structure is accompanied by corresponding change of the Brillouin zone and the electronic bandstructure. Since the rhombohedral IV-VI compounds crystallize in the same structure as the elemental group V elements, e.g. Bi, it can be assumed that the corresponding bandstructure has some similarities with the Bi-bandstructure: i.e. the four equivalent L-valleys in the conduction band will be split into singlet T-states and triplet L-states below T_c (Fig.1) Thus measurements of properties related to the electronic bandstructure also give direct information on the phase transition. Already measurements of the temperature dependence of the electrical conductivity σ can be used to determine T_c [1]. Enhanced scattering of conduction electrons in the temperature region around T_c and a corresponding anomaly in $\sigma(T)$ were used by several authors (e.g. [13,15]).

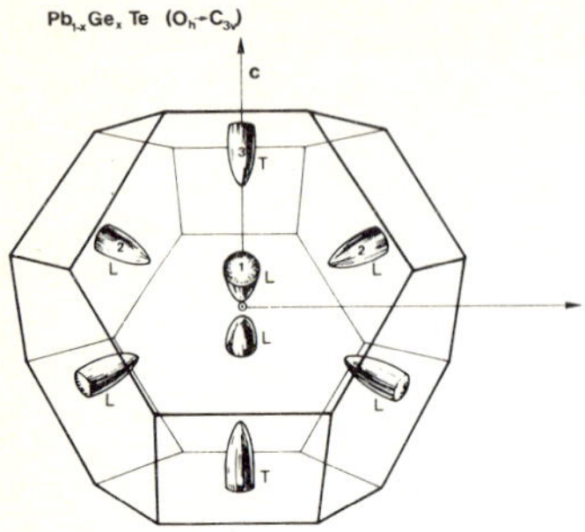

Fig.1: Brillouin zone of PbGeTe for $T<T_c$.

Shubnikov de Haas effect has been investigated in $Pb_{1-x}Sn_xTe$ and $Pb_{1-x}Ge_xTe$ by Takaoka et al [15,16] above and below T_c to get information on the change of the fermi surface below T_c.

More direct methods to get information on effective masses, their anisotropies, and effective g-factors are magnetooptical investigations. Lewis et al [17], Kawamura et al [18] have reported intraband magnetooptical properties above and below T_c in $Pb_{1-x}Ge_xTe$ and Fantner et al [19] investigated interband transitions in the same material.

No magnetooptical investigations of the $Pb_{1-x}Sn_xTe$ and $Sn_{1-x}Ge_xTe$ systems have been reported so far in the rhombohedral phase due to problems with crystal perfection. $Pb_{1-x}Ge_xTe$ with Ge content smaller than 1.5% is ideal for these investigations since both in the rhombohedral as well as in the cubic phase magnetooptical experiments can be performed at temperatures below about 80 K. (Fig.2)

In addition, the crystalline perfection and relatively low carrier concentrations of less than 5×10^{17} cm^{-3} (especially in epitaxially grown films) allow intraband magnetooptical experiments with sufficient resolution.

It is the purpose of this paper to present inter- and intraband magnetooptical measurements on PbGeTe, both in the O_h as well as in the C_{3v} phase. The data are analysed within a $\vec{k}\cdot\vec{p}$ model developed by Bangert [20] starting with the Mitchell-Wallis [21] scheme for PbTe, and considering for the C_{3v} phase the reduced symmetry at the T- and L-points. The lack of inversion symmetry induces k-linear terms and lifts the Kramers degeneracy of the bands.

II. Experimental

Epitaxial samples of $Pb_{1-x}Ge_xTe$ on BaF_2 [111] substrates grown by "hot-wall" epitaxy [22] were used. For the intraband magnetooptical experiments optically pumped molecular gas lasers were used as a light source. The transmission experiments were performed in the temperature region 2... 70 K in magnetic fields 0... 19 Tesla. All far infrared magnetooptical experiments were performed in Faraday-geometry, B||k||[111]. For an analysis of the intraband data, information on the dielectric function is necessary since the data are taken in the frequency region $\omega_{TO}<\omega<\omega_L+$ between the TO- and coupled LO-phonon-plasmon-mode frequencies. For the analysis of the free carrier resonances an extended oscillator model was used taking into account carriers in the T- and L-valleys. The data were analysed by taking into account multiple reflection and interference effects within the film and substrate according to the procedure as described in Ref.23.

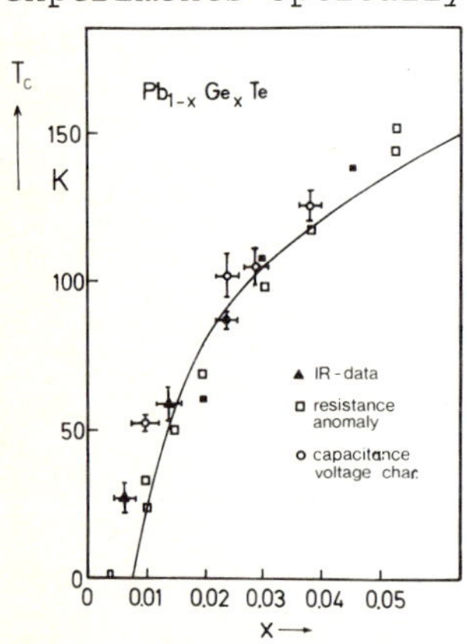

Fig.2: Transition temperature as a function of conposition x, determined by various methods; see main text.

For interband magnetooptical transmission experiments CO-laser lines with $\hbar\omega$=195... 240 meV were used. Experiments were performed in fields up to 6 T, with

1,6 K<T<55K. The data were taken in Faraday geometry B||k|| [111] using CdS quarter wave plates for circular polarization, as well as in Voigt geometry B||E, B⊥E with B|| [110], k|| [111].[19,24]

III. Theory

Above the phase transition temperature the symmetry at the L-points of the Brillouin zone is D_{3d}. For the calculation of the Landau levels of cubic PbGeTe the $\vec{k}.\vec{p}$ treatment according to Mitchell and Wallis [21] is used as given in Ref.23, 25. Below T_c, the symmetry is reduced and a rhombohedral Brillouin zone with T^c and L-points results. The symmetry of the T-singlet states is C_{3v} and of the triplet L-states C_s. In contrast to the elemental V semiconductors, inversion symmetry is lost, and thus k-linear terms are induced and the Kramers degeneracy of the bands is lifted. A $\vec{k}.\vec{p}$ band structure calculation for the C_{3v}-phase has been performed by Bangert [20], which will be followed. The structure of the lowest conduction and the highest valence bands at the T-point can be obtained from that of the cubic one electron hamiltonian by adding a lattice periodic potential V' of C_{3v} symmetry. This leads to addidional matrix elements in the 4x4 matrix hamiltonian on the main diagonal as well as new off diagonal matrix elements. The conduction band ($T_{\alpha,\beta}$) and valence band states ($T_{\alpha,\beta}$) are mixtures of the $L_{6\alpha}^{\pm}$, $L_{6\beta}^{\pm}$ states and the C_{3v} matrix hamiltonian is given by [20]:

$$
\begin{array}{ccccc}
 & T^c_\alpha & T^c_\beta & T^v_\alpha & T^v_\beta \\
H = & \left(\begin{array}{cccc}
Ak_\perp^2 + Bk_z^2 & iS(k_x - ik_y) & Qk_z & L(k_x - ik_y) \\
-iS(k_x + ik_y) & Ak_\perp^2 + Bk_z^2 & -L(k_x + ik_y) & Qk_z \\
Qk_z & -L(k_x - ik_y) & Ck_\perp^2 + Dk_z^2 - \tilde{E}_G & iT(k_x - ik_y) \\
L(k_x + ik_y) & Qk_z & -iT(k_x + ik_y) & Ck_\perp^2 + Dk_z^2 - \tilde{E}_G
\end{array}\right) &&&
\end{array}
\qquad (1)
$$

The coefficients L and Q reflect the direct two-band valence-conduction interaction, the coefficients A, B, C, D are due to the remote bands. The k-linear terms S and T are new and demonstrate the lack of inversion symmetry due to the reduction of symmetry $D_{3d} \to C_{3v}$.

The calculation of the Landau levels for the T-valleys for a magnetic field B|| to the rhombohedral c-axis is described in Ref.20 and for this particular configuration good Landau quantum numbers hold.

As far as the L-points of the Brillouin zone are concerned, the low symmetry of the group of the k-vector, C_s, yields a very complex matrix hamiltonian which was obtained by Bangert by the method of invariants. Only in a special case, this matrix is equivalent to the one of Eq. 1, with the z-coordinate oblique to the c-axis. For explicit calculations this matrix was used. However, even for magnetic field || c, this oblique orientation leads to a situation where no good Landau quantum numbers exist and larger matrices (24x24) have to be diagonalized to calculate the first few Landau states.

It is well known, that in the rhombohedral phase domains occur since each of the cubic <111> directions can be transformed in a rhombohedral c-axis [1,13,14]. For B|| [111], this leads to a situation where B is either || to a rhombohedral c-axis (c-domain) or oriented under an angle of about 70° to it (c'-domains). A T-point of a c-domain corresponds that to one of the L-points of a c'-domains and as a result there are non-equivilent oblique L-valleys in the c- and c'-domains. For the calculation of the Landau- ladder systems within the Bangert-model, the Landau- ladder system at a T-point of a c-domain is identical to the corresponding Landau -ladder of the interchanged L-point of a c'-domain. For the obliquely oriented valleys, for B||c also just one system of Landau-ladders occurs. For B|| [110] there are 3 different classes of domains.

Below T_c, the surfaces of constant energy are not simple ellipsoids. Due to the k-linear terms, a cross section of the surfaces of constant energy in the k_z-$k_\perp$-plane (z-direction ||c-axis) consists of two intersecting ellipses, their splitting being proportional to $k_o \propto \frac{S}{2L} \times (E+\tilde{E}g)/L$ (for the conduction band) in a two band approximation.

IV. Interband Transitions

Fig.3 shows as an example magneto-transmission in Faraday geometry B||k|| [111] for a CO-laser line ($\tilde{\nu}$=1794.4 cm^{-1}) for temperatures above and below the phase transition temperature. In the cubic phase, the transitions are better resolved, in the rhombohedral phase the overall transmission decreases by approximately an order of magnitude and the modulation due to magnetooptical interband transitions gets weaker. For the Voigt geometry a recording is shown for B|| [110], for B||E as well as for B⊥E (Fig.4)

Fig.5 shows as an example experimentally observed transmission minima as a function of magnetic field for various CO-laser lines for a PbGeTe sample with x=0.01 at T=17K. A comparison between the experimental and calculated transition energies is made. The dominant features in the experimental as well as calculated data are due to transitions between the T-Landau ladder systems where according to Ref.20 also the strongest dipole transition matrix elements occur (Δn=0, ↑→↓ and ↓→↑ for σ± radiation, respectively for $\vec{B}||c||\vec{k}$).

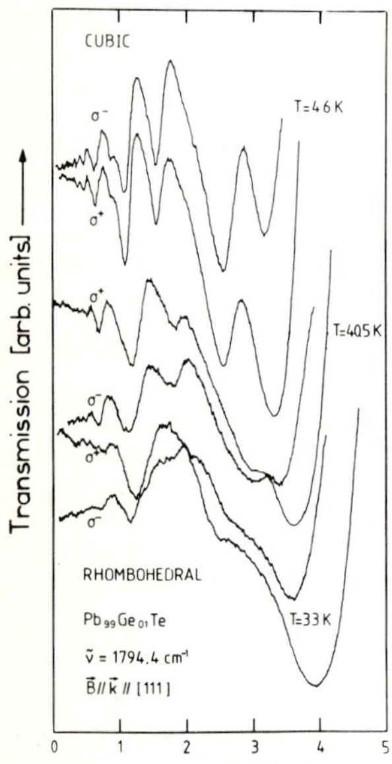

Fig.3: Transmission vs B in Faraday geometry, below and above phase transition temperature [19].

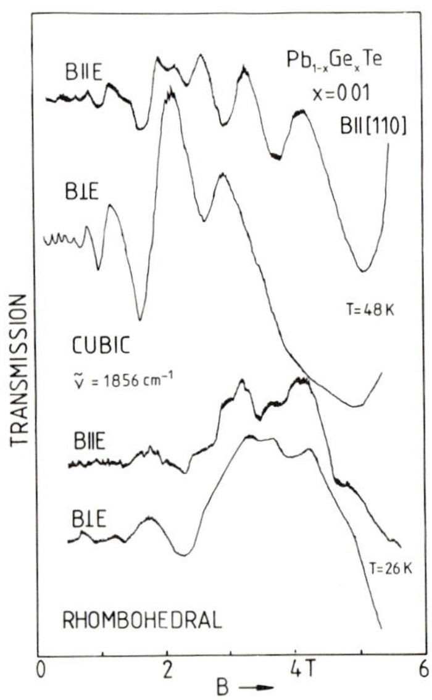

Fig.4: Transmission vs B in Voigt geometry B|| [110], B||E and B⊥E below and above phase transition temperature [24].

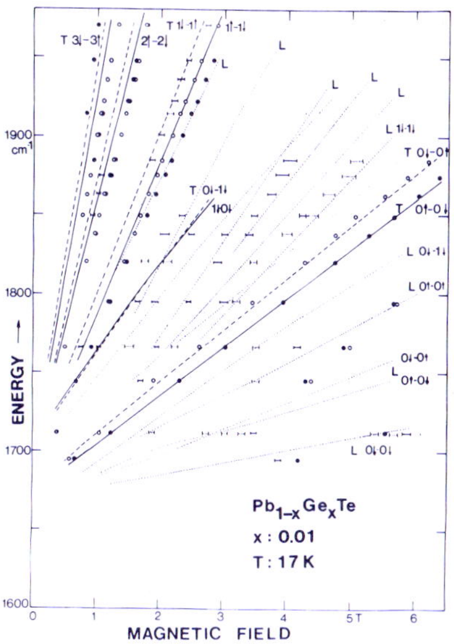

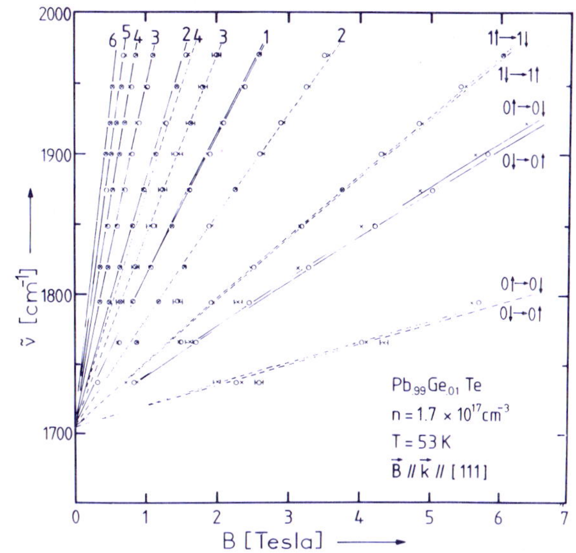

Fig.5: Interband transitions in rhombohedral PbGeTe. Experimental data: σ^+ (●), σ^- (o); weak transitions: σ^+ (⊢—⊣), σ^- (⊢--⊣) [19] calculated values: T-valley: σ^+ (—) σ^- (---); L-valley: $\sigma^\pm$ (...) after [20].

Fig.6: Interband transitions in cubic PbGeTe; experimental data σ^+:o; σ^-:x; calculated data: (—) within [111] Landau ladder system (---) within oblique <$\bar{1}$11> systems [19].

The tendency apparent from the data is the fact that transitions involving carriers in the obliquely oriented valleys are much weaker below T_c than above this temperature. The reason is the following: the selection rules are relaxed, the oscillator strength of several closely lying transitions are approximately equal. Due to broadening effects most of these transitions cannot be resolved and mostly contribute to the enhanced background absorption. In addition the oscillator strength of various transitions is magnetic field dependent and therefore difficult to identify unambiguosly. As a function of temperature, a small shift of energy gap below T_c is observed in agreement with earlier absorption data without magnetic field, [13,26]. In addition we observe a splitting between the two most prominent features for σ^+- and σ^--radiation (0↑→0↓, 0↓→0↑, 1↑→1↓, 1↑→1↓) which increases with decreasing temperature.

The data indicate, that this splitting is already observable at relatively low magnetic fields and an extrapolation for the σ^+- and σ^--transitions to B→0 does not yield the same apparent energy gap. The transmission for the σ^+ polarization for the 0→0 transition of the valley with its main axis parallel to the surface normal is much weaker by about a factor of 3, than for the σ^- polarization throughout all laser photon energies used. Also for the higher transitions (1→1, 2→2) radiation with the same photon energy but different sense of polarization is not absorbed equally.

In the cubic PbGeTe as shown in Fig.6 the interband magnetooptical transitions are described by the Mitchell and Wallis-model. Transitions within the Landau ladder systems oriented || to B ([111]) as well as oblique to B (<111>) are observed for σ^+- and σ^--radiation with the following selection rules: n↓→n↑ and n↑→n↓, respectively [27].

V. Intraband Transitions

Magnetooptical intraband transitions below and above the phase transition temperature have been studied by various groups [17,18, 28,29]. Fig.7 shows as a representative example, the temperature dependence of transmission versus B taken at $\lambda=118.8\mu m$ in Faraday geometry $B||k||$ [111] [29].

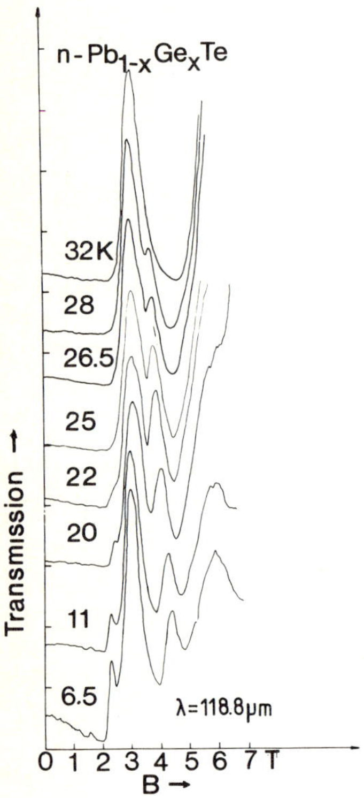

Fig.7: Intraband-transmission vs B in PbGeTe x=0.0084 as a function of temperature. $T_c \sim 32K$. [29].

Above T_c the structure is due to the two cyclotron resonances, in the || oriented L-valley and the three obliquely oriented L-valleys with respect to B:

$$m_{c1}=m_t;\ m_{c2}^{-1}=\sqrt{\frac{1}{9m_t}\left(\frac{8}{m_l}+\frac{1}{m_t}\right)}$$

Below T_c, additional structures appear and a small shift is observed. The transmission minima cannot be directly related to the cyclotron resonance positions, since for the chosen laser wavelength the realpart of background lattice dielectric[+] function is negative ($\omega_{TO}<\omega_L<\omega_{LO}$) and thus the data have to be fitted to an oscillator model.

In the cubic situation ($T\geq 32K$ in the case of Fig.7), the results and their analysis are very similar to PbTe apart from some what larger masses, mainly due to the increase in energy gap.

Below T_c, Fig.7 shows that with decreasing temperature additional resonances occur. The reduced symmetry splits considerably the transition energies between the up- (↑) and down - states (↓) for a given Landau quantum number n. Thus instead of the relatively closely lying 0↑→1↑,0↓→1↓ spin split transitions in the cubic structure, which are not resolved but contribute just to one broader absorption, in the rhombohedral phase the up-up(n↑→n+1↑) and down-down (n↓→n+1↓) are split. Thus one expects at least four resonance structures. Two of these originate from the valley oriented with its c-axis || to surface normal and B(T-valley in c-domain and one of the L-valleys per c'-domain). The other two structures result from cyclotron resonance transitions in the obliquely oriented valleys (3 L-valleys in a c-domain, 2 L-valleys and a T-valley in the c'-domains, respectively).

Fig.7 shows, that at low temperatures there are definitely more resonances than four. Thus the experiments were extended to higher magnetic fields. Fig.8 shows the dependence of the transmission of n-$Pb_{1-x}Ge_xTe$ on B up to 19 Tesla for three different laser lines. The data indicate, that apparently apart from the four resonances identified as cyclotron resonances more structures are present below T_c.

For the number of carriers contributing to the various resonances, the position of the Fermi level with respect to band edges and therefore also the relative shift of the conduction band edges (n-type samples) or the valence band edges (p-type samples) is of importance. For the analysis, a complete dielectric function, with contributions from ε_∞, the infrared active phonon (classical oscillator) and the free carriers were used. Instead of a single plasma frequency, the number of carriers in a given valley was calculated, and from the

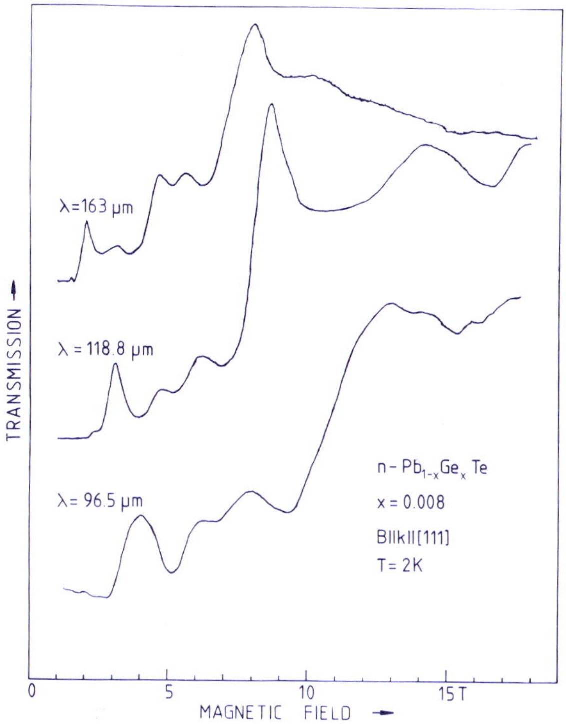

Fig.8: Transmission versus B for n-PbGeTe in the rhombohedral phase (T=2K, T_c∿31K) for three laser wavelengths.

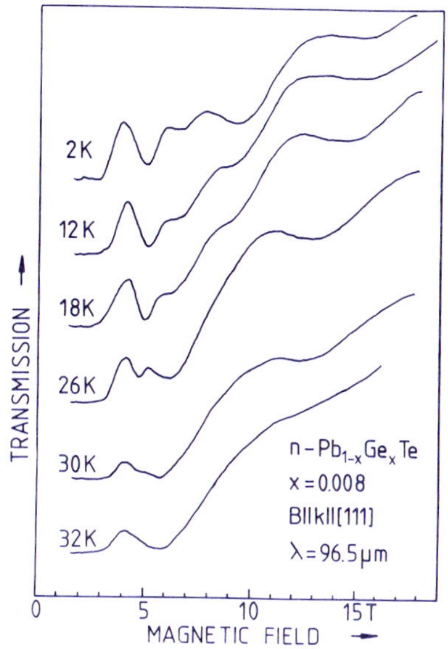

Fig.9: Cubic-rhombohedral phase transition as evidenced by temperature dependence of transmission as a function of B for λ=96.5μm. Curve at highest temperature represents cubic phase.

matrix elements calculated by Bangert [24] the oscillators strengths were derived and the results compared with experimental data. The complete analysis will be published elsewhere [24].

According to the Bangert-model [20, 24] apart from normal cyclotron resonance transitions combined resonance transitions (0↓→1↑) as well as "spin"-flip transitions (0↓→0↑) in T- and L-valleys are allowed in the rhombohedral phase.

To get more information on normal cyclotron transitions as well as the "spin"-flip transitions, the measurements on the influence of temperature on far infrared laser transmission were expended to higher fields and various wavelengths. Figs.9 and 10 show as an example data for λ=96.5μm and λ=118.8μm up to 19T.

The phase transition manifests itself very clearly in the change of the structure of the dielectric anomalies associated with the various resonances as evidenced from Figs.9 and 10.

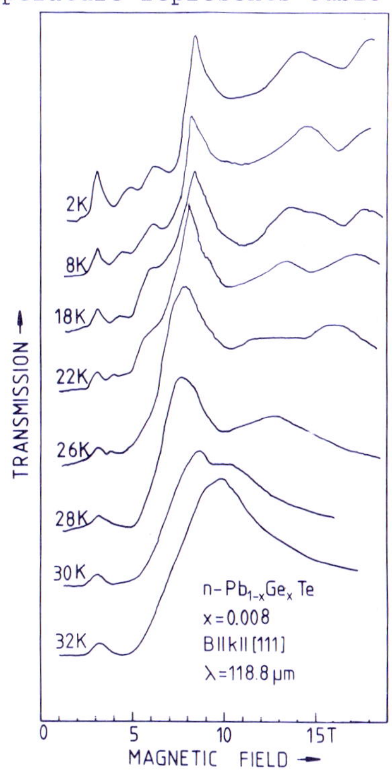

Fig.10: As Fig.9 but for laser wavelength λ=118.8μm.

Investigations on p-type samples have been reported too [17,18,28, 29]. The overall behaviour is very similar to the conduction band, the T- and L-valley $n\uparrow\rightarrow(n+1)\uparrow$ transitions and $n\downarrow\rightarrow(n+1)\downarrow$ transitions are considerably split and additional transitions occur. A T-valley (B||c) combined resonance transition has been detected already in [17,28].

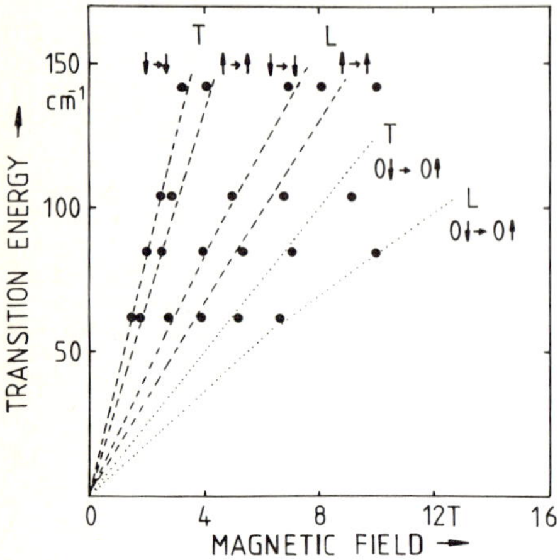

Fig.11: Transition energies vs B in Faraday geometry B|| [111] in the rhombohedral phase: T=10K n-PbGeTe, x=0.008; experimental data: o; theoretical data: cyclotron transitions: - - - ; spin-flip transitions: ....; T_c = 31K.

The experimentally deduced transition energies from the different laser-transmission experiments are plotted in Fig.11 as a function of magnetic field for the data shown in Figs.7-10. Apart from the split cyclotron-resonance transitions of the valley oriented parallel to surface normal (for brevity indicated as T in Fig.11) and the cyclotron transitions of the obliquely oriented valleys with respect to B (indicated as L in Fig.11) additional relatively strong transitions occur. These are identified as spin flip transition in the T- and L-valleys. From the splitting of the normal cyclotron resonance transition, which increases with decreasing temperature (see Figs.7,9 and 10) below the phase transition temperature, the "rhombohedral" band parameters S and T and their temperature dependence can be deduced. The broken lines and the dotted lines in Fig.11 were calculated with the Bangert-model (Eq.1) using the band parameters for $T\sim 10K$ as given in Table I:

Table I: Band parameters

(rhombohedral phase, x = 0.008, T = 10 K)		
two band:	$L = 4.3\times10^{-8}$ eV cm $Q = 1.53\times10^{-8}$ eV cm	far band: $A = 7.2\times10^{-15}$ eV cm² $B = 1.5\times10^{-15}$ eV cm² $C = -4.5\times10^{-15}$ eV cm² $D = -0.2\times10^{-15}$ eV cm² $g_1^c = -3$
"rhombohedral"	$S = 0.35\times10^{-8}$ eV cm² $T = 0.22\times10^{-8}$ eV cm²	

For this low Ge-concentration x, (x=0.008) the apparent gap $\tilde{E}_g$=210meV has been used. The S-parameter determines mainly the influence of the reduction of symmetry $D_{3d}\rightarrow C_{3v}$ in the conduction band, the T-parameter the corresponding change in the valence band. For $T>T_c$ these two parameters disappear and the matrix hamiltonian (Eq.1) is then completely equivalent to the cubic Mitchell-Wallis hamiltonian. Since the phase transition is second order, the band parameters exhibit a dependence on temperature for $T<T_c$, even in the case $\tilde{E}_g\sim$ const., as found for small Ge contents below phase transition temperature [26,1].

The analysis presented above does not take into account the effect of stresses between the films and substrate at low temperatures [5, 16,28]. Its influence is discussed in the next section.

VI. Discussion

As far as the cubic phase of $Pb_{1-x}Ge_xTe$ ($x < 0.015$) is concerned, intra- and interband magnetooptical data are described satisfatorily by sets of band parameters obtanined within the Mitchell-Wallis band - model. In comparison to PbTe it is not only necessary to increase the energy gap, but also to change the two-band as well as the far-band parameters [19,24].

Below T_c, the k.p model by Bangert has been used for a comparison between theory and experiment. A main feature of this model is the fact, that the reduced symmetry (C_{3v} at T-point, C_s at L-point of a c-domain) leads to a considerable splitting of moderately split $\uparrow \rightarrow \uparrow$, $\downarrow \rightarrow \downarrow$ intraband transitions and $\uparrow \rightarrow \downarrow$, $\downarrow \rightarrow \uparrow$ interband transitions in the cubic phase and to additional selection rules. For a quantitative description of the data several difficulties are encountered:

(i) In the Bangert-model the transitions within the Landau-ladder systems of various domains (c- and c'-domains for $B||c$) are treated under the simplifying assumption that for the L-points not the complex k.p hamiltonian for general C_s-symmetry is used but just a special case of that matrix with z-coordinate oblique to the rhombohedral c-axis. Thus in cyclotron resonance for $B||c$ just four transitions ($n\downarrow \rightarrow (n+1)\downarrow$, $n\uparrow \rightarrow (n+1)\uparrow$) for T- and L-valleys occur. All other observed transitions are thus explained in this model as resulting from transitions involving spin-flip, with transition-probabilities about an order of magnitude smaller as compared to cyclotron-resonance. The very complex structure , found in magnetoplasma -experiments by Kawamura et al. [18] cannot easily be explained by this model, whereas the data presented in this work (Fig.11) and earlier far-infrared transmission experiments [17,28,29] are satisfactorily described.

(ii) Below T_c, the energy gaps at the T- and L-points of the Brillouin-zone (BZ) are not identical. $Pb_{1-x}Ge_xTe$ is in the rhombohedral phase an indirect semiconductor with a minimum gap between the T-valence and L-conduction band states (c-domain). The relative shift between the T - and L-edges depends on composition x which determines the energy shifts due to strain (ε) and relative displacement ($\vec{u}$). For the Ge-compositions used in magnetooptical experiments ($x<0.015$) an energy difference between the T-and L-band edges of less than 5 meV results [16].
However, both the differences in the gaps at the L-and T-points of the BZ, as well as the relative shifts of the band edges in the conduction as well as the valence bands are important. For carrier concentrations of the order of $10^{17}cm^{-3}$, the Fermi-energies are of the order of 10 meV. Population effects thus influence the intra-band transitions and blocking effects the interband transitions.
Increments of the gaps at T- and L - extrema were invoked also by Drabkin et al. [30] for an explanation of changes of magnetic susceptibility.

From the temperature dependence of the "rhombohedral" parameters S and T in the Bangert-model one could in principle deduce also the change of the relative sublattice displacement $\vec{u}$, the ordering parameter with temperature below T_c for a given composition x, if the relevant deformation potentials were known. Takaoka et al. [16] have used PbTe deformation potentials for the calculation of energy shifts, using a Landau-expression u^2 = const (T_c-T) and obtained an optical deformation potential of 12 eV for $x = 0.015$.

The S- and T-parameters are, in addition to the two-and far-band contributions, e.g. responsible for the splitting of the ($n\uparrow \rightarrow n\downarrow$, $n\downarrow \rightarrow n\uparrow$) transitions, the magnitude of which being dependent on quantum number n. The magnitude of S and T and their signs determine whether the σ^+ or σ^- - interband transitions occur at a given B at higher or lower energies.

Table II: Splitting of σ^+-σ^- interband transitions in rhombohedral phase, B||c, $Pb_{1-x}Ge_xTe$, x = 0.01.

Temperature	Δ {O↑ → O↓ ; O↓ → O↑}		
	B = 2T	B = 3T	B = 4T
1.6 K	16cm^{-1}	19 cm^{-1}	21cm^{-1}
18.5 K	14	15	16.5
31 K	9	10	11

In Table II, the temperature dependence of these energy differences is shown. From the data in Fig.5 we conclude, that the σ^+ and σ^--interband transitions do not extrapolate to the same energy gap. Experimentally, we observe transitions not only within the T-valence and conduction states of the c-domain but also in the L-states of c'-domains oriented with their main axis ||B.In n-type samples the relative shift of the T- and L- band edges blocks more effectively the σ^--transitions of these L-states. Transitions whithin the T-ladders will be less affected and the σ^+-transitions will occur both in the T- and L-Landau-ladder systems. In Fig. 5 the T-L band gap split varies from about 1 meV at 31 K to 2 meV at 1.6K. The observed differences in absorption constant for σ^+- and σ^--polarized radiation indicate in addition the relative band edge shift between the T- and L-conduction band edges, effects of blocking due to population and also is in agreement with a domain ratio of c'/c -domains of approximately 3:1.

The intraband data (Figs.7 - 10) show that for $T < T_c$ with decreasing temperature the first new cyclotron transitions are due to the splitting of the n↑ → (n+1)↑ and n↓ → (n+1)↓ transitions in the obliquely oriented valleys with respect to B. This results both from the domain ratio as well as from population effects.When the temperature is reduced further, then also the valleys oriented with their main axis || B show the splitting. All the data indicate, that the shift of the resonance positions with decreasing temperature is rather small. There is only a small increase of the resonance field position difference of ↑ → ↑ and ↓ → ↓ transitions. Thus, the population of the T- and L-states and their change with T for $T < T_c$ dominate the far-infrared magnetotransmission. Also in p-type samples, there is practically only a minor shift of resonance positions below T_c [28].

In Table III cyclotron masses for the T- and L valleys are given.

Table III: Cyclotron resonance masses (n-$Pb_{1-x}Ge_xTe$, T=2K)

B||c	T- transitions		L- transitions		
	↓ → ↓	↑ → ↑	↓ → ↓	↑ → ↑	
m_c/m_o	0.021	0.027	0.045	0.057	x = 0.008
m_c/m_o	0.016	0.027	0.049	0.061	x = 0.01

All the data presented in this paper were obtained with epitaxially grown films on insulating substrates. In the $Pb_{1-x}Ge_xTe$ system grown on BaF_2 substrates, the lattice constant mismatch decreases with increasing Ge-content [5]. For contents x > 0.047, Hohnke et al. [5] have observed a change of the coefficient of thermal expansion at $T = T_c$. Since the mobility of carriers in samples with x > 0.015 is too small, only for relatively low Ge-contents magnetooptical investigations are possible. X-ray investigations on the change of lattice constant with T were not yet performed for $x<0.02$, and also no information on the T-dependence of thermal expansion coefficient exists. However, for low Ge-contents of PbGeTe with transition temperatures $T_c < 50$ K, the effects of strains due to differences in thermal expansion and lattice constants will be similar to the PbTe/BaF_2 system. A slight distortion of the films in this case causes a downwards shift in energy of the L-valley oriented with its main axis || to the [111] surface normal. With $T < T_c$ the opposite trend in the relative energy shift of the corresponding T-state (c-domain) will occur. In our case the relevant shifts are of the order or less than 5 meV.

VII. Conclusions

The cubic-rhombohedral phase transition was investigated in $Pb_{1-x}Ge_xTe$ by magnetooptical methods. The results are the following:

a) The four equivalent L-states of the O_h-structure are split into singlet T- and triplet L-states in the C_{3v}-structure.
b) Below T_c, the inversion symmetry is lost, thus k-linear terms occur in the dispersion and the Kramers-doublets are split.
c) The extrema in the E(k) relationship in direction perpendicular to the rhombohedral c-axis are shifted away from T-and L-points of the BZ. The surfaces of constant energy are not simple ellipsoids of revolution.
d) The reduction in symmetry causes new selection rules for inter- and intraband magnetooptical transitions.
e) The minimum energy gap is indirect for $T < T_c$. The gap is different at L-and T-points and this difference increases with decreasing temperature.

Acknowledgments:

I am grateful to E.Bangert, E.J.Fantner, and H.Pascher for their collaboration in this work. All samples were prepared by A.Lopez-Otero. High field experiments were performed at the High Magnetic Field Laboratory of the Max Planck Institut für Festkörperforschung and the CNRS, Grenoble, France, in collaboration with M.von Ortenberg, J.Thielemann, and W.Staguhn, and G. Belle.
This work was supported by Fonds zur Förderung der wissenschaftlichen Forschung, Vienna, Austria.

References:

1) H. Kawamura in Proc.Int.Conf.Phys. Narrow Gap Semiconductors, ed. J. Rauluszkiewicz (PWN Warszawa 1978) p.7 and in "Narrow Gap Semiconductors" ed. W. Zawadzki, Lecture Notes in Physics 133 (1980) 470.
2) R.T. Bate, D.L. Carter, J.S. Wrobel, Phys.Rev. Lett. 29 (1970) 159.
3) R.S. Allgaier, SPIE-Infrared Detector Materials 285 (1981) 3.
4) W.Jantsch, G.Bauer, A.Lopez-Otero in Proc. 14th Int.Conf. Phys. Semicond. ed. B.L.H. Wilson (Edinburgh 1978) p.445.
5) D.K. Hohnke, H.Holloway, S.Kaiser, J.Phys.Chem.Solids 33,(1972) 2053.
6) R. Clarke, Phys.Rev. B 18 (1978) 4920.
7) G.S. Pawley, W.Cochran, R.A.Cowley, G.Dolling, Phys.Rev.Lett.17 (1966) 759;
G.Dolling, W.J.L. Buyers, J.Nonmetals 1 (1973) 159;
W.J.Daughton, C.W.Tompson, E.Gurmen, J.Phys.C: Solid State Physics 11 (1978) 1573.
8) G.Bauer, H.Burkhard, W.Jantsch, F.Unterleitner, A.Lopez-Otero, in Proc. Int.Conf. Lattice Dynamics; ed. M.Balkanski (Flammarion, Paris 1977) p.669;
W.Jantsch, G.Bauer, A. Krost, A.Lopez-Otero, Ferroelectrics 38 (1981) 905.
9) W.Jantsch, H.Mitter, A.Lopez-Otero, Z.Physik B 41 (1981) 287.
10) S.Nishi, H.Kawamura, K.Murase, phys.stat.sol. (b) 97 (1980)581.
11) E.F.Steigmeier and G.Harbeke, Solid State Commun. 8 (1970) 1275.
12) S.Sugai, K.Murase, S. Katayama, S.Takaoka, S. Nishi, H.Kawamura, Solid State Commun. 24 (1977) 407.
13) K.Murase, J.Phys.Soc. Japan 49 Suppl.A (1980) 725.
14) W.Jantsch in Physics of Narrow Gap Semiconductors, ed. E.Gornik, Lecture Notes in Physics 152 (1982) 221.
15) S.Takaoka, K. Murase, Phys.Rev. B20 (1979) 2823.
16) S.Takaoka, K.Murase, J.Phys.Soc. Japan 51 (1982) 1857.
17) A.V.Lewis, R.J.Nicholas, J.C. Ramage, G.Bauer, A.Lopez-Otero, R.A.Stradling, J.Phys. C13 (1980) 561; C13 (1980) L443.
18) H. Kawamura, S.Nishikawa, K.Murase in Proc.Int.Conf. on Appl. High Magnetic Fields in Semicond. Physics, ed. J.F. Ryan (Oxford 1978) p.170.
19) E.J.Fantner, H. Pascher, G.Bauer, R.Danzer, A.Lopez-Otero, J.Phys. Soc.Japan 49 Suppl.A (1980) 741.
20) E.Bangert, in Physics of Narrow Gap Semiconductors, ed.E.Gornik, Lecture Notes of Physics 152 (1982) 216.
21) D.L.Mitchell and R.F.Wallis, Phys.Rev. 151 (1966) 581.
22) A.Lopez-Otero, Thin Solid Films 49 (1978) 1.
23) H.Burkhard, G.Bauer, W.Zawadzki, Phys.Rev. B19 (1979) 5149.
24) E.Bangert et al., to be published
25) G.Bauer, in Narrow Gap Semiconductors, ed. W.Zawadzki, Lecture Notes in Physics 133 (1980) 427.
26) W.Jantsch, Z.Phys. B 40 (1980) 193.
27) R.Grisar, H. Burkhard, G.Bauer, W.Zawadzki, in Proc.3rd Int.Conf. Phys.Narrow Gap Semiconductors, ed. Rauluszkiewicz (PWN, Warszawa (1977) p.115.
28) A.V. Lewis, G.Bauer, R.J.Nicholas, J.C. Ramage, Int.J.Infrared and Millimeter Waves 1 (1980) 485.
29) G.Belle, E.J.Fantner, G.Bauer, A.Lopez-Otero, E.Bangert, Verhandl. DPG (VI) 16 (1981) 190;
G.Belle, Diploma Thesis, Montanuniversität Leoben, (1981);unpublished.
30) I.A.Drabkin, T.B.Zhukova, M.A.Kvantov, O.E. Kvyatkovskii, L.M. Sysoeva, Sov.Phys.Semicond. 15 (1981) 1163.

TRANSPORT PROPERTIES IN THE ACTIVE LAYER OF A DOUBLE HETEROSTRUCTURE INJECTION LASER MEASURED BY A STRONG MAGNETIC FIELD

H.J.A. Bluyssen and A.van Etteger
Katholieke Universiteit van Nijmegen, Nijmegen, The Netherlands

L.J.van Ruyven
Philips Research Laboratories, Eindhoven, The Netherlands

abstract

The transport properties of the active layer of a double heterostructure injection laser can be investigated by experimental observation of the wavelength shift and the threshold current as a function of the magnetic field. The effect of the magnetic field is twofold i.e. a transformation of the energy bands into Landau levels as well as localized states into Zeeman states and on the other hand the Lorentz force acting on the carrier transport and therefore on the threshold current.

The energy of the stimulated emission coincides with the energy at maximum gain, which is determined by the matrixelement of the optical transition, the Fermi-Dirac distribution function and the density of states function of the Landau levels which includes the electronic scattering time. Both the matrixelement and the electronic scattering time are independent of the magnetic field.

From measurements of the energy shift of the peak of the laser emission as a function of the magnetic field, the scattering time and the magnetic field dependence of the carrier density at threshold can be determined. Although the threshold current must be proportional to the carrier density at threshold, measurements show a significant reduction in the threshold current for both directions of the magnetic field with respect to the p-n junction layer. The magnitude of this reduction is assumed to depend on the geometry of the laser structure and may be attributed to the spreading in the injection current. Lasers which have narrow stripes and low resistive cladding layers have an appreciable current spreading and therefore should show a significant reduction in the threshold current. On the other hand lasers with a braod stripe and a high resistive cladding layer have practically no current spreading and therefore the threshold current should follow the carrier density at threshold as a function of the magnetic field.

QUANTUM OSCILLATIONS OF INDIRECT INTERBAND MAGNETOABSORPTION BELOW THE BURSTEIN EDGE

J. Mycielski(a, b), G. Bastard(a), J.K. Furdyna(c), C. Rigaux(a), H. Serre(a), A. Simiczyjew(b)

(a) Groupe de Physique des Solides de l'École Normale Supérieure, 24 rue Lhomond, 75231 Paris Cedex 05, France

(b) Institute of Theoretical Physics, Warsaw University, 00-681 Warsaw, Poland

(c) Physics Department, Purdue University, West Lafayette, Ind. 47907, USA

For a simple semiconductor the indirect interband magnetoabsorption below the Burstein edge, due to the electron-ionized impurity interaction, is calculated. The magnetoabsorption coefficient shows quantum oscillations as a function of reciprocal magnetic field. Oscillations of this type and amplitude were observed recently by the present authors.

I. Introduction

It is well known that in a degenerate n-type semiconductor the direct absorption or magnetoabsorption gap is shifted into higher energies /"Burstein shift"/ . The direct transitions between the valence band states and empty conduction band states occur for the photon energies $\hbar\omega > E_B$, where /for parabolic bands/ the Burstein edge E_B is

$$E_B = E_g + [1 + (m_c/m_v)]\, E_F. \tag{1}$$

Here E_g, E_F, m_c, and m_v are the energy gap, Fermi energy, and the effective masses of electron and hole, respectively. It should be noted, however, that the energy difference between the highest valence band state and the lowest empty conduction band state is smaller than E_B and equals

$$E_{IND} = E_g + E_F + (1/2)\hbar\omega_v, \tag{2}$$

where ω_v is the hole cyclotron frequency. Interband absorption or magnetoabsorption may, therefore, appear also for photon energies in the range

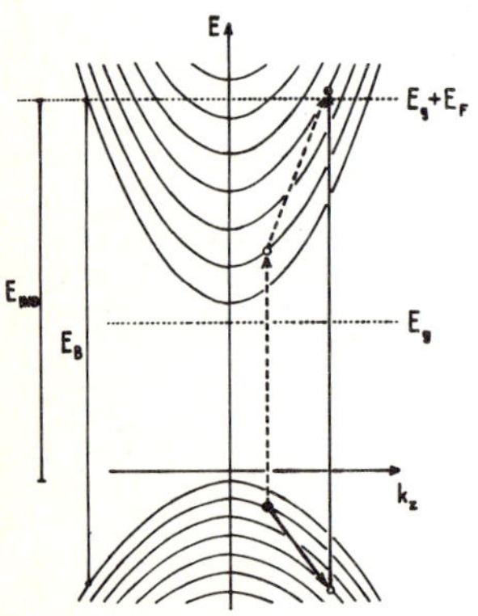

Fig. 1. Indirect interband magnetoabsorption processes in a degenerate n-type semiconductor.

$$E_{IND} < \hbar\omega < E_B, \tag{3}$$

due to indirect transitions of electrons interacting with crystal imperfections. This is shown in Fig. 1.

A strong indirect interband absorption and magnetoabsorption below the Burstein edge in both longitudinal and transverse polarizations was observed recently in zero-gap strongly degenerate n-type $Hg_{1-x}Fe_xSe$ /$x \simeq 0.01$/ at helium temperatures and in magnetic field $H \leq 5$ T [1,2 The absorption coefficient gradually increases with the photon energy in the range 240 to 280 meV. This range corresponds to (3) in the investigated samples.

In the magnetic field, the magnetoabsorption coefficient in the range (3) shows small /up to 2 %/ oscillations periodic in 1/H. The position of maxima and minima are independent of ω. It was concluded from the period that these oscillations correspond to crossing the Fermi level by Landau levels in the conduction band.

The observed effect is not the intraband Shubnikov-de Haas one as the photon energy in the range (3) is much higher than the cyclotron energy $\hbar\omega_c$ in the conduction band. Thus, the free-carrier absorption cannot occur between states both close to the Fermi level. Oscillations of the intraband /free-carrier/ absorption were calculated recently [3] and it was shown that their period depends on ω, opposite to what was observed in [1, 2]. Close to the plasma edge, small changes of the plasma frequency ω_p may produce an important change of the reflectivity and, hence, of the transmission. However, the range (3) was well above the plasma frequency in [1, 2] /$\hbar\omega_p \simeq 130$ meV/. Moreover, the effect of plasma frequency oscillations /as a function of 1/H/ calculated in [4, 5], followed from neglecting the dependence of E_F on H. In the zero-gap semiconductor, the dependence of E_F on H may be a reason for a very weak oscillations of ω_p, through the dependence of the effective mass on energy in non-parabolic conduction band. One may also expect some oscillations of the refractive index due to virtual interband transitions. However, both effects are at least two orders of magnitude to weak to explain the experimental findings.

The purpose of the present paper is to calculate the indirect interband magnetoabsorption below the Burstein edge and to show that it oscillates with a period independent of ω and an amplitude comparable to that found experimentally.

II. Calculation of magnetoabsorption

We consider a model semiconductor with simple parabolic and spherical energy bands. Spin splitting in magnetic field and spin-orbit coupling are neglected. We assume a non-vanishing interband momentum matrix element. Assuming that the electrons interact with the /unscreened/ Coulomb potentials of the ionized impurities /defects/ , we will calculate the indirect interband magnetoabsorption for the longitudinal polarization of radiation.

We take $\underline{H}$ in z-direction and use the Landau gauge for the vector potential: $\underline{A} = (0, Hx, 0)$. The wave functions in the valence /v/ and conduction /c/ bands are

$$|v(c)\ k_y\ k_z\ n\rangle = (L_yL_z)^{-1/2}\exp[i(k_y y + k_z z)]\, f_n(x - x_o)\,|v(c)\rangle, \quad (4)$$

where $|v(c)\rangle$ is the Bloch function /normalized in unit volume/ at $\underline{k} = 0$ in the valence /conduction/ band, f_n is the nth harmonic-oscillator function of the characteristic length $\lambda = (c\hbar/eH)^{1/2}$, $x_o = -\lambda^2 k_y$, L_y and L_z are the dimensions of the periodicity box, and k_y, k_z have values allowed by this box.

The radiation perturbation in the Hamiltonian has the form

$$H_r = -i(\hbar e/m\omega)\mathrm{Im}[E_z\exp(-i\omega t)]\ \partial/\partial z\ , \quad (5)$$

where m is the free-electron mass, and E_z is the amplitude of electric field. From (5) , the time-independent radiation perturbation operator is [6]

$$H_+ = -(\hbar e/m\omega)E_z\ \partial/\partial z\ . \quad (6)$$

The perturbation due to the fluctuations of the electrostatic potential in space is of the form

$$H_f = \sum_{\underline{q}}{}' U(\underline{q})\exp(i\underline{q}.\underline{r})\ , \quad (7)$$

where $\underline{q}$ runs over all $\underline{q} \neq 0$ allowed by the periodicity box, and $U(-\underline{q}) = U^*(\underline{q})$. The interband matrix element of H_f vanishes. The intraband elements are [6]

$$\langle v(c)\ k_y + q_y\ k_z + q_z\ n'\,|\,H_f\,|\,v(c)\ k_y\ k_z\ n\rangle$$
$$= \sum_{q_x}{}'' \exp(-i\lambda^2 k_y q_x)U(\underline{q})F(n', n, q_y, q_x)\ , \quad (8)$$

where summing over q_x we omit $q_y = 0$ if $q_y = q_z = 0$. F is defined as

$$F(n', n, q_y, q_x) = \int f_{n'}(x + \lambda^2 q_y)\exp(iq_x x)f_n(x)dx . \qquad (9)$$

The interband matrix element of H_+ is

$$\langle c\cdot k_y + q_y\ k_z + q_z\ n' | H_+ |\ v\ k_y\ k_z\ n\rangle$$
$$= (eP/\hbar\omega)E_z\ \delta_{0q_y}\ \delta_{0q_z}\ \delta_{nn'} , \qquad (10)$$

where $P = -(\hbar^2/m)\ \langle c|\ \partial/\partial z\ | v\rangle$. Itegration in P is over unit volume.

The two second-order perturbation contributions to the indirect interband transition rate are shown in Fig. 1. Using the standard secon-order-perturbation formula for this rate and the explicite expression for $F(n', n, q_y, q_x)$ [6], we obtain the transition rate from the state $|v\ k_y\ k_z\ n\rangle$ to the state $|c\ k_y'\ k_z'\ n'\rangle$.

Assuming S types of impurity ions /defects/ of charges $Z_l e$ and concentrations N_l /l = 1, ... , S/ , the neutrality condition is

$$\sum_{l=1}^{S} Z_l N_l = N , \qquad (11)$$

where N is the electron concentration. For a perfectly random distribution of ions the average square modulus of $U(\underline{q})$ is [7]

$$\langle |U(\underline{q})|^2\rangle_{av} = 16\pi^2 e^4 DN/L_x L_y L_z \varkappa^2 q^4 , \qquad (12)$$

where $\varkappa$ is the static dielectric constant and

$$D = \sum_{l=1}^{S} Z_l^2 N_l/N . \qquad (13)$$

Summing the transition rate over the initial and final states and using (12), we obtain the indirect interband absorption coefficient K. Integrating analitically over three variables, we have finally

$$K = (e^6 |P|^2 m_c^{5/2} D/c\hbar^7 \varkappa^{5/2} E_F^{1/2}\ \omega)$$
$$\times f[m_c/m_v, E_F/\hbar\omega_c, (\hbar\omega - E_g)/E_F] , \qquad (14)$$

where the dimensionless function f(u, w, z) is a two-fold sum over n, n' of two-fold integrals over $q_\perp = (q_x^2 + q_y^2)^{1/2}$ and k_z.

In a similar way one can also calculate the indirect interband absorption of our model semiconductor in the absence of magnetic field, i. e. for $E_F/\hbar\omega_c = \infty$. One obtains $K(H = 0) = K_\infty$ in the form (14) , with $f(u, \infty, z) = f_\infty(u, z)$ being an analitic function.

III. Numerical results and discussion

Using the numerical integration method [8], we have calculated an example with the parameters close to those of the samples in [1, 2]: $E_g = 0$, $m_c = 0.04m$, $m_c/m_v = 0.125$, $N = 6\times10^{18}$ cm^{-3}. With these parameters, the range (3) is 300 meV < $\hbar\omega$ < 340 meV. The functions f and f_∞ are shown in Fig. 2, together with the magnetoabsorption coefficient K, calculated with the parameters $|P| = 7.2\times10^{-8}$eVcm, D = 1, $\varkappa$ = 16. It should be noted that this value of |P| corresponds to that for transitions $\Gamma_6 \rightarrow \Gamma_8$ in [1, 2]. Of course, transitions from the heavy hole Γ_8-band to the electron Γ_8-band in a zero-gap semiconductor have a vanishing matrix element at the Γ-point, increasing with k, different than in our model semiconductor.

The magnetoabsorption coefficient K oscillates with w /i. e. with 1/H/ , the positions of maxima being 1/2 + n /Landau level crossing Fermi level/ independent of ω. Oscillations are due to the fact that the transitions to the states close to the Fermi level are privileged by both the H_f matrix elements and the energy denominators.

The absolute amplitude of the oscillations of K is of the order of that observed in [1, 2]. However, K_∞ is an order of magnitude lower than that observed. This may be due to the presence of a higher /and non-oscillating/ contribution from short-range potential scattering /alloy scattering?/.

IV. Acknowledgement

The research of one of the authors /J. M. / was supported in part by the Institute of Physics, Polish Academy of Sciences.

References

1. H. Serre: Thèse, Université de Paris VI 1981.
2. H. Serre, G. Bastard, C. Rigaux, J. Mycielski, and J.K. Furdyna: Physics of Narrow Gap Semiconductors, Proc. 4th Int. Conf. Linz 1981, ed. E. Gornik, H. Heinrich, and L. Palmetshofer /Springer, Berlin 1982/, p. 321.
3. H. Adamska and H.N. Spector: Bull. Am. Phys. Soc. 27 (1982) 143.
4. M.S. Dresselhaus and G. Dresselhaus: Phys. Rev. 125 (1962) 499.
5. M.S. Dresselhaus and J.G. Mavroides: Solid State Comm. 2 (1964) 297.
6. J. Mycielski, G. Bastard, and C. Rigaux: Phys. Rev. B 16 (1977) 1675.
7. G. Bastard, J. Mycielski, and C. Rigaux: Phys. Rev. B 18 (1978) 6990.
8. A. Dydek and M. Pindor: Preprint IFT/11/78 (1978), Institute of Theoretical Physics, Warsaw University.

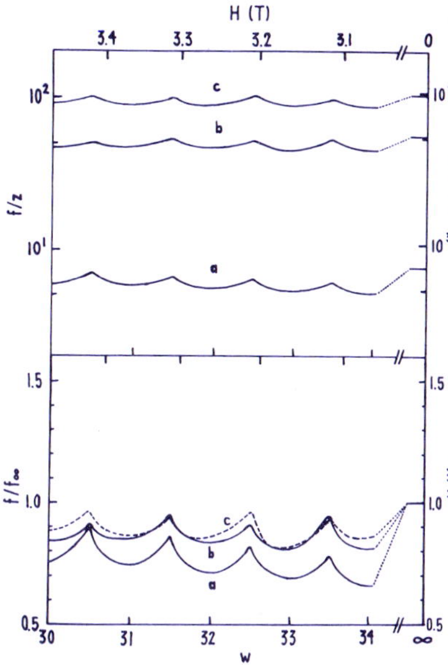

Fig. 2. Indirect interband magnetoabsorption coefficient /a numerical example/ at:
a) $z = 1.020$ /$\hbar\omega = 307$ meV/ ,
b) $z = 1.047$ /$\hbar\omega = 315$ meV/ ,
c) $z = 1.057$ /$\hbar\omega = 318$ meV/ ,
for the material parameters given in the text.

NEUTRAL BOUND EXCITONS IN A MAGNETIC FIELD

B. Stébé and F. Dujardin
Laboratoire de Physique des Milieux Condensés, Université de Metz,
Ile du Saulcy, 57045 Metz, France

G. Munschy
Laboratoire de Spectroscopie et d'Optique du Corps Solide,
(associé au C.N.R.S. n° 232), Université Louis Pasteur
5, rue de l'Université, 67084 Strasbourg Cedex, France

We are studying the influence of an external uniform magnetic field on the energy of the ground state of an exciton bound to a neutral impurity. In the low field limit, we treat the diamagnetic operator as a perturbation using our previously obtained Page and Fraser type wave function. In the high field limit, we have solved the Schrödinger equation by means of the adiabatic method. We compare our results with the observed diamagnetic shift in GaAs.

1. INTRODUCTION

The application of a magnetic field is particularly useful in the optical study of neutral bound exciton states, but apart from the analysis of the Zeeman splitting, we dispose of very few theoretical information predicting their behaviour in a magnetic field. In fact, it is only very recently that even in the absence of a magnetic field, it has been possible to prove definitively the stable binding[1,2] of an exciton with a neutral donor or a neutral acceptor, and that sufficiently accurate wave functions have become available. This lack of theoretical results[3] does not allow one at the present time, to take full advantage of all informations resulting from the magneto optical experiments. Only approximate models have been proposed in the past in order to describe some particular situations, by using the analogy with simple excitons, donors and acceptors.

In the phenomenological model of Willmann et al.[4,5] the interaction between an exciton and a neutral impurity is represented by an effective short range potential, so that the binding energy does not depend on the magnetic field, and the transition energy shifts of free and neutral bound excitons should be equal. In the pseudodonor model of the (A^o,X)-complex, the energy is just the sum of those of a neutral donor and a positive charged acceptor A^+. The transition energy shift should here be equal to that of a neutral donor increased by the difference of the energy shifts between the positive charged acceptor A^+ and the neutral acceptor ground state. If the latter contribution can be neglected, the transition energy shift reduces to that of a neutral donor[6,7] and should not be comparable to that of a free exciton[7]. Furthermore, the experimental situation seems not to be quite clear. For instance, measurements of the diamagnetic shift in CdS [8], GaSb [9] and GaAs [4,5] reveal that the magnitudes of the transition shifts of free and bound excitons are approximately equal, whereas more recent results on GaSb [6] and GaAs [7] show that the transition shift is almost identical to the energy shift of an effective mass donor, as already reported earlier[10]. It seems therefore desirable to dispose of more precise predictions of the magnetic properties of neutral bound excitons. The purpose of this paper is to present a first attempt in this direction in the two limits of low and high effective magnetic fields.

2. EFFECTIVE HAMILTON OPERATOR

We investigate here the magnetic field dependence of the exciton-neutral donor complex (D^o,X) and the exciton-neutral acceptor complex (A^o,X) in the effective mass approximation and a simple non degenerate isotropic two-band model. By interchanging the electrons and the holes, the studies of the two complexes are equivalent. We choose to consider the case of the (A^o,X)-complex which consists of the holes 1 and 2 and the electron e bound to the acceptor centre A, in the presence of an external uniform magnetic field $\vec{B}$ directed along the z-axis, with the Lorentz gauge $\vec{A} = (\vec{B}\times\vec{r})/2$, $\vec{A}$ being the vector potential describing the magnetic field. By introducing the usual atomic units $a_A = \varepsilon\hbar^2/m_h^* e^2$ of length, $2|E_A| = m_h^* e^4/\varepsilon^2\hbar^2$ of energy and the dimensionless effective magnetic field parameter $\gamma_A = \hbar\omega_c/2|E_A|$, where $\omega_c = eB/m_h^* c$ is the effective hole cyclotron frequency, ε an appropriate dielectric constant, the envelope Hamiltonian becomes by neglecting the spins as well as the electron-hole exchange interaction :

$$H = T_h + \sigma^{-1} T_e + V + H_1' + H_2' \tag{1}$$

where

$$T_h = -(\nabla_1^2 + \nabla_2^2)/2 , \qquad T_e = -\sigma^{-1} \nabla_e^2/2 \tag{2}$$

$$V = -1/r_1 - 1/r_2 + 1/r_e + 1/r_{12} - 1/r_{1e} - 1/r_{2e} \tag{3}$$

$$H_1' = \gamma_A (-L_{1z} - L_{2z} + \sigma^{-1} L_{ez})/2 \tag{4}$$

$$H_2' = \gamma_A^2 (\rho_1^2 + \rho_2^2 + \sigma^{-1} \rho_e^2)/8 \tag{5}$$

with $\sigma = m_e^*/m_h^*$, $\rho_i^2 = x_i^2 + y_i^2$

L_{iz} are the projections of the angular momentum operator. H_1' denotes the linear Zeeman term whereas H_2' is the quadratic diamagnetic operator. The energies of the two complexes are related by :

$$E_{(A^o,X)}(\sigma,\gamma_A)/E_A = E_{(D^o,X)}(\sigma^{-1},\gamma_D = \gamma_A)/E_D \tag{6}$$

with $\gamma_D = \gamma_A/\sigma^2$ for a fixed magnetic field. The importance of the magnetic terms H_1' and H_2' depends on the values of the effective magnetic field γ which may be considerably different for donor or acceptor impurities in a given magnetic field. For acceptors the effective field γ may often be one order of magnitude smaller than for donors. Therefore, in order to investigate the (D^o,X) and (A^o,X) complexes in magnetic fields up to 10 Teslas, it is necessary to consider a large scale of γ values ranging from about 10^{-3} to 10^{+2}.

3. LOW FIELD APPROXIMATION

For small enough values of the effective field γ, we can treat the magnetic terms H_1' and H_2' as a perturbation with respect to the zero field Hamiltonian in (1), the Schrödinger equation of which we have solved previously[1] using a 35-term Page and Fraser type wave function :

$$\phi(r_{ij}) = \psi(kr_{ij})$$

$$\psi(r_{ij}) = \sum_{mnpq} c_{mnpq} (1+P_{12})|mnpq>$$

$$|mnpq> = r_1^m r_{1e}^n r_e^p r_{12}^q \exp(-\alpha r_1 - \beta r_2 - r_e) \tag{7}$$

$(m,n,p,q$: positive integers ; $m+n+p+q \leq 3)$

Because this function does not depend on the angles, the linear term (4) does not give any contribution to the ground state, so that the magnetic shift arises only from the quadratic diamagnetic term (5). In the same low field approximation, the acceptor and exciton diamagnetic shifts are given by $\Delta E_A = \gamma_A^2|E_A|/2$ and $\Delta E_X = \gamma_X^2|E_X|/2$, with $\gamma_X = \gamma_A(1+\sigma^{-1})^2$. Our results reported in figure 1 show that the energy shift ΔE of bound excitons may be rather different from that of a free exciton. Furthermore, this shift is in general different from the sum of that of an acceptor impurity and a free exciton. Therefore, the binding energy $W = E-E_X-E_A$ depends on the magnetic field in contradiction to the result of Willmann et al.[4,5]. In a low magnetic field, the (A^o,X)-complex becomes more stable for σ less than 0.69 and less

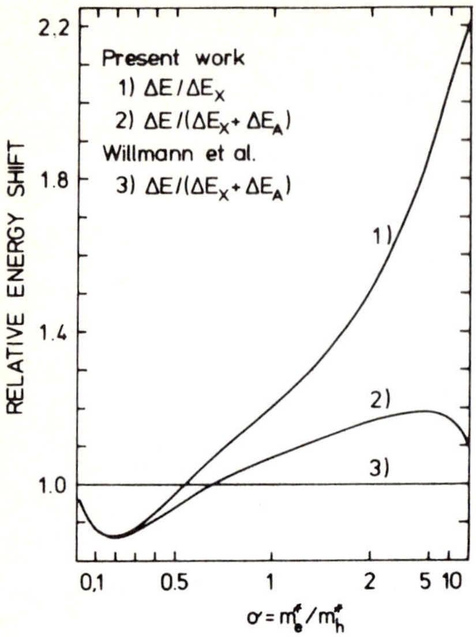

Fig. 1. Energy shift of the (A^0,X)-complex obtained by the perturbation theory, 1) and 2), and comparison with the result 3) of Willmann et al.[5].

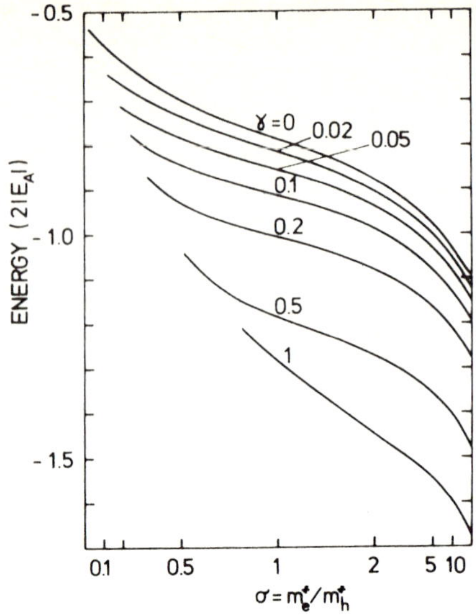

Fig. 2. Energy of the (A^0,X)-complex obtained by the variation method. The zero of energy is the first Landau level.

stable for higher values of σ. However, these conclusions are valid only for small enough effective magnetic fields. For example, the condition $\Delta E/\ E\ < 0.1$ is only realized if $\gamma_A < 0.1$ for $\sigma = 0.5$ and if $\gamma_A < 0.4$ for $\sigma = 10$, whereas the similar condition $\Delta E/|E_A| < 0.1$ requires that $\gamma_A < 0.447$.

We have extended the validity of our perturbation results to some higher values of the effective magnetic field by means of a variational calculation with the same wave function (7), but by varying only the linear variational parameters c_{mnpq}. In figure 2 we have reported the energies obtained for effective fields up to unity with reference to the first Landau level $E_L/|E_A| = \gamma_A\ (2+\sigma^{-1})$ of the complex. We have also determined the transition energy shift $\Delta h\nu = \Delta E - \Delta E_A$ and compared it to the analog exciton transition shift $\Delta h\nu_X = \Delta E_X$. Our results are reported in figure 3 together with those relative to the phenomenological model of Willmann et al.[4,5], and to the pseudodonor model for which $\Delta h\nu = \Delta E_{A^+} + \Delta E_D - \Delta E_A$. The energy shifts of the exciton, the donor and the acceptor impurities have been obtained after the highly accurate results of Cabib et al.[11] whereas that of the positively charged acceptor A^+ are due to Larsen[12]. As a result, the transition energy shifts of the neutral bound excitons are not equal to that of the exciton, although they have the same order of magnitude. The shift increases with the effective magnetic field and takes higher values for the (D^0,X)-complex than for the (A^0,X)-complex. The pseudodonor model, on the other hand, leads to somewhat different results. In this model, the bound exciton line shifts less than the exciton line.

4. HIGH FIELD APPROXIMATION

When the magnetic energy is of the order of the Coulomb energy (γ=1), the previous procedure breaks down. In the absence of the Coulomb term (3), the motion of the three particles is free along the direction of the field and circular in the plane perpendicular to it. The ground state energy reduces then to that of the lowest Landau level. For large enough values of γ, we can solve the Schrödinger equation by means of the adiabatic approximation[13,14], by separating still the two motions and

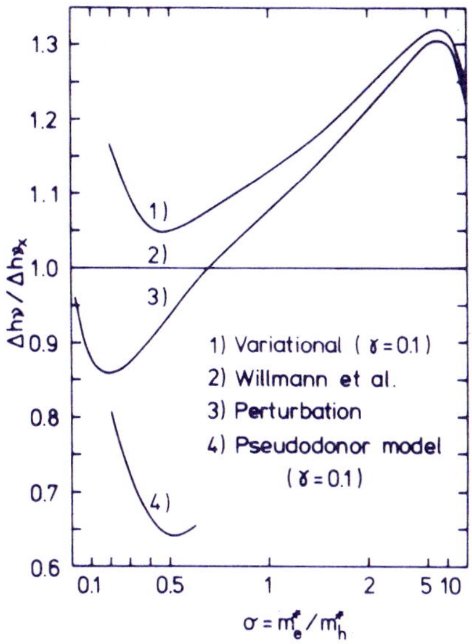

Fig. 3. Comparison between the transition energies of the neutral acceptor bound exciton and the exciton obtained by different methods.

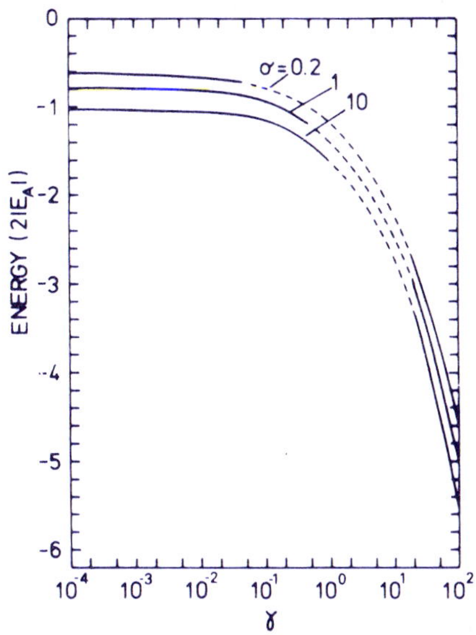

Fig. 4. Energies of the (A^O,X)-complex in the low and high field limits (solid lines). The dashed lines represent the expected energies in the intermediate field region. The zero of energy is the first Landau level.

by treating the Coulomb potential as a perturbation with respect to the fundamental Landau level. The mean value of the Coulomb potential has been approximated using the method of Elliott and Loudon[15]. It gives rise to an attractive potential for the z- dependent equation, which we have solved within the variational method by using a four term trial function of the type :

$$Z = \sum_{ijk} c_{ijk}\,\phi_{ijk}\,, \qquad (c_{ijk} = \pm\, c_{jik})$$

$$\phi_{ijk} = \exp\,(-\,a_i z_1^2 - a_j z_2^2 - b_k z_e^2)\,, \qquad (8)$$

a_i, a_j, b_k and c_{ijk} being variational parameters. In figure 4, we have reported the results obtained for some values of σ. The function (8) is better adapted for small values of σ because it does not take explicite account of the correlations between the different particles. Nevertheless, it appears at the present stage, that the low and high field calculations do not connect too badly. A study for the intermediate field region is in progress.

5. COMPARISON WITH EXPERIMENT

In the past, different attempts[4,5,7] have been made to interpret the observed transition shifts in Sn-doped GaAs. In this case, the comparison between the experimental and the theoretical transition shifts appears to be questionable because the effective mass theory is not adapted to deep impurity states, although the shifts observed with different impurities seem to be identical[5]. On the other hand, the valence band of GaAs is degenerated at the centre of the Brillouin zone and cannot be accurately described by a simple isotropic hole mass parameter. Nevertheless, it may be interesting to compare our results with those obtained previously. To this purpose, we use the following parameters[16] : $\varepsilon = 12.56$, $m_e^* = 0.0665\ m_o$ and the "isotropic" hole mass $m_h^* = m_o/\gamma_1 = 0.131\ m_o$. Then for $\gamma_A = 0.1$, i.e. B = 2.55 T, we obtain a theoretical shift of 1.12 meV, whereas the model of Willman et al. leads to

1.07 meV and the pseudodonor model to 0.69 meV. All these values are higher than the observed shift of 0.4 meV, which on the other hand is remarkably well fitted[7] by the neutral donor energy shift ΔE_D = 0.4 meV. But this latter cannot be identified with the pseudodonor transition energy shift because the contribution $\Delta E_{A^+} - \Delta E_A$ = 0.29 meV is not negligible. The discrepancy between the experimental and the theoretical transition shifts may arise from the fact that in the effective mass approximation the mean radii of the orbitals are larger than for deep impurities. As a consequence, the theoretical diamagnetic shifts are probably overestimated for GaAs : Sn.

REFERENCES

1 B. Stébé and G. Munschy, Solid State Comm. 35, 557 (1980).
2 J. Adamowski, S. Bednarek and M. Suffczinski, J. Phys. C12, L325 (1979).
3 A. Baldereschi, in Theoretical Aspects and New Developments in Magneto-Optics, J.T. Devreese, ed. (Plenum Press, New York and London, 1980), p.291.
4 W. Dreybrodt, F. Willmann, M. Bettini and E. Bauser, Solid State Comm. 12, 1217 (1973).
5 F. Willmann, W. Dreybrodt and M. Bettini, Phys. Rev. B8, 2891 (1973).
6 W. Rühle and D. Bimberg, Phys. Rev. B12, 2385 (1975).
7 W. Schairer, D. Bimberg, W. Kottler, K. Cho and M. Schmidt, Phys. Rev. B13, 3452 (1976).
8 D.G. Thomas and J.J. Hopfield, Phys. Rev. 128, 2135 (1962).
9 E.J. Johnson and H.Y. Fan, Phys. Rev. 139, A1991 (1965).
10 A.M. White, J. Hinchcliffe and P.J. Dean, Solid State Comm. 10, 497 (1972).
11 D. Cabib, E. Fabri and G. Fiorio, Solid State Comm. 9, 1517 (1971).
12 D. Larsen, Phys. Rev. B20, 5217 (1979).
13 L.I. Schiff and H. Snyder, Phys. Rev. 55, 59 (1939).
14 Y. Yafet, R.W. Keyes and E.N. Adams, J. Phys. Chem. Solids, 1, 137 (1956).
15 R.J. Elliott and R. Loudon, J. Phys. Chem. Solids, 8, 382 (1959).
16 D.D. Sell, S.E. Stokowski, R. Dingle and J.V. Dilorenzo, Phys. Rev. B7, 4568 (1973).

MAGNETO-OSCILLATION OF RELAXATION TIME IN ELECTRON HOLE-DROPS

H. Nakata
Department of Physics, College of General Education
Osaka University, Toyonaka, Osaka 560, Japan

Magneto-oscillating behavior of the relaxation time is shown to exist for the carriers in electron-hole drops created in Ge by analyzing the far-infrared magneto-absorption results. The oscillation is discussed in terms of the far-infrared photon absorbing electron-hole scattering across the Fermi level.

Shubnikov-de Haas type magneto-oscillations have been observed for electron-hole drops (EHD) in Ge.[1] It is true that the main part of magneto-oscillations seems to originate in the oscillation of the carrier pair density.[2] In our last paper, indeed we reported that the magneto-oscillation of far-infrared (FIR) absorption is due to the oscillatory carrier pair density like luminescence.[1] But FIR absorption seems to be more complicated than luminescence. The luminescence intensity is influenced very little by the relaxation time of carriers in EHD, whereas FIR absorption seems to be strongly connected with that. We now believe that the magneto-oscillation in FIR absorption is caused not only by the variation of the carrier pair density but by the oscillatory electron-hole scattering relaxation time as well.

The relaxation time of carriers in EHD can be obtained from two sources: one from the absorption intensity for the photon energy, lower than the resonance energy of 8.7 meV, and the other from the linewidth of the electric dipole resonance. The two methods are complementary to each other.

A pure germanium sample with the size of 1×3×4 mm^3 is used in this experiment. It is placed in a superconducting magnet capable to produce 7 T. The temperature has been kept at 1.7 K and the magnetic field is directed along the [100] axis. The sample is excited by 3 W argon ion laser, the beam of which is chopped by an acousto-optical modulator. The chopping frequency is chosen 90 Hz and the pulse width of the chopped beam is 200 μs.

The source of FIR radiation is an optically pumped FIR laser that is internally chopped at the frequency of 180 Hz, producing the pulse width of 1 ms. A multi-channel time resolution system with microcomputer-connected digital instruments is used for data accumulation.[3]

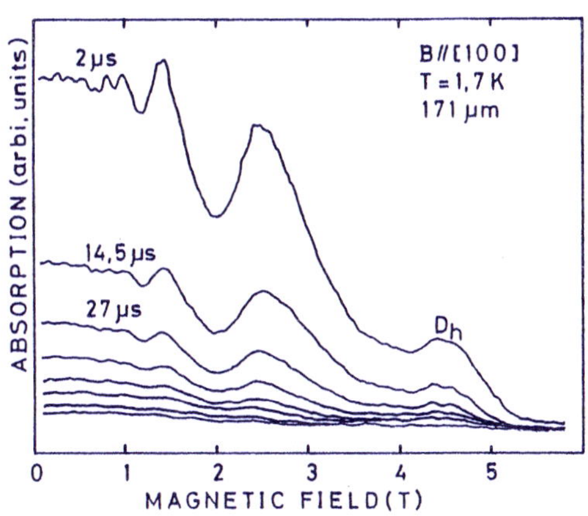

Fig. 1 Magneto-absorption of EHD after the end of the excitation pulse.

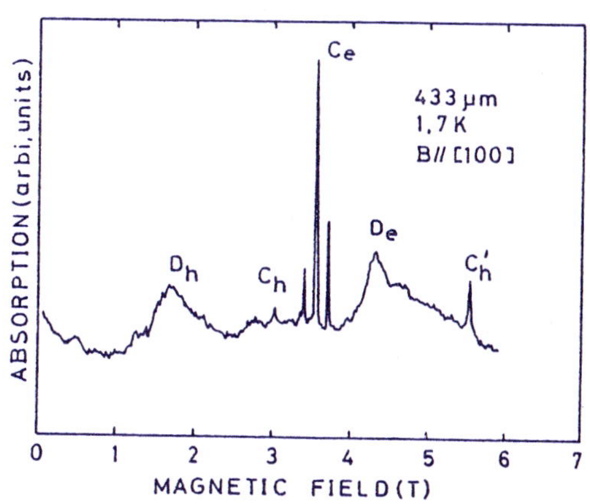

Fig. 2 FIR magneto-absorption of EHD, electrons and holes during exposure to the excitation pulse.

Magneto-absorption of FIR radiation by photo-excited carriers in Ge is shown in Fig. 1 and Fig. 2. Figure 1 shows the transient profiles of the 171 μm magneto-absorption after the photo-pulse. Several traces for different delay-times as indicated are taken in a single sweep of magnetic field. Magneto-absorptions for 433 μm during excitation are presented in Fig. 2. Magneto-oscillations of EHD are observed for 163 to 255 μm in the magnetic fields lower than 3 T. Especially for 171 and 184 μm, prominent oscillations are observed.

The peak near 4.5 T for 171 μm is electric dipole resonance. It is denoted by D_h because of its connection with hole quantum cyclotron resonance. It is also observed for 184 to 433 μm. Three sharp absorption peaks indicated C_e are cyclotron resonances of electron. The peak D_e observed at 4.3 T for the wavelength of 433 μm is electric dipole resonance associated with electron cyclotron resonance. The two sharp peaks observed at 3.1 and 5.6 T for 433 μm are hole quantum cyclotron resonances.

The following equation is used to calculate the relaxation time from magneto-absorption intensity. The FIR absorption coefficient at the frequency ω is given by

$$\alpha = (4\pi\omega/c)R^3N_d\varepsilon_0^{1/2}\mathrm{Im}[(\varepsilon^* - 1)/(\varepsilon^* + 2)] \ . \tag{1}$$

Here ε* is to denote the relative dielectric function in EHD, R is the radius of EHD, N_d is the number of EHD in unit volume. We assume that all the generated electron-hole pairs condense into EHD and that the pair generation rate does not depend on the magnetic field. The volume fraction occupied by EHD is

$$(4/3)\pi R^3N_d \propto 1/n \ . \tag{2}$$

By using Eq.(2), Eq.(1) becomes

$$\alpha \propto (1/n)\mathrm{Im}[(\varepsilon^* - 1)/(\varepsilon^* + 2)] \ . \tag{3}$$

For the dielectric function ε*, we use zero magnetic field value; i.e.,

$$\varepsilon^* = 1 - \omega_p^2/[\omega(\omega + i/\tau)] \ ; \tag{4}$$

where ω_p is the plasma frequency of EHD and τ is the relaxation time by electron-hole scatterings.

In the above approximation, the variation of relaxation time can be calculated only if one knows the magnetic field dependence of the carrier pair density. We borrow the data from luminescence experiments for the carrier pair density in the presence of a magnetic field.[2)]

If the radius of EHD is too large to satisfy the condition kR << 1, where k is the wave number of FIR in EHD, use of Eq. (1) should be avoided. The decay curve of the absorption is no longer expected to be a simple exponential. We analyze the data taken 27 μs after the end of the photopulse. The relaxation times calculated from 171 μm magneto-absorption are given by solid line in Fig. 3.

The relaxation time is also calculated from the linewidth of the electric dipole resonance D_h in the following manner. The dielectric function ε* has several poles at the magnetic fields of cyclotron resonance. We treat the peak D_h, which is associated with the pole of the

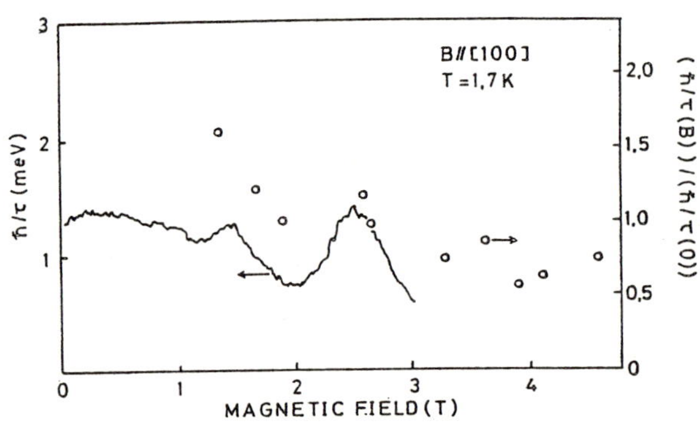

Fig. 3 The relaxation time of carriers in EHD obtained from the linewidth and absorption intensity.

light hole. In the vicinity of this resonance, we can neglect the influence from other poles. The linewidth ΔB of the peak D_h is connected with the relaxation time by the equation

$$1/\tau = (e\Delta B)/(2m^*_{\ell h}m_0 c) \qquad (5)$$

The calculated relaxation time as a function of magnetic field is shown by circles in Fig. 3. We adopt Zarate et al.'s result as the zero field value $\hbar/\tau(0)$.[4] The ratio of the damping constant $\hbar/\tau(B)$ to the zero field value $\hbar/\tau(0)$ decreases gradually in magnetic fields higher than 2.5 T.

The relaxation time by carrier-carrier scatterings in a Fermi liquid was calculated by Landau.[5] For low photon energy ($\hbar\omega << E_F$), the inverse relaxation time is proportional to ω^2. For electron-hole scatterings in EHD, Zarate et al.[4] calculated the relaxation time by the theory of Tzoar and Platzman[6], who used the temperature Green function approach and considered the correlation of carriers in EHD. In our case, the relaxation time has to be calculated in the magnetic field. The calculation by Zarate et al., accordingly, must be extended to non-zero magnetic field. Instead of actually carrying out the procedure, we have tried to calculate the relaxation time by electron-hole scattering in a simple way. Several crude approximations are taken: The correlation between carriers is neglected. The interaction between electron and hole is replaced by the delta function. The transfer of transverse momentum is neglected. The result of such a crude calculation can in no way be compared with experiment quantitatively. Certain qualitative information, however, can still be obtained. From the investigation of the matrix elements, two properties can be deduced. One is that the scattering with large momentum change is difficult to occur, and the other is that the scattering to higher subband is also difficult to occur. Almost all of holes are heavy holes whose Landau levels lie closely. If a heavy hole absorbs a photon, it should undergo a transition jumping over many Landau levels. Such a transition is rather hard to occur. Thus an electron is more likely to absorb a photon than a hole.

The Fermi energy and Landau levels of electron are shown in Fig. 4. The electron Fermi energy E_F^e is calculated from the carrier pair density by the relation

$$n = (2\sqrt{2}eHm_d^{*3/2}\sqrt{m_0}/\pi^2\hbar^2 cm^*_{ce})\sum_\ell [E_F^e - (\ell + \tfrac{1}{2})\hbar\omega_{ce} \pm \tfrac{1}{2}g^*\mu_B H]^{1/2} \; ; \qquad (6)$$

where m_d^* is the density-of-states mass of electron, g^* the effective g value of electron and μ_B the Bohr magneton. The crossing fields of the Fermi level with Landau levels 1^-, 1^+, 2^- and 2^+ are 2.35, 2.02, 1.29 and 1.22 T, respectively. The peaks of the inverse relaxation time $\hbar/\tau$ occur at the magnetic fields of 2.49 and 1.46 T. One thus finds that the maxima of $\hbar/\tau$ lie at magnetic fields somewhat higher than the crossing fields.

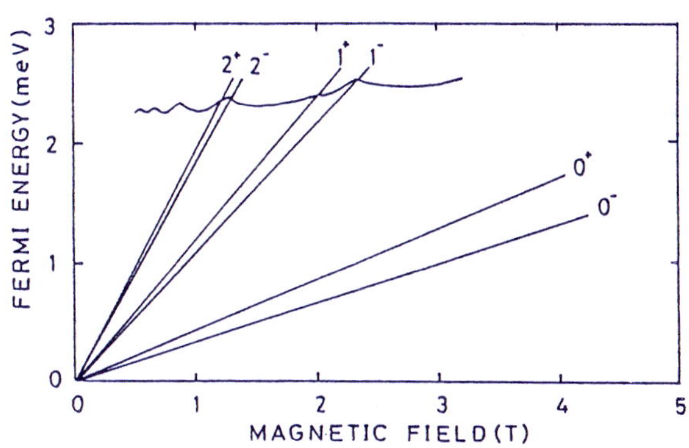

Fig. 4 The Fermi energy and Landau levels of electrons against the magnetic field.

In our experiment, the electron Fermi energy E_F^e is about 2.5 meV and the photon energy $\hbar\omega$ of FIR is about 7 meV, so that $\hbar\omega >> E_F^e$. For $\hbar\omega << E_F^e$, the magneto-oscillation behaves like the DC Shubnikov-de Haas effect. Such a situation is realized in the ultrasonic attenuation by a large EHD.[7] In that case, the magnetic fields corresponding to the maxima of $\hbar/\tau$ occurred near the crossing field. The difference of the two cases may be accounted for by difference in scattering mechanism. In the case of the DC Shubnikov-de Haas case, scatterings occur near the Fermi level.

In the AC Shubnikov-de Haas case, on the other hand, the carrier absorbs a photon and is excited to the virtual state with higher energy. The scattering occurs between the high energy carrier and a carrier near the Fermi energy. The difference in phase between the DC Shubnikov-de Haas and the magneto-oscillation of the relaxation time may be explained as follows. In the DC-SdH measurement, the electron-hole scattering occurs near the Fermi level as shown in Fig. 5(a). As the electron having an energy of $(\ell + 1/2)\hbar\omega_{ce}$ sees so large a density of states, that the scattering is easy to occur when the Landau level crosses the Fermi energy. In the AC Shubnikov-de Haas case, the matrix element for electron hole scattering should also be considered in addition to the density of states. The matrix element is large for small momentum transfer. The electron in the initial state with smaller k_z requires larger momentum transfer when it undergoes inter-subband transition absorbing a photon as indicated in Fig. 5(b). In this way the momentum transfer dependence of the matrix element may induce the phase difference in magneto-oscillation.

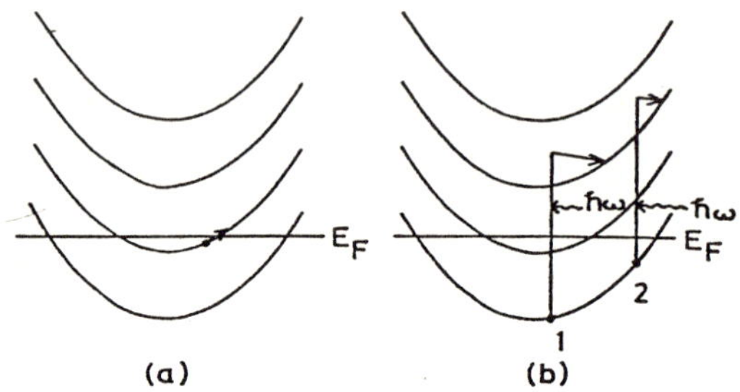

Fig. 5 Electron-hole scattering in a magnetic field. (a) DC Shubnikov-de Haas (b) AC Shubnikov-de Haas

The decrease of relaxation time at magnetic fields higher than 3 T may be explained by decrease of the screening length in the interaction potential. As the magnetic field becomes higher, the carrier pair density in EHD increases. It induces then a larger screening. If the Coulomb potential is strongly screened, electron-hole scattering may become difficult to occur.

To summarize, the magneto-oscillation of FIR absorption cannot be explained only by the oscillation of the carrier pair density. The relaxation time also has to oscillate in magnetic fields.

I thank Prof. E. Otsuka and Dr. T. Ohyama for helpful discussions. Thanks are also due to Mr. T. Yodo for experimental assistance and discussions.

This work has been supported by Grant-in-Aid for Scientific Research of the Ministry of Education, Science and Culture.

[Reference]

1) H. Nakata and E. Otsuka, Phys. Rev. B23 5428 (1981); other references are cited therein.
2) K. Betzler, B. G. Zhurkin and A. L. Karuzskii, Solid State Commun. 17 577 (1975).
3) H. Nakata and E. Otsuka, Appl. Phys. B27 207 (1982).
4) H. G. Zarate, J. P. Carbotte and T. Timusk, Phys. Rev. B21 4649 (1980).
5) L. D. Landau, Soviet Physics JETP 5, 101 (1957).
6) N. Tzoar and P. M. Platzman, in Linear and Non-linear Electron Transport in Solids, edited by J. T. DeVreese and V. E. Van Doren (Plenum, New York, 1976) p. 199.
7) T. Ohyama, A. D. A. Hansen and J. L. Turney, Solid State Commun. 19 1083 (1976).

EXPERIMENTAL VERIFICATION OF NODAL SURFACE NUMBER CONSERVATION RULE IN MAGNETO-OPTICAL EFFECTS OF HYDROGENIC EXCITON

N. Kuroda, Y. Nishina, H. Hori* and M. Date*

The Research Institute for Iron, Steel and Other Metals,
Tohoku University, Sendai 980, Japan

*Department of Physics, Faculty of Science,
Osaka University, Toyonaka 560, Japan

The correspondence relationship between low and high field levels of a hydrogen atom has been examined through the analogy of the magneto-absorption spectra of the A-exciton in InSe up to the field of 39.5 T at 4.2 K. Our result shows that the nodal surface number conservation rule as proposed by Shinada et al. gives more appropriate criterion of correspondence than the noncrossing rule of Boyle and Howard.

1. INTRODUCTION

The electronic structure of the hydrogen atom in magnetic fields has been a subject of extensive studies since 1930's. Despite the long history, the correspondence between discrete (n ℓ m) levels of their energy eigenvalues $E_{n\ell m}$'s

$$E_{n\ell m} = -R/n^2, \quad n = 1, 2, \cdots, \infty, \tag{1}$$

at zero field, and (N M i) levels of E_{NMi} [1],

$$E_{NMi} = (N + \tfrac{1}{2})\hbar\omega_c - R/\nu^2, \quad N,\nu = 0, 1, 2, \cdots, \infty, \tag{2}$$

at a very high field, has never been clarified, where $i = 2\nu$ for states with even parity under inversion of the coordinate along the field and $i = 2\nu - 1$ for states with odd parity; R is the Rydberg constant and $\hbar\omega_c$ the cyclotron energy. Boyle and Howard (BH) have proposed that the correspondence can be found from the noncrossing rule based on the conservation of magnetic quantum number (m = M) and parity.[2] Kleiner has claimed that the number of nodal surfaces of wavefunction can also be a good quantum number, and presented a different correspondence relationship.[3] Shinada, Akimoto, Hasagawa and Tanaka (SAHT) have modified Kleiner's conjecture into a more refined form.[4]

Magneto-spectroscopic experiments on donors in semiconductors have been performed by a number of groups.[5] Their results have disproved Kleiners conjecture, but have failed to suggest which of the BH and SAHT schemes is valid. The present experimental problem, therefore, is to verify which of these schemes is more reasonable. From this view point, we report the results of our experimental study of the magneto-absorption of the A-exciton in the layer compound InSe. The conduction and valence bands which are responsible for the direct interband transition consists of In-s and Se-p Bloch states, respectively.[6] The A-valence band consists of $p_1\alpha$ and $p_{-1}\beta$ at its extremum, where α and β are spin functions. The effective Rydberg constant R* is known to be 15 meV, which is more than an order of magnitude smaller than the splittings of valence bands. Thus the Coulomb interaction gives negligible mixing between the A-valence band and the split-off B- and C-bands. In this sense, the single-conduction-band-single-valence-band model holds very well in InSe. Consequently, only the exciton states with m = M = 0 are allowed for optical transition.

2. THEORETICAL BACKGROUNDS

2.1 The Noncrossing Rule of BH

According to BH, a pair of states with a common magnetic quantum number and parity can not be degenerate in any magnetic field. In this scheme, a low field state is to connect to a high field state associated with the Landau edge of $N = (M + |M|)/2 = (m + |m|)/2$. For these levels, the connection between both extremes of H should be made between a pair of states which have the same order of energy. Results are listed in Table 1. Baldereschi and Bassani [7] and Lee, Larsen and Lax [8] have calculated energy eigenvalues of several even parity $m = 0$ states by means of variational methods. Their results are consistent with the noncrossing rule of BH.

2.2 Nodal Surface Number Conservation Rules

Kleiner has assumed that numbers of nodes along ρ and z directions of cylindrical coordinates are individually conserved for the field parallel to the z direction. This assumption leads to the occurrence of a number of level crossings in finite fields. Being aware of discrepancy between experiments and Kleiner's hypothetical rule, SAHT [4] have proposed that the total number, S, of nodal surfaces would be conserved. The states with the same set of S, m and parity repel each other in magnetic fields because of the noncrossing rule. The results for s and p states are very similar to the BH scheme as shown in Table 1. The significant difference from the BH scheme is seen for $N > (M + |M|)/2$ states. Each of these high field states is connected by a low field state, whereas none of them is connected by a low field states in the BH scheme. In deriving the relationship in Table 1, we have found that the magneto-optical experiments on shallow donors would not discriminate the essential difference between BH and SAHT on account of the selection rule being different from that for exciton-formation [9]. Praddaude [10] has calculated energy eigenvalues of the first 14 states of n = 1, 2 and 3 by means of expansion of wavefunctions in Laguerre polynomials. He has obtained an empirical correspondence relationship which is identical with that of SAHT.

Table 1. Correspondence relationship of BH, Kleiner and SAHT for $m = M = 0$ states.

(N 0 i)	BH	Kleiner	SAHT
(2 0 3)	—	6f	7h
(2 0 2)	—	5d	6g
(2 0 1)	—	4p	6h
(2 0 0)	—	3s	5g
(1 0 3)	—	5f	5f
(1 0 2)	—	4d	4d
(1 0 1)	—	3p	4f
(1 0 0)	—	2s	3d
(0 0 3)	3p	4f	3p
(0 0 2)	2s	3d	2s
(0 0 1)	2p	2p	2p
(0 0 0)	1s	1s	1s

3. EXPERIMENTAL RESULTS AND DISCUSSION

Magneto-absorption spectrum has been measured in pulsed fields up to 39.5 T at 4.2 K. Experimental arrangements have been described elsewhere.[9][11] Figure 1 shows the field dependences of magneto-absorption peaks measured in the Faraday configuration with $\mathbf{H} \parallel \mathbf{c}$ and $\mathbf{E} \perp \mathbf{c}$, where **H** is the magnetic field, **E** the electric field of light and **c** the c-axis of the hexagonal unit cell. The orbit of electron-hole relative

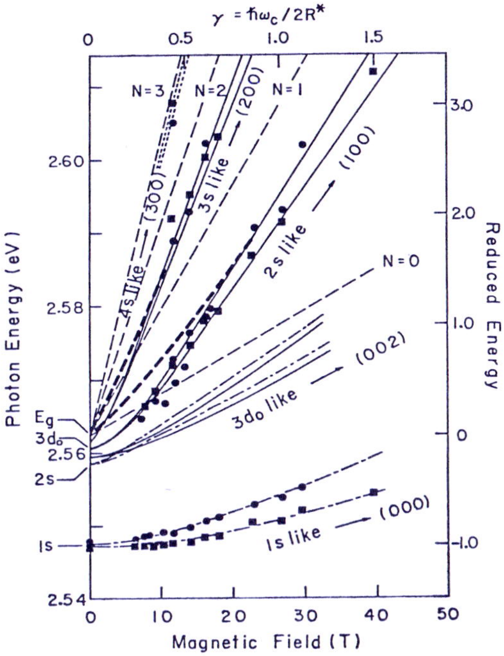

Fig. 1 Magnetic field dependences of energies of absorption peaks measured with right-circularly-polarized (circles) and left-circularly-polarized (squares) light. Dashed lines are Landau edges. Solid and dash-dotted lines are excitonic levels calculated by Tanaka and Shinada [13] and Cabib et al. [12], respectively. Dotted lines are energies of the (300) state evaluated from the formula of Zilich and Monozon.[14] Assignments denoted as 1s-like, $3d_o$-like etc. represent dominant characters of levels in the indicated region of fields. The bold dotted lines are given by qualitative prediction in the BH scheme.[2]

motion is almost isotropic for the A-exicton in InSe.[5] Thus one can compare the above experimental results with numerical calculations for the hydrogen atom if energy and field are scaled in terms of the effective Rydberg constant R* and $\gamma = \hbar\omega_c/2R^* = e\hbar H/2\mu c R^*$, respectively, where μ is the reduced effective mass of electron and hole. Exact eigenvalues have been calculated by Cabib et al. for (000) and (002) states.[12] Tanaka and Shinada have obtained solutions for (000), (002), (100) and (200) states in an adiabatic approximation.[13] The theoretical curves for R* = 15 meV and $\mu = 0.10_2\ m_o$ are shown in Fig. 1. The linear Zeeman splitting corresponding to the effective g-value of 2.2 is taken into account for all the states. The energy of the (300) state has been evaluated from the approximated analytical solution obtained by Zilich and Monozon.[14] The result is also shown in Fig. 1. All the theoretical curves agree very well with our experimental results.

The BH scheme predicts that high field (N 0 0) states with N ≧ 1 originate from the ionization edge and lie in energy between the N and N − 1 Landau edges at any field as shown by bold dotted lines in Fig. 1. The SAHT scheme, on the other hand, claims that these states originate from $3d_o$, $4g_o$ etc. as shown in Table 1. For example, the (100) state is connected by the $3d_o$ state. Thus it should cross the N = 0 Landau edge at a certain field. The adiabatic approximation breaks down in the low field region, so that the agreement of the theoretical curve of the (100) state with the $3d_o$ level at zero field should be regarded as fortuitous. Our experimental points of the (100) state, however, cross the N = 0 Landau edge near 10 T. This result agrees with the SAHT scheme and not with the BH. Furthermore, the dipole transition to create a hydrogenic exciton is allowed for states containing the s-character. According to the SAHT scheme, ns states do not connect in energy to (N 0 0) states except for (000). They couple with d_o, g_o etc. in finite fields. The couplings result in interchanges in the orbital character and the oscillator strength between states. Thus, the (N 0 0) state looks as if it originates from the ns state with n = N + 1. The field dependences of oscillator strengths thus expected explain why (100), (200) and (300) states become observable only above 7 T as shown in Fig. 1. This oscillator strength variation has never been calculated in the BH scheme.

The absorption intensities of experimental peaks exhibit remarkable increases with increasing field. Figure 2 shows the field dependence of oscillator strength of the (100) peak as an example. The result agrees with the prediction of the SAHT scheme on the basis of calculation of Elliott and Loudon.[1] No magneto-absorption peak has been observed for the (0 0 i) (i even) states such as (002). It merges with the continuum absorption band at any field strength. This result is also consistent with the nodal surface number conservation rule.

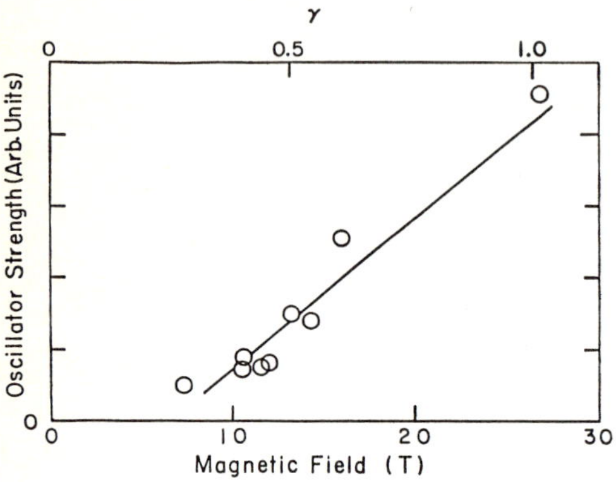

Fig. 2 Magnetic field dependence of oscillator strength of the (100) absorption peak. One division of the ordinate corresponds roughly to 10 % of the oscillator strength of the 1s absorption peak at zero field.

In conclusion, our experimental results show that the SAHT scheme gives a reasonable description of the field dependence of the excitonic energy levels. The variation in the oscillator strength, however, may be understood rather well in terms of Kleiner's scheme in the high field region where one may nearly discard the effects of level repulsion due to the noncrossing rule employed by SAHT.

This work was supported by the Grant-in-Aid from the Ministry of Education in Japan.

REFERENCES

[1] R.J. Elliott and R. Loudon, J. Phys. Chem. Solids 15, 196 (1960).
[2] W.S. Boyle and R.E. Howard, J. Phys. Chem. Solids 19, 181 (1961).
[3] W.H. Kleiner, MIT Lincoln Lab. Prog. Rep. Feb., 1953 (unpublished) p.36. See also the reference [1].
[4] M. Shinada, O. Akimoto, H. Hasegawa and K. Tanaka, J. Phys. Soc. Jpn. 28, 975 (1970).
[5] R.J. Wagner and B.D. McCombe, Phys. Stat. Sol. (b) 64, 205 (1974). See also the references therein.
[6] N. Kuroda, I. Munakata and Y. Nishina, Solid State Commun. 33, 687 (1980).
[7] A. Baldereschi and F. Bassani, Proc. 10th Int. Conf. Phys. Semicond. edited by S.P. Keller, J.C. Hensel and F. Stern (USAEC Division of Technical Information, Washington, D.C., 1970) p.191.
[8] N. Lee, D.M. Larsen and B. Lax, J. Phys. Chem. Solids 34, 1059 (1973).
[9] N. Kuroda and Y. Nishina, Proc. Int. Symposium on High Field Magnetism, edited by M. Date (North-Holland, 1982) to be published.
[10] H.C. Praddaude, Phys. Rev. A 6, 1321 (1972).
[11] Y. Sasaki, N. Kuroda, H. Hori, M. Shinoda and M. Date, Phys. in High Magnetic Fields, edited by S. Chikazumi and N. Miura (Springer-Verlag, Berlin 1981) p.195.
[12] D. Cabib, E. Fabri and G. Fiorio, Solid State Commun. 9, 1517 (1971), and Nuovo Cimento B 10, 185 (1972).
[13] K. Tanaka and M. Shinada, J. Phys. Soc. Jpn. 34, 108 (1973).
[14] A.J. Zilich and B.S. Monozon, Sov. Phys.-Solid State 8, 2846 (1967).

Identification of Contaminating Donors in III-V Compounds by Far-Infrared Laser Magneto-Optical Studies.

C.J.Armistead, A.M.Davidson, P.Knowles, S.P.Najda, R.A.Stradling
Department of Physics,
University of St. Andrews,
North Haugh, St. Andrews,
Fife, Scotland, KY16, 9SS, UK.

R.J.Nicholas, and S.J.Sessions
Clarendon Laboratory,
University of Oxford,
Oxford, OX1, 3PU, UK.

Abstract

The chemical shifts of the 1s-2p Zeeman transitions between donor states in n-GaAs and n-InP have been studied in samples from numerous sources. In the best samples of InP eleven separate peaks were observed. The approximate locations of the S and Si components have been established using back-doped samples. The only radically different set of residual donors in InP is produced by bulk growth.

Introduction

The chemical shift or central cell effect observed on transitions originating on the 1s state of hydrogenic impurities in semiconductors is the result of the nonvanishing wavefunction of this state at the impurity site. Variations in the impurity potential close to the core (ie in the central cell) from one impurity species to another can thus produce shifts in the level of the 1s state.

In the III-V compound semiconductors with direct energy gaps these shifts from the effective mass value of the 1s donor states are of the order of 0.5 cm^{-1}. Thus central cell splitting of transitions between the 1s and 2p levels is only just resolved at zero magnetic field even with the very highest purity GaAs. However when a magnetic field of the order of a Tesla is applied the transitions to the Zeeman split components of the 2p level narrow dramatically allowing this structure to be observed in more heavily doped samples.

The experiments described in this paper were carried out using the technique of Far Infrared Laser Spectroscopy. This technique offers greater resolution and signal to noise ratio than the standard Fourier method (1) but has the disadvantage that the magnetic field must be swept to obtain a spectrum since the source frequency is fixed. Band gap illumination was used on many of the samples in order to neutralise some of the ionized impurities and to narrow the lines. Reducing the temperature from 4 to 2K also narrowed the lines in some cases.

Discussion of GaAs

Figure 1 shows the principle results obtained for the 1s-$2p_+$ transition in GaAs. The bottom trace is the sample RR98B (μ_{77} = 1.6 X 10^5 cm^2/Vs), grown at the Wright-Patterson Air Force Base, and shows a fairly typical vapour phase epitaxy (VPE) signature. We label the components according to Low's convention (2). The donor X3, which is characteristically present in VPE material, is attributed to Ge, though C may also give a peak in this region. RR39CA is a sample from the same source but with a much lower mobility of 9.0 x 10^4 cm^2/Vs. The peaks are much broader and it is likely that they are X_1 and X_2, although some shifts occur with respect to RR98B, probably because inhomogeneous broadening has moved the line centres. This is more apparent in the structure on the 1s-2p_ transition discussed later. The integrated intensities of the lines suggests that the improvement in mobility has been achieved by a reduction of X_2. E351 and R137 are high mobility liquid phase epitaxy (LPE) grown samples from the Max-Planck Institute in Stuttgart. The principle difference is that E351 was grown when the lab was located in the centre of Stuttgart, while R137 was grown after the lab was moved to a rural site. A substantial improvement in purity has been achieved as can be seen in the recordings. The two high field peaks of E351 have reduced in intensity to the extent that a fourth donor X_1, very rarely seen in LPE material, is now resolved between them. The central donor of E351 has been previously identified as Sn, while Wolfe (3) has labelled the high field donor as Pb. This result suggests that metal pollution in the city atmosphere was responsible for a significant fraction of the contamination in E351.

Figure 2 shows the results obtained from spectra of the 1s-2p_ transition at a wavelength of 302.8 microns. The 2p_ level is not broadened as much as the $2p_+$ and so the resolution is much better. Furthermore, the 1s-2p_ transition is moving very slowly with respect to field with the result that the experimental spectrum is well spaced and high reproducibility and accuracy are readily achieved. Note that the laser energy is less than the minimum transition energy for the X_3 donor which consequently does not appear. Thus the lowest field transition is that due to X_2. However the resolution is much better than Fourier spectroscopy and a fifth donor is apparent slightly deeper than X_1. It is possible that this is the Sn donor observed in LPE samples and also seen in some other VPE samples. The shoulder on the $1s\text{-}2p_+$ X_2 transition resolves as a separate sharp peak and a much broader peak. This result reopens the question whether X_2 is attributable to sulphur (2). Structure on the high field side of the sharp peak may be due to complexes (4). Most interesting, however, is that on reducing the temperature to 1.8K a further transition is resolved between X1 and Pb. The topmost trace of Figure 2 shows RR39CA. The intensity ratio of the two peaks appears different from the $1s\text{-}2p_+$ spectrum and the high field peak appears to coincide with this new peak in RR98B. It is not entirely clear whether it is this peak or the one on the low field side of it in RR98B which is the donor X_1. Two molecular beam epitaxy samples were studied and both showed the doublet structure shown in the top recording of figure 1. These impurities are likely to be the donors X_1 and X_2.

Discussion of n-InP

The spectra in Figure 3 were all taken at a temperature of 1.8K with band gap illumination. The samples are all grown by the VPE method and have mobilities in the range 90000 to 120000 cm^2/Vs at 77K. The samples generally show three or four components and in the case of KV308 there are seven components. With the exception of CR276, all four samples have the main rather broad peak at 3.661 T dominant. Later KV samples (468 and 469) and the NAG samples in Figure 4 show this strong donor and a second equally strong component at 3.715 T. Experiments with samples thinned by etching suggests that the donors responsible for the lower field peak may be accumulating near to the interface with the substrate.

Three back-doped samples have been run so far. The first sample was only very lightly doped with S while the second and third were lightly and more heavily doped with Si. The sulphur peak was broad and centred at 3.669 Tesla which is close to the dominant peak, reproducing an earlier result (4). Both the silicon-doped samples have their peak in the region 3.71 T. The position in the more lightly doped sample is about 10 mT below the strong component observed with the NAG samples and KV468 and 469, suggesting that silicon may be a less common contaminant in VPE material than had previously been thought. However the multiplicity of lines present in the undoped samples, eleven in total, and the presence of asymmetric broadening caused by ionized donors, makes it impossible to unambiguously associate lines in the undoped samples with lines in back-doped samples.

Studies of VPE samples grown with water vapour present were found to have only the sulphur donor present above the 10% level. Water vapour is believed to suppress silicon donors through the oxygen-silicon association. Some alkyl grown samples were also run, as well as a high purity bulk grown sample. In the alkyl samples (eg Fig. 4) the sulphur donor was a strong component, with the second strong line lying near the silicon position. The bulk sample was markedly different with four components resolved at 4K. Although two of the strong lines could be attributed to sulphur and silicon, the two strongest components were substantially shallower in energy. From a comparison of the chemical trends in GaAs, the donors lead, tin and selenium would be present in this region. Bulk growth is then the only technique to produce a radically different set of residual donors in InP.

We should like to thank the following individuals for the gift of samples: D.Ashen, S.J.Bass, E.Bauser, P.Blood, B.Cockayne, M.J.Cardwell, P.Colter, P.Giles, M.Hailes, C.W.Litton, D.S.Robertson , I.R.Saunders and L.Taylor. We also gratefully acknowledge useful discussions with P.J.Dean.

References

1. R.A.Cooke, et.al., J.Phys.D. 11, 945(1978)
2. T.S.Low, et.al., Appl.Phys.Lett. 41, 183(1982)
3. C.M.Wolfe, et.al., Proc.Sym.on GaAs, St.Louis,(1976) pp 120-128.
4. P.A.Fedders, Phys.Rev. B 25, 3846(1982)
5. R.F.Kirkman, D.Phil. Thesis, 1975 (Univ. of Oxford).

FIGURE 1: N-GAAS 1S - $2P_+$ TRANSITION

λ = 118·8 MICRONS; T = 4·2K

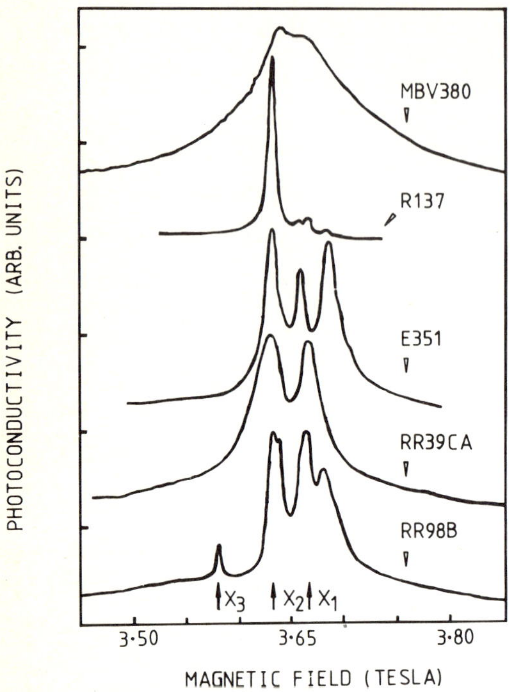

FIGURE 2: N-GAAS 1S - $2P_-$ TRANSITION

λ = 302·8 MICRONS;

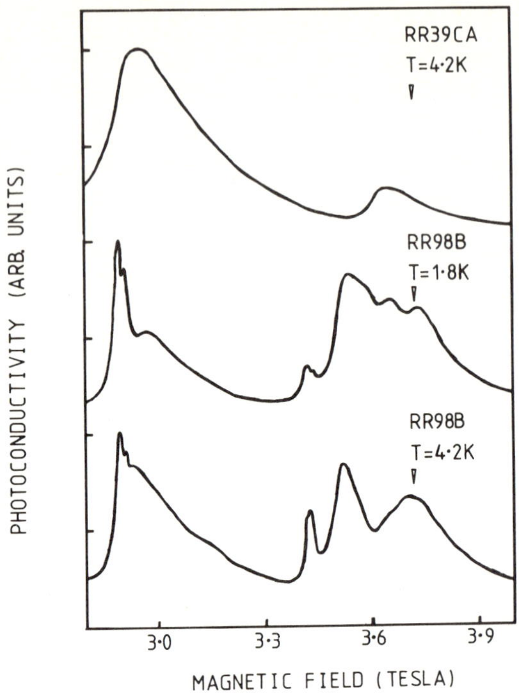

FIGURE 3: N-INP 1S - $2P_+$ TRANSITION

λ = 118·8 MICRONS; T = 1·8K

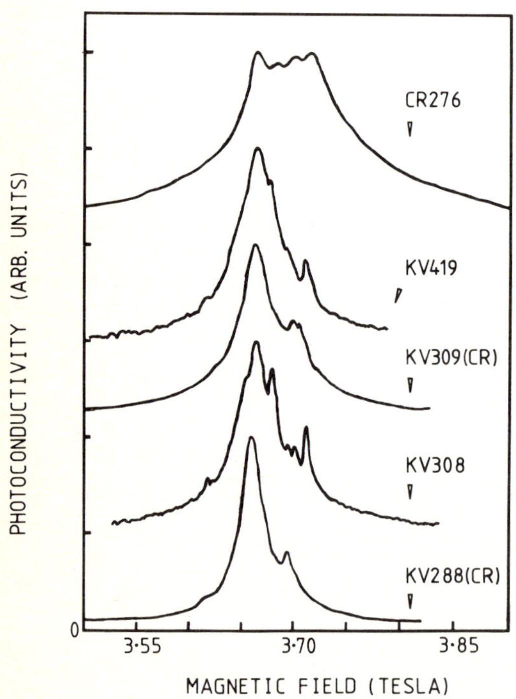

FIGURE 4: N-INP 1S - $2P_+$ TRANSITION

λ = 118·8 MICRONS; T = 4·2K

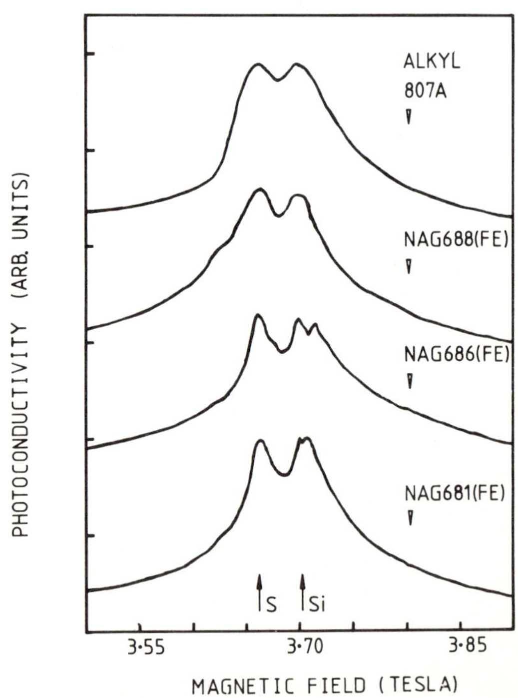

MAGNETOPLASMA OSCILLATIONS IN A SMALL CONDUCTING SPHERE

J.K. Furdyna
Physics Department, Purdue University, West Lafayette, Indiana 47907, USA
S. Goettig
Institute of Physics, Technical University of Warsaw, 00-661 Warsaw, Poland
J. Mycielski
Institute of Theoretical Physics, Warsaw University, 00-681 Warsaw, Poland
W. Trzeciakowski
High Pressure Research Center "Unipress", Polish Academy of Sciences, 01-142 Warsaw, Poland

We study the eigenmodes of a free-carrier lossless magnetoplasma in a small conducting sphere. We determine the frequencies of the oscillations, the motion of the carriers, charge densities etc. The modes fall into three classes of bulk-, surface-, and helicon-like magnetoplasmons. The interaction of the modes with an external electromagnetic field is calculated and the expressions for the power absorption are given.

1. Introduction

The investigation of magnetoplasma resonances has proved to be a useful method for determining basic parameters of a semiconductor like the carrier effective mass, concentration or the dielectric constant of the lattice /see [1] and references therein/. The experiments are usually performed in microwave resonant cavities, where the sample is placed in a static magnetic field and the plasma oscillations are excited by the harmonically varying electric and magnetic fields [2,3]. Changing the orientation of the static field and the position of the sample in the cavity /i.e. changing the radiation field/ one may observe various modes of excitation of magnetoplasma in a sample. The helicon-like mode [4] at the frequency of the order of cyclotron frequency is particularly interesting as the resonance conditions can be easily achieved by changing the static magnetic field.

All resonances depend strongly on the size and the shape of the samples [2]. Here we present a theoretical study of the eigenmodes of a free-carrier lossless magnetoplasma in a small conducting sphere. Then we determine the interaction of the modes with the external electromagnetic field. This approach is more complete and gives a better physical insight than the study of magnetoplasma oscillations forced by the external radiation [1,5]. Many eigenmodes cannot be excited by the radiation field of a simple form, but could be observable in characteristic energy-loss experiments.

2. Assumptions and Basic Equations

Let us consider a sphere of radius r_0 filled with an ideal electron /or hole/ plasma placed in a uniform compensating background having a dielectric constant ε and unit permeability. The carriers of charge q and equilibrium concentration n are assumed to have isotropic, energy-independent effective mass m^*. The system is placed in a static uniform magnetic field $\vec{B}_0$.

In the following we shall consider small collective oscillations of the spherical plasma, using the jelly model. The position of an oscillating carrier can then be written as the sum of the equilibrium position $\vec{r}$ and a small displacement vector $\vec{\xi}(\vec{r},t)$ which is a continuous function of $\vec{r}$. In all equations we shall keep only terms linear in $\vec{\xi}$. The interaction between carriers will be described in terms of a self-consistent electric field $\vec{E}(\vec{r},t)$ and magnetic field $\vec{B}(\vec{r},t)$. As these fields vanish for vanishing $\vec{\xi}$, they are at least linear in $\vec{\xi}$.

The linearized equation of motion for a carrier in a lossless plasma is

$$-m^* \omega^2 \vec{\xi}(\vec{r}) = q\vec{E}^-(\vec{r}) - (i\omega q/c)\vec{\xi}(\vec{r}) \times \vec{B}_0 \quad , \qquad (1)$$

where the harmonic time-dependence $\vec{\xi}(\vec{r},t)=\vec{\xi}(\vec{r})\,e^{-i\omega t}$, $E(\vec{r},t)=E(\vec{r})\,e^{-i\omega t}$ was assumed. The electric field inside and outside the sphere fulfills the Maxwell equations

$$\varepsilon\nabla\cdot\vec{E}^-(\vec{r}) = 4\pi q n\nabla\cdot\vec{\xi}(\vec{r}) \qquad (r<r_0), \tag{2}$$

$$\nabla\cdot\vec{E}^+(\vec{r}) = 0 \qquad (r>r_0), \tag{3}$$

$$\nabla\times\vec{E}^\pm(\vec{r}) = 0 \qquad (r<r_0 \text{ and } r>r_0), \tag{4}$$

where the $(-)$ and $(+)$ indices denote the fields inside and outside the sphere, respectively. Equation (4) is valid if we neglect terms proportional to

$$\varepsilon\,(r_0\,\omega_p/c)^2 \ll 1, \tag{5}$$

where $\omega_p=[4\pi nq^2/m^*\varepsilon]^{1/2}$. This means that the radius of the sphere must be small as compared to the wavelength of radiation of the plasma frequency. In this case the equations for the electric and magnetic fields decouple, i.e. we use the quasistatic approximation.

We complete our equations with the boundary conditions on the surface of the sphere

$$\hat{\vec{r}}_0\times\vec{E}^+(\vec{r}_0) = \hat{\vec{r}}_0\times\vec{E}^-(\vec{r}_0), \tag{6}$$

$$\hat{\vec{r}}_0\cdot\vec{E}^+(\vec{r}_0) = \varepsilon\,\hat{\vec{r}}_0\cdot\vec{E}^-(\vec{r}_0) + 4\pi\sigma(\vec{r}_0), \tag{7}$$

where

$$\sigma(\vec{r}_0) = nq\,\hat{\vec{r}}_0\cdot\vec{\xi}(\vec{r}_0), \tag{8}$$

is the surface charge density and $\hat{\vec{r}}_0 = \vec{r}_0/r_0$.

From Eq. (1) we can express $\vec{\xi}(\vec{r})$ by $E^-(\vec{r})$, Eq. (4) allows us to introduce the electric potentials $\Phi^\pm(\vec{r})$. Therefore Eqs. (2),(3)(6) and (7) become, correspondingly

$$(\omega^2-\omega_p^2-\omega_c^2)\,\Delta\Phi^-(\vec{r}) + (\omega_p\,\omega_c/\omega)^2\,\frac{\partial^2\Phi^-(\vec{r})}{\partial z^2} = 0, \tag{9}$$

$$\Delta\Phi^+(\vec{r}) = 0, \tag{10}$$

$$\Phi^+(\vec{r}_0) = \Phi^-(\vec{r}_0), \tag{11}$$

$$(\omega^2-\omega_c^2)\,\vec{r}_0\cdot\nabla\Phi^+(\vec{r}_0) = \varepsilon(\omega^2-\omega_p^2-\omega_c^2)\,\vec{r}_0\cdot\nabla\Phi^-(\vec{r}_0) + \varepsilon(\omega_p\omega_c/\omega)^2\left[z\frac{\partial\Phi^-}{\partial z}\right]\Bigg|_{\vec{r}=\vec{r}_0} + i\varepsilon(\omega_p^2\omega_c/\omega)\left[y\frac{\partial\Phi^-}{\partial x} - {} + x\frac{\partial\Phi^-}{\partial y}\right]\Bigg|_{\vec{r}=\vec{r}_0}, \tag{12}$$

where $\omega_c=(qB_0/m^*c)$ is the cyclotron frequency and $\vec{B}_0$ was chosen paralell to the z-axis. Equations (9)-(12) form a complete set determining the potentials and eigenfrequencies. We shall now study some special classes of solutions of these equations.

3. Zero-Magnetic-Field Modes

In the simplest case of $B_0=0$ /i.e. $\omega_c=0$/ Eqs. (10),(11) remain unchanged while Eqs. (9) and (12) become, respectively

$$(\omega^2-\omega_p^2)\,\Delta\Phi^-(\vec{r}) = 0, \tag{13}$$

$$\frac{\partial\Phi^+(\vec{r})}{\partial r}\bigg|_{\vec{r}=\vec{r}_0} = \varepsilon\left[1-(\omega_p/\omega)^2\right]\frac{\partial\Phi^-(\vec{r})}{\partial r}\bigg|_{\vec{r}=\vec{r}_0}. \tag{14}$$

In the case $\omega\neq\omega_p$ the regular solutions of the Laplace equations (10) and (13) are

$$\Phi^-(\vec{r}) = (q/r_0)\sum_{l=0}^{\infty}\sum_{m=-l}^{l} C_{lm}^-\,(r/r_0)^l\,Y_{lm}(\Theta,\varphi), \tag{15}$$

$$\Phi^+(\vec{r}) = (q/r_0)\sum_{l=0}^{\infty}\sum_{m=-l}^{l} C_{lm}^+\,(r/r_0)^{-(l+1)}\,Y_{lm}(\Theta,\varphi), \tag{16}$$

where $C_{lm}^\pm$ are arbitrary complex dimensionless coefficients and $Y_{lm}(\Theta,\varphi)$ are orthonormal spherical harmonics. Inserting (15) and (16) into (11) and (14) we obtain /for every l,m/ two equations for $C_{lm}^\pm$ which for $l=0$ give $C_{00}^\pm=0$ and which for $l\geqslant1$ give non-zero $C_{lm}^\pm$ only if

$$\omega = \omega_{pl} = \left\{\frac{4\pi nq^2}{m^*[\varepsilon+(l+1)/l]}\right\}^{1/2}. \tag{17}$$

The corresponding potentials are given by (15) and (16) with the summation over l abandoned. Thus to every eigenfrequency there correspond $(2l+1)$ independent modes of motion i.e. each frequency is $(2l+1)$ fold degenerate. It is worth noting that due to Eq. (13) the space-charge density for these modes vanishes, and so the only source of the electric field is the surface-charge density.

Let us now consider the $\omega = \omega_p$ case. Equation (10) together with (14) gives

$$\Phi^+(\vec{r}) \equiv 0, \tag{18}$$

and then (11) leads to

$$\Phi^-(\vec{r}_0)=0, \qquad (19)$$

Therefore the oscillations with the frequency ω_p give a vanishing potential /and the field/ outside the sphere, while the potential inside the sphere is arbitrary provided it vanishes on the surface. Both space and surface charge densities are in general non-zero.

The external magnetic field $\vec{B}_0$ applied to the system can either modify the zero-field modes considered above or give rise to new, helicon-like modes. In the low magnetic field range /$|\omega_c| \ll \omega_p$/ we shall then expect two types of solutions of Eqs. (9) - (12): high-frequency, i.e. with $\omega \sim \omega_{pl}$ or ω_p and low-frequency, i.e. with $\omega \sim |\omega_c|$.

4. Vanishing Space-Charge-Density Modes

In the presence of an arbitrary magnetic field it is difficult to find exact general solutions of Eqs. (9) - (12). However, in the previous section we found that an important class of oscillations /the ω_{pl} modes/ was characterized by a vanishing space-charge density. Let us therefore make such an additional assumption in (9) -(12) i.e. let us demand that both components of (9) vanish separately. The potentials inside and outside the sphere are then given by (15) and (16) respectively, and the condition $\partial^2\Phi^-(\vec{r})/\partial z^2=0$ is satisfied only for the modes with m=±1, ±(l-1). The continuity condition (11) gives $C^+_{lm}=C^-_{lm}$ and then the second boundary condition (12) leads to the equation determining the eigenfrequencies. For m=±1, l≥1, we obtain

$$\omega_{lm1}=\omega_{pl}\left[\sqrt{1+(\omega_c/2\omega_{pl})^2}-(\omega_c/2\omega_{pl})\,\mathrm{sgn}\,m\right]. \qquad (20)$$

For m=±(l-1) we get $\omega_{lm1}=\omega_{pl}$ for l=1 and for l≥2 the expressions for the frequencies are cumbersome so that we shall write down only their approximate forms in the case of weak magnetic field i.e. $|\omega_c| \ll \omega_p$. Up to the second order in (ω_c/ω_p) we obtain the high-frequency solutions /subscript 1/ in the form

$$\omega_{lm1}\cong\omega_{pl}-m/2l\cdot\omega_c+\frac{(l-1)(l+3)}{8l^2}\,\frac{\omega_c^2}{\omega_{pl}}\quad, l\geq 2\quad, m=\pm(l-1), \qquad (21)$$

and for m=-(l-1) sgn q the low-frequency solutions /subscript 2/

$$\omega_{lm2}\cong\frac{|\omega_c|}{l}\left[1-\frac{l-1}{l^2}(\omega_c/\omega_{pl})^2\right]\quad, l\geq 2\quad, m=-(l-1)\,\mathrm{sgn}\,q\,. \qquad (22)$$

The potentials, corresponding to any of the frequencies given by Eqs. (21),(22),and (20) are of the form

$$\Phi^-_{lm}(\vec{r})=(q/r_0)(r/r_0)^l\,Y_{lm}(\Theta,\varphi)\quad, \qquad (23)$$

$$\Phi^+_{lm}(\vec{r})=(q/r_0)(r/r_0)^{-(l+1)}\,Y_{lm}(\Theta,\varphi)\quad, \qquad (24)$$

so that every (l,m) mode posesses an oscillating 2^l-th electric multipole moment.

It is interesting to analyze the simplest high-frequency mode /l=1,m=0,±1/. If we calculate the displacement $\vec{\xi}(\vec{r})$/or the current $\vec{j}(\vec{r})$/ it turns out that all carriers move in the same way: circularly in the plane perpendicular to $\vec{B}_0$ for m=±1 and linearly along $\vec{B}_0$ for m=0 /electric dipole/. The simplest low-frequency /helicon-like/ mode is the one with l=2, m=-sgn q. In that case the current density $\vec{j}(\vec{r})$ looks as if a uniformly charged sphere was rotating around the vector which in turn performs the rotation with frequency $|\omega_c|/2$ in the plane perpendicular to $\vec{B}_0$. This mode possesses a rotating magnetic dipole moment.

5. High-Frequency Modes

Another situation when Eq. (9) can be simplified arises in the case of a weak magnetic field /$|\omega_c| \ll \omega_p$/ and $\omega\approx\omega_p$ or $\omega\approx\omega_{pl}$ if we retain only terms up to the first order in ω_c/ω_p. Let us first consider the case $\omega=\omega_{pl}+\omega'$, where ω' is a small correction of the order of ω_c. Equation (9) reduces into a Laplace equation so that $\Phi^-(\vec{r})$ and $\Phi^+(\vec{r})$ are given by (15) and (16), respectively. Inserting them into the boundary conditions we obtain $C^\pm_{00}=0$ and for l≥1 we have non-zero $C^\pm_{lm}$ only if

$$\omega=\omega_{lm1}=\omega_{pl}-(m/2l)\,\omega_c\quad, \quad l\geq 1\,,\ |m|\leq l\,, \qquad (25)$$

The corresponding potentials are given by Eqs. (23) and (24). This means that the degeneracy of the ω_{pl} modes occuring in $\vec{B}_0=0$ case has been lifted by the magnetic field. The space-charge density for these modes is at least of the second order in (ω_c/ω_p) and in some cases /for m=±1,±(l-1)/ it vanishes exactly /see section 4/. We shall therefore call them surface magnetoplasmons.

The second type of high-frequency mode arises from the ω_p-modes in $\vec{B}_0=0$ case. Setting $\omega=\omega_p+\omega''$ we obtain from our basic system (9)-(12) that $\omega''=0$ and the potentials are still determined by the conditions (18) and (19) i.e. up to the first order in

(ω_c/ω_p) both the frequencies and the potentials remain unchanged in the presence of the magnetic field. Both the space and surface charge densities are non-zero for these modes and so we call them bulk magnetoplasmons. They will be discussed in detail in a separate paper.

We shall now outline the calculation of the absorption /of electromagnetic radiation/ connected with generation of the magnetoplasma $\omega_{lm\tau}$-modes considered in Sec.4,5.

6. Hamiltonian and Absorption

The classical energy of the plasma can be written as

$$H_o = (n/2)m^* \cdot \int_{r\leq r_o} d^3r \left[\dot{\vec{\xi}}_R(\vec{r},t)\right]^2 + (nq/2) \int_{r\leq r_o} d^3r\, \vec{\xi}_R(\vec{r},t) \nabla\phi_R^-(\vec{r},t), \quad (26)$$

where the subscript R denotes the real part and the dot denotes the time derivative. One can write down H_o in terms of canonical variables $\vec{\xi}_R(\vec{r},t)$ and $\vec{p}_R(\vec{r},t)$. If we then introduce an external radiation field $\vec{A}(\vec{r},t)=\vec{A}(\vec{r})e^{-i\omega t}$ in the vanishing scalar potential gauge, we obtain the Hamiltonian in the form $H = H_o + H_1$, where

$$H_1 = -(qn/c) \cdot \int_{r\leq r_o} d^3r\, \vec{\xi}_R(\vec{r},t) \cdot \mathrm{Re}\left\{\vec{A}(\vec{r})\, e^{-i\omega t}\right\}. \quad (27)$$

If we insert into (26) the potential and the displacement being a linear combination of the eigenmodes potentials and displacements /with $C_{lm\tau}^{(t)}$ amplitudes/ we obtain H_o in terms of normal coordinates $b_{lm\tau}(t) = C_{lm\tau}(t)/\gamma_{lm\tau}$

$$H_o = \sum_{lm\tau} \hbar\omega_{lm\tau}\, b^*_{lm\tau}(t)\, b_{lm\tau}(t), \quad (28)$$

where $\gamma_{lm\tau}$ are normalization coefficients. The perturbation Hamiltonian H_1 can also be expressed by $b_{lm\tau}(t)$ and $b^*_{lm\tau}(t)$. The system can then be quantized by replacing $b_{lm\tau}(t)$ and $b^*_{lm\tau}(t)$ amplitudes by the anihilation and creation operators $b_{lm\tau}$ and $b^+_{lm\tau}$

The vector potential $\vec{A}(\vec{r})$ in (27) can be expanded around the center of the sphere in terms of the dimensionless parameter $(r_o\omega/c)$. Due to the assumption (5) only first two terms of this expansion should be retained. $A(\vec{r})$ acting on carriers can be expressed by the radiation fields $\vec{E}^r$ and $\vec{B}^r$ in the absence of the sphere, taken at $\vec{r}=0$. In this way we obtain

$$A_\mu(\vec{r}) = -\frac{3}{(\varepsilon+2)}\frac{ic}{\omega}E^r_\mu - \frac{1}{2}\varepsilon_{\mu\nu\kappa} r_\nu B^r_\kappa - \frac{5}{(2\varepsilon+3)}\frac{ic}{2\omega}\left[E^r_{\mu,\nu} + E^r_{\nu,\mu}\right] r_\nu, \quad (29)$$

where $\varepsilon_{\mu\nu\kappa}$ is the totally antisymmetric Levi-Civita tensor and comma denotes differentiation. Inserting (29) into (27) it turns out that only $b_{lm\tau}$ and $b^*_{lm\tau}$ with l=1,2 appear in H_1 i.e. only ω_{1m1} and $\omega_{2m\tau}$ modes can be excited by the external electromagnetic radiation.

The first order perturbation calculation in H_1 shows that the absorption occurs only if the radiation frequency is equal to one of the $\omega_{lm\tau}$ frequencies. The corresponding net /i.e. with stimulated emission substracted/ power absorbed by the sphere $P_{lm\tau}$ is non-zero only for l=1 and l=2 and is given by /up to a multiplicative factor/

$$P_{1m\tau}(\omega) \propto \delta(\omega-\omega_{1m\tau})\,\delta_{\tau 1}\left[2\left(1-\frac{\omega_c^2}{\omega_{1m1}^2}\right)^2 \left|E^r_z\right|^2 \delta_{m0} + \sum_{\alpha=\pm1}\left(1-\frac{\alpha\omega_c}{\omega_{1m1}}\right)^2 \left|E^r_x - i\alpha E^r_y\right|^2 \delta_{m\alpha}\right], \quad (30)$$

$$P_{2m\tau}(\omega) \propto \delta(\omega-\omega_{2m\tau})\left[1-\delta_{\tau 2}(1-\delta_{m,-\mathrm{sgn}\,q})\right]\Big(\frac{8}{3}\Big|\Big(3-\frac{2\omega_c^2}{\omega_{2m\tau}^2}\Big)E^r_{z,z} + \frac{(2\varepsilon+3)}{5}\frac{\omega_c}{c}B^r_z\Big|^2\delta_{m0} + \sum_{\alpha=\pm1}\Big\{\Big(1-\frac{\alpha\omega_c}{\omega_{2m\tau}}\Big)^2\Big|\Big(2+\frac{\alpha\omega_c}{\omega_{2m\tau}}\Big)\cdot$$
$$\cdot\left[E^r_{y,z}+E^r_{z,y}+i\alpha(E^r_{x,z}+E^r_{z,x})\right] + \frac{(2\varepsilon+3)}{5}\frac{\omega_c}{c}(B^r_y + i\alpha B^r_x)\Big|^2\delta_{m\alpha} + 4\Big(1-\frac{\alpha\omega_c}{\omega_{2m\tau}}\Big)^2\Big|E^r_{x,y}+E^r_{y,x}+i\alpha(E^r_{x,x}-E^r_{y,y})\Big|^2\delta_{m2\alpha}\Big\}\Big), \quad (31)$$

The result is temperature independent. It could be obtained by a classical calculation as it does not depend on $\hbar$. It is easy to see from (30) and (31) which modes shall be excited by a particular polarization of the radiation field. For instance, the electric field paralell to $\vec{B}_o$ excites only the ω_{101} mode.

It turns out that for $|\omega_c| \ll \omega_p$, $(P_{2m\tau}/P_{2m1}) \sim 10^{-3}$ so that the absorption for the frequencies with l=2 is much weaker than for those with l=1.

REFERENCES

1 G.W. Ford, J.K. Furdyna, S.A. Werner, Phys.Rev.B 12, 1452 /1975/
2 T.A. Evans, J.K. Furdyna, Phys.Rev.B 8, 1461 /1973/
3 J.R. Dixon, J.K. Furdyna, Phys.Lett. 54A, 59 /1975/
4 J.K. Furdyna, J. Mycielski, Bull.Am.Phys.Soc. 19, 246 /1974/
5 G.W. Ford, S.A. Werner, Phys.Rev.B 8, 3702 /1973/

Two-Photon Spectroscopy in InSb at High Magnetic Fields*

D. G. Seiler, M. W. Goodwin†, and S. W. McClure
Center for Applied Quantum Electronics
Department of Physics
North Texas State University
Denton, Texas 76203

and

L. A. Veilleux††
Francis Bitter National Magnet Laboratory
Massachusetts Institute of Technology
Cambridge, MA 02139

ABSTRACT

High resolution two-photon magneto-absorption (TPMA) spectra for n-InSb have been investigated for magnetic fields up to 15 T using cw CO_2 lasers and photoconductivity techniques. A modified Pidgeon-Brown energy band model using spherical and nonspherical selection rules is used to interpret the data. Anisotropic effects are observed for the first time in the TPMA photoconductive spectra. TPMA structure is also observed for the first time in the photovoltaic response.

INTRODUCTION

Magneto-optical studies have proven capable of determining energy band parameters of semiconductors because of the optical transitions that occur between magnetically quantized electronic or impurity states. Many studies during the past 30 years have used one-photon absorption (OPA) techniques with conventional photon sources. However, the increasing availability of laser sources now opens the possibility of carrying out nonlinear magneto-optical studies in semiconductors. In particular, two-photon absorption (TPA) techniques offer unique spectroscopic advantages over the traditional OPA techniques. TPA has a small cross section for absorption in many semiconductors and therefore has a large penetration depth resulting in uniform bulk absorption. This reduces problems associated with surface effects often encountered in OPA measurements (OPA has a large absorption cross section and thus a small penetration depth, usually on the order of a micron). Another advantage of TPA is that different states of the energy bands can be accessed. For an applied magnetic field not only different but more selection rules between Landau levels are allowed. Thus, TPA and TPMA (two-photon magneto-absorption) measurements can yield significantly more information about the eigenstates of a material. Recently[1], we have shown that TPMA is easily observable in n-InSb using only cw CO_2 lasers operating at even milliwatt powers. This result opens up a new dimension for two-photon spectroscopy using cw lasers only.

Two-photon absorption experiments on InSb divide basically into two categories; those without an applied magnetic field and those with an applied magnetic field. Although TPA in a magnetic field is the main subject of this paper much of the concepts and theories developed for the no magnetic field case are extremely relevant.

The purpose of this investigation was to examine the TPMA process in more detail than has previously been possible (The historical background on TPMA effects in InSb was recently reviewed by Goodwin et al.[2]). With high resolution experimental techniques many transitions in the TPMA spectra have been observed. In past studies there has been much ambiguity in the identification of the TPMA transitions. However, with more observable transitions and better resolution, a systematic investigation of this complex process has now been performed using polarized light and high magnetic fields. Most of the transitions can be explained by the usual spherical approximation two-photon selection rules. However, there are some transitions which can not be described with these selection rules. Possible explanations for these transitions are warping and inversion asymmetry-induced effects due to the band structure of InSb. For the first time anisotropic effects are observed and shown to be related to the anisotropy of the initial hole states.

THEORY

The interpretation of results of any magneto-optical experiment depends strongly on the model used for the energy band structure of the material. One of the most useful energy band models was developed by Pidgeon and Brown.[3] Their 8 x 8 Hamiltonian gives the energies of the Γ_6 conduction band, Γ_8 light and heavy hole valence bands, and the Γ_7 spin-orbit split-off band and includes the effects of higher bands to order k^2. They also extended their results to yield the Landau level energies when a magnetic field B is applied.

The most recent extensive calculations of the energy bands of InSb have come from Weiler.[4,5] The double group representation of the intermediate states in the second order matrix elements of the $\vec{k} \cdot \vec{p}$ Hamiltonian were used instead of the single group as in Pidgeon and Brown's. As a result of the spin-orbit splitting of the higher bands, this analysis yields other parameters N_1, N_2, and N_3 (in Weiler's notation). These results were used to explain the presence of certain anamolous magneto-optical effects in InSb, such as spin-flip and cyclotron harmonic transitions. By neglecting the contributions of small terms and certain parameters, but including the presence of a magnetic field, the 8 x 8 Hamiltonian was reduced to two 4 x 4 matrices, one for the a-set (spin-up, +1/2) and one for the b-set (spin-down, -1/2) energies, which can then be solved numerically. However, calculations of the eigenvalues and eigenvectors are not sufficient for a good description of TPMA processes; a proper choice of selection rules is also extremely important.

Zawadzki and Wlasak[6] gave a summary of the two-photon selection rules for the different light polarizations (σ_L, σ_R, σ, and π) for the case of $\vec{B} \parallel$ <001>. Extra transitions were shown to occur because of warping and inversion-asymmetry effects. More recently, Goodwin et al.[2] gave more complete two-photon selection rules for the same light polarizations and for $\vec{B} \parallel$ <111> and <110> directions, as well as the <001> direction. The selection rules in the spherical approximation governing transitions between initial and final states for the different light polarizations are for σ_L: $\Delta n = 2$, $\Delta s = 0$; σ_R: $\Delta n = -2$, $\Delta s = 0$; σ: $\Delta n = 0$, $\Delta s = 0$; π: $\Delta n = 0$, $\Delta s = 0$. Here Δn is the final minus initial state Landau level number and Δs is the final minus the initial state spin number (a-set spin up +1/2 and b-set spin-down -1/2). In some of the previous papers the authors renumbered the valence band Landau levels (to eliminate n = -1 states) which caused their selection rules for n to differ from ours by 1.

Once the matrix elements have been calculated and the selection rules are known, the two photon magneto-absorption coefficients are straight forward to calculate. Following second-order perturbation theory, the transition rate[7,8] for a two-photon process in c.g.s. units is

$$T = \frac{2\pi}{\hbar} \frac{4\pi^2 e^4 I^2}{n_p^2 c^2 m^4 \omega^4} \sum_f \int \left| \sum_t \frac{\langle f|\vec{\varepsilon}\cdot\vec{p}|t\rangle\langle t|\vec{\varepsilon}\cdot\vec{p}|i\rangle}{E_t - E_i - \hbar\omega} \right|^2 \delta(E_f - E_i - 2\hbar\omega) \frac{d\vec{k}}{(2\pi)^3} \qquad (1)$$

for both magnetic field and no magnetic field cases. The symbols in this equation are n_p which is the index of refraction , I is the incident intensity, $\vec{\varepsilon}$ is the polarization vector in the direction of the electric field, and i, t, and f, are the initial, intermediate, and final states, respectively. The delta function is a statement of energy conservation between the initial and final states and the two photons. The two-photon absorption coefficient is

$$K_2 = (2\hbar\omega/I^2)T \quad (2)$$

Although these expressions were derived for two-photon absorption in the absense of a magnetic field, they can be easily adapted to the magnetic field case. The principal differences between the two cases are in the matrix elements and the integration over $\vec{k}$, which is related to the joint density of states. In general the magnetic field density of states in the parabolic approximation[9] has the form

$$g(E) = \frac{1}{8\pi^2}\left(\frac{2m^*}{\hbar^2}\right)^{3/2} \sum_{n=0}^{n_{max}} \frac{\hbar\omega_c}{[E - (n + 1/2)\hbar\omega_c]^{1/2}} \quad (3)$$

where n_{max} is the largest positive integer for which the summand is real, m* is a cyclotron effective mass, and ω_c is the cyclotron frequency (= eB/m*). Immediately one sees that the density of states diverges whenever the energy E equals the energy of a Landau level. This is completely different from the case without a magnetic field where the density of states is proportional to $\sqrt{E}$ and no divergence is observed. This divergence in g(E) is responsible for the two-photon resonances observed in the conductivity. To remove the infinite divergence whenever $E = (n + 1/2)\hbar\omega_c$, it has been customary to introduce a broadening term Γ into the energy of the Landau levels.[10] This broadening factor is a result of collisions of electrons with other particles (i.e. other electrons, phonons, or impurities). Kubo et al.[11] has shown that Γ can be expressed as $\Gamma = \hbar/(2\tau)$ where τ is a characteristic life-time of the state.

Including broadening into the transition rate and assuming all transitions occur at $\vec{k} = 0$, we can write a general expression for the two-photon magneto-absorption coefficient as follows:

$$K_2^{i,j} = \frac{e^5 \hbar\, 4\pi}{n_p^2 c^2 m^4} \frac{B}{(\hbar\omega)^3}\left(\frac{2\tau\mu}{\hbar}\right)^{1/2} \left| \sum_t \frac{\langle f|\vec{\varepsilon}\cdot\vec{p}|t\rangle\langle t|\vec{\varepsilon}\cdot\vec{p}|i\rangle}{E_t - E_i - \hbar\omega} \right|^2 \quad (4)$$

where μ is an effective mass for an electron and light or heavy hole system ($1/\mu = 1/m_e + 1/m_h$). The final state is some Landau level in the conduction band and the initial state is a Landau level of either the light or heavy hole band. The intermediate state t can be a Landau level in any one of the four bands (conduction, heavy or light hole, or spin-orbit split-off band). Notice that $K_2^{i,f}$ is dependent on the initial and final states. Therefore each transition between specific Landau levels in the valence and conduction bands has a characteristic two-photon absorption coefficient, which depends upon magnetic field B and photon energy $\hbar\omega$. We have restricted the sum over intermediate states to the four bands just mentioned; the contribution of higher bands is expected to be small, because of the large energy difference in the denomimators. The prescription for calculating $K_2^{i,f}$ is to numerically solve for the eigenvectors of the two 4 x 4 Hamiltonians and then use these eigenvectors to calculate the matrix elements. In Figure 1 the splitting of the bands into Landau levels is shown along with a representative two-photon transition for σ_L. Figure 2 schematically shows the two-photon transitions for the different light polarizations according to the spherical selection rules.

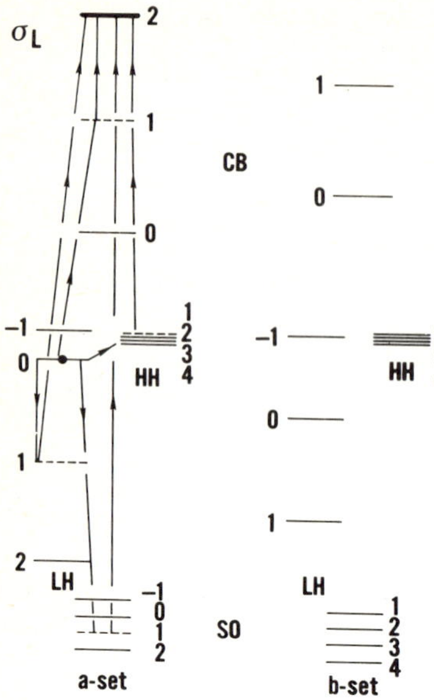

Figure 1. Splitting of the bands into Landau levels. a-set are the spin up levels (+1/2) and b-set are the spin down levels (-1/2). (CB-conduction band, HH-heavy hole band, LH-Light hole band, SO-spin orbit split off band). A two-photon transition ($a^+(0) \rightarrow a^c(2)$ for σ_L) is shown with all the intermediate states. (——•—— initial state, ----- intermediate state, ——— final state).

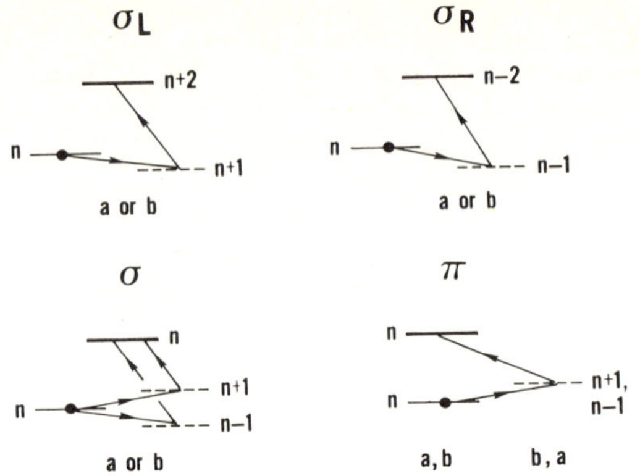

Figure 2. Schematic diagram of the initial, intermediate and final states of the two-photon process for the different light polarizations. For $\vec{e} \perp \vec{B}$ (σ_R, σ_L, σ) all transitions are within the a or b set. For $\vec{e} \parallel \vec{B}$ (π) the initial and final states are of the same spin while the intermediate state is of opposite spin.

Although there has been extensive theoretical work on TPA in the absence of a magnetic field, there has been only two major attempts to calculate TPMA coefficients. The first attempt was by Bassani and Girlanda.[12] Using second-order perturbation theory, they considered the absorption of two photons each with different energies and different polarizations. They also used the parabolic approximation to the Landau levels and considered both inter-inter and inter-intra band transitions between one valence band and two conduction bands. In addition broadening was not included in their model; K_2 therefore exhibited infinite divergences at the resonance points. No comparison with experiment was ever made with their results. The other major attempt was by Nguyen, Strnad and Yafet,[13] who also used second-order perturbation theory. In the sum over intermediate states they included the conduction, light, and heavy hole bands and did not include the spin-orbit split-off band, which is usually only a small contribution. However, they were able to obtain analytical expressions for K_2, because they used first-order perturbation wave functions in calculating the matrix elements. Their theoretical results compared favorably with their experiment.

EXPERIMENTAL WORK

Oriented samples of high purity n-InSb ($\approx 9 \times 10^{13}$ cm^{-3}) were mounted in a variable temperature dewar for the lower magnetic field work (0 - 2 T) carried out at NTSU using a Varian electromagnet. High field (0 - 15 T) studies were carried out at the Francis Bitter National Magnet Laboratory with the samples immersed directly in pumped liquid helium. Light from grating tunable cw CO_2 lasers was mechanically chopped by a rotating slotted wheel with a small duty cycle to minimize sample heating effects. In the low field region the photoconductive measurements were carried out using either boxcar integrator techniques or a combination of sampling oscilloscope and magnetic field modulation techniques. At high fields only the boxcar was used. Field positions, amplitudes and line widths could then be directly measured and studied as a function of laser intensity, lattice temperature, etc. The addition of modulation techniques significantly improves the sensitivity and resolution of these magneto-optical experiments allowing a more accurate determination of the resonant field positions and observation of weak structure or resolution of closely spaced doublets.

Figure 3 shows a reproduction of the output of the lock-in detector versus magnetic field for sample dc bias currents of I = 1 mA (photoconductive response) and I = 0 mA (photovoltaic response). The detector response is proportional to the second derivative of the photo-induced magnetoresistance or photovoltaic effect. The observed amplitude of the resonant structure contains the effects of the modulation method. To our knowledge this is the first time that two-photon absorption effects have been observed in the photovoltaic response of a semiconductor. In all subsequent experiments the photovoltaic response was minimized by keeping the laser beam spot positioned between the sample contacts. Thus, in most cases two-photon effects were not discernible in the photovoltaic response, while the photoconductive response showed excellent two-photon resonances.

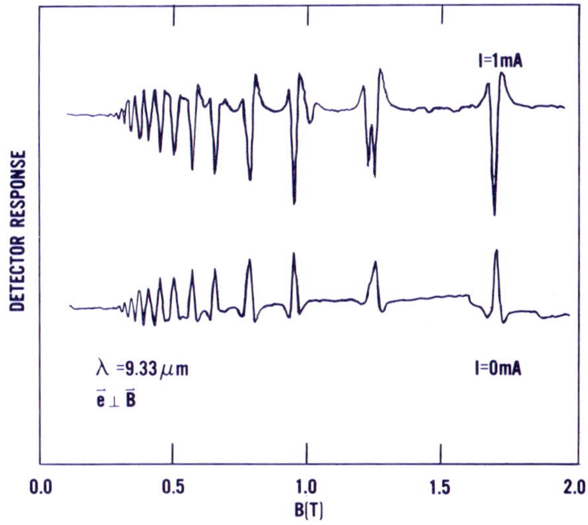

Figure 3. TPMA effects in the photo-voltaic response (I = 0 mA trace). The top trace contains both the photoconductivity and the photovoltaic signals.

RESULTS

Experiments were performed in two principal magnetic field ranges: 0 - 2 T and 0 - 15 T. Most of the results, analysis, and interpretation are focused on the 0 - 2 T range. However, the higher magnetic fields are important for the observation of certain transitions near the TPMA band edge, which are not predicted by the spherical selection rules. In addition, at the higher fields the hole Landau levels are split further apart allowing clear resolution of unresolved transitions at the lower magnetic fields.

Figure 4 shows the wavelength dependence of the TPMA structure. The numbers correspond to distinct transitions that have been identified from all the data that has been taken and not just this figure. For TPMA each transition corresponds to an increase in the conductivity or a decrease in the magnetoresistance. That is, as the magnetic field is increased to some value where the energy between two allowed Landau levels equal twice the photon energy, electrons in the initial valence band Landau level will be photo-excited to the final conduction band Landau level. Transitions can originate from either the light or heavy hole Landau level. This process is repeated at every magnetic field where a two-photon transition is allowed. The increase in amplitude of the structure with higher magnetic fields is partly due to the increase with field of the joint density of states. In addition, the sampling and lock-in technique produces a Bessel function envelope to the amplitudes. Figure 4 dramatically illustrates the movement of the TPMA structure towards lower magnetic fields as the wavelength is increased. The usefulness of the CO_2 laser in studying TPMA in InSb is also demonstrated in that its spectral output can be tuned below the TPMA threshold (the energy gap equal to twice the photon energy) or to higher photon energies where much complex structure can be observed since electrons can now be excited high into the conduction band. To resolve the transitions close to the TPMA band edge, these experiments were extended to 15.0 T using the boxcar averager technique. Figure 5 shows the results of those experiments where the TPMA band edge transitions (1, 2, and 3) are clearly resolved. The detector response in this figure is proportional to the photo-induced change in magnetoresistance. At the higher fields, transitions 2 and 3 are well resolved, whereas at the lower magnetic fields, they are not resolved even

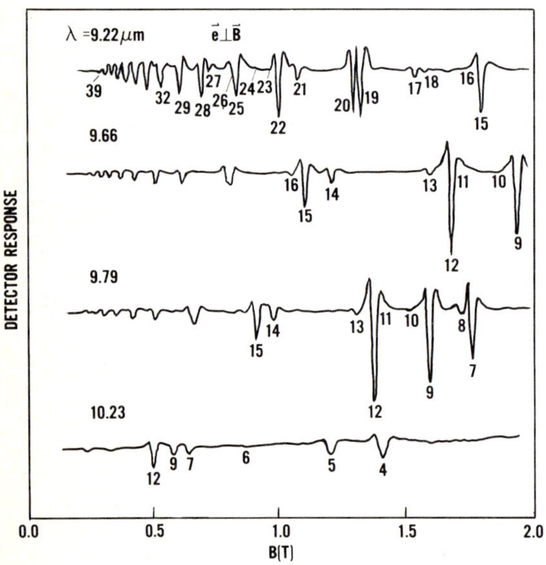

Figure 4. Wavelength dependence of the TPMA structure for $\vec{B} \parallel$ <110> and for 0 - 2 T using sampling oscilloscope and field modulation techniques. The numbers correspond to distinct transitions.

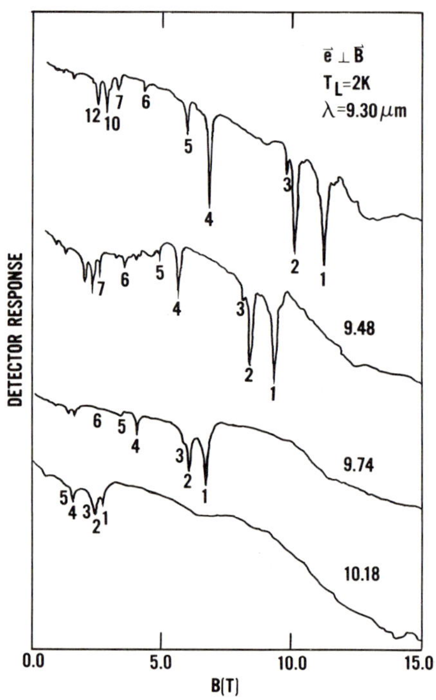

Figure 5. Wavelength dependence of TPMA for $\vec{B} \parallel$ <211> and for 0 - 15 T. The data were taken with the boxcar averager technique and shows directly the TPMA band edge.

with the sampling and lock-in technique. Figure 5 also shows the growth in amplitude of the structure as it moves toward higher magnetic fields with higher photon energies. Although use of the boxcar averager technique eliminates the Bessel function envelope to the structure, there is present a background photo-induced magnetoresistance due to non-resonant absorption processes. For our experiments TPMA was not observed for wavelengths above 10.35 μm.

The results of polarization studies for $\vec{e} \perp \vec{B}$ and $\vec{e} \parallel \vec{B}$ in the Voigt configuration have recently been presented.[2] Here we present additional results using circularly polarized light (obtained by using a Fresnel Rhomb) in the Faraday geometry. The TPMA structure obtained for σ_R, σ_L, and $\pi(\vec{e} \parallel \vec{B})$ polarizations are shown in Figure 6. The magnetic field positions of the structures for σ_R and σ_L are basically the same. Figure 6 also shows additional structure between transitions 16 and 17 in the σ_R polarization which have not been labeled. However, there are some noticeable differences in the strengths of some of the transitions. For instance, transitions 11, 13, 14, 17, and 21 appear to be stronger for σ_R, while transitions 12, 16, and 20 appear to be dominant for σ_L. Some transitions such as 15, 19, and those higher than 22 have no significant difference between σ_R and σ_L. These transitions are the result of σ polarization, which is the sum of σ_R and σ_L. The reason why these transitions are present in both σ_R and σ_L is that each polarization, σ_R or σ_L, contains some residual polarization of the opposite circularly polarized wave. That is, rather than having pure circular polarization, the wave is slightly elliptical. We estimate from a polarization analysis of the beam that perhaps as much as 10 percent of the opposite circular polarized electric field is present for either of the σ_R or σ_L polarizations. The transitions in the π polarization should match magnetic field positions for the σ transitions, because they both obey the same selection rules for transitions between initial and final states. However, the strengths of the σ transitions are different from those of the π transitions, because the intermedite states are different.

Specific transitions are assigned to the structure in Figure 7 by comparing theoretical calculations of K_2, the TPMA coefficient, from Eq. (4) with the amplitude and magnetic field positions for the structure obtained using σ_L and σ_R polarizations. Each line for K_2 represents the strength of a specific transition. The magnetic field position of each transition was estimated from an energy band calculation of the Landau levels. Only those transitions which obey the spherical selection rules are plotted. The amplitude of the structure can only be compared qualitatively with the calculations for K_2 because (1) the detector response signal is a second-derivative which may be slightly different from the actual signal and (2) the detector response signal contains a Bessel function type envelope to the transitions. As can be seen, some of the structure, certainly at the lower magnetic fields, are actually the sum of several transitions which are not resolved. The observed resonant structure in Figure 7 was identified with only the strongest calculated transitions. The results of the identification are given in Table I. We have only made comparisons of the $\vec{e} \perp \vec{B}$ transitions with the calculated K_2 values. In principle the same method can be applied for the $\vec{e} \parallel \vec{B}$ or π transitions. However, the structure is much less complicated making the identification of the transitions easier. Therefore for the purpose of identification of transitions K_2 values were not calculated.

After the structure in the data has been tentatively identified, the energy band Landau level calculations can be used to determine a set of band parameters. This involves a considerable amount of computer usage in fitting the theoretical calculations to the data. The two-photon transition energies as a function of magnetic field are plotted in the usual "fan chart" method and compared to the experimental data. The results of our fitting procedure for both $\vec{e} \perp \vec{B}$ and $\vec{e} \parallel \vec{B}$, are shown in Figures 8-10, where the data are the experimental transition energies (twice the photon energy) versus resonant magnetic field positions and the lines (solid and dashed) are the calculated results for $\vec{B} \parallel$ <110>. There are many more transitions plotted than have been found from the earlier studies using pulsed lasers. Interband transitions involved in magneto-absorption experiments are usually considered as being to exciton levels. However, there is at present no

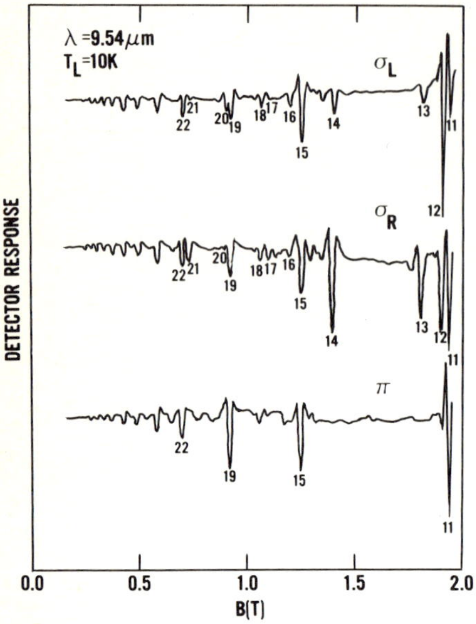

Figure 6. Polarization dependence of TPMA for $\vec{B} \parallel$ <110>.

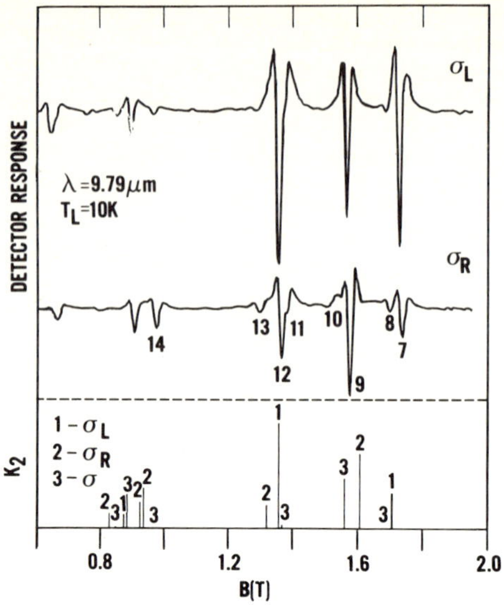

Figure 7. Comparison of the amplitude of the TPMA structure for σ_L and σ_R and for $\vec{B} \parallel$ <110> with theoretically calculated values of the TPMA coefficient for the possible two-photon transitions.

adequate exciton theory for TPMA in semiconductors with complex coupled energy bands. Here we adopt the approach of not correcting for exciton effects and of treating the Landau-level theory as accurately as possible in calculating transition energies. We then look for differences between the calculated and observed energies for the lowest transitions where exciton effects should be most important. The theoretically calculated solid lines shown in Figures 8-10 result from using only the spherical selection rules. It appears that most of the low field experimental data can be adequately described (but with a new set of band parameters) with these rules without using exciton corrections. A good fit of these theoretical transitions to the low field data results from using the following band parameters: E_g = 235.2 meV, E_p = 23.2 eV, Δ = 0.803 eV, γ_1 = 3.25, γ_2 = -0.2, γ_3 = 0.9, κ = -1.3, F = -0.2, q = 0.0, N_1 = -0.55. No attempt was made to determine a value of Δ from our two-photon data. Instead we have used the stress modulated magnetoreflectance results of Δ = 0.803 ± 0.005 eV obtained by Aggarwal.[14] There are some noticeable differences, however, between the theory and data in Figure 10 where the transitions have been plotted out to 15 T. If excitons are present, the transition energies would be slightly altered from the no-exciton case, which the data clearly indicates for transitions 7 and lower. The deviation between the exciton and no-exciton theories should also be more evident with increasing magnetic fields. An adequate TPMA exciton theory is thus needed to improve the fit for these transitions.

Additional structure labeled (1,2), (6), (17,18), and (24) in Figure 8, (1,2) and (6) in Figure 10, and (18) in Figure 9, appears not to be explainable by the spherical selection rules. These transitions could be described by nonspherical selection rules allowed because of inversion asymmetry and warping effects.

Table I. Theoretical two-photon transition assignments for each distinct experimental transition number (Exp. Trans. No.) of the observed resonant structure in the photoconductivity for $\vec{e} \perp \vec{B}$ polarization (σ_R, σ_L, $\sigma_R + \sigma_L = \sigma$) in the Voigt geometry. The * marks indicate those transitions which can only be explained by nonspherical or extra transitions.

Exp. Trans. No.	Polarization	Theoretical Assignment	Exp. Trans. No.	Polarization	Theoretical Assignment
*1,*2		$a^-(1)\ a^c(0)$	25	σ_R	$a^-(7)\ a^c(5)$
		$a^+(-1)\ a^c(0)$		σ	$b^+(2)\ b^c(2)$
		$b^-(2)\ a^c(0)$	26	σ_L	$b^+(1)\ b^c(3)$
		$b^-(1)\ a^c(0)$	27	σ_R	$a^+(4)\ a^c(2)$
		$b^+(-1)\ a^c(0)$	28	σ_R	$a^-(8)\ a^c(6)$
3	σ_R	$a^-(2)\ a^c(0)$		σ	$a^+(3)\ a^c(3)$
4	σ	$a^+(0)\ a^c(0)$	29	σ_R	$a^-(9)\ a^c(7)$
5	σ_R	$b^-(2)\ b^c(0)$		σ	$b^+(3)\ b^c(3)$
*6		$a^+(0)\ b^c(0)$	30	σ_L	$b^+(2)\ b^c(4)$
7	σ_L	$a^+(-1)\ a^c(1)$	31	σ_R	$a^+(5)\ a^c(3)$
8	σ	$a^-(1)\ a^c(1)$	32	σ_R	$a^-(10)\ a^c(8)$
9	σ	$b^+(0)\ b^c(0)$		σ	$a^+(4)\ a^c(4)$
10	σ_R	$a^-(3)\ a^c(1)$	33	σ_L	$b^+(3)\ b^c(5)$
11	σ	$b^-(1)\ b^c(1)$		σ_R	$a^-(11)\ a^c(9)$
12	σ_L	$b^+(-1)\ b^c(1)$		σ	$b^+(4)\ b^c(4)$
13	σ_R	$b^-(3)\ b^c(1)$	34	σ_R	$a^+(6)\ a^c(4)$
14	σ_R	$a^+(2)\ a^c(0)$		σ_R	$a^-(12)\ a^c(10)$
15	σ_R	$a^-(4)\ a^c(2)$		σ	$a^+(5)\ a^c(5)$
	σ	$a^+(1)\ a^c(1)$	35	σ_L	$b^+(4)\ b^c(6)$
16	σ_R	$b^-(4)\ b^c(2)$		σ_R	$a^-(13)\ a^c(11)$
*17		$a^+(2)\ b^c(0)$		σ	$b^+(5)\ b^c(5)$
		$a^-(4)\ b^c(2)$	36	σ_R	$a^+(7)\ a^c(5)$
*18		$a^+(1)\ b^c(1)$		σ_R	$a^-(14)\ a^c(12)$
		$a^+(0)\ b^c(2)$		σ	$a^+(6)\ a^c(6)$
	σ_R	$b^+(2)\ b^c(0)$	37	σ_L	$b^+(5)\ b^c(7)$
19	σ_R	$a^-(5)\ a^c(3)$		σ_R	$a^-(15)\ a^c(13)$
	σ	$b^+(1)\ b^c(1)$		σ	$b^+(6)\ b^c(6)$
20	σ_L	$b^+(0)\ b^c(2)$	38	σ_R	$a^+(8)\ a^c(6)$
21	σ_R	$a^+(3)\ a^c(1)$		σ_R	$a^-(16)\ a^c(14)$
22	σ_R	$a^-(6)\ a^c(4)$		σ	$a^+(7)\ a^c(7)$
	σ	$a^+(2)\ a^c(2)$	39	σ_L	$b^+(6)\ b^c(8)$
23	σ_R	$b^-(6)\ b^c(4)$		σ_R	$a^-(17)\ a^c(15)$
*24		$a^+(2)\ b^c(2)$		σ	$b^+(7)\ b^c(7)$

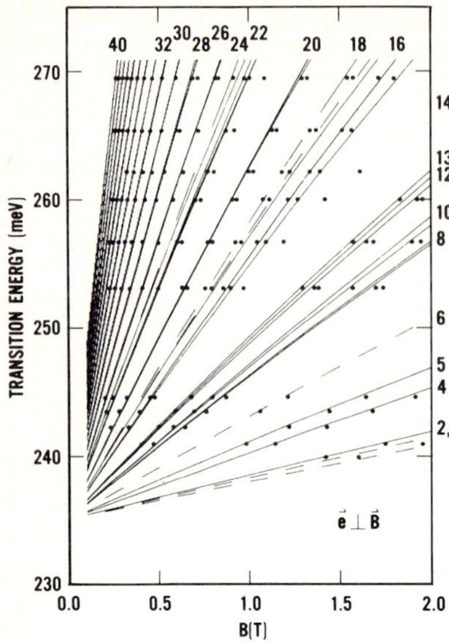

Figure 8. Fan chart of TPMA transitions energies ($\vec{e} \perp \vec{B} \parallel$<110>). Solid lines represent theoretical calculations with selection rules $\Delta s = 0$, $\Delta n = 0, \pm 2$. Dashed lines, nonspherical selection rules. Dots are the data.

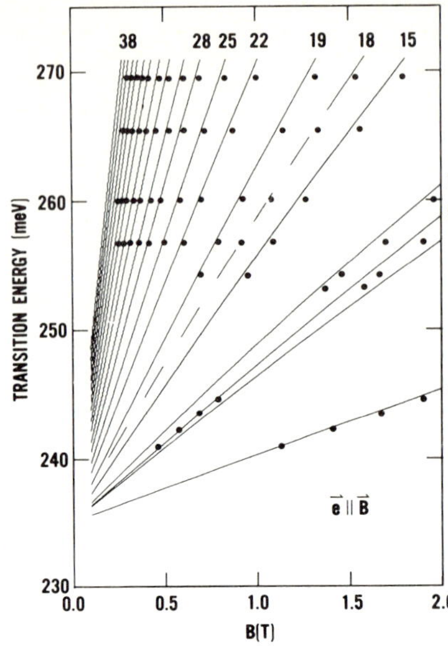

Figure 9. Fan chart of TPMA transition energies $\vec{e} \parallel \vec{B} \parallel$ <110>.

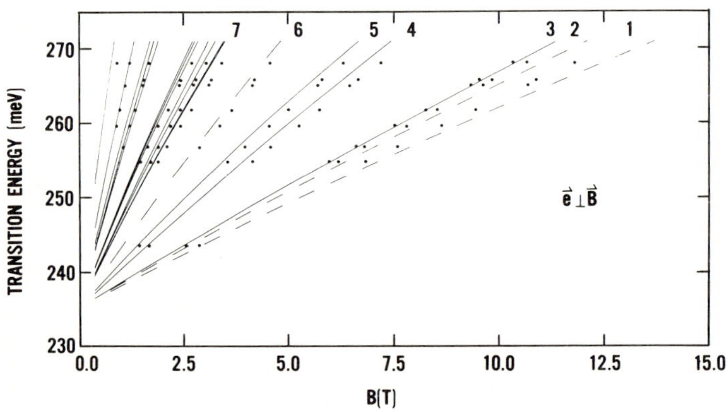

Figure 10. Fan chart of TPMA transition energies $\vec{e} \perp \vec{B} \parallel$<211> for the magnetic field range 0 - 15 T.

In Figure 11 TPMA spectra are shown for the major crystal directions. Manlief and Palik[15] reported that they observed no difference in the TPMA spectra for the three crystal directions. From Figure 11 one can see that for much of the structure, there is no appreciable differences in the spectra between the three crystal directions. However, one striking feature seen is the clear resolution of the doublet nature to transitions 11 and 12 for $\vec{B} \parallel$ <111>. This has not been previously reported in any TPMA study. There are also some definite magnetic field shifts for some of the weaker transitions (8, 10, 11, 13) observable at the higher magnetic fields for $\lambda = 9.79$ µm. The observed shift in these structures is due to the difference in energy of the heavy hole Landau levels for the three crystal directions. Calculations of the heavy hole Landau level energies give good agreement with the observed shifts. This is further proof for our heavy hole identification of these transitions. From cyclotron resonance measurements of the heavy hole masses the heavy hole mass $m^-[111]$ is the largest, while the $m^-[100]$ mass is the smallest ($m^-[111] > m^-[110] > m^-[100]$). Since the Landau level energies are inversely proportional to the effective masses of a particular band ($E \simeq n\hbar eB/m^*$ in the parabolic approximation), the heavy hole energies will be somewhat smaller for the <111> direction than the <100> direction, thus causing a shift in the heavy hole transitions toward higher magnetic fields for the <111> direction. The light hole transitions do not show a significant shift in magnetic field position, because there is not a large anisotropy associated with the light holes.

We have calculated the TPMA spectra as a function of magnetic field. In Figure 12 we show an experimental spectrum (top curve) and a calculated spectrum (bottom curve). The experimental trace was obtained using the boxcar averager technique and is proportional to the photo-induced magnetoresistance. The theoretical spectrum was calculated assuming a parabolic magnetic field dependent joint density of states function, similar to Eq. (3), for each transition with broadening of the energy levels included. The energy of the Landau levels as a function of magnetic field in the joint density of states function was calculated from the nonparabolic 8 x 8 Hamiltonian matrix. The spectrum at each magnetic field point was obtained by summing the contributions of all the transitions in the spectrum at that point. Thus, transitions in close proximity will be significantly influenced by each other, while transitions further away in magnetic field will have a much less influence, depending on the strength of the transitions. The theoretical spectrum is plotted as $1/K_2$ since the experimental spectrum is $\propto 1/n$ or $1/K_2$.

The theoretical spectrum is in qualitative agreement with the experimental spectrum. The two high field sets of transitions (9, 10, and 11, 12) show a sharp rise on the high magnetic field side of a transition while a much more gradual decrease is evident on the low field side of a transition because of the density of states. The magnetic field width of a transition depends on the amount of broadening of the energy levels. For this spectrum an energy width of $\Gamma = 0.1$ meV was used corresponding to a collision broadening time $\tau = 1 \times 10^{-12}$ sec. Although the two large transitions at the high magnetic fields are at a slightly lower magnetic field than the experimental ones, the theoretical spectrum reproduces transitions 9-12 of the experimental spectrum very well. However, the theoretical spectrum in the vicinity of transitions 14 and 15 shows a complex multiplet of structure while the experimental shows one seemingly large transition. Either certain transitions are stronger and dominate the spectrum or a slightly different set of band parameters would perhaps reproduce the spectra more accurately. The theoretical spectrum does show however, that some of the transitions, certainly at the lower magnetic fields, are actually the sum of several unresolved transitions. Also the experimental spectrum contains a noticeable magnetic field background in addition to the resonant structure.

CONCLUSIONS

The TPMA spectra in InSb is shown to be very complex, exhibiting characteristic polarization, crystalline anisotropy, and magnetic field dependences. Many aspects of these spectral features are presently understood with calculations using a modified Pidgeon-Brown energy band model and the appropriate selection rules. However, a real quantitative understanding of the role of excitons in the TPMA spectra is lacking at the present time.

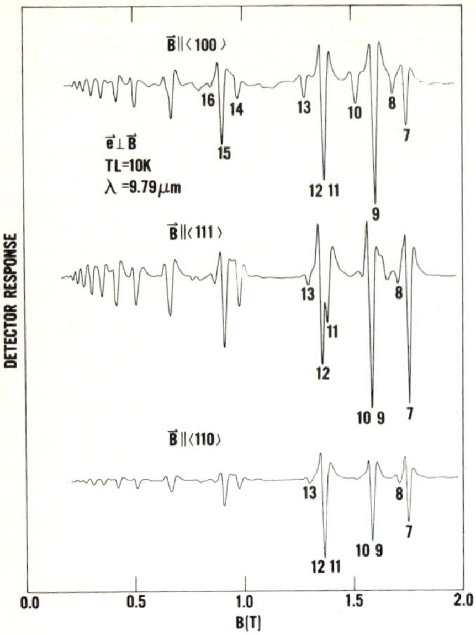

Figure 11. Crystal orientation dependence of TPMA.

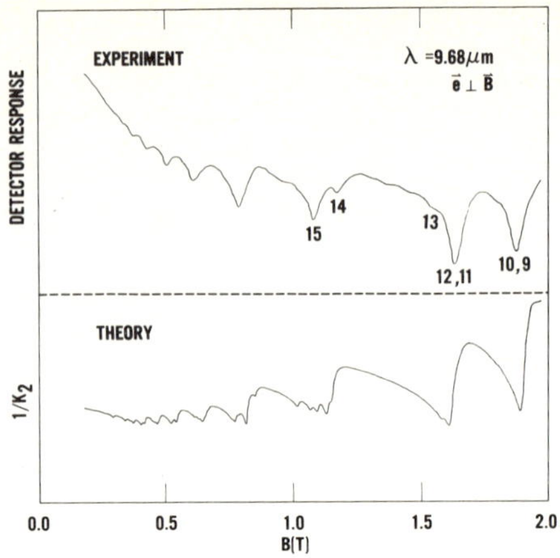

Figure 12. Comparison of an experimental TPMA spectrum with a theoretically calculated value of $1/K_2$.

REFERENCES

*Work supported in part by the Office of Naval Research and a Faculty Research Grant from North Texas State University.
†Present address: Texas Instruments, Inc., Dallas, Texas 75265.
††Also, Department of Electrical Engineering.

1. D. G. Seiler, M. W. Goodwin, and M. H. Weiler, Phys. Rev. B 23, 6806 (1981).
2. M. W. Goodwin, D. G. Seiler, and M. H. Weiler, Phys. Rev. B 25, 6300 (1982).
3. C. R. Pidgeon and R. N. Brown, Phys. Rev. 146, 575 (1966).
4. M. H. Weiler, R. L. Aggarwal, and B. Lax, Phys. Rev. B 17, 3269 (1978).
5. M. H. Weiler, Ph.D. Thesis, Massachusetts Institute of Technology (1977).
6. W. Zawadzki and J. Wlasak, J. Phys. C: Solid State Physics 9, L663 (1976).
7. C. C. Lee and H. Y. Fan, Phys. Rev. B 9, 3502 (1974).
8. C. R. Pidgeon, in Theoretical Aspects and New Developments in Magneto-Optics, edited by J. T. Devresse (Plenum Press, New York, 1980), p. 255.
9. A. H. Kahn and H. P. R. Frederikse, in Solid State Physics, edited by F. Seitz and D. Turnbull (Academic Press, New York, 1959), Vol. 9, p. 257.
10. L. M. Roth and P. N. Argyres, in Semiconductors and Semimetals, edited by R. K. Willardson and A. C. Beer (Academic Press, New York, 1966), Vol. 1, . 159.
11. R. Kubo, N. Hasitsume, and S. J. Miyake, in Solid State Physics, edited by F. Seitz and D. Turnbull (Academic Press, New York, 1965), Vol. 17, p. 269.
12. F. Bassani and R. Girlanda, Opt. Comm. 1, 359 (1970).
13. Van-Tran Nguyen, A. R. Strnad, and Y. Yafet, Phys. Rev. Lett. 26, 1170 (1971).
14. R. L. Aggarwal, in Semiconductors and Semimetals, edited by R. K. Willardson and A. C. Beer, (Academic Press, New York, 1972), Vol. 9, p. 151.
15. S. K. Manlief and E. D. Palik, Sol. State Commun. 12, 1071 (1973).

ELECTRON SPIN RESONANCE IN InSb UNDER UNIAXIAL STRESS

F. Kuchar
Institute for Solid State Physics, University of Vienna and Ludwig Boltzmann Institute, Vienna, Austria

M. Kriechbaum
Institute for Theoretical Physics, University, Graz, Austria

I. Introduction

Intraband magnetooptical experiments have been successfully used to determine band parameters of semiconductors. In particular, in narrow gap semiconductors the influence of the interaction of e.g. the conduction band with other bands can be studied. This interaction leads to a mixing of electronic states. The application of uniaxial stress can induce an additional strain interaction. The magnetooptical transition between the Landau subbands 0^+ and 0^-, the spin resonance (SR), is particularly sensitive to these interactions. The conduction band-valence-band interaction causes the nonparabolicity of the bands. The Landau subband wavefunctions 0^+, 0^- are a mixture of conduction and valence band states and are not spin eigenstates. However the electric dipole spin resonance matrix element $\langle 0^+|p_\perp|0^-\rangle$ is proportional to $k_\parallel$ and $\langle 0^+|p_\parallel|0^-\rangle = 0$ [1]. In high magnetic fields, magnetic freeze-out causes the matrix element to be very small because of its $k_\parallel$ dependence. In crystals like InSb, as a consequence of the inversion asymmetry, two additional terms appear in the conduction band energy dispersion relation: (1) k^3 terms due to the interaction with higher bands (strength determined by the constant B) and (2) in uniaxially stressed crystals, k linear terms due to the strain interaction between conduction and valence band (constant C_2). This means that the inversion asymmetry leads to an additional mixing of the 0^+ and 0^- states and thus to an enhancement of the matrix element $\langle 0^+|p|0^-\rangle$. Experimentally, this effect could be observed in a SR photoconductivity experiment [3]. It was explained in terms of a six band model taking into account the reduced symmetry of the stressed crystal. With transmission experiments [4] we determined the absorption coefficient and the deformation potential C_2 which governs the strength of the stress enhanced spin resonance.

In this paper, we present an experimental study using Fourier transform and laser spectroscopy as well as a theoretical treatment of the problem based on an exact solution of the eight band model (conduction band, valence band, split off band) with higher bands included by perturbation theory. We also discuss the consequences for the interpretation of other experiments, i.e. Shubnikov-de Haas and [5], magnetophonon [6] and intra-valence-band magnetooptical experiment [7].

II. Experimental Arrangement

The experiments were performed using either a Fourier transform spectrometer (FTS) or an optically pumped far-infrared laser system (FIRL, 163μm, 118.8μm, 96.5μm). Data were taken in Faraday geometry with linearly polarized (FIRL) or unpolarized radiation (FTS). The details of the FTS stress apparatus at the sample chamber are shown in Fig.1. It fits in the axial bore of a superconducting magnet. The radiation is falling onto the sample from the top via a 0.75 inch light pipe. The force is applied via a steel wire. The maximum force which can be applied is about 30 kp, e.g. yielding a stress of 5 kbar in a sample with 0.6 mm^2 cross section. Further details are described in Ref.[8]. The laser experiments were performed in a split-coil magnet. The uniaxial stress was applied in a similar way, besides the fact

Table I: Sample characteristics

Sample No.	N_D-N_A (cm^{-3})	μ_{77} ($cm^2/Vsec$)	stress direction
1	8×10^{13}	6.5×10^5	[110]
2	6×10^{13}	6.7×10^5	[110]
3,4	1×10^{14}	6.3×10^5	[110]
5	1×10^{14}	6.3×10^5	[100]
6	2×10^{14}	5.5×10^5	[110]

that the pulleys O (Fig.1) were not neccessary because of the use of the radial bore of the superconducting magnet.

Measuring temperatures were 4.2 K and below. High purity InSb samples (Cominco) were used. Sample characteristics are given in Table I. Sample no.6 had a slightly higher carrier concentration obtained by nuclear transmutation doping [9]. Differently orientied samples were used, the stress direction being either [110] or[100].

Spectra were recorded using transmission as well as photoconductivity (P.C) techniques. The first one was only applied in the laser

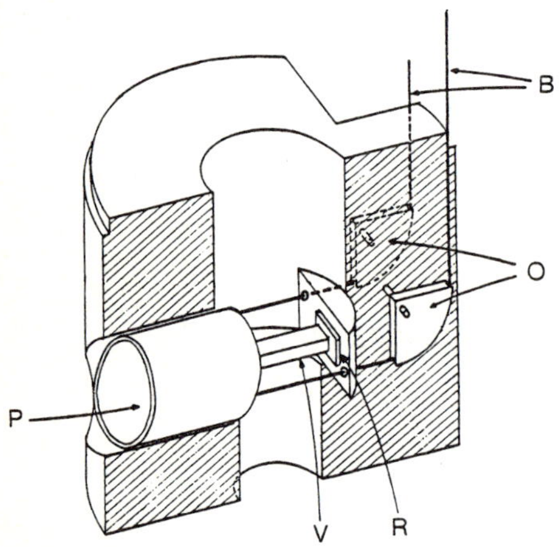

Fig.1: Sectional view of the sample chamber in the FTS experiment. B...steel string, O... pulleys, P...moveable piston, R...Mylar, V...sample.

experiments where a Ge bolometer was used as the detector. In photoconductivity the spin resonance can be easily detected by applying stress of the order of 1 kbar. The absorption at these stress values is onlyof the order of a few percent (λ=118.8μm). Although this needs a very high stability of the laser system it has the big advantage of being able to determine the absorption coefficient and the deformation potential C_2. Because of the weak absorption sample thicknesses had to be at least 1 mm in the transmission experiment. Typical cross sections of the samples were about 1 mm^2. The irradiated length of the samples is approximately 2 mm.

III. Experimental Results

Figure 2 shows P.C. spectra (FTS) of two different high purity samples a T = 4.2 K. The sharp peak at 54 cm^{-1} is within 1% of the position where the electron spin resonance (SR) transition $0^+ \rightarrow 0^-$ is expected to occur according to theoretical results [10] and experimental data obtained with FIR lasers at zero stress [2,11].The broad background is

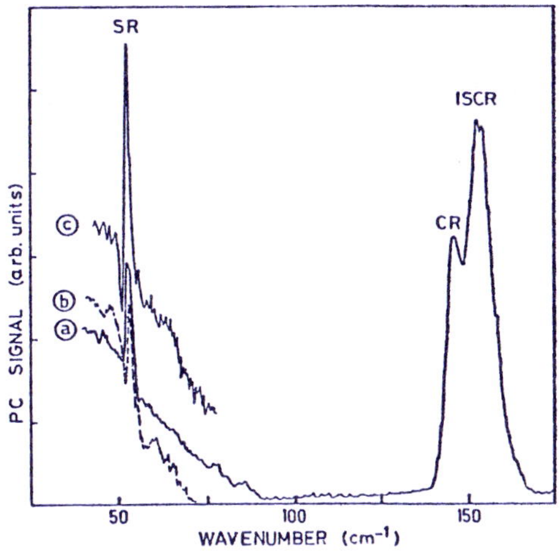

Fig.2: FTS photoconductivity (P.C.) spectrum of sample 1, X=1 kbar (a): sample 2, X=1.15 kbar (b)und X=1.9 kbar (c). B=2.5 T, T=4.2 K, X||[110], B||[1$\bar{1}$2].

due to the ionizing transition from the impurity ground state (000) to the Landau continuum. The cyclotron resonance line is split into the free carrier transition $0^+ \rightarrow 1^+$ (CR) and the impurity shifted cyclotron resonance $000^+ \rightarrow 110^+$ (ISCR). Already at the relatively low stress values of 1-2 kbar the peak intensity of the SR P.C. signal becomes comparable to that of the CR. In zero-stress transmission experiments the SR was observed to be at least three orders of magnitude weaker than the CR [2]. The strength of the CR and ISCR peaks remain unchanged under stress. In the low resolution spectra of Fig.2 a single SR line is observed. There is only a single free carrier transition ($0^+ \rightarrow 0^-$) possible which can contribute to the signal. From the laser experiments at zero stress [2,11], however, it is known that up to three narrow lines ($\Delta B \sim 2x10^{-3}T$) with a spacing of about $3x10^{-2}T$ can be observed at 4.1 T. Experimentally [2] the central line can be shown to be

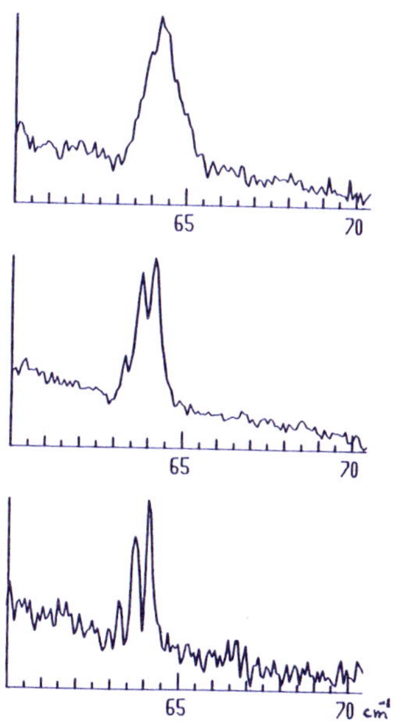

Fig.3: Stress dependence of the spin resonance lines at B = 3 T and T=4.2 K (sample 3). From top to bottom : 4.4 kbar, 3.4 kbar, 2.1 kbar, X||[110], B||[1$\bar{1}$2].

the free carrier SR transition and the two others to be impurity related. With high resolution (0.2 cm^{-1}) these three lines can also be resolved in our FTS experiment. Fig.3 shows the stress dependence of the P.C. spectrum at B = 3 T up to 4.4 kbar. The strength of the P.C. signal increases with stress up to 4.4 kbar. Below 2 kbar the shape of the three lines remains unchanged. Above 2 kbar broadening of the lines occurs which makes the resolution of the three peaks impossible at the highest achieved stress value. The broadening cannot be caused by inhomogenous stress in the sample, since the position of the lines does not change within the accuracy of our FTS experiment.

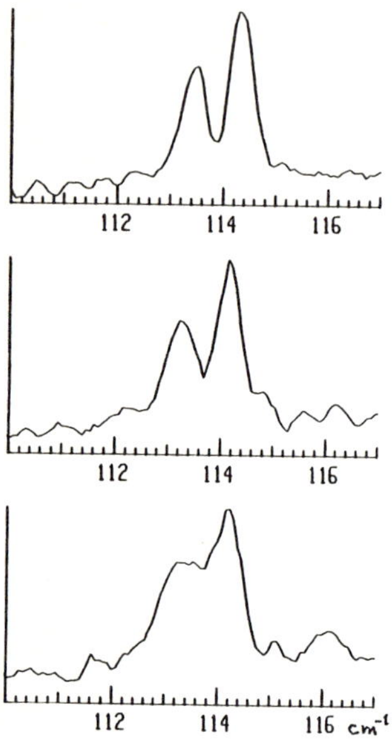

Fig.4: Temperature dependence of the spin resonance at B=6 T and X=1.9 kbar (sample 2). From top to bottom: 4.2 K, 3.4 K, 2.2 K, X||[110], B||[1$\bar{1}$2].

At B = 6 T two lines are resolved. The temperature dependence is shown in Fig.4. The essential effect of decreasing temperature is a broadening of the low frequency line. From comparison of the relative intensities with our 3 T data and our laser data as well as with the zero stress data [2,11] we conclude that the low frequency line corresponds to the free carrier transition $0^+ \rightarrow 0^-$. The FTS spectra shown in Figs. 3 and 4 were normalized to the strongest peak. Since the sensitivity of the photoconductivity technique can change with temperature, stress, and magnetic field no attempt was made to compare the absolute intensities. The following results were obtained using the optically pumped laser system. Figure 5 shows the anisotropy of the strength of the photoconductivity signal at the spin resonance for the two stress directions [110] (sample 4) and [100] (sample 5). The position of the three [110] peaks at about 4.12, 4.15, and 4.19 T are in accordance with the zerostress data [2,11] The weakest line could not be resolved with X||[100]. As pointed out above, the central line is the free carrier transition $0^+ \rightarrow 0^-$ and the other lines are impurity related. The dependence of the ESR signal on stress for X|| [100] is shown in Fig.6. At low stress values the P.C. signal increases as for X||[110]. The decrease at higher stress values remains unexplained at present. It might have its origin in the photoconductivity mechanism. Results of P.C. and transmission experiments with X||[110] are shown in Fig.7 for sample 6. Due to the higher impurity concentration the P.C. signal is weaker than in the case of sample 4. The weakest line is not observed

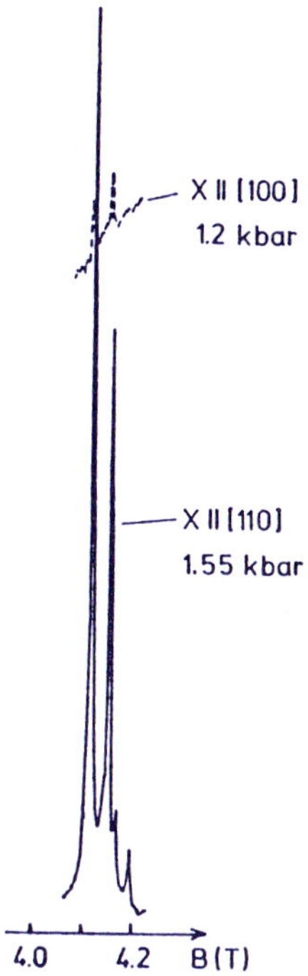

Fig.5: Photoconductivity spectra at $\lambda=118\ \mu m$.
Sample 4: $X||[110]$, $B||[1\bar{1}2]$;
sample 5: $X||[100]$, $B||[011]$.

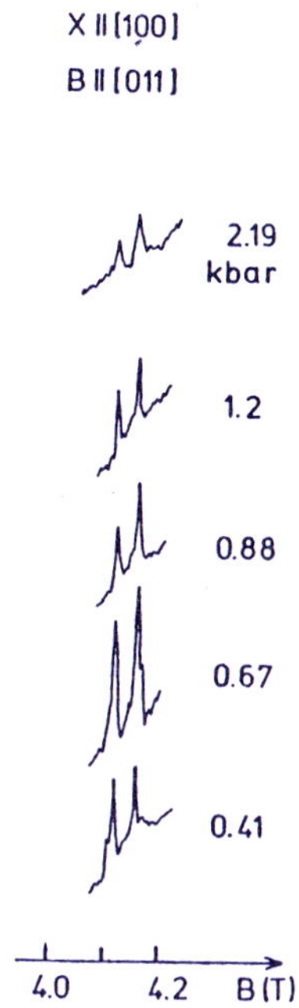

Fig.6: P.C. spectra of sample 2 at $\lambda=118.8\ \mu m$ for various stress values. Background signal is the same for all curves (2.19 kbar spectrum: no zero offset).

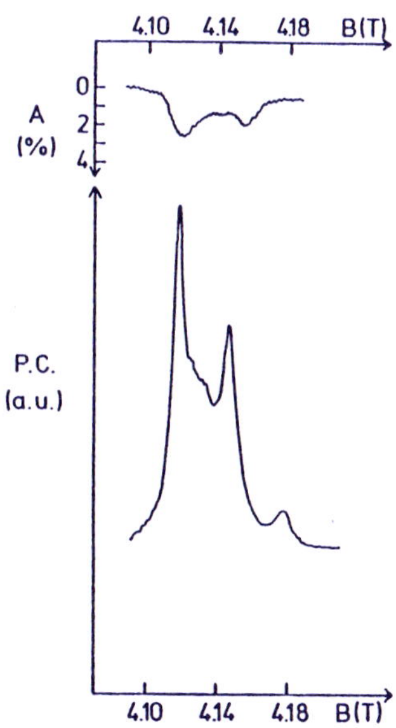

Fig.7: P.C. and absorption (A) spectra of sample 6 at $\lambda=118.8\ \mu m$ and 1.56 kbar, $X||[110]$, $B||[1\bar{1}2]$.

in transmission. From the comparison with the P.C. spectrum it is obvious that the high-field absorption line corresponds to the central P.C. line and consequently to the free carrier transition $0^+ \to 0^-$.
The spectra of Figures 5 - 7 were taken with a laser wavelength of 118.8 µm. Using the 163 µm line, less carriers are frozen out into impurity states due to the lower resonance magnetic field. This enhances the spin resonance signal, the three lines, however, cannot be separately resolved. Since the same behavior of band electron and shallow impurity transitions with stress can be expected, the 163 µm transmission data can be used to determine the stress dependence of the

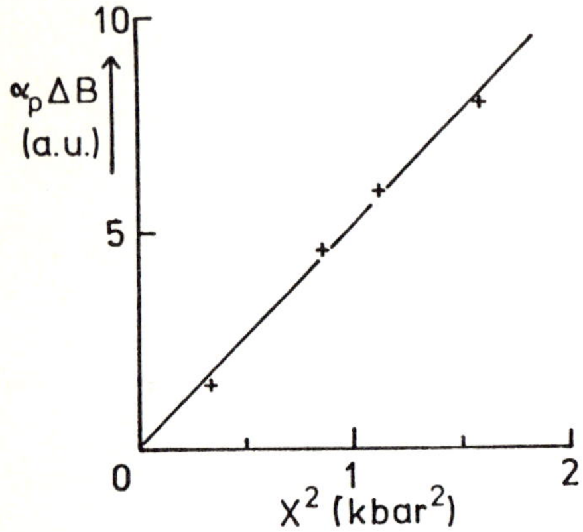

Fig.8: Dependence of the integrated intensity $\alpha_p \Delta B$ on the square of the stress X. α_p... peak absorption coefficient, ΔB ...linewidth. λ= 163 µm, sample 6, X||[110], B||[1$\bar{1}$2].

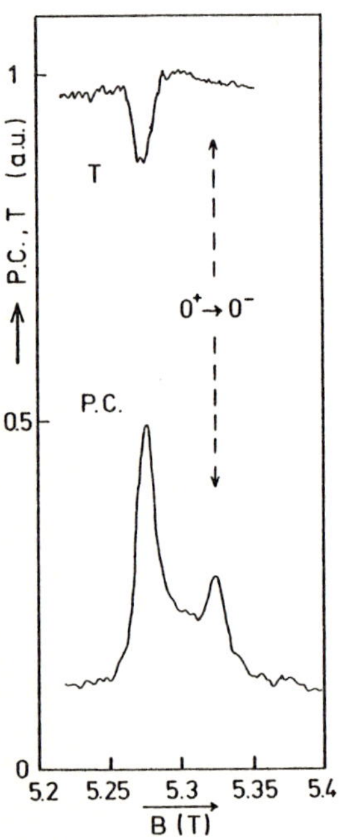

Fig.9: Transmission and P.C. spectra of sample 6 at λ= 96.5 µm. X = 0.93 kbar, X||[110], B||[1$\bar{1}$2].

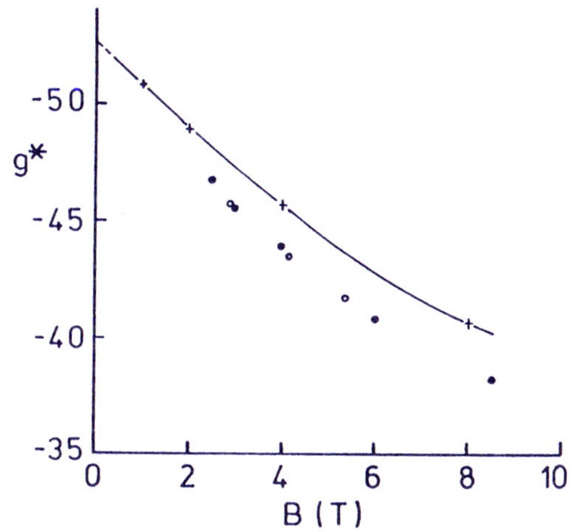

Fig.10: Magnetic field dependence of g^*. Experimental: • (FTS), samples 2,3), o (laser, samples 4,5). Theoretical: +.

SR absorption coefficient. Results of the integrated absorption vs the square of the stress are shown in Figure 8.

The influence of the magnetic freeze-out is again demonstrated in Fig.9. With the 96.5 μm laser line the resonance field is at about 5.3 T. The stronger one of the impurity related lines is clearly observable in transmission whereas the $0^+ \rightarrow 0^-$ transition is too weak due to the enhanced freeze-out. The fact that it is observable in photoconductivity demonstrates again the sensitivity of this technique and shows that the photoconductivity mechanisms are presumable of different sensitivity for the bound and free carrier transitions.

IV. Theory

The experimental data are analyzed within the framework of effective mass theory taking into account the Γ_6 conduction band and the spin orbit split Γ_8 and Γ_7 valence bands. The effects of higher bands on these three bands are included by perturbation theory. The Hamiltonian acting on the envelope functions is then entirely determined by symmetry and is given by the method of invariants [12] and is explicitly written in Table IV of Trebin et al [13]. It has the block form

$$\begin{bmatrix} H_{cc} & H_{cv} & H_{cs} \\ H_{vc} & H_{vv} & H_{vs} \\ H_{sc} & H_{sv} & H_{ss} \end{bmatrix} \tag{1}$$

including terms up to second order in the kinetic momentum K and first order in the strain tensor ε. The magnetic field H causes the commutation relation

$$K \times K = \frac{eH}{i\hbar c} \equiv \frac{1}{il^2} \tag{2}$$

with the cyclotron radius l.
For analysing conduction band features it suffices in a first approximation to retain only H_{cc}, H_{cv} and H_{cs}. They are, after introducing dimensionless quantities by dividing through the free electron cyclotron energy $\hbar\omega_c$ and replacing K by k=Kl

$$h_{cc} = \frac{a'k^2}{2} + \frac{g_s}{2}\sigma \cdot h + c_1 \operatorname{tr} \varepsilon$$

$$h_{cv} = \sqrt{3}\,[\, k \cdot T + b\, i(T_x\{k_y k_z\} + cp) + i\, c_2(T_x \varepsilon_{yz} + cp)]$$

$$h_{cs} = -\frac{1}{\sqrt{3}}[\, pk \cdot \rho + b\, i\,(\rho_x\{k_y k_z\} + cp) + ic_2(\rho_x \varepsilon_{yz} + cp)] \tag{3}$$

$$\{k_x k_y\} = (k_x k_y + k_y k_x)/2$$

h is the unit vector in the direction of magnetic field. T and ρ are the cross space matrices and obey

$$T_i T_i^+ = 2/9 \qquad T_i T_j^+ = -\frac{i}{9}\varepsilon_{ijk}\sigma_k. \tag{4}$$

σ are the Pauli spin matrices.
They transform like the radius vector under proper rotations. The constants are all real and are given by

$$p=P/(\hbar\omega_c\, l), \quad b= B/(\hbar\omega_c\, l^2), \; c_2=C_2/\hbar\omega_c \tag{5}$$

$$\text{with } P = \langle S|\frac{\hbar P_z}{m}|\, Z\rangle \tag{6}$$

$$iB = 2\,(\frac{\hbar}{m})^2 \sum_j \frac{\langle S|P_x|j\rangle\langle j|P_y|Z\rangle}{(E_c+E_v)/2-E_j} \tag{7}$$

$$iC_2 = \langle S|D_{xy} + D_{yx}|Z\rangle \tag{8}$$

where the sum in (7) extends over all bands with Γ_5 symmetry save the valence band and D_{xy} is the strain operator [13] . The eight dimensional problem $(h-e)\psi=0$ may now be reduced to a 2 x 2 Hamiltonian for the conduction band envelope functions

$$\begin{aligned} h_{eff} &= \frac{a'k^2}{2} + \frac{g_s}{2}\sigma \cdot h + c_1 \operatorname{tr} \varepsilon \\ &+ \frac{p^2}{3}\left(\frac{2}{e+e_g} + \frac{1}{e+e_g+\delta}\right)k^2 - \frac{p^2}{3}\left(\frac{1}{e+e_g} - \frac{1}{e+e_g+\delta}\right)\sigma \cdot h \\ &+ \frac{2}{3}pb\left(\frac{1}{e+e_g} - \frac{1}{e+e_g+\delta}\right)[\sigma_x\{k_x(k_z^2-k_y^2)\}+ cp] \\ &+ \frac{2}{3}pc_2\left(\frac{1}{e+e_g} - \frac{1}{e+e_g+\delta}\right)[\sigma_x(k_z\varepsilon_{zx}-k_y\varepsilon_{xy})+cp] \end{aligned} \tag{9}$$

The term proportional to b causes cubic terms in the energy dispersion relation, whereas the c_2 term causes k-linear terms in the energy of a uniaxially stressed crystal in the case of vanishing magnetic field. These terms will be treated in first order perturbation theory. Restricting for simplicity to H||[001] the unperturbed solutions are the Landau subbands with wave functions Φ_{N+}, Φ_{N-}. The perturbing terms cause a mixing between these states proportional to b and c_2. The wavefunctions become now (only spin flip terms included and terms proportional to k_z are neglected as $k_z<g^*m^*/2m$ for electrons below the 0 band).

$$\begin{aligned} |0+\rangle &= \Phi_{0+} - ic_2\frac{2}{3}p\left(\frac{1}{e_g} - \frac{1}{e_g+\delta}\right)\sqrt{2}\,\frac{1}{e_{0+}-e_{1-}}\varepsilon_{xy}\Phi_{1-} \\ |0-\rangle &= \Phi_{0-} + b\frac{2}{3}p\left(\frac{1}{e_g} - \frac{1}{e_g+\delta}\right)\frac{1}{\sqrt{2}}\frac{1}{e_{0-}-e_{1+}}\Phi_{1+} \end{aligned} \tag{10}$$

The electric dipole spin flip transition $\langle 0+|\vec{p}|0-\rangle$ will be enhanced by this mixing. After some calculation one obtains

$$\begin{aligned} \langle 0+|\frac{k_x+ik_y}{\sqrt{2}}|0-\rangle &= \frac{\sqrt{2}}{3}p^2k_z\left(\frac{1}{e_g+e_{0-}} - \frac{1}{e_g+e_{0+}} + \frac{1}{e_g+\delta+e_{0-}} - \frac{1}{e_g+\delta+e_{0+}}\right) \\ &+ i\frac{2\sqrt{2}}{9}p^3c_2\varepsilon_{xy}\left(\frac{1}{e_g} - \frac{1}{e_g+\delta}\right)\left(\frac{4}{e_g} + \frac{2}{e_g+\delta}\right)\frac{1}{e_{0+}-e_{1-}} \\ \langle 0+|\frac{k_x-ik_y}{\sqrt{2}}|0-\rangle &= \frac{\sqrt{2}}{9}p^3b\left(\frac{1}{e_g} - \frac{1}{e_g+\delta}\right)\left(\frac{4}{e_g} + \frac{2}{e_g+\delta}\right)\frac{1}{e_{0-}-e_{1+}} \\ \langle 0+|k_z|0-\rangle &= 0 \end{aligned} \tag{11}$$

In (11) the first term is the spin flip term of the unstressed crystal in the 3-band model caused by nonparabolicity. It is also present if the crystal possesses inversion symmetry. It is proportional to the wavevector in the direction of magnetic field. In the other terms only the dominant contributions were retained. Terms proportional to k_z were neglected. Both are proportional to the spin orbit splitting and would vanish in a crystal with inversion symmetry.
The linear term (c_2) yields an absorption for cyclotron inactive polarisation, the cubic term (b) for cyclotron active polarization. The matrix element for the three band contribution (k_z) and the cubic term

(b) are proportional to the magnetic field, whereas the stress term is independent of it.

The k-linear terms in the conduction band energy could cause a shift of the minimum of energy out of the Γ point if the two Landau subbands are sufficiently close. It was shown however [3] that the splitting of the Landausubbands is sufficient for all experimental conditions (B>0.5T, X<10 kbar) to keep the minimum at $k_z=0$.

A further contribution to the spin flip matrix element arises from the valence band Hamiltonians H_{kl}^{vv} (k-linear terms), $H_{\varepsilon k}^{vv}$ and $H_{\varepsilon k}^{vs}$ (mixed strainmagnetic interaction) (see Table IV of [13] for their explicit form). In the following an estimate of these contributions is presented for magnetic field of 1 Tesla. For a stress of 1 kbar $\varepsilon \sim 10^{-3}$, $P \sim 10^{-7}$ ev cm, $C \sim 10^{-10}$ eV cm, $C_4 \sim 10^{-7}$ eV cm, $C_2 \sim 1.6$ eV and observing that the valence band part of the eight component wavefunction is of order p/e_g one obtains in dimensionless quantities for the spin flip

$$
\begin{aligned}
&\text{k-linear terms of (C)} && \sim \frac{p^2}{e_g}\left(\frac{p}{e_g}\right)^2 \frac{c}{\Delta e} \sim 0.1 \\
&\text{mixed terms } (C_4, C_5) && \sim \frac{p^2}{e_g}\left(\frac{p}{e_g}\right)^2 \frac{C_4\varepsilon}{\Delta e} \sim 0.1 \\
&\text{Equ. (11)}(C_2) && \sim \frac{p^2}{e_g}\,\frac{pC_2}{e_g}\,\frac{\varepsilon}{\Delta e} \sim 9 \\
&\text{Equ. (11) (B)} && \sim \frac{p^2}{e_g}\,\frac{pb}{e_g \Delta e} \ll 9
\end{aligned}
\tag{12}
$$

The calculation of these terms in first order perturbation theory is straightforward but lengthy and will not be reproduced here.

V. Discussion

Two terms contribute to the spin flip matrix element. From the absence of an ESR signal in the zero stress experiments an upper limit for B can be deduced as $B < 10^{-16}$ eV cm^2. This is by a factor of 10 smaller than the value given by Seiler [5]. In that paper B and C_2 were extracted from beating effects in the Shubinikov de Haas effect of uniaxially stressed InSb. However, the data presented in [5] do not seem to be sensitive to B at all.

The results shown in Fig.4 show a pronounced dependence on the direction of the applied stress. This feature can be understood from Equ.11. For X ∥ [100] there is no shear component for zincblende structure. As already stated in [3,4] the ESR, Equ.11, becomes for X ∥ [110] and B ⊥ X and θ the angle between B and [001]

$$
\left\langle 0+\left|\frac{p}{m}\right|0-\right\rangle = -\frac{1}{\hbar\omega_{sc}}\,\frac{id}{l}\,\frac{\hbar}{lm^*}\,e_{CRI} \tag{13}
$$

with

$$
d = \frac{2}{3}\,PC_2\left(\frac{1}{E_g+E} - \frac{1}{E_g+\Delta+E}\right)\frac{\sqrt{2}}{4}\cos\theta\; s_{44}\,X, \tag{14}
$$

$$
\omega_{sc} = \omega_s + \omega_c .
$$

The absorption constant is related to the matrix element as

$$\alpha = \frac{8\pi e^2 N}{n\omega c\ \Gamma} \quad |\langle 0+|\frac{p}{m}|0-\rangle|^2 \qquad (15)$$

with Γ the experimental line width. From this a value $|C_2|$=1.6±0.5 eV is deduced. This value of C_2 is comparable to the one obtained by Seiler [5] (C_2=2 eV) from the beating effect in SdH. However, as the evaluation of his experimental results was done with a quasiclassical model, the result for C_2 may be erranous by an order of magnitude compared to an exact quantum transport treatment [14]. C_2 affects also intravalence band magneto optics. However, it occurs in the valence band Hamiltonian only in the combination [13]

$$C_5' = C_5 - \frac{2}{3} C_2\ P/E_g \qquad (16)$$

so that by adjusting C_5 the intravalence band data [7] remain unaltered. Finally, it should be mentioned that the oscillator strength depends quadratically on applied stress as it was also observed in the experiments (Fig.8).

An exact numeric diagonilisation of the eight band Hamiltonian including effects of higher bands has been performed yielding the g^* values given in Fig.10. The parameter values used were the same as given in Ref.[4]. The calculated stress dependence of g^* turns out to be negligible as corresponds to the experimental finding. No numerical fitting for the parameter values was attempted to decrease the discrepancy in the magnetic field dependence of g^* as this would affect also the intravalence and interband spectroscopic data.

Acknowledgement

This work was supported bei "Fonds zur Förderung der wissenschaftlichen Forschung in Österreich," project no. S-22 and by Ludwig Boltzmann Gesellschaft. We thank Profs. K.Seeger and K.Baumann for their interest in this work. The FTS experiments were performed together with R.Meisels, the laser experiments together with Dr.E.Fantner at the Institute of Physics, University of Leoben, Austria.

References

[1] V.I.Sheka, Fiz.Tverd.Tela 6, 3099 (1964)[Sov.Phys.Solid State 6, 2470 (1965)].
[2] B.D.McCombe, Int.Conf.Appl.High.Magn.Fields, Würzburg 1974; B.D.McCombe, R.J.Wagner, Proc.11th Int.Conf.Phys.Semicond.,Warsaw 1972, p.321.
[3] F.Kuchar, R.Meisels, M.Kriechbaum, Proc.4th Int.Conf.Phys.Narrow Gap Semicond., Linz 1981 (Springer 1982), p.197.
[4] M.Kriechbaum, R.Meisels, F.Kuchar, E.Fantner, Proc.16th Int.Conf. Phys.Semicond.,Montpellier 1982 (in print).
[5] D.G.Seiler, B.D.Bajaj, A.E.Stephens, Phys.Rev. B16, 2822 (1977).
[6] D.G.Seiler, D.L.Alsup, R.Muthukrishnan, Solid State Commun.10, 865 (1972).
[7] R.Ranvaud, H.R.Trebin, U.Rössler, F.H.Pollak, Phys.Rev.B20, 701 (1979).
[8] F.Kuchar, K.Veigl, E.J.Fantner, Rev.Sci.Instrum.50, 345 (1979).
[9] F.Kuchar, E.Fantner, G.Bauer, phys.stat.sol.(a)24, 513 (1974).
[10] W.Zawadzki, Lecture Notes in Physics No.133, Springer 1980, p.85.
[11] R.J.Wagner, B.D.McCombe, F.Kuchar (unpublished).
[12] G.L.Bir, G.E.Pikus, Symmetry and Strain Induced Effects in Semiconductors (Wiley, 1974).
[13] H.R.Trebin, U.Rössler, R.Ranvaud, Phys.Rev. B20, 686 (1979).
[14] D.G.Seiler (private communication).

FAR-INFRARED CYCLOTRON RESONANCE OF HOT CARRIERS IN InSb

Eizo Otsuka

Department of Physics, College of General Education
Osaka University, Toyonaka, Osaka 560, Japan

Far-infrared laser cyclotron resonance has been applied to InSb for the study of the hot carrier system. The hot carriers are produced at 4.2 K or below either by pulsed electric field or by band-gap photoexcitation. Joint application of electric field and photoexcitation is also put on trial. Electrical excitation yields normal orbital electron temperature analysis, whereas optical excitation evokes the spin temperature concept of the conduction electron and enables us to find its spin relaxation time.

I. Introduction

Application of the far-infrared laser cyclotron resonance to n-type InSb saw a considerable success in the study of the hot electron system at liquid helium temperatures. Initially, use was made of the pulsed-electric field modulation method, that helped us to find out origins of some unidentified signals and also to count the number of electrons transfered from a state to another.[1] Later, based on the information obtained by this modulation technique, direct absorption measurements were amply carried out for widening our knowledge of the hot electron system.[2] Recently, it has been found that hot carriers can be produced also by band-gap illumination.[3,4] In contrast to the electrical excitation case, that employed only n-type materials, the optical excitation yields hot electrons even in p-type materials. While the electrical excitation sees various properties of the orbital electron temperature, the optical excitation urges us to think of the spin temperature analysis of the conduction electrons. In this treatise, we shall put emphasis on the newly developing feature of the photoexcited cyclotron resonance.

II. Experimental

A. Samples

Since the experiment is essentially cyclotron resonance, only very pure samples were employed, all of which are nearly intrinsic. They were taken from four pieces of ingots, two n-types and two p-types. The average impurity concentrations of these ingots are given in Table 1. Each sample was so shaped as to have the typical dimension of 4 x 4 x (0.15 - 0.50) mm^3. The flat face is always (111).

Table 1. List of InSb samples

Ingot No.	N_D(cm^{-3})	N_A(cm^{-3})
N1	1.9 x 10^{14}	1.8 x 10^{14}
N2	3.8 x 10^{14}	1.7 x 10^{14}
P1	1.0 x 10^{14}	1.1 x 10^{14}
P2	1.3 x 10^{14}	2.6 x 10^{14}

B. Far-Infrared Source and Detector

The discharge-type far-infrared (FIR) molecular laser containing H_2O or D_2O vapour could yield various wavelengths; 84, 119, 172, 220 μm etc.. Main use was made of 84 μm, with which the resolution was optimum. The laser oscillated in pulses of ∿40 μs at 20 Hz. Transmitted laser beam was detected by an n-type InSb Putley unit. Detection of the FIR cyclotron emission, on the other hand, was done mainly by Sb-doped Ge detector.

C. Method of Photoexcitation

Electrical or optical excitation was made at 10 Hz, in every other turn synchronisation with FIR pulses. Electric pulses had the duration of 5 μs, with various heights to adjust the electron temperature. Optical excitation was mostly by xenon flash lamp, having the width of 1 μs. The light was led to the sample in the He dewar through a special self-focusing dual core glass light-guide. The maximum photoexcitation power just before entering the sample was 3 mW/cm^2. For observing cyclotron emission in optical excitation case, a tungsten lamp, a YAG laser and further an argon-ion laser were also employed.

D. Magnetic Field and Geometry

The magnetic field was always applied along the <111> direction of the sample, vertical to the flat sample surface. In the case of electrical excitation the FIR beam was mostly in the Voigt configuration with respect to the magnetic field, while in the case of optical excitation it was always in the Faraday configuration.

III. Electrical Excitation —— Summary of the Past

In n-type InSb embedded in liquid helium, say at 4.2 K, both conduction electrons and donor electrons are present. They yield, in the absence of an electric field, the lowest energy cyclotron transition in the conduction band, or specifically $0^+ \rightarrow 1^+$, and the so-called impurity cyclotron resonance (ICR).[1,5] Here the numerals stand for the Landau quantum numbers and the + sign indicates the spin. On application of the electric field, electrons residing at the donor impurities are ejected into the conduction band, thus reducing the intensity of ICR. Conduction electrons are accelerated by the electric field, resulting in the cyclotron transitions between higher Landau subbands, $0^- \rightarrow 1^-$, $1^+ \rightarrow 2^+$, $1^- \rightarrow 2^-$ etc., which show up at higher magnetic fields on account of the slightly nonparabolic nature of the conduction band. The relative intensities of these quantum cyclotron transition lines yield the electron temperature, to be more exact, the so-called intersubband electron temperature. It is defined in effect by[1,2]

$$C_2/C_1 = \exp[-\{E_-(0) - E_+(0)\}/k_B T_E^{inter}] \quad . \tag{1}$$

Here C_1 and C_2 are the intensities of the transitions $0^+ \rightarrow 1^+$ and $0^- \rightarrow 1^-$, respectively. $E_+(0)$ and $E_-(0)$ are the energies of the 0^+ and 0^- subbands at $k_z = 0$, where k_z is the electron wave number along the magnetic field. If the electron distribution is Maxwellian, the same T_E^{inter} should be derived also from C_3/C_1, C_4/C_1 and so on, where C_3 and C_4 are to denote the intensities of the $1^+ \rightarrow 2^+$ and $1^- \rightarrow 2^-$ transitions, respectively. As a matter of fact, however, this is not the case. Deviation from the Maxwellian distribution becomes more and more noticeable as the applied electric field is intensified. An interesting feature is the directional effect; difference in electron temperature is observed between two geometries $E \perp B$ and $E \parallel B$, details of which are already described elsewhere.[2] The electron distribution at $k_z = 0$ is derived from the line-shape, which, in turn, gives the intrasubband electron temperature T_E^{intra}. Determination of T_E^{intra} is a kind of self-consistent parameter fitting.[2] Disagreement between T_E^{inter} and T_E^{intra} is observed at each applied electric field. Geometrical difference is also observed for T_E^{intra}.

On top of these two electron temperatures, one can further define an impurity-associated electron temperature, that is determined by the electron population under an electric field at the donor states, in comparison with the thermal equilibrium donor electron population under no electric field. Interesting enough, the geometrical effect is again observed.[6]

Electrical excitation further gives rise to cyclotron emission.[7,8] That is a reverse process of cyclotron absorption and considered mainly to arise from the transition $1^+ \rightarrow 0^+$. On changing the magnetic field, an emission peak appears when the cyclotron energy meets the maximum photoresponse of the infrared detector. If one changes the detector, the emission peak shifts in the magnetic field. On closer examination, it is found that the emission peak consists of more than one cyclotron transition. Even harmonics, possibly reflecting the transitions, say $2^+ \rightarrow 0^+$ etc., appear at the lower magnetic fields. It is not to speak of the geometrical effect that occurs also for the cyclotron emission. Of further interest has been the polarisation characteristics of the emitted light. In the "Faraday" configuration, in which the average direction of the emitted light is along the magnetic field, the emitted light is found to be circularly polarised, while in the "Voigt" configuration, it is linearly polarised.[9] Since the cyclotron emission is a tunable far-infrared source, device application has been tried for quite some time. No report of brilliant success, however, has arrived at the author's ear so far.

IV. Optical Excitation

Optical excitation by band-gap light has also been found effective in making the carrier system hot in semiconductors. A pioneer work was carried out by Ulbrich (1973) for GaAs.[10] Similar feature of optical hot carriers is observed in InSb.[3,4] In cyclotron resonance, the behaviour is more distinct and interesting. Typical cyclotron resonance traces in time resolution from an optically excited n-type InSb (sample from N2) is shown in Fig. 1 for the FIR laser wavelength of 84 μm.

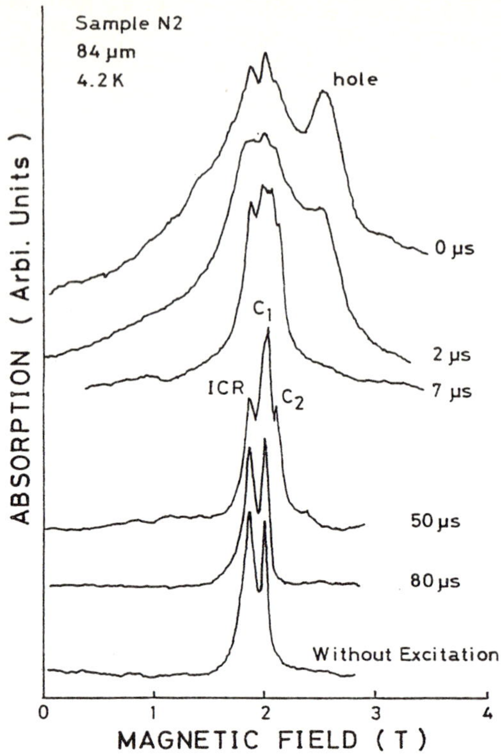

Fig. 1 Time resolution series of the photoexcited cyclotron resonance of an n-type InSb.

The photoexcitation is made by xenon flash-lamp, having the pulse width of ∿1 μs. Time variation of the signals ICR, C_1 and C_2 is pursued for electrons, in both n- and p-type materials. Here we have written C_1 and C_2 for the transitions $0^+ \rightarrow 1^+$ and $0^- \rightarrow 1^-$, respectively. Similar procedure is taken also for holes. A general description of the photocarriers is as follows: After excitation by xenon flash-lamp, the produced photocarriers start to recombine on one hand and also start to neutralise the ionised impurities on the other. The intensities of the cyclotron resonance signals decline with time. The signals of holes, in particular, become unobservable after the lapse of a few microseconds both in n- and in p-type materials. The electron signals in n-type material persist and arrive at their equilibrium state. Even in p-type material, electron signals survive much longer than hole signals. The reason why hole signals die down so fast is considered due to the deepness of the acceptor level. An acceptor, once capturing a hole, would not emit it back into the valence band at low temperatures. On the other hand, a donor has an extremely shallow level for an electron. The electron can easily be reemitted into the conductron band. Electron capture and reemission processes by shallow donors are seemingly responsible for the long life of the electron signals. The lifetime of electrons as a whole, both for the conduction electrons and for the donor electrons, changes from sample to sample and is not necessarily definable in terms of an exponential decay process. In an n-type material, as long as a few millisecond time has been observed for the eventual decay of the photoexcited electron signal.[4]

The time constant is more clearly defined in the case of p-type material. The decay characteristic of electrons can be more or less approximated by an exponential function. The electron signals in p-type material are thoroughly photoexcited signals, while those in n-type material are mixtures of photoexcited and thermal equilibrium ones. Thus analysis of photoelectrons becomes much easier in p-type material. For this reason, emphasis will be given to the electronic behaviour in p-type material.

V. Optically Excited Electron System in p-InSb

In p-type InSb, the electron signals, whenever observed at liquid helium temperatures, are of non-thermal equilibrium. A typical time-resolved cyclotron resonance series obtained also at 84 µm is given in Fig. 2 for a sample from P1. After a long lapse of time, all the signals die down to the zero level. The eventual flat ground line, indicated by "without excitation", thus corresponds to thermal equilibrium. Since the intensity of photoexcitation is adjusted to be moderate in this example, one cannot see hole signals even right after the photopulse. In the course of decay, electron signals C_1, C_2 and ICR are all observed. It should again be emphasised that ICR is observable in p-type material. Existence of ICR is telling us the least rise of lattice temperature.

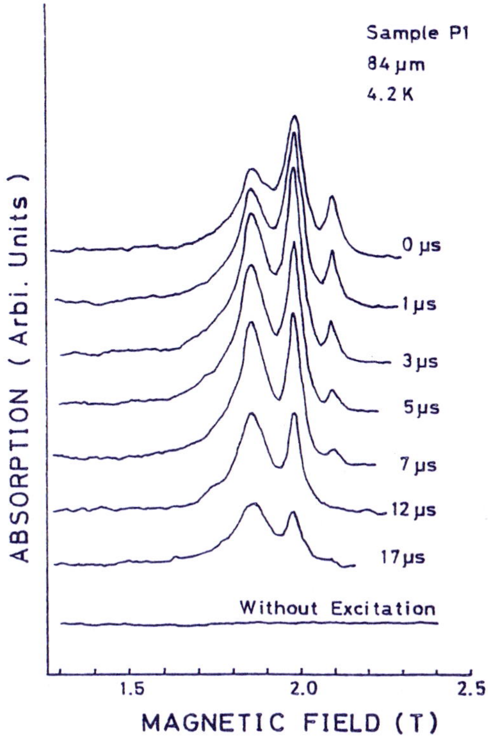

Fig. 2 Time resolution series of the photoexcited cyclotron resonance of a p-type InSb

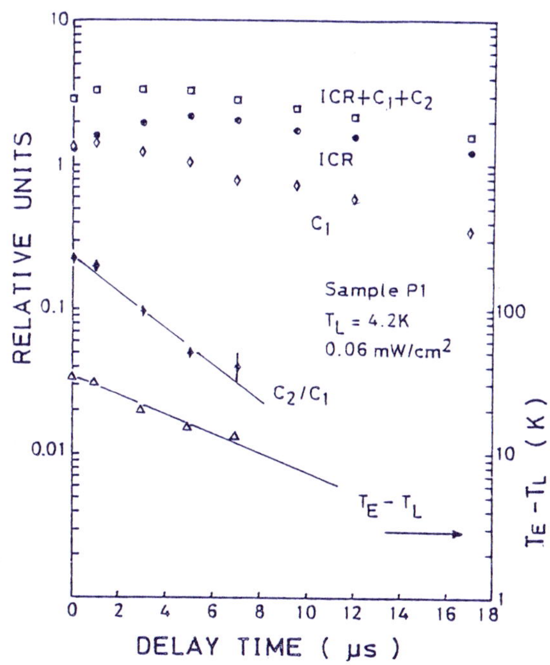

Fig. 3 Time variation of several quantities from a photoexcited p-type InSb

In Fig. 3, we plot the time variation of C_1, ICR etc. for our sample from P1. The intensity ratio C_2/C_1 and the intersubband electron temperature derived from the relation (1) are also given, the latter being in the form of difference from the lattice temperature, which is kept 4.2 K. Fitting the data with Newton's cooling law

$$-d(T_E - T_L)/dt = (T_E - T_L)/\tau_c, \quad (2)$$

where T_L is the lattice temperature and τ_c is the cooling time constant, one finds $\tau_c \sim 6.5$ µs. On lowering T_L to 1.7 K, we see a very slight increase of τ_c, almost within the limits of error. The time constant for C_2/C_1 is ~ 3.0 µs at $T_L = 4.2$ K. It is seen that the duration of C_1 and ICR signals are longer than 10 µs.

General behaviour of other p-type InSb samples is more or less the same. One thing to be noted is that the decay rate of the intensity ratio C_2/C_1 as well as the temperature difference $T_E - T_L$ has some dependence on the strength of photoexcitation. The qualitative feature is shown in Fig. 4 for a sample from P2, and details will be discussed in VIII.

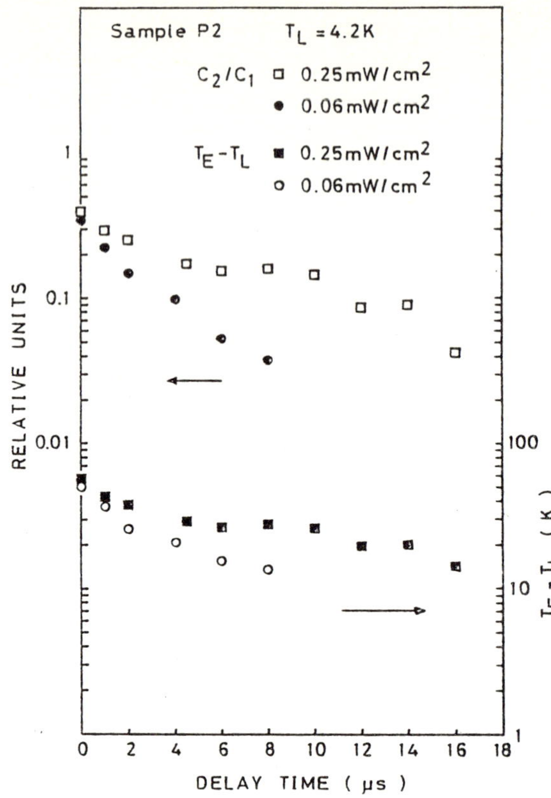

Fig. 4 Excitation dependence of the time variations for C_1/C_2 and $T_E - T_L$.

VI. Energy Dissipation of Hot Carriers

If the hot carrier system loses its energy by acoustical phonon emissions, it has been suggested by Shockley[11] that the energy loss rate of electron is given by $(ms^2/\tau)(T_E/T_L)$, where s is the sound velocity and τ is the collision time of electron. Straight application of this mechanism to Eq.(2) leads to the definition of the cooling time constant

$$\tau_c = \tau(k_B T_L/ms^2) \quad . \qquad (3)$$

Since τ is the electron-phonon scattering time, it depends both on T_E and T_L; *i.e.*, on the electron velocity and on the phonon density. In other words, τ_c cannot be tractable as a plain constant during the entire course of the cooling process. As a matter of fact, however, a large deviation from an exponential does not show up in the cooling process of $T_E - T_L$.

The above mechanism of energy dissipation, however, loses its ground as one substitutes numerical values into (3). Putting T_E = 40 K and T_L = 4.2 K, and taking only the deformation potential scattering, one finds $\tau_c = 1.3 \times 10^{-7}$ s. This is considerably shorter than the observed time constant, indeed by a facter of 50. A more exact treatment, inclusion of the piezo-electric scattering for example, would give a further small value of τ_c. It thus seems inadequate, accordingly, to give credit to the Shockley mechanism. A different idea, making the carrier relaxation slow through screening the electron phonon scattering by frequent shallow impurity scatterings, was raised by the author.[12] It is an impurity version of the screening by electron-hole plasma. Before obtaining a substantial support for this idea, however, we have come across another more plausible interpretation of the slow relaxation phenomenon, which is associated with the spin flipping within the conduction electron system.

VII. Spin Temperature Model

Though we have defined the intersubband electron temperature by the electron populations at 0^+ and 0^- subbands, we have implicitly been assuming that electrons are also residing at higher Landau subbands 1^+, 1^- and so on. In the electrical excitation case, this assumption has been consistent with our observation of C_3, C_4, ... lines. In the case of optical excitation, however, we have been unable so far to detect the C_3 line, not to speak of C_4, C_5, ... An essential difference, which we have to emphasise, between the two cases is that in the former the resonance is observed during the application of electrical excitation, pulsed as it is, while in the latter the resonance is observed only after the photopulse. The electron distribution showing up in the resonance lines should then be different. For the carrier-carrier relaxation process will still be going on in the former while it is very likely completed in the latter. As one comes to think over this, he may start understanding the complete absence of the C_3, C_4, $\cdots$ signals in the optical case. It would not be merely the matter of signal-to-noise ratio, but possibly is a fact.

It would be suggested that T_E^{inter} as defined by Eq. (1) could also be regarded as the spin temperature within the zeroth Landau subband. In other words, one may rewrite Eq. (1)

$$C_1/C_2 = \exp[-\{E_-(0) - E_+(0)\}/k_B T_S] \ , \tag{4}$$

where T_S is to denote the spin temperature. The durable observation of the C_2 line may be telling the long spin relaxation time.

Based on this picture, the relaxation story of the optical hot electron system can be rewritten as follows. First, the electron system assumes a very high temperature, its distribution extending to the subbands 1^+, 1^-, 2^+, 2^-, $\cdots$. Owing to the energy dissipation to the lattice, due either to optical or to acoustical phonon emissions, the system rapidly comes down to the lowest states, 0^+ and 0^-. Here is waiting the bottleneck for the energy relaxation. If the spin-flipping from - to + states takes a long time, the durable observation of C_2 is understandable. What then causes the spin to flip? One may recall the spin relaxation mechanism of the conduction electron suggested by Elliott.[13] The mechanism owes to a strong spin-orbit as well as interband coupling for causing the spin of an electron to change by interacting with a purely electric potential. Indeed Boguslawski and Zawadzki carried out actual calculation for InSb,[14] by using a screened Coulomb potential for an ionised impurity centre. A slightly extended calculation for spin-relaxation caused by screened Coulomb potentials, not by immobile ionised impurity centres but by moving conduction electrons, was also made by Bogusławski.[15] Both of these calculations will be applicable to the present case. One has to be careful, however, not to forget another kind of scattering centres; namely, neutralised impurities. In our experiment, photoexcited carriers are to neutralise a considerable portion, if not the whole, of initially ionised impurity centres and these are to join in scattering electrons. Thus we may count on three kinds of scatterers to cause the spin of an electron to flip: i) ionised impurities, ii) free carriers and iii) neutralised impurities. The free carriers are almost entirely electrons, since holes are no longer free when we are watching the decay process of the ratio C_2/C_1.

The scattering potential corresponding to i) as well as to ii) is of the screened Coulomb type

$$V(r) = \pm(e^2/\kappa r)\exp(-r/r_c) \ ; \tag{5}$$

where r_c is the screening radius and κ is the static dielectric constant. The scattering potential corresponding to iii), on the other hand, may as well be approximated by the Hartree potential for the hydrogen atom; namely,

$$V(r) = \pm(e^2/\kappa)(1/r + 1/a_B^*)\exp(-r/a_B^*) \ ; \tag{6}$$

where a_B^* is the effective Bohr radius for the impurity atom. The double signs in Eqs. (5) and (6) correspond to acceptor and donor centres in the case of impurity scattering, either ionised or neutral. In the case of carrier-carrier scattering, one can forget the negative sign, since only electrons are scattering an electron.

The spin relaxation time T_1 is obtainable by calculating the transition probability from 0^- to 0^+.[14] It can be written, using the Fourier component $v(k)$ of the scattering potential :

$$1/T_1 = (2\pi/\hbar)(NC^2/V)\sum_k \ v(k)\ ^2|J_{01}|^2\ \delta(E_+ - E_-) \ ; \tag{7}$$

where

$$c = \hbar k_z(\hbar\omega_c)^{1/2}(2m_e^* D_+(k_z)D_-(k_z))^{-1/2}\ [b_+c_-/\sqrt{2} - b_-(b_+ - \sqrt{2}c_+)/2] \tag{8}$$

with

$$D\pm(k_z) = \hbar\omega_c/2 + E_z \pm g_e{}^*\mu_B H/2, \tag{9}$$

$$b_\pm{}^2 = \beta^2 E_\pm(k_z)/3(E_g + 2E_\pm(k_z)), \tag{10}$$

$$c_\pm{}^2 = 2\gamma^2 E_\pm(k_z)/3(E_g + 2E_\pm(k_z)), \tag{11}$$

$$\beta^2 = \Delta^2/(\Delta + E_g)(\Delta + 3E_g/2), \tag{12}$$

$$\gamma^2 = (\Delta + 3E_g/2)/(\Delta + E_g), \tag{13}$$

$$E_z = \hbar^2 k_z{}^2/2m_e{}^* , \tag{14}$$

$$|J_{01}|^2 = \zeta e^{-\zeta}, \quad \zeta = \ell_H{}^2(k_x{}^2 + k_y{}^2)/2, \quad \ell_H = (c\hbar/eH)^{1/2} ; \tag{15}$$

In the above, N is the density of scattering centres of the given type, V is the volume of the crystal, Δ is the spin-orbit splitting, E_g the energy gap, H the applied magnetic field in oersted, $g_e{}^*$ the electronic g-facter, $m_e{}^*$ the electronic effective mass, ω_c the cyclotron frequency for electron, and k the wave number of electron with k_z being its component along the megnetic field. The energies at the subbands in question, $E_+(k_z)$ and $E_-(k_z)$, can further be written

$$E_\pm(k_z) = - E_g/2 + [(E_g/2)^2 + E_g D_\pm(k_z)]^{1/2} . \tag{16}$$

The Fourier component of the scattering potential (5) is given by

$$v(k) = \pm 4\pi e^2/\kappa(k^2 + k_s{}^2) ; \tag{17}$$

where $k_s = 1/r_c$. Similarly the Fourier component of the potential (6) may be written

$$v(k) = \pm 8\pi e^2/\kappa a_B{}^{*2}(k^2 + a_B{}^{*-2})^2 . \tag{18}$$

The relative importance of the scattering potentials changes in accordance with the degree of excitation. It should further change in the passage of time. If the excitation is strong enough to neutralise all the impurities, one may think that the potential form (6) dominates at first and then, after a reasonable lapse of time, will be replaced by the form (5). The form (5), of course, cannot entirely be neglected even shortly after the photopulse, since we have a considerable number of conduction electrons. The contribution from electron-electron scatterings, however, could be put aside for the moment on account of two reasons. First, the scattering potential is repulsive and its cross-section is expected rather small in comparison with that by impurities, which no doubt contain attractive forces. Second, the density of conduction electrons at the decay measurement stage seems not to depend very much on the excitation level, that definitely changes the relative importance of ionised and neutral impurities. In other words, the mechanism ii) merely adds an almost fixed contribution on the competing ground of the mechanisms i) and iii). We shall, accordingly, be concerned primarily with the role of impurities. In order to get a rough idea of what is happening, we have carried out some qualitative experiment to supplement the already mentioned time-resolution measurement. Results will be described in the next section.

VIII. Dependence on Excitation Intensity

One can expect, at least qualitatively, a change in relative concentration of ionised and neutralised impurities by varying the intensity of photoexcitation. This change in turn will be reflected on the value of the observed T_1. A fresh p-type sample from P2, the impurity concentrations of which are $N_D = 1.3 \times 10^{14}$ cm^{-3} and $N_A = 2.6 \times 10^{14}$ cm^{-3}, has been employed. In Fig. 4, we plot the time variation of C_1/C_2 as well as of $T_S - T_L$ for two typical intensities of excitation, weak (0.06 mW/cm^2) and strong (0.25 mW/cm^2). The lattice temperature is held at 4.2 K. One can see that for this sample the time constant for C_1/C_2 as well as for $T_S - T_L$ is not so well defined. That is not too surprising if one thinks of the time-dependent mechanism of electron scattering during the course of relaxation. All we can say from

Fig. 6 is that the cooling rate of the spin temperature is faster when the excitation is weaker. This information, however, helps us a great deal in building up a conjecture. The sole reasoning will be that the Coulomb scattering by ionised impurities, even if screened, is more effective than the neutral impurity scattering. The time rate of change both in C_1/C_2 and in $T_S - T_L$ is larger for weaker excitation, when more ionised impurities are expected to exist. In order to distinguish the two time constants, we shall write one for C_1/C_2 as τ_{21} and the other for $T_S - T_L$ as τ_c. The constant τ_{21} is more directly connected with T_1. Because of the somewhat nonexponential character of the decay process, it is indeed difficult to define those time constants very clearly. Nevertheless, we give rough values extending over the time interval given in Fig. 6 as follows; τ_{21} = 3.5, (6 ± 1) and (9 ± 2) μs for excitation intensities of 0.06, 0.17 and 0.25 mW/cm^2, respectively, while τ_c = 6, (13 ± 3) and (14 ± 3) μs for the same excitation intensities. It seems that the cooling rate of the spin temperature becomes slower as one increases the intensity of excitation and tends to saturate to a value which is characteristic to the neutral impurity scattering.

We tentatively compare our experimental value of τ_{21} with the theoretical T_1. We shall consult with the Boguslawski-Zawadzki model in which the ionised impurity scattering is playing the main role. Crudely we put $N = 10^{14}$ cm^{-3} and r_c = 500 Å, taking the Debye-Hückel screening model. We shall further assume that the electron temperature T_E, now rather intrasubband than inter, to be ∿10 K, somewhat higher than T_L (= 4.2 K) to emphasise a non-equilibrium state. Using these values of parameters, one finds $T_1 \underset{=}{\sim}$ 6 μs. This is apprently a quite reasonable value in comparison with our experimental observation of τ_{21}. Of course this comparison contains a problem, since we have to examine the effects of conduction electrons and neutralised impurities. On critical examination, we further face a strictly limited validity of the above calculation. The theory is essentially based on the Born Approximation, which does not distinguish between donors and acceptors, whether ionised or not. As a matter of fact, a big difference is expected between donor and acceptor scatterings.[16-19] Though not explicitly calculated, one has to be careful about the neutral impurity scattering, too. The potential form (6) should not be given too much credit, since the Hartree potential is just a distant view of an isolated atom. The actual view of a neutral impurity in InSb cannot be that of an isolated atom because of the large Bohr orbits of the donor impurities, which are next to filling up the crytal. An exact treatment of the Elliott process will thus become rather complicated. But the above estimate using the ionised impurity scattering mechanism is nevertheless considered to give a model study of the phenomenon, good enough to predict an order of magnitude of the spin relaxation time.

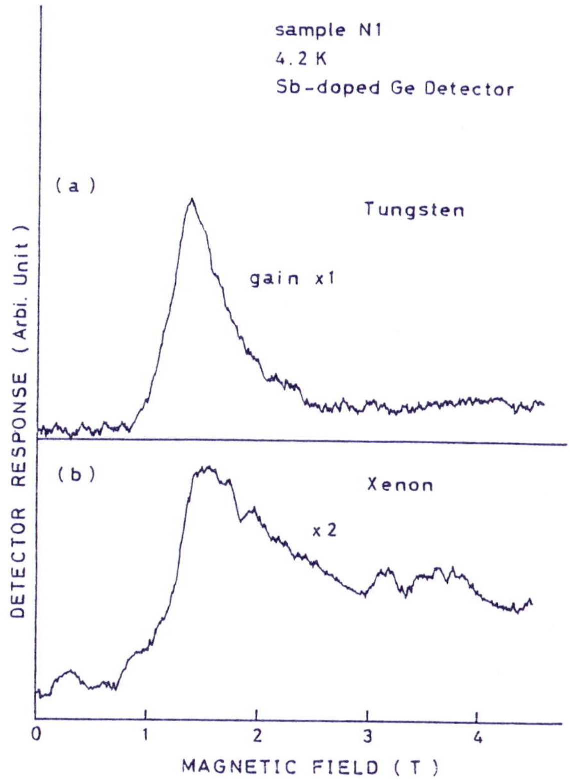

Fig. 5 Cyclotron emission signals from a photo-excited n-type InSb :(a) tungsten lamp-excited (b) xenon flash lamp-excited.

IX. Cyclotron Emission

Details of the cyclotron emission phenomena from electrically excited InSb are given elsewhere.[1,3,6,7,20] We shall describe here some additional results in the optical excitation case.[9,10] We tried four kinds of photoexcitation source; a xenon flash-lamp, a tungsten lamp, a YAG laser and an argon-ion laser. A typical emission signal, obtained from an n-type sample (N1) under photoexcitation by chopped light of a tungsten lamp (1 ms, 30 Hz), is shown in Fig. 5(a). Similar emission signals were obtained for optical excitation by YAG as well as argon-ion laser. Excitation by xenon flash lamp yielded considerably poorer signal-to-noise ratio, followed by a broad tail on the higher magnetic field side [Fig. 5(b)]. Since the main part of the cyclotron emission arises from the $1^+ \rightarrow 0^+$ transition, its observation is an evidence of the electron existence in the 1^+ subband. The long tail in the case of xenon flash-lamp excitation may even indicate the transition from higher subbands, say $1^- \rightarrow 0^-$, $2^+ \rightarrow 1^+$ etc.. Such a situation is in contrast with the statement made in the spin temperature argument, VII, where we were almost entirely engaged with the 0^+ and 0^- electrons. Two important things are to be pointed out. First, the cyclotron emission is not observed in time-resolution, but in a steady-state or next to steady-state optical excitation. Second, so far we fail to observe cyclotron emission from a p-type InSb. The first thing indicates the possibility that when we are measuring the electron temperature (or spin temperature) in the time resolved experiment, we have electrons only at o^+ and 0^- levels, though we had them at higher subbands during as well as right after the photopulse. As a matter of fact, we have never been able to observe C_3, C_4, ··· lines in the time-resolved optical excitation measurement. The second thing offers another support for the spin temperature model in the case of p-type material. In p-type material, photoexcited electrons may relax faster to 0^+ and 0^- states than in n-type material. A similar difference between n- and p-type materials has been noticed for $In_{0.53}Ga_{0.47}As$, as for the occurrence of the negative differential resistance.[21,22]

X. Photo- and Electric Field Excitation

Another indication of a striking feature in this study would be the joint application of optical and electrical excitations. The experimental procedure is as follows. We first apply an optical pulse to a p-type InSb. A considerable, if not the whole, portion of the donor impurities are neutralised. Then, before the donor electrons are transfered to acceptor levels, we apply an electrical pulse with duration of 5 μs. The delay-time is chosen also to be 5 μs. The rest of the procedure is like the electrical excitation experiment for an n-type InSb, except that we perform a crude time-resolution within the 5 μs electrical pulse. In a sense, we are conducting electrical excitation for an artificial "n-type" InSb. This type of experiment can be performed by virtue of the sufficiently long life of the optically neutralised donors.

The act of applying an electrical pulse causes a drastic effect on this artificially made " n-type" InSb. The intensity of the C_1 line, and hence the lifetime of the photoelectrons in the conduction band, is

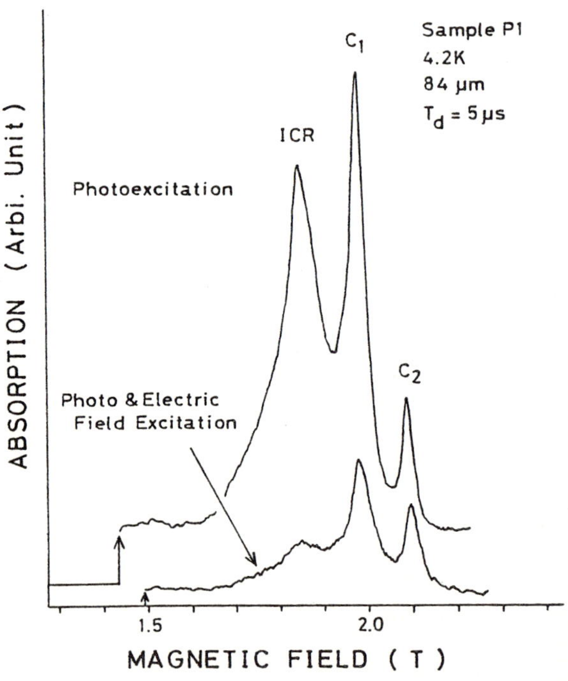

Fig. 6 Photoexcited cyclotron resonance from a p-type InSb, after 5 μs followed by electric field excitation.

decreased, not to speak of the ICR. This behaviour is examined by shifting the 1 μs gate of time-resolution within the 5 μs electrical pulse. Evidently the electrical pulse is helping the electron transfer to the acceptor sites bý ejecting donor electrons into the conduction band, where electrons migrate to the sites of acceptor impurities. In the absence of the electrical pulse, the donor electrons can be transported to the acceptor sites only after waiting for thermal ionisation. Difference in time constant is then quite understandable. One more point of interest would be the apparent line-broadening of the C_1 as well as the C_2 line on application of the electric field. The electric field no doubt is raising the intrasubband electron temperature.

Some features of the photo- and electric field excitation obtained from a P1 sample are given in Fig. 6. One may note tha overall sinking down of the electron signals, corresponding to the prompted transfer of electrons to the acceptor levels. Especially that of ICR is remarkable, being all the more prompted by impact ionisation. On top of the line-broadening mentioned above, one finds noticeable peak shift both for C_1 and for C_2 lines. These are characteristics observed for electrically excited n-type InSb.[2] Further details of the analysis will be presented in a later paper.

XI. Other Features

Under electrical excitation, the hot electron feature was observable not only in the relative intensities of the electron cyclotron transitions, but also in the line-width or line-shape of the resonance signal. That was indeed a basis for defining the so-called intrasubband electron temperature. An obvious characteristic was an asymmetry of the resonance. The line-shape in the case of optical excitation, however, is very close to Lorentzian in all cases, including that of the hole resonance. In other words, little asymmetry has been observable. The asymmetry in the case of electrical excitation was reflecting mainly the k_z-broadening. Absence of the asymmetry, accordingly, means very small k_z-broadening. In other words, the rise of the intrasubband electron temperature seems to be lower in the time-resolved optical excitation case than in the steady-state electrical excitation case. It would not, however, be so low as the lattice temperature. This is already obvious in the time-resolved measurement for the n-type InSb (Fig. 1). One finds there that the linewidth of the C_1 line is broader for the smaller delay-time. After a long lapse of time, one arrives at the thermal equilibrium signals of C_1 and ICR, the linewidth of the former being smaller than the nonequilibrium value at a small delay-time.

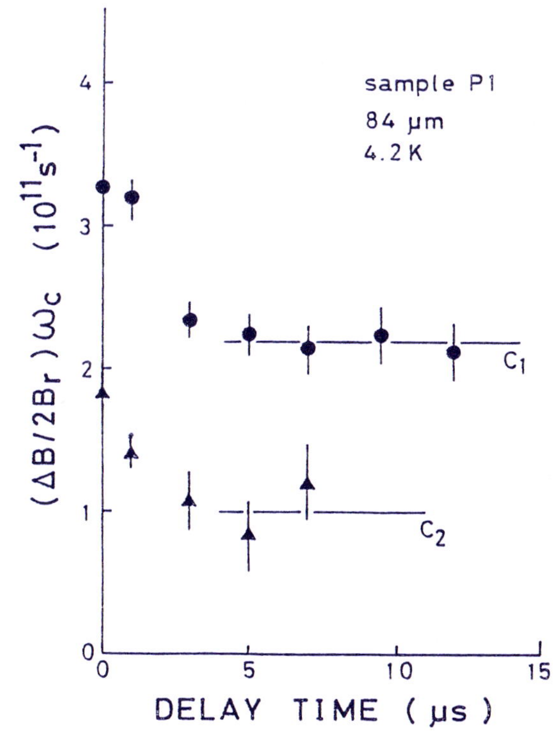

Fig. 7 Time variation of the linewidth for C_1 and C_2.

There remains still a considerable amount of mystery about the linewidth or line-shape problem of the nonequilibrium resonance. Analysis of the photoexcited carrier behaviour is easier for p-type material than for n-type, since all the signals observable are under excitation. Thus we give some more details of linewidth analysis for the P1 sample. Time variation of the linewidth is given in Fig. 7 for both C_1 and C_2 lines. The linewidth decreases at first but tends to a fixed value in each case after a short lapse of time. Assuming the simple relation

$$\omega_c \tau = 2B_r/\Delta B \ , \tag{19}$$

where B_r is the resonance field and ΔB is the half-width, one can obtain the apparent collision time τ. From the final linewidth we find

$$1/\tau = (2.2 \pm 0.1) \times 10^{11}\ s^{-1} \qquad \text{for } C_1 \tag{20}$$

and

$$1/\tau = (1.0 \pm 0.2) \times 10^{11}\ s^{-1} \qquad \text{for } C_2\ . \tag{21}$$

The reason why the C_1 line becomes broader than the C_2 line is not clear. An existing interpretation based on the spin polarised electron scattering by neutral donor does not account for the difference; nay, it predicts the C_2 line to be broader.[23]

Now some mention should be made of the impurity states. In the case of electrical excitation, the electron population at the donor level gave a kind of electron temperature. It is harder to deploy a similar argument in the case of optical excitation. All we can say is that donor electrons will be exhausted at a certain lattice temperature. The stable appearance of ICR in large amount offers an evidence that the lattice temperature does not undergo a substantial rise by photoexcitation. The steady decay of the ICR signal is due to the electron transfer from donor to acceptor sites.[24,25] The transfer rate depends on concentration, distribution and degree of compensation of the impurities, and thus changes sample to sample. In an n-type sample from N1, the eventual time constant of ICR tending to its thermal equilibrium value amounts to as long as a few milliseconds at liquid helium temperatures.

An interesting feature of ICR appears for both n- and p-materials, when one applies an excessive photoexcitation. One can see that partially in Fig. 1. For the delay-times 0 and 2 μs, when we have a high density of electrons and holes, the ICR signal is not resolved so well from C_1. It further has a long tail on the lower magnetic field side. This is a part of the optical hot electron effect involving the impurity states. It is not clear yet if the long tail is really connected with the proper ICR signal itself. For strong excitation, the two lines ICR and C_1 are apparently merged almost into one. In other words, the donor states could be regarded as a part of the conduction band. It would then be more appropriate to associate the long tail with the conduction electron system that is showing an excessive line broadening due to carrier-carrier interaction. For a convincing elucidation, however, one is required to make a further careful study.

XII. Conclusions

Cyclotron resonance of the hot electron system in InSb has been studied both for electrically excited and for optically excited cases. In the case of optical excitation, the electron temperature concept seems to be more likely replaced by the spin temperature concept, except for the case of cyclotron emission which occurs under steady-state excitation. The apparent spin relaxation time is measured for the conduction electron in p-type samples. Its comparison with the theoretical estimate by Boguslawski and Zawadzki shows a reasonable agreement, thus giving a strong credit to the spin temperature model. It may then be said that we have seen an unexpected use of cyclotron resonance in finding out the spin relaxation time of the conduction electrons in InSb.

Acknowledgements: This review owes to the work by T. Ohyama, K. Fujii, K. L. I. Kobayashi and O. Matsuda. The reviewer is further indebted to N. Nakata for fruitful discussions. We have been supported by Grant-in-Aid for the Scientific Research by the Ministry of Education, Science and Culture.

References

1. K. L. I. Kobayashi and E. Otsuka: J. Phys. Chem. Solids 35 (1974) 839.
2. O. Matsuda and E. Otsuka: J. Phys. Chem. Solids 40 (1979) 819.
3. E. Otsuka, T. Ohyama and K. Fujii: J. de Phys. 42 (1981) C7-393.
4. T. Ohyama, K. Fujii and E. Otsuka: Jpn. J. Appl. Phys. 21 (1982) 865.
5. O. Matsuda and E. Otsuka: J. Phys. Chem. Solids 40 (1979) 809.
6. O. Matsuda and E. Otsuka: Solid State Commun. 26 (1978) 925.
7. E. Gornik: Phys. Rev. Lett. 29 (1972) 595.
8. K. L. I. Kobayashi, K. F. Komatsubara and E. Otsuka: Phys. Rev. Lett. 30 (1973) 702.
9. E. Otsuka: Infrared and Millimeter Waves, Vol. 3, ed. K. J. Button (Academic, New York 1980) 347.
10. R. Ulbrich: Phys. Rev. B8 (1973) 5719.
11. W. Shockley: Bell System Tech. J. 30 (1951) 990.
12. E. Otsuka, T. Ohyama and K. Fujii: Jpn. J. Appl. Phys. 20 (1981) L885.
13. R. J. Elliott: Phys. Rev. 96 (1954) 266.
14. P. Boguslawski and W. Zawadzki: J. Phys. C13 (1980) 3933.
15. P. Boguslawski: Solid State Commun. 33 (1980) 389.
16. F. J. Blatt: J. Phys. Chem. Solids 1 (1957) 262.
17. E. Otsuka, K. Murase and J. Iseki: J. Phys. Soc. Jpn. 21 (1966) 1104.
18. E. Otsuka, K. Fujii and K. L. I. Kobayashi: Jpn. J. Appl. Phys. 12 (1973) 1600.
19. E. Otsuka: J. Phys. Soc. Jpn. 50 (1981) 189.
20. K. L. I. Kobayashi, K. F. Komatsubara, E. Otsuka, T. Ohyama and O. Matsuda: Proc. 15th Int. Conf. on Solid State Devices, Tokyo 1973, J. Jpn. Soc. Appl. Phys. 43 Supplement (1974) 301.
21. J. Degani, R. F. Leheny, R. E. Nahory and J. P. Heritage: Appl. Phys. Lett. 39 (1981) 569.
22. Jagdeep Shah, R. E. Nahory, R. F. Leheny, J. Degani: Appl. Phys. Lett. 40 (1982) 505.
23. T. Ohyama, K. Murase and E. Otsuka: J. Phys. Soc. Jpn. 29 (1970) 912.
24. D. G. Thomas, J. J. Hopfield and W. M. Augustyniak: Phys. Rev. 140 (1965) A202.
25. T. Sanada, K. Matsushita, T. Ohyama and E. Otsuka: J. Phys. Soc. Jpn. 45 (1978) 501.

MAGNETO-THERMAL CONDUCTIVITY OSCILLATIONS IN N-TYPE InSb

R. Mansfield and P. Fozooni
Bedford College (London University)

Regent's Park, London, NW1 4NS. England.

Measurements of the thermal conductivity of n-type InSb have been made at low temperatures [1,2,3]. Below 10 K phonon-phonon scattering becomes negligible and the scattering of phonons is mainly by the specimen boundaries and by isotopes of different masses. For samples with a free electron concentration greater than 10^{16} cm^{-3} electron-phonon scattering also occurs. For this level of concentration InSb is quasi-metallic and the electrons are degenerate. The theory of phonon-electron scattering under such conditions has been developed by Ziman [4]. Kosarev et al [3] found that the electrons primarily scatter the longitudinal phonons and that for a concentration of 4×10^{18} cm^{-3} the effect due to the electrons is to reduce the thermal conductivity by a temperature independent term. Confirmation that the electrons couple only with longitudinal phonons has been provided in a direct and convincing way using a heat pulse technique [5,6]. Little work has been done on the effect of a magnetic field on the thermal conductivity. Challis et al [1] could detect no change in conductivity of their specimens for fields up to 9 kG. The electrical conductivity, however, is much affected by a field which causes a large amplitude oscillatory de Haas-Shubnikov effect. The amplitude is large, for quite modest fields, because the Fermi level crosses Landau levels close to the extreme quantum limit with n = 1, 2, 3 etc. A similar large amplitude oscillation should also occur in the electron contribution to the thermal conductivity but this contribution is negligible compared to the phonon contribution. However, the results described here show a small amplitude oscillation having the same period as the de Haas-Shubnikov effect and it will be proposed that the phonon contribution to the heat current may oscillate with magnetic field due to an oscillation in the phonon-electron scattering in a similar manner to that observed in bismuth [7].

Measurements have been made on four tellurium doped n-type samples. The table gives the electron concentrations.

Sample	B	C	D	E
Electron concentration	0.35	4.4	35	120 x 10^{15} cm^{-3}

A special microprocessor based control system was developed for the measurement of the thermal conductivity [8]. The system was fully automated and could determine the zero field conductivity to within 2%. The change in thermal conductivity with magnetic field is much less than 2% and was measured by a differential method. The Allen-Bradley resistors used to measure the temperature gradient were situated outside the super-conducting coil producing the field with the residual magnetic field shielded by wrapping the resistors with niobium foil. The resistors were thermally anchored to two positions along the sample using copper wires as thermal links. The resistors were in a bridge which was balanced with a constant heat flow along the sample. The magnetic field was then swept up and the out of balance current determined using a lock-in amplifier and monitored on an X-Y recorder. The field was also swept down to check that the recorder retraced the initial curve. With this arrangement a precision of better than 0.2% was possible in the relative value of the thermal conductivity in the magnetic field. The magnetic field was parallel to the heat current. Figure 1 shows the experimental results for typical runs on samples B and D. Oscillations similar to de Haas-Shubnikov oscillations, but of much smaller amplitude are clearly visible for sample D.

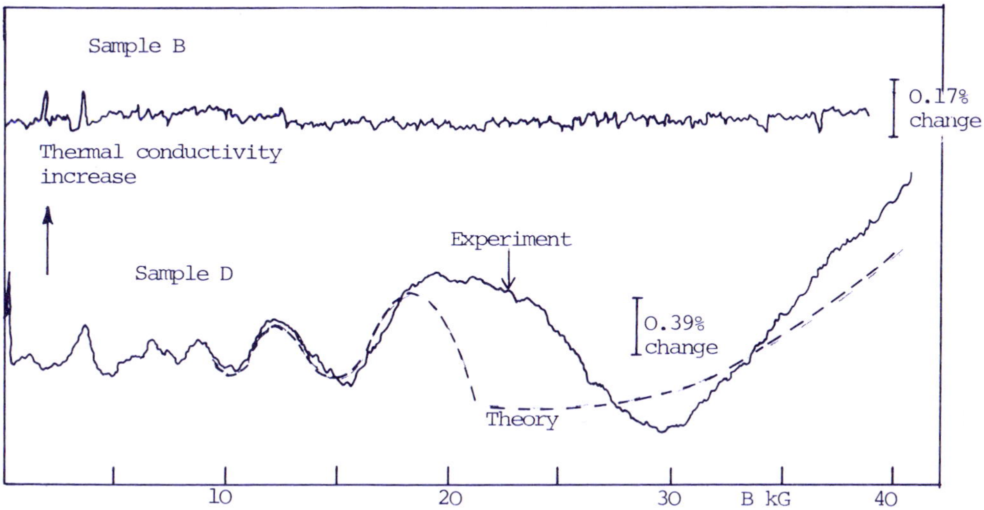

Figure 1. Thermal conductivity of samples B and D as a function of magnetic field at 1.3 K.

The results can be analysed using Callaway's expression for the thermal conductivity [9].

$$K(T) = \frac{k^4T^3}{6\pi^2\hbar^3}\sum_{t=1}^{3}\frac{1}{v_t}\int_0^\infty \frac{x^4e^x}{(e^x-1)^2}\,\tau\,dx \tag{1}$$

Here t represents the phonon mode with velocity v_t, k is the Boltzmann constant, $x = (\hbar\omega/kT)$ and T is the temperature. Assuming the principle of additivity of scattering frequencies the relaxation time τ is given by

$$\tau^{-1} = \tau_b^{-1} + \tau_i^{-1} + \tau_{el-ph}^{-1} \tag{2}$$

where the scattering due to boundaries (τ_b), isotopes (τ_i) and charge carriers (τ_{el-ph}) are included and all other mechanisms have been neglected. When a magnetic field is applied the only term in (2) which is affected is τ_{el-ph}. The way in which τ_{el-ph} is affected can be determined using the theory developed to account for measurements of the absorption of ballistic heat pulses in bismuth in a magnetic field [7]. Spike like quantum oscillations are found in the absorption of longitudinal heat pulses propagating in bismuth in a magnetic field. They are due to electron transitions within a Landau level being restricted to electrons close to the Fermi level and then only when the level is just above the Landau level. The expression for the rate of phonons q being absorbed is

$$W(q) = \sum_{n\,m\,s} W^s_{nm}(q) \simeq \tau^{-1}_{el-ph} \tag{3}$$

where $$W^s_{nm}(q) = \frac{L^3}{4\pi^2\lambda^2}\int W(q,n,k_z,s \rightarrow o,m,k_z+q_z,s)\,f_{n,k_z,s}(1-f_{m,k_z+q_z,s})\,dk_z \tag{4}$$

$f_{n,k_z,s}$ is the Fermi distribution function for the electron (n,k_y,k_z,s) and

$$W(q,n,k_z,s \rightarrow 0,m,k_z+q_z,s) = (2\pi/\hbar)(D^2\hbar q^2/2L^3\rho\omega_q)F_{nm}(t^2)\delta(\hbar\omega_q+E(n,k_z,s)-E(m,k_z+q_z,s)) \quad (5)$$

where $$F_{nm}(x) = e^{-x}x^{n-m}(m!/n!)\left[L_m^{n-m}(x)\right]^2. \quad (n\geq m) \quad (6)$$

D is the deformation potential parameter, L the dimension of the sample and L_m^n is the Laguerre bi-polynomial. In these expressions n,m and s refer to quantum numbers assosicated with Landau levels. The type of transitions which are allowed are shown in Figure (2) where intra- and inter-Landau level transitions are shown as unprimed (1,2) and primed (1',2') respectively.

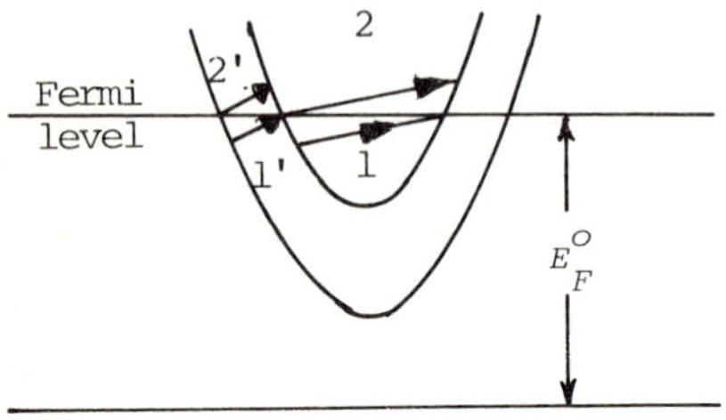

Figure 2

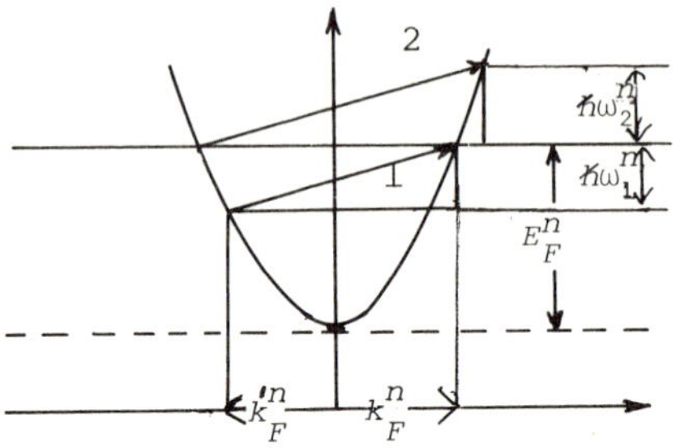

Figure 3

Transitions within a Landau level are restricted by the Pauli principle to processes which lie within a band extending from kT below process 1 to kT above process 2. Denoting the wavenumbers of the phonons associated with processes 1 and 2 by q_1^n and q_2^n then from Figure 3

$$k_F^n = (2m^*/\hbar^2)^{\frac{1}{2}}(E_F^n)^{\frac{1}{2}} \quad (7)$$

and $$k'^n_F = (2m^*/\hbar^2)^{\frac{1}{2}}(E_F^n - \hbar\omega_1^n)^{\frac{1}{2}} \quad (8)$$

hence $$q_1^n = k_F^n + (2m^*/\hbar^2)^{\frac{1}{2}}(E_F^n - \hbar\omega_1^n)^{\frac{1}{2}} \quad (9)$$

where $$E_F^n = E_F^O - (n+\tfrac{1}{2})\hbar\omega_c \text{ and } \hbar\omega_1^n = \hbar q_1^n v\ .$$

Then $$q_1^n = 2(2m^*/\hbar^2)^{\frac{1}{2}}\left[E_F^O - (n+\tfrac{1}{2})\hbar\omega_c\right]^{\frac{1}{2}} - 2m^*v/\hbar \quad (10)$$

Similarly $$q_2^n = 2(2m^*/\hbar^2)^{\frac{1}{2}}\left[E_F^O - (n+\tfrac{1}{2})\hbar\omega_c\right]^{\frac{1}{2}} + 2m^*v/\hbar\ . \quad (11)$$

The term $2m^*v/\hbar$ is very small compared to q_1^n. For example for specimen D with $N_D - N_A = 3.5\times10^{16}\ cm^{-3}$ and $B = 18$ kG

$$q_1^1 = (0.903 - 0.009) \times 10^6\ cm^{-1}\ . \quad (12)$$

As a consequence only a narrow band of phonon frequencies is absorbed due to electron-phonon scattering. In addition the function $F_{nm}(x)$ decreases rapidly as the argument becomes larger than unity. The physical meaning of this is that phonons with wave-numbers larger than the inverse of the cyclotron radius are not absorbed effectively. There is therefore an additional restriction on the wave-number of the phonons which are absorbed which is given by

$$q < q_{cr}^{nm} \simeq \lambda^{-1} \quad (13)$$

where $\lambda^{-1} = (eB/\hbar c)^{\frac{1}{2}} =$ (magnetic length)$^{-1}$.

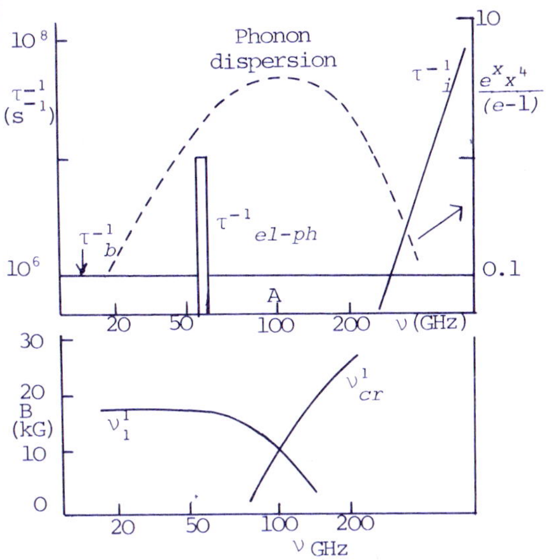

Figure 4. Contributions to the scattering frequency from equation (2) for a particular magnetic field. Also shown is the phonon spectrum at $T = 1.3$ K. The lower part of the figure shows the magnetic field dependence of $\gamma_{1 \atop 2} = q^{nm}_{cr} v_\ell$ and $\gamma_{1 \atop 2} = q^{nm}_{cr} v_\ell$. At A and for lower fields the band of frequencies when τ^{1}_{el-ph} is finite disappears.

Figure 4 shows the frequency dependencies of the inverse of the various relaxation times. The electron-phonon term has a finite value of τ^{-1}_{el-ph} only for a narrow band of frequencies. As the magnetic field is decreased the band moves to higher frequencies until condition (13) is no longer satisfied, which occurs near A on the figure. The amplitude of τ^{-1}_{el-ph} then rapidly decreases but at the same time another band appears due to the next higher Landau level crossing the Fermi level so that electrons in this next level causes scattering. The sequence is repeated as the field is continuously decreased.

The procedure for analysing the results was as follows. Starting with the zero magnetic field thermal conductivity, the specularity constant in boundary scattering and the constant in the Rayleigh term for isotope scattering were determined in the usual way by fitting equation (1) to experiment in the temperature range 1-10 K assuming that electron-phonon scattering was negligible [8]. The thermal conductivity in a magnetic field was then calculated by deriving τ_{el-ph} using (3) to (6) for the allowed values of m and n appropriate to the field. Spin effects were neglected. For example when the Fermi level is just above the Landau level with $n = 2$ then there are three bands of frequencies for which τ_{el-ph} is finite. These are $m = n = 1$, $n = m = 2$ and $n = 1$ $m = 2$. Equation (1) was then numerically integrated over the phonon spectrum introducing τ_{el-ph} for each of the bands of energies when electron-phonon scattering occurs assuming that only the longitudinal phonons are scattered by electrons. The theoretically derived thermal conductivity is compared with experiment in Figure 1. Good agreement is obtained and is also obtained for other temperatures.

Acknowledgements

We are grateful to Dr. H. Tokumoto for helpful discussions, Mr. A.K. Betts for assistance with the development of the microprocessor system, and the Central Research Fund of the University of London for a grant.

References

1 L.J.Challis, J.D.N.Cheeke and J.B.Harness, Phil.Mag.7 (1962) 1941.
2 M.G.Holland, Phys.Rev. 134 (1969) A 471.
3 V.V. Kosarev, P.V.Tamarin and S.S.Shalyt, Phys.Stat.Sol.(b) 44 (1971) 525.
4 J.M.Ziman, Phil.Mag. 1 (1956) 191.
5 J.P.Maneval, A.Zylbersztejn and D.Huet, Phys.Rev.Lett. 27 (1971) 1375.
6 J.P.Maneval and D.Huet, Physics Letters 48A (1974) 463.
7 T.Ishiguro, K.Kajimura, S.Kagoshima, H.Tokumoto and J.Kondo, J.Phys.Soc. Japan 39 (1975) 1547.
8 P.Fozooni, Ph.D. Thesis, London University (1980).
9 J.Callaway, Phys.Rev. 113 (1959) 1046.

INFRARED EMISSION DUE TO THE MAGNETOELECTRIC-PHOTO EFFECT IN InSb

T. Morimoto and M. Chiba

Institute of Atomic Energy, Kyoto University, Uji, Kyoto, Japan

Nonlinear infrared emission from n-InSb due to the magneto-electric-photo effect, having a central wavelength nearly equal to the energy gap, has been observed in the temperature range from 80 to 300 K.

In the transverse configuration ($J \perp H$) of current density J and magnetic field H, as shown in Fig.1, remarkable nonlinear infrared emission from n-InSb has been observed in the temperature range from 80 to 300 K [1]. This phenomenon is different from the well known cyclotron emission at lower temperatures [2] in its nature. It originates from the radiative recombination of excess electron-hole pairs produced near the surface by the J x H force (see Fig.1), and so the emission would be expected to have a central wavelength,

$$\lambda_o \cong ch/(E_g + kT), \tag{1}$$

corresponding to the value of the direct energy gap E_g. Here, c is the light velocity, h Planck's constant, k the Boltzmann constant, and T the temperature. In eq.(1) the effect of the carrier distribution is taken into account, since at relatively high temperatures it can not be neglected with respect to the value of the energy gap E_g for narrow-gap semiconductors, such as InSb and $Hg_{1-x}Cd_xTe$. The probability of radiative recombination for indirect-transition materials, such as Ge and Si, is too small to expect remarkable emission, but large values can be expected for direct-transition, narrow-gap semiconductors.

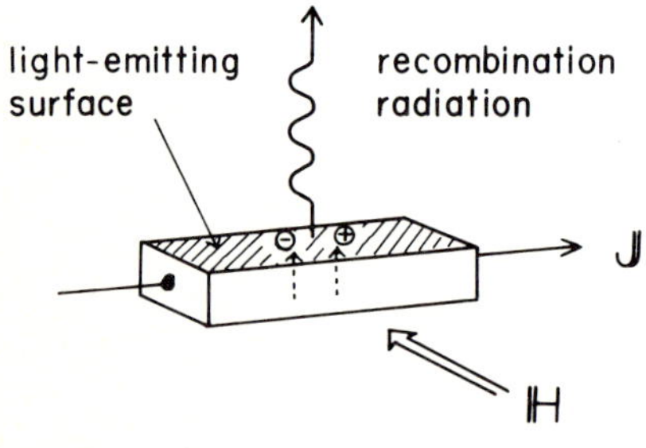

Fig.1. The motion of electrons and holes at high magnetic fields, and the recombination radiation, where ⊖ means an electron, and ⊕ a hole.

The experiments were carried out using n-InSb crystals from which ∿5.3 μ m infrared emission could be expected. The block diagram of the apparatus is shown in Fig.2. The typical sample #F' used here is 14.86 mm long, 4.0 mm wide and 0.88 mm thick. The values of the electron concentration n_o and the Hall mobility μ at 77 K were 5×10^{15} cm^{-3} and 1.66×10^5 $cm^2/V.s$, respectively, and the values at 294 K were 2.38×10^{16} cm^{-3} and 6.41×10^4 $cm^2/V.s$, respectively. The light-emitting surface was slightly polished using #4/0 emery paper to increase the emission efficiency after etching by CP-4A solution. Both end areas of the light-emitting surface were masked by black tape to prevent inclusion of possible injection emission of light near the electrodes.

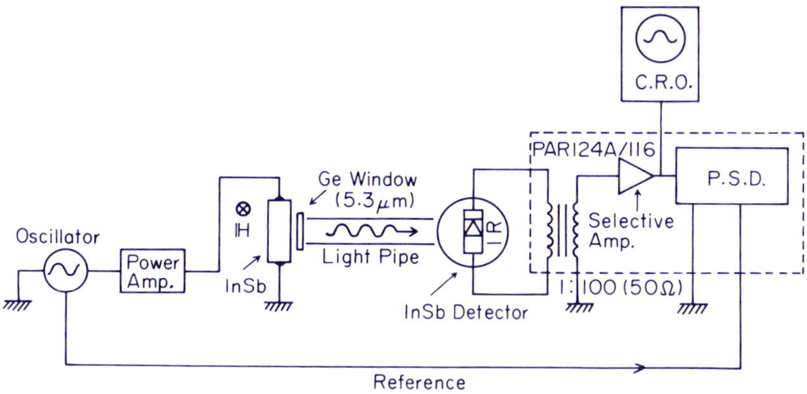

Fig.2. Block diagram of the apparatus for IR emission measurements.

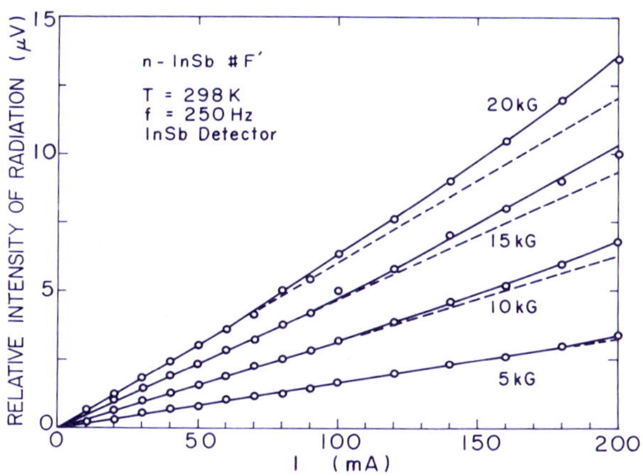

Fig.3. Relative intensity of radiation vs current characteristics at 298 K. H = 5, 10, 15 and 20 kG.

The sample was mounted in a metal dewar having a Ge window (5.3 μ m band pass filter) of 9 mm effective diameter. An ac current up to 200 mA at 250 Hz was passed through the sample across the magnetic field applied by an electromagnet. The emission from the sample was detected using a narrowband infrared detector, a photovoltaic InSb detector, mounted in another dewar being connected by a light pipe of 50 cm length and 10 mm inner diameter. The detector operates at 80 K in the range of the wavelength from 2 to 5.7 μ m and has maximum spectral detectivity $D^* = 1.3 \times 10^{11}$ cm $Hz^{1/2}W^{-1}$ at 5.1 μ m. The photosignal from the detector modulated at a frequency of 250 Hz was synchronously detected (PSD) using a PAR Model 124A/116 lock-in amplifier operated in a differential mode with 50 Ω input impedance.

Figure 3 shows the emitted intensity versus current curves at room temperature for given transverse magnetic fields, H = 5, 10, 15 and 20 kG. It will be seen that these curves indicate linear dependence (dotted lines in the figure) on I x H for lower values, while deviations from the linear relation are observed at higher values of I x H, from which we can estimate the value of the parameter γ in eq.(3) to be

3.12×10^{-29} cm^3 s. The detected PSD voltage, which corresponds to the quantity (R_c - R) in eq.(3) and is referred to as "relative intensity of radiation", changes its sign associated with the polarity reversal of the applied magnetic field, indicating a phase reversal of the emitted light of 180° (see Fig.4). This comes

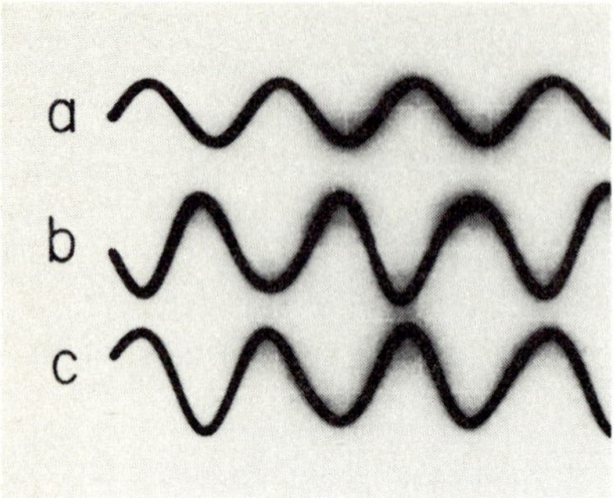

Fig.4. Oscilloscope traces of the waveforms of the current and the photosignals. f = 250 Hz, I = 100 mA, and T = 295 K. (a): the current waveform. (b): waveform of the photosignal for H = 20 kG. (c): waveform of the photosignal for H = -20 kG. Note the phase differnce of 180° between the waveforms b and c.

from the phase reversal of the excess carrier concentration near the light-emitting surface being modulated by the J x H force at a frequency of 250 Hz through the polarity reversal of the applied magnetic field. The result clearly indicates that remarkable radiative recombination takes place at or near the light-emitting surface. In view of the fact that a narrowband InSb detector is utilized it will be apparent that the emitted light has wavelength near 5.3 μ m as suggested in eq.(1).

Figure 5 shows the detector signal versus current curves at 80 K for given transverse magnetic fields, H = 20 kG and H = -20 kG. Here, it is to be noted that the PSD voltage is referred to as "relative intensity of radiation", so that it can take negative values associated with the phase reversal. At low temperatures sign reversal of the signal and remarkable nonlinear emission are observed at relatively high currents, while at lower currents the response of the detector signals to the polarity reversal of the magnetic field is different from that of the room temperature sample and the emission intensity is lowered by several times compared with it. This seems to suggest the necessity of taking into account any other recombination mechanisms [3,4] at low temperatures [1].

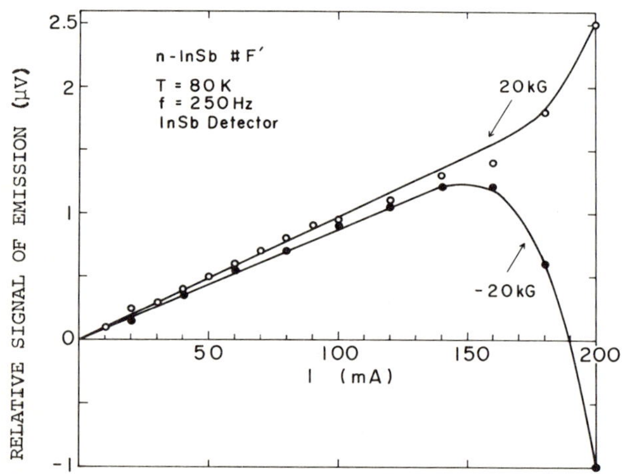

Fig.5. Relative signal of radiation vs current characteristics at 80 K. H = 20 kG (open circles), and H = -20 kG (solid circles).

Now, the observed results might be understood as follows by taking into account the action of the J x H force on charge carriers [1] . In a system where electrons and holes are coexisting under high magnetic fields, the J x H force causes a collective flow of electrons and holes, as shown by the dotted lines in Fig.1, resulting in a population inversion of charge carriers, particularly in the vicinity of the light-emitting surface; $np > n_i^2$, where n and p are the concentrations of electrons and holes, respectively and n_i is the intrinsic carrier concentration. Thus, near the surface, radiative recombination of excess electron-hole pairs must take place with a rate R_c;

$$R_c = Bnp = (np/n_i^2)R, \tag{2}$$

with $R = B\, n_i^2$, where R is the rate of radiative recombination in thermal equilibrium and B is the probability of radiative recombination [5,6]. Here, it should be noted that the rate R_c takes a form similar to the rectifying equation for a diode;

$$R_c = R\exp(\Delta E/kT) \simeq R\exp(\gamma JH/kT), \tag{3}$$

where ΔE is the action of the Lorentz force on charge carriers approximated by γ JH with $\gamma \simeq d/cn_o$, d being the sample thickness [7,8], so that a linear relationship with respect to the product JH is expected only when $\Delta E \ll kT$. Assuming eq.(3), we can estimate the value of the parameter γ for the room temperature InSb sample to be 3.12×10^{-29} cm^3 s, which seems to be a reasonable value for this room temperature sample. At lower temperatures, because of a large contribution from the exponential factor in eq.(3), remarkable nonlinearity might be expected for the emission intensity in spite of the small value of n_i^2 at lower temperatures.

At the relatively low current densities examined here, no current oscillations have been observed. However, the present experiment clearly indicates that a part of the input electric power is taken out of the sample in a form of infrared emission. This must be closely related to the negative conduction and associated phenomena in InSb [9,10], from the viewpoint of the activity of the sample arising from the nonequilibrium state caused by application of high magnetic fields. Finally, it should be noted that the emission from the room temperature sample is clear and intense enough so that it can be utilized for realizing a magneto-infrared emitting diode (or laser diode) especially operating at room temperature.

The authors would like to thank Dr.S.Akai for supplying the sample and for helpful advise, Professors M.Umeno and S.Narita for valuable advice and discussions.

References

1. T.Morimoto and M.Chiba, Phys. Lett. 85A, 395 (1981).
2. See, for example, E.Gornik, J. Magn. Magn. Mater. 11, 39 (1979), Suppl. Proc. Intern. Conf. on Solids and plasmas in high magnetic fields (Cambridge, MA, USA).
3. W.Shockley and W.T.Reed Jr., Phys. Rev. 87, 835 (1952).
4. R.N.Zitter, A.J.Strauss and A.E.Attard, Phys. Rev. 115, 266 (1959).
5. W.van Roosbroek and W.Shockley, Phys. Rev. 94, 1558 (1954).
6. J.I.Pankove, Optical processes in semiconductors, Chaps. 6 and 9 (Prentice-Hall, NJ, 1971).
7. T.Morimoto, Bull. Inst. Atomic Energy, Kyoto Univ. 59, 32 (1981).
8. See also M.Bernard, J. Phys. Chem. Solids 8, 332 (1959).
9. T.Morimoto, M.Yamamoto and J.Kawamoto, Phys. Lett. 49A, 261 (1974).
10. T.Morimoto, Phys. Lett. 50A, 183 (1974).

OPTICALLY PUMPED RECOMBINATION RADIATION IN ZINCBLENDE SEMICONDUCTORS UNDER CROSSED MAGNETIC AND STRAIN FIELDS

B. Wolfstädter[+], H.-R. Trebin[++], H. Pascher[+]
[+]Physik. Institut Universität Würzburg, D-8700 Würzburg, FRG
[++]Institut für Theoret. Physik Universität Regensburg, D-8400 Regensburg, FRG

Stimulated recombination radiation (SRR) and spin-flip Raman scattering (SFR) has been observed in n-InSb. Uniaxial stress T was applied perpendicular to a magnetic field $\underline{B}$ (T// [100], $\underline{B}$// [001]). For this geometry a $\underline{k}\cdot\underline{p}$ Hamiltonian was established within an 8x8 Kane model and was diagonalized exactly. Energy levels, transition energies and corresponding oscillator strengths were calculated and compared to the experimental data.

1. Introduction

In zincblende-type semiconductors the valence band Landau levels are anomalously spaced due to the degeneracy of heavy and light hole band at $k = 0$. Uniaxial stress lifts the degeneracy and allows an interpretation of the optical spectra. From intraband quantum cyclotron resonance transitions in the valence bands of uniaxially stressed InSb a set of band parameters was deduced /1/, which, however, disagrees with data derived from interband absorption experiments and intra-conduction band transitions. In an attempt to resolve the discrepancies we observed stimulated recombination radiation (SRR) and spin-flip Raman scattering (SFR) under uniaxial stress T, which was applied parallel to the [100]-direction and perpendicular to the magnetic field $\underline{B}$ || [001].
In contrast to Ge, for InSb this geometry is not inferior to the one T// $\underline{B}$// [001] used previously /1/, since for both the symmetry group of the $\underline{k}\cdot\underline{p}$ Hamiltonian is the two-element group C_2, when all symmetry breaking effects are taken regard of (tetrahedral crystal point group, magnetic and strain field, longitudinal momentum k_z).

2. Computational methods

For the calculation of the magnetic states we used a $\underline{k}\cdot\underline{p}$ Hamiltonian /1/ acting in the eight-dimensional space of the Γ_6 conduction band and the spin-orbit-split Γ_8 and Γ_7 valence band states at the zone center:

$$H = H^{cc} + H^{vv} + H^{ss} + \left[(H^{cv} + H^{cs} + H^{vs}) + \text{h.c.}\right].$$

The influence of the higher spin-orbit-split Γ_5 conduction band states, separated by about 3 eV from the band edge, was treated carefully in second-order perturbation theory. In H^{cc} these states modify bare electron mass and g-factor, and in H^{cv} they add additional terms of tetrahedral symmetry. The Hamiltonian matrix was diagonalized exactly up to the desired numerical accuracy.

3. Stress dependence of the Landau levels (theory)

In Figure 1 for both T// $\underline{B}$ and T⊥$\underline{B}$ the energy of the band-edge Landau levels is plotted versus stress at $B = 20$ kG and $k_z = 0$. The band parameters used are those of Ref. 1; for the correction terms of the Γ_5 conduction band, parameters were taken from Weiler et al. /2/. The energy scale is inverted, and all energies are referred to the lowest conduction-band level. For recombination radiation from this level the diagram directly displays the frequency. At $k_z = 0$ the symmetry group for T// $\underline{B}$// [001] is S_4 of four good quantum numbers $Q = 0,\dots,3$. For T⊥$\underline{B}$ a twofold symmetry axis gives rise to the two irreducible representations labeled by $P = 0,1$. For T// $\underline{B}$ strong anticrossing effects appear at 1980 cm^{-1}. They arise from the final state of the fundamental transition, which in Ge is parallel to the state (0,-1/2) (Q=3), but in InSb is perturbed by stress-induced k-linear terms. No trace is seen of this structure in the other geometry. When going from T// B to T⊥B the valence band levels (0,+1/2) and (0,-1/2) are interchanged and split apart strongly by strain. In the conduction band the g-factor decreases with stress for T// B, but remains constant (49.5) for T⊥B. Due to stress-induced k_z-linear terms, off center extrema appear in the k_z-dependence of the valence-band energies for T⊥B. In recombination radiation these will not cause any noncentral transitions since the dispersion relations of the conduction band Landau levels remain strongly parabolic at $k_z = 0$.

4. Characteristic features of interband transitions

We now discuss the case T⊥B only. According to the figure 1b electrons, which recombine from conduction band level (0,+1/2), will essentially emit two lines, of final states (0,+1/2) and (0,-1/2), respectively, in the valence band. The first is polarized parallel to the magnetic field and declines monotonically in frequency, the second, of polarization E⊥B, rises to a frequency maximum before decreasing. Recombination radiation from CB level (0,-1/2) will show the same frequency dependence, but shifted by 46 cm^{-1} to higher energies; the polarization type of the two lines is exchanged. Thus two series of lines will be observed, each one copy of the other and resembling the stress behaviour of the valence band levels. Inter- and intraband features appear simultaneously in the spectra.

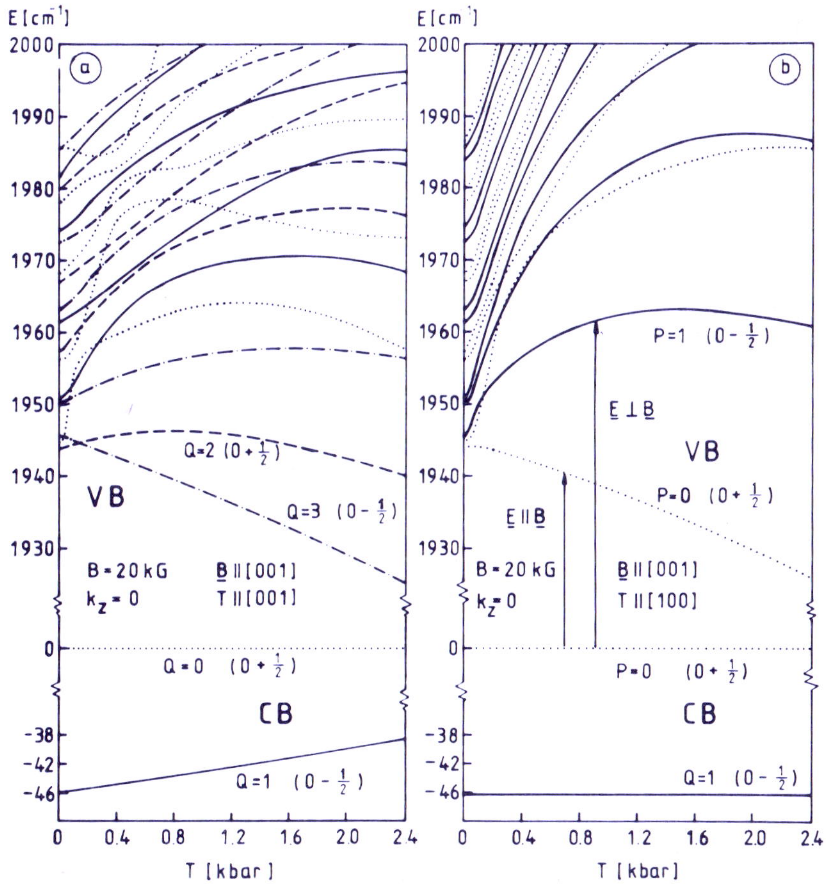

Fig.1 Landau levels versus stress at B = 20 kG and $k_z = 0$.

5. Experimental Methods

Two kinds of experiments were carried out in order to observe spin-flip transitions in the conduction band and interband transitions. The effective g-values were determined by spin-flip Raman gain measurements and four-wave mixing spectroscopy, respectively, with the experimental set-ups described in /3,4/. The interband transitions were measured by optically pumped stimulated recombination radiation. The experimental arrangement for these experiments is described in /5/. In all experiments the sample (n-InSb; $n = 9\cdot10^{14}/cm^3$) was mounted in Voigt-configuration ($k_L \perp B$; k_L : wave vector of optical radiation) in a split coil superconducting magnet at the temperature of pumped liquid helium. Uniaxial stress T up to 4 kbar was applied perpendicular to the magnetic field (T// [100] ; B// [001] ; k_L// [010]).

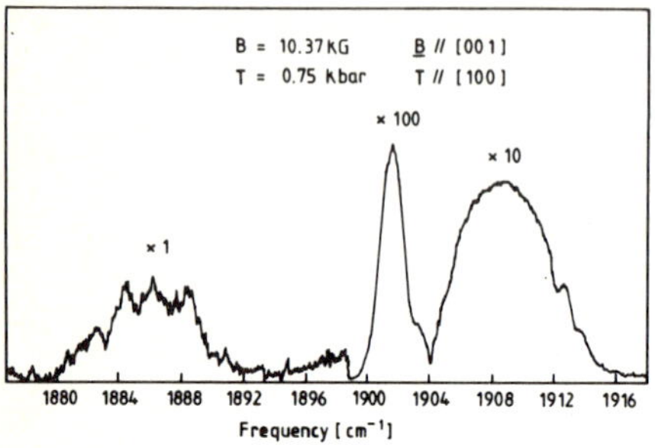

Fig.2 Spectrum of the stimulated recombination radiation at a magnetic field of 10.37 kG and an uniaxial stress of 0.75 kbar.

Figure 2 shows a spectrum of the emitted SRR taken at a magnetic field B = 10.37 kG and a stress T = 0.75 kbar. Three transitions with different intensities and linewidths can be seen. The frequencies of the peaks are evaluated and either plotted versus magnetic field with the stress held fixed or versus strain at constant magnetic field.

6. Results and Discussion

The effective g-value of the conduction band turned out to be independent of stress. In uniaxial stress up to 3 kbar the absolute value of the g-factor decreased by less than 0.5 %, compared to zero strain. This result is also confirmed by our theoretical calculations.

In order to obtain information on the band structure from SRR measurements one needs a model for the generation of the stimulated recombination radiation. Due to the high concentration of photocreated carriers (about 10^{16} cm^{-3} in our experiment) one has to take into account modifications of the band structure by exchange and correlation effects. We assume exchange and correlation energy to be independent of strain and magnetic field. Therefore the only manybody effect is a reduction of the energy gap of the order of ΔE_g= 4 meV. The intensity of the SRR is proportional to $\exp(\gamma-\alpha)x$ where γ is the gain factor, α is the absorption coefficient for the SRR, and x is the position in the sample. Due to the high mass of the heavy hole bands γ is proportional to the density of occupied states, and α to the density of unoccupied states in the conduction band only. Thus at a given magnetic field the frequency of the emitted radiation can be calculated from the maximum of $\gamma-\alpha$, the energy gap (reduced by manybody effects) and the energy of the final states in the valence band. Due to the narrow linewidth of the stimulated emission we think this method to be more exact than cyclotron resonance for the investigation of the valence bands.

At a magnetic field of about 10 kG only the two spin energy states of the n = 0 conduction band Landau level are occupied. The calculations of oscillator strengths show that the strongest transitions starting from these levels are: $c(0,\pm1/2) \rightarrow v(0,\pm1/2)$. In figure 3 the theoretically calculated strain dependence of the energies for these transitions is compared to experimental values corrected with respect to ΔE_g. In figure 3a the magnetic field is held fixed at 10.37 kG. Figure 3b shows the same for 17.89 kG. At this higher magnetic field only the c(0,+1/2) level is populated.

The shape of the tuning curves and the slopes $\Delta E/\Delta T$ are described very well by the theory using the band parameters as indicated above. The energy differences between tuning curves with the same final states $c(0,-1/2) \rightarrow v(0,-1/2)$, $c(0,+1/2) \rightarrow v(0,-1/2)$, or $c(0,-1/2) \rightarrow v(0,+1/2)$, $c(0,+1/2) \rightarrow v(0,+1/2)$ are in accordance to $g^*\cdot\mu_B\cdot B$ with g^* measured by SFR. The relative intensities of the different lines correspond well to the occupation probability of the conduction band-levels and the oscillator strengths calculated.

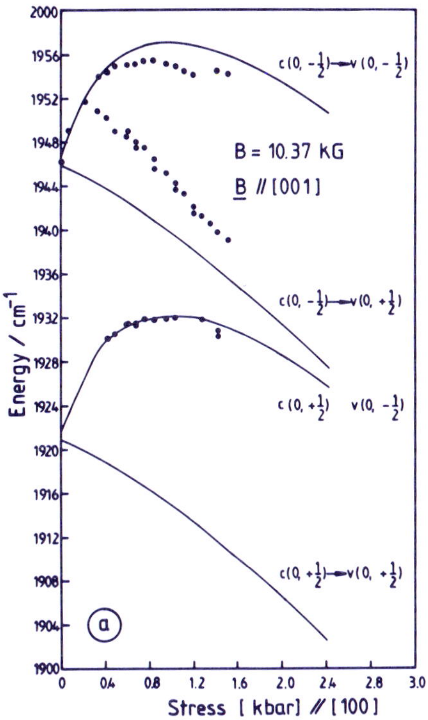

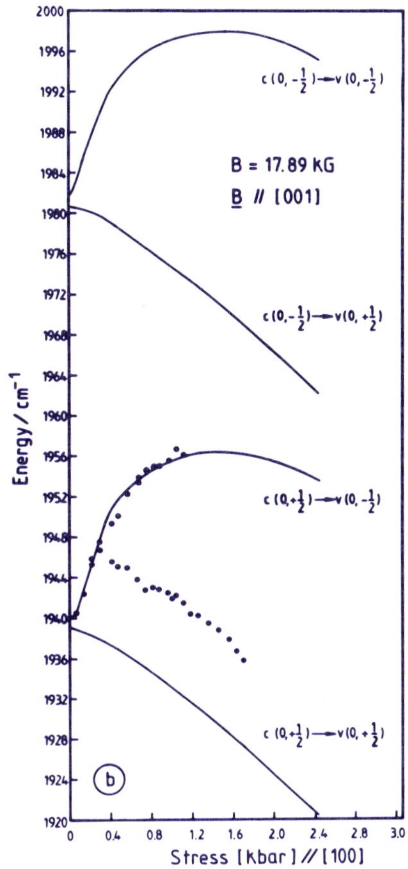

Fig.3 Stress dependence of the recombination frequencies at two magnetic fields

But there is one discrepancy between experiment and theory. The transitions to the v(0,+1/2) levels have a higher energy than calculated (8 cm^{-1} at 17.89 kG). At 0.3 kbar the v(0,+1/2) and v(0,-1/2) levels cross each other in contradiction to the calculations and to the cyclotron resonance data by Ranvaud /1/.

Due to symmetry such a level crossing is not forbidden since the two states belong to different representations. But in the framework of the present theory there is no set of bandparameters which describes these observations. Since there is no indication for experimental errors this discrepancy remains to be solved.

References

1. H.-R. Trebin, R. Ranvaud, U. Rössler, Phys. Rev. B 19, 686 (1979)
   R. Ranvaud, H.-R. Trebin, U. Rössler, Fred H. Pollak, Phys. Rev. B 19, 701 (1979)
2. M.H. Weiler, R.L. Aggarwal, B. Lax, Phys. Rev. B 17, 3269 (1978)
3. H. Pascher, G. Appold, R. Ebert, H.G. Häfele, Appl. Phys. 15, 53 (1978)
4. H. Pascher, Opt. Commun. 41, 106 (1982)
5. H. Pascher, H.G. Häfele, Appl. Phys. 20, 75 (1979)

THE METAL — NON METAL TRANSITION IN N-TYPE InSb IN HIGH MAGNETIC FIELDS

A. RAYMOND

Centre d'Etudes d'Electronique des Solides, associé au C.N.R.S.
Université des Sciences et Techniques du Languedoc ,

Place E. Bataillon, 34060 - MONTPELLIER- CEDEX , France

§-1 INTRODUCTION

The problem of the metal-non metal transition (M-NMT) i.e. the transition from a conducting to a non conducting state is one of the basic problems of the condensed matter physics. The doped semiconductor is an excellent system for studying the N-NMT : for example it provides a possibility of varying the doping and the compensation and consequently allows to induce activated or non activated conductivity. On the other hand this system is simple and well studied. Shallow donors and acceptors act as hydrogen-like centers embedded in a host crystal with the dielectric constant χ and the localized electron has the effective mass m^*. Nevertheless, the theoretical and experimental advantages should not conceal the problems arising from the random distribution of impurities, A doped semiconductor is a disordered system because of the randomness of the impurity distribution. The effect of the disorder on the electronic properties, and particularly on the M-NMT , is a great importance. Indeed as we shall see (§-2) we can classify different theoretical studies of the M-NMT in two categories : the first, which treats the M-NMT as being mainly due to electron-electron (e - e) interactions, and the second, which assumes that the disorder can induce a transition of the Anderson type.

In this paper we will give a short review (§-2) of the different theoretical aspects of the M-NMT in doped semiconductors and we will discuss (§-3) the problem of the M-NMT in n-type InSb induced by magnetic field, without or under a hydrostatic pressure. The problem of the minimum metallic conductivity and of the ε_2 process will also be discussed. Finally, we will present a new experimental method of investigating the M.NMT in n-type InSb. The entire study shows the preponderant role of the correlation effects for the M-NMT in n-type InSb.

§-2 SOME THEORETICAL ASPECTS OF THE N-NM TRANSITION*

The idea of impurity conduction was first expressed by Gudden and Schottky (1). These authors proposed a mechanism of conduction in semiconductors relying on the fact that a carrier has a finite probability of tunneling from an occupied impurity center to

**For review of the M-NM Transition we propose :*

1. *N.F. Mott "Metal-Insulator transition", London 1974, Taylor and Francis, London*
2. *J. Physique 37, C4 (1976), Int. Proc. Coll. International du C.N.R.S. Autrans, France*
3. *The Metal Non-Metal Transition in disordered systems (1978) Ed. L.R. Friedman and D.P. Tunstall (Proc. 9th Scottish Univ. Summerschool in Physics)*
4. *Impurity bands in semiconductors, Int. Conf. Wurzburg (1979) Phil. Mag. B Vol. 42, n° 6 (1980)*

an equivalent, but non occupied one. Several authors have pointed out that, at large impurity concentrations, the overlap of the wave functions of adjacent impurities become so strong that carriers are no longer localized around individual atoms, and conduction proceed via the impurity band. Mott (2) suggested that this transition from the tunneling to the metallic type of conduction might occur abruptly at a certain critical concentration n_c such that

$$n_c^{1/3} a_H = 0.25 \qquad (1.1)$$

where a_H is the effective Bohr radius of the impurity center :

$$a_H = (\chi \hbar^2) / m^* e^2 \qquad (1.2)$$

This criterion (called the Mott criterion) is obtained assuming that the Thomas-Fermi screening constant is equal to the Bohr radius. As it has been said in the introduction, the different theories of the M-NMT can be classified in two essential categories. We shall briefly discuss them now in the context of the doped semiconductors.

2.1 Localization due to disorder. The Anderson transition

In 1958 Anderson (3) published a paper on "the absence of diffusion in certain random lattices". The absence of diffusion is a consequence of the "localization" of the wave functions, which takes place when the disorder parameter becomes large enough (4). An electron can then move from one site to another only by exchanging energy with phonons (thermally activated hopping). This is the Anderson localization and the transition from the extended to the localized states is the Anderson transition. Mott (5) suggested that, even when the "Anderson disorder" is not sufficient to allow localization, localised states would exist for energies near the band edge, and that an energy E_c must separate the localized and the extended states. This energy E_c, has been named later "mobility edge" (6). If we denote the Fermi energy at zero temperature by E_F, two important cases arise :

- if $E_F > E_c$ the system is "metallic" and the conductivity σ would tend to a finite value as $T \to 0$.

- if $E_F < E_c$, the electrons can only move by thermally activated hopping and the conductivity σ tends to zero as $T \to 0$; such a system has been called by Anderson (7) a "Fermi glass".

In consequence if E_F and E_c can be made to vary in such a way that $E_F - E_c$ changes sign, a metal-insulator transition will occur.

Mott (8) noticed that if E_F lies above E_c, σ shows a weak T dependence and, as $E_F \to E_c$, approaches a finite value that he called the "minimum metallic conductivity" σ_{min}, which can be written as :

$$\sigma_{min} \simeq 0.026\ (e^2/\hbar\ d_c) \qquad (1.3)$$

where d_c is the average distance between impurity centers for n_c defined by :

$$3/4\ \Pi\ d_c^3\ n_c = 1$$

According to the Mott criterion $n_c^{1/3} a_H = 0.25$ one obtains $d_c \simeq 2.5\ a_H$.

Thus, when the states are localized, two forms of conduction are possible and generally coexist :

a- conduction involving electrons excited to the extended states at a mobility edge, which gives the following contribution to the conductivity :

$$\sigma_{ext} = \sigma_{min} \exp \{ - (E_c - E_F) /kT \} \qquad (1.4)$$

b- conduction by hopping near the Fermi energy E_F. For the hopping transition between nearest neighbour :

$$\sigma_{hop} = \sigma_3 \exp(-\varepsilon_3/kT) \qquad 1.5)$$

2.2 The M-NM Transition due to correlations : Hubbard model and Mott transition

The most important step in understanding the role of the e-e interactions in the M-NMT we owe to Hubbard (11). He showed that a model which includes only the intra-atomic energy U, leads to a continuous M-NMT as a function of the carrier density. U can be defined as the energy required to transfer an electron from one atom to another so that :

U = (ionization energy) minus (electron affinity)

The application of the Hubbard model to doped semiconductors is certainly justified in the insulating region. A neutral impurity D° can capture a second electron forming a D^- state. The D^- states form a band as the concentration increases. This band can be considered as the "upper Hubbard band" separated from the "lower Hubbard band" D° by the correlation energy U.

In 1958 Fritzsche (12) proposed an activated conduction process, with the activation energy ε_2 which corresponded to the excitation of electrons from the D° donor states to the D^- donor states. Fritzsche (12) established a strong decrease of ε_2 when the concentration was increased. Several authors (13,14) obtained similar results and showed that at the M-NMT ε_2 approaches zero.

By considering the intra-atomic interaction only, Mott (8) suggested that a Metal-Insulator transition occurs when $B \simeq U$, (or $\varepsilon_2 = B - U \simeq 0$) where B is the bandwidth This transition is sometimes called the Mott transition. Neglecting the disordered arrangement of the centers, this leads to the formula :

$$n^{1/3} a_H \simeq 0.2 \qquad (1.6)$$

2.3 Disorder and correlations : "Mott-Hubbard-Anderson" model

There exist few studies of simultaneous effects of disorder and correlations on the M-NMT . Mott (15) proposed, for the impurity states, a model combining both the effects of correlations (Hubbard model) and of disorder (Anderson model). We can summarize this model in the following way : in the low concentration region, the Hubbard hamiltonian describes both the ground state of the neutral donor D° and the D^- state with two electrons, the energy separation between these two states being the correlation energy U. When the concentration increases the states are displaced and finally are formed two bands D° and D^-, called the lower and upper Hubbard bands. Mott argued that, because of the disorder, the M-NMT does not occur when the D° and D^- bands begin to overlap (as explained in chapter 2.2) ; the band edge states are "Anderson localized" and the Fermi energy E_F is below the mobility edge E_c. The transition occurs only when E_F crosses E_c. Consequently, in this model the energy ε_2 is equal to $E_c - E_F$.

In Fig. 1 we sketched the density of states for a doped semiconductor assuming a very small compensation. The dashed regions correspond to localized states. The neutral donor states D° are at an energy ε_1 below the conduction band edge, equal to the effective Rydberg :

$$\varepsilon_1 = R_y = (m^* e^4)/2\hbar^2 \chi^2 \qquad (1.7)$$

The singlet D^- states are very weakly bound. They form a broader band because the extent of their wave functions is about 3 to 4 times that of the D° states. Since each donor atom has one electron, the lower band is completly filled and therefore

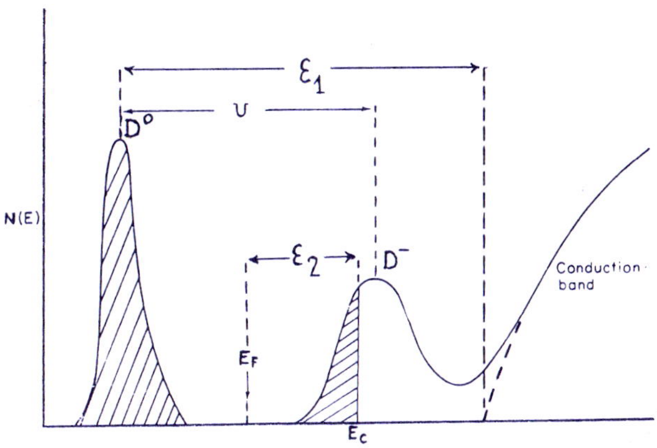

Fig. 1 Sketch of the density of states in two Hubbard bands for a doped semiconductor. Localized states are darked.

the conduction occurs when electrons are excited to the upper Hubbard band at the mobility edge E_c (energy ε_2).For concentrations $n << n_c$ the electrons excited to energy E_c yield an activated conduction behaviour as defined in chapter 2.1

$$\sigma = \sigma_2 \exp(-\varepsilon_2/kT)$$

where $\varepsilon_2 = E_c - E_F$ and σ_2 is the minimum metallic conductivity σ_{min}.

At higher impurity concentrations, the D° and D^- bands broaden and begin to merge because of disorder and stronger wave function overlap. The energy ε_2 decreases but the density is sufficiently high to yield hopping at very low temperature.

At $n = n_c$, the states near the Fermi level become extended and the M.NMT occurs ($\varepsilon_2 = 0$).

However, most of the theoretical studies of the M-NMT in doped semiconductors show the essential role of the electronic interactions in this transition (16,17).

In 1978, Ghazali and Leroux-Hugon (18) described the Metal-Insulator transition in doped semiconductors by using a model which incorporates e-e interactions through the Kohn-Sham density functional scheme within the cellular approximation. These authors predict the coexistence of the localized and the extended states at the Fermi energy, in a narrow range of concentrations and temperatures. Recently (19) Ghazali and Serre showed that, as the impurity concentration in semiconductors is decreased from high values, "an erosion appears between the main body of the conduction band and its tail, leading to the formation of an impurity band separated from the conduction band by an energy gap which widens out". They obtained this result by using the multiple-scattering theory of Klauder (20).

So these two recent papers show that the electronic interactions are essential for inducing the M-NMT in doped semiconductors.

§-3 THE METAL-NON METAL TRANSITION IN N-TYPE InSb

Following Fritzsche (12) in the activated conduction regime, the conductivity can be generally written as,

$$\sigma = \sigma_1 \exp(-\varepsilon_1/kT) + \sigma_2 \exp(-\varepsilon_2/kT) + \sigma_3 \exp(-\varepsilon_3/kT) \quad (3.1)$$

In the metallic phase the conductivity is approximatively temperature-independent. Up to now three methods have been used to induce the M-NMT. They involve,

a- changing of the overlap of neighbouring donor states by changing the concentration of impurities or by adding compensating impurities,

b- reduction of the overlap when a magnetic field is applied

c- reduction of the overlap caused by hydrostatic pressure.

The first method requires experiments on different samples with different values of doping and compensation and as a result different degrees of disorder. Variation of the disorder is difficult to assess and additionally it changes from sample to sample. In the other methods the transisition is studied on the same sample, thus disorder is being unchanged. The basic effect of magnetic field is to shrink the Bohr radius. The first experimental study of this problem was by Keyes and Sladek (21) and the theoretical study by Yafet, Keyes and Adams (22). Recently the problem was studied for InSb-type semiconductors by Zawadzki and Wlasak (23). The effect is particularly strong when the effective mass in the conduction band is small and the dielectric constant is high. All these reasons justify the choice of InSb to study the magnetic field induced M-NMT.

3.1 Review of the different studies of the Metal-Non Metal Transition in InSb

The metallic behaviour of InSb at zero magnetic field has been attributed to various causes. Some authors, as Sandercock (24), and Ghazali and Leroux-Hugon (18,25), suggest that the impurity states, in the purest available crystals, merge with the conduction band. Others, as Miyazawa and Ikoma (26), Ikowa (27), and Ferré et al (28) argue that the metallic behaviour is governed by an impurity band slightly separated from the conduction band. It should be pointed out that Ferré et al (28), Ishida and Otsuka (29) and also Gershenzon et al (30) neglect the temperature dependence of the conduction-band mobility as a possible explanation for the conductivity variations and the metallic behaviour in zero magnetic field. Ferré et al (28) presented results of magneto-transport experiments (up to 5 T) on three samples with excess donor density near the critical concentration $n_c \simeq 6 \times 10^{13}$ cm^{-3}. The authors analyse their data by postulating the existence of the activation energy ε_1 at the zero magnetic field even for the most heavily doped sample (3.9×10^{14} cm^{-3}) at high temperatures (T > 2.5 K). In the low temperature range (T < 2.5 K) they conclude that :

a- there exists an activation energy ε_2 in zero magnetic field for the least doped sample (5.7×10^{13} cm^{-3}); this energy has been found practically equal to ε_1,

b- there exists a unique impurity band completely separated from the conduction band at the zero magnetic field even for the sample with the highest doping.

For low temperatures, they assumed the presence of the metallic conduction involving the impurity band. According to Mott et al (4) "a value for the minimum metallic conductivity σ_{min} can be estimated either from the values of σ at which an activation energy disappears, or form the extrapolation to 1/T = 0 of the curves, plotting ln σ against 1/T". Ferré et al. obtained a common intercept of the curves ln ρ versus 1/T at 1/T = 0 and used it to postulate the existence of a minimum metallic conductivity σ_2.

This fact is rather convincing and seems to assert the existence of the ε_2 regime. However, it should be pointed out that Ferré et al. (28) deduce the activation energies from the analysis of the resistivity, whereas the theoretical predictions concern only the conductivity.

Ishida and Otsuka (29) have studied the galvanomagnetic properties of n-type InSb samples with excess donor concentrations between 1×10^{13} and 2×10^{15} cm^{-3} in vanishing and weak magnetic fields. They concluded that an impurity conduction of metallic type exists at concentrations in excess of 4.1×10^{14} cm^{-3}and suggested that a magnetic-field-induced M-NMT occurs in the impurity band. From an analysis of the curves of σ_{xx} for different magnetic fields, as functions of the donor concentration N_d, they concluded that a minimum metallic conductivity exists in the neighbourhood of the transition.

In 1977, Walton and Dutt (31) discussed these conclusions and saw no evidence for an ionization energy at doping levels above 10^{14} cm^{-3} in zero magnetic field. They suggested that the temperature dependence of the conductivity might reflect the variation of the electron mobility in the conduction band.

In 1978 Mansfield and Kusztelan (32) examined the validity of the procedure of Ferré et al. (28) and questioned the use of the resistivity in the M-NMT analysis. They also suggested that a common intercept of ln ρ versus 1/T lines at 1/T = 0 could be explained as a saturation of the resistivity when all the electrons are excited into the conduction band. They also pointed out that the cyclotron resonance studies and the galvanomagnetic measurements indicate that the mobility of electrons corresponds to that of electrons in the conduction band.

Similar results had been reported in 1970 by Oka and Narita (33). In 1979, after having analysed the previous results on the M-NMT in InSb, the Montpellier group, Robert, Raymond, Aulombard and Bousquet (34) concluded a nonexistence of an impurity ionisation energy in the absence of a magnetic field and suggested metallic conductivity in the conduction band. The Montpellier group proposed a study of the M-NMT by using simultaneously the magnetic field and the hydrostatic pressure.

3.2 A study of the Metal-Non Metal transition in InSb by transport experiments in high magnetic field and under hydrostatic pressure

We performed galvanomagnetic measurements on several samples with doping between 10^{14} and 2.5×10^{16} cm^{-3} (see Table I). We used magnetic fields up to 20 T and hydrostatic pressure up to 7.5 kbars to separate the different conduction processes and analyse them.

TABLE I
Characteristics of the investigated samples

Samples	$N_D - N_A$ (cm^{-3})	μ (cm^2/v.s) (T = 4.2°K)	Threshold field (T)
Fd - a	7.5×10^{13}	70000	< 0.2
547-25 d	10^{14}	94000	0.18
547-3-a	4.5×10^{14}	110000	0.7
547-25-c	10^{15}	106000	1.5
536-23-a	5.4×10^{15}	107000	5.8
536-21-c	5.8×10^{15}	106000	6
550-2-C	1.2×10^{16}	100000	10
536-27-a	2.14×10^{16}	92000	17.7

For the less doped samples ($N_D - N_A \simeq 10^{14}$ cm^{-3}) the temperature dependence of the resistivity with and without hydrostatic pressure is explained by the variation of the conduction band mobility without the use of the activation energy concept. This implies that the impurity band is merged with the conduction band in zero magnetic field for carrier concentrations larger than 10^{14} cm^{-3}. This result has been corroborated (35) on a very low doped and compensated sample $N_D \simeq 7.5 \times 10^{13}$, $N_A \simeq 6.10^{12}$ cm^{-3} as it will be shown later. Moreover, the transport experiments we performed in presence of a magnetic field (34) show that a common intercept at $1/T = 0$ of the ln σ_{xx} v.s. $1/T$ curves is not obtained and we cannot thus deduce the existence of the minimum metallic conductivity nor the presence of the ε_2 process.

3.2.1 The conduction processes in the neighbourhood of the magnetic field-induced Metal-Non Metal transition

As indicated above, the common intercept of the ln ρ v.s. $1/T$ curves at $1/T = 0$ could be treated as an evidence for the existence of the minimum metallic conductivity and the ε_2 process. In agreement with the observations of Ferré et al (28) we have found (34) that for sample 547-25-d, comparable to sample B of Ferré et al (28) the ln ρ v.s. $1/T$ curves had a common intercept at $1/T = 0$ (fig. 2), in the magnetic field range 0.27 - 1.1 T. However, this cannot be used as an evidence for the existence of the minimum metallic conductivity and the ε_2 process, because a common intercept is not observed when a similar analysis is used for the σ_{xx} component (fig.3).

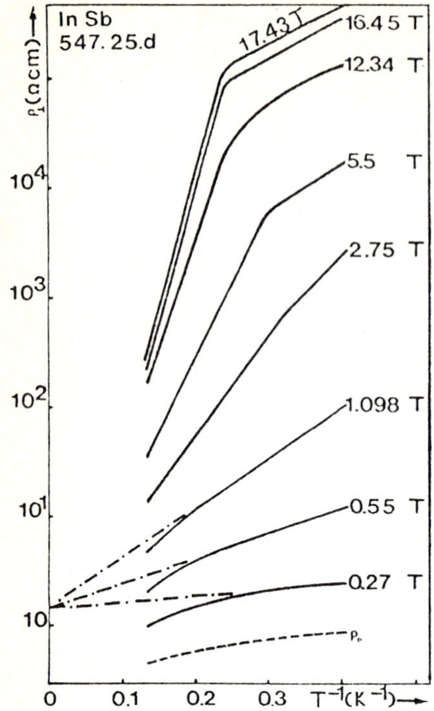

Figure 2
Temperature dependence of the transverse magnetoresistivity of sample 547-25-d.

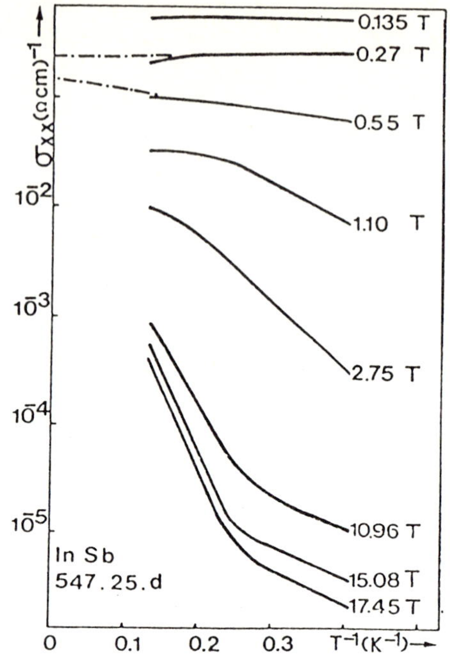

Figure 3
Temperature dependence of the σ_{xx} component of sample 547-25-d

Ishida and Otsuka (29) claimed the existence of the minimum metallic conductivity varying between 0.3 and 0.6 $(\Omega.\mathrm{cm})^{-1}$ as a function of doping. We obtained a comparable value for σ_{min} equal to 0.7 $(\Omega.\mathrm{cm})^{-1}$ for sample 536-23-a. For the non activated behaviour, however the common intercept at $1/T = 0$ was not found (fig. 4). Nevertheless if we assume that this value 0.7 $(\Omega.\mathrm{cm})^{-1}$ is the minimum metallic conductivity from (1.3) it follows that it should be pressure independent. We have however observed that σ_{min} varies from 0.7 to 1 $(\Omega.\mathrm{cm})^{-1}$ when the hydrostatic pressure increases up to 7.5 kbar (fig. 4).

Figures (5) and (6) show ln $(nT^{-1/2})$ v.s. $1/T$ functions at $p = 0$ and at $p = 7.5$kbar.

It can be readily seen that the activated regime i.e. the M-NMT appears for lower magnetic field when the pressure is applied. Similar results have been obtained on the most havily doped sample. Furthermore, we observe a similar behaviour for the σ_{xx} component at the threshold magnetic field : σ_{xx} varies from approximately 0.8 to 1.2 $(\Omega.\mathrm{cm})^{-1}$ when the pressure increases from 0 to 7.5 kbar. Also in this case we do not observe a common intercept at $1/T = 0$. It seems as if we were dealing with a conductivity process which does not correspond to the ε_2 process.

This is not really surprising : the activation energy of the D- band is very small even at high fields. Its value has been recently determined by Larsen (36) who obtained approximatly 0.9 meV at 14.5 T. The upper Hubbard band D^- seems thus to overlap with the conduction band in n-type InSb.

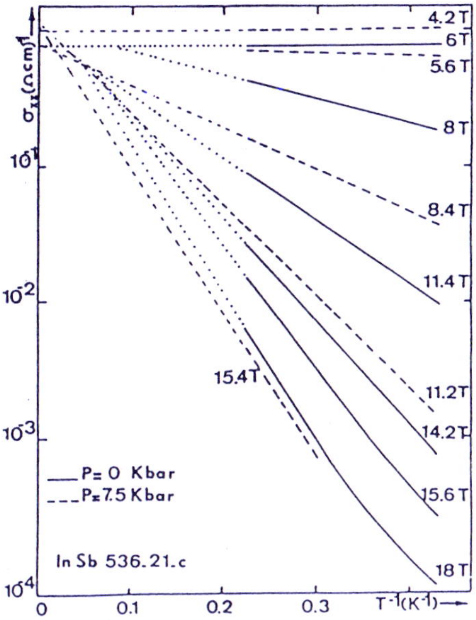

Figure 4

Temperature dependence of the σ_{xx} component of sample 536-21-c with and without pressure.

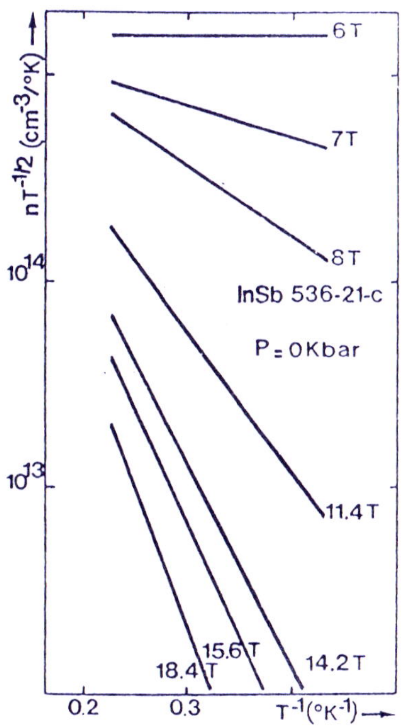

Figure 5

Plot of the curve ln n $T^{-1/2}$ versus 1/T for sample 536-21-c without pressure

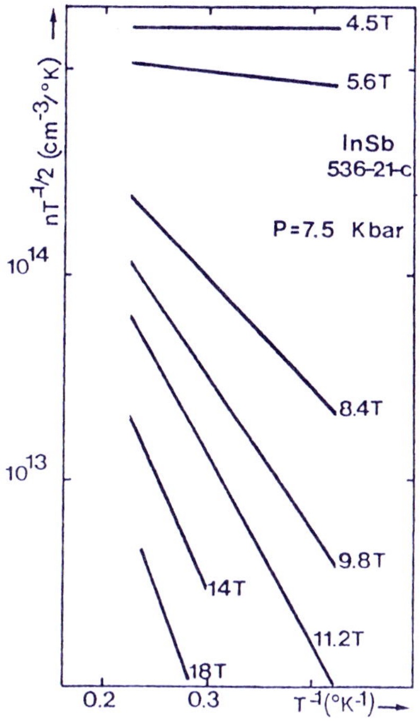

Figure 6

Plot of the curve ln n $T^{-1/2}$ versus 1/T for sample 536-21-c under hydrostatic pressure

If we attribute the activated regime to the ε_1 process, the dependence of the critical field on pressure is easily explained by an increase of the effective Rydberg (Eq.(1-7)). This hypothesis requires that the conduction occurs in the conduction band and the experimental values (table 1) of the mobility can be thus easily explained. In this case, the carrier concentration is related to the activation energy ε_1 by the well known formula

$$\frac{n(N_A + n)}{N_D - N_A - n} = N_B \exp(-\varepsilon_1/kT) \qquad (3\text{-}2)$$

To determine the concentration n we used the procedure described in reference (34). We have calculated (34) the position of the Fermi level at which the transition occurs. We observed that, for all the investigated samples, the transition occurs when the reduced Fermi energy η is nearly equal to zero i.e, when the Fermi level coincides with the last Landau sublevel 0^+ (g value is negative).

Figure (7) shows the magnetic field dependence of the free carrier density at different temperatures for sample 536-27-a, with and without pressure. For this sample we have

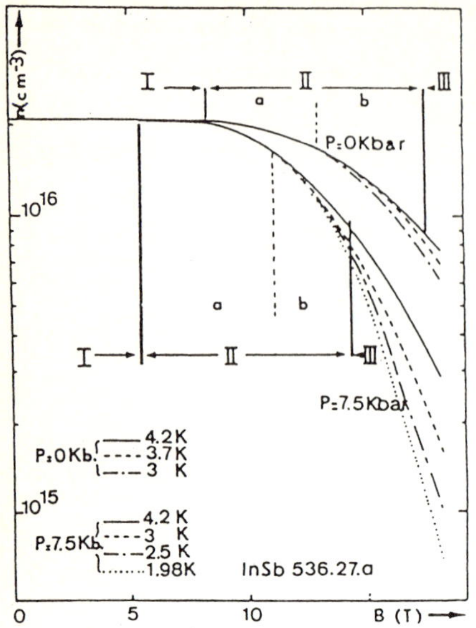

Figure 7

Magnetic field dependence of the free carrier density at different temperatures of sample 536-27-a with and without pressure

observed the Shubnikov-de Haas effect in the low magnetic field range, which corresponds to a degenerate statistics and a constant free carrier density (region I). When the magnetic field increases still further, a part of the electrons is trapped in the impurity band, although the electron gas in the conduction band is still degenerate. (The contribution to the σ_{xy} conponent of the trapped electrons can be neglected because of their low mobility). The free carrier density thus decreases without any activated process (region II). When the magnetic field increases still further, the impurity band and the last sublevel 0^+ are well separated and an activation energy which we associate to the ε_1 process appears (region III). This experimental result and its interpretation agree well with the theoretical results of Ghazali and Leroux-Hugon (18) who described the M-NMT solely in terms of the electron-electron interactions.

These results demonstrate the power of the simultaneous use of a magnetic field and a hydrostatic pressure in studying the nature of the metallic conductivity at the transition. In these experiments we have never used pressures in excess of 7 kbar [x] because of the presence of resonant states which could modify the electron trapping process (see Appendix 1). However, as suggested by Mott (37), it is difficult to ignore in our analysis the fact that the symetry of the wave functions is destroyed by the application of the magnetic field. So, it would be desirable to use a method which permits to avoid this problem. Such a method involves the use of metastable states associated with the X minimum in n-type.

3.2.2. New method of studying the M-NM transition in n-type InSb

By using the properties of the impurity levels associated with the X minimum of the conduction band, the free carrier density in InSb can be lowered to the record value of $n \simeq 10^{11}$ cm^{-3} at 4.2 K (Kadri et al 35). The concentration can be also varied, in the same sample, from 10^{11} cm^{-3} to 75 x 10^{13} cm^{-3} at temperatures from 4.2 to 95 K. It should be emphasised that the experiments with these extremely low free electron concentrations are performed in InSb free of any external stress.

As we previously said the "X impurity state" is metastable and lies about 140 meV above the Γ point at zero pressure. The main feature of this state is that a large potential barrier separates it from the conduction band states. If hydrostatic pressure exceeding 13 kbar is applied at a temperature higher than 140 K, electrons of the conduction band are trapped on this level in such a way that the decrease of the temperature below 100 K prevents any reexcitation of the carriers, even when the hydrostatic pressure is removed. If temperature is then raised to 100 - 110 K, a slow electron relaxation is observed, giving different "states" for the same sample characterized by different free carrier density.

We performed this experiment (35) on the purest sample (InSb Fd a) with an initial free carrier density of 7.5 10^{13} cm^{-3} at low temperature and a compensation ratio of about 5% (this value and the magnetic field dependence of ε_1 have been determined by

[x] In the first experiments that we performed using hydrostatic pressure, the pressure values were overestimated.

a graphic resolution of the neutrality equation for each magnetic field strength). From the Hall coefficient at low magnetic field (0.01-0.5 T), we obtained (Fig. 8), on the same sample, the temperature dependence of the free electron concentration for five different "states" (a,b,c,d,e,). Line f corresponds to intrinsic carrier concentration. In Fig. 9 we show the temperature dependence of the resistivity for the five "states". These two figures provide evidence for the M-NMT.

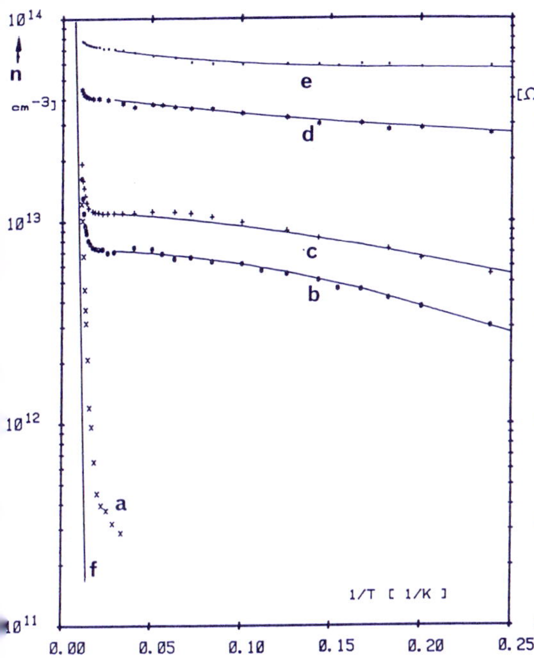

Figure 8

Electron concentration versus 1/T. Points are experimental, solid lines are calculated including compensation (≃ 5%)

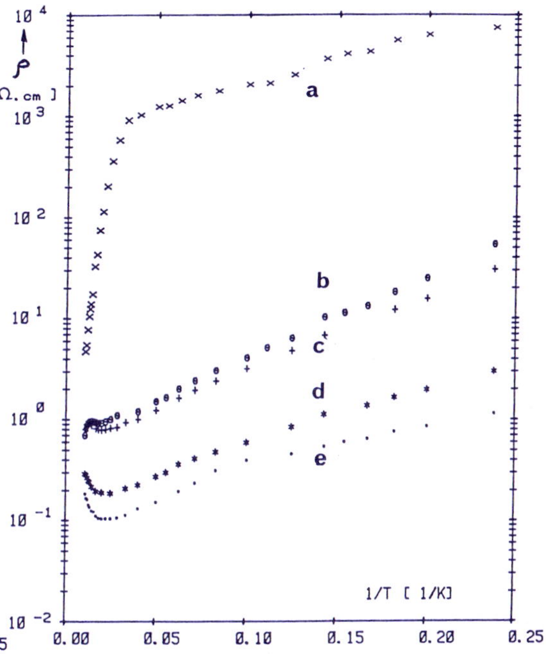

Figure 9

Variations of experimental resistivity versus 1/T

A detailed analysis of the data presented in Fig. 8 has been performed on the basis of the neutrality equation. This indicated the existence of three levels : a hydrogenic one, a deep level (55 meV) and one or several acceptor levels which are always ionized. The hydrogenic donor level is active at temperatures from 4.2 to 15 K. For the "state" 1b (Fig. 8) calculations in this temperature range give an activation energy of about 0.6 meV. This value is close to the effective Rydberg (ε_1 process) which is calculated to be 0.63 meV in n-type InSb. In this "state" the free electronic concentration is much smaller than the critical value ($n_c \simeq 6.10^{13}$ cm^{-3}). Thus the impurity band corresponding to the hydrogenic level is well separated from the bottom of the conduction band. For "state" (1c) in the same temperature range, the calculated activation energy is about 0.4 meV. This can be understood if we assume that the increase of the electron concentration broadens the impurity band moving it slightly towards the conduction band. On the contrary, calculations performed for "states" (1d) and (1e), give no activation energy in the whole temperature range. For both these cases, the free electron concentration is very close to the critical value n_c. Therefore in this situation, the impurity band should merge with the conduction band. These results agree with the theoretical works of Ghazali and Leroux-Hugon (18), and Ghazali and Serre (19) and show that, in InSb, the M-NMT is governed by e-e interactions.

§-4 CONCLUSION

We conclude that in "natural" n-type InSb studied up to date the D° and D^- states merge with the conduction band at zero magnetic field, even for the purest available samples. There is no evidence of M-NMT involving the D^- band. This seems natural since the ionisation energy of the D^- state in InSb is negliglibly small at zero field and does not exceed 1 meV at 20 T. By freezing out the conduction electrons into the donor state associated with the X band minimum, with the use of hydrostatic pressure we observed the temperature activated conductivity corresponding to the donor ionisation energy of 0.6 meV, in agreement with the theoretical extimations. This seems to be the first direct observation of the shallow donor in InSb in absence of external fields.

APPENDIX 1

In 1972, Konczykowski et al (38) have shown the existence of two donor impurity levels (probably oxygen impurity) which the authors associated with L and X minima. Later the metastable character of the impurity state associated with the X minimum has been demonstrated (39,40,41). The bottom of the conduction band at the Γ point and the two impurity levels associated with the X and L minima of the conduction band are so far apart that they cannot be brought together by the magnetic field alone . The distance between these levels can be reduced by the application of a hydrostatic pressure. Then the small energy difference which corresponds to the crossing of the levels, can be reduced to zero by the magnetic field. Porowski et al (41) used this method to observe the magnetic freeze-out involving the subsidiary minima and to induce a M-NMT at high temperature (50-140 K). We used the same experimental method (42) using a pressure at which the oxygen level (L) was still resonant with the Γ conduction band at zero magnetic field. The experiments have been made at low temperatures so that only the L level could be active ("X level" is a slow relaxation level). For lower magnetic fields range the classical magnetic freeze-out is observed involving the hydrogenic level associated with Γ minimum (ε_1 process). For the highest magnetic fields the freeze-out becomes much stronger. The interpretation of this experiment is based on the model of interaction of two hydrogenic states associated with two nonequivalent minima Γ and L when they cross each other. We have observed this kind of magnetic freeze-out and M-NMT on $Hg_{1-x}Cd_xTe$ (43). In this material the resonant level is not too far apart from the Γ minimum which allows the interaction at the absence of pressure.

Acknowledgements :

This article reviews the work of our group: Dr R.L. Aulombard, C. Bousquet, A. Kadri Dr A. Raymond and Pr J.L. Robert. I would like to acknowledge the cooperation with the Polish Physicists : Dr M. Baj, Dr L. Konczewicz and Pr S. Porowski. I am grateful to Pr W. Zawadzki for elucidating discussions. I thank Mrs J.C. Picoche and P. Rub of Service National des Champs Intenses (C.N.R.S.) in Grenoble for technical assistance during our experiments. Dr J. Chroboczek contributed to improve the form of the manuscript.

References

1. B. Gudden and W. Schottky, J. Tech. Phys. 16 323 (1935)
2. N.F. Mott, Proc. Phys. Soc. (London) 62, 416 (1949)
Phil. Mag. 6, 287 (1961) Adv. Phys. 16, 49 (1967)
3. P.W. Anderson, Phys. Rev. 109, 1492 (1958)
4. S.N. Mott, M. Pepper, S. Pollitt, R.E. Wallis and C.J. Adkins, Proc. R. Soc. Lond. A. 345, 169-205 (1975)
5. N.F. Mott, Adv. Phys. 16 49, (1967)
6. M.H. Cohen, H. Fritzsche and S.R. Owskinsky , Phys. Rev. Lett. 22, 1065 (1969)
7. P.W. Anderson, Comments on Solid State Phys. 2, 193 (1970)
8. Mott N.F., Phil. Mag. 26 1015 (1972) "Metal-Insulator Transitions" London : Taylor and Francis (1974)
9. A. Miller and F. Abrahams, Phys. Rev. 120 745 (1960)
10. N.F. Mott, J. Non Cryst. Solids 1, 1, (1968)
11. J. Hubbard, Proc. R. Soc. A277 237, (1964) Proc. R. Soc. A 281, 401 (1964)
12. H. Fritzsche, J. Phys. Chem. Solids 6, 69 (1958)
13. E.A. Davis and W.D. Compton, Phys. Rev. 140, A 2183 (1965)
14. C. Yamanouchi, K. Mizuguchi and W. Sazaki, J. Phys. Soc. Japan 22 859 (1967)
15. N.F. Mott, Adv. Phys. 21, 785 (1972)
16. F. Yonezawa, M. Watabe, M. Nakamura and Y. Ishida, Phys. Rev. 10 2332 (1974)
17. P. Antoniou and E. Economou, Phys. Rev. B16, 3768 (1977)
18. A. Ghazali and P. Leroux-Hugon, Phys. Rev. Lett. 41, 1569 (1978)
19. A. Ghazali and J. Serre, Phys. Rev. Lett. 48 886 (1982)
20. J.R. Klauder Ann. Phys. (N.Y.) 14, 43 (1961)
21. R.W. Keyes and R.J. Sladek, J. Phys. Chem. Solids 1, 143 (1956)
22. Y. Yafet, R.W. Keyes and Adams, J. Phys. Chem. Solids 1, 137 (1956)
23. W. Zawadzki and J. Wlasak, 14th Int. Conf. Physics Semiconductors, Edinburgh (1978) Ed. Wilson. The Institute of Physics, Bristol and London
24. J.R. Sandercock, Solid St. Com. 7 721 (1969)
25. P. Leroux-Hugon and A. Ghazali, Phys. Rev. B 14 602 (1976)
26. H. Miyazawa and H. Ikoma, J. Phys. Soc. Japan 23, 290 (1967)
27. H. Ikowa, J. Phys. Soc. Japan 30, 1509 (1971)
28. D. Ferré, H. Dubois and G. Biskubski, Phys. Stat. Sol. (b) 70 81 (1975)
29. S. Ishida and E. Otsuka, J. Phys. Soc. Japan 42, 542 (1977a) Ibid, 43, 124 (1977b)
30. E.M. Gershenzon, I.N. Kurilenko and Litvak-Gorskaya, L.B. Soviet. Phys. Semicond. 8, 689 (1974)
31. A.K. Walton and J.C. Dutt, J. Phys. C 10, L 29 (1977)
32. R. Mansfield and L. Kusztelan, J. Phys. C 11 1 (1978)
33. Y. Oka and S. Narita, J. Phys. Soc. Japan 28, 674 (1970)
34. J.L. Robert, A. Raymond, R.L. Aulombard and C. Bousquet, Int. Conf. "Impurity bands in semiconductors" Wurzburg (1979). Phil. Mag. B 42 1003 (1980)
45. A. Kadri, R.L. Aulombard, C. Bousquet, A. Raymond, J.L. Robert, S. Porowski and L. Konczewicz presented at 16th Int. Conf. Phys. of Semicon. Montpellier 1982
36. D.M. Larsen, Physics in High Magnetic Fields, Hakone 1980, Springer series in Solid State Sciences 24 (1981)
37. N.F. Mott, Phil. Mag. B 44, 265 (1981)
38. M. Konczykowski, S. Porowski and J. Chroboczek, Proc. XI Inter. Conf. Phys. Semicond. Warsaw, Vol. 2 p. 1050 (1972)
39. M. Konczykowski, S. Porowski and J. Chroboczek, High Temp. High Press. Vol. 6 p. 111 (1974)
40. E. Litwin-Staszewska, W. Szymanska, R. Piotrzkowski, Phys. Nar. Gap Semicond. Ed. Springer Verlag, Proc. Linz (1981) p. 387
41. S. Porowski, L.Konczewicz, M.Konczykowski, R.L.Aulombard and J.L.Robert 15th Inter. Conf. Physics of Semicond. Kyoto, 1980, J.Phys. Soc.Japan 49 Suppl. A 271 (1980)
42. S.Porowski, L.Konczewicz, A.Raymond, R.L.Aulombard, J.L.Robert and M.Baj, This conference : "The application of high magnetic fields in semicond.Physics, Grenoble (1982) "
43. A. Raymond, J.L. Robert, D.A. Kyriakos and M. Royer, presented at 16th Int. Conf. Physics of Semicond. Montpellier 1982

SHUBNIKOV-DE HAAS EFFECT IN HEAVILY DOPED n-InSb UNDER PRESSURE

L.Konczewicz, E.Litwin-Staszewska, S.Porowski
High Pressure Center "Unipress", Polish Academy of Sciences, 01-142 Warsaw, Poland

R.Aulombard
Université de Montpellier II, C.E.E.S. associé au C.N.R.S. 34060 Montpellier-Cédex

Abstract

The transverse magnetoresistivity and Hall effect have been investigated in n-InSb heavily doped with sulphur and tellurium. The measurements have been performed under hydrostatic pressures up to 2500 MPa, magnetic fields up to 20 T and in the temperature range from 4.2 K to 120 K.

The distinct oscillations of both magnetoresistivity and Hall effect have been observed. From the analysis the amplitude of SdH oscillations, the value of Dingle temperature T_D has been determined. For the first time the pressure dependence of Dingle temperature for different impurities in InSb has been reported up to 2500 MPa. In the low pressure range /n(p)=const./ the experimental T_D/p/ dependence agrees with the theoretical calculations which take into account the scattering on both the ionized impurities and on the short range potential centers. At the highest pressure range the distinct difference between sulphur and tellurium doped samples is observed. For Te-doped samples the amplitude of SdH oscillation softly increases with pressure and for the S-doped ones a sharp increase both in the amplitude and the period of the oscillations take place. Such effect for S-doped material has been attributed to the transfer of electrons between conduction band and resonant states of S(1). However, the observed abrupt decrease of the damping of the oscillations is too strong to be explained by the decrease of the free carrier concentration and the improvement of the homogeneity of the sample when partial deionization of resonant S-states take place and suggest that direct interaction between free and resonant electrons is responsible for this effect.

1. E.Konczewicz, E.Litwin-Staszewska, S.Porowski, Proc.Int.Conf.on Phys.Narrow Gap Semicond., Warsaw 1977

INTERACTION BETWEEN Γ AND L MINIMA ASSOCIATED IMPURITY STATES IN n-InSb**

S. Porowski, L. Konczewicz,

High Pressure Research Center "Unipress", Polish Academy of Sciences Warsaw, Poland

A. Raymond, R.L. Aulombard, J.L. Robert, M. Baj*

University of Montpellier, France

Introduction

Due to the small effective mass in the Γ conduction minimum of InSb the effect of magnetic field on shallow donors is very strong and has been a subject of very intensive studies /1,2/. Recently it was shown that typical donors /Te, Se, S and O/ besides the shallow states give also the nonhydrogenic /strongly localized/ resonant states lying above the bottom of the conduction band /3/. The effect of magnetic field on their energy relative to L and X minima is not very strong, however the shift of the bottom of Γ minimum with magnetic field causes the substantial decrease of their energy relative to Γ minimum /4/. Also the pressure decreases their energy relative to minimum Γ and two lowest resonant states, which are probably due to oxygen, can be pushed into the forbidden gap. As a consequence - at pressures above 800 MPa when nonhydrogenic states become deeper than the shallow ones, magnetic field causes strong freeze-out of the free carriers into these states /4/. In this paper we present the measurements of the magnetic freeze-out effect in the pressure and magnetic field range when energies of both hydrogenic and nonhydrogenic levels of oxygen are very close one to the other and freeze-out into both types of levels has to be taken into account. For this energetic configuration one can also expect anomalous effects due to the interaction between the levels. Such effects have been recently observed in magneto-optical experiments /5/.

Experimental results

Two samples of high purity n-InSb with the free carrier concentration /at T=77 K/ : $n_1 = 7.5\ 10^{13}\ cm^{-3}$ and $n_2 = 1.5\ 10^{14}\ cm^{-3}$ were studied. The Hall coefficient R_H and transversal magnetoresistivity ρ as a function of magnetic field up to 18 T in the temperature range 4.2K - 14K and pressure range 600 MPa $\leqslant P \leqslant$ 800 MPa were measured. The CuBe clamped cell was used as a pressure vessel /6/. The cryostat with the cell was placed inside the 10 MWatt Bitter type solenoid /SNCI Grenoble/. In fig. 1 the experimental n/B/ dependences at high pressure (for the sample 1) for different temperatures are presented. Two regions of magnetic field can be distinguished. For the lower magnetic field range the free carrier concentration n slightly decreases with magnetic field and then the effect becomes much stronger. The results for the sample 2 are presented in fig. 2 and fig. 3. Fig. 2 shows n/B/ dependence at T=10K for several pressures. The curves are step-like in character and the region of sharp decrease of n/B/ shifts with pressure toward lower magnetic field. In Fig. 3 the n/B/ dependences at P=720 MPa for different temperatures are presented. Like for the sample 1 with decreasing temperature the decrease of n/B/ becomes more abrupt.

Analysis of the Experimental Results

Infrared magneto-optical spectra of high purity InSb show well defined central cell structure. The 1s-2p_ transitions are split into 4 lines A,B,C and D /5/ line. It is believed that the line A corresponds to oxygen impurities and the lines B,C and D to other donors.

* On leave from University of Warsaw, Poland

** All the experiments have been performed at the SNCI-CNRS Grenoble France

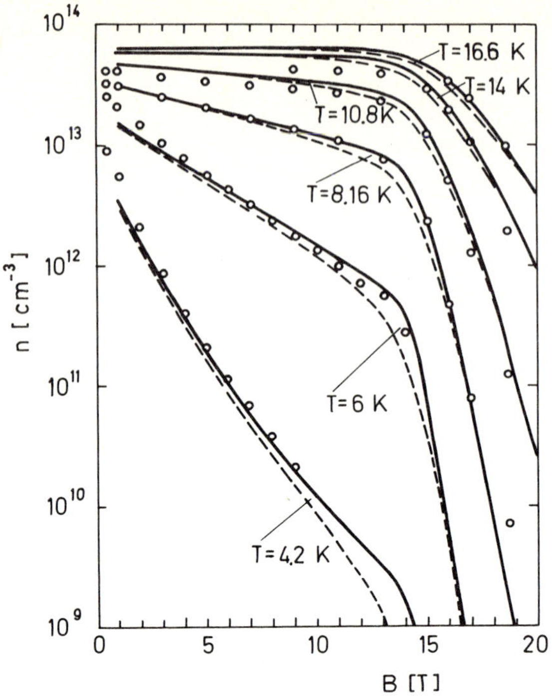

Fig. 1 - The n/B/ dependences at P = 553 MPa for several temperatures. Results for the sample no.1. The theoretical curves : ——— for the interaction parameter 2B = 0 /no interaction/ ; — — — 2B = 2 meV

The oxygen centers /A/ are characterized by the concentration N^o and three energies E^o_Γ, E^o_L and E^o_X /energies relative to the Γ minimum/. The pressure coefficients of E^o_L and E^o_X are -10.5 meV/100 MPa and -20 meV/100 MPa respectively /6-7/. The pressure dependence of E^o_Γ level which is a shallow one is about two orders of magnitude weaker /8/ and has to be found separetely for each sample because due to the interactions between impurities it depends on donor concentration in a sample. At low temperatures the E^o_X states are metastable due to its strong lattice relaxation /7/ and their population does not change with magnetic field. For the donors which correspond to the lines B, C and D the energy separation between their ground states is so small that for analysis of the transport data they can be characterized by the only one energy E^h and the total concentration N^h. The existence of nonhydrogenic states for these donors can be neglected because their energy is much higher than E^h and they are never populated.

The effect of magnetic field on E_Γ and E_L states is similar to that of pressure - magnetic field increases the energy of Γ-minimum associated states faster than L-ones and therefore can also cause the crossing of these levels. To describe the magnetic field dependences of E_Γ and E_L in the crossing range it is necessary to take into account the interaction between them. In our analysis the Altarelli model /9/ describing the pressure induced crossing of two hydrogenic Γ and L minima

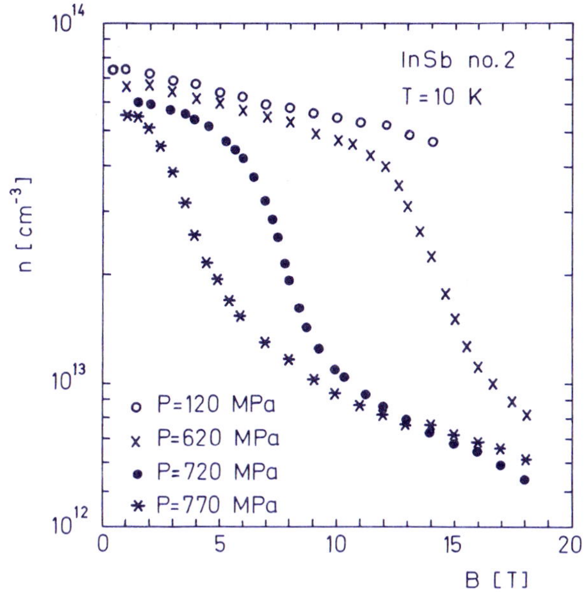

Fig. 2 - The n/B/ dependences at T=10K for several pressures. Results for the sample no. 2.

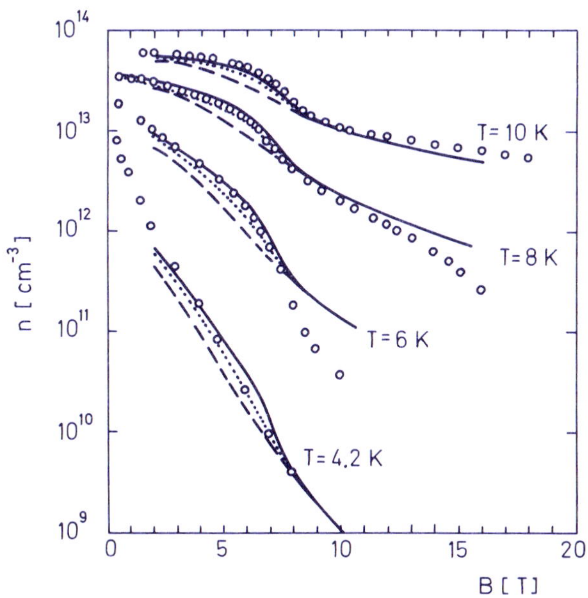

Fig. 3 - The n/B/ dependences at P=720 MPa for several temperatures. The results for the sample no.2. The theoretical curves : ——— for the interaction parameter 2B=0 /no interaction/ ; ··········· 2B=2 meV ; — — — 2B=4meV.

associated states was adopted. We make a simplifying assumption that pressure and magnetic field shifts of level energies are equivalent. We disregard the fact that the pressure and magnetic field effects on the wave functions are different. The model predicts that 4 fold degenerate level associated with L minimum is split by interaction of 4 equivalent L minima into $E_{L,\Gamma 1}$ singlet and E_{L,Γ_4} triplet /without spin/. When two levels with Γ_1 symetry get closer they repulse each other due to the noncrossing rule. Γ_4 level is not affected by this interaction /see fig 4a/. According to /8/ the interaction between both Γ_1 levels is described by the formula :

$$E'_{\Gamma,L} = \frac{1}{2} \left(E_\Gamma + E_L \pm \sqrt{(E_\Gamma - E_L)^2 + 4B^2}\right)$$

where E_Γ and E_L are the nonperturbed by interaction energies of Γ and L minima associated state with Γ_1 symmetry and 2B is the minimum energy separation of two Γ_1 states /see fig. 4b/. In the numerical calculations it was assumed that :

- the magnetic field dependence of oxygen E^o state out of the interaction region is the same as for shallow B,C and D states : $E^o_\Gamma/B/ = E^h_\Gamma\ /B/$
- the positions of L minimum and of E_L states relative to L minimum are not affected by the magnetic field and as a consequence the $E^o_L/B/$ dependence is described only by the shift of the lowest spin-up Landau subband of Γ minimum with the magnetic field : $E^o_L(B) - E^o_L(0) = 1/2\ (\hbar\,\omega_c - g^* \mu_B . B)$
- the interaction takes place between spin-up Landau subband shallow level E^o_Γ and the spin-up $E^o_{L,\Gamma 1}$ level / at high magnetic field the spin-down Γ states have considerably higher energies/

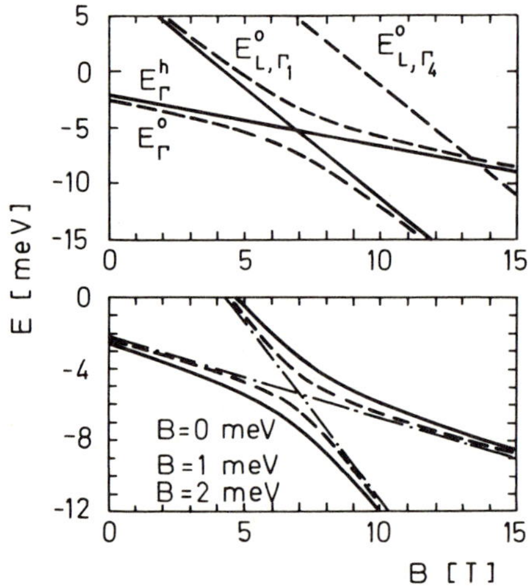

Fig. 4 - Theoretical model of interaction of Γ and L minima associated states.

The neutrality equation is described in the form :

$$N_{tot} = n_\Gamma/B/ + n^h/B/ + n^o/B/$$

where N_{tot} is the total carrier concentration determined at p=0 and T=77K. n_Γ is the carrier concentration in conduction band. It was calculated according to the formula for nonparabolic band at high magnetic field /10/ with the following band parameters : $E_g/p=0/=0.2357$ eV, dE_g/dp = 15 meV/100 MPa, $m^*_o = 0.0415$, $g^* = -50$. n^h and n^o are the concentrations of electrons occupying hydrogenic and oxygen states respectively and they were calculated using classical formula for the excited impurity states.

The model describes well the experimental data. The solid and dashed lines in fig. 1 and fig. 3 are calculated from this model. The low magnetic field n/B/ dependences correspond to classical freeze-out into the shallow levels, the steepest part of the n/B/ dependence is due to the freeze-out into the L associated level. The intermediate part is determined by the deionization of both interacting states. The different character of n/B/ dependences for sample 2 is caused by the fact that for this sample $N^o < N_{tot}$. At sufficiently high magnetic field all E_L^o states are deionized and the curve corresponds again to the frezee-out into the shallow donors. The fitting has been done for the interaction energies equal 2B=0, 2 meV, 4 meV. The satisfactory agreement is obtained for two smallest value of 2B therefore we can conclude that the interaction energy is about 2 meV /or lower/. The small value of the interaction parameter makes its precise determination difficult because thermal spread out of the distribution function is of the same order. Unfortunatelly the lowering of the temperature does not improve much the precision of the energies determination. It is due to the strong increase of the classical freeze-out into shallow donors what makes the R_H measurements difficult.
From the other hand at low magnetic field when freeze-out is not yet very strong and the R_H measurements are possible the effect of level interaction is additionaly masked by interaction between impurities due to large Bohr radius.
Recently another theoretical model has been developed by Trzeciakowski and Krupski /11/ which also predicts the anomalous deepening of the shallow donors /in InSb/ and even the change of its character from Γ to L under pressure. The deepening arises in this case from the increase of the chemical shift due to the compression of the wavefunction in the range of the local potential of the donor. However this effect is much less rapid in comparison with the observed data. In the presented experiments the observed range of Γ minimum displacement, at which the change from Γ to L level character takes place, is less then 10 meV. To observe the similar effect the model of Trzeciakowski and Krupski predicts the value of at least one order of magnitude higher. Although the interaction proposed by Trzeciakowski model cannot be excluded, the predominant character of the observed effect must be due to the crossing of the levels.

References

/1/ J.L. Robert, A. Raymond, R.L. Aulombard, C. Bousquet.
Phil. Mag. B (G.B.), 42, 6, (1980).
/2/ R. Kaplan
Proc. Int. Summer School Narrow Gap Semidond., Nimes, 1979.
/3/ S. Porowski
Proc. 4 th Int. Conf. Phys. Narrow Gap Semicond., Linz, 1981.
/4/ S. Porowski, L. Konczewicz, M. Konczykowski, R.L. Aulombard, J.L. Robert
Proc. 15 th Int. Conf. Phys. Semicond., Kyoto, 1980.
/5/ Z. Wasilewski, A.M. Dawison, P. Knowles, S. Porowski, R.A. Stradling
Proc. 4 th Int. Conf. Phys. Narrow Gap Semicond., Linz, 1981.
/6/ S. Porowski, L. Konczewicz, J. Kowalski, R.L. Aulombard, J.L. Robert
Phys. Stat. Sol. (b), 104, p. 657, (1981).
/7/ L. Dmowski, H. Konczykowski, R. Piotrzkowski, S. Porowski
Phys. Stat. Sol. (B), 73, (1976)
/8/ R.L. Aulombard, A. Raymond, L. Konczewicz, J.L. Robert, S. Porowski
Int. Sem. Appl. High Magnetic Fields in Phys. Semicond., Hakone, 1980
/9/ M. Altarelli, G. Iadonisi
Il Nuovo Cimento, 5, (1971).
/10/ D. Dobrowolski
Phys. Stat. Sol. (6), 70, K63, (1975).
/11/ W. Trzeciakowski, J. Krupski
Will be published in Sol. St. Comm.

ANOMALOUS NMR RELAXATION PEAK IN InSb IN THE EXTREME QUANTUM LIMIT

R E Harper and D P Tunstall
Department of Physics, University of St Andrews, St Andrews, Fife, U.K.

The extreme quantum limit in doped InSb shows up a strong nmr relaxation peak as the external magnetic field is swept. This peak is explored at helium temperatures in fields up to 11T and doping densities up to $10^{23}m^{-3}$. A beat between the plasma zero-point motion and the electronic Larmor frequency is conjectured as an explanation of the effect.

I. Introduction

Some years ago Bridges and Clark[1,2] (BC) published papers on nuclear relaxation in transmutation-doped InSb samples, covering a wide range of both concentration and temperature. Their investigations highlighted, amongst other things, an anomalous peak at low temperature in the relaxation rate of all 3 nuclear (In^{115}, Sb^{121}, Sb^{123}) species in the region of the magnetic field just before magnetic freeze-out occurs. This peak was labelled the A peak, and it exhibited a $B_A \propto n_D^{\frac{1}{2}}$ dependence, where n_D is the donor density, over a density range $8 \times 10^{20} < n_D < 30\ 10^{20} m^{-3}$, and for fields ranging up to 2.0 Tesla. All previous, and subsequent, studies have placed the onset of magnetic freeze-out at fields greater than the observed B_A. For fields below B_A, BC detected standard relaxation peaks associated with Landau levels and electron-nuclear flip-flops via the Fermi contact interaction as the dominant nuclear relaxation mechanism. Further work[3,4,5] failed to provide any convincing reason for the appearance of this strong relaxation peak in the region of field defined as the extreme quantum limit.

The extreme quantum limit in doped semiconductors, where all conduction electrons are in a single sub-Landau level, has been the subject in recent years of some experimental and theoretical interest[6,7,8], with the main focus concentrating on the possibilities of observing a Wigner transition or charge density wave in this regime. InSb provides a semiconductor system where the extreme quantum limit is defined by the $B \propto n_D^{0.66}$ dependence of the so-called 0^--line in magneto-transport (where the field is just sufficiently large to accommodate all the n_D conduction electrons in the lowest sub-Landau level), as a lower limit, and by the $B \propto n_D^{0.83}$ dependence[9] of the onset of magnetic freeze-out, as the upper limit.

The nuclear spin-lattice relaxation rate $1/T_1$ is sensitive to magnetic fluctuations at the relevant nuclear Larmor frequency. In a metallic system these fluctuations are normally generated by a cancellation between the time dependence of the transverse electron spin precession at the electronic Larmor frequency (huge in InSb) and the change of time dependence of the phase factor $e^{-iE_k t}$ of the spatial wavefunction describing an electron of energy E_k and wave-vector k at the Fermi surface.[11] An important element in our study is the fact that this usual nuclear spin-lattice process becomes forbidden in the extreme quantum limit, because the electronic Larmor frequency becomes larger than any possible change in the time dependence of $e^{-iE_k t}$ i.e. $(g\beta B)_{electronic} > E_F$, the Fermi level measured from the bottom of the only occupied Landau sub-band.

II. Experimental Details

A. Samples

The samples were cut from high purity InSb($8 \times 10^{19} m^{-3}$) in the shape of rectangular bars ~15x1x1 mm, and a range of doping densities from $1 \times 10^{21} m^{-3}$ to $1 \times 10^{23} m^{-3}$. was obtained by inserting the bars for different times into the neutron flux of the Scottish Universities Research Reactor (SURRC); the thermal to fast ratio was approximately two for these samples.[10] Two other bars were irradiated in the PLUTO reactor at Harwell, where the thermal to fast ratio of the neutron flux was 800:1. All bars were wrapped in indium foil in order to remove the 1.4ev resonance neutrons from the beam. A check experiment was also run on a Te melt doped sample, supplied by Mining and Chemical Products.

After doping, and an appropriate period for cooling of the radioactivity, samples were annealed at 350°C for 1 hour[10] and etched in CP4.

B. T_1 measurements

Most of the data was obtained with a fixed frequency nmr pulsed spectrometer operating at 50MHz and, in order to obtain the magnetic field variation of T_1, 'field-cycling' was employed; this is a technique whereby the relevant nuclear polarisation is first saturated at the field corresponding to resonance at the 50MHz frequency of the spectrometer, then the field is quickly switched to the value at which a measurement of T_1 is required. Nuclear polarisation recovery proceeds via spin-lattice interaction for a pre-determined time τ, and then the field is switched quickly back to the 50MHz initial value for a measurement of how much nuclear polarisation has recovered. Because of the enormously long T_1's involved in these samples, the demands placed on the magnet by the requirement of fast switching between fields were minimal, and were easily satisfied by the performance of our 11T superconducting magnet, in which zero to full field sweep times of as little as 3 minutes could be obtained.

III. Experimental Results

One of the main aims of this study was to follow the density dependence of the BC work up to higher fields and densities. The results on this aspect are shown in Figure 1.

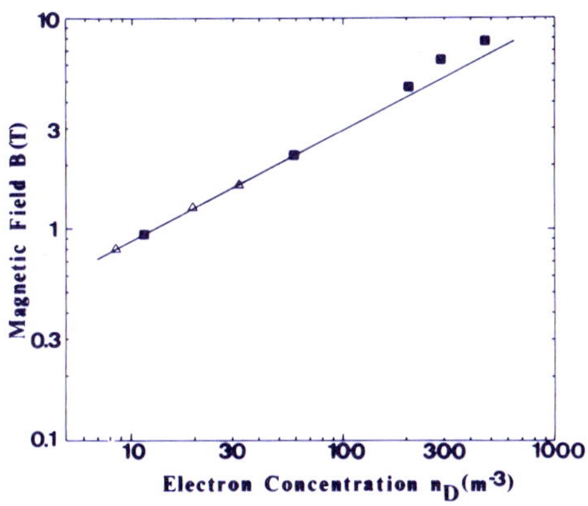

Figure 1. A plot of the field B_A at which $1/T_1$ is a maximum, in the extreme quantum limit, against doping density n_D(in m^{-3}). The full line represents $B_A \propto n_D^{\frac{1}{2}}$ drawn through, as a best fit, the low density data. The open triangles are BC data, squares represent our data.

Several other important features emerged; (i) all three types of sample studied (SURRC, PLUTO, melt) exhibited A-peaks of the same general characteristics, and at fields B_A such that it is reasonable to suppose that Figure 1 is a true representation of the density dependence for all sample fabrication methods.

(ii) Throughout the density range B_A takes the same value in a given sample for all three isotopic species.

(iii) At low densities the amplitude of the peak, i.e. the value of $1/T_1$ at B_A, is greater for Sb^{123} than for Sb^{121}; at the higher densities, $1/T_1$ at B_A for Sb^{123} is less than for Sb^{121}.

(iv) the temperature dependence of the peak position B_A becomes very weak at high density, whereas BC reported 10% variations in B_A on cooling from 4.2 to 1.5K.

(v) the peak amplitude of the higher end of the density range $1/T_1$ at B_A, seems to vary with temperature T in the liquid helium range approximately in a manner consistent with T_1T = constant.

IV. Interpretation

To understand the context of the observations above, some additional information is necessary. Table 1 lists nuclear data:

TABLE 1

Nucleus	I	Abundance(%)	NMR Frequency in 1T(MHz)
In^{115}	9/2	96.0	9.3
Sb^{121}	5/2	57.3	10.2
Sb^{123}	7/2	42.7	5.5

Figure 1 supports in large measure the initial, low-density, diagnosis of BC that $B_A \propto n_D^{2}$. Point (i) above reinforces the belief that the A-peak phenomenon is a property of doped InSb and not an effect associated with impurities. Point (ii) above, where B_A is the same for Sb^{121}, Sb^{123} and In^{115} in a given sample, indicates strongly that the peak phenomenon must be associated with some modification of the conduction electron states induced by the magnetic field, whereas point (iii) indicates that this modification interacts more strongly, in the low density samples, with the Sb^{123} nuclei than with Sb^{121}, in spite of the low value of the Sb^{123} magnetic moment. A low value of this moment usually means that nuclear relaxation via magnetic coupling with the conduction electrons proceeds more slowly than for isotopes with larger moments. In the region of field below the extreme quantum limit, $1/T_1$ for Sb^{123} is indeed substantially less than for Sb^{121}. In the present case, the difference must be that, at a given B_A, Sb^{123} has a substantially smaller nuclear Larmor frequency than Sb^{121} and that the relevant fluctuations causing nuclear relaxation are much stronger at the lower Larmor frequency.

Points (iv) and (v) above emphasize that as the electronic system is pushed towards stronger degeneracy by increasing the density the temperature effects become characteristic of normal metals.[11]

These nmr results as outlined above define a boundary within which any interpretation must operate. One of the most striking connections that has already been noticed in this A-peak phenomenon[4] is that between ω_p, the electronic plasma frequency, and $g\beta B_A/\hbar$ the electronic Larmor frequency. Our measurements (Fig.1) reinforce this connection. One can surmise that the deviation from $B_A \propto n_D^{2}$ at high density is due to the reduction in the electronic g-value as n_D increases,[12] so that matching of the $\omega_p = g\beta B_A/\hbar$ condition requires a higher B_A.

We believe then that these experiments indicate that as the electron spin resonance frequency is swept through the plasma frequency, some low frequency fluctuations appear at the nuclear sites; at low density the width in frequency of these low frequency fluctuations is rather narrow, sufficiently so as to make their spectral density much stronger at Sb^{123} frequencies than at those of Sb^{121}. This width in frequency of the fluctuations, perhaps due to interactions between the zero-point plasma motion and the electronic Larmor precession, must then broaden as the doping density increases, so permitting, at high density, $1/T_1$ at the Sb^{121} frequency to become once again bigger than $1/T_1$ at the Sb^{123} frequency.

References

1. F Bridges and W G Clark, Phys.Rev. 164 288 1967.
2. F Bridges and W G Clark, ibid 182 463 1969.
3. G A Miranda and W G Clark Phys.Rev. B9 495 1974.
4. L Gunther, M Revzen and A Ron, Physics 3 115 1967.
5. G Benford and N Rostoker, Phys.Rev. 182 375 1969.
6. G Nimtz and B Schlicht, Advances in Solid State XX 369 1980.
7. W G Kleppman and R J Elliott, J.Phys. C8 2729 1975
8. Y Kuramoto, J.Phys.Soc. Japan 44 1572 1978.
9. J L Robert, A Raymond, R L Aulombard and C Bousquet, Phil.Mag.B 42 1003 1980.
10. W G Clark and R A Isaacson, J.Appl.Phys. 38 2284 1967.
11. see C P Slichter, Principles of Magnetic Resonance, Springer-Verlag, 1978.
12. N G Gluzman, L D Sabirzyanova, I N Tsidil 'kovskii, Sov.Phys.Semicon. 13(3) 275 1979.

Dielectric Function of Zero-Gap HgCdTe and HgMnTe in a Magnetic Field

R.Dornhaus and R.Faymonville, I.Physikalisches Institut der RWTH Aachen, D-5100 Aachen, F.R.Germany
G.Bauer, Institut für Physik, Montanuniversität Leoben, A-8700 Leoben, Austria
R.R.Galazka, Institute of Physics, Polish Academy of Sciences, Warsaw, Poland

In zincblende zero-gap materials the dielectric tensor contains contributions of both Γ_8-intra and Γ_8-Γ_8-interband transitions in the same frequency region and, hence, represents a frequency- and magnetic-field dependent function. Magnetooptical investigations have played an essential role in the determination of the detailed bandstructure of HgCdTe alloys (for two recent reviews see /1,2/). In most cases magnetic spectroscopy at fixed frequency and variable magnetic field has been used, resulting in elaborated Fan-charts displaying transition energies versus magnetic field, but little information on the complete dielectric function has been obtained. Also the theoretical determination of the interband contribution of the dynamical dielectric function in a magnetic field has proven to be very difficult even in the collisionless case considered up to now /3,4/.

We have studied far infrared ($20cm^{-1} < \omega/2\pi c_o < 700cm^{-1}$) magnetoreflectivity of $Hg_{1-x}Cd_xTe$ ($0.14<x<0.18$) and $Hg_{1-x}Mn_xTe$ ($0.04<x<0.10$) close to the transition from zero-gap to narrow-gap conditions in magnetic fields up to 5T using Fourier transform spectroscopy in Faraday and Voigt ($E\parallel B$ and $E\perp B$) configuration. The data were analysed by Kramers-Kronig relations in the case of Voigt geometry. In Faraday geometry no circular polarized measurements were possible and, hence, such an analysis is not valid.

Magnetooptical intraband and interband transitions were observed below, between and above the reststrahlen bands in both materials.

An example of measured reflectivities and corresponding $Im(\varepsilon_\perp) = Im(\varepsilon_{xx}+\varepsilon_{xy}^2/\varepsilon_{xx})$ for $Hg_{.845}Cd_{.155}Te$ in Voigt ($E\perp B$) geometry is shown in Fig.1. Besides the plasma-cyclotron-resonance two interband transitions near $3\omega_c/4$ and $5\omega_c/4$ may easily be identified in reasonable agreement with calculations /3/ and established selection rules /5/.

Magnetooptical investigations of $Hg_{1-x}Mn_xTe$ were performed so far both in zero-gap ($x<0.075$, T=5K) as well as in the semiconducting range (positive energy gap). The investigation of interband magnetoabsorption ($\Gamma_6 \rightarrow \Gamma_8$) in zero-gap /6,7/ and positive gap ($\Gamma_8 \rightarrow \Gamma_6$) /7,8/ alloys has emphasized the strong influence of spin-spin exchange interaction between the localized magnetic moments and band electrons on the spin splitting of Landau levels in these materials.

The influence of exchange interaction on the $\Gamma_8 \rightarrow \Gamma_8$ magnetoabsorption in zero-gap ($x<0.02$) $Hg_{1-x}Mn_xTe$ was investigated by Pastor et.al. /9/ using far infrared laser lines.

For the $Hg_{1-x}Mn_xTe$ samples investigated here the interaction gap E_o is about -100 meV for x=0.04 and -40meV for x=0.06 (T=5K) /7,8,10,11/.

In contrast to zero-gap $Hg_{1-x}Cd_xTe$, where an applied magnetic field always opens the gap, in $Hg_{1-x}Mn_xTe$ with an inverted band structure the situation is more complicated. The exchange interaction pushes some of the heavy hole states upwards /10,12/ particularly the $b_v(-1)$ level which overlaps with the first electronic level a(0) (using the notation of ref. /12/).

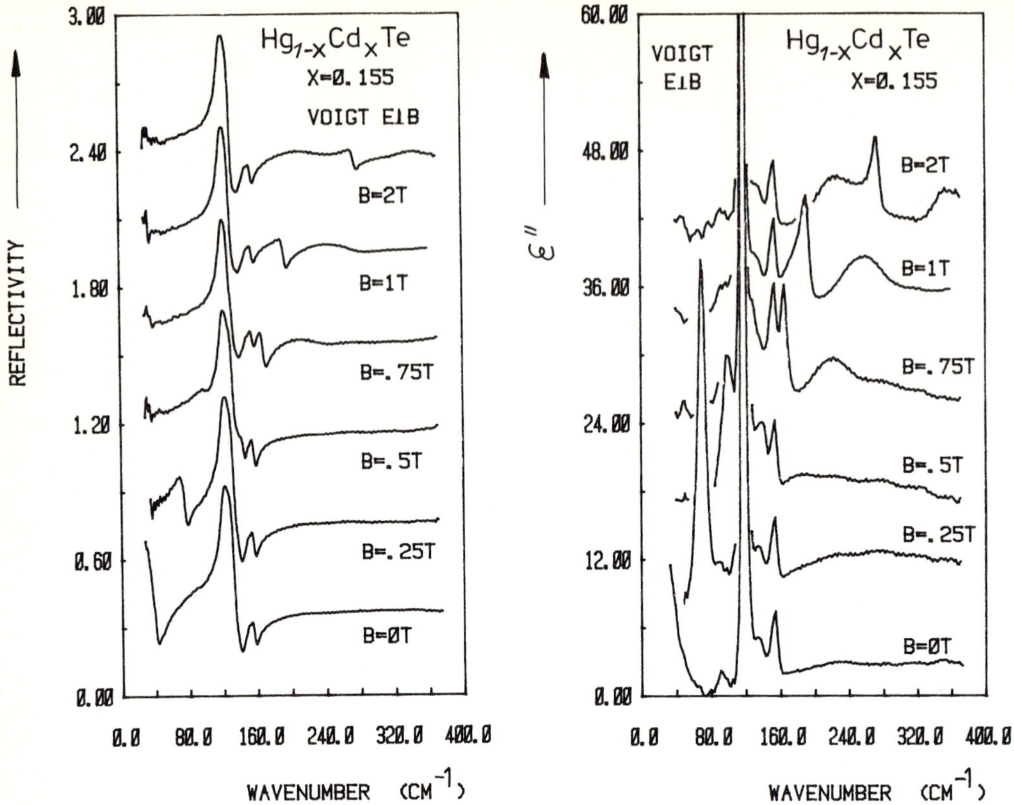

Fig. 1: Reflectivity of $Hg_{.845}Cd_{.155}Te$ in Voigt-configuration (E⊥B) at T=6K for various magnetic fields between 0 and 2 T (left hand side) and $Im(\varepsilon(\tilde{\nu},B))$ obtained from reflectivity curves by Kramers-Kronig analysis (right hand side).

Depending on composition and temperature, in higher fields, the magnetic field induced gap opens when the exchange terms in the Hamiltonian become smaller than the usual terms, linear in B /12/. In low fields, a semimetallic situation exists due to the overlap of the bands, direcly proved by magnetotransport experiments for samples with x up to 0.05 /12,13/.

Our experiments were performed just in this range. Fig. 2 shows reflectivity spectra as a function of $\tilde{\nu}$ for various fields for E∥B as well as for E⊥B. In Voigt-configuration these data can be analysed by Kramers-Kronig relations yielding the dielectric function in a magnetic field.

In Fig. 3 $Im(\varepsilon(\tilde{\nu},B))$ is shown for various fields for another HgMnTe sample with x=0.04. For B=0 the structures in $Im(\varepsilon)$ are due to the HgTe- (117 cm^{-1}) and MnTe- (190 cm^{-1}) like TO phonon modes and an additional optical mode (96 cm^{-1}) , probably resulting from Mn-clustering effects. With the application of a magnetic field several magnetooptical ($\Gamma_8 \rightarrow \Gamma_8$) transitions appear with different selection rules for E∥B and E⊥B /9/.

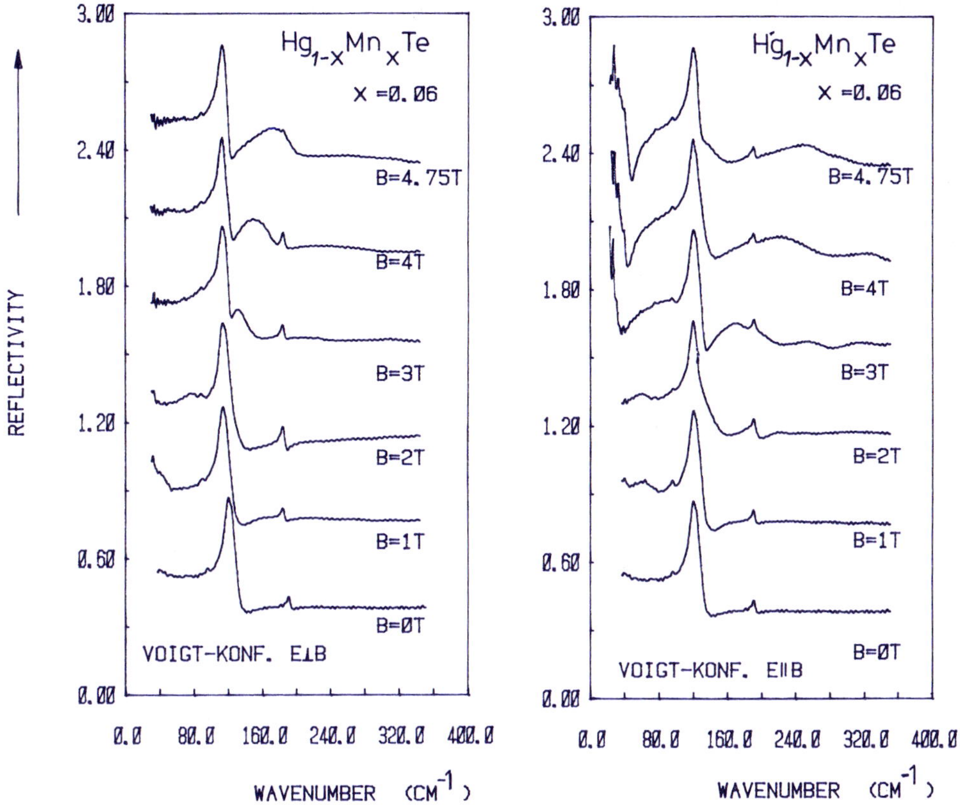

Fig. 2: Reflectivity of $Hg_{.94}Mn_{.06}Te$ in Voigt-configuration with E⊥B (left hand side) and E∥B (right hand side) at T=6K for various magnetic fields in the range between 0 and 4.75T.

It is interesting, that especially in the E⊥B configuration pinning phenomena are observed as can be directly seen from the dielectric function (Fig.3). These phenomena are particularly strong with the HgTe-like phonon mode (the mode with the largest oscillator strength). As evidenced in Fig. 3 in the E∥B configuration these pinning phenomena are less pronounced.

Our experimental results indicate, that both inter- and intraband contributions to the dielectric function have a pronounced influence on the dependence of the reflectivity on frequency in a magnetic field in zero gap materials like HgCdTe and HgMnTe. These contributions can be evaluated from measurements in Voigt-configuration by Kramers-Kronig analysis and yield information not only on transition energies and selection rules, and hence the bandstructure in a magnetic field, but also on the complete dynamical dielectric function. For the analysis of our reflectivity measurements in Faraday configuration model calculations would be necessary including the influence of finite temperatur and damping on the interband part of the dielectric function, which are not yet available.

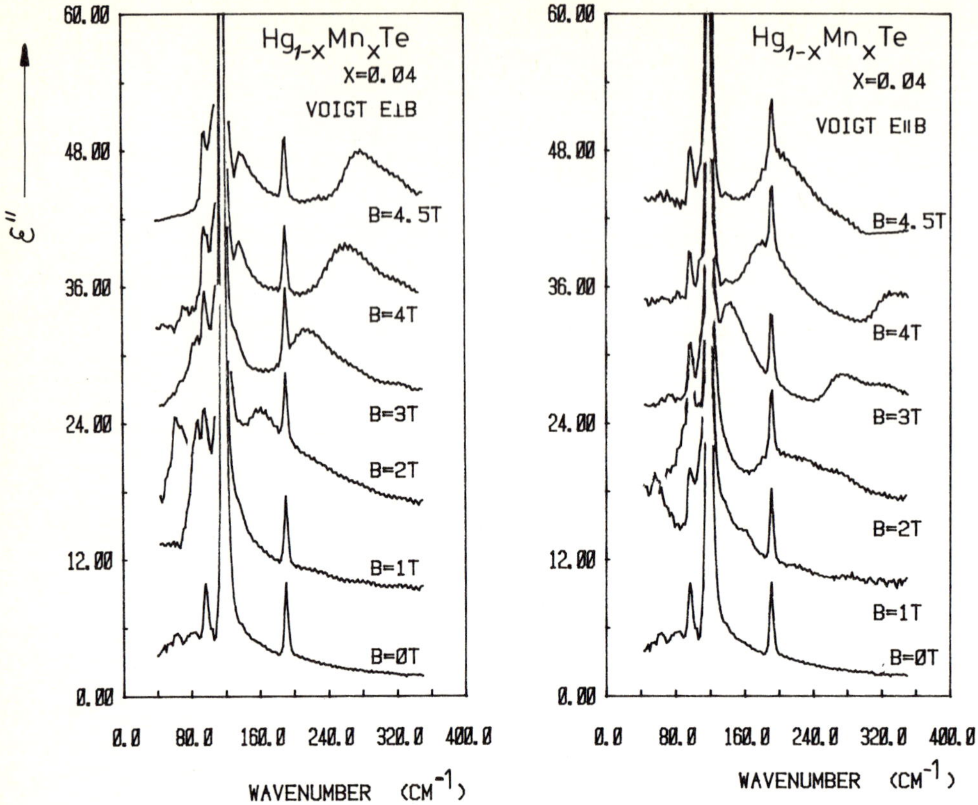

Fig. 3: Im(ε($\tilde{\nu}$,B)) for $Hg_{.96}Mn_{.04}Te$ in Voigt-configuration with E⊥B (left hand side) and E∥B (right hand side) at T=6K for magnetic fields between 0 and 4.5T obtained from reflectivity curves by KK-analysis.

References:

/1/ M.H.Weiler in: Semiconductors and Semimetals, ed. by R.K.Willardson, A.C.Beer (Academic, N.Y. 1981), Vol.16,p.119.
/2/ R.Dornhaus, G.Nimtz in: Springer Tracts in Modern Physics, ed. by G.Höhler, E.A.Niekisch (Springer, Berlin 1982)
/3/ P.Kacman, J.Blinowski, M.Grynberg, Phys.Stat.Sol.(b) 89,521(1978)
/4/ W.Leung, L.Liu, Phys.Rev. B8,3811(1973)
/5/ R.Stepniewski, K.Pastor, M.Grynberg, J.Phys.C 13,5783(1980)
/6/ G.Bastard, C.Rigaux, Y.Guldner, J.Mycielski, A.Mycielski, Journal de Physique 39,87(1978)
/7/ M.Dobrowolska, W.Dobrowolski, J.Phys.C 14,5689(1981)
/8/ G.Bastard, C.Rigaux, Y.Guldner, A.Mycielski, J.K.Furdyna, D.P.Mullin, Phys.Rev. B24,1961(1981)
/9/ K.Pastor, M.Grynberg, R.R.Galazka, Solid State Commun. 29,739(79)
/10/ R.R.Galazka in:Proc.Int.Conf.Phys.Semic.(Edinburgh 1978), p.133
/11/ S.W.McKnight, P.M.Amirtharaj, S.Perkowitz, Solid State Commun. 25,357(1978)
/12/ R.R.Galazka, J.Kossut, Lecture Notes in Physics 133,245(1979)
/13/ A.M.Sandauer, P.Byszewski, Phys.Stat.Sol.(b) 109,167(1982)

FERMI SURFACE PROPERTIES OF $Pb_{.71}Sn_{.29}Te$ ABOVE AND BELOW THE STRUCTURAL PHASE TRANSITION

Yusheng He* and A.D.C. Grassie†, Physics Division,
University of Sussex, Brighton, U.K.

Below the structural phase transition temperature (T_c) in $Pb_{1-x}Sn_xTe$ the symmetry of the conduction and valence bands is reduced from D_{3d} to C_{3v} for the T points of the rhombohedral Brillouin zone (BZ). Bangert (1) has extended the k.p model for the cubic phase of IV-VI compounds by adding a periodic lattice potential of C_{3v} symmetry to the cubic one-electron Hamiltonian and finds that the four equivalent valleys at the L points split into a single T valley and three equivalent L valleys, with two pieces of Fermi surface associated with each valley on account of the lifting of spin degeneracy due to the absence of inversion symmetry. The form of the Fermi surfaces are reminiscent of a slightly axially elongated apple and its core and we will refer to them as T^+, T^-, L^+ and L^-.

The angular dependence of the Shubnikov-de Haas oscillations (SdH) in the distorted phase has been measured and the frequency spectrum of the oscillations has been obtained from Fourier analysis, using a window function such that the side lobes are less than 1% of the fundamental peak.

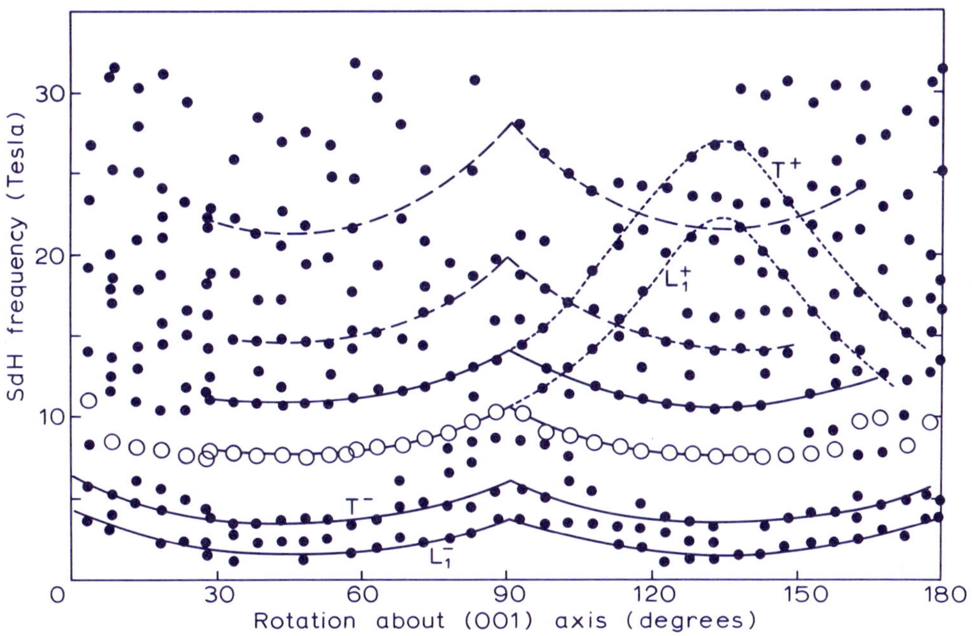

Figure 1. SdH oscillation frequencies for B rotation in (001) plane. The larger dots are the maximum amplitude oscillations and the dotted curves are harmonics

Figure 1 shows the frequencies obtained for rotation of B in the (001) plane of the cubic phase. The frequency with the largest amplitude at each orientation is indicated by a open circle and its harmonic is picked out by a dashed curve. In addition, we can trace a frequency branch running parallel to the large amplitude curve and its related harmonic. Since we expect from Bangert's band structure that the Fermi surfaces are close to ellipsoids (2), we can trace dotted extensions of these curves with maximum frequencies at 135°. We associate these frequencies with ellipsoids at the T and L points (L_1). Furthermore, considering the very low frequencies in figure 1 and bearing in mind the effect of lifting spin degeneracy, we can trace evidence of similar minimum cross sections of other smaller ellipsoids (T^- and L_1^-).

Returning to consideration of the maximum amplitude curve, we can see evidence of a splitting near 0° and 180° and also a small additional branch below the cusp at 90°. These areas are impossible to explain if the L ellipsoids are symmetrically disposed with their major axes perpendicular to the BZ face and we are forced to consider the possibility of tilted L ellipsoids.

In figure 2, we have used the observed maximum area of the L_1 ellipsoid at 135° to allow us to redraw the dotted curves to include the split-off data and many other frequencies. The resulting curves correspond to L_2 and L_3 ellipsoids with maximum areas in this field traverse separated by 34°. The corresponding tilt from the normal to the BZ face is 14°. The assignment of the exact angular separation is carried out using the established angular variation of the L_1 ellipsoid.

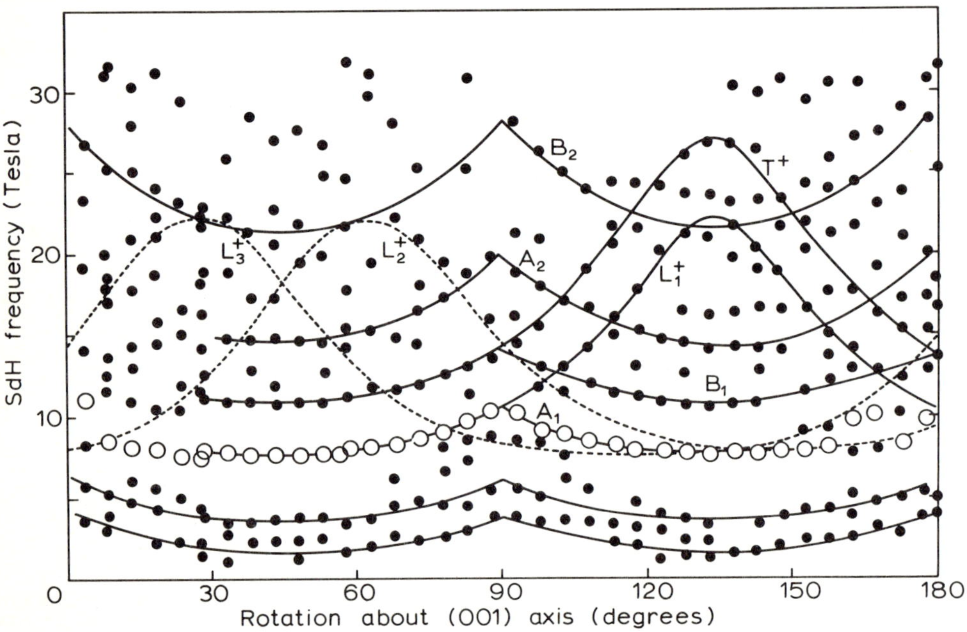

Figure 2. SdH oscillation frequencies for B rotation in (001) plane. The dashed curves are the tilted L ellipsoids.

However, one important element of the maximum amplitude data, the branch between 90° and 110° and its harmonic, labelled A_1 and A_2 in figure 2, still remains unexplained, as also does the branch between 120° and 160° and its harmonic, labelled B_1 and B_2. At this point, we recall that a domain structure is expected for these compounds below T_c and has indeed been observed (3). We would therefore expect to see evidence of T and L_1 ellipsoids from the other domains and this is indeed the case. The resulting complex fit is shown in figure 3 and is consistent with the distortions taking place along each of the <111> directions in the cubic phase, giving two mirror image patterns of identical frequency data for rotation of B in the (001) plane of the cubic phase.

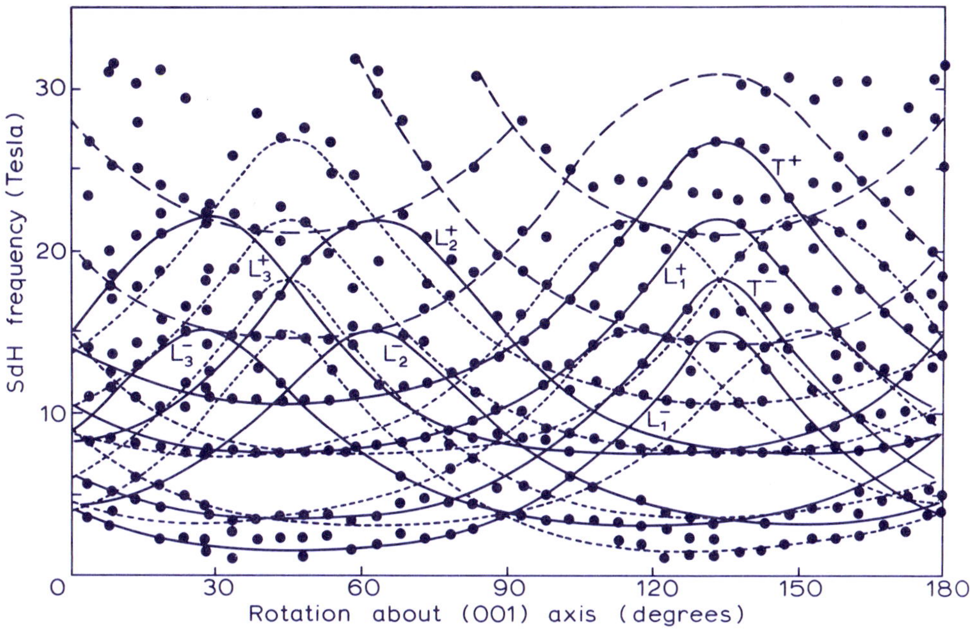

Figure 3. SdH oscillation frequencies for B rotation in (001) plane. The dotted curves are additional frequencies due to domain structure in the sample.

Confirmation of this identification is obtained at an even more complex level for the Fermi surface areas observed for B rotation in the (110) plane of the cubic phase. The two samples concerned were cut from the same crystal.

Applying a simple ellipsoidal model as an approximation to the true band structure, the carrier redistribution in the distorted phase can be deduced and is shown in table 1.

TABLE 1

Carrier Redistribution in the Distorted Phase, in Units of $10^{17} cm^{-3}$

B rotation	$p(T^+)$	$p(T^-)$	$p(L^+)$	$p(L^-)$	total p	p from Hall effect
(001) plane	2.5±.1	1.0±.1	1.7±.1	0.5±.1	10±1	
(110) plane	2.6±.1	1.3±.1	1.7±.1	0.6±.1	11±1	10±1

Since the SdH oscillations are dominated by the L^+ minimum area at low fields, values for the effective masses on this surface can be obtained for various B orientations. The anisotropy ratio for this surface (defined as the square of the ratio of maximum to minimum cross-sectional areas) is found to be 8 and the angular variation of the effective mass is consistent with this anisotropy ratio, yielding a value of $m^* = (0.035 \pm 0.005)\, m_o$ for the minimum area orbit of the L^+ surface. Using this value of the anisotropy constant, we can deduce a value for the parameter r in the Hall effect expression $p = {}^{r}/_{R_H e}$. The effect of the multi-valley ellipsoids is to set $r = \frac{3K(K+2)}{(2K+1)^2}$, where K is defined as the ratio of the effective masses at maximum and minimum areas (4). The statistical factor related to the scattering mechanism has value 1 for degenerate statistics (4). The resulting carrier concentration deduced from measurements of the Hall effect at 10K is compared in table 1 with that deduced from the simple ellipsoidal model and satisfactory agreement is obtained.

A further sample cut from the same crystal was used for measurements of the SdH oscillations over an extended temperature range from 2K to 46K. Four orientations were studied with B parallel to each of the <111> directions in the cubic phase. Since the current was restrained to flow along the $[1\bar{1}0]$ direction, two of the measurements were of pure transverse magnetoresistance and the other two had a predominantly longitudinal configuration.

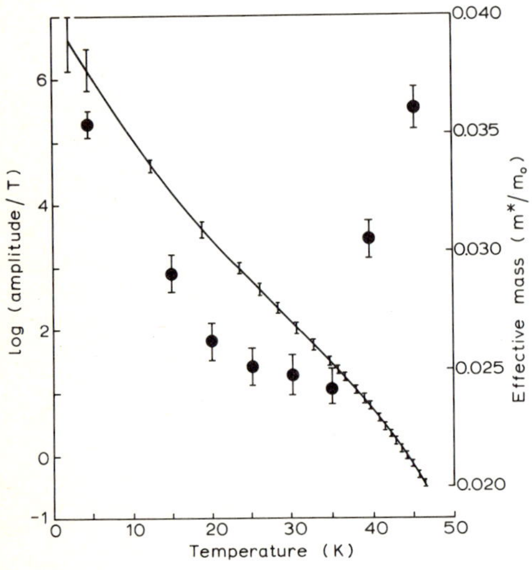

Figure 4. Temperature dependence of log (A/T), where A is SdH amplitude, and of derived effective mass.

Figure 4 shows the temperature dependence of the logarithm of the SdH oscillation amplitude divided by temperature, the slope being proportional to the effective mass. The frequency concerned is that associated with the minimum area of an L ellipsoid and the amplitude has been corrected for departures of the function $\sinh(\alpha T m^*/B)$, where α is a numerical constant, from a simple exponential and also for the contribution of other oscillations to the observed amplitude, both of these corrections only being relevant for temperatures less than 20K. Values of m^* deduced at specific temperatures are also shown in figure 4. The change in m^* is consistent with what would be expected from the change of the average energy gap through T_c reported for the $Pb_{1-x}Ge_xTe$ system (5). A much smaller change in effective mass through T_c was observed in the configuration dominated by the longitudinal magneto-resistance.

We wish to express our gratitude to K.L.I. Kobayashi for the samples used in this work and to E. Bangert for the band structure information.

*Permanent Address: Department of Physics Pekin University, China.

†Present Address: Philips Research Laboratories, Redhill, U.K.

References

(1) Bangert, E. "Physics of Narrow Gap Semiconductors" Proceedings Linz 1981, p.216 (Springer Verlag 1981).

(2) Bangert, E. Private communication 1982.

(3) Jantsch, W. et al. Ferroelectrics 38, 905 (1981).

(4) Allgaier, R.A. The Hall Effect and Its Applications p.389 ed C.L. Chien and C.R. Westgate (Plenum Press, New York and London 1981).

(5) Murase, E. et al. Inst. Phys. Conf. Series 43 p.437, 1979.

EVIDENCE OF FERMI LEVEL PINNING IN $Pb_{1-x}Sn_xTe/In$ AT HIGH MAGNETIC FIELDS

K. Murase, S. Takaoka, T. Itoga and S. Ishida

Department of Physics, Faculty of Science, Osaka University,
1-1 Machikane-yama, Toyonaka, 560 Japan

1. Introduction

In this decade, since the first report on PbTe/In [1], many papers have been concerned with unusual properties of IV-VI compounds doped with group III elements at low temperatures and/or high pressures, such as metal-semiconductor transitions, and slow relaxation of electrical resistivity and Hall coefficient associated with the process of carrier flow between the impurity and the extended band. Among them the $Pb_{1-x}Sn_xTe$ /In-system has been most extensively studied in galvanomagnetic effects, transient photoconductivity and magnetic properties [1-12], yet the microscopic mechanisms for these properties are open to investigation. It has been considered that there are three charged states of indium impurities; In^{3+}, In^{2+} and In^{+} [2,4,8]. Magnetic studies in PbTe/In [5] suggested that most indium impurities are split into two charged states; In^{3+} and In^{+}. Such an autocompensation had been shortly discussed with the ionic lattice polarization around the charged impurities located at the Pb^{2+} sites [2]. Recently an alternative explanation with chemical bondings and local lattice distortions for the origin of the autocompensation has been proposed to describe the temperature dependence of the Hall coefficient for $Pb_{1-x}Sn_xTe/In$ with various alloy composition x [10].

We made a systematic investigation of metal-insulator transitions, and determined the composition and temperature dependence of the chemical potential in $Pb_{1-x}Sn_xTe$ ($x=0.2 \sim 0.35$) doped with In. It is unambiguously shown in p-$Pb_{1-x}Sn_xTe$ ($x=0.35$, $\sim 1\%$ In) that oscillatory Hall coefficient of large amplitudes with magnetic fields can be analyzed by change of valence band hole concentrations originating from the Fermi level pinning by the impurities. The decay time of photoconductivity is confirmed to be "infinite" only in insulating n-type specimens at 4.2K, though it is still very long in p-type specimens. It is discussed that the various unusual effects can be understood by mechanisms based on the model of the three-charged-state impurity with different local lattice distortions.

2. Results and Discussions

Single crystals $Pb_{1-x}Sn_xTe$ were grown by a vapor transport method. The slices of ~ 1 mm thickness were doped with indium by the isothermal diffusion, and then step-annealed in indium vapor, with argon gas to avoid sublimation. The specimens saturated with $\sim 1\%$ indium atoms gave temperature dependences of the resistivity as shown in Figure 1 and 2. The corresponding Hall coefficients are shown in Figure 3 and 4. The specimens with tin compositions $0.25 \leq x \leq 0.30$ are semi-insulating at low temperatures, resistivity measurements were difficult to make due to a non-linearity and some instability especially below 20K. For $0.21 \leq x \leq 0.27$, specimens give n-type conduction in the whole temperature range, while those with $0.30 \leq x \leq 0.35$ exhibit the n-type at high temperatures and the p-type at low temperatures as determined by Hall measurements at 9.95kG. The chemical potential was determined so as to fit the standard Hall coefficient formula including both conduction electron and valence band hole contributions, and the conductivity expression based on the Drude model, to the experimental Hall coefficient at 9.95kG and the resistivity. To calculate the carrier concentrations n (or p), we used the Fermi distribution function with a chemical potential as an adjustable parameter and the density of states of the two

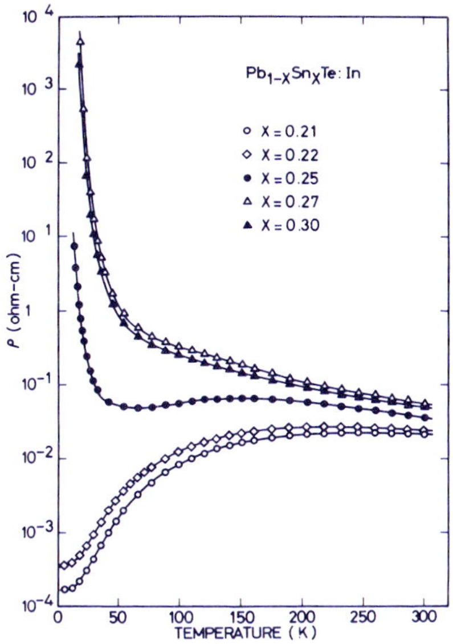

Fig.1 Resistivity ρ of $Pb_{1-x}Sn_xTe/In$ with different tin composition x (0.21 ∿ 0.30) at temperatures between 4.2 ∿ 300K

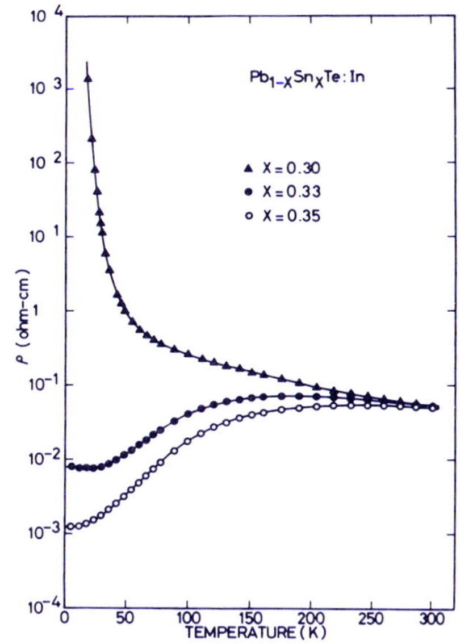

Fig.2 Resistivity ρ of $Pb_{1-x}Sn_xTe/In$ with different tin composition x (0.30 ∿ 0.35) at temperatures between 4.2 ∿ 300K

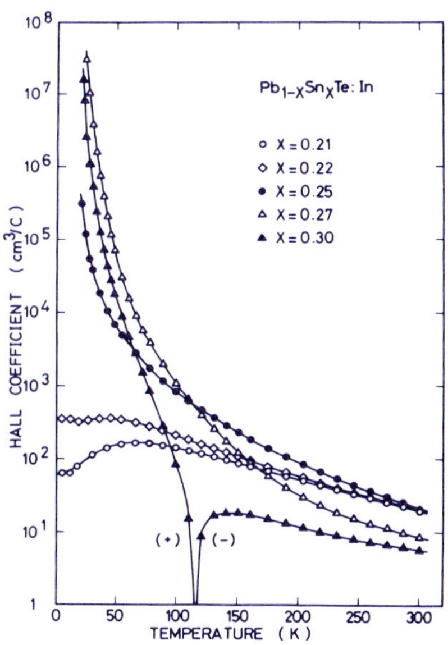

Fig.3 Hall coefficient at 9.95kG of $Pb_{1-x}Sn_xTe/In$ corresponding to Fig.1

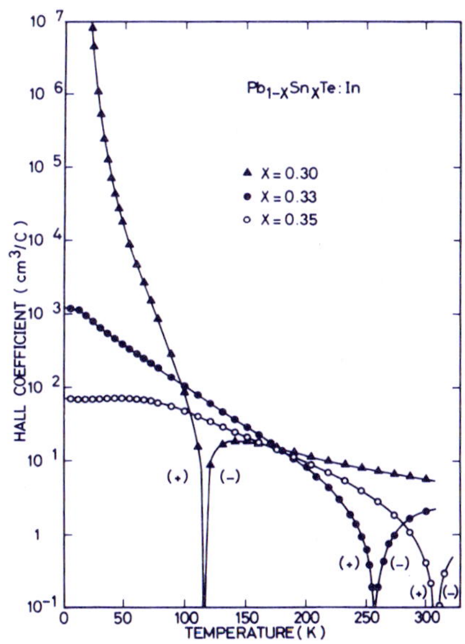

Fig.4 Hall coefficient at 9.95kG of $Pb_{1-x}Sn_xTe/In$ corresponding to Fig.2

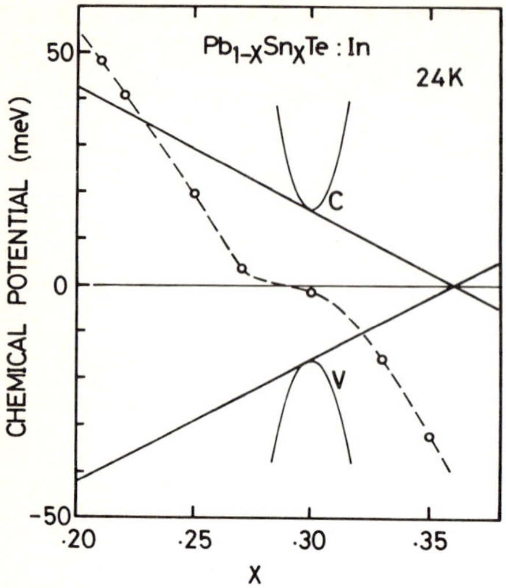

Fig.5 Chemical potential of $Pb_{1-x}Sn_xTe$/In at 24K derived from the data shown in Fig.1 - 4. A brief description for the derivation is presented in the text

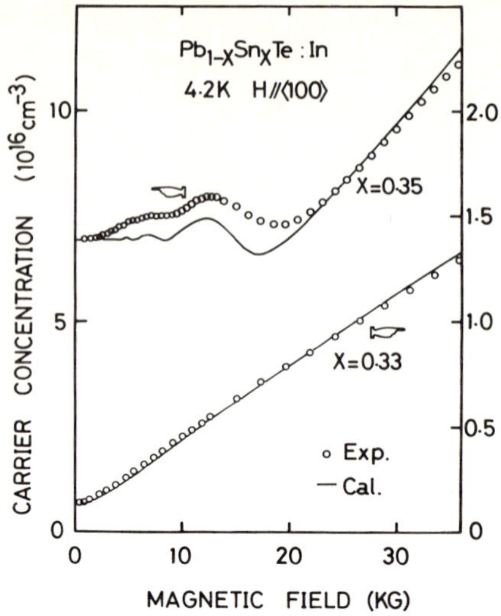

Fig.6 Open circles: Magnetic field dependent "carrier concentrations" for Hall measurements. Solid lines: calculations based on the Fermi level pinning by indium impurities

band model. Since the band structure is practically unchanged by ∿1% indium impurities, the composition and temperature dependence of the band gap for $Pb_{1-x}Sn_xTe$: $E_g=|0.46T + 182 - 543x + 20x^2|$ meV was used [13]. The mobility ratio μ_e/μ_h is assumed to be 1, This ratio is not very sensitive to the determination of the chemical potential. Thus obtained chemical potentials for 24K and at 9.95kG are shown in Figure 5. The effect of the magnetic field is small in this case.

For the semi-insulating specimen (x=0.25) below 21K, the photoconductive effect is prominent. The resistivity drops by 4∿5 orders of magnitude under room-temperature radiation or an illumination of light by a gallium arsenide photo-diode. The photoconductivity is practically unchanged for several hours after shutting off the irradiation as reported in [3]. While the resistivity of the specimens of x=0.27 and 0.30 recovered to the dark value within 1 minute. The photo-Hall measurement with the photodiode revealed excess electrons for n-type materials and holes for p-type materials.

Shubnikov-de Haas oscillations were clearly observed in the metallic specimens (x=0.21, 0.22 and 0.35) at low temperatures. The x=0.35 specimen revealed an oscillatory Hall coefficient with large amplitudes at high magnetic fields. The origin of the oscillation cannot be attributed to the influence of the transverse magnetoresistance $\rho_\perp$ which is estimated to be ∿3 x 10^{-5}, B=10kG. The other possibility has been discussed for the oscillatory Hall coefficient in PbTe/In, as an oscillatory change of carrier concentrations by the magnetic quantization of the extended band on the likely assumption that the Fermi level is stabilized by the indium impurities [9]. In the case of the PbTe/In, however, the oscillation of the Hall coefficient could not be identified unambiguously [9]. In Figure 6, "the hole concentrations" derived from the Hall coefficient are plotted for the specimens with X = 0.33 and 0.35. The magnetic field dependent carrier concentrations are well reproduced by the Fermi level pinning [2] as shown in Figure 6 for x=0.33 and 0.35 specimens. The Fermi energy was set to the value at B=0 which was determined from the carrier concentration by using the two band model. The carrier concentrations at magnetic fields were obtained by the calculation of the number of states of the valence band above E_F. The broadening of the Landau levels was taken into account for the both specimens. The magnetic field was swept very slowly to ensure the carrier transfer, though the sweep time is still inadequate for the x=0.35 specimen. In the case of the n-type specimens, the relaxation time is too long to get the equilibrium at magnetic fields, as demonstrated in [8]. The reason why the PbTe/In did not show the large amplitude of the oscillation [9] is probably due to the high carrier concentrations and the very long relaxation time [10]. Judging from the changes in the lattice parameter by In in PbTe, it is probable that indium atoms substitute for lead in PbTe [14]. Then, the apparently weak donor action [15] is due to the autocompensation among the substitutional indium atoms [2,4] rather than the compensation by an equal occupation of two kinds of completely different sites [15]. It may be noted that "In^{+}" can gain much energy by chemical bonding with some p components to the neighboring three tellurium atoms and is at slightly off-center position along a <111> direction, as discussed for Ge atoms in PbTe [16]. A far-infrared transmission dip was observed at 160 cm^{-1} in $Pb_{0.73}Sn_{0.27}Te$/In, and at 187 cm^{-1} in $Pb_{0.78}Sn_{0.22}Te$/Ga, respectively. These could be assigned to higher frequency F_{1u} localized modes. Since Ga^{+} has much smaller ionic radius than that of In^{+}, it will be more stable at an off-center position. In gallium doped $Pb_{1-x}Sn_xTe$, larger energy separation among different charged states are expected in comparison with the indium impurities. The energy separation between Ga^{3+} and Ga^{+} would be comparable with the band gap. Then, one may understand a rapid change of chemical potential in the forbidden gap as a function of pressures in $Pb_{1-x}Sn_xTe$/Ga [17]. There is, however, no assurance that the group III impurities enhance the ferroelectric instability [18] because of the lack of p-electrons. The occurrence of the large photoconductance below 21K can be explained by the configurational coordinate model[8], where the barrier to the impurity level, intercepts the retreat of the conduction electrons associated with the temperature shift of the band gap. The anomaly at the mid-gap in Fig.5 might be due to a reduction of the conduction band states by excited p-electron states of impurities.

The authors would like to thank Dr. T.Fukunaga for discussions and S.Shimomura for the infrared experiments. One of them (K.M.) was supported by the 22th Science and Technology Grants of Toray Science Foundation.

References

1. A.A.Averkin, V.I.Kaidanov and R.B.Mel'nik, Sov. Phys. Semicond. 5, 75 (1971).
2. Yu.V.Andreev, K.I.Geiman, I.A.Drabkin, A.V.Matveenko, E.A.Mozhaev and B.Ya.Moizhes, Sov. Phys. Semicond. 9, 1235 (1975).
3. B.M.Vul, I.D.Voronova, G.A.Kalyuzhnaya, T.S.Mamedov and T.Sh.Ragimove, JETP Lett. 29, 20 (1979).
4. K.Weiser, A.Klein and M.Ainhom, Appl. Phys. Lett. 34, 607 (1979).
5. I.A.Drabkin, M.A.Kvanto and V.V.Kompaniets, Sov. Phys. Semicond. 13, 1206 (1979).
6. B.A.Akimov, L.I.Ryabova, O.B.Yatsenko and S.M.Chudinov, Sov. Phys. Semicond. 13, 441 (1979).
7. B.A.Akimov, N.B.Brandt, S.A.Bogoslovskii, L.I.Ryabova and S.M.Chudinov, JETP Lett. 29, 10 (1979).
8. Claude M.Penchina, A.Klein and K.Weiser, Proc. 15th Int. Conf. Physics of Semiconductors, Kyoto, 1980; J. Phys. Soc. Japan 49, suppl.A 783 (1980).
9. S.N.Lykov and I.A.Chernik, Sov. Phys. Semicond. 14, 25 (1980).
10. S.N.Lykov and I.A.Chernik, Sov. Phys. Semicond. 14, 730 (1980).
11. T.D.Aitikeeva, K.I.Geiman, A.I.Levedev, A.V.Matveenko and A.E.Yunovich, JETP Lett. 32, 122 (1980).
12. B.A.Akimov, N.B.Brandt, V.I.Stafeev, V.N.Nikiforov and O.B.Yatsenko, JETP Lett. 32, 128 (1980).
13. C.R.Hewes, M.S.Adler and S.D.Senturia, Phys. Rev. B7, 5195 (1973).
14. K.Weiser, Phys. Rev. B15, 2741 (1981).
15. A.J.Rosenberg and F.Wald, J. Phys. Chem. Solids 26, 1079 (1965).
16. S.Katayama and K.Murase, Solid State Communs. 36, 707 (1980).
17. B.A.Akimov, N.B.Brandt, L.I.Ryabova, D.R.Khokhlov, S.M.Chudinov and O.B.Yatsenko, JETP Lett. 31, 280 (1980).
18. K.Murase, Ferroelectrics 35, 67 (1981), see references.

MAGNETOTRANSPORT PROPERTIES OF $Hg_{0.8}Cd_{0.2}Te$ IN PULSED FIELDS UP TO 35 T.

Godelieve De Vos and Fritz Herlach
Katholieke Universiteit Leuven
Celestijnenlaan 200 D, B- 3030 Leuven (Belgium)

The longitudinal and transverse magnetoresistance and the Hall voltage of $Hg_{0.8}Cd_{0.2}Te$ have been measured in the temperature range 1.5 K - 4.2 K in pulsed magnetic fields up to 35 T. The results can be interpreted in terms of magnetic freeze-out at magnetic fields below 10 T. At higher fields and sample currents above 10 μA, an impact ionization process is activated by the high electric field associated with the high magnetoresistance.

I. Introduction

The peculiar magnetotransport properties of the narrow gap semiconductor $Hg_{1-x}Cd_xTe$ have given rise to different interpretations [1-3]. In the n-type material magnetic freeze-out is not expected theoretically because the donor levels are smeared out in the conduction band [4].

At fields below 10 T, the magnetoresistance follows a power law as a function of the magnetic field. As this is in agreement with the theory of conduction by free electrons in a magnetic field, the occurence of magnetic freeze-out is considered to be unlikely.
At fields above 10 T, a break in the magnetoresistance curves occurs at the lower temperatures (2 K). If magnetic freeze-out does not occur, this anomaly can hardly be due to hot electron effects. An interesting alternative was proposed by Nimtz et al. [1] i.e. that at these low temperatures and high magnetic fields, the electrons condense into a Wigner lattice. Our previous data provided support for this hypothesis [3].

However, an extended study of the temperature dependence of the magnetoresistance and Hall voltage in the lower field range have revealed magnetic freeze-out in our sample. The longitudinal magnetoresistance and the carrier concentration depend exponentionally on 1/T, and from the slope of the curves the activation energy has been determined at several magnetic fields.
The dependence of the transport data on the sample current indicates that the anomaly at higher fields and low temperatures is a hot carrier effect. Unfortunately, we could not yet measure at sample currents below 10 μA, so we cannot arrive at a conclusion about the anomaly in the Ohmic region. We cannot directly compare our measurements with those of Nimtz et al., which were done with much lower sample currents.

II. Experiments

The $Hg_{0.8}Cd_{0.2}Te$ sample was obtained from Cominco (Canada), and has the following characterics at 4.2 K :
$n = 2.4 \times 10^{20} m^{-3}$, $\mu_H = 95\ m^2V^{-1}s^{-1}$, $Eg = 70$ meV and $m^* = 0.0048\ m_0$.
The experimental technique has been outlined in a previous publication [3].

a) The low field region.

The longitudinal magnetoresistance depends exponentionally on 1/T at temperatures below 4.2 K (Fig. 1). The carrier concentration, obtained from the Hall coefficient by the equation $n = R_H B^2 / e(\rho_\perp^2 + R_H^2 B^2)$, follows the same temperature dependence,

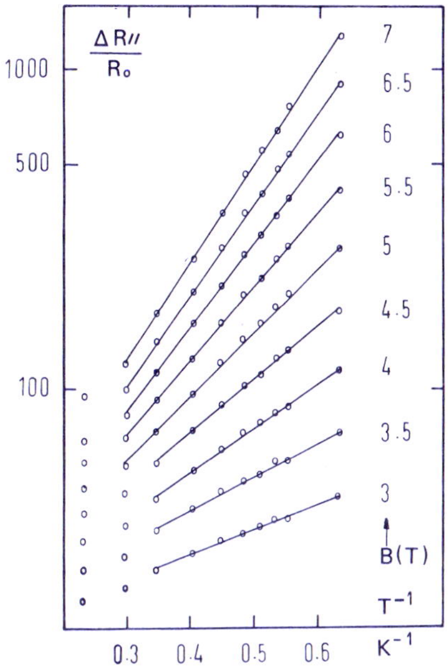

Fig. 1 Longitudinal magnetoresistance as a function of T^{-1} in the low field region at 50 μA.

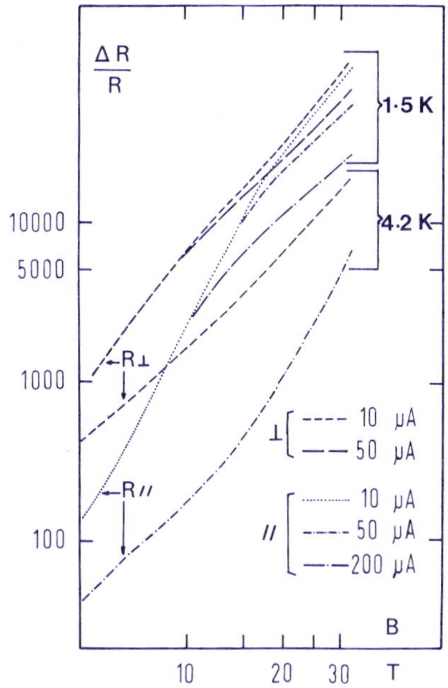

Fig. 2 Transverse and longitudinal magnetoresistance as a function of magnetic field at several sample currents.

within the limits of experimental error. We preferred this equation because the simple relation $n = (eR_H)^{-1}$ gives an almost temperature independent carrier concentration, which is at variance with the magnetoresistance data. We conclude that magnetic freeze-out occurs at these low magnetic fields. The activation energy is obtained from the slope of the curves, and is shown in Table I for several magnetic fields. The difference between the activation energy obtained from the $\rho_{//}$ (1/T) and n(1/T) curves is within experimental error. The activation energy increases with increasing magnetic field, which is expected when magnetic freeze-out takes place. In this field range our experiments are in agreement with those of Raymond et al. [2], who explain the freeze-out at those low magnetic fields by assuming the existence of a resonant level located above the Fermi energy at zero field. Freeze-out occurs due to the shifting of the lowest Landeau level with respect to the resonant level in a magnetic field. This result does not exclude the formation of a Wigner lattice at higher fields, in fact it is rather in favour of this because it lowers the carrier concentration.

B(T)	3	3.5	4	4.5	5	5.5	6	6.5	7
E_{act} (10^{-2} meV)	16.5	22.1	28.5	35.2	39.3	45.8	52.9	56.6	60.9

Table 1 : Ionization energy obtained from the slope of the $\rho_{//}(\frac{1}{T})$-curves at various magnetic fields.

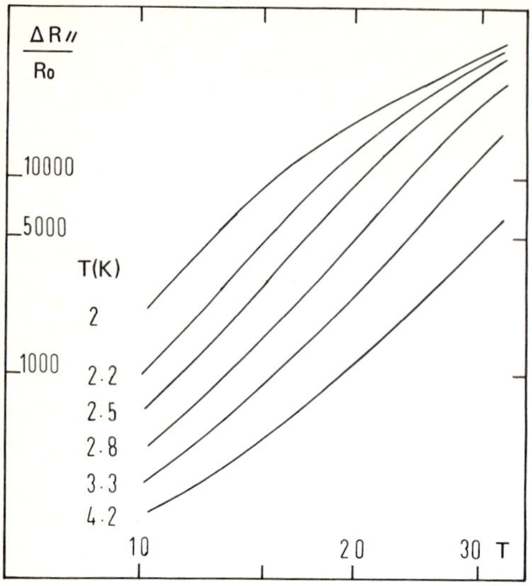

Fig. 3 Longitudinal magnetoresistance as a function of magnetic field at several temperatures. The sample current is 50 μA.

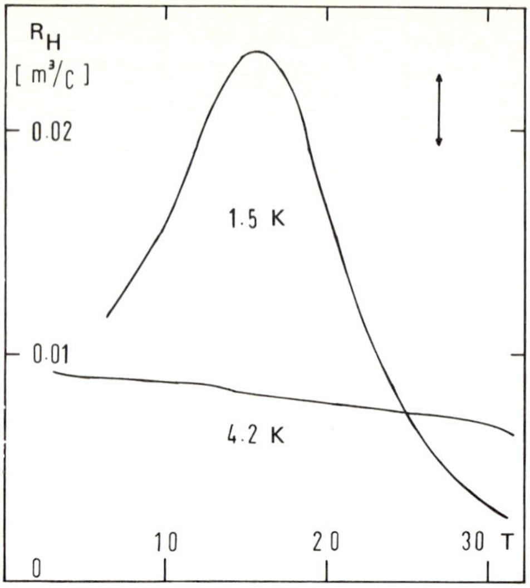

Fig. 4 Hall coefficient as a function of the magnetic field at 4.2 K and 1.5 K. The arrow in the upper right corner indicates the error limit. The sample current is 50 μA.

b) the high field region

At 4.2 K and fields above 15 T, the transverse and the longitudinal magnetoresistance deviate slightly from the assumed power law. However, the Hall coefficient does not change at these fields; if anything, it decreases.
To check if the break in the magnetoresistance curves at the lowest temperatures is a hot carrier effect, measurements with sample currents between 200 μA and 10 μA were done (Fig. 2). At 4.2. K, the magnetoresistance is found to be independent of the current, in contrast to the 1.5 K measurement. The break in the magnetoresistance curves becomes less pronouned at lower sample currents. For the longitudinal magnetoresistance, it is evident that the break shifts to lower field values with higher currents. On basis of this strong current dependence, we conclude that the presumed anomaly is a hot carrier effect. It would be most desirable to study this anomaly in the Ohmic region. Unfortunately, we were not able to do measurements with lower sample currents due to vibrations of the wires in the pulsed magnetic field, which induce pick-up voltages comparable to the magnetoresistance signal. It is likely that we can further improve our detection system in order to overcome this difficulty.
Fig. 3 shows the longitudinal magnetoresistance at various temperatures. The break in the curves shifts to lower field values when the temperature is lowered.
At 1.5 K the Hall coefficient first increases with the magnetic field and at the position of the break, a sudden decrease in the Hall coefficient is observed (Fig.4). The observed dependence on the sample current does not agree with the behaviour expected for Wigner condensation. In that case, the break would be more pronounced when the sample current is lowered; this is at variance with out measurement.
All these phenomena at high fields and low temperatures can be explained by an electric field activated impact ionization process, which excites the frozen-out electrons into the conduction band.

III. Conclusion

In the low field range, the magnetotransport data can be interpreted in terms of magnetic freeze-out. The activation energy, as calculated from $\rho_{/\!/}$ (1/T) and n(1/T) increases with the magnetic field. Our data are not yet sufficient to compare the magnetic field dependence of the activation energy with theoretical calculations. In the high field and low temperature region, the high electric field on the sample (0.4 V/cm at 10 T and I_S = 50 μA) activates an impact ionization process, which induces a break in the magnetoresistance curves and a sudden decrease in the Hall coefficient. Our present data cannot be used to support the hypothesis of the formation of a Wigner lattice.

REFERENCES

[1] G. Nimtz and B. Schlicht : Festkörperprobleme ed. J. Treusch, Vieweg, Braunschweig (Advances in Solid State Physics XX, 1980) 368
J. Phys. Soc. Japan 49, suppl. 1 (1980) 313.

[2] A. Raymond et al. : physics of Narrow Gap Semiconductors, eds. E. Gornik, H. Heinrich and L. Palmetshofer (Springer Lecture notes in physics 152, 1982) 387.

[3] G. De Vos and F. Herlach : physics of narrow gap semiconductors, eds. E Gornik, H. Heinrich and L. Palmetshofer (Springer Lectures notes in physics 152, 1982) 378.

[4] G.W. Pratt, J. Nonmetals 1, 103 (1973).

INFLUENCE OF AN ACCEPTOR STATE ON TRANSPORT IN ZERO-GAP $Hg_{1-x}Mn_xTe$

M. Sawicki, T. Dietl, W. Plesiewicz, P. Sękowski, L. Śniadower

Institute of Physics, Polish Academy of Sciences
Al. Lotników 32/46, 02-668 Warsaw, Poland

M. Baj

Institute of Experimental Physics, Warsaw University
00-681 Warsaw, Poland

L. Dmowski

High Pressure Research Center "Unipress"
01-142 Warsaw, Poland

1. Introduction

In this paper we report our results of transport measurements performed on a zero-gap p-type ($N_A > N_D$) HgMnTe. Results are discussed in view of the present knowledge on:

(i) acceptor states degenerate with the conduction band continuum |1,2,3|,
(ii) intrinsic scattering mechanisms in zero-gap materials |4|,
(iii) behavior of bands |5,6,7| and impurity states |8| in semimagnetic semiconductors

2. Experimental

The experiments were performed on $Hg_{1-x}Mn_xTe$ samples with molar composition $x=0.06\pm0.01$. Samples of this composition have a small negative energy gap $\mathcal{E}_0=E_{\Gamma_6}-E_{\Gamma_8}\simeq-50$meV |5,6| and are therefore particulary sensitive to temperature, magnetic field and hydrostatic pressure. The samples have been annealed for 12 months at 473K in saturated mercury vapour prior to the measurements. The Hall coefficient R_H and the restivity ρ were measured by a.c. technique in the standard Hall geometry with six In contacts. The measurements below 1.5K were performed in a dilution refrigerator while keeping the power dissipated in the sample below 10^{-13}W. Transport measurements as a function of hydrostatic pressure were made using helium as a pressure transmitting medium.

3. Influence of an acceptor state on the transport properties

Our experimental results demonstrate in a very direct way that our samples are p-type (in a sense that $N_A > N_D$). More important, they show how an acceptor state degenerate with the conduction band affects various transport properties of the material. The relevant results and their brief description follow.

3.1. Magnetic field dependence of R_H and ρ at 77K.

Magnetic field dependence of the conductivity tensor components show that a high temperatures ($T \gtrsim 77$K) the hole concentration exceeds that of electrons (see Fig.1). This indicates that in our samples $N_A > N_D$ and enables to estimate $N_A - N_D \simeq 2\cdot10^{16}\,cm^{-3}$.

3.2. Contribution of heavy electrons.

An accurate description of σ_{xx} and σ_{xy} is possible if a contribution of heavy electrons is taken into account (Fig.1.). Their presence is usually assigned to the impurity conduction in the acceptor band |3|, although no satisfactory model has been proposed up to now.

3.3. Pressure-induced change of sign of R_H.

As it is known, an application of hydrostatic pressure to a zero-gap material leads to the openning of the gap |9| and therefore to an abrupt reduction of the intrinsic carrier concentration. Hence, the positive sign of R_H under pressure again indicates that $N_A > N_D$ (Fig.2). Furthermore, R_H at the highest pressure gives $N_A - N_D = 2\cdot 10^{16}\,cm^{-3}$ in an agreement with the value obtained from the field dependence of σ_{xx} and σ_{xy}.

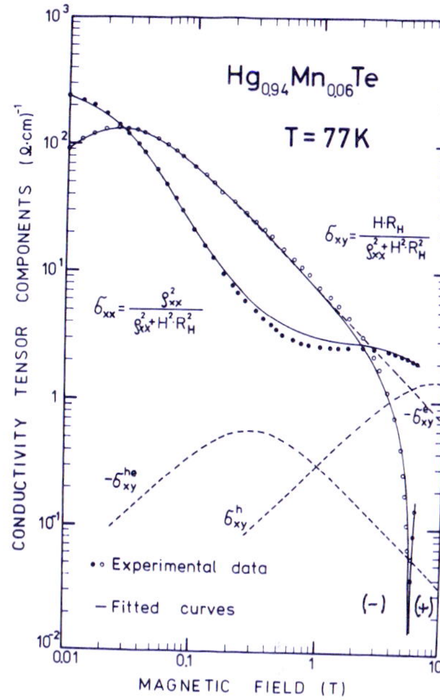

Fig.1. Conductivity tensor components vs. magnetic field at 77K. Solid lines are calculated assuming contribution from electrons, heavy electrons and holes with concentrations (in cm^{-3}) and mobilities (in cm^2/Vs), respectively: $n=4.5\cdot 10^{15}$, $\mu_e=3.7\cdot 10^5$; $n_{he}=2\cdot 10^{14}$, $\mu_{he}=3.5\cdot 10^4$; $p=1.6\cdot 10^{16}$, $\mu_h=1.1\cdot 10^3$.

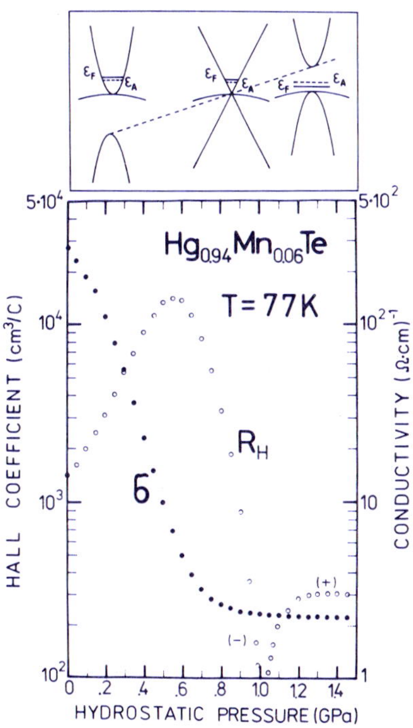

Fig.2. Hall coefficient and electric conductivity vs. hydrostatic pressure at 77K. Change of sign of R_H is connected with the zero-gap to open-gap transition, leading (when $N_A > N_D$) to freeze-out of electrons as sketched in the upper part of the figure.

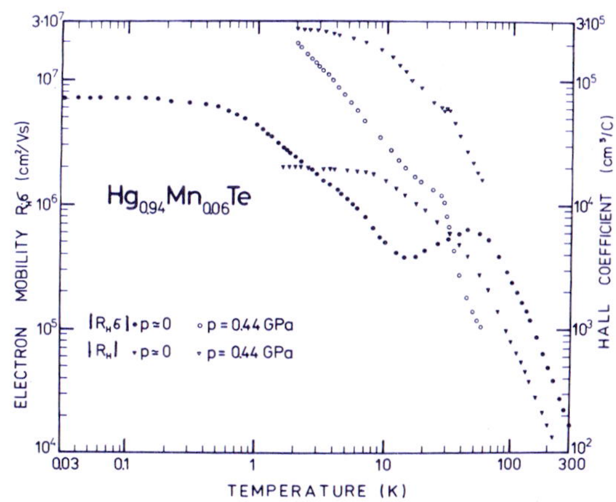

Fig.3. Hall coefficient and electron mobility vs. temperature at pressure set to zero and 0.44GPa, (H≃0). Note: the mobility minimum at 15K attributed to the resonant scattering and a peak value of mobility $2\cdot 10^7\,cm^2/Vs$.

3.4. Electron concentration and low-temperature pinning of the Fermi level

As it is known, in zero-gap HgTe-type materials a donor does not form bound states wheras an acceptor does. Therefore at high temperatures p>n because of the presence of thermally activated holes from acceptor states. At low temperatures, however, n >p since a freeze-out of holes occurs.

If the number of states N_o lying below the acceptor level in the conduction band is smaller than N_D, the Fermi level at low temperatures is pinned by the acceptor state, i.e. $\varepsilon_F = \varepsilon_A$. Under these circumstances the free electron concentration as measured by the Hall effect, is equal rather to N_o than to N_D. From the value of R_H below 10K (Fig.3) we get $N_D \gtrsim N_o = (3\pm0.2)\cdot10^{14}\mathrm{cm}^{-3}$ at zero pressure and $3\cdot10^{13}\mathrm{cm}^{-3}$ at 0.44GPa.

3.5. Temperature dependence of electron mobility.

High values of the electron mobility at low temperatures ($\mu \simeq 10^7 \mathrm{cm}^2/\mathrm{Vs}$, see Fig.3), observed previously by Morrissy |10| and Jaczyński |11| have so far been regarded as an outstanding property of HgMnTe |6|. In fact, such a high value has been reported also for HgCdTe |2|. In both cases it is due to the small effective mass, related to the vicinity of zero-gap to open-gap transition and the pinning of the Fermi level by the acceptor state. On the other hand, the decrease of mobility above 1K is not easy to understand theoretically. In order to describe it one has to assume that the efficiency of resonant scattering increases with temperature. This may be caused by an increase of the number of electrons having energies corresponding with energetic position of the peak of density of states of the acceptor band.

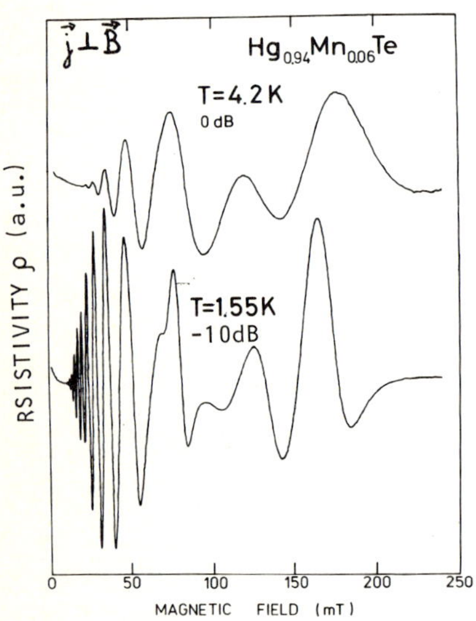

Fig.4. Recorder trace of the SdH oscillations measured at the second harmonics of the modulation field (B_M=0.45mT) at 4.2K and 1.55K. The scale for 1.55K is reduced by the factor 3.16.

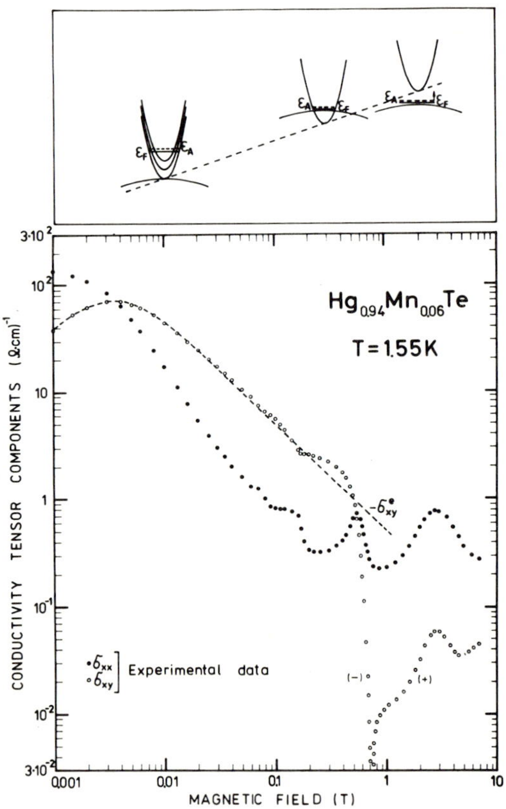

Fig.5. Conductivity tensor components vs. magnetic field at 1.55K. Note the ShdH oscillations of both σ_{xx} and σ_{xy} below 0.2T, and maxima of σ_{xx} at 0.5T and 3T of the origin discussed in the text.

3.6. Shubnikov-de Haas oscillations of σ_{xy}.

A careful analysis of SdH oscillations (Figs.4 and 5) shows that oscillation of both σ_{xx} and σ_{xy} contribute to the oscillations of ρ. The oscillations of σ_{xy} once more demonstrate that the Fermi level is pinned by the acceptor state. Indeed, the oscillations of σ_{xy} may be reproduced theoretically |1| by replacing the usual condition $n(B)$=const. by $\varepsilon_F(B)=\varepsilon_A$=const. From the oscillation period we get $N_o=(3.1\pm0.1)\cdot10^{14}$ cm^{-3} in a good agreement with the Hall effect data. Furthermore from the temperature dependence of the amplitude (under approximately fulfilled assumptions that the Dingle temperature and the effective $\bar{g}$-factor are temperature independent), we obtain $m^*/m_o=0.004$. Having N_o and m^* we may estimate ε_A, $\varepsilon_A=\hbar^2(3\pi^2N_o)^{2/3}/2m^*\simeq4$meV.

4. Effects of the exchange interaction.

The results of the previous section were given to stress the similarities between high mobility samples of HgMnTe and HgCdTe. Now we focus on the properties of HgMnTe which are not encountered in nonmagnetic zero-gap materials and, as we will suggest, are caused by the p-d exchange interaction. Comparing the ShdH oscillations (Fig.4) with those observed in a similar sample of HgCdTe |1| we note the presence of splittings of oscillations in HgCdTe and their absence (or presence in a residual form) in HgMnTe. This indicates that in HgMnTe the spin-splitting energy $\hbar\omega_1$ is not equal to half of the cyclotron energy $\hbar\omega_c$. This is because the exchange interaction affects $\hbar\omega_1$ and in the case considered here makes it roughly equal to $\hbar\omega_c$ |12|.
Turning now to the quantum limit region (B>0.2T) we may note in Fig.5 two distinct maxima on $\sigma_{xx}(B)$ curve. The increase of σ_{xx} between 0.2T and 0.5T we attribute to the exchange-induced upward shift of the valence band. In weak fields this shift exceeds the diamagnetic shift of the conduction band edge and leads to an overlap of Γ_8 bands |12|. The overlap increases the carrier concentration and thus σ_{xx}. At higher fields, however, a thermal gap appears and the freeze-out of electrons takes place. Consequently σ_{xx} decreases and σ_{xy} changes its sign. In fields B >1T σ_{xx} starts to rise again. This we assign to exchange induced lowering of the acceptor ionization energy ε_A |8|. At the highest fields the diamagnetic increase of ε_A takes place and σ_{xx} starts to decrease |13|.

References

|1| A.Kozacki, S.Otmezguine, G.Weill, C.Vérié, Proc. 13th Int. Conf. Physics of Semiconductors, Rome 1976, ed. F.G. Fumi, p.467.

|2| G.Weill, A.Kozacki and C.Vérié, Proc. 3rd Int. Conf. Narrow-Gap Semiconductors, ed. J.Raułuszkiewicz, M.Górska, E.Kaczmarek (PWN, Warsaw 1978) p.173.

|3| K.Szlenk, Z.Dziuba and R.R.Gałązka, phys. stat. sol. (b) 91, 255 (1979) and references therin.

|4| J.Dubowski, T.Dietl, W.Szymańska and R.R.Gałązka, J. Phys. Chem. Sol. 42, 351 (1981).

|5| G.Bastard, C.Rigaux, Y.Guldner, J.Mycielski and A.Mycielski, J. Physique 39, 87 (1978).

|6| M.Jaczyński, J.Kossut and R.R.Gałązka, phys. stat. sol. (b) 88, 73 (1978).

|7| B.B.Ponikarov, I.M.Tsidilkovskii, N.G.Shelushina, Sov. Phys. Semiconductor 15, 170 (1981).

|8| for the review, see, J.Mycielski, this conference proceedings.

|9| J.Stankiewicz, W.Giriat and M.V.Bien, phys. stat. sol. (b) 68, 485 (1975).

|10| J.H.Morrissy, Ph. D. thesis, Oxford 1971, unpublished.

|11| M.Jaczyński, Ph. D. thesis, Warsaw 1978, unpublished.

|12| A.Sandauer and P.Byszewski, phys. stat. sol. (b) 109, 167 (1982).

|13| P.Byszewski, K.Szlenk and R.R.Gałązka, in ref. |2| p.191.

SHUBNIKOV-DE HAAS EFFECT IN AMORPHOUS Cd_3As_2

L. ZDANOWICZ, J.C. PORTAL*, W. ZDANOWICZ
*Département de Physique I.N.S.A., Toulouse, France
Institute of Solid State Physics, Polish Academy of Sciences, Zabrze, Poland

I. INTRODUCTION

Shubnikov-de Haas oscillations (SdH) have been observed in single crystals and sometimes in bulk polycrystalline samples for a number of semiconductors of III-V, II-VI and other groups with sufficiently high magnetic fields (usually up to about 5 Tesla) when the well known conditions are fulfilled: $\omega\tau > 1$; $\hbar\omega$ (or E_F) $>kT$, where $\omega = eB/m_c^*$ is cyclotron frequency, τ is the collision time, B is the magnetic field, m_c^* is the cyclotron effective mass, E_F is the Fermi energy.

One especially promising material for the observation of the SdH oscillations is cadmium arsenide - Cd_3As_2, a degenerate semiconductor of the II-V group with an electron concentration in the range of $5x10^{17}$ - 10^{19} cm^{-3} and with extremaly high electron mobility of the order of 10000 to 25000 cm^2/Vs at room temperature and up to 280000 cm^2/Vs at liquid helium temperatures and very small effective mass of the electrons (1). Shubnikov-de Haas oscillations of magnetoresistance were investigated for the first time by Rosenman (2) on single crystals of Cd_3As_2 with an electron concentration in the range $1\text{-}8x10^{18}$ cm^{-3} and then by Aubin et al (3), and also by Blom et al (4) for bulk crystalline samples of lower concentration, down to $n \simeq 2x10^{17}$ cm^{-3}.

Also the de Haas von Alphen effect (dHvA) was examined by Doi et al (5). Rosenman (2) showed that the Fermi surface of Cd_3As_2 consists of only one ellipsoid of revolution elongated in the fourfold c-axis direction with an anisotropy ratio of the axes of about 1.18. He found also some anisotropy of the cyclotron effective mass and its dependence on the electron concentration (the anisotropy increases with decreasing concentration **n** especially below the "characteristic" $n \simeq 2x10^{18}cm^{-3}$). The very accurate experimental SdH data obtained by Rosenman (2) and dHvA results by Doi (5) were the basis for the theoretical $\bar{k}.\bar{p}$ calculations of the energy band structure of this tetragonal semiconductor (a=b= 12.6461 A, c=25.4378 Å) made by Bodnar (6).

Bodnar generalized the three-level Kane model for a tetragonal semiconductor with a narrow gap by taking into account the tetragonal field splitting (by adding an anisotropic crystal potential to the Hamiltonian and incorporation of the anisotropy of the interband momentum matrix element). According to the band model proposed by Bodnar, Cd_3As_2 has an inverted band structure of the HgTe - type with some anisotropy. It should be pointed that Cd_3As_2 has not only interesting semiconductor properties, but also a number of interesting technical applications (1).

Very distinct oscillations of magnetoresistance and Hall voltage were obtained for thin vacuum deposited crystalline films of Cd_3As_2 in our laboratories (7,8). It was probably the first case of observation of the SdH effect in vacuum deposited polycrystalline films at all. The Hall mobility of electrons in these films was of 0.5-0.7 m^2/Vs at room temperature and was not significantly changed down to helium temperatures. So, the quantization conditions were nearly fulfilled for $B \simeq 5$ Tesla. The calculated oscillation periods and effective masses were in good agreement with those obtained for the bulk crystals. No special attention was given to the preparation conditions of the investigated films (7). For some of the films, which showed texture, also very accurate anisotropy of the oscillation periods and effective masses was established giving anisotropy factors in good accordance with those calculated by Rosenman for single crystals (8).

However, most surprising and unexpected were very well defined series of SdH oscillations of the magnetoresistance and Hall voltage in thin vacuum deposited amorphous films of Cd_3As_2 (9). The present paper is devoted to the investigation of the SdH effect in these amorphous films (a-films) of cadmium arsenide.

Thin amorphous films of Cd_3As_2 - structure and properties.

Thin films of cadmium arsenide for SdH investigations have been prepared at Zabrze by direct thermal vacuum deposition of bulk Cd_3As_2 on cleaved mica substrates by the method described in our earlier works (see for ex.11-13). Films thicknesses were in the range 0.8-10μm. For electrical measurements silver-gold electrodes were vacuum deposited.

Depending on the temperature of the substrate T_s during the deposition process it was possible to prepare crystalline films above $T_s \simeq 400$ K or amorphous films below this temperature (11,13,14). Structure as well as electrical and optical properties of amorphous films were thoroughly investigated mainly for a-films prepared on the substrates held at room temperature ($T_s \sim 300$ K) (see for ex. 14-18). Depending on the rate of deposition two distinct groups of a-films can be prepared (15-17):

- high resistance films with resistivity of 10^{-2} - 10 ohm cm and low electron mobility of $\mu_H \simeq 1 - 100$ cm²/Vs obtained at high deposition rates ($v \sim 100 - 500$ A/s), (group A) and

- low resistance films (slowly deposited) with resistivity of the order of $10^{-2} - 10^{-3}$ ohm cm and very high Hall mobility μ_H 200 - 500 cm^2/Vs not observed yet for any other amorphous material but considerably lower than those for single crystal and even for polycrystalline films (group B).

Structural testing of the investigated films was performed by X-ray and electron diffraction methods (11,12). A typical halo-type diffractogram confirming the amorphous character of the films is shown in Fig. 1.

The microstructure of the a-films,investigated by electron microscopy in transmission or by replica method, consists of "clusters" which are typically in the range 1000 - 2000 A for the a-films prepared at room temperature. The Fourier analysis of the radial distribution function for the a-film of group A (20) showed that there is no significant coordination for an interatomic distance above about 15 Å and that the coordination numbers of Cd and As atoms in the a-films are different from those in Cd_3As_2 single crystals. The X-ray microprobe testing revealed some excess of arsenic in the obtained a-films. This excess of As is stronger for the group A films of high resistance and is significantly weaker for the low resistivity group B (see table . This deviation of stoichiometry seems to be the reason of the amorphous

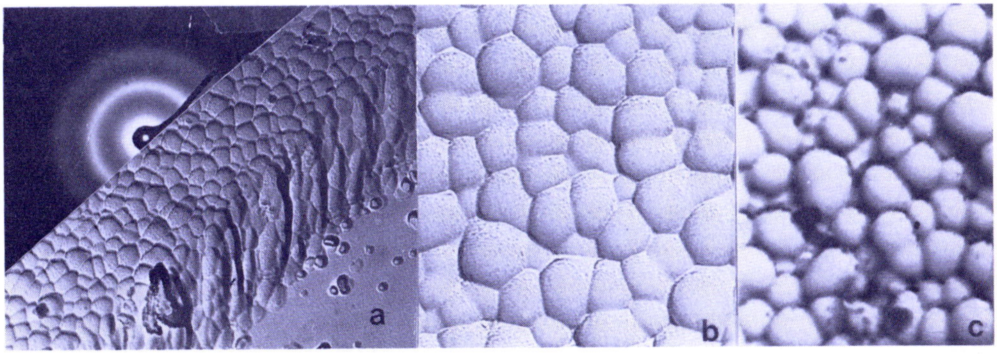

Fig. 1. Typical structure of amorphous cadmium arsenide films;
a-surface and cross-section, a and b - low resistance films (group B), c - high resistance films (group A). x20000 (9).

character of the films (17). However, the question in which manner the excess As-atoms are incorporated in the films' cluster-like microstructure is not resolved as yet. Probably, the interior of the clusters is less perturbed by disorder than the outside.

Table 1. The chemical composition and resistivity for a-films of Cd_3As_2 T_s - 300 K (17).

Sample no	% at Cd	% at As	Resistivity ohm m	
a - 145	49.5	50.5	high-ohmic	
a - 143	49.6	50.4	2.49×10^{-2}	group A
a - 149	49.9	50.1	5.45×10^{-2}	group A
a - 147	53.2	46.8	2.77×10^{-2}	group A
a - 144	54.7	45.3	2.68×10^{-3}	group B
a - 39	55.9	44.1	1.02×10^{-4}	group B
a - K20	59.1	40.9	1.75×10^{-5}	group B
bulk	60	40	10^{-6}	

The typical parameters for a number of the investigated a-films are shown in Table 2. The typical temperature dependences for both group A and group B a-films are presented in Fig. 2 in comparison with those for crystalline films and for single crystal of Cd_3As_2 (15,16). It is clearly seen the temperature dependences of resistivity are significantly different for the both types of a-films. Also temperature dependences for Hall mobility and for thermoelectric power distinctly separate the both mentioned groups showing different transport properties in each group (16,17). For the discussion of the transport mechanisms in these a-films the Mott - CFO model in connection heterogeneous chaotic band model due to disorder and granular structure of the film have been applied. In these models the electronic states at the band extremities are localized. For the a-films of group A the carrier transport through localized states below the mobility edge E_c at low temperatures has been postulated, while for the low-resistance films of group B the high electron concentration ($n \simeq 10^{18}$ cm^{-3}), the very high electron mobility and its non-activated character at low temperature suggest that conduction in the a-films of this group is through the extended states (above the critical energy E_c) (16,17).

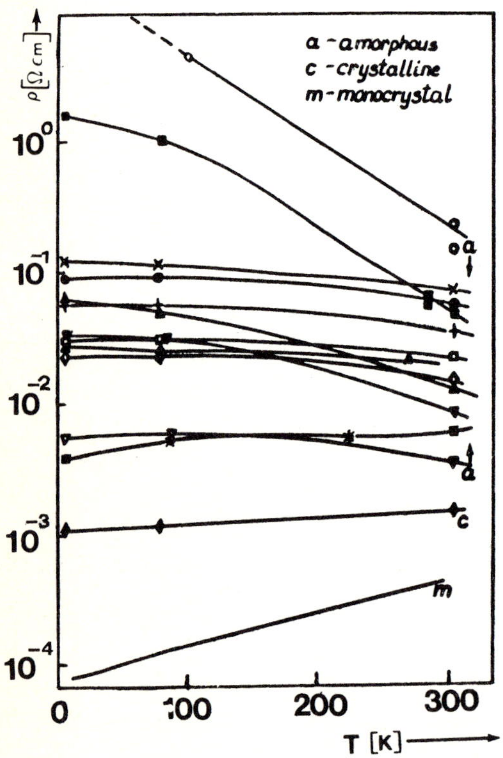

Fig. 2. Resistivity versus temperature for Cd_3As_2 a-films. (16).

It is the latter group for which SdH oscillations have been observed by us (9-10). For the high-resistivity films of group A oscillations have not appeared (see Table 2).

Shubnikov-de Haas oscillations in a-films of cadmium arsenide

The Shubnikov-de Haas oscillations have been studied both in the magnetoresistance and the Hall voltage using the Toulouse pulsed magnetic field facility at temperatures of 4.2K and 20K.

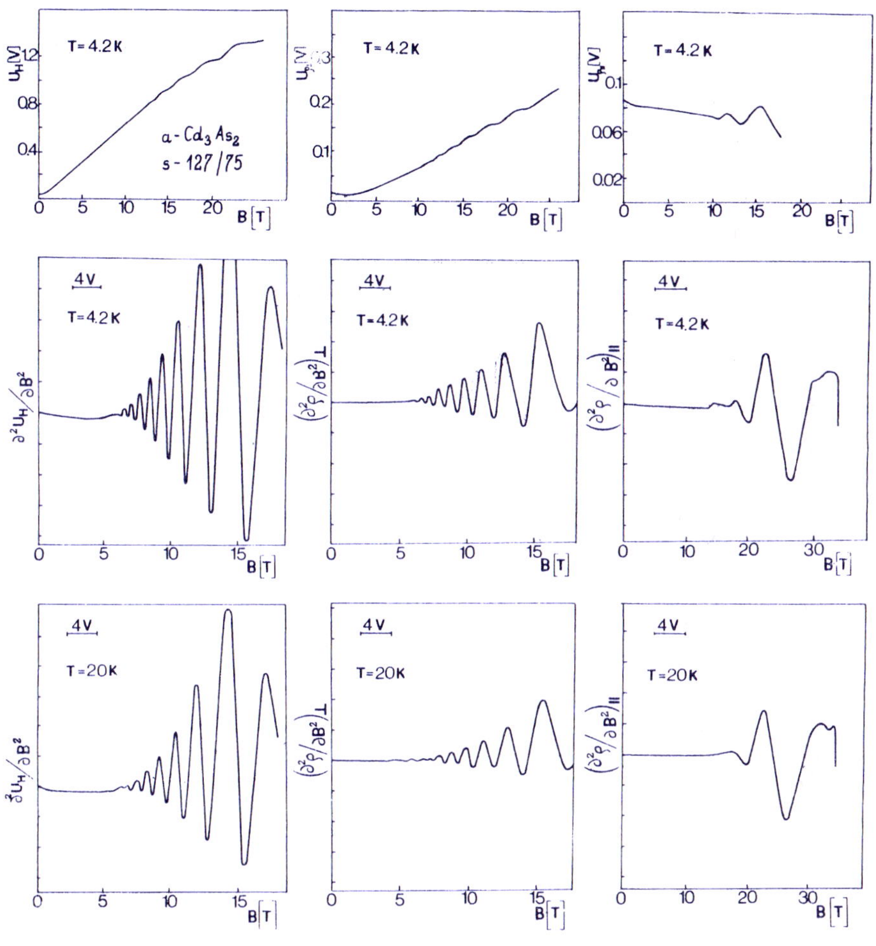

Fig. 3. SdH oscillations for a thin a-film of Cd_3As_2 (sample 127/75). The second derivatives of the Hall voltage, the transverse magnetoresistance and the longitudinal magnetoresistance at 4.2 K and 20 K are shown.

Strong pulsed magnetic fields up to B = 38 T were produced by discharging a 20 mF capacitor bank onto a nitrogen-cooled solenoid (7). The investigated sample was immersed in liquid helium and d.c. (i_x = 2 - 10 mA) or an a.c. current was passed through the sample. Longitudinal and two transverse configurations have been used. To extract the oscillations' spectrum in greater detail either a technique of second differentiation or linear compensation of the non-oscillatory part with the field were used. The dependences on the magnetic field of the Hall voltage U_H, of the longitudinal magnetoresistance $(\Delta\rho / \rho_o)_{\parallel}$ and the transverse magnetoresistances $(\Delta\rho/\rho_o)_{\perp}$ and $(\Delta\rho/\rho_o)_{\perp\!\perp}$ for the two configurations with respect to the magnetic field B (8) and of the second derivatives of these variables were measured for a number of low-ohmic Cd_3As_2 a-films (Table 2). Typical results are presented in Fig. 3 and 4. The traces are taken directly from the oscillograms. The observed oscillations series which appear for the magnetic field B ⩾ 10 T in both Hall voltage and resistivity are very surprising and quite comparable with those observed for polycrystalline films with Hall mobility of more than one order in magnitude higher than for single crystals in which the low-field mobility at helium temperatures is even three orders in magnitude higher (9,10).

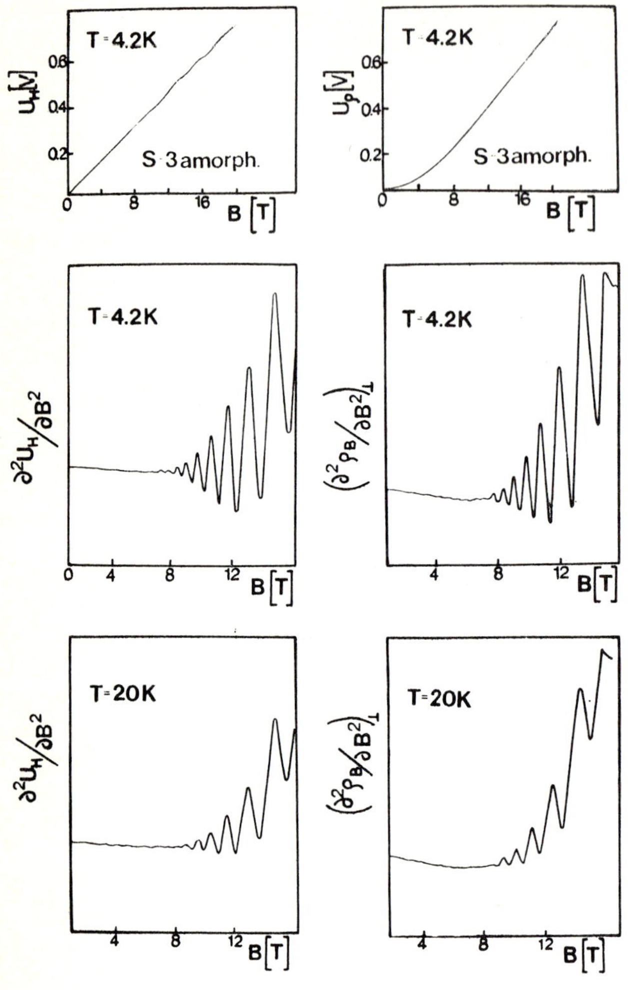

Fig. 4. SdH oscillations for a thin a-film of Cd_3As_2 (sample 3). The second derivatives of the Hall voltage and the transverse magnetoresistance at 4.2 K and 20 K are shown.

Period of oscillations

The oscillations shown in Figs. 3 and 4 are exactly periodic with respect to the reciprocal magnetic field B and calculated periods oscillations $P_n = \Delta(1/B)$ were found to be identical for all orientations of the sample contrary to the crystalline samples where distinct anisotropy of P_n was established. From the comparison of P_n values for amorphous and crystalline films in Fig. 5 it follows that also the magnitude of $P_n = 0.091\ T^{-1}$ for a-films differs considerably from that for crystalline films

The electron concentration calculated from SdH oscillations is equal to $6.6\times10^{18}\ cm^{-3}$ for the presented sample no 127/75 (9).

This value is about two orders of magnitude higher than that obtained from the Hall data at the same temperature ($n_H \simeq 0.05\times10^{18}\ cm^{-3}$ at 4.2 K). The values of n_{SdH} calculated for the other samples summarized in Table 2 are of the same order of magnitude ($5\text{-}7\times10^{18}$ cm^{-3}) (9).

Cyclotron effective mass

The temperature and field dependence of the amplitude A of the oscillations of magnetoresistance is expressed by the well known classical relation :

$$A \sim \frac{\Delta\rho}{\rho_o} \sim B^{1/2} \frac{2\pi^2 kT/\hbar\omega_c}{\sinh(2\pi^2 kT/\hbar\omega_c)} \exp\left(-\frac{2\pi^2 kT_D}{\hbar\omega_c}\right) \qquad |1|$$

where T is the temperature of the sample and T_D - is the Dingle temperature.

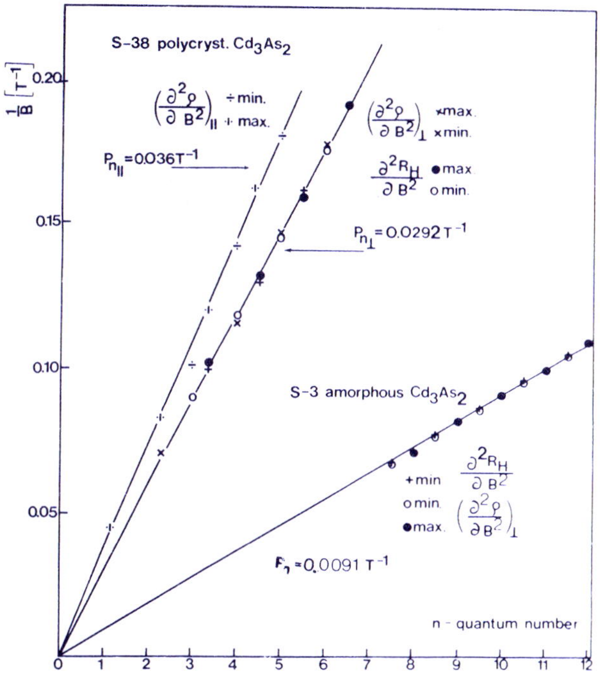

Fig. 5. Periods of oscillations for crystalline and amorphous thin films of Cd_3As_2.

To a good approximation equation |1| can be written as

$$\frac{A_o}{A_T} = \frac{\sinh U_T}{U_T} \qquad |2| \qquad \text{where } U_T = \frac{2\Pi^2 kT}{\hbar\omega_c}$$

From the experimental values of the amplitude A_o at T_o = 4.2 K and A_T at 20 K measured from the oscillograms for the identical values of the magnetic field B we have calculated U_T and then the cyclotron effective mass m^*_c from eqn. |2| (8,21)

$$\frac{m^*_c}{m_o} = \frac{eh}{2\pi^2 k}\,\frac{BU_T}{T} = \frac{BU_T}{14.775\ T} \qquad |3|$$

where B is in teslas.

The cyclotron effective masses m^*_c calculated in this manner for the three orientations of the samples, i.e. m^*_H for ($\Delta U/U_o$), $m^*_{\parallel}$ for $(\Delta\rho/\rho_o)_{\parallel}$ as well as $m^*_{\perp}$ for $(\Delta\rho/\rho_o)_{\perp}$ are presented in Table 3 (9) together with data obtained by compensation method.

Unlike the case of crystalline Cd_3As_2, the obtained effective mass m^* for amorphous samples is found to be completely isotropic i.e., independent on the direction of the magnetic field B with respect to the sample current for all the a-films investigated and was also independent of the magnetic field B through the whole range examined. The average value was found to be $m_c^* \simeq 0.1\ m_o$ which is about three times larger than that for crystalline samples at the same concentration ($m_1^* = m_2^* = 0.027\ m_o$, $m^* = 0.031\ m_o$) (9).

The a-films for which the calculated value of m_c^* was lower than $\sim 0.1\ m_o$ we found to be partly crystallized near the electrodes after room - and low-temperature electrical measurements (see Table 2).

Final remarks

The appearance of so clear oscillation series in amorphous film was very surprising and unexpected. Such oscillations have been observed for amorphous material for the first time at all. The low-field Hall mobility for a-film of Cd_3As_2 measured at helium temperatures did not usually exceed 200 - 700 cm^2/Vs. This value is extremely high for amorphous film but is really too small to fulfil the quantization conditions of the Landau levels. At 4.2 K it is more than two orders in magnitude less than the Hall mobility in single crystal of Cd_3As_2 and more than one order of magnitude less than that for crystalline films of Cd_3As_2.

The possible explanation of this phenomenon seems to be connected with the observed granular structure of the investigated a-films, with grain dimensions typically of 1000 Å (12).

The small quantity of excess As atoms seems to be built-up into the lattice network in such a manner that interior of the clusters may be less perturbed and thus higher electron mobility inside grains can be expected, while the low Hall mobility and low film conductivity are due to scattering at grain boundaries.

At 4.2 K the relation |1| can be recast in the form (9) :

$$\frac{A_n}{A_{n+1}} = \left(\frac{B_n}{B_{n+1}}\right)^{1/2} \exp\left\{\frac{2\pi^2 k m^*}{h\ e} T_D \left(\frac{1}{B_{n+1}} - \frac{1}{B_n}\right)\right\} = \left(\frac{2n+3}{2n+1}\right)^{1/2} \exp \frac{\pi}{\mu_c} p_n \qquad |1a|$$

for two different magnetic fields B_1 and B_2 corresponding to the relative amplitudes A_n and A_{n+1} of the two successive oscillation peaks where T_D is the Dingle temperature and $\mu_c = e\ \tau_F/m_c^*$ is the apparent microscopic mobility, called by us "cyclotron" mobility, τ_F being the relaxation time of the carriers. The relation |1 a| has been applied to crystalline materials at very low-temperatures and strong magnetic fields |1,1a| . We used this equation to deduce the variation of cyclotron microscopic mobility μ_c with the magnetic field B despite the fact that it must be treated in this case as an approximation only. We have obtained completely unexpected result (9). The data calculated by us for sample 127/75 are presented in Fig. 6 together with the values of $\mu_c = f(B)$ for a crystalline Cd_3As_2 film. The monotonous decrease of μ_c for crystalline film is similar to that reported for heavily doped GaAs at the oscillation region (20) and is probably due to the scattering processes at so strong magnetic field. The rapid rise of μ_c at $B > 10$ T for a-film has quite different picture from that found for polycrystalline film. However, this ensures the satisfying of the quantization conditions necessary to the appearance of the SdH oscillations. This behaviour of μ_c is probably governed by various factors related to the granular amorphous structure of the film and particularly to local order, inhomogeneities, grain boundaries and some difficulty of forming cyclotron orbits inside the grains.

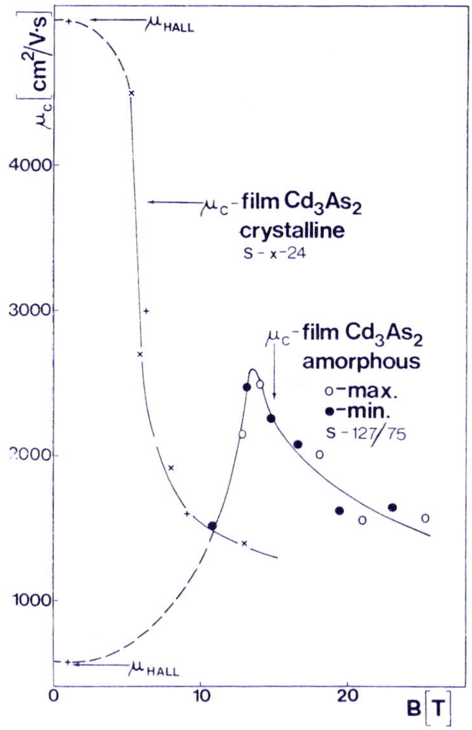

Fig. 6. Apparent cyclotron mobility μ_c as a function of the magnetic field B calculated for Cd_3As_2 films.

The physical parameter describing in some sense the inhomogeneity of the sample is the Dingle temperature T_D, which represents the nonthermal damping factor due to collisions (mainly inhomogeneities, impurities, grain boundaries). It is clearly related to the relaxation time τ_F :

$$T_D = \frac{\hbar}{2\pi k} \cdot \frac{1}{\tau_F} = \frac{\hbar}{2\ \pi k} \cdot \frac{e}{\mu\ m^*} \qquad |4|$$

For an amorphous structure composed of clusters it is reasonable to write (9) :

$$T_{D,ef.} = \frac{\hbar}{2\pi\ k}\ \frac{1}{\tau_{ef}} = \frac{\hbar}{2\pi\ k}\ (\frac{1}{\tau_{int.}} + \frac{1}{\tau_{dimens.}}) \qquad |5|$$

where τ_{int} is the relaxation time of carriers due to relaxation processes inside the clusters connected with disorder or colisions that are not affected by dimensional and boundary effect, and τ_{dimens} due only to the dimensional effect caused by scattering from the cluster boundaries. The value of T_D calculated by us for a number of examined a-films varied from 30 K up to 35 K. This value is considerably higher than that calculated for crystalline films, $T_{D,av.}$ = 22.7 K (8) or for Cd_3As_2 single crystal, where T_D was not more than 7.2 K (2).

It is reasonable, that cyclotron orbits can only be formed inside the clusters, then in order to produce the SdH oscillations the condition $2r_c < D$ must be satisfied, where D is the amorphous grain diameter and r_c is the classical cyclotron radius defined as

$$r_c = r_o\ \{2n+1\}^{1/2} = \frac{260}{B^{1/2}}\ \{2n+1\}^{1/2}\ [\text{Å}] \qquad |6|$$

The magnetic field B (in teslas) must be then significantly larger of some minimum value, $B > B_{min}$.

Thus at 10 T, where we are concerned with the 12th Landau level, the calculated radius of the cyclotron orbit is ≃ 450 Å ; at higher fields it decreases as $B^{1/2}$ and the broadening of the Landau level is strongly reduced in the direction of the grain centre.

This calculation confirms our suggestion that the cyclotron orbit can be formed when its radius is sufficiently small compared to the grain size (∿1000 Å), since there is less perturbation by the boundary regions (with some local order).

This seems to be the reasonable physical explanation for the rapid increase observed in the microscopic cyclotron mobility μ_c for B greater than 10 T which satisfies the quantization condition in the above mentioned sense. The observed slope in the increasing part of $\mu_c = f(B)$ curve can be due to statistical appearance of oscillations in successive cluster for which the radius of the cyclotron orbit satisfies the condition $2r_c < D$. Then, the value of B_{min} depends on the microstructure of the real film, i.e. on the average cluster dimensions and the distribution of the grain sizes as well as on the internal local order.

The proposed mechanism for the production of the SdH oscillations is also supported by the very high cyclotron electron concentration ($n_{SdH} = 5\text{-}7 \times 10^{18}$ cm^{-3}) different from the measured Hall concentration n_H (see Table 2). The cyclotron concentration can be assigned to the interior of the clusters, where as the Hall concentration is always an average value that includes the boundary regions where there are a number of broken bonds and, of course, some excess of the As atoms.

It is interesting to add, that from our investigations of a-films of the Hall mobility as a function of electron concentration we have obtained some maximum value of Hall mobility for $n \simeq 6 \times 10^{18}$ cm^{-3} (22) while for the crystalline material we have always $n_H \simeq n_{SdH}$ (2,7,8). The existence of the quantized states in an amorphous material suggests that the electron states can be treated here in the effective mass approximation (10). As suggested by Kivelson and Gelott (23) this can follow from the assumption that the wave function can be treated as the product of a rapidly varying "molecular" functions and a slowly varying envelope function which satisfies a Landau equation with an effective mass depending on the molecular function. This can be possible in the amorphous material like amorphous cadmium arsenide, where local order exists. However, no theoretical treatment of this problem exists up to now.

Finally we note, that we have shown for the first time the possibility of producing of the quantum phenomena in an amorphous semiconductor. Our results suggest that long-range order is not essential for sharp Landau quantization, though we believe that local order may be.

Acknowledgements

We wish to thank Professor J.L. ROBERT, Dr. M. DUSSEAU, Dr. J. CISOWSKI for useful discussions and S. ABD-ELMOATY for valuable experimental contribution.

Literature references

1. W. ZDANOWICZ, L. ZDANOWICZ, Ann. Rev. Mater.Sci., 5/1975/301
2. I. ROSENMAN, J. Phys. Chem. Solids, 30/1966/1385
3. M.J. AUBIN, A.RAMBO, J.C.PORTAL, C. HOULBERT, S.ASKENZAZY, P.R. WALLACE Proc. 3rd Int. Conf. on Physics of Narrow-gap Semiconductors, Warsaw 1977, ed.PWN Warsaw 1978, p. 317
4. F.A.P. BLOM, M.J. GELTEN, Int. Conf. on Physics of Semicond., Edinburgh 1978; Proc. on Phys. of Narrow-gap Semiconductors, Warsaw 1977, ed. PWN Warsaw 1978, p. 257
5. H. DOI, T. FUKUROI, T. FUKASE, Y. MUTO, K. TANAKA, Sci. Rep. Inst. Tohodu Univ. Ser. A 20/1969/190
6. J. BODNAR, Proc. 3rd Int. Conf. on Phys. of Narrow-gap Semiconductors, Warsaw 1977, ed. PWN Warsaw 1978, p. 311
7. W. ZDANOWICZ, L. ZDANOWICZ, J.C. PORTAL, S. ASKENAZY, Thin Solid Films 61/1979/41

8. J.C. PORTAL, M.A. MOATY, W. ZDANOWICZ, L. ZDANOWICZ, Thin Solid Films 76/1981/391
9. W. ZDANOWICZ, L. ZDANOWICZ, J.C. PORTAL, S. ASKENAZY, M.A. MOATY, Thin Solid Films, 81/1981/101
10. J.C.PORTAL, P.R. WALLACE, W. ZDANOWICZ, L. ZDANOWICZ, Proc. 15th Int. Conf. Physics of Semiconductors, Kyoto 1980, J. Phys. Soc. Japan 49/1980/1167
11. L. ZDANOWICZ, Acta Physica Polonica, 31/1967/1021
12. J. JURUSIK, L. ZDANOWICZ, Thin Solid Films, 67/1980/285
13. L. ZDANOWICZ, Influence of deposition condictions on the structure and properties of Cd_3As_2 films, Habilitation, ed. PWN Warsaw-Wroclaw, 1979/in/Polish
14. L. ZDANOWICZ, S. MIOTKOWSKA, Thin Solid Films, 48/1980/397; J. Crystal Growth, 48/1980/387
15. Cz. WECLEWICZ, L. ZDANOWICZ, Proc. Int. Conf. "Amorphous Semiconductors-78", Pardubice, Czechoslovakia 1978, p. 340; "Amorphous Semiconductors-76" Balatonfüred, Hungary 1976, ed. Adademiai Kiado, Budapest 1976, p. 437
16. Cz. WECLEWICZ, L. ZDANOWICZ, J. JURUSIK, Proc. 1st Int. Symp. on Phys. and Chemistry of II-V Compounds, Mogilany, Poland, Sept. 80, ed. M.J. GELTEN, L. ZDANOWICZ, Eindhoven University, p. 125
17. L. ZDANOWICZ, W. ZDANOWICZ, Cz. WECLEWICZ, J.C. PORTAL, Journal de Physique Colloque C4, suppl. 10,V42/1981/C4-1069
18. L. ZDANOWICZ, T. KWIECIEN, Vacuum, 27/1977/409
19. A. BURIAN, J. AULEYTNER, as in ref. 16, p. 119
20. B.M. VUL, N.V. KOTELNIKOVA, E.J. ZAVARITSKAYA, J.D. VORONOVA, Proc. 13th Int. Conf. Physics of Semiconductors, Rome 1976, North-Holland, Amsterdam 1976, p. 1188
21. S. ASKENAZY, in "New Developments in Semiconductors" V.19, part IV, Noordhoff, Leydon, p. 333
22. L. ZDANOWICZ, Cz. WECLEWICZ, to be published
23. S. KIVELSON, C.D. GELOTT, Phys. Rev., 319,10/1979/5160

Table 2. Electrical parameters for some Cd_3As_2 a-films/see also (9)/.

Sample	d	$R_x (\Omega)$		$n \times 10^{18} (cm^{-3})$		$\mu_H (cm^2/Vs)$		Hall effect		$(\Delta\rho/\rho_0)\perp$		$(\Delta\rho/\rho_0)\parallel$	
								from derivative oscillations					
	μm	300 K	4.2 K	300 K	4.2 K	300 K	4.2 K	$n \times 10^{18}$	m^*/m_o	$n \times 10^{18}$	m^*/m_o	$n \times 10^{18}$	m^*/m_o
118/27	0.82	9000	14×10^6	0.33	-	39	-	no oscillations					
119/13	2.3	905	21000	1.13	0.07	95	-	no oscillations					
129/56	1.07	1210	44250	0.76	0.2	190	210	-	-	2.30	0.045	-	-
129/55	1.07	1080	2500	1.08	0.4	150	170	-	-	1.85	0.050	-	-
7	0.54	366	1600	3.28	0.51	300	430	4.6	0.08	4.2	0.082	2.25	0.080
3	0.54	358	317	2.3	1.62	420	620	6.50	0.102	6.59	0.104	2.8	0.1
140/77	0.91	470	2100	0.52	0.38	84	256	5.1	0.102	5.67	0.101	-	-
127/75	8	590	900	0.137	0.05	290	500	6.5	0.095	6.5	0.095	6.0	0.118
127/77	8.5	720	1100	0.11	0.035	280	863	7.1	0.101	7.35	0.098	6.0	0.095
128/63	11	1030	1600	0.087	0.039	190	297	5.16	0.095	-	-	-	-

Table 3. Effective mass of electrons in Cd_3As_2 a-films. (Sample 127/75 ; T_o = 4.2K, T = 20 K)

Property	Parameter	Maxima										Average
$\frac{\partial^2 U_H}{\partial B^2}$	B(T)	1.6	13.94	12.37	11.14	10.11	17.142	17.771	13	4.71	10.62	-
	m^*/m_o	0.0964	0.0991	0.0977	0.0956	0.0954	0.0933	0.0932	0.0968	0.0941	0.0945	0.0954
$\frac{\partial^2 \rho_B}{\partial B^2}\perp$	B(T)	17.23	14.86	13.06	11.71	-	18.74	15.86	13.86	12.29	11	-
	m^*/m_o	0.094	0.099	0.098	0.1	-	0.092	0.092	0.096	0.097	0.096	0.095
$\frac{\partial^2 \rho_B}{\partial B^2}\parallel$	B(T)	20.333	26.9	-	-	-	23.33	-	-	-	-	-
	m^*/m_o	0.0939	0.0938	-	-	-	0.1189	-	-	-	-	0.1005
$\frac{\Delta U_H}{U_H}$	B(T)	20.76	17.80	15.52	13.76	12.38	22.76	19.28	16.66	14.66	13.09	
	m^*/m_o	0.126	0.113	0.113	0.106	0.113	0.116	0.118	0.117	0.113	0.110	0.114

HOPPING CONDUCTION IN THE MAGNETIC FIELD

J.A. Chroboczek

Max-Planck-Institut für Festkörperforschung, Hochfeld-Magnetlabor
166X, F-38042 Grenoble Cédex, France
and Institute of Physics; Pol. Acad. Sci.; Warsaw, Poland

This paper reviews the effects of the magnetic field upon the phonon-assisted tunneling (hopping) [1]-[3] and covers the following topics: DC and AC conductivity in the magnetic field, electron-electron correlations and their modification by the magnetic field and the Hall effect in hopping conduction.

FEASIBILITY OF THE EFFECTS

In order to discuss the effect of the magnetic field on shallow impurities in semiconductors several authors [4]-[6] have made use of the parameter $\gamma = h\omega/2Ry^*$, where $h\omega/2$ is the energy of the lowest Landau state and Ry^* is the effective Rydberg. The magnitude of γ provides a measure of the degree of perturbation of a hydrogenic state by the magnetic field. When discussing the magnetic field effect on hopping it is instructive to rewrite γ as a^{*2}/λ^2, where a^* is the effective Bohr radius of the ground state wave function and λ is the amplitude of the first cyclotron orbit. Table I lists some values of the effective Bohr radius for several impurities in semiconductors, next to the field strengths at which the parameter λ assumes values comparable with a^*.

Percolation theory arguments [3] indicate that for the evaluation of the effect of the magnetic field upon the hopping conduction, involving hydrogenic states, one should consider another parameter, $\delta = a^* N^{-1/3}/\lambda^2 = \gamma N^{-1/3}/a^*$, since the average separation of the impurities of the concentration N is about $(4/3\ \pi N)^{-1/3}$. The expression for δ implies that the effect of the magnetic field on the hopping conduction is more pronounced in more dilute samples. This has also been born out by experiment [9],[16].

TABLE I

Magnetic field strength H (T)	$(\frac{ch}{eH})^{1/2}$ λ (Å)	Bohr radius a^* (Å)	Impurity/ Semiconductor	Reference
0.2	570	600	donor in InSb	[7] and [8]
10	81	99	Sn in GaAs	[9]
		87	Ga in Ge *)	[10]
		84	donor in InP	[11]
30	47	41	Sb in Ge *)	[12]
100	26	26	P in Si	[8]
		16	P in Si	[13]
200	18	21	Al in Si *)	[14]
		17	B in Si *)	[14]
5000	3.6	3.6	charged soliton in $(CH)_x$	[15]

*) Asymptotic value

As is known [1], [12], the average separation of the majority impurities at which the pair hopping becomes the predominant mechanism of conduction is larger than $5a^*$, i.e. in the expression for δ one should replace $N^{-1/3}$ by $5a^*$ at the onset of the hopping conduction mechanism. Consequently, the values of the field which has a pronounced effect on the hopping conduction is at least 5 times smaller than the field strengths at which $\gamma = 1$. Indeed, large magnetoresistance effect in semiconductors has been reported by numerous authors, at relatively weak fields. To give a few representative papers, InSb: references [7,17]-[20]; InP: references [20]-[23]; GaAs: references [9,24]-[26]; and Ge: references [15,27]-[29]. It should be pointed out that the hopping conduction in n-type InSb cannot be observed without the application of a magnetic field. The absence of data for Si follows from the fact that $\delta = 1$ requires fields of about 20 T or employment of very dilute samples, which have very high resistance, both providing serious difficulties for the experimentalist. As Table I implies hopping conduction in polyacetylene, $(CH)_x$, may be affected by the magnetic fields totally non-realisable today. It has been verified experimentally [30] that a magnetic field of 15 T has no effect whatsoever on the AC and DC conductance of $(CH)_x$ at liquid N_2 temperatures, where hopping is supposed to be the dominant conduction mechanism. This result is in agreement with the estimate of the spatial extent of the wave function of the charged soliton.

DC MAGNETORESISTANCE

Several authors [3,31]-[34] evaluated the effect of the magnetic field on the hop conduction, however, all these theoretical treatments applied either to the case when $\delta \ll 1$ (low field) or $\delta \gg 1$ (high field). Recently Ioselevich [35] following the procedure proposed by Shklovskii (described in [3]) calculated the effect of the magnetic field upon hopping between spherically symmetric impurity states at an arbitrary magnetic field. Since the wave functions of the localized states are exponentially small in the barrier that separates the states, the quasiclassical apporoximation was used. The action integral in classically forbidden regions was computed by the analytic continuation, the imaginary time, method. Figure 1 summarizes some theoretical and experimental results for hopping magnetoresistance involving spherically symmetry impurities. Here H^*, the normalized magnetic field, is H/H_c where $H_c = 3.47\, N^{1/3}\, c\hbar/ea^*$; ξ^* is the critical exponent for percolation, also normalized by the factor $\xi_o = 1.73/N^{1/3}\, a^*$. The magnetoresistance is related to ξ^* through the expression, $\ln(\rho(H)/\rho(0)) = \xi_o(\xi^* - 1)$.

The experimental results of Kahlert et al. [9] (GaAs) follow the universal ξ^* vs. H^* dependence, consistent with the Ioselevich theory, however, the experimental data points lie above line 4, that is given by theory [35]. Note that the data points for the 105 InP sample [21] follow the same dependence as the data points for GaAs, however, the data points obtained for the other (202) InP sample do not. A possible reason for this is an error in establishing the concentration of impurities in InP samples. It should be pointed out that this requires the knowledge of the compensation, which in InP is difficult to establish.

It is worth emphasizing that a similar underestimate of the magnetoresistance effect in n-type Ge [16] has been reported by Movaghar and Sauer [34], who used the Green's function method for calculating the change of resistivity in the magnetic field.

AC CONDUCTIVITY

The AC conductivity measurements of hopping conduction in semiconductors have been done by several authors [36]-[40]. Papers [39] and [40]involve the use of a magnetic field. As is known that the real and the imaginary parts of the conductance follow the power dependence,

$$(\sigma - \sigma_{DC}) \propto \omega^s \quad \text{and} \quad \omega(C - C_o) \propto \omega^s$$

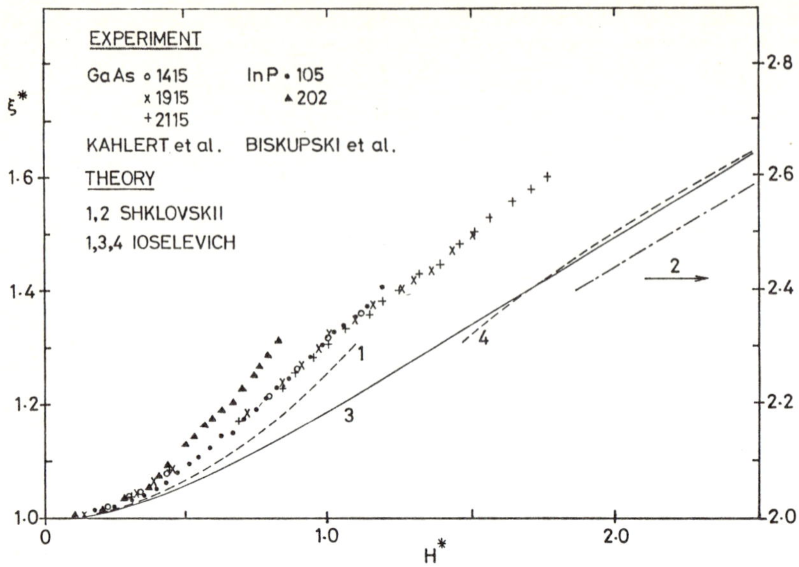

Figure 1: Variation of the critical exponent for percolation, ξ^* with the normalized magnetic field $H^* = H/3.47N^{1/3}c\hbar/ea^*$. Line 1 (dashed) $\xi^* = 1 + H^{*2}/4$ (Shklovskii [3], low field limit); Line 2 (dash-dot) $\xi^* = 1 + H^{*1/2}$ (Shklovskii [3] high field limit), note that the scale (right hands side) is shifted by unity, however, the slope of lines 2 (dash-dot) and 3 (solid) are very close; Line 3 (solid) present theory (Ioselevich, [35]) Line 4 (dashed) high field limit of [35] (see Eq. (17) of [35]). The experimental data of GaAs have been provided by H. Kahlert, and were published in ref. [9]. The data for InP were provided by G. Biskupski (also published in [23].

where C_o is the capacitance of the host lattice (a constant). The exponent s is found to be about 0.8, however, in samples with higher impurity concentration, where multiple hops are possible, the value of s can be as small as 0.2 (Amitay and Pollak, [41]. In such cases the magnetic field was reported [39] to increase towards 0.8, and shift the values of σ downwards (see Fig. 2, reproduced from paper [39]). The increase of the slope has been explained in [39] by the disruption of the multiple hop chains by strong magnetic fields at the low frequency limit. The strong magnetic field also affects the impurities surrounding the impurity pair between which the actual hop is executed, and reduces the e-e correlation effects. Since the correlations increase the impedance of the hop [42], their elimination should increase the conductance. However, since the conductivity decreases with the magnetic field, the reduction of e-e correlation has probably a secondary effect. Recent work of Mansfield et al. [40] reports that AC conductance can be lower for higher impurity concentrations, although the DC conductance behaves inversely. The ffect is attributed to the presence of strong correlations that entails polarisation loss as discussed by Jonscher [43, 44] in his theory of universal dielectric response.

CORRELATION EFFECTS

Pollak and Knotek [42] showed that at concentrations at which the activation energy ε_3 falls off the linear dependence, if plotted versus $1/r_{av}$, $r_{av} = (4/3\pi N)^{-1/3}$,

(Miller and Abrahams pair approximation), the preexponential ρ_3 exhibits a distinct increase. They attributed this effect to the onset of multiple, correlated transitions. The magnetic field decreases the spatial extent of the wave functions and decreases,in consequence, the e-e correlations. If the concentration of impurities is such that correlations are not very strong the magnetic field may eliminate them completely increasing ε_3 to the value it should have in the regime of validity of the pair hopping mechanism. This effect was observed by Knotek [29], who also proposed the explanation of that phenomenon in terms of the elimination of multiple hops. The increase of ε_3 was also observed by Biskupski [20] in InP and also in Ge by Chroboczek et al. [45]. The latter proposed that the magnetic - field induced spin - flip hopping transitions are responsible for the increase of the activation energy. The explanation of the effect through the elimination of multiple hopping transitions seems much more plausible, particularly because the phenomenon has been observed in several materials in which spin interactions are difficult to imagine.

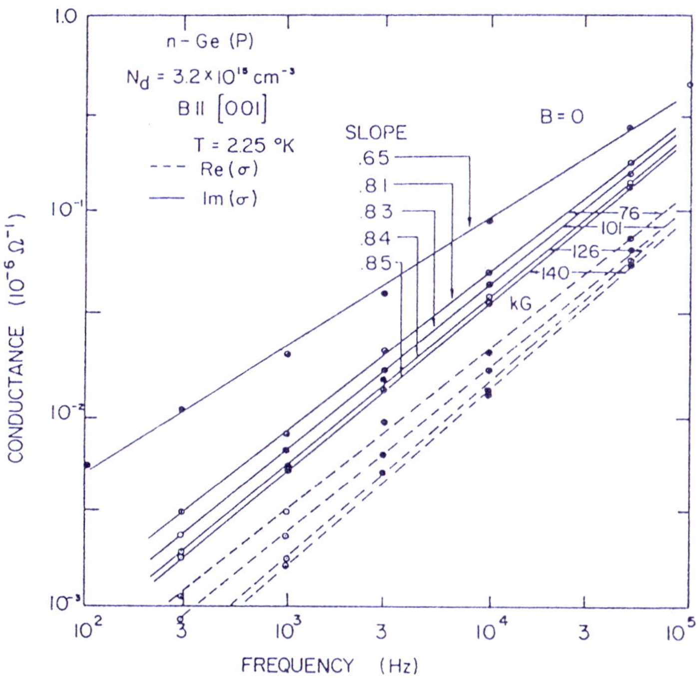

Figure 2: (reproduced from [39]) The real and imaginary parts of the conductance plotted as a function of frequency in a log-log scale. The ω^s dependence is depicted as a straight line with the slope s. The Re(σ) line for B=0 had been eliminated for clarity. The values Re(σ) and Im(σ) are defined in the text.

HALL EFFECT

The Hall effect in hopping conduction is related to the change in the phases of the impurity wave functions as the field is increased (Mikoshiba and Gonda [31] and Holstein [46]). Consequently, the transition rates in a three - site arrangement of impurities is altered and the transition rate component in the direction perpendicular to the field is augmented. Extensive theorectical studies of the Hall effect in hopping are given by Holstein [46] and Pollak and Friedman [47], [48]. The AC Hall effect was studied experimentally by Amitay and Pollak [41] and Klein and Pollak [49]. The experimental accuracy achieved in [49] should have permitted the detection of Hall voltages of the level predicted by Holstein's theory, however, no effect was observed.

Butcher and Kumar [50] and Butcher and McInnes [51] have calculated the DC mobility using the linearized rate equation in the zero frequency limit. Recently Biskupski and Dubois [20] showed that their data obtained on several InP samples comply with the predictions of the DC Hall effect theory, [50], [51]. It should be pointed out that the magnetoresistance data presented in Fig. 1 were taken on two of the samples used for the Hall measurements. Since the values of the impurity concentrations in these samples may be incorrect, as discussed above, reexamination of the assumption of the predominance of the hopping conduction in the InP samples used in the Hall effect studies seems, therefore, desirable.

ACKNOWLEDGEMENTS

I am indebted to Prof. Kahlert and Dr. Biskupski for making their original data available to me. I am very thankful to Dr. McInnes for his help in computing the results presented in Fig. 1. This work was partially supported by the Science and Engineering Research Council, Great Britain.

REFERENCES

[1] A. Miller and E. Abrahams, Phys. Rev. 120, 745 (1960)
[2] N.F. Mott and E.A. Davis, Electronic Processes in Non-Crystalline Semiconductors, Second Edition, Oxford, Clarendon Press, 1979
[3] B.I. Shklovski and A.L. Efros, Electronic Properties of Doped Semiconductors, Moscow, Nauka, 1979 (English translation to appear shortly, Springer Verlag)
[4] Y. Yafet, R.W. Keyes, and E.N. Adams, J. Phys. Chem. Solids 1, 137 (1956)
[5] H. Hasegawa and R.E. Howard, J. Phys. Chem. Solids 21, 179 (1961)
[6] D.M. Larsen, J. Phys. Chem. Solids, 29, 271 (1968)
[7] G. Biskupski, H. Dubois, and D. Ferré, Phil. Mag. 43, 177 (1981)
[8] K.F. Berggren and B.E. Sernelius, Phil. Mag. 43, 115 (1981)
[9] H. Kahlert, G. Landwehr, A. Schlachetzki, and H. Salow, Z. für Physik B 24, 361 (1976)
[10] M. Pollak, J. Non-cryst. Solids 11, 1 (1972)
[11] M. Cardona, p. 125 in Semiconductors and Semimetals, Vol. 3 (1966)
[12] H. Fritzsche, J. Phys. Chem. Solids 6, 69 (1958)
[13] T.G. Castner, Phil. Mag. B 42, 873 (1981)
[14] R.K. Ray and H.Y. Fan, Phys. Rev. 121, 768 (1961)
[15] S. Kivelson, Phys. Rev. B 25, 3798 (1982)
[16] J.A. Chroboczek and R.J. Sladek, Phys. Rev. 151, 595 (1966)
[17] R. Mansfield, and L. Kusztelan, J. Phys. C 11, 4157 (1978)
[18] H. Tokumoto, R. Mansfield and M.J. Lea, Phil. Mag. B 46, 93 (1982)
[19] J.L. Robert, A. Raymond, R.L. Aulombard, and C. Bousquet, Phil. Mag. B 42, 1003 (1980)
[20] G. Biskupski, Thèse, Univ. des Sci. et Techniques de Lille, 1982 (unpublished)
[21] G. Biskupski, and H. Dubois, Phil. Mag. B 42, 19 (1980)
[22] O.V. Emelyanenko, V.G. Masagutov, P.N. Nasledov, and D.N. Timchenko, Fiz. Tekh. Poluprov. 9, 503 (1975), Sov. Phys., - Semicond. 9, 330 (1975)
[23] G. Biskupski and H. Dubois, pp. 29-31 in Recent Developments in Condensed Matter Physics, Vol. 3, Plenum Press 1981, J.T. Devreese, Edit.
[24] L. Halbo and R.J. Sladek, Phys. Rev. 173, 794 (1968)
[25] O.V. Emelyanenko, T.S. Lagunova, D.N. Nasledov, L.D. Nedolgo, and I.N. Timchenko, Fiz. Tekh. Poluprov. 7, 1919 (1973), Sov. Phys.- Semicond. 7, 1280 (1974)
[26] G.A. Matveev, Fiz. Tekh. Poluprov. 15, 2333 (1981), Sov. Phys. - Semicond. 15, 1355 (1981)
[27] A.R. Gadziev and I.S. Shlimak, Fiz. Tekh. Poluprov. 6, 1582 (1972) Sov. Phys. - Semicond. 6, 1364 (1973)
[28] J.A. Chroboczek, A. Klokocki, and K. Kopalko, J. Phys. C 7, 3042 (1974)

[29] M.L. Knotek, Phys. Rev. B 16, 2629 (1977)
[30] J.A. Chroboczek, M. Nechtschein, and J.-P. Travers in Polymers Electroactifs ; TCP457, CNRS, Autrans, 1982 (unpublished)
[31] N. Mikoshiba and S. Gonda, Phys. Rev. 127, 1954 (1962), and N. Mikoshiba, Phys. Rev. 127, 1962 (1962)
[32] B.I. Shklovskii, Fiz. Tekh. Poluprov. 8, 416 (1973), Sov. Phys. - Semicond. 8, 268 (1974)
[33] B.I. Shklovskii, Zh. Eks. Teoret. Fiz. 61, 2033 (1971), Soviet Phys. - JETP 34, 1084 (1972)
[34] B. Movaghar and G.W. Sauer, Solid State Commun. 35, 841 (1980)
[35] A.S.Iosolevich, Fiz. Tekh. Poluprov. 15, 2373 (1981), Sov. Phys. - Semicond. 15, 1378 (1981)
[36] M. Pollak and T.H. Geballe, Phys. Rev. 122, 1745 (1961)
[37] S. Golin, Phys. Rev. 132, 178 (1963)
[38] H. Kahlert, J. Phys. C, 9, 491 (1976)
[39] J.A. Chroboczek, E.W. Prohofski, and R.J. Sladek, p. 653 in Proc. IX Intern. Conf. Phys. Semicond., Leningrad, Nauka, 1968; S.M. Ryvkin Edit.
[40] R. Mansfield, Chen Zhonggan, and P. Fozoni, presented at XVI Intern. Conf. Phys. Semicond., Montpellier 1982
[41] M. Amitay and M. Pollak, p. 549 in Proc. VIII Internat. Conf. Phys. Semicond., Kyoto 1966, J. Phys. Soc. Japan, 21, Supplement 1966
[42] M. Pollak and M.L. Knotek, J. Non-crystalline Solids, 32, 141 (1979)
[43] A.K. Jonscher, p. 22 in Inst. Phys. Conf. Series No. 58, Phys. Diel. Solids, 1980
[44] R.M. Hill and A.K. Jonscher, J. Non-crystalline Solids, 32, 53 (1979)
[45] J.A. Chroboczek, E. W. Prohofsky, and R.J. Sladek, Phys. Rev. 169, 593 (1968)
[46] T. Holstein, Phys. Rev. 124, 1329 (1961)
[47] L. Friedman and T. Holstein, Phys. Rev. 165, 1019 (1968)
[48] L. Friedman and M. Pollak, Phil. Mag. B 38, 173 (1978)
[49] R. Klein and M. Pollak, private communication
[50] P.N. Butcher and A.A. Kumar, Phil. Mag. B 42, 201 (1980)
[51] P.N. Butcher and J.A. McInnes, Phil. Mag. B, 44, 595 (1981)

EFFECT OF DILUTE IONIZED IMPURITIES ON THE MAGNETORESISTIVITY OF BISMUTH IN HIGH MAGNETIC FIELDS

Hiroshi Bando, Kokichi Oshima, Masahiko Manabe and Mitsuru Suzuki
Cryogenic Centre, University of Tokyo
Bunkyo-ku, Tokyo, 113 JAPAN

In crossed electric-(E) and magnetic-(H) fields where the cyclotron motion of the carriers is well-defined, the cycloidal drift of the carriers drives the current along the ExH direction, whereas scattering processes may deflect the current in every direction - including that parallel to E - and this causes the Shubnikov-de Haas(SdH) oscillations. In order to detect the SdH oscillations it is necessary to apply strong magnetic fields to make the Landau level spacing $h\omega_c$ much larger than the broadening h/τ of each state, where ω_c and τ denote the cyclotron frequency and the mean lifetime. However, this condition is insufficient to observe the oscillations up to the highest harmonics [1]. The wave function of the carriers under strong magnetic fields has the form of an harmonic oscillator in the plane perpendicular to H and of a plane wave with wave number up to $k_z(H)$ in the direction along H; it is composed of wave packets with the size of the order L and $1/k_z(H)$ in each direction, where L denotes the Larmor radius $2.56\times10^{-6}/\sqrt{H}$ cm and $k_z(H)$ is an oscillatory function of H. If the size of this wave packet is so large or the concentration of scatterers is so high that the carriers interact with more than two scatterers at the same time, the broadening of each level grows rapidly and the magnetoconductivity ceases to reflect the density of states of carriers at $E_F \pm kT$, where E_F denotes the Fermi energy. Since the maximum of $k_z(H)$ is determined from the total carrier density through the Fermi distribution, at least the density of the scatterers should be kept lower than the carrier density in order to study the SdH effect in detail. This condition is quite limiting for the bulk system of doped semiconductors containing the same number of impurities and carriers. On the other hand, bismuth has intrinsic carriers with small mass and is fairly easily purified and crystalized. Moreover, one has the possibility to control the impurity concentration by doping up to 10 ppm, the order of the intrinsic carrier density. However, not much work has been done in this direction.

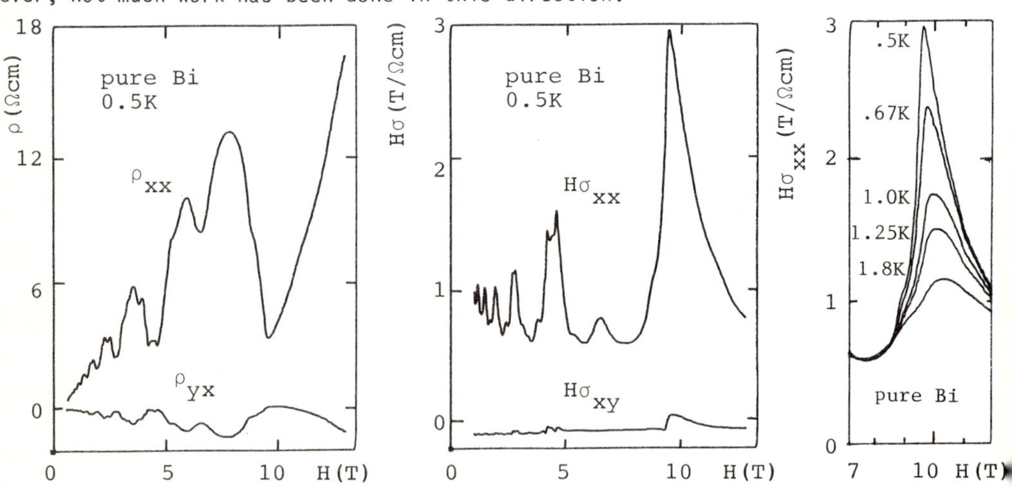

Figure 1: (a) ρ_{xx} and ρ_{xy} for a pure bismuth single crystal (H ||trigonal axis) as a function of H.
(b) $H\sigma_{xx}$ and $H\sigma_{xy}$, obtained from the data shown left, as function of H
(c) Amplitude of the last $H\sigma_{xx}$ maximum as function of H.

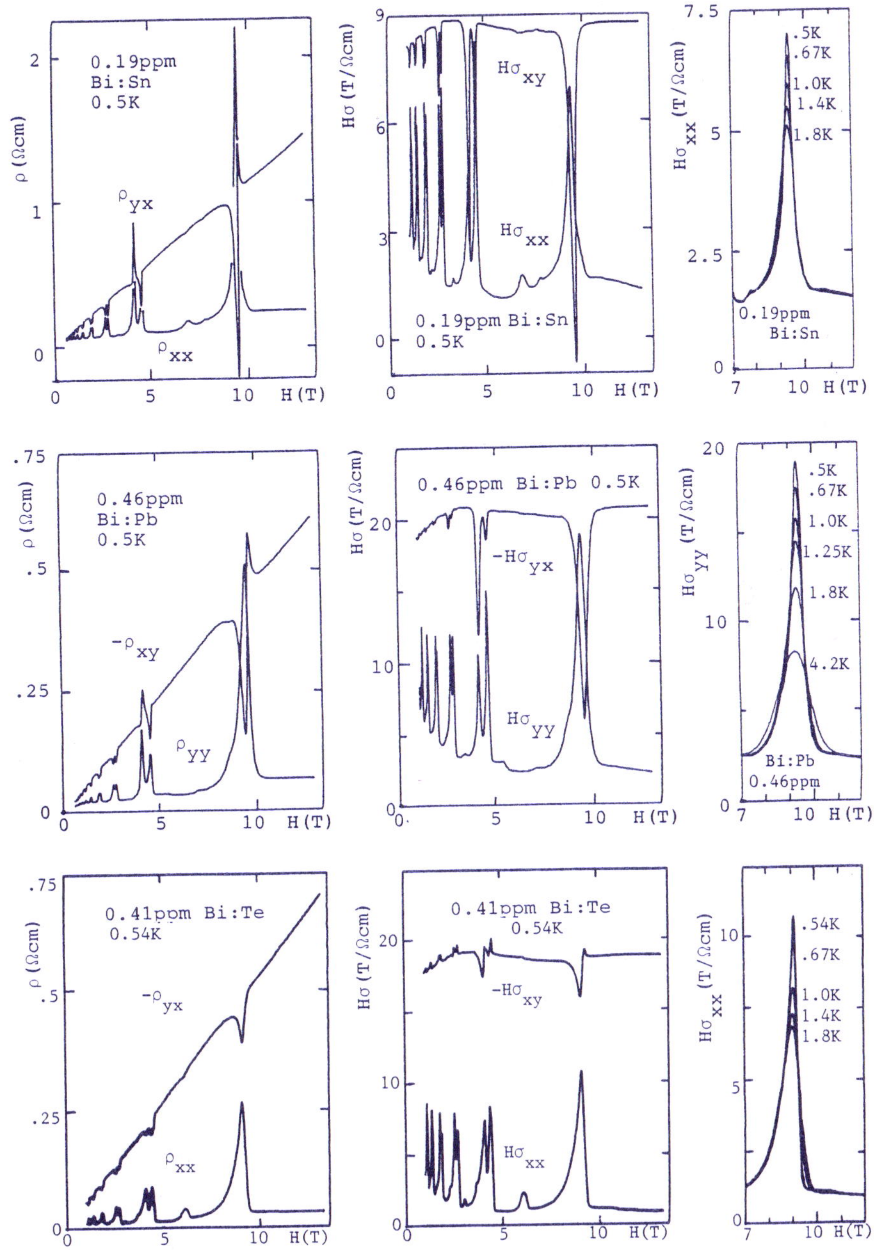

Fig.2: (a) Experimental results of magnetoresistivity tensor components. H//trigonal axis. (b) Magnetoconductivity tensor components multiplied by H, transformed from plot(a) of the figure. (c) Temperature dependence of $H\sigma_{ii}$ in large fields.

In this study single crystals of lightly doped Bi:Sn, Bi:Te or Bi:Pb, besides pure Bi, were prepared by the Czochralski method.[2] Samples were pulled along the bisectrix(y) or the binary(x) axis and carefully cut so that the segregation of impurities during crystal growth did not cause inhomogeneity.[3] Measurements of magnetoresistivity tensor components ρ_{xx} and ρ_{yx} were performed in pumped liquid ^{3}He using a 13.5T superconducting magnet. For the convenience of theoretical analysis, the conductivity tensor components σ_{xx} and σ_{xy} were transformed from the experimental data recorded digitally. The level of impurity density was determined from the non-oscillatory part of the Hall magnetoconductivity $\sigma_{xy} \cong (p-n)e/H$, the values of which ranged between 0.11 and 2.3ppm.

Data for undoped samples obtained at T=0.5K are shown in figure 1. On the left hand side ρ_{xx} and ρ_{xy} have been plotted . The results are essentially the same as those which have been obtained before. In the middle part σ_{xx} and σ_{xy} multiplied by H are shown, and on the right hand side the temperature dependence of the last SdH-maximum.

The data obtained for various doped specimens are shown in figure 2a-2c. The dopants and their concentrations are marked in the plots. It is obvious that the magneto-resistance has decreased considerably in the doped samples and that the σ_{xx}- peaks have reversed their direction. Surprinsingly the SdH peaks in the doped crystals are much sharper than in the undoped specimens. The lineshape of the oscillations and the temperature dependence of the last peak, which is substantial at temperatures as low as 0.5 K, suggest that in the pure samples acoustic phonon scattering dominates. In the doped crystals elastic scattering by ionized impurities should be of major importance, and the density of states at the Fermi surface should be reflected by the sharpness of the magnetoconductivity peaks. With increasing doping, the average conductivity becomes larger and the SdH oscillations are enhanced at first, though the temperature dependence of the peak height becomes smaller. Damping due to collision broadening, which is usually expected, seems to appear above impurity concentrations of 1ppm. The dependence of the maximum values of $H\sigma_{ii}$, obtained at 0.5K, has been plotted against the impurity concentration for various doping in fig.3.

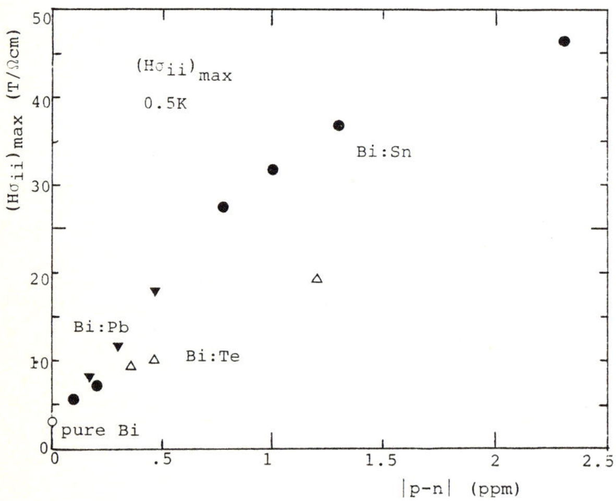

Fig.3: Maximum values of $H\sigma_{ii}$, plotted against the impurity density $|p-n|$, at 0.5 K.
(○ — pure Bi, ● — Bi:Sn, ▼ — Bi:Pb, △ — Bi:Te.)

The Hall resistivity of doped samples is based on the classical part proportional to H, which is cancelled in the pure specimens. In addition, quantum oscillations are seen. Especially ρ_{yx} of Bi:Sn and Bi:Pb samples with various impurity densities have an anomalous shape, resembling a capital letter N (or a dispersion curve), ρ_{xy} even crosses zero at 0.5 K in the most dilute samples. This feature appears to be due to the quantum Hall effect:[5,6]. The anomalous shape in ρ_{xy} reduces to a sharp but simple oscillation in σ_{xy}. In detail, the resonance in σ_{xy} appears at slightly stronger H values than in σ_{xx}. Remembering that the total Hall conductivity is determined by both holes and electrons, and that σ_{xy} in pure Bi also changes sign, it can be concluded that the change in sign is not essential. The oscillation is due to the interaction between resonating holes and anions mainly screened by holes themselves. On the other hand, the above mentioned anomaly is not seen in the raw data of Bi:Te samples with any impurity density.

Instead the resonance of holes reveals a complicated shape in σ_{xy}: Quantum oscillations of the Hall conductivity due to electrons with positive amplitude seem to overlap these due to holes with negative amplitude and different line shape. Both are due to the scattering by cations mainly screened by electrons. Though the problem of the quantum Hall effect in Bi:Te is delicate, considering the experimental method used, one can conclude from the present results that the magneto-conductivity in lightly doped bismuth strongly depends on the sign of the potential of the screened ionized impurities. This speculation is also consistent with the experimental result that the SdH oscillations caused by electrons are especially enhanced in Bi:Te samples and depend on the relative direction of current and main direction of each pocket. The difference in size of Bi cores and dopant ions seems not essential in this problem since there is little difference between the results in Bi:Sn and Bi:Pb samples.

In summary, light doping of bismuth up to 1ppm causes enhancement instead of damping of the SdH oscillations.

We are grateful to Dr.S.Ikehata for valuable discussions and the staff of the Cryogenic Centre, University of Tokyo for their experimental assistance.

REFERENCES

[1] R.Kubo,S.J.Miyake and N.Hashitsume: Solid State Phys. 17(1965)265.
[2] H.Bando,K.Oshima,M.Manabe and M.Suzuki: Solid State Commun. 43(1982)35.
[3] J.M.Noothoven van Goor: Philips Res.Repts.Suppl. 1971, No.4.
[4] M.Suzuki,H.Bando and K.Oshima: Jpn.J.Appl.Phys. 20(1981)1323.
[5] H.Shiba,K.Kanada,H.Hasegawa and H.Fukuyama: J.Phys.Soc.Jpn. 30(1971)972.
[6] T.Ando,Y.Matsumoto and Y.Uemura: J.Phys.Soc.Jpn. 39(1975)279.

A MAGNETIC FIELD INDUCED TWO-DIMENSIONAL STRIPPING OF THE POLARON

F.M. Peeters[+] and J.T. Devreese[*]
Universitaire Instelling Antwerpen, Departement Natuurkunde, Universiteitsplein 1, B-2610 WILRIJK, Belgium

A polaron in a magnetic field shows, as function of the magnetic field, two regimes: 1) for small magnetic field strength a dressed polaron moves under the influence of the magnetic field while 2) for relatively large magnetic field strength the polaron shakes off its polarization cloud in the direction perpendicular to the magnetic field. The influence of this stripping effect on the magneto-absorption spectrum is discussed.

1. Introduction

A conduction electron in a polar semiconductor or ionic crystal polarizes its surroundings. If the electron moves slowly through the crystal this polarization cloud will move with it. The quasi-particle consisting of the electron together with its surrounding polarization cloud is called *a polaron*.

Mathematically this problem is described by the Fröhlich Hamiltonian. Without any external fields and at zero temperature Feynman [1] has obtained an accurate approximation for the polaron ground state energy for all values of the electron-phonon coupling constant (α). Recently we generalized [2] the Feynman result to arbitrary magnetic field strength ($\mathcal{H}$) and arbitrary temperature (T).

We found that the thermodynamic quantities exhibit discontinuities for certain values of α, T and ω_c ($= e\mathcal{H}/mc$). The discontinuities are a consequence of a two-dimensional stripping of the polaron in the direction perpendicular to the magnetic field and correspond to an instability. In this paper we discuss the consequences of this stripping to the magneto-optical absorption spectrum.

2. The Feynman polaron theory generalized to arbitrary magnetic field strength

In [1] Feynman used the path-integral technique to calculate the polaron ground state energy (and polaron mass) at T=o and ω_c=o. The power of his approach is due to the fact that by using path-integrals he was able to eliminate the phonon variables exactly. Thus the problem is reduced to a one electron problem which was calculated approximately by using the Feynman inequality (this is the path-integral analogon of the Bogolubov inequality [3]). In doing so a trial action has to be chosen. Feynman took a trial action which can be derived from a two-particle model: an electron quadratically coupled to a second particle. The latter one is called "fictitious particle" because its coordinates have to be eliminated to obtain the trial action.

In the case that a magnetic field is present this two particle model (also called Feynman polaron model) is generalized (see also [4]) in the sense that we have to take into account the specific symmetry of the physical problem. Using the axial symmetry around the direction of the magnetic field we have introduced the following generalization of the Feynman polaron model

$$H_F = H_\parallel + H_\perp \tag{1}$$

with

$$H_\parallel = \frac{p_z^2}{2m} + \frac{p_z'^2}{2m'} + \frac{1}{2}\kappa_\parallel (z-z')^2 \tag{2}$$

and

$$H_\perp = \frac{1}{2m}(p_x - \frac{m\omega_c}{2}y)^2 + \frac{1}{2m}(p_y + \frac{m\omega_c}{2}x)^2 + \\ + \sum_{i=x,y}\{\frac{p_i'^2}{2m_i'} + \frac{1}{2}\kappa_\perp (r_i - r_i')^2\} \tag{3}$$

where $(\vec{r},\vec{p})$ are the electron variables, m the electron mass, $(\vec{r}',\vec{p}')$ the variables of the second particle which has masses $m_\parallel'$ parallel to the magnetic field and $m_\perp'$ perpendicular to the magnetic field. The quadratic coupling parallel (perpendicular) to the magnetic field between the electron and the fictitious particle is related to $\kappa_\parallel$ ($\kappa_\perp$). For convenience we introduced four new parameters: $w_i^2 = \kappa_i/m_i'$ and $v_i^2 = \kappa_i(1+m_i')/m_i'$; $i=\perp,\parallel$ (we have taken m=1 and ω_o=1, ω_o is the longitudinal optical phonon frequency).

With the model Hamiltonian H_F (see Eqs. 1-3) a trial action can be constructed. The resulting approximation to the free energy depends on three physical parameters (α,ω_c,T) and four variational parameters ($v_\perp$, $w_\perp$, $v_\parallel$, $w_\parallel$). The latter are determined such that the free energy attains its minimal value. From this free energy the magnetization, the susceptibility, the internal energy, the entropy and the specific energy were calculated [2]. It is this variational calculation which gives rise to the above mentioned polaron instability which is closely related to the two-dimensional stripping of the polaron (see [2]).

3. The magneto-optical absorption spectrum: an estimation of the oscillator strength and position of the peaks

In [5] the optical absorption spectrum of polarons is calculated in the absence of a magnetic field. If a magnetic field is applied a similar calculation can be performed. At present we limit ourselves to the calculation of the zeroth-order term which is given by the magneto-optical absorption resulting from H_F. The quantity which one has to calculate (see e.g. [6]) is

$$\sigma_{\pm}(\omega) = -\frac{e^2}{\omega}\int_0^{\infty} dt\, e^{i\omega t} < [A_{\pm}(t)^+, A_{\pm}(o)] > \tag{4}$$

the Fourier transform of the (complex) velocity auto-correlation function ($A_{\pm}(t) = \dot{x}(o) \pm i\,\dot{y}(t)$). For the Hamiltonian H_F one finds for the absorptive part

$$\mathrm{Re}\ \sigma_+(\omega)/P_o = T_1\,\delta(\omega - s_1) + T_2\,\delta(\omega + s_2) + T_3\,\delta(\omega - s_3) \tag{5}$$

with $P_o = 2\pi e^2$ and $\mathrm{Re}\ \sigma_-(\omega) = \mathrm{Re}\ \sigma_+(-\omega)$. The frequencies $s_1 < s_2 < s_3$

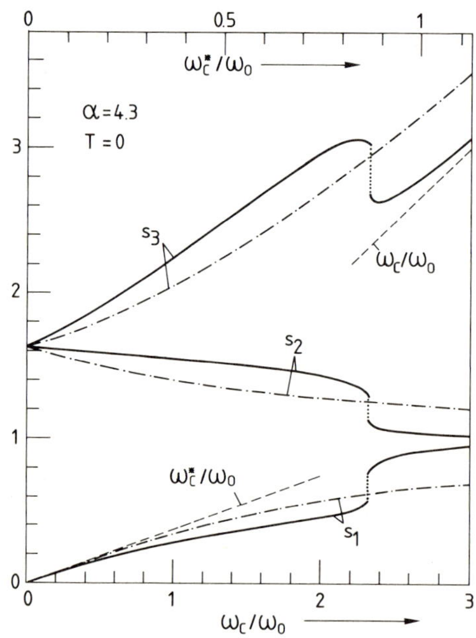

Fig. 1. Zeroth-order approximation to the peak position of the absorption peaks of polarons in the magneto-optical absorption spectrum (full lines). For comparison the peak positions (–.–.–) are also shown in the case that the polaron mass is magnetic field independent. $\omega_c = e\mathcal{H}/mc$ is the cyclotron mass of a bare electron while $\omega_c^{\star} = e\mathcal{H}/m^{\star}c$ is the cyclotron resonance frequency with $m^{\star} = M_{\perp}$ taken at zero magnetic field.

are given by the positive roots of

$$s^2(s^2-v_\perp^2)^2 - \omega_c^2(s^2-w_\perp^2)^2 = 0 \qquad (6)$$

and the oscillator strengths are equal to

$$T_i = \frac{s_i^2-w_\perp^2}{3s_i^2+2(-)^i\omega_c s_i-v_\perp^2} \quad ; \; i=1,2,3. \qquad (7)$$

In Fig. 1 we plotted the frequencies s_i as a function of the magnetic field for $\alpha=4.3$ and T=o. s_1 and s_3 are the resonant frequencies in the Faraday configuration while s_2 is the resonant frequency in the Voigt configuration. For the calculation of s_i we have considered the physical Feynman polaron model which means that $w_\perp = 1$ and $v_\perp = \sqrt{M_\perp}$. The mass $M_\perp$ of the Feynman polaron‡ model was calculated in [2]. In Fig. 2 the oscillator-strength of the different peaks is shown. To show the effect of the magnetic field dependence of the polaron effective mass we also plotted s_i and T_i in the case that the polaron mass is magnetic field independent.

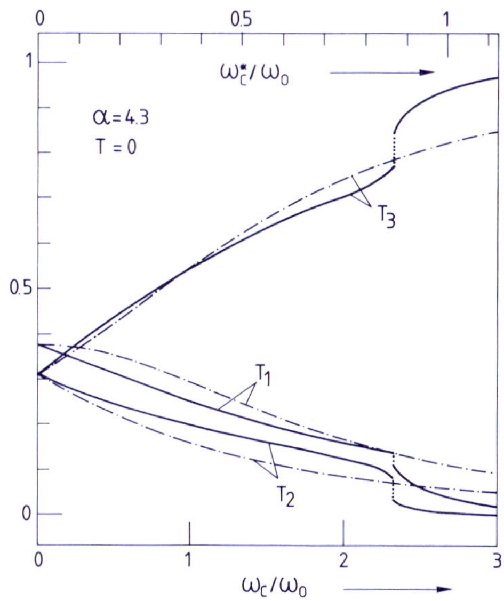

Fig.2..The oscillator strength of the different peaks calculated in zeroth order (full lines). The (–.–.–) line shows the oscillator strength in the case that the polaron mass is chosen as magnetic field independent.

‡ Strictly speaking $M_\perp$ is the variational parameter which, as $\mathcal{H}\to 0$, tends to the (Feynman) polaron mass.

4. Conclusion

The magnetic field dependence of the Feynman polaron mass $M_\perp$ alters considerably the behaviour of s_i and T_i. At the instability point the polaron shakes off its polarization cloud in the direction perpendicular to the magnetic field. This leads to a decrease in s_2 and s_3 and in an increase of s_1. The frequency s_3 approaches asymptotically the cyclotron resonance frequency of a bare electron. The sudden decrease of the oscillator strengths T_1 and T_2 at the instability point is understood in terms of a sudden decrease of the effective electron-phonon interaction.

As is well-known,the Feynman polaron model does not exhibit dissipation which implies that the dissipative part of Eq. (4), as given by Eq. (5), consists of a delta function. It is the first-order term, in a series expansion in the difference between the exact action and the trial action, which takes into account the dissipative effects.

Acknowledgement: This work was supported by F.K.F.O., Belgium, project nr. 2.0072.80.

References

+ **Research assistent of the National Fund for Scientific Research, Belgium**

* Also at R.U.C.A., Antwerpen (Belgium) and Technische Hogeschool Eindhoven, the Netherlands

1. R.P. Feynman, Phys. Rev. B7, 660 (1955).
2. F.M. Peeters and J.T. Devreese, Phys. Rev. B25, 7281 (1982), ibid. 7302 (1982).
3. S.V. Tyablikov, *Methods in the Quantum Theory of Magnetism*, (Plenum Press, New York, 1967), p. 180.
4. J.T. Devreese, E.P. Kartheuser, R. Evrard and A. Baldereschi, Phys. Stat. Sol. (b) 59, 629 (1973).
5. R. Feynman, R. Hellwarth, C. Iddings and P. Platzman, Phys. Rev. 127, 1004 (1962). See J.T. Devreese, J. De Sitter and M. Goovaerts, Phys. Rev. B5, 2367 (1972) for a calculation of the optical absorption.
6. J.T. Devreese, in *Theoretical Aspects and New Developments in Magneto-Optics*, ed. J.T. Devreese (Plenum, New York, 1980), p. 217.

VARIABLE RANGE HOPPING IN INDIUM PHOSPHIDE NEAR THE METAL NON METAL TRANSITION IN PRESENCE OF A MAGNETIC FIELD

BISKUPSKI G.*, DUBOIS H.*, LABORDE O.**

* L.S.H., Université de Lille I, 59655 VILLENEUVE D'ASCQ (F)
** C.R.T.B.T., S.N.C.I. (C.N.R.S.), BP 166 X, 38042 GRENOBLE CEDEX (F)

Introduction

In lightly doped n type InP, an impurity band of donors levels separated from the conduction band by an energy of a few meV is observed. These donor states are quite well described by hydrogenic wave functions having an envelope characterized by the exponential $e^{-\alpha R}$.

The properties of the material depend on the overlap between nearby states and a convenient parameter for the description of physical properties is the ratio d/a_H between the average distance between the impurity centres, d, and the effective Bohr radius a_H (84 Å).

Some experiments involving the InP samples in which d/a_H was greater than 3 [1] have been already carried out. On the insulating side of the metal non metal [M N M] transition, the three processes ε_1 ε_2 ε_3 described in the literature [2] were observed in these samples. By increasing the ratio d/a_H with the magnetic field we could affect the activation energies ε_1 ε_2 ε_3 and also provide evidence for existence of the minimum metallic conductivity.

In the present experiment we have studied the conductivity of a sample which is metallic at zero magnetic field. Some characteristics of this sample are listed in Table 1.

TABLE 1

Sample	$N_D - N_A$ (cm^{-3})	N_D (cm^{-3})	N_A (cm^{-3})	K	d/a_H ($N_D - N_A$)	d/a_H (N_D)
InP 1002	$4.1\ 10^{16}$	$8.84\ 10^{16}$	$4.74\ 10^{16}$	0.54	2.03	1.7

Since $d/a_H = 1.7$, the sample is on the metallic side of the MNM transition, and the Fermi level, E_F, is in the region of non localized states of the impurity band. When the magnetic field is increased, the sign of $E_c - E_F$ (where E_c is a mobility edge) is reversed and the conduction by excitation to a mobility edge appears for moderate temperatures while variable range hopping is expected in the lowest temperature region.

Experimental details

The resistivity has been measured at temperatures between 50 K and 1.6 K and at magnetic fields up to 7.6 Tesla as in our previous works. The value of $N_D - N_A$ was obtained from the hall effect at room temperature and the compensation K was estimated from the hall mobility versus temperature curve with the BROOKS HERRING analysis, in a temperature range where the scattering by ionised impurities is predominant.

Results and discussion

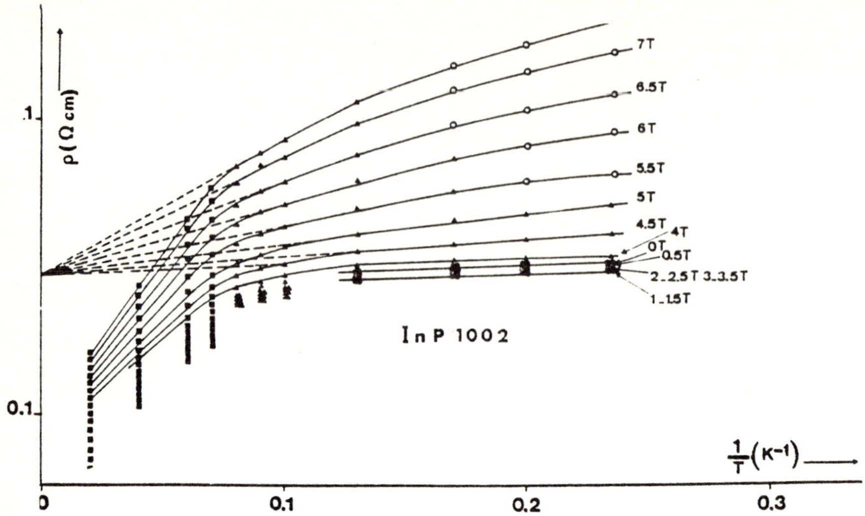

Fig. 1 : Variation of the resistivity versus inverse temperature for different values of the magnetic field.

Figure 1 shows the variations of the resistivity versus inverse temperature for different magnetic fields. Three different conduction regions may be distinguished. Region I corresponds to the ε_1 process which appears for temperatures greater than 12.5 K

In Region II, between 12.5 K and 8 K the resistivity is a linear function of 1/T and lines for different B have a common intercept at 1/T = 0. Thus in Region II, the conduction process may be identified as the ε_2 process, described by the equation,

$$\sigma_2 = \sigma_{02} \exp - \left(\frac{\varepsilon_2}{kT} \right)$$

Region III, for temperatures lower than 8 K cannot be identified with the ε_3 process the activation energy depends on the magnetic field and its value is too high to account for the ε_3 process.

A remarquable feature of the results is the existence of a common intercept ρ_o = 0.3 Ωcm. This value is compared in the Table 2 with the MOTT minimum metallic conductivity value.

TABLE 2

Sample	σ_{min} experimental $(\Omega cm)^{-1}$	σ_{min} calculated with N_D	σ_{min} calculated with $N_D - N_A$
InP 1002	3.33	4.35	3.39

The agreement is satisfactory as it has been already observed in previous works [1, 4] involving several samples of InP and InSb. Similar conclusions apply for samples with concentrations varying over several decades. This provides a crucial argument for the validity of the concept of the minimum metallic conductivity.

For magnetic fields up to 4.5 Tesla , the energy ε_2 is zero. For magnetic fields greater than 4.5 Tesla , ε_2 has a positive value increasing with the magnetic field. The behaviour of ρ and ε_2 is characteristic for the Anderson transition ; the factor which induces the transition is the magnetic field which modifies the value of d/a_H while all other parameters remain constant.

In Region III we have not been able to interpret the results by the ε_3 process. Therefore we examined the applicability of the variable range hopping mechasnism near the MNM transition.

There exist different expressions giving the $T^{-1/4}$ variation of the resistivity [5]. they depend on different initial approximations and methods of derivation [5] .

It should be pointed out that the choice between these expressions is difficult. In each case the plot of experimental results is reasonably accounted for by theory. So, for the first three forms we have calculated the physical parameters as : the localisation length $a = \alpha^{-1}$, the density of states at the fermi level $N(E_F)$, and the carrier density obtained from $N(E_F)$ and the width of the band which is assumed equivalent to ε_2.

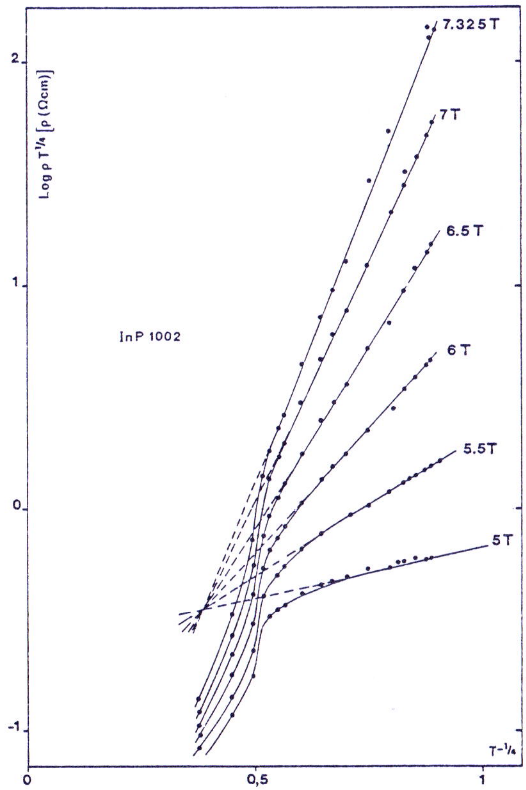

Fig. 2 : Variation of the resistivity in the variable range hopping conduction regime.

The best results were obtained when the following expression [5] was used :

$$\sigma = \frac{e^2\, E_1^2}{12\,\Pi\rho s^5\,\hbar^4} \left[\frac{8\Pi N k T \alpha}{9}\right] \left[\frac{e^2\,\alpha}{2\varepsilon\varepsilon_o}\right]^2 \exp - 2.063 \left[\frac{\alpha^3}{N\,k\,T}\right]^{1/4}$$

The corresponding experimental results appear in Fig. 2, the values of a, $N(E_F)$ and n are showed in Table 3.

TABLE 3

B(T)	$\varepsilon_2(10^{-5}$ eV)	a(Å)	$N(E_F)$ (cm^{-3} eV^{-1})	n(10^{16} cm^{-3})
5	21	1 573	6.49	13.63
5.5	34	1 058	0.56	1.90
6	48	800	0.166	0.79
6.5	62	592	0.078	0.48
7	77	460	0.051	0.39
7.5	-	385	0.043	-

Note that the localisation length decreases as the magnetic field is increased, while it is always several times greater than the mean distance between sites. This is compatible with the variable range hopping process and the values of $N(E_F)$ are acceptable.

The experimental results are satisfactorily accounted for by the expression derived f the variable range hopping. The decrease of $N(E_F)$ with the magnetic field is compatible with a model of two overlapping Hubbard bands. The entire picture is consistent with a model proposed by MOTT, [6] , in which the mobility edge lies in a D^- band which is broader than the D° band.

References

[1] G. BISKUPSKI, H. DUBOIS, O. LABORDE, X. ZOTOS, Phil. Mag. B. 42, 19 (1980).

[2] N.F. MOTT, E.A. DAVIS, *Electronic Processes in non crystalline materials*, 2nd edition (Oxford University press) (1979).

[3] A. MILLER, E. ABRAHAMS, Phys. Rev., 120, 745 (1960).

[4] G. BISKUPSKI, Thesis Lille (1982).

[5] F.R. ALLEN, G.L. ADKINS, Phil. Mag. 26, 1027 (1972).

[6] N.F. MOTT, J.H. DAVIES, Phil. Mag. 42, 845 (1980).

A Kinetic Approximation for Quantum Magneto-Transport

J. Hajdu and Ursula Paulus

Institut für Theoretische Physik
Universität zu Köln

D-5000 Köln 41, W. Germany

1. The Problem

The method for calculating the dynamic conductivity developed by Götze and Wölfle [1] (which can be considered as a simplified version of the Mori formalism [2]) has proved to be useful in the quantum limit magnetoresistance and cyclotron resonance theory [3] . A weakness of this method is that, when defining the relaxation kernel $M(\omega)$ by

$$\sigma(\omega) = \frac{ie^2 n/m}{\omega - \omega_c + M(\omega)} \tag{1}$$

(where $\sigma(\omega)$ is the complex transverse conductivity) and using the simplest "lowest order correlation approximation" (LOCA) for $M(\omega)$,

$$\mathrm{Im}\, M \sim U^2 A^2 (U) \tag{2}$$

where A is the single electron spectral function and U is - in the case considered here - the electron-impurity potential, for the Hall coefficient the method always yields the classical constant value $1/enc$. This contradicts both the observation and the kinetic theory which is exact in the weak coupling limit (van Hove [4]). Also the results for the transverse and longitudinal magnetoresistances deviate, in the Shubnikov - de Haas regime, from experiment and the expressions derived by kinetic theory. Formally, the trivial result for the Hall coefficient is due to the fact that, in LOCA, $\mathrm{Re}\, M(\omega)$ vanishes in the zero frequency limit.

2. Kinetic Approximation

In order to avoid the contradictions mentioned above we suggest, instead of (1), a more general matrix formulation. For zero temperature

$$\sigma(\omega) = \sum_{\nu} \sigma_\nu(\omega) = \sum_{\nu\mu} \sigma_{\nu\mu}(\omega) \tag{3}$$

$$\sum_{\lambda} \left\{ (\omega - \omega_c)\delta_{\nu\lambda} + M_{\nu\lambda}(\omega) \right\} \sigma_{\lambda\mu} = i\alpha_{\nu\mu} \tag{4}$$

Here $\mu, \nu, .. = 0,1,2,..$ refer to Landau levels. The transitions to finite temperatures can be executed by using the "Greenwood formula"

$$\sigma(\omega, T) = \int \left(-\frac{\partial f}{\partial \varepsilon} \right) \sigma(\omega, \varepsilon_F = \varepsilon)\, d\varepsilon \tag{5}$$

where f is the Fermi distribution function .

We can easily find interpolating expressions for $\sigma_\nu(\omega)$ by solving eq. (4) in the "kinetic approximation"

$$\sigma_\mu/\alpha_\mu \simeq \sigma_\nu/\alpha_\nu , \tag{6}$$

$\alpha_\nu = \sum_\mu \alpha_{\nu\mu}$, which yields

$$\sigma_\nu(\omega) = \frac{i\alpha_\nu}{\omega - \omega_c + M_\nu(\omega)} \tag{7}$$

where

$$M_\nu = \frac{1}{\alpha_\nu} \sum_\mu M_{\nu\mu} \alpha_\mu \tag{8}$$

$M_{\mu\nu}$ and $\alpha_{\mu\nu}$ can be calculated by comparing (4) with the exact formula

$$\sigma_{\mu\nu}(\omega) = \frac{1}{i\omega} \left\{ \chi_{\mu\nu}(\omega) - \chi_{\mu\nu}(0) \right\} \tag{9}$$

$$\chi_{\mu\nu}(\omega) = -\frac{1}{Vol} \frac{1}{\hbar} \langle\langle j^+_\mu ; j_\nu \rangle\rangle_\omega \tag{10}$$

$$\langle\langle A;B \rangle\rangle_z = \int_{-\infty}^{\infty} dt\, (\mp i)\, \theta(\pm t) \langle [A(t), B] \rangle e^{izt} , \quad Im\, z \gtrless 0 \tag{11}$$

For an electron - impurity system with Hamiltonian

$$H = \sum_\alpha \varepsilon_\alpha a^+_\alpha a_\alpha + \sum_{\alpha\beta} V_{\alpha\beta} a^+_\alpha a_\beta \tag{12}$$

$\alpha = (\nu, k_y, k_z)$

$$j_\mu = -ie \sqrt{\frac{\hbar\omega_c}{m}} \sum_{k_y k_z} \sqrt{\nu+1}\, a^+_{\alpha+1} a_\alpha \tag{13}$$

in LOCA

$$\alpha_\mu = \frac{e^2 \hbar}{m} (\mu+1) \sum_{k_y k_z} \left(f_\alpha - f_{\alpha+1} \right) \tag{14}$$

and $\operatorname{Im} M_{\mu\nu}(\omega)$ is linear in U^2 and in

$$\int d\omega' A_\alpha(\omega) A_\beta(\omega'-\omega) \frac{f(\omega'-\omega)-f(\omega)}{\omega} \tag{15}$$

α_μ obeys the sum role $\sum_\mu \alpha_\mu = e^2 n/m$ and $\operatorname{Re} M_{\mu\nu}(\omega)$ can be obtained from $\operatorname{Im} M_{\mu\nu}(\omega)$ by using the Kramers-Kronig relations.

3. Results

These are formulae for the conductivity components in LOCA. For $\omega = 0$ but at arbitrary temperature

$$\sigma_{xx} = \frac{e^2 n}{m} \left\langle \frac{\operatorname{Im} M}{\omega_c^2 + (\operatorname{Im} M)^2} \right\rangle \tag{16}$$

$$\sigma_{xy} = \frac{e^2 n}{m} \left\langle \frac{-\omega_c}{\omega_c^2 + (\operatorname{Im} M)^2} \right\rangle \tag{17}$$

$$\sigma_{zz} = \frac{e^2 n}{m} \left\langle \frac{1}{\operatorname{Im} M} \right\rangle \tag{18}$$

$$\langle A \rangle = \frac{m}{e^2 n} \int \left(-\frac{\partial f}{\partial \varepsilon} \right) \sum_\mu \alpha_\mu A_\mu(\varepsilon)\, d\varepsilon \tag{19}$$

In general eqs. (16) and (17) yield corrections to the classical Hall constant.

Inspections of eqs. (16) - (19) show that our "kinetic approximation" is a generalization of the usual simple kinetic theory. Since it avoids, however, the solution of a kinetic equation it is easy to take into account (i) impurity damping effects (in a self-consistent way), (ii) realistic electron-impurity potentials (including the spin orbit interaction) and (iii) band structure effects (like non-parabolicity).

Concrete results for InSb are in preparation .

We thank Professor Wolfgang Götze for numerous valuable discussions .

References

1. W. Götze and P. Wölfle, J. Low Temp. Phys. 5, 575 (1971 ; Phys. Rev. B 6, 1226 (1972)

2. H. Mori, Progr. Theor. Phys. 34, 399 (1966)

3. W. Götze and J. Hajdu , J. Phys. C 11, 3993 (1978) ; Solid State Comm. 29, 89 (1979)

4. L. van Hove, Physica 21, 517 (1955)

SCATTERING MECHANISMS IN n-CdTe UNDER CLASSICAL AND QUANTUM LIMIT CONDITIONS

A.P.Radliński, K.Pastor and M.Baj
Warsaw University, Institute of Experimental Physics,
00-681 Warszawa, ul.Hoża 69, Poland

The relaxation time of electrons in high-purity n-CdTe was studied by means of weak-field Hall and conductivity measurements and far-infrared cyclotron resonance (CR) transmission measurements. The transport properties were studied with a low frequency ac experimental arrangement (ν = 43 Hz, B = 1.5 mT) in the temperature range 15 K - 300 K. Additionally, dependence of the Hall coefficient on the strength of the magnetic field was measured at several temperatures with a classical static-field Hall arrangement (B up to 1.5 T). Results of the ac measurement are presented in Fig.1. The Hall (concentration) data were analysed according to single level statistics and nondegenerate carrier densities [1]:

$$\frac{n(N_A + n)}{N_D - N_A - n} = \frac{1}{2} N_c \exp(-E_d/kT) \qquad (1)$$

where $N_c = 2(m^* kT/2\pi\hbar^2)^{3/2}$.

To calculate the concentration from our Hall results we used values of r = 1.0 above 80 K and r = 1.4 in the range 15 K - 20 K, r being the Hall scattering factor. These values of r were determined in our static-field experiment. We found the following values characteristic of our sample: donor concentration $N_D = 5.7 \times 10^{13}$ cm^{-3}, acceptor concentration $N_A = 1.9 \times 10^{13}$ cm^{-3}, donor ionization energy $E_d = 14.4 \pm 0.5$ meV. The numerical value of E_d found here remains in very good agreement with the effective rydberg extracted from the magneto-optical data [2].

In our analysis of the mobility we took into account scattering by optical phonons [3], deformation potential [4], piezoelectric effect [5] and screened ionized impurity potential [6]. The total scattering lifetime τ_t has been calculated according to Matthiessen formula $\tau_t^{-1} = \sum_i \tau_i^{-1}$.

The result of our analysis is shown in Fig.2. It is interesting to note the importance of the deformation potential scattering in the intermediate temperature region, not seen that clearly for less pure samples [1, 7]. Thus, the parameter E_1 determined here can be considered as reliable : $E_1 = 15 \pm 2$ eV (we assumed $u_l = 3.11 \times 10^5$ cm/sec, u_l being velocity of sound in CdTe [8]). The value of $\alpha = 0.40$ found in our analysis of scattering by LO phonons remains in very good agreement with the previous magnetooptical result [4]. Finally, we can conclude that at temperatures below 20 K the dominant scattering mechanism is due to ionized impurities.

The CR spectra were investigated in the temperature range 2 - 40 K with the 337 μm and 311 μm radiation of far infrared HCN laser (CRA polarization was used in the Faraday configuration). Both sets of data coincide very closely with each other. The position of the resonance peaks remains in very good agreement with the previous results

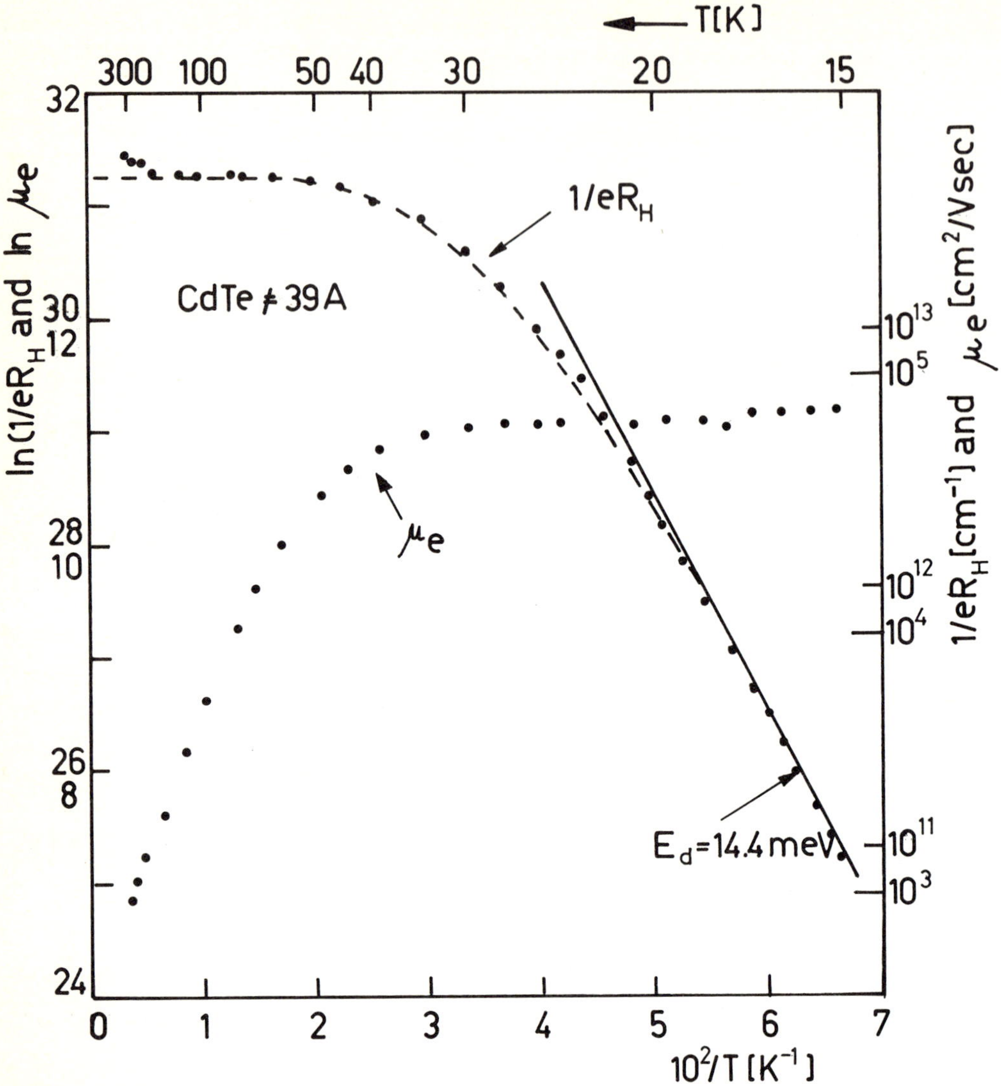

Figure 1. Mobility and Hall coefficient for the sample 39 A. Broken line represents a theoretical fit for $r = 1$, $N_D = 6.5 \times 10^{13}$ cm^{-3}, $N_A = 2.7 \times 10^{13}$ cm^{-3}, $E_d = 14.4$ meV. For correct values of r, N_D, and N_A see text.

(see [9] and literature cited therein). The experimental values of the CR line halfwidth ($\lambda = 337$ µm) versus temperature, transposed into relaxation times, are shown in Fig.2. At temperatures below 15 K screening of the ionized impurity potential by free electrons is negligible. Therefore, the screening length λ is approximately equal to 3×10^{-5} cm, an average distance between ionized impurities. On the other hand, the cyclotron radius ($l = (\hbar/eB)^{1/2}$ for $\hbar\omega_c \gg kT$, $\hbar\omega_c \cong 4$ meV) is then 1.3×10^{-6} cm, so $l \ll \lambda$. Under these conditions two main CR line broadening mechanisms are active: an adiabatic one,

MAGNETOPHONON AND MAGNETOIMPURITY RESONANCES IN N-GERMANIUM

C. Hamaguchi, Y. Hirose and K. Shimomae
Department of Electronics
Osaka University
Suita, Osaka 565, Japan

Oscillatory structures of the magnetoresistance in n-Ge are investigated in three different temperature regions, above 77K, from 10 to 20K and below 10K. Intervalley magnetophonon series due to the degenerate (LA,LO) phonons at the X point is observed at T > 77K in addition to the intravalley LO phonon series, and the deformation potential ratio of the intervalley scattering to the intravalley scattering is determined. Magnetophonon resonance due to the intervalley TA phonons at the X point is observed for the first time at the intermediate temperature range. Magnetoimpurity resonance is observed at T < 10K and a tentative assignment of the observed five series is given.

1. Introduction

A study of oscillatory magnetoresistance in semiconductors provides a very important information such as effective mass, phonon energy, impurity state and energy loss mechnisms of hot electrons. The oscillatory structure of magnetoresistance has been shown to arise from resonant scattering of electrons in Landau subbands and the observed resonances are classified in ordinary magnetophonon series, 2TA phonon series, intervalley magnetophonon series, impurity assisted magnetophonon series (impurity series) and magnetoimpurity resonance series [1,3]. Except the simplest case where only one type of resonance series is involved, the oscillatory structure is very complicated and therefore its analysis is made with the help of Fourier transform [4]. One of the main purposes of the present work is to investigate magnetophonon resonance due to the intervalley phonon scattering. Although group theoretical analysis shows that two types of intervalley phonon scatterings are allowed in Ge [5], only magnetophonon resonance due to the degenerate (LA,LO) phonons at the X point is observed [4,6,7] and magnetophonon resonance due to the intervalley TA phonons at the X point has not been reported. Since our experimental method is based on the magnetic field modulation technique, the observed second derivative signals of the magnetoresistance with a high S/N enable us to resolve the complicated structure when we use Fourier transform analysis.

2. Experimental Procedure

We employed the magnetic field modulation technique to eliminate the structureless background of the magnetoresistance and to observe the oscillatory components with a high S/N. A static magnetic field up to 8T was produced by a superconducting solenoid and the sample temperature was controlled by a small heater from 4.2 to 300K. A small modulating coil was installed in the sample room. The experimental data were processed digitally by using HP-9825A or HP-9826A personal computer and WX4675 plotter. Fourier transform analysis was performed using the method described elsewhere [8,9]. Samples used in the experiments at temperatures T > 10K are high purity n-Ge (∿40Ωcm at 300K) and those used for magnetoimpurity resonance at T < 10K are Sb doped n-Ge (2∿3Ωcm at 300K). Although the experiments were carried out in the magnetic fields of <100>, <110> and <111> directions, the results of the <100> are mainly presented.

3. Results and Discussion

3.1 High Temperature Range (T > 77K)

Experiments at T > 77K were performed to investigate magnetophonon resonance due to

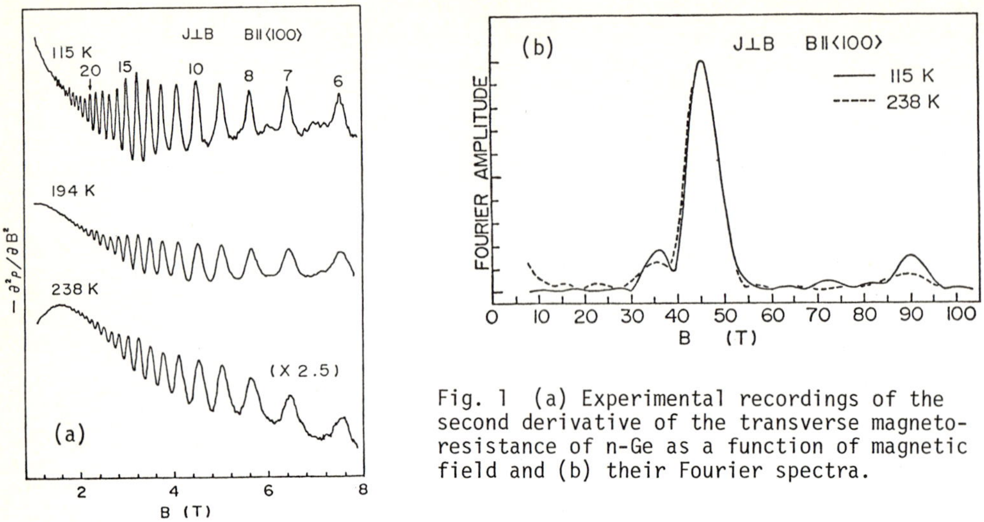

Fig. 1 (a) Experimental recordings of the second derivative of the transverse magnetoresistance of n-Ge as a function of magnetic field and (b) their Fourier spectra.

the intervalley phonon scattering and to determine the deformation potential ratio of the intervalley phonon scattering to the intravalley LO phonon scattering. Typical examples of the experimental recordings are presented in Fig. 1(a), where the second derivative of the transverse magnetoresistance is shown as a function of magnetic field applied in the <100> direction. We can see a beating of oscillations in Fig. 1 (a) which is caused by an interference of the high and low frequency components. The components are resolved by Fourier spectra as shown in Fig. 1(b), where we find a fundamental field at 45T and its second harmonic at 90T which arise from the magnetophonon resonance due to the intravalley LO phonon scattering. In addition to these two peaks we find a weak peak at 36T which is assigned to the magnetophonon resonance due to the intervalley phonon scattering and ascribed to the low frequency component observed as the beating in Fig. 1(a). Using the LO phonon energy reported by Nilsson and Nelin [10] we obtain the effective masses $m_t = 0.0823m$ and $m_l = 1.59m$ from the present work [7]. These values provide the intervalley phonon energy 349K which is in good agreement with the degenerate (LA,LO)(X) phonons of 346K (at the lattice temperature 80K) [10].

As shown in Fig. 1(b) the intervalley phonon peak at 36T is very weak but we can see it clearly, and therefore we tried to determine the deformation potential ratio D_X/D of the intervalley to intravalley phonon scatterings. We used the relation [7]

$$\frac{a_X}{a} = \frac{|D_X|^2}{|D|^2}\frac{\omega_{LO}}{\omega_X}\exp[\hbar(\omega_{LO} - \omega_X)/kT] \qquad (1)$$

where a_X and a correspond to the amplitude coefficient of the first term of the expression derived by Barker [11], and $\hbar\omega_{LO}$ and $\hbar\omega_X$ are the LO phonon and intervalley phonon energies, respectively. We determined the ratio by fitting both the second derivative of the experimental magnetoresistance and its Fourier spectra with those calculated from Barker's expression. From this analysis we obtained $D_X/D = 0.28 \pm 0.07$. Using the intervalley scattering rate determined by Weinreich et al.[12] from the acoustoelectric effect which gives $D_X = 1.6\times10^8$eV/cm, we obtain $D = 5.71 \pm 1.8\times10^8$ eV/cm for the deformation potential of the intravalley LO phonon scattering.

3.2 Intermediate Temperature Range (10K < T < 20K)

At lower temperatures (T < 40K) ordinary ohmic magnetophonon resonance disappears but an application of a high electric field produces magnetophonon resonance due to oscillatory behavior of energy relaxation time of hot electrons [13]. Experiments were

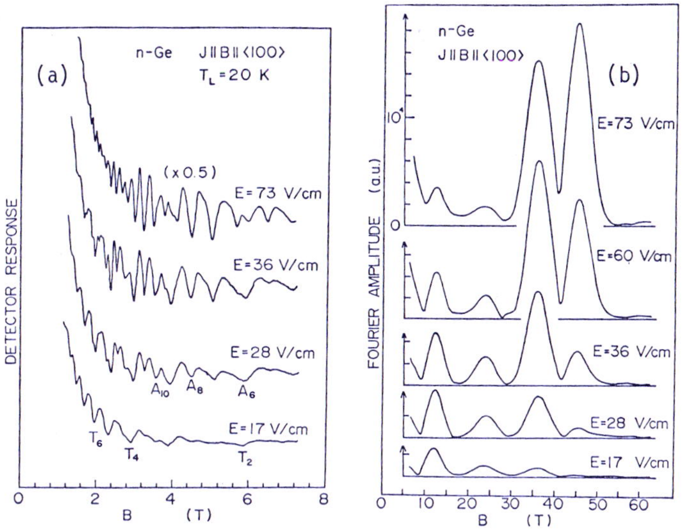

Fig. 2 (a) Experimental recordings of the second derivative of the longitudinal magnetoconductivity at 20K as a function of magnetic field in the <100> direction for several applied electric fields and (b) their Fourier spectra.

carried out by measuring magnetoconductance at temperatures from 10 to 20K in longitudinal magnetic fields applied in the <100>, <110> and <111> directions. Experimental recordings of the second derivative of the magnetoconductivity at T = 20K and their Fourier spectra are shown in Fig. 2(a) and (b), where the applied electric field was changed from 17 to 73V/cm. The Fourier spectra consist of four peaks of which fundamental fields are 12, 23, 36 and 45T. The fundamental fields of 36 and 45T are the same as those observed at the higher temperature range (T > 77K) and assigned to the resonance series due to, respectively, the intervalley phonons of 346K at the X point and the intravalley LO phonons of 438K at the Γ point. The other two peaks have not been reported yet. The fundamental field of 12T is assigned to the magnetophonon resonance due to the intervalley TA phonons of 115.5K at the X point because the observed extrema T_N in Fig. 2(a) are in good agreement with the calculated peak positions. Group theoretical analysis shows that the intervalley phonon scatterings by the TA phonons and the degenerate (LA,LO) phonons at the X point are allowed, but the intervalley magnetophonon series due to the TA phonons has not been reported. The series of the fundamental field 23T is ascribed to the second harmonic of the intervalley TA phonon series or to the 2TA phonon series in which simultaneous emission of the TA phonons of 115.5K at the X point is involved.

The above assignment is confirmed by the analysis of the experimental results obtained in the magnetic fields of the <110> and <111> directions [14]. In the magnetic field of the <111> direction we observed impurity series of the intervalley (LA,LO) phonons at the X point in which an electron in a Landau subband makes a transition to 1s donor state associated with another valley [14].

3.3 Low Temperature Range (T < 10K)

In this temperature region magnetoimpurity resonance is observed, in which the resonant interaction between electrons in Landau subbands and in impurity states results in an oscillatory structure of magnetoconductivity in the presence of a magnetic field B. The resonance condition is given by

$$N\hbar\omega_c = E_2(B) - E_1(B) \tag{2}$$

where N = 1,2,3,...., $\omega_c = eB/m^*$ is the cyclotron angular frequency, and $E_1(B)$ and $E_2(B)$ are the energies of two impurity states. Such a resonance in n-Ge was first observed by Instone et al. [15] in the photomagnetic EMF at 4.2K, where the impurity states involved are 1s and $2p_-$ states. The magnetoimpurity resonance effects in other semiconductors are reviewed by Eaves and Portal [3]. It is very interesting to point out that magnetoimpurity resonance in n-InP involves a process in which the shallow

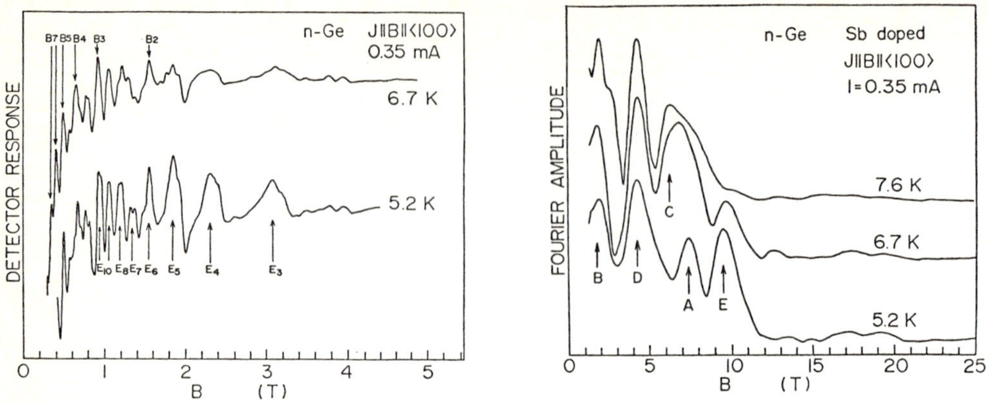

Fig. 3(a) Magnetoimpurity resonance in Sb doped n-Ge at 5.2 and 6.7K.

Fig. 3(b) Fourier spectra of magnetoimpurity resonance in Sb doped n-Ge at T = 5.2, 6.7 and 7.6K.

donor is excited from the 1s ground state to the bottom of the conduction band continuum. Therefore $E_2(B)$ in n-InP is the N=0 Landau level close to $k_B = 0$. In the present work we observed magnetoimpurity resonance due to transition of an electron between an excited state of the donor and the conduction band continuum, in addition to the magnetoimpurity resonance series observed in n-Ge by Instone et al. Figures 3(a) and 3(b) represent experimental recordings of the second derivative of the longitudinal magnetoresistance and their Fourier transform. In Fig. 3(b) we find five series of which fundamental fields are 2.0, 4.3, 6.4, 7.4 and 9.6T. The series E_N (N = 3,4,5,...) in Fig. 3(a) and the peak E at 9.6T in Fig. 3(b) are assigned to the magnetoimpurity resonance in which electrons in Landau subbands interact with shallow donors causing transition between the 1s and $2p_-$ states. From the value of the fundamental field 9.6T we obtain the resonance energy $E_2 - E_1$ = 8.2meV which is in good agreement with 8.5 ± 0.3meV reported by Instone et al. [15] and 8.46meV calculated by Faulkner [16]. From the fundamental fields 2.0, 4.3, 6.4 and 7.4T we obtain the resonance energies 1.7, 3.7, 5.4 and 6.3meV, respectively. Such small energies are not found in the phonon energies at the zone edges and therefore are assigned to the excitation energies of electrons at the shallow donor states. The most probable assignment at the present stage is B(C.B.- $2p_-$), D(C.B. - 2s), C(C.B. - $2p_0$) and A(2s - 1s), where the calculated energies from the values of Faulkner are respectively 1.73, 3.52, 4.74 and 6.67meV. The assignment was made from the fitting of energies only.

1. P.G. Harper, J.W. Hodby and R.A. Stradling: Rep. Prog. Phys. 36, 1 (1973)
2. R.L. Peterson: *Semiconductors and Semimetals*, ed. by R.K. Willardson abd A.C. Beer, vol. 10 (Academic Press, New York, 1975), p. 221
3. L. Eaves and J.C. Portal: J. Phys. C12, 2809 (1979)
4. L. Eaves, R.A. Hoult, R.A. Stradling, R.J. Tidey, J.C. Portal and S. Askenazy: J. Phys. C8, 1034 (1975)
5. H.W. Streitwolf: phys. stat. sol. 37, K47 (1970)
6. N.G. Glutzman and I.M. Tidil'kovskii: Sov. Phys.-Solid State 10, 2469 (1969)
7. Y. Hirose, K. Shimomae and C. Hamaguchi: J. Phys. Soc. Jpn. 51, 2226 (1982)
8. K. Kasai, T. Shirakawa and C. Hamaguchi: J. Phys. Soc. Jpn. 44, 216 (1978)
9. K. Senda, K. Shimomae, K. Kasai and C. Hamaguchi: J. Phys. Soc. Jpn. 47, 551 (1979)
10. G. Nilsson and G. Nelin: Phys. Rev. B3, 364 (1971)
11. J.R. Barker: J. Phys. C5, 1657 (1972)
12. G. Weinreich, T.M. Sanders, Jr. and H.G. White: Phys. Rev. 114 33 (1959)
13. R.A. Stradling and R.A. Wood: Solid State Commun. 6, 701 (1968)
14. Y. Hirose and C. Hamaguchi: unpublished
15. T. Instone, L. Eaves, J.C. Portal, C. Houlbert, P. Perrier and S. Askenazy: J. Phys. C10, L585 (1977)
16. R.A. Faulkner: Phys. Rev. 184, 713 (1969)

CLASSICAL TRANSPORT EFFECTS REVISITED : A BASIS FOR NEW APPLICATIONS OF HIGH MAGNETIC FIELDS

S. Cristoloveanu
Institut National Polytechnique, Lab. Physique des Composants à Semiconducteurs, ERA CNRS 659
ENSERG, 23, rue des Martyrs - 38031 GRENOBLE Cedex FRANCE

1 - INTRODUCTION

High magnetic fields are applied in semiconductor physics in order to observe and to study quantum effects. The aim of this paper is to outline that measurements performed in a wide range of magnetic fields can qualitatively help to improve the efficiency of classical effects and assist in the characterization of transport- and recombination properties of semiconductor materials. We consider several phenomena (magnetoresistance, magnetoconcentration-, magnetodiodes and photomagnetoelectric effects) for which extensive experimental data have been obtained at the Service National des Champs Intenses - Grenoble in the range B = 0 to 15 Teslas.

2 - TRANSVERSE MAGNETORESISTANCE (MR) IN MIXED SEMICONDUCTORS

In near-intrinsic semiconductors the existence of opposite Hall fields for electrons and holes yields an important transverse MR effect (Fig.1). The MR coefficient ρ_B/ρ_0 is given (for isotropic and parabolic bands) by [1] :

$$\frac{\rho_B}{\rho_0} = \frac{n\mu_{n_0}+p\mu_{p_0}}{n\mu_{n_B}+p\mu_{p_B}}\,(1+\mu_H^{*2}B^2)^{-1}, \quad \mu_H^* = \frac{p\mu^2_{p_B}-n\mu^2_{n_B}}{p\mu_{p_B}+n\mu_{n_B}} \qquad (1)$$

where μ_H^* is the ambipolar Hall mobility, n,p are electron and hole concentrations and $\mu_B^{1,2}$ are usual energy averages of the elementary mobilities.

The use of a wide range of magnetic fields allows to simplify Eq.(1) by considering monokinetic carriers [1] : $\mu_B^{1,2} = \mu_0^{1,2}(1+\mu_0^2B^2)^{-1}$. Fitting the experimental results of Fig.1 (and referring to ρ_0 and $\mu_H^*(B\to 0)$ values or fitting also $\mu_H^*(B)$ curves), we have accurately determined the electron and hole concentrations and mobilities as well as the intrinsic density [1]. This method, applicable to each sample and valid for all temperatures, appears to be much more versatile than the conventional approach (Dunlap, etc.) where the doping or the temperature were taken as experimental parameters instead of B.

Our method has been successfully tested on Ge and InSb[1] and in addition been used for the study of the influence of cadmium doping on the mobilities in PbSnTe[2]. It was also used for the characterization of semi-insulating GaAs[3]. The only limit to this technique is that the ambipolar effects must substantially exceed the effects of scattering, anisotropy, non-parabolicity or quantum transport. However, in *extrinsic* semiconductors the ambipolar MR of *photogenerated* carriers can be used for the determination of mobilities and compensation mechanisms.

Finally, this method has been recently extended to arbitrary shaped samples by generalizing the Van der Pauw theory to the case of high magnetic fields [4].

3 - LONGITUDINAL MAGNETORESISTANCE (LMR)

In anisotropic semiconductors like Silicon On Sapphire (SOS) thin films, the LMR exhibits a strong variation with B [5]. Scattering mechanisms can be ignored with confidence and computer modelling gives an opportunity to deduce the key parameters : drift mobility and anisotropy.

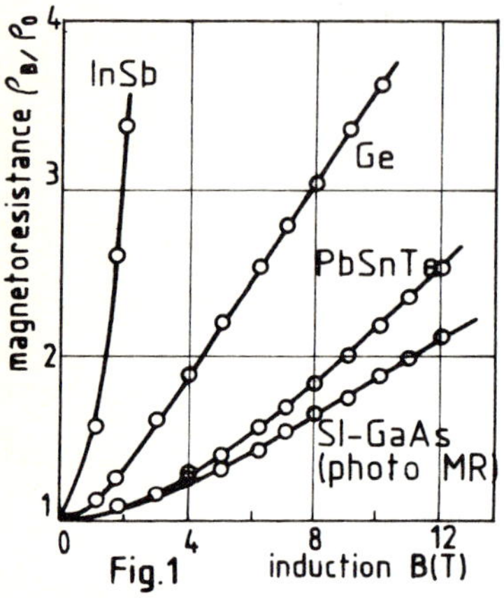

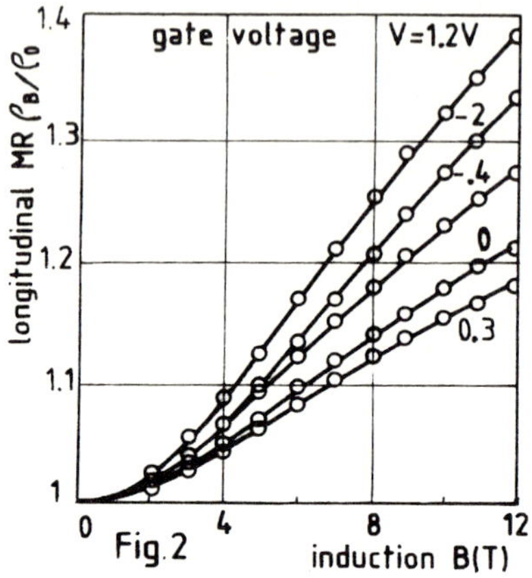

Fig.1 : Experimental transverse MR coefficient ρ_B/ρ_0 against magnetic induction B in several semiconductors at room temperature.

Fig.2 : Longitudinal MR coefficient $(\rho_B/\rho_0)^{110}_{110}$ versus magnetic induction in MOS-Hall devices with gate voltage as an experimental parameter (0.65 µm thin n-SOS film, T = 293 K).

Moreover, in gate-controlled SOS samples (MOS-Hall devices) measurements of LMR versus B have been performed with the gate voltage as parameter (Fig.2), i.e. by progressively depleting the film. The differentiation of these curves allowed us to find the transport coefficients in planes inside the film [6]. Resulting profiles of the anisotropy coefficient, the drift and Hall mobilities are of practical interest for the investigation of internal stress and transport degradation along the epitaxial direction in SOS films.

4 - MAGNETOCONCENTRATION (MC) EFFECT

This effect occurs in near intrinsic semiconductor slabs subjected to crossed electric E_x and magnetic B_z fields. A non-homogeneous carrier distribution n(y) is induced by Lorentz forces if at least one of the active surfaces is provided with non-infinite recombination rate. It has been shown that the magnetic field plays a more complex role than the electric field ; in particular, there is an optimum induction B for which the carrier non-equilibrium has a maximum.

The carrier gradient involves modifications of average density and surface density as well as the generation of a transverse diffusion current J_D. The MC effect is experimentally by an anomalous behaviour of thermo-galvanomagnetic effects (MR, Hall-, Ettingshausen-, etc.). We have pointed out that high magnetic fields cause the diffusion current J_D to dominate the transport [7]. At very low electric fields, J_D gives rise to partial or whole suppression of the ambipolar MR [7]. For moderate fields E_x, the apparent MR can be positive ($\rho_B > \rho_0$) or negative ($\rho_B < \rho_0$), while in the region of high electric fields the MR is always an increasing function with E_x and B_z, as illustrated in Fig.3. For instance, the MR coefficient measured in Ge at

12 Tesla and 20 volts is 500 % higher than the usual value (without MC effect).

The fitting of such experimental curves (Ge, InSb, etc.) is a powerful method for the simultaneous determination of bulk- and surface recombination parameters.

5 - MAGNETODIODE (MD) EFFECT

This effect is the extension of MC to double injecting structures (p^+nn^+ diodes) [8]. Magnetodiodes realized on Ge and SOS are ultrasensitive magnetic sensors [8, 9]; they are also useful tools for the characterization of recombination processes in *extrinsic* semiconductors where the MC effect is no more efficient.

The general theory of the MD effect shows that during the double-injection *semiconductor* regime, the current-voltage I(V) law is expressed by [8]:

$$I \sim \tau_{eff} V^2 \qquad (2)$$

where τ_{eff} is the carrier *effective* lifetime, depending on mobilities, E_x, $\vec{B}_z$, bulk and surface recombination rates, etc.

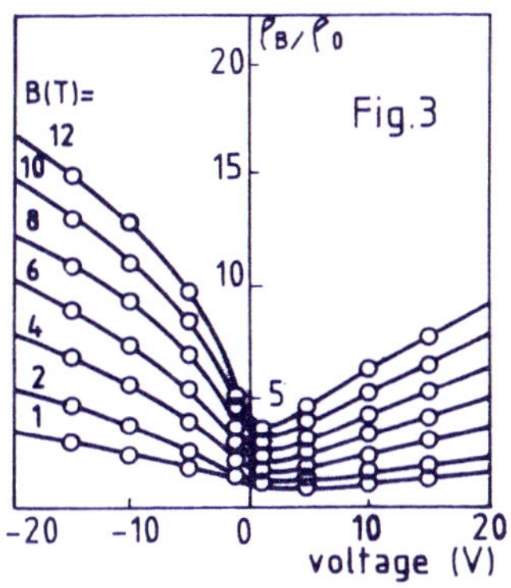

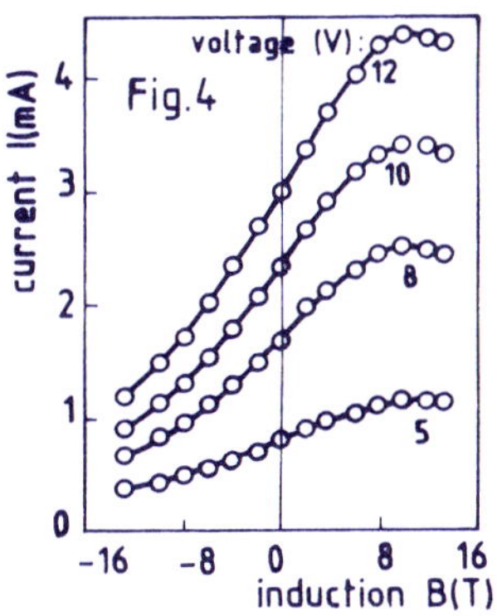

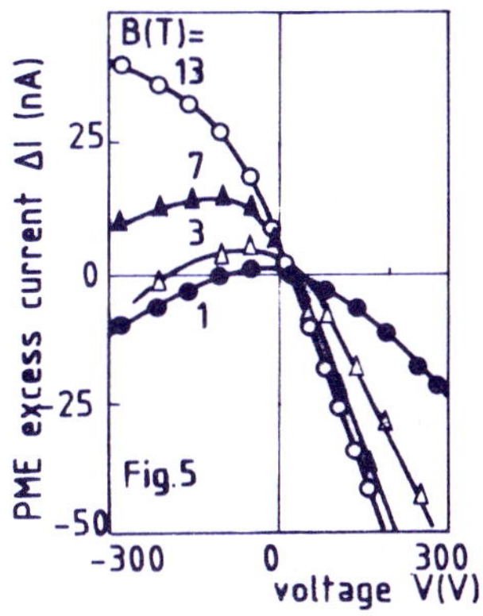

Fig. 3 : MC effect in near-intrinsic Ge : experimental MR coefficient ρ_B/ρ_0 against voltage with the magnetic induction as parameter.

Fig. 4 : Experimental I(B) characteristics in a SOS p^+nn^+ magnetodiode (length = 10 μm, thickness = 0. 65 μm, doping $\simeq 10^{22}m^{-3}$).

Fig. 5 : Experiment of the "field-assisted" PME effect in SI GaAs : excess current $\Delta I = I(B) - I(0)$ against longitudinal voltage V.

Typical experimental characteristics of SOS micro-magnetodiodes are given in Fig.4. At low magnetic field ($B < 2T$) I and τ_{eff} are proportional to B. A more interesting feature is observed in the region of higher fields where the I(B) characteristics become non-linear. For $B < 0$ carriers are deflected towards the highly recombining interface $Si\text{-}Al_2O_3$, so that the current is reduced and a *quasi-ohmic* law is observed: $I \sim V$, i.e. $\tau_{eff} \sim V^{-1}$ in Eq.2. For $B > 0$ the good quality of the accumulated interface $Si\text{-}SiO_2$ is responsible for the current increase ; the I(V) characteristic is now quadratic as expected from Eq.2. The current maximum, observed at about 10 Tesla, suggests that τ_{eff} has reached the upper limit, which is the carrier bulk lifetime.

Furthermore, computer simulations of the data give the recombination velocities of $Si\text{-}SiO_2$ and $Si\text{-}Al_2O_3$ interfaces and also indicate that in 0.6 μm thin SOS films the carrier lifetime (10^{-8} s) is reduced by one order of magnitude from the Si-SiO_2 interface towards the substrate [10].

6 - PHOTOMAGNETOELECTRIC (PME) EFFECT

It is well-known that a short-circuit current I_{PME} is induced in illuminated semiconductor slabs by application of a magnetic field B_z perpendicular to the light flux. The classical PME method consists in analysing I_{PME} and the photoconductivity. Obviously, this experiment is not sufficient for the evaluation of all PME effect parameters : electron and hole lifetimes and mobilities, recombination velocities, compensation rate, etc. A possible solution, usually adopted, is to make additional assumptions. Another method is to increase the number of experimental parameters by performing, in a first step, measurements of both photo-MR (Fig.1) and I_{PME} in high magnetic fields. For example I_{PME} is no longer proportional to B and gives additional information in comparison with the conventional low-field PME effect.

In a second step we propose a modification of the PME effect by applying also high electric fields E_x :as a consequence of Lorentz forces the photogenerated carrier distribution n(y) will be modified. This phenomenon, called "field assisted PME effect" [3], is therefore a superposition of MC and PME effects. It is characterized by non-linear $I_x(V_x)$ relations depending on the magnetic field intensity and direction. As already noted for the MC effect, the total current is governed in high fields B by the diffusion component I_{PME}. For the first time typical experiments performed on chromium-doped semi-insulating GaAs are shown in Fig.5, which fully support our theoretical model.

7 - CONCLUSION

Several galvanomagnetic effects have been reconsidered in this paper in order to demonstrate the interesting role of high magnetic fields in classical transport. This role is based on three prominent features :

- magnetoresistive reduction of carrier mobilities and diffusion coefficients,
- enhancement of Lorentz forces,
- increase of the influence of transverse diffusion currents.

The application of high magnetic fields is not only of fundamental interest but can also be successfully exploited for characterization purposes. Transport and recombination properties have been deduced for different semiconductors : extrinsic or near-intrinsic bulk materials (Ge, InSb, GaAs, PbSnTe,...) as well as thin SOS layers with low crystalline quality.

REFERENCES

1 S. Lakeou, S. Cristoloveanu, A. Chovet : Phys. Stat. Sol. (a), *43*, 213 (1977)
2 F.F. Sizov, S. Cristoloveanu, M.M. Kondratenko : Phys. Stat. Sol. (a), *58*, K 185 (1980)
3 K.N. Kang, S. Cristoloveanu, A. Chovet : 2nd Conf. of Semi-insulating III-V Materials, Evian, France (1982)
4 D. Marion, S. Cristoloveanu, A. Chovet : IEE Proceed part I : Solid State and Electron Device (1982), in press
5 S. Cristoloveanu, J.H. Lee, A. Chovet : Rev. Phys. Appl. *15*, 725 (1980)
6 J.H. Lee, S. Cristoloveanu, A. Chovet : Solid-State Electron, *25* (1982) in press
7 S. Cristoloveanu, A. Chovet : J. Phys. C : Solid State Phys., *11*, 3981 (1978)
8 S. Cristoloveanu : Phys. Stat. Sol. (a), *64*, 683 (1981) and *65*, 281 (1981)
9 A. Mohaghegh, S. Cristoloveanu, J. de Pontcharra : IEEE Trans. Electron Devices, *28*, 237 (1981)
10 S. Cristoloveanu, A. Chovet, G. Kamarinos : Solid State Electron, *21*, 1563 (1978)

IMPURITIES IN SEMIMAGNETIC SEMICONDUCTORS

Jerzy Mycielski
Institute of Theoretical Physics
Warsaw University
00-681 Warsaw, Poland

A review of information on donor, acceptor, and resonant states in semimagnetic semiconductors $Cd_{1-x}Mn_xTe$, $Cd_{1-x}Mn_xSe$, and $Hg_{1-x}Mn_xTe$ is given. Theory of the impurity states is discussed on the basis of the band theory of semimagnetic semiconductors. The results of magnetooptical and transport experiments are presented.

I. Introduction

The so-called "semimagnetic semiconductors" /or "diluted magnetic semiconductors"/ are alloys of non-magnetic and magnetic semiconductors. After several years of intense studies, the probably best known semimagnetic semiconductors are now $Hg_{1-x}Mn_xTe$, $Cd_{1-x}Mn_xTe$, and $Cd_{1-x}Mn_xSe$ /see the review articles [1-7] /. All of them are of $A^{II}B^{VI}$ type. The first and the second are of the zinc-blende cubic structure, and the third is of the wurtzite hexagonal structure. All have wide conduction and valence bands. However, $Cd_{1-x}Mn_xTe$ and $Cd_{1-x}Mn_xSe$ are wide--gap semiconductors while $Hg_{1-x}Mn_xTe$ is a narrow- or zero-gap semiconductor /for $x > 0.075$ and $x < 0.075$, respectively/. In the following we will discuss only these three materials.

The purpose of this paper is to review on the impurity and defect electronic states. We will discuss only shallow states and ignore more complex systems like, e.g., excitons bound to impurities.

The specific features of the semimagnetic semiconductors follow from the exchange interactions of the magnetic ions with each other, and of these ions with conduction and valence band electrons. The former interaction determines the magnetic properties of the alloy /see, e.g., [7]/. The latter gives rise to peculiarities of the band structure [8-10] and affects the impurity states. It should be stressed, however, that there exists a feedback: not only the localized spins of the magnetic ions may influence the carrier on an impurity state but also this carrier may produce some ordering of the surrounding localized spins. The latter effect - formation of the bound magnetic polaron - will not be discussed here, at least in principle, as it is the topic of another paper at this conference [11].

In the following we shall review briefly on the band structure of wide-, narrow-, and zero-gap semimagnetic semiconductors. Then we will discuss the effect of the exchange interaction with magnetic ions on the carriers bound to donors or acceptors in open-gap semimagnetic semiconductors. Some information on resonant states in zero-gap materials will also be presented.

II. Band structure

We will discuss the electron energy levels in semimagnetic semiconductors within the virtual crystal approximation, i.e. assuming that the one-electron microscopic potential of the alloy is a weighted average of the microscopic potentials of the constituents. In this approximation the microscopic potential of the alloy is, of course, periodic. Suppose for a while that there is no external magnetic field and the average component of the spin of magnetic atoms vanishes for all directions. The materials under consideration behave then very similarly to non-magnetic semiconducting alloys, like $Hg_{1-x}Cd_xTe$. They have a direct band gap /increasing with x/ at the Γ-point of the Brillouin zone. In the open-gap <u>zinc-blende</u>-type semiconductors /T_d point group/ the symmetry at the Γ-point of the conduction band and upper and lower valence bands is Γ_6 , Γ_8 , and Γ_7 , respectively. In the zero-gap case /$Hg_{1-x}Mn_xTe$ for low x/ the positions of Γ_6 and Γ_8 levels are reversed and the light-hole band Γ_8 of the open-gap structure becomes the conduction band. The basis for the Γ_6 , Γ_8 , and Γ_7-level is, respectively,

$$\Psi_{1/2,\ 1/2} = |S\uparrow\rangle\ , \qquad \Psi_{1/2,\ -1/2} = |S\downarrow\rangle\ , \tag{1}$$

$$\begin{aligned}\Phi_{3/2,\ 3/2} &= 2^{-1/2}|(X + iY)\uparrow\rangle\ , \quad \Phi_{3/2,\ 1/2} = i6^{-1/2}|(X + iY)\downarrow - 2Z\uparrow\rangle\ ,\\ \Phi_{3/2,\ -1/2} &= 6^{-1/2}|(X - iY)\uparrow + 2Z\downarrow\rangle\ , \quad \Phi_{3/2,\ -3/2} = i2^{-1/2}|(X - iY)\downarrow\rangle\ ,\end{aligned} \tag{2}$$

$$\Phi_{1/2,\ 1/2} = 3^{-1/2}|(X+iY)\downarrow + Z\uparrow\rangle\ , \quad \Phi_{1/2,\ -1/2} = i3^{-1/2}|-(X-iY)\uparrow + Z\downarrow\rangle\ , \tag{3}$$

where the position-dependent functions S and X, Y, Z transform under the operations of T_d point group as s- and p-type functions, respectively.

Using the molecular-field and virtual-crystal approximations, we can write the exchange interaction of the band electron with the $3d^5$-electrons of the Mn^{2+} ions in the form [8,9]

$$H_{exch} = \underline{s}\cdot\langle\underline{S}\rangle_{av}\ x\sum_{\underline{R}} J(\underline{r} - \underline{R})\ , \tag{4}$$

which has the periodicity of the crystal lattice, as the summation is over all metal lattice sites. J is a short-range function of the distance. $\underline{s}$ is the spin operator of the band electron, and $\langle\underline{S}\rangle_{av}$ is the manganese spin averaged thermally and over all manganese ions /spins are in units of $\hbar$/. In the absence of an external magnetic field $\langle\underline{S}\rangle_{av}$ vanishes. In the presence of the field $\underline{H}$ parallel to the z-direction $\langle\underline{S}\rangle_{av}$ has a non-vanishing z-component $\langle\overline{S}_z\rangle_{av}$ which tends to −5/2 for strong fields and low temperatures.

The exchange Hamiltonian (4) written in the basis (1)-(3) has the form /for $\langle\underline{S}\rangle_{av} \parallel z$/

$$H_{exch} = \begin{Vmatrix} 3A & 0 & 0 & 0 & 0 & 0 & 0 & 0\\ 0 & -3A & 0 & 0 & 0 & 0 & 0 & 0\\ 0 & 0 & 3B & 0 & 0 & 0 & 0 & 0\\ 0 & 0 & 0 & B & 0 & 0 & i\sqrt{8}B & 0\\ 0 & 0 & 0 & 0 & -B & 0 & 0 & -i\sqrt{8}B\\ 0 & 0 & 0 & 0 & 0 & -3B & 0 & 0\\ 0 & 0 & 0 & -i\sqrt{8}B & 0 & 0 & -B & 0\\ 0 & 0 & 0 & 0 & i\sqrt{8}B & 0 & 0 & B \end{Vmatrix} \tag{5}$$

$$A = (1/6)xN_0\alpha\langle S_z\rangle_{av}\ , \qquad B = (1/6)xN_0\beta\langle S_z\rangle_{av}\ , \tag{6}$$

$$\alpha = \langle S|J|S\rangle\ , \qquad \beta = \langle Z|J|Z\rangle\ . \tag{7}$$

N_0 is the number of unit cells in unit volume.

Mixing of Γ_8 and Γ_7 functions due to the off-diagonal elements of (5) can usually be neglected, as the spin-orbit splitting of Γ_8 and Γ_7 levels is large. Thus, the only effect of exchange interaction consist in splitting the Γ_6 , Γ_8 , and Γ_7 levels into singlets. This splitting depends - through $\langle S_z\rangle_{av}$ - on the external magnetic field and temperature, and is of the order of 10 meV. The $\underline{k}.\underline{p}$ calculation yields then the following results [10]. The splitting of Γ_6 band changes only slow with $\underline{k}$ and can be described by an effective g-factor. As $\alpha < 0$ for all known materials /i.e. $A > 0$/ and $N_0\alpha$ is of the order −0.2 to −1 eV, thi g-factor is very high and positive even in narrow-gap semimagnetic semiconductors, in spite of the fact that in narrow-gap non-magnetic semiconductors the g-factor of conduction electrons is negative. For the Γ bands, the splitting depends dramatically on both the direction and the absolute value of $\underline{k}$. This is shown schematicly for an open-gap band structure in Fig. 1. The reduced variables used here are $\varepsilon = E/B$, $\varkappa = \hbar k/(2Bm_{hh})^{1/2}$, where E is the electron energy and m_{hh} is the heavy-hole mass. Of particular interest are the dispersion relations close t the Γ-point, i.e. for $\hbar^2k^2/2m_{lh} \ll B$, where m_{lh} is the light-hole mass The four non-degenerate valence bands are

$$E_{\pm 3/2}(\underline{k}) = -(1/2)\hbar^2[(3/4m_{lh} + 1/4m_{hh})(k_x^2 + k_y^2) + (1/m_{hh})k_z^2] \pm 3B, \quad (8)$$

$$E_{\pm 1/2}(\underline{k}) = -(1/2)\hbar^2[(1/4m_{lh} + 3/4m_{hh})(k_x^2 + k_y^2) + (1/m_{lh})k_z^2] \pm B . \quad (9)$$

Thus, the constant energy surfaces are rotational ellipsoids, cucumber-like for the highest and the lowest bands /±3/2/ and disc-like for the intermediate bands /±1/2/. For $m_{hh} \gg m_{lh}$, the mass anisotropy is particularly strong for the bands ±3/2, being equal to $3m_{hh}/4m_{lh}$. At $\underline{k}$ = 0, the Bloch functions of the bands $E_{3/2}(\underline{k})$ to $E_{-3/2}(\underline{k})$ are given by $\varphi_{3/2,\ 3/2}$ to $\varphi_{3/2,\ -3/2}$ (2), respectively. As $\beta > 0$ for all known materials /i.e. $B < 0$/, $N_0\beta$ being of the order 0.5 to 1.5 eV, $E_{-3/2}(\underline{k})$ is the highest valence band.

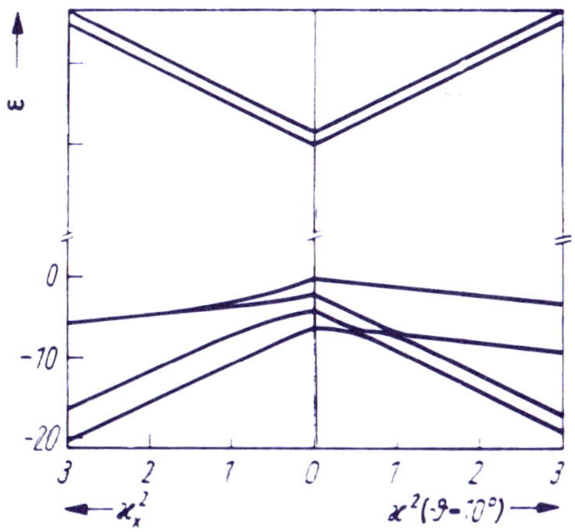

Fig. 1. Splitting of Γ_6 and Γ_8 bands due to exchange in an open-gap semimagnetic semiconductor for two values of angle ϑ between $\underline{k}$ and $\underline{H}$: 90° and 10° [10].

So far we have neglected the direct effect of the external magnetic field on the orbital motion of the carrier, as well as the intrinsic spin splitting, taking into account only the ordering of manganese spins by the field. The direct effect can be neglected for rather weak magnetic fields in wide-gap semiconductors, like $Cd_{1-x}Mn_xTe$, but not in narrow- or zero-gap materials, like $Hg_{1-x}Mn_xTe$, with low effective masses and high cyclotron frequencies. In the latter case the Landau splittings are comparable to those due to the exchange interaction. Thus, both effects have to be treated together. This is done by adding the exchange interaction (5) to the Pidgeon-Brown $\underline{k}.\underline{p}$ Hamiltonian [12] of a zinc-blende-type semiconductor in a magnetic field. The numerical solution of the modified Pidgeon-Brown Hamiltonian [8,9] fit very well the observed positions of the Landau levels in narrow- and zero-gap semimagnetic semiconductors.

Of particular interest is the level called $b_v(-1)$ and originating from Γ_8 , for which we have an analitic formula /warping of the valence band being neglected here/

$$E_{b_v(-1)} = -(1/2)\hbar\omega_{\perp 3/2} + (3/2)\hbar\omega_0\kappa - 3B , \quad (10)$$

where κ is a Luttinger parameter of the valence band [13], ω_0 is the free-electron cyclotron frequency, and $\omega_{\perp 3/2}$ is the cyclotron frequency corresponding to the transverse effective mass of the bands $E_{\pm 3/2}(\underline{k})$ given by (8). The wavefunction is /in the Landau gauge, and up to a normalization factor/ $\exp(ik_y y)\exp[-(x+L^2k_y)^2/2L^2]\varphi_{3/2,\ -3/2}$, where $L^2 = \hbar c/eH$. For the usual values of parameters, the state $b_v(-1)$ is the highest valence band level in both the open- and zero-gap case. It is the top of a heavy-hole Landau subband. From (10) and the wave function, one can see that $b_v(-1)$ is the first Landau level originating from $E_{-3/2}(\underline{k})$ valence band. Because of the -3B term in (10), $E_{b_v(-1)}$ increases with the magnetic field up to the saturation of $\langle \underline{S} \rangle_{av}$ /at few T/ and only then starts to decrease slowly.

In an open-gap wurtzite-type semiconductor /C_{6v} point group/, like $Cd_{1-x}Mn_xSe$, in the absence of exchange interaction, the symmetry at the Γ-point of the conduction band and upper, middle, and lower valence bands is Γ_7 , Γ_9 , Γ_7 , and Γ_7 , respectively. Each band is doubly degenerate at the Γ-point. Because of the uniaxial symmetry and the lack of inversion symmetry, the spin-orbit interaction yields a term linear in $\underline{k}$ in the effective-mass Hamiltonian of a Γ_7-band, of the form [14,15

$$\lambda\, \underline{c}\cdot(\underline{s}\times\underline{k}) \quad . \qquad (11)$$

λ is a constant and $\underline{c}$ is the unit vector along the c-axis. The effect of the exchange interaction of the manganese ions with the band electrons on the energy levels at the Γ-point consists in splitting all the levels Γ_7 and Γ_9 .

III. Donors in open gap

As we have mentioned, the effect of exchange interaction on the conduction band consists in a very large Zeeman-type splitting dependent /through $\langle \underline{S}\rangle_{av}$/ on the magnetic field and temperature.

In wide-gap semiconductors like $Cd_{1-x}Mn_xTe$ and $Cd_{1-x}Mn_xSe$, the effective mass in the conduction band is rather high. Therefore, the cyclotron energy is much lower than the donor binding energy, up to the magnetic fields of few T, and the direct effect of the external magnetic field on the orbital motion of the donor's electron can be neglected However, the exchange-induced splitting is comparable to the donor binding energy.

The exchange-induced splitting of donor states was observed in $Cd_{0.9}$ $Mn_{0.05}Se$ by spin-flip Raman scattering [16,17]. At 1.6K, the Zeeman-type splitting reaches 9.3 meV already in H = 1 T and nearly saturates in H = 2 T, with the value of about 14 meV /the donor binding energy being about 20 meV/. With increasing temperature the splitting decrease rapidly. This behaviour is in a good agreement with the expected dependence of $\langle \underline{S}\rangle_{av}$ on H and T. However, the splitting does not vanish in H = 0. This is interpreted as due to the bound magnetic polaron formation and to fluctuations of magnetization of the system of magnetic ions [11,18,19].

The exchange-induced splitting of donor states in $Cd_{1-x}Mn_xSe$ was observed also in another interesting experiment. A strong electric dipole spin resonance was found in far infrared magnetotransmission of the n-type samples with x = 0.1 [20]. This resonance was interpreted as the spin-flip transition on the 1s ground state of the donor, allowed by the term (11) in the Hamiltonian of a wurtzite-type crystal. The measured values of the effective g-factor are positive and very high. For example, in H = 1.1 T at 4.2K the Zeeman-type splitting of the 1s state is 7.2 meV which corresponds to the g-factor equal to 111. The splitting increases with magnetic field and decreases rapidly with increasing temperature. It does not extrapolate to zero for $H\to 0$ which is interpreted by the bound magnetic polaron effect.

An effect of the exchange interaction in n-$Cd_{1-x}Mn_xSe$ was observed in dc magnetoconductivity measurements [21]. In the extrinsic range of conductivity there is no much effect of the magnetic field. In the hopping conductivity range a strong negative and isotropic magnetoresistance is observed in fields of few T. In this range the resistivity depends on temperature as $\rho_\infty \exp(\varepsilon_a/k_BT)$, where ε_a is the activation energy of conductivity. ε_a decreases strongly in the magnetic field in $Cd_{1-x}Mn_xSe$ /with no much change in ρ_∞/ while for CdSe it does not depend on H. The effect is interpreted by the destroying action of the magnetic field on the bound magnetic polaron and on fluctuations of magnetization. However, the observed effect is stronger than the theoretical predictions for higher concentrations of donors which suggests a correlated /multi-electron/ hopping process /see, e.g., [22]/.

In narrow-gap semimagnetic semiconductors the effective mass in the conduction band is low. Thus, the cyclotron energy is comparable to the exchange-induced splitting and can easily be much higher than the donor binding energy. As in a non-magnetic narrow-gap semiconductor, below each Landau level there is a set of bound or resonant donor states. In a semimagnetic semiconductor, however, these states are strongly shifted together with the corresponding Landau levels, due to the exchange interaction.

This was observed recently in an intraband absorption and photoconductivity experiment on indium-doped n-$Hg_{0.89}Mn_{0.11}Te$ /energy gap 165 meV/ in magnetic field up to 18T, at helium temperatures [23]. Starting from 8T, the freezing of conduction electrons on donor states occurs and transitions from the donor ground state below the lowest Landau level to states below higher Landau levels and/or to these Landau levels were observed. In the magnetic field the donor binding energy increases and is observed to be about 4 meV for the state associated with the lowest Landau level n=0 and about 11 meV for n=3, at H = 15 T. The difference of these binding energies is probably due to non-parabolicity of the conduction band, higher states corresponding to higher effective masses. The positions of the Landau levels fit well with the predictions of the modified Pidgeon-Brown Hamiltonian including exchange [8,9].

IV. Acceptors in open gap

We have mentioned already that in wide-gap materials the direct effect of the magnetic field on the orbital motion of the carrier can be neglected /for not too strong magnetic fields/. Because of high effective masses of holes, the acceptor binding energy E_I in the absence of exchange is much higher than the exchange-induced splitting:

$$E_I \gg 6\,|B|\,. \tag{12}$$

The exchange-induced splitting is, in turn, much higher than the Landau splitting. Therefore, it is justified to treat the exchange interaction as a perturbation in the acceptor problem in wide-gap semimagnetic semiconductors, neglecting the direct effect of magnetic field on the orbit.

The wave functions of the four-fold degenerate ground state of a shallow acceptor in zinc-blende structure have the form

$$\Psi_i(\underline{r}) = \sum_{j=-3/2}^{3/2} \varphi_{3/2,\,j}(\underline{r})\, F_{ji}(\underline{r})\;, \quad (i = -3/2,\ -1/2,\ 1/2,\ 3/2) \tag{13}$$

where $\varphi_{3/2,\,j}$ are given by (2), and F_{ji} are slowly-varying envelope functions [24,25]. Using the envelope functions and parameters determined by the Schechter's variational procedure [24] one can show that the functions Ψ_i are not mixed by the exchange interaction (5), but the quadruplet ground state is splitt into four singlets with the wave functions Ψ_i and energies [5,26]

$$E_{I\,\pm 3/2} = E_I \pm [3 - 2d_{zz} - (8/3)d_{xy}]B\;, \tag{14}$$

$$E_{I\,\pm 1/2} = E_I \pm (1 - 2d_{zz})B\;. \tag{15}$$

d_{zz} and d_{xy} are norms of the two d-type parts of the Schechter's envelope four-function, depending on x^2, y^2, z^2, and on xy, yz, zx, respectively. The predominant s-type part of F_{ji} vanishes for j = i but the d-parts do not. Therefore, they are responsible for mixing different $\varphi_{3/2,\,j}$ in a given Ψ_i. This mixing reduces the acceptor ground-state splitting (14),(15) as compared with that of free holes at the Γ-point, and the splitt levels are not equidistant. For the parameters of CdTe [25], on the right-hand side of (14) we have about ±2.3B /instead of ±3B for free holes/, and on the right-hand side of (15) about ±0.8B /instead of ±B/.

In the absence of magnetic field, the theoretical ionization energy of a shallow acceptor in CdTe is 87.4 meV [25], and should be the same in $Cd_{1-x}Mn_xTe$ /in the absence of exchange/, as the parameters of the valence band [27] and the static dielectric constant [28] were found to be composition independent. In a photoluminescence experiment values 57.8, 58.8, 108, and 147 meV of the ionization energies were found in CdTe and interpreted as corresponding to the substitutional impurities Li, Na, Ag, and Cu, respectively [29]. In $Cd_{1-x}Mn_xTe$ the absorption due to acceptor photoionization was measured and analized [30]. The photoionization spectrum was fitted using a formula derived by the quantum defect method [31]. This gave a rather high quantum defect for the ground state, and E_I equal to 165 and 170 meV, for x being 0.05 and 0.10, respectively. Thus, the acceptor is not very shallow.

The exchange contribution to the binding energy of acceptors in $Cd_{1-x}Mn_xTe$ was deduced recently from the temperature dependence of the extrinsic conductivity of p-type samples /in the absence of magnetic field/, for different compositions [32]. The activation energy ε_a of conductivity increases with increasing x for both samples with shallow and deeper acceptors. As we have mentioned already, ε_a should be independent of x in the absence of exchange. However, with increasing x the bound magnetic polaron and the magnetization fluctuation effects will appear and ε_a will increase. The calculated dependence of ε_a on x is in a reasonable agreement with the experimental data.

In narrow-gap semimagnetic semiconductors, like $Hg_{1-x}Mn_xTe$ /$x > 0.075$ the hole effective masses are low and, therefore, the direct effect of the external magnetic field on the orbital motion of the hole /Landau splitting/ is strong and comparable to the splitting due to exchange interaction. Moreover, it may be comparable or larger than the low acceptor ionization energy E_I in the absence of exchange. This makes the theory of acceptors states more involved than in the case of wide-gap materials.

Nevertheless, it is still easy to calculate the lowest acceptor states in the case of large exchange and Landau splittings, i.e. for

$$2\,|B|,\quad \hbar\omega_{\perp 3/2} \gg E_I \;. \tag{16}$$

As can be seen from (8)-(10), one can use then the ultra-quantum limit approach, i.e. construct the acceptor lowest states' wave functions from the functions of the highest Landau valence subband. Using the symmetric gauge we have an acceptor wave function in the form

$$\Psi(\underline{r}) = (2\pi L^2)^{-1/2}\exp[-(x^2+y^2)/4L^2]F(z)\Phi_{3/2,-3/2}(\underline{r}) \;, \tag{17}$$

where the normalized envelope function F(z) fulfils the equation

$$\{-(1/2)(d^2/d\xi^2)-(2\pi L^2/a_{hh}^2)^{1/2}\exp(\xi^2)[1-\Phi(|\xi|)]+2E/\hbar\omega_{hh}\}F(\xi) = 0 \;. \tag{18}$$

Here $\xi = z/2^{1/2}L$, $a_{hh} = h^2\varepsilon_s/e^2m_{hh}$, $\omega_{hh} = eH/cm_{hh}$, and

$$\Phi(x) = (2/\pi^{1/2})\int_0^x \exp(-t^2)dt \;. \tag{19}$$

$-E$ is the hole energy in respect to $b_v(-1)$ level.

In the vacuum or in non-magnetic semiconductors the ultra-quantum limit corresponds to the magnetic length much smaller than the Bohr radius. Equation (18) can be then solved numerically /see, e.g., [33]). For the acceptor in a semimagnetic semiconductor, due to the anisotropy of the $E_{-3/2}(\underline{k})$ band (8), the ultra-quantum limit condition (16) does not imply $L < a_{hh}$. It exists a broad range of magnetic field intensities in which there is

$$L \gg a_{hh} \;. \tag{20}$$

If (20) is fulfilled, the "potential" in the equation (18) can be expanded assuming $|\xi| \ll 1$ [34,35], and the resulting equation can be solved in Bessel and Hankel functions [36]. One obtains a series of symmetric and antisymmetric eigenstates. The ground state is symmetric and the ionization energy is

$$E_I = (\pi/2)^{1/2}(e^2/\varepsilon_s L)[1 - 0.73(a_{hh}/L)^{1/3}] \;. \tag{21}$$

Because of (20), the ionization energy (21) is lower than the ionization energy in the absence of magnetic field, and the acceptor transverse radius /L/ is larger than that in the absence of magnetic field. This behaviour is very unusual as compared with that known in vacuum and non-magnetic semiconductors and is due to the exchange-induced anisotropy of the band $E_{-3/2}(\underline{k})$. It should be noted that the decrease of ionization energy and the increase of transverse radius occur before reaching the ultra-quantum limit - in the ultra-quantum limit the ionization energy increases and the transverse radius decreases with increasing magnetic field.

One can expect that the strange behaviour of the acceptor in magnetic field should reflect on the low temperature magnetoresistance. In fact, the resistivity of $Hg_{1-x}Mn_xTe$ drops strongly in magnetic field at low temperatures [37,38]. A very strong negative magnetoresistance was investigated at few T in p-type narrow-gap $Hg_{1-x}Mn_xTe$-samples with the net acceptor concentration of the order of $10^{16}cm^{-3}$, in both extrinsic and hopping conductivity range [39,40]. Some of the results are shown in Fig. 2.

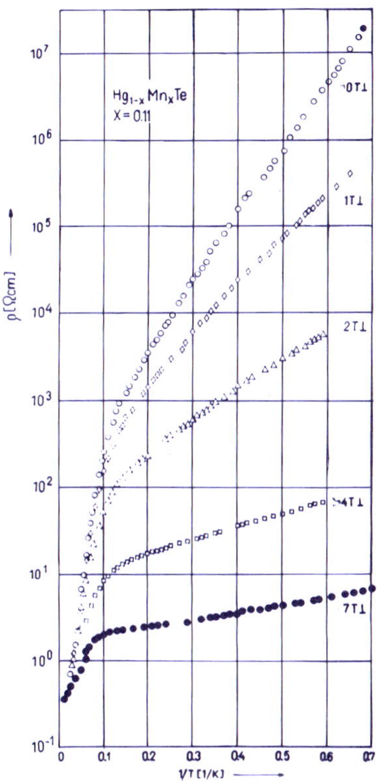

Fig. 2. Transverse resistivity in p-$Hg_{1-x}Mn_xTe$ /x=0.11/ as the function of temperature and magnetic field [40].

In the extrinsic conductivity range the negative magnetoresistance is due to a decrease of the ionization energy with increasing magnetic field /"boil-off" effect/. E.g., for x = 0.155, the values of E_I observed at H = 0 and 3.5T are 8.2 and 5.6 meV, respectively [39]. The value of E_I at H = 0 calculated by the Schechter's method [24] is 9 meV, and E_I at H = 3.5T calculated from (21) is 5.5 meV. In the hopping conductivity range both ρ_∞ and ε_a decrease strongly with increasing magnetic field /Fig. 2/. ρ_∞ depends on the overlap of the wave-functions of the neighbouring acceptors and, therefore, should decrease with increasing radius of the wave function, i.e. when approaching the ultra-quantum limit. Because of the anisotropy of the acceptor wave function, the conductivity should be higher in the transverse direction, which was confirmed by experiment. The increase of the overlap may also cause the transition from non-metallic to metallic type of conductivity, which was observed in one of the samples in [40]. Well in the ultra-quantum limit, the acceptor radius starts to decrease with increasing magnetic field and the resistivity should start to increase slowly, which seems to be confirmed by the experimental results for low temperatures.

The decrease of ε_a in the magnetic field may be caused by approaching the nonmetal-metal transition or may correspond to the destroying action of the magnetic field on the bound magnetic polaron and on fluctuation of magnetization [40]. The decrease of ε_a with increasing acceptor radius may be also due to the increasing role of the correlated /multi-hole/ hopping [22]. If a part of ε_a is given by fluctuations of the composition x of the alloy in space [41,42], then ε_a will decrease because of the fact that with increasing wave-function's radius the fluctuations are averaged over a bigger volume and their effect decreases.

For a narrow-gap semimagnetic semiconductor in a moderate magnetic field /~1T/, the problem of calculating the acceptor states is very involved, as (16) is not fulfilled and $2|B|$, $\hbar\omega_{3/2}$, and E_I may all be comparable. The problem of the ground state in the whole range of magnetic field intensity was treated [43] including the exchange interaction (5) into the valence band effective mass Hamiltonian in magnetic field and using the spherical tensor operators and reduced matrix elements technique [25,44]. In the variation procedure the envelope functions /see (13)/ were assumed to be sums of products of spherical harmonics and exponentials of distance. The results show the development of four levels originating from the accept ground state quadruplet, over a broad rang of magnetic field.

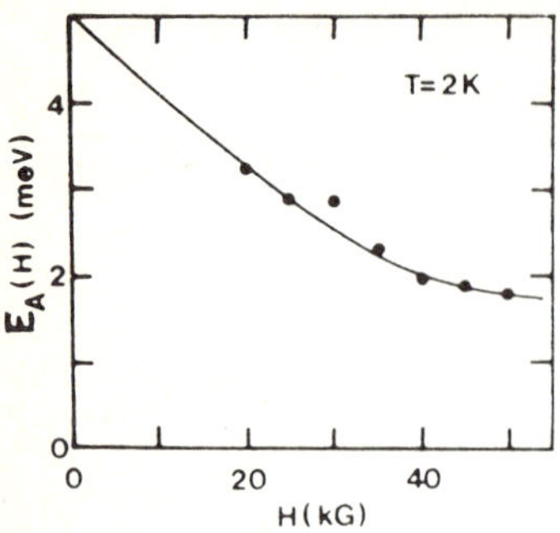

Fig. 3. Magnetic field dependence of the acceptor level observed in p-$Hg_{1-x}Mn_xTe$ /x=0.128, energy gap 219 meV/ [46].

Magnetotransmission measurements were performed on narrow-gap p-$Hg_{1-x}Mn_xTe$ at helium temperatures [45,46]. A line corresponding to the transition from an accepto level to $b_c(1)$ Landau state in the conduction band was observed. The acceptor level was found to decrease with increasing magnetic field, in respect to the top of the valence band at H = 0 /Fig. 3/. The transmission measurement in the far infrared /a H = 0/ has shown an absorption line at abo 5 meV which may correspond to the transition from the valence band to this accepto level [45]. At 4T, the level is about 12 meV below $b_v(-1)$ Landau level Thus, it is not the acceptor ground state but probably one of the excited states.

V. Resonant states in zero gap

In zero-gap semiconductors, donors and acceptors may only exist as resonant states degenerate with the valence and conduction band, respe tively. Because of a very high density of states in the heavy-hole ban quasi-discrete donor resonant states do not exist. However, the densit of states in the conduction band is low and the resonant acceptor stat are observed.

In non-magnetic semiconductors, a lot of experimental and theoretic work was done on the resonant acceptors /see, e.g., [47] and the revie paper [48]/. The effect of the magnetic field on resonance states was also discussed [49].

In semimagnetic zero-gap $Hg_{1-x}Mn_xTe$ the magnetoabsorption experimen were performed [8,50,51]. Transition from b(0) Landau level in Γ_6 valence band to a resonant acceptor state was observed. The zero-field energy of the resonant state was found to be about 1 meV [8,50] or 1.5 meV [51]. In [50] a strong quenching of the acceptor g-factor was also found: the acceptor state does not split or shift in the magnetic field. However, such a shift was observed in [8,51]. In [50], the submillimeter resonance experiment was performed and Γ_8-Landau levels to acceptor, and acceptor to Γ_8-Landau levels transitions were observed. Two acceptor levels were found: 0.9 meV at very low x, and 2.2 meV for $x \geq 0.008$. They were identified as A_0 /impurity/ and A_1 /mercury vacanc levels, respectively, observed also in non-magnetic zero-gap $Hg_{1-x}Cd_xT$ [52]. The energy of A_1 was found to be nearly independent of magnetic field.

Resonant acceptor states were also investigated by magnetotransport measurements on zero-gap p-$Hg_{1-x}Mn_xTe$. Temperatures down to 30mK and pressures up to 10 Kbar were used [53]. It was shown [54-56] that in very strong magnetic fields / >10T/ the energy gap opens and the nonresonant acceptor states can be observed. Hopping conductivity over these states decreases strongly with increasing magnetic field, due to shrinking of the acceptor wave function, and the transverse conductivi is the lower one [55,56]. Conductivity in the impurity band of resonan acceptors /i.e. for closed gap/ was also observed and shown to increas strongly at the fields 1 to 2T, in particular in the transverse direction [55-57]. This effect was interpreted as the decrease of spin-disorder scattering in the impurity band. However, it may also have an explanation similar to that presented in the previous Section, for the strong negative hopping magnetoresistance in open-gap materials.

Acknowledgement

This work was supported by the Institute of Physics, Polish Academy of Sciences.

References

1) R.R. Galazka: Physics of Semiconductors, 1978, ed. B.L.H. Wilson /The Institute of Physics, Bristol 1979/, Conf. Ser. 43, p. 133.
2) R.R. Galazka and J. Kossut: Proc. Int. School Narrow-Gąp Semicond., Nimes, France 1979, p. 245.
3) J.A. Gaj: Proc. 15th Int. Conf. Phys. Semicond. Kyoto (1980), ed. S. Tanaka and Y. Toyozawa, J. Phys. Soc. Japan 49 (1980) Suppl. A, p. 797.
4) G. Bastard, J.A. Gaj, R. Planel, and C. Rigaux: J. de Physique C 5 (1980) 247.
5) J. Mycielski: Recent Developments in Condensed Matter Physics, Vol. 1, ed. J.T. Devreese /Plenum Press, New York 1981/, p. 725.
6) T. Dietl: Physics in High Magnetic Fields, ed. M. Miura, Springer Ser. Solid State Sciences, Vol. 24 /Springer, Berlin 1981/, p. 344.
7) R.R. Galazka: Physics of Narrow Gap Semiconductors, Proc. 4th Int. Conf. Linz 1981, ed. E. Gornik, H. Heinrich, and L. Palmetshofer /Springer, Berlin 1982/, p. 294.
8) G. Bastard, C. Rigaux, Y. Guldner, J. Mycielski, and A. Mycięlski: J. de Physique 39 (1978) 87.
9) M. Jaczyński, J. Kossut, and R.R. Gałązka: phys. stat. sol. (b) 88 (1978) 73.
10) J.A. Gaj, J. Ginter, and R.R. Gałązka: phys. stat. sol. (b) 89 (1978) 655.
11) R. Planel: this conference.
12) C.R. Pidgeon and R.N. Brown: Phys. Rev. 146 (1966) 575.
13) J.M. Luttinger: Phys. Rev. 102 (1956) 1030.
14) R.C. Casella: Phys. Rev. Lett. 5 (1960) 371.
15) R. Romestain, S. Geschwind, and G.E. Devlin: Phys. Rev. Lett. 39 (1977) 1583.
16) M. Nawrocki, R. Planel, G. Fishman, and R. Galazka: Proc. 15th Int. Conf. Phys. Semicond. Kyoto (1980), ed. S. Tanaka and Y. Toyozawa, J. Phys. Soc. Japan 49 (1980) Suppl. A, p. 823.
17) M. Nawrocki, R. Planel, G. Fishman, and R.R. Galazka: Phys. Rev. Lett. 46 (1981) 735.
18) T. Dietl and J. Spałek: Phys. Rev. Lett. 48 (1982) 355.
19) P.A. Wolff: Bull. Am. Phys. Soc. 27 (1982) 143.
20) M. Dobrowolska, H.D. Drew, J.K. Furdyna, T. Ichiguchi, A. Witowski, and P.A. Wolff: to be published.
21) T. Dietl, J. Antoszewski, and L. Świerkowski: Proc. 16th Int. Conf. Phys. Semicond. Montpellier (1982), to be published.
22) M. Pollak: Phil. Mag. B 42 (1980) 781.
23) M. Grynberg, G. Martinez, and L.C. Brunel: Solid State Comm., to be published.
24) D. Schechter: J. Phys. Chem. Solids 23 (1962) 237.
25) A. Baldereschi and N.O. Lipari: Phys. Rev. B 8 (1973) 2697.
26) J. Mycielski and C. Rigaux: to be published.
27) A. Twardowski and J. Ginter: phys. stat. sol. (b) 110 (1982) 47.
28) W. Gębicki and W. Nazarewicz: phys. stat. sol. (a) 61 (1980) K135.
29) E. Molva, J.P. Chamonal, and J.L. Pautrat: phys. stat. sol. (b) 109 (1982) 635.
30) R. Wojtal, A. Golnik, and J.A. Gaj: phys. stat. sol. (b) 92 (1979) 241.
31) H.B. Bebb: Phys. Rev. 185 (1969) 1116.

32) J. Jaroszyński, T. Dietl, M. Sawicki, and E. Janik: Proc. 16th Int. Conf. Phys. Semicond. Montpellier (1982), to be published.
33) M. von Ortenberg: J. Phys. Chem. Solids 34 (1973) 397.
34) S.D. Beneslavskii, N.B. Brandt, E.M.Golyamina, S.M. Chudinov, and G.D. Yakovlev: Zh. Eksp. Teor. Fiz. Pisma 19 (1974) 256.
35) S.D. Beneslavskii and E. Entral'go: Zh. Eksp. Teor. Fiz. 68 (1975) 2271.
36) J. Mycielski: to be published.
37) R.T. Delves: Proc. Phys. Soc. 87 (1966) 809.
38) J.H. Morrissy: Ph.D. Thesis, Oxford University (1971).
39) A. Mycielski and J. Mycielski: Proc. 15th Int. Conf. Phys. Semicond. Kyoto (1980), ed. S. Tanaka and Y. Toyozawa, J. Phys. Soc. Japan 49 (1980) Suppl. A, p. 807.
40) T. Wojtowicz and A. Mycielski: Proc. 16th Int. Conf. Phys. Semicond. Montpellier (1982), to be published.
41) B.L. Gelmont, A.R. Gadzhiev, B.I. Shklovskii, I.S. Shlimak, and A.L. Efros: Fiz. Tekh. Poluprov. 8 (1974) 2377.
42) I.S. Shlimak, A.L. Efros, and I.Ya. Yantchev: Fiz. Tekh. Poluprov. 11 (1977) 257.
43) T.R. Gawron and J. Trylski: Physics of Narrow Gap Semiconductors, Proc. 4th Int. Conf. Linz 1981, ed. E. Gornik, H. Heinrich, and L. Palmetshofer /Springer, Berlin 1982/, p. 312.
44) N.O. Lipari and M. Altarelli: Solid State Comm. 33 (1980) 47.
45) C. Rigaux, G. Bastard, Y. Guldner, G. Rebmann, A. Mycielski, J.K. Furdyna, and D.P. Mullin: Proc. 15th Int. Conf. Phys. Semicond. Kyoto (1980), ed. S. Tanaka and Y. Toyozawa, J. Phys. Soc. Japan 49 (1980) Suppl. A, p. 811.
46) G. Bastard, C. Rigaux, Y. Guldner, A. Mycielski, J.K. Furdyna, and D.P. Mullin: Phys. Rev. B 24 (1981) 1961.
47) W. Trzeciakowski: J. Phys. C: Solid State Phys. 15 (1982) 1199.
48) G. Bastard: Physics of Semiconductors, 1978, ed. B.L.H. Wilson /The Institute of Physics, Bristol 1979/, Conf. Ser. 43, p. 231.
49) Z.W. Gortel, J. Szymański, and L. Świerkowski: Physics of Semiconductors, 1978, ed. B.L.H. Wilson /The Institute of Physics, Bristol 1979/, Conf. Ser. 43, p. 261.
50) G. Bastard, C. Rigaux, Y. Couder, H. Thome, and A. Mycielski: phys. stat. sol. (b) 94 (1979) 205.
51) M. Dobrowolska, W. Dobrowolski, M. Otto, T. Dietl, and R.R. Galazka: Proc. 15th Int. Conf. Phys. Semicond. Kyoto (1980), ed. S. Tanaka and Y. Toyozawa, J. Phys. Soc. Japan 49 (1980) Suppl. A, p. 815.
52) Y. Guldner, C. Rigaux, A. Mycielski, and Y. Couder: phys. stat. sol. (b) 81 (1977) 615.
53) M. Sawicki, T. Dietl, W. Plesiewicz, P. Sękowski, L. Śniadower, M. Baj, and Ł. Dmowski: this conference.
54) P. Byszewski, K. Szlenk, J. Kossut, and R.R. Gałązka: phys. stat. sol. (b) 95 (1979) 359.
55) B.B. Ponikarov, I.M. Tsidilkovskii, and N.G. Shelushinina: Fiz. Tekh. Poluprov. 15 (1981) 296.
56) A.B. Davydov, B.B. Ponikarov, and I.M. Tsidilkovskii: Fiz. Tekh. Poluprov. 15 (1981) 881.
57) A.B. Davydov, B.B. Ponikarov, and I.M. Tsidilkovskii: phys. stat. sol. (b) 101 (1980) 127.

BOUND MAGNETIC POLARONS IN SEMIMAGNETIC SEMICONDUCTORS

R. PLANEL

Groupe de Physique des solides de l'Ecole Normale Supérieure
Université Paris 7, Tour 23, 2 Place Jussieu 75251 Paris Cédex 05
France

Semimagnetic semiconductors provide a link between semiconductor physcis and magnetism. This is especially true for the problem of magnetic polarons bound to neutral donors or acceptors, which have recently been found in these compounds. Spectroscopy provides accurate information about the magnetization around impurity centers with small dimensions in the bound carrier orbit. The experimental access to the exchange energy differs for donors and acceptors, and also the orders of magnitude of the obtained quantities (average magnetization, mean number of localized moments involved, exchange energy) are different. Due to the small size of the impurity, magnetic fluctuation effects are noticeable, especially in the high temperature range where the exchange energy does not vanish due to these fluctuations.

1. Introduction

"Semimagnetic semiconductors" (SMSC) named by R. Galazka (1), are diluted magnetic alloys. Up to now, the most studied are the II-VI compounds containing manganese. The interest in these mixed crystals has grown considerably in the recent years, due to the availability of a well known basic compound (such as CdTe) and to the feasibility of gradually introducing magnetism in the problem. Cd_{1-x} Mn_x Te alloys may cover a quasi-continuous composition range from x = 0 to x = 0.7. Thus, it is possible to benefit from powerful experimental techniques and physical concepts elaborated for standard semiconductors physics, mainly of the accurate spectroscopy and transport measurements [2]. With respect to other magnetic semiconductors, SMSC form a more suitable link between semiconductor physics and magnetism. Thus, they appeared of great interest for the study of bound magnetic polarons (BMP), a concept which was introduced much earlier to account for anomalous transport properties in europium chalcogenides [3]. As a matter of fact, the use of optical spectroscopy in the visible range provides an accurate access to magnetization averaged over various scales, and thus reveals to be a powerful tool to detect and study BMP in wide gap SMSC. The purpose of this paper is to give a rather detailed account of these works. It is to be noted that the formation of BMP has also been observed in narrow gap SMSC (such as HgMnTe alloys), especially by transport measurements. These data will be presented by J. Mycielski during this conference.

In the next section, we briefly remind of the way to include the exchange interaction in the frame work of semiconductor physics, which gives us the possibility to dwell on spin flip Raman scattering experiments (SFRS) in sec.3. These experiments provide reliable data on donor-BMP, and point out the importance of magnetic fluctuations (sec.4). In sec. 5, we deal with acceptor-BMP which is found qualitatively different and requires different experimental approaches. Then, we present some other experimental results which shed light on various aspects of the physical situation (sec.6).

2. Exchange-induced spin-splitting of extended electronic states

The band structure of CdMnTe or CdMnSe alloys is quite similar to the one of CdTe or CdSe, respectively, at least around the center of the Brillouin zone. In the low and intermediate ranges of composition ($x \lesssim 0.3$) it has been clearly established that they are direct gap materials where the usual concepts of semiconductor physics (excitons, shallow impurities,...) remain useful. Extended electronic states, which may be described within the framework of the effective mass approximation, originate from the orbitals of s-(cations) and p-(anions) type; they are responsible for the semiconductor properties. On the contrary, half filled $3d^5$ shells of Mn^{++}ions are responsible for magnetic properties.

The exchange interaction between these two types of electronic states is described by the Heisenberg term

$$H_{exc} = - J(\vec{r} - \vec{R})\ \vec{\sigma}\cdot\vec{S}$$

where $(\vec{r},\vec{\sigma})$ (resp. $[\vec{R},\vec{S}]$) are the space and spin variables of the extended (resp. d-shell) states. Further simplification is obtained by neglecting the Mn^{++} excited states: thus Mn^{++} ions are considered as having rigid 5/2 spins, with g = 2 since the orbital momentum of the $3d^5$ shells vanishes.

Treating the short-range exchange operator as a δ-function $J(\vec{r} - \vec{R}) = I\delta\ (\vec{r}-\vec{R})$ the one-electron Hamiltonian describing the extended states may be written as:

$$(1) \qquad H_{exc} = -I\vec{\sigma} \cdot (\sum_i \vec{S}_i |\phi(\vec{R}_i)|^2)$$

where $|\phi(\vec{R}_i)|^2$ is the electronic density at cation site $\vec{R}_i$, $\vec{S}_i$ is the ith rigid spin operator; the summation is running over all Mn^{++} ions in the wave function. The exchange parameter I depends on the periodic part of the wave function, and is usually denoted α for the s-like conduction band, and β for the p-like valence bands. It is worth noting that our definition of exchange parameters (and of the related quantity N_o) is something unlike the original one [4]. It is more suggestive in the case of BMP but is not as general and causes problems if there is more than one cation site in each primitive lattice cell Ω. In such a case (as in wurtzite structure CdMnSe) I should be considered as a mean value per cation site, averaged over Ω.

The exchange term (eq.1) can generally be approximated as

$$(2) \qquad H_{exc} = -N_o x I\vec{\sigma}\cdot\int |\phi(\vec{R})|^2\ \vec{S}(\vec{R})\ d^3 R$$

where N_o is the number of cation sites per unit volume, X is the Mn fraction, and the Mn^{++} average spin value is considered as a continuous function $\vec{S}(\vec{R})$.

In the case of a constant magnetization $N_o x g\mu_B$ $<\vec{S}>$ eq. (2) reduces to

$$(3) \qquad H_{exc} = - N_o x I\vec{\sigma} \cdot <\vec{S}>$$

This exchange term is usually obtained in the framework of mean-field approximation and first order perturbation theory. It results in a huge enhancement of the band splittings in a magnetic field and at low temperatures, when <S> is high. In wide gap materials, the ordinary g-factor is small and may often be neglected. This enhancement results in several giant magnetooptical effects, such as Faraday rotation [5] and exciton "Zeeman like" effects [6] (see Fig. 1). In wide gap compounds, typical exchange parameters are $N_o\alpha \sim 0.2$ eV and $N_o\beta \sim 0.9$ eV (7). It has been shown that magnetooptical experiments could be considered as optical measurements of magnetization with a good precision (5, 8).

The detailed magnetic behaviour of SMSC shows a $Mn^{++} - Mn^{++}$ interaction of antiferromagnetic sign. Due to the random distribution of Mn^{++} ions in the crystal it behaves as a frustrated system and spin-glass properties have been observed in $x > 0.2$ compounds (9). In samples with smaller x or in high fields, the magnetization may be approximated by a Curie-Weiss law (8):

$$(4) \qquad M = N_o x\ g\mu_B\ S_o B_{5/2} \left(\frac{5/2\ g\ _B\ H}{k(T + T_o)}\right)$$

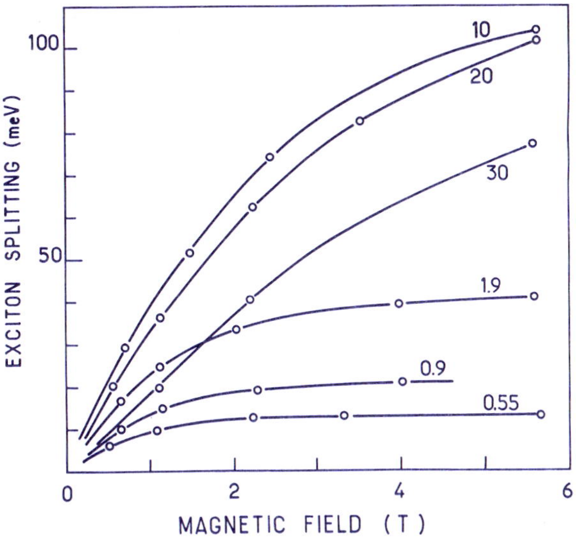

Fig. 1 Giant exciton Zeeman-like splitting in CdMnTe alloys; T = 1,6 K. Note the saturation of splittings with Mn^{++} magnetization. [after ref. (6)]

where $B_{5/2}$ is the Brillouin function for 5/2 spins and the departure from a pure paramagnetic law is accounted by $S_o < 5/2$ effective maximum spin value and by T_o being a positive parameter. It is worth noting that in an insulating medium, the indirect exchange interaction via paramagnetic electrons (the RKKY mechanism in metals) does occur. The dominant interaction in SMSC is believed (10) to happen via virtual interband transitions (Bloembergen-Rowland mechanism) which results in weak- and short range coupling, especially in wide gap materials (11).

3.Spin-Flip Raman Scattering from neutral donors

Let us call that SFRS became a classical technique in wide gap semiconductors since the work of Thomas and Hopfield in CdS (12). The Stokes shift $\Delta(H)$ is a direct measure of the spin splitting of paramagnetic electrons (i.e. unpaired electrons) in external fields. Typical $\Delta(H)$ values are of the order of 1 cm^{-1}/Tesla and temperature independent. At low temperature, electrons are trapped at donors and, in the framework of the effective mass approximation, g-factors do not differ substantially. As the temperature is increased, an anti-Stokes line reveals the existence of donors in the excited spin-state. The linewidth of SFRS essentially reveals, in homogeneous materials, the donor state spin relaxation time. It is usually smaller than $\Delta(H)$ in insulating samples.

The same experiment has been performed in n-CdMnSe (13) and n-CdMnS (14) alloys. Typical results are shown in Fig. 2. At first sight, results in the magnetic field range 0.1 to 2 Tesla merely illustrate, through a non-resonant technique, the giant spin-splitting of conduction band- and related states. On the contrary, in the low-field range, the donor electron spin-splitting does not vanish in contrast to the bulk magnetization deduced from exciton splitting. This reveals a local non-zero magnetization in the donor orbit, which is interpreted as due to the formation of a donor BMP.

The existence of an unpaired electron with non-vanishing electronic density $|\phi(\vec{r})|^2$ in the region surrounding an impurity causes each ith Mn^{++} ion to be influenced by exchange energy $H_i = -\alpha|\phi(\vec{R}_i)|^2\, \vec{\sigma}\cdot\vec{S}_i$. It results in a mean magnetization which, in turn, induces a non-zero donor electron spin splitting. Obiviously, the BMP energy E_p essentially depends on α, x, T, and donor Bohr radius a_D. From these easily estimated quantities, one obtains thermal mean values $< S(\vec{R}_i) >$ and, from Eq. 2, an estimation of E_p, the mean value of H_{exc}. But a much more lucid approach is obtained by taking into account the effect of magnetic fluctuations on the donor orbit, as pointed out by Dietl and Spaek [15].

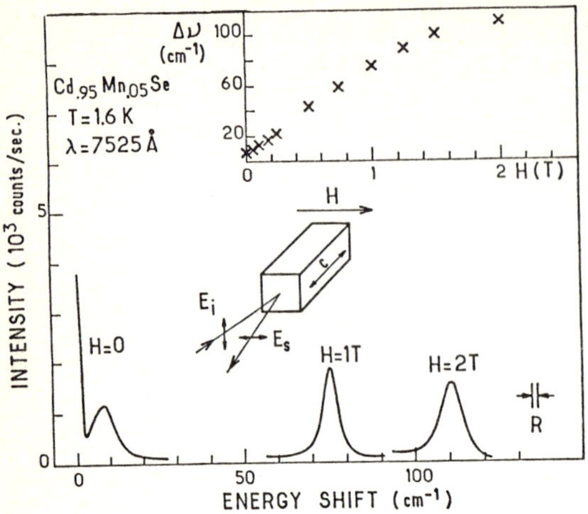

Fig. 2 Typical SFRS spectra from neutral donors.
Insert: the Stokes shift as a function of magnetic field.
Note the existence of a zero field magnetization in the donor orbit. [after ref. (13)]

4. Fluctuations in BMP

Actually, the number N of Mn^{++} ions involved in a BMP is far from being macroscopic ($N \sim 200$ in the case of $Cd_{0.95}Mn_{0.05}Se$). This has several consequences

a) SFRS by donors is more sensitive to composition fluctuations than, say, free exciton spin splitting. Free excitons integrate magnetization over their wave function volume, which is, practically, reduced by their mean free path. Thus, inhomogeneous broadening is expected for donors, growing with the Stokes shift and of the order $(\sqrt{N}/N)\,\Delta(H)$. It is observed in experimental results (see Fig. 2) but a 5 cm^{-1} zero field linewidth must be explained otherwise

b) Neglecting composition fluctuations the magnetization from one BMP to another fluctuates around its thermal average: due to the reduced number of Mn^{++} involved, the free energy of the coupled M^{++}-carrier system is not strongly peaked around its minimum value. This accounts for a zero-field linewidth, of the order of $(E_pkT)^{1/2}$ [15], *which remains constant in the high temperature range.*

c) The Mn^{++} ions in the BMP are coupled to the impurity carrier via an isotropic exchange Hamiltonian. In the simplest case of purely paramagnetic ions, and of a "square" wave function ($|\phi(R)|^2 = 1/\pi a^3$ inside a volume πa^3), it reduces to

$$(5)\quad H_{exc} = -\frac{\alpha}{\pi a^3}\vec{\sigma}\cdot\sum_{i=i}^{N}\vec{S}_i = -\frac{\alpha}{2\pi a^3}\left[(\vec{\sigma}+\Sigma\vec{S}_i)^2 - (\Sigma\vec{S}_i)^2 - \sigma^2\right]$$

This formulation implies that the energy states may be labelled by the $(\Sigma S_i)^2$ quantum number J; the density of states as a function of energy, which is easily calculated for small N, shows a minimum at zero energy. In the classical limit, it is possible to calculate the magnetic entropy as a function of magnetization for greater N: this problem reduces to a random walk problem in 3 dimensional space. According to the number of magnetic ion states with $|\Sigma\vec{S}_i| = s$ is

$$P(s) \propto (s^2/S^2)\exp(-3s^2/2NS^2)$$

which is valid for $s/SN \ll 1$

As a function of magnetization $\vec{m} = g\mu_B\vec{s}$, one obtains

$$(6)\quad P(m) \propto m^2 \exp(-m^2/2kT\chi)$$

where χ is the low field susceptibility: $\chi = Ng^2\mu_B^2S^2/3kT$. Note that P(m) vanishes at zero magnetization. This suggests how the classical gaussian approximation has to be modified in the case where the variable of interest is $|\vec{M}|$ rather than M_z as in a Zeeman Hamiltonian. This results in a mean non-zero carrier spin-splitting even in the high temperature range $kT > E_p$. It is found also to be of the order $[E_pkT]^{1/2}$ [15].

Coming back to SFRS experiments, it is worth noting that we must introduce an adiabatic approximation neglecting all changes in the Mn^{++} magnetization during the spin-flip. We also note that we get a Stokes or anti-Stokes shifts depending on the

σ quantum number with respect to $\Sigma \vec{S}_i$, which completes the determination of $(\vec{\sigma} + \Sigma \vec{S}_i)^2$ eigenstates (see Fig. 3). Thus, the SFRS line should exactly reproduce the energy distribution of the coupled Mn^{++} -electron system, if the BMP can be considered to be in thermal equilibrium and if non-diagonal terms of the electron spin density matrix may be neglected (which means that the electron spin is well quantized along $\Sigma \vec{S}_i$).

In the more realistic case of an hydrogenic donor wave function, we may, following (15), parametrize the local magnetization by

(7) $m(\vec{r}) = \vec{\eta} \exp(-2r/a)$ far from saturation. After straight forward integration, the SFRS zero-field line shape, which is twice the exchange induced donor energy, comes out as

(8) $P(\Delta,\sigma) \propto \Delta^2 \exp(-\Delta^2/8 E_p kT) \exp(\sigma\Delta/kT)$

where Δ is the Raman shift and $\sigma = \pm 1/2$ for the Stokes and anti-Stokes part of the line, respectively. On Figure 4, comparison is made with a "high" temperature experiment, below the donor ionization temperature. It is shown that the effect of magnetic

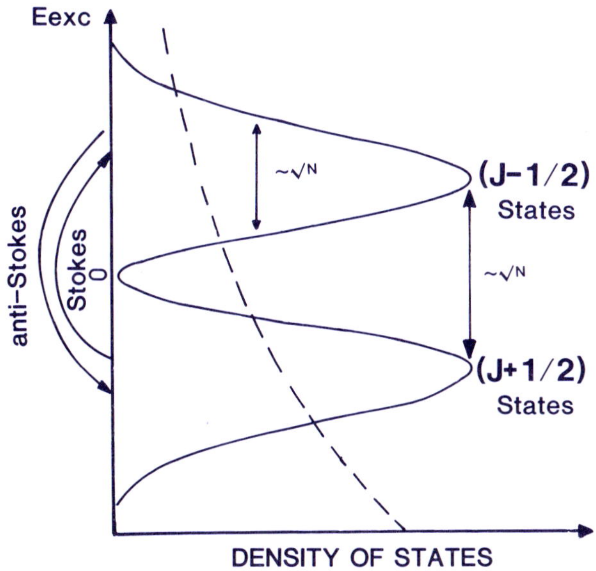

Figure 3 The density of states of the exchange hamiltonian $H_{exc} = -\vec{\sigma}\cdot\Sigma \vec{S}$ (continuous line). Energies E_{exc} depends on J and J $\pm$ 1/2 quantum numbers. In the classical limit, J and J + 1 become proportional to $|\vec{M}|$ (see eq. 5 and 6 in the text). These states are populated through a Boltzmann factor exp $(-E_{exc}/kT)$ (dashed line). Stokes and anti-Stokes transitions are suggested on the left part of the figure.

fluctuations (finite linewidth and finite mean spin splitting at high temperature) are accessible by SFRS experiments in the case of donors. The only parameter used for the fitting procedure is the high temperature constant E_p. The Mn^{++} mean spin value in the donor orbit is of the order of 10^{-2}. In material with 5 % Manganese content the number of Mn^{++} involved in the BMP is of the order of 10^2. Thus, one can imagine a donor-BMP as consisting of a few hundred loosely coupled Mn^{++}ions.

5. Acceptor-BMP

Due to heavier effective masses in the valence bands, the physical situation is quite different in the case of neutral acceptors: the typical Bohr radius ratio a_D/a_A being of the order of 5 in II-VI compounds, one expects that the saturation may be reached in the bound hole orbit, and that the fluctuation effects, which are related to the average number of Mn ions involved, are much more important than in the donor case.

The experimental determination of acceptor-BMP energies is actually more difficult. To our knowledge, SFRS experiments in p-type SMSC have up to now not been successful. This may partly be due to huge fluctuation broadening, and also due to lack of tetrahedral symmetry of A° centers, a situation frequently suspected in pure II-VI compounds [18] as well as in alloys [19]. However, the acceptor ground state energy has been

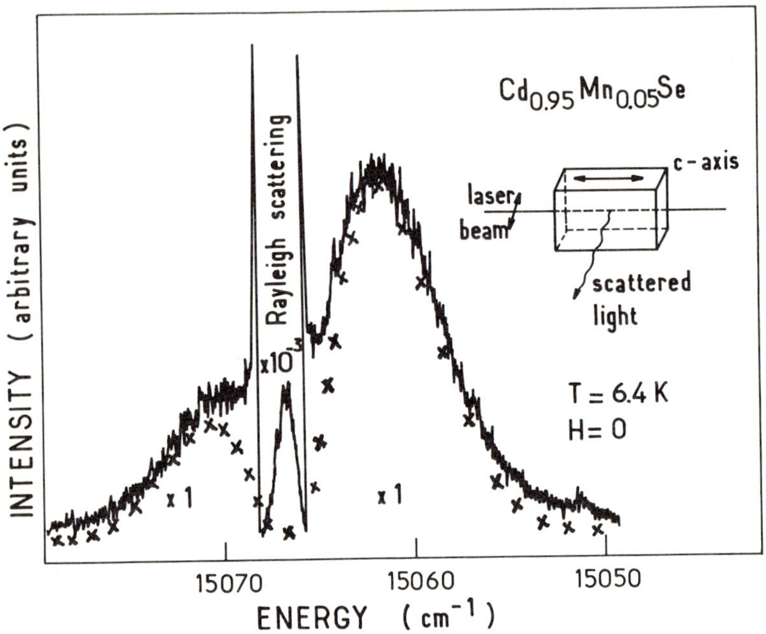

Figure 4 Experimental SFRS spectra in CdMnSe at "high" temperature, when the spin-splitting is close to its high temperature limit $(8\ E_p\ kT)^{1/2}$. Laser energy is 15070 cm^{-1}. The theoretical fit (x) is obtained from eq. (8) in the text. [after ref. 17].

measured, in p-type CdMnTe alloys, through analysis of donor-acceptor pair recombination (20). This technique is sensitive to composition fluctuations through the energy gap dependence on composition and to random pair distribution. Thus, inhomogeneous broadening of the acceptor energy is more important and the attempt to get precisely the exchange-induced energy distribution is hopeless. However, the experimental mean values, which are shown in Figure 5, may be fairly well accounted for by the theoretical approach proposed in eq. (6) and (8) providing the magnetization is parametrized in a way different from eq. (7). As a matter of fact, saturation of magnetization is expected, and it leads to the crude assumption that $m(\vec{r})$ is saturated in a volume $\pi\gamma^3 a_A{}^3$ around the impurity, then proportional to the hole density in the outer volume. In addition, it seems uncertain whether the bound hole wave function is well accounted for by the effective mass theory, and whether the local magnetization is described correctly by a continuous function. As a consequence, we may conclude that our knowledge of acceptor-BMP is poorer than the one of donor-BMP, from both experimental and theoretical points of view.

However, several features may be clearly stated: - Saturation of magnetization is reached, at low temperature, in the whole Bohr orbit; the effect of increasing temperature is to reduce the saturation volume. Thus, we may describe an acceptor-BMP as a ferromagnetic bubble in a paramagnetic bulk, which includes typically 8 Mn^{++} in x=5% compounds, at 1,6 K.

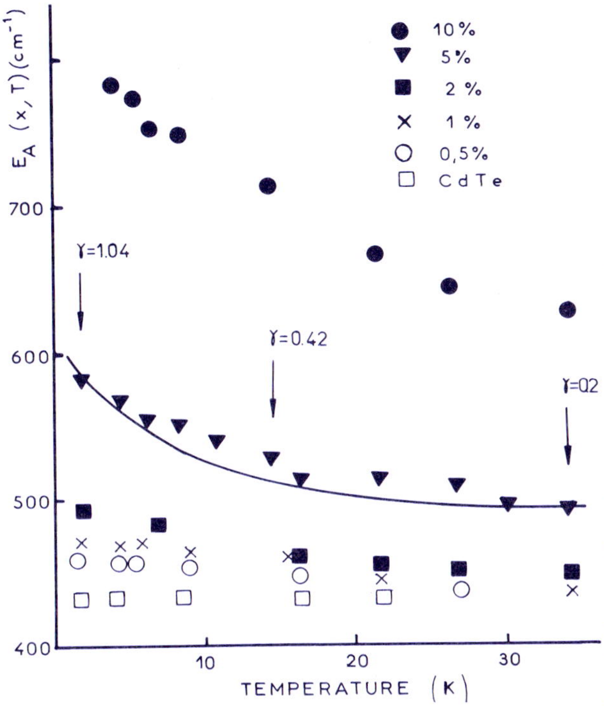

Figure 5 Acceptor binding energy in CdMnTe as a function of temperature (10 % results are poorly reliable), obtained from donor-acceptor recombination. The continuous line is a theoretical fit. Typical values of γ are pointed by arrows (see text) [after ref. 20]

- Fluctuation effects are quite significant, leading to large exchange energies at high temperature (some 40 cm^{-1} in $Cd_{0.95}Mn_{0.05}Te$).

- At low temperature, the BMP energy may reach 50 % of the orbital binding energy (in x = 5 % compounds). This raises the question of a temperature dependant electronic wave function, which has been considered in ferromagnetic semiconductors (3) as well as in SMSC (15). Up to now, there is no experimental evidence for such a situation in SMSC. Actually, the indirect exchange coupling between Mn ions, which is of antiferromagnetic sign, probably reduces the chances to observe this effect in SMSC.

6. Giant spin effects

The BMP may be considered as a localized "bubble", with weak (donor case) or strong (acceptor case) coupling between localized moments. In both situations it carries a giant moment, typically 10 times larger than an isolated ion. Although somehow qualitative, this approach enlightens the physical meaning of various experiments.

In $Cd_{0.95}Mn_{0.05}Te$ alloys, the presence of an acceptor-BMP has been originally detected (21) by an unusual temperature shift of the bound exciton luminescence line A°X and by its polarization behaviour in weak external fields (see Fig. 6).

Although the free exciton line X splits into two Zeeman-like components σ^+ and σ^-, as expected, the A°X line behaviour is better understood: Bound excitons are thermalized along the acceptor-BMP directions, which are randomly distributed in zero external field. A weak field does not significantly modify this picture, but aligns the giant spin associated to A°-BMP, which explains the observed polarization differential spectrum of A°X line.

More quantitative results may be obtained in the case of donors, owing to the use of SFRS selection rules [17]. They are strongly dependent on the excitation energy, through the relative influence of virtual intermediate states. Close to the resonance with excitons formed from the highest valence bands, it is possible to get a SFRS

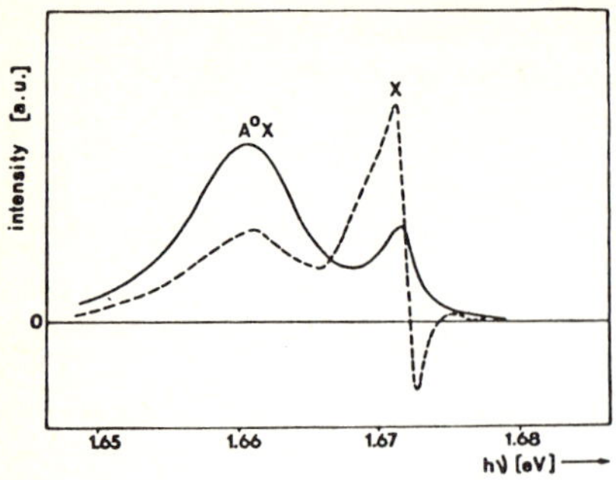

Figure 6 Luminescence spectrum and differential polarization spectrum $(\sigma^+ - \sigma^-)$ of $Cd_{0.95}Mn_{0.05}Te$ at 15 K and in a 0.1 T magnetic field. Note the different behaviour of X (free exciton) and A°X (bound exciton) lines. [after ref. 21]

strongly dependant on circular polarization in high fields, when all BMP are presumably aligned along the external field: a 75 % degree of circular polarization of the scattered light is reached. As a consequence, the SFRS degree of circular polarization measures the average alignement of BMP in weaker fields. Experimental results are shown in Figure 7.

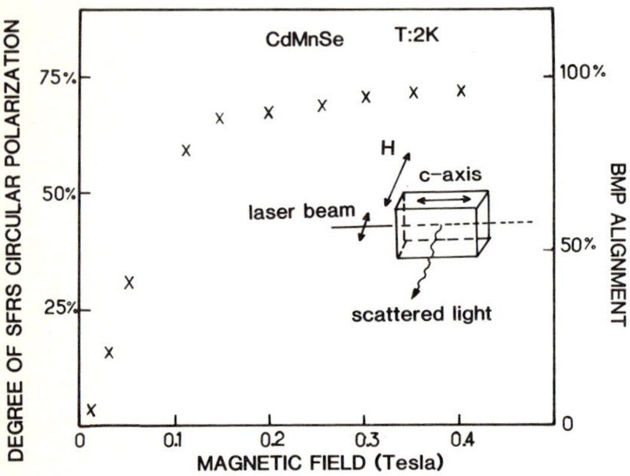

Figure 7 Degree of circular polarization of SFRS in $Cd_{0.95}Mn_{0.05}Se$ at T = 2K in weak external field. Laser energy 15070 cm^{-1}. From high field data, a 75 % degree of polarization is assumed to correspond to 100 % BMP alignment. Note the magnetic field scale [after ref. 17]

The theoretical treatment of this problem has been using the statistical approach of eq. (8), taking into account the external field as in [15]. It is thus easily shown that the degree of BMP alignment follows a Langevin function $L(x) = \mathrm{ctnh}(x) - \frac{1}{x}$:

$$< \cos\phi >= L(g\mu_B\ S_T\ H/kT)$$

where S_T is the BMP total spin.

Taking into account additional selection rules (such as SFRS is forbidden if $S_T /\!/$ c-axis) somehow complicates the result, but the essential feature remains: saturation of these giant spins is attained at typically, 0.1 Tesla at helium temperature, which is one order of magnitude smaller than for isolated paramagnetic ions.

7. Concluding remarks

In wide gap SMSC, optical spectroscopy is a useful tool for the study of BMP. This contrasts with the situation in ferromagnetic semiconductors [3] where the existence of BMP accounts for surprising transport properties. In this latter case, the exchange induced binding energy plays a dominant role in the formation of this complex with respect to the binding on impurities.

Transport measurements can also reveal the influence of BMP in wide-gap SMSC, as shown by Antoszewski et al. [22]. Such works should provide interesting information in the future, especially in the case of the acceptor BMP which remains noticeable in the intermediate temperature range (T ∿ 50 K), where spectroscopic techniques are less accurate.

Acknowledgements

Many works presented in the present paper result from a long standing collaboration between the author's laboratory on one hand, the University of Warsaw and the Polish Academy of Science on the other hand. I would like to thank all colleagues for fruitful cooperation. For the most recent period, enlightening discussions are acknowledged with C. Benoit à la Guillaume, G. Fishman and T. Dietl.

References

[1] R. GALAZKA, Proc. XIV Int. Conf. Phys. Semicond. Edinburg (1978). Inst. Phys. Ser. N° 43 (1979) p. 133

[2] General recent reviews on SMSC are by
- J. GAJ, Proc. XVth Int. Conf. Phys. Semicond. Kyoto 1980 J. Phys. Soc. Jpn. Suppl. A 49 (1980) p. 797
- M. GRYNBERG, Proc. XVIth Int. Conf. Phys. Semicond. Montpellier 1982. In press.

[3] - T. KASUYA and A. YANASE, Rev. Mod. Phys. 40, 684 (1968)
- P. LEROUX-HUGON: Proc. Disc. Meet. Magn. Semicond., Ed. W. Zinn North-Holland Publ. Co Amsterdam (1976)
- P. KUIVALAINEN, J. SINKKONEN, K. KASKI and T. STUBB, Phys. Stat. Sol. (b) 94, 181 (1979)

[4] - J. KOSSUT, Phys. Stat. Sol. (b) 78, 537 (1976)
- G. BASTARD, C. RIGAUX and A. MYCIELSKI, Phys. Stat. Sol. (b) 79, 585 (1977)

[5] - J. GAJ, R. GALAZKA and M. NAWROCKI, Sol. Stat. Comm. 25, 73 (1978)
- H. KETT, W. GEBHARDT, U. KREY and J.K. FURDYNA, J. Magn. Magn. Mat. 25, 215 (1981)

[6] - J. GAJ, P. BYSZEWSKI, M. CIEPLAK, G. FISHMAN, R. GALAZKA, J. GINTER, M. NAWROCKI, NGUYEN THE KHOI, R. PLANEL, R. RANVAUD and A. TWARDOWSKI, in ref. 1 p. 1113

[7] See the review by J. GAY in Ref. 2

[8] J. GAJ, R. PLANEL and G. FISHMAN, Sol. State Comm. 29, 435 (1979)

[9] - R. GALAZKA, S. NAGATA and P.H. KEESOM, Phys. Rev. B22, 3344 (1980)
- M. ESCORNE and A. MAUGER, Phys. Rev. B 25, 4674 (1982)

[10] C. LEWINER and G. BASTARD, J. Phys. C13, 2347 (1980)

[11] C. LEWINER, J. GAJ and G. BASTARD, J. Phys. C5, 289 (1980)

[12] D.G. THOMAS and J.J. HOPFIELD, Phys. Rev. 175, 1021 (1968)

[13] M. NAWROCKI, R. PLANEL, G. FISHMAN and R. GALAZKA, Phys. Rev. Lett 46, 735 (1981)

[14] D.D. ALOV, S.I. GUBAREV, V.B. TIMOFEEV and B.N. SHEPEL', Pis'ma Zh ETF 34, 76 (1981) [JETP Lett. 34, 71 (1981]

[15] T. DIETL and J. SPALEK, Phys. Rev. Lett. 48, 355 (1982)

[16] This elementary result may be found in Dr. WATSON: Theory of Bessel functions ed. Cambridge Univ. Press (1966) p. 419

[17] R. PLANEL, TRAN HONG NHUNG, M. NAWROCKI and G. FISHMAN, to be published

[18] J. DESSUS, LE-SI-DANG, A. NAHMANI and R. ROMESTAIN, Sol. State. Com. 37, 689 (1981)

[19] C.E.T. GONCALVES DA SILVA, Rev. Brasileira Fisica, 11, 231 (1981)

[20] TRAN HONG NHUNG and R. PLANEL, Proc. XVIth Int. Conf. Phys. Semicond. Montpellier 1982 - In press

[21] A. GOLNIK, J. GAJ, M. NAWROCKI, R. PLANEL and C. BENOIT A LA GUILLAUME Proc. XVth Int. Conf. Phys. Semicond. Kyoto 1980, J. Phys. Soc. Jpn Suppl. A49, 814 (1980)

[22] J. ANTOSZEWSKI, U. DEBSUA, M. DIETL, E. JANIK, J. JAROSZYNSKI and M. SAWICKI, Proc. 10th Polish Sem. Semicond. Comp. Jaszowiec (1980)

"SPIN-DOPING" IN SEMICONDUCTOR PHYSICS

Michael von Ortenberg
Max-Planck-Institut für Festkörperforschung
Hochfeldmagnetlabor Grenoble

F-38042 Grenoble, France

1. INTRODUCTION:

The development of semimagnetic semiconductors gave rise to the observation of many interesting phenomena, because for the first time the usually rigid energy-band structure had become tunable by externally controlled parameters as temperature and magnetic field [1]. This unique possibility was achieved by the solid solution of paramagnetic ions M especially in II-VI-compounds $A_{1-x}M_xB$. The localized spin of the paramagnetic impurity couples strongly with the spin of the quasi-free carriers by the exchange interaction, which can be described by the following effective Heisenberg hamiltonian [2] :

$$\mathcal{H} = xJs_z\langle S_z\rangle \qquad (1)$$

Here x is the molar fraction of the paramagnetic material in the ternary mixed-crystal system, J is the exchange coupling constant and describes in the case of Mn as paramagnetic ion the exchange interaction of the localized d-electrons with the quasi-free carriers. s_z is the quasi-free carrier spin, whereas $\langle S_z\rangle$ indicates the thermodynamical expectation value of the localized spins, which is usually an extremely sensitive function of both, temperature and external magnetic field. As result the energy levels of the quasi-free carriers are modified in such a way, that the electron g-factor depends now explicitely on the above exchange interaction [2] :

$$g = g_o + \frac{xJ\langle S_z\rangle}{\mu_B H} \qquad (2)$$

where g_o is the g-factor of the pure material AB without admixture of paramagnetic ions. H and μ_B are the external magnetic field and the Bohr magneton, respectively.

As a matter of fact, this modification of the g-factor is generally not the only change involved by addition of paramagnetic ions, but there is also a more or less pronounced change of the other energy-band parameters, as gaps and momentum matrix elements due to the alteration in the effective periodic potential. Especially the energy gaps are very sensitive to changes in the composition [3,4].

There are, however, special cases as discussed below in detail, where only the change in the g-factor with respect to the reference material becomes effective and all other energy-band parameters can be treated to very good approximation as being unchanged. In these special cases we call the admixture of paramagnetic ions "spin-doping". In the following sections we discuss effiency and possibilities of "spin-doping" in semiconductor physics.

2. "SPIN -DOPING" AS A NEW TOOL IN THE ENERGY-BAND-STRUCTURE DETERMINATION

The special effiency of "spin-doping" as a new tool in the energy-band-structure determination applying submillimeter-magneto spectroscopy was pointed out by us in a recent communication [5]. This kind of magneto spectroscopy provides detailed information on the intraband-energy separation of the Landau levels in the presence of an external magnetic field. For the interpretation, however, it is absolutely necessary, that the initial and final state of the resonance transition are unambiguously identified. This identification is often somewhat arbitrary and for more complicated band structures there is generally no way to discriminate between a combined-spin flip and a pure spin flip as schematically indicated in fig. 1. Polarization selection rules are easily broken and also the study of the temperature dependence of the resonance lines will usually not succeed in proper identification, since both, cyclotron- and spin-split energy exhibit in canonical semiconductors the same temperature behaviour. In such a situation "spin-doping" can be applied to overcome the problems. In this case we mean by "spin-doping" the substitution of only a very small amount of crystal constituents of the material in question by paramagnetic ions. For very small values of the composition parameter x only the g-factor of the material is affected and the influence on the other characteristic quantities of the energy-band structure is negligible. For the mole fraction x of paramagnetic ions M in the ternary mixed crystal $A_{1-x}M_xB$ of the order $x \approx 10^{-3}$ the energy gaps and momentum matrix elements are to very good approximation those of the undoped reference material AB [3]. Since the quantity $\langle S_z \rangle$ in equ. (2) for the g-factor is usually a very sensitive function of the temperature, the temperature control provides a direct means of controlling the spin-splitting. It should be noted that for very high temperature the average polarization $\langle S_z \rangle$ vanishes, so that in this limit the band structure merges into that of the undoped material.

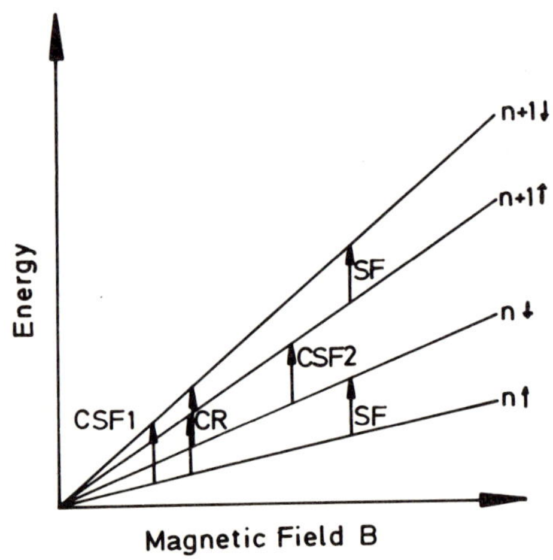

FIG. 1 The scheme for different magnetoresonances shows that the pure spin-flip (SF) and combined spin-flip transitions (CFS2) may have about the same resonance position and are then difficult to distinguish.

Due to the temperature dependence of the effective g-factor the three different intraband-resonance transitions exhibit a characteristic temperature behaviour: the cyclotron resonance should in this approximation be not affected by the temperature, whereas the CSF2-transition should exhibit the opposite sign in the temperature shift in comparison with the CSF1- and SF-transition.

As an example of the effiency of the "spin-doping" method we refer to the results obtained on HgSe ↔ $Hg_{1-x}Mn_xSe$ respectively. For HgSe there was some discussion concerning the correct assignment of spin- and combined spin-flip transition observed in submillimeter-magneto spectroscopy [6,7]. We investigated therefore $Hg_{1-x}Mn_xSe$ for x = 0.0015, that means for only a very small Mn-composition. The experimental set-up applied the strip-line technique in submillimeter-magneto spectroscopy [8]. Utilizing several wavelengths varying from 70.6 μm to 570 μm we were not only able to follow the frequency dependence of the observed resonance lines, but also to record in the long-wavelength limit the optical Shubnikov-de Haas effect, which provides the necessary information for the carrier concentration and hence the Fermi energy in the sample [8]. The key experiment, however, was the investigation of the temperature dependence of the observed resonance lines as demonstrated for 170.6 μm - wavelength radiation in fig. 2. The opposite shift of the α- and β-lines proves de-

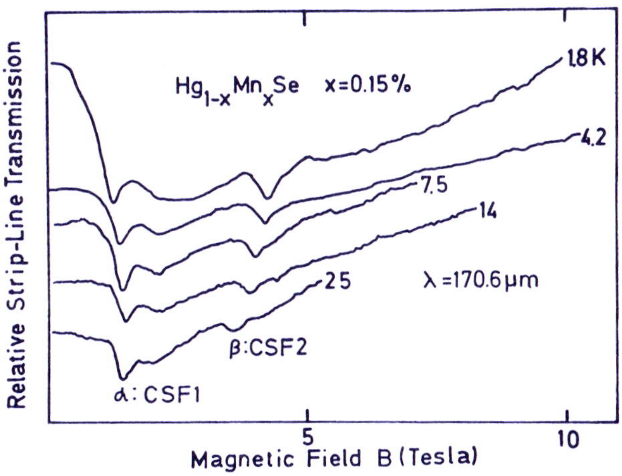

Fig. 2 The temperature dependence of the observed resonances in $Hg_{1-x}Mn_xSe$, for x = 0.0015, proves clearly that the α and β lines are caused by the CSF1 and CSF2 transitions, respectively. (after ref. [5])

finitely that the β-line is caused by a CSF2-transition: $n\uparrow \rightarrow n+1\downarrow$, whereas the α-line is due to a CSF1-transition: $n\downarrow \rightarrow n+1\uparrow$. Actually by raising the temperature from 1.8 K to 25 K the effective g-factor has been decreased from 42 to 31, which is about a 25% effect despite the very small composition parameter of only x = 0.0015. This fact demonstrates the importance of the interaction term of equ. (1) in the context of the $\vec{k}\cdot\vec{p}$-hamiltonian. We have applied a modified Pidgeon-Brown model to calculate the scheme of Landau levels in $Hg_{1-x}Mn_xSe$. The results are plotted in fig. 3. The levels were calculated for Mn-doped material (x = 0.0015 solid curves) and pure HgSe (x = 0 broken curves) using the parameters of Dobrowolska et al. for a temperature of 4.2 K [9]. The solid arrows represent the highest possible observed transitions near the Fermi energy (dashed horizontal line) for 170.6 μm and 163 μm wavelength radiation, respectively. The open arrows indicate the magnetic field position of the corresponding spin-flip transition, which is astonishingly not obser-

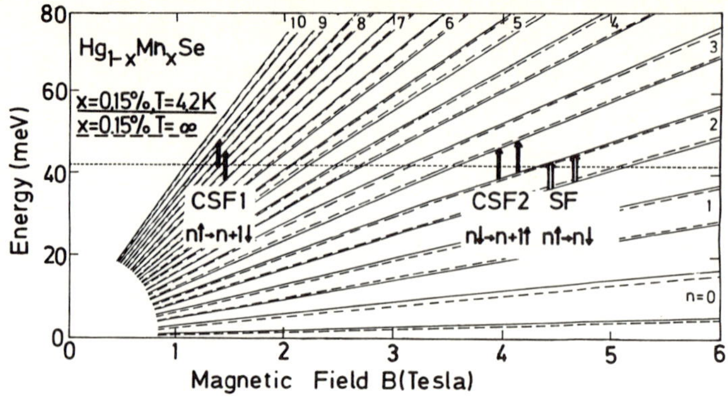

Fig. 3 The scheme of Landau levels fo $Hg_{1-x}Mn_xSe$ (x = 0.0015, T = 4.2 K)as plotted by the solid curves has only a slightly larger spin-splitting than that for pure HgSe represented by the broken curves. The full arrows indicate the observed transitions of 170.6 µm and 163 µm wavelength radiation near the Fermi energy of ζ= 43 meV. The pure spin-flip, as indicated by the open arrows, was not observed in our experimental results. (after ref. [5])

ved in our experiment. It should be noted that both, SF- and CSF2-transition, are not allowed in a simple isotropic model, but require an additional perturbation mechanism, which breaks the angular momentum selection rule.

3. THE "SPIN-HETERO JUNCTION"

The incorporation of paramagnetic ions in a mixed crystal system allows to vary the g-factor of the quasi-free carriers by external control. The range of this variation is, of course, proportional to the composition parameter x (equ. (2)). On the one hand it is desirable to have large and pronounced effects due to the exchange interaction of equ. (1), on the other hand secondary effects as non-negligible changes of the band gaps will enter, if the composition parameter is too much increased. This shortcoming, however, can be overcome if the reference material is chosen in such a way, that it is compatible with the unavoidable changes in the band gaps. For "spin-doped" mixed crystals $A_{1-x}M_xB$ the suitable reference material is $A_{1-y}D_yB$, where D indicates a nonmagnetic dopant. Such complementary materials are for example mercury-manganese-selenide and mercury-cadmium-selenide. In fig. 4 we have plotted the dominant energy gap of these materials as function of the composition parameter [3,4]. It is quite evident that manganese as dopant is about twice effective in changing the energy gap in comparison with cadmium. Actually for equal energy gap E_o we obtain the condition:

$$y \approx 2.08x \tag{3}$$

if x and y are the composition parameters for mercury-manganese-selenide and mercury-cadmium-selenide, respectively. Complementary materials in the sence of "spin-doping" are then $Hg_{0.99}Mn_{0.01}Se$ and $Hg_{0.98}Cd_{0.02}Se$, where only manganese acts as "spin-dopant", but cadmium has no influence on the electronic g-factor of the system. If we compare the physical properties of these two materials, we notice only the effect of "spin-doping".

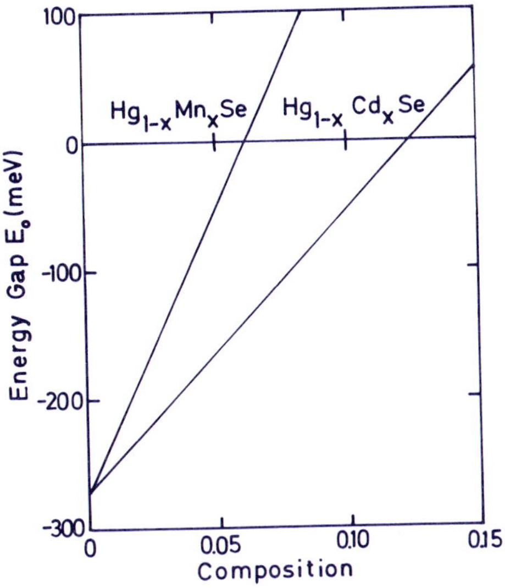

Fig. 4 The change in the negative energy gap E_o in $Hg_{1-x}Mn_xSe$ is about twice as pronounced than in $Hg_{1-y}Cd_ySe$ for varying composition parameter x. (after ref. [3] and [4])

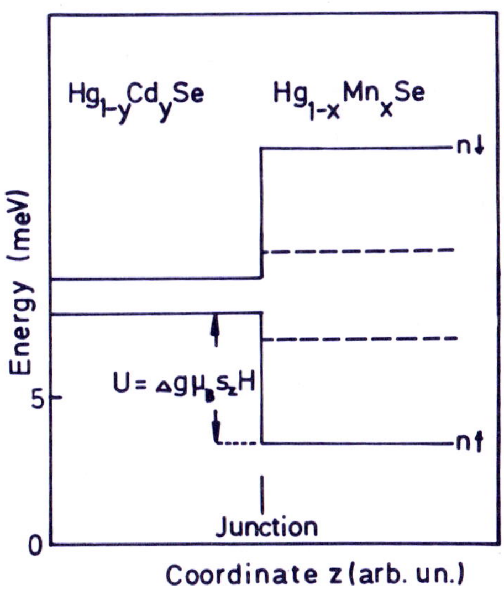

Fig. 5 At the "spin-hetero junction" of $Hg_{1-x}Mn_xSe$: $Hg_{1-y}Cd_ySe$ interface with x = 0.01 and y = 0.02 a potential barrier with $U = \Delta g\mu_B s_z H$ becomes effective. Due to the pronounced temperature dependence of the g-factor the potential step can be tuned by temperature from U = 4.4 meV at T = 1.8 K to U = 0.9 meV at T = 25 K and an external magnetic field of B = 4 Tesla.

The combination of such two complementary materials at an internal semiconductor interface leads to a completely novel kind of hetero junction. For an external magnetic field H perpendicular to the interface the effective hamiltonian for the motion parallel to the field in z-direction is given by the following expression:

$$\mathcal{H} = E_o + \hbar\omega(n+1/2) + \mu_B g(z) s_z H + p_z^2/2m \qquad (4)$$

where the effective g-factor is a function of the coordinate z. We like to point out again, that the band edge energy E_o as well the effective mass values have been set equal for both components of the hetero junction. The effective potential for the motion in z-direction is step-like and its amplitude is determined only by the variation of the g-factor:

$$U = g\mu_B s_z H = xJs_z\langle S_z\rangle \qquad (5)$$

Since only the g-factors differ in both materials, but the energy gaps and hence the mass values are equal, we have an externally tunable "spin-hetero junction" as shown schematically in fig. 5. We have assumed an external magnetic field of B = 4 Tesla and a temperature of T = 1.8 K (solid lines) resulting in a potential step of U = 4.4 meV as extrapolated from our previous data [5]. For larger "spin-doping" the potential amplitude should be correspondingly increased. Raising the temperature from t = 1.8 K to T = 25 K the amplitude is considerably decreased (broken lines) and finally vanishes for very high temperatures. This means that in this high temperature limit we have a homogeneous medium with respect to the electron transport properties. It should be noted that the potential step across the junction depends on the spin polarity and is opposite for "spin-up" and "spin-down" states. This fact implies a spin-density modulation across the junction, which is, however, not accompanied by a charge-density modulation.

4. THE "SPIN-SUPER LATTICE"

Superlattices are periodically arranged layers of semiconductor materials having different energy-band parameters with a periodicity much greater than the fundamental lattice constant [10]. The superlattice periodicity leads to the formation of different subbands separated by mini gaps. The energy position of the subbands and the width of these gaps depend directly on the effective lattice potential. Up to now the parameters of this potential were essentially set during the growing process of the superlattice, usually by molecular-beam epitaxy. Afterwards only minor changes could be obtained by space charge effects. The possibility of a tunable "spin-hetero junction" leads consequently to a tunable "spin-super lattice".

We have proposed a periodic arrangement of $A_{1-x}M_xB$: $A_{1-y}D_yB$ double layers in a superlattice, where only the electron g-factor has a spatial periodic modulation, essentially a "spin-super lattice" [11]. In fig. 6 we have plotted in consequent extrapolation of the above concept using the same parameters the effective potential U(z) attached to the different spin-split Landau levels for electrons in a $Hg_{0.99}Mn_{0.01}Se$: $Hg_{0.98}Cd_{0.02}Se$ superlattice. In a microscopic picture the dimensions of the superlattice constant a should be large in comparison with the average Mn - Mn distance, otherwise the effectice potential looses its step-like character.It should be noted again that for "spin-up" and "spin-down" states the effective potential has reversed modulation. For such a superlattice with layers of equal thickness and a superlattice constant a $\approx$ 97 Å - equivalent to 16 fundamental lattice cells - we have plotted in fig. 7 the Kronig-Penney energy-wavenumber relation near the boundary of the superlattice Brillouin zone as indicated by the vertical arrow at about $k_{Br} = 3.2\cdot10^6$ cm^{-1} [12]. At the zone boundary we obtain clearly a mini gap, which can be tuned externally by temperature. The value of the mini gap is of that order of magnitude, that submillimeter spectroscopy becomes applicable for the excitation of inter-mini gap transitions. For this purpose the external magnetic field should be tuned in such a way, that the Fermi energy lies within one of the mini gaps attached to the different Landau levels.

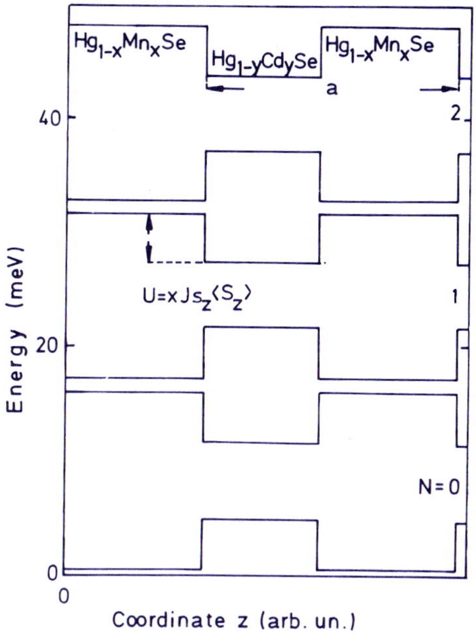

Fig. 6 In a $Hg_{1-x}Mn_xSe$: $Hg_{1-y}Cd_ySe$ superlattice the motion perpendicular to the layers (z-direction) is affected only by the spatial variation of the g-factor, so that the amplitude U of the effective potential can be tuned by an external magnetic field and temperature. The parameters used are B = 4 Tesla and T = 1.8 K. (after ref. [11])

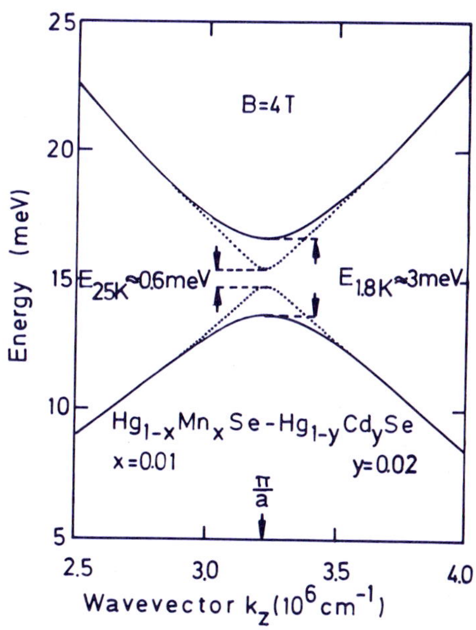

Fig. 7 The mini gap at the zone boundary of a $Hg_{0.99}Mn_{0.01}Se$: $Hg_{0.98}Cd_{0.02}Se$ spin-super lattice can be tuned from 0.6 meV to 3 meV decreasing the temperature from 25 K to 1.8 K, respectively. (after ref. [11])

From the technical point of view the growing of "spin-super lattices" using semimagnetic materials should not envisage too large difficulties, because the lattice constant of both semiconductors $Hg_{0.99}Mn_{0.01}Se$ and $Hg_{0.98}Cd_{0.02}Se$ match within 10^{-2}% [13]. We hope that the experimental realization of the "spin-super lattice" by molecular-beam epitaxy will be achieved in the near future, so that the interesting features of this new kind of superlattice can be investigated by many different experimental techniques.

5. SUMMARY

"Spin-doping" provides externally controllable variation of the electron g-factor of the quasi-free carriers in semiconductors without affecting the other energy-band parameters. This method is not only a useful tool in the analysis of optical transitions, but opens new aspects in spin physics, which were not possible before.

6. REFERENCES

1 see for references in the review articles by
R.R. Galazka and J. Kossut
Lecture Notes in Physics 133 "Narrow Gap Semiconductor Physics and Applications", Proceedings Nîmes 1979, Springer Verlag Berlin Heidelberg New York, p. 245 (1980)
T. Dietl
Springer Series in Solid State Sciences 24 "Physics in High Magnetic Fields" ed. S.Chikazumi and N.Miura, p. 334 (1981)

2 J. Kossut
Solid State Commun. 27, 1237 (1978)

3 M. Dobrowolska, W. Dobrowolski, R.R. Galazka and A. Mycielski
phys. stat. sol. b105, 477 (1981)

4 A. Mycielski, J. Kossut, M. Dobrowolska and W. Dobrowolski
J. Phys. C Solid State Phys. 15, 3293 (1982)

5 W. Dobrowolski, M. von Ortenberg, J. Thielemann and R.R. Galazka
Phys. Rev. Letters 47, 541 (1981)

6 K. Schwarzbeck and M. von Ortenberg
Proc. "14th Int. Conf. of Semiconductors", Edinburgh 1978, ed. BLH Wilson, Conf. Series No 43, p. 1125

7 K. Pastor, M. Jaczyinski and J.K. Furdyna, Proc. "15th Int. Conf. of Semiconductors", Kyoto 1980, J. Phys. Soc. Japan 49 Suppl., p. 770 (1980)

8 M. von Ortenberg
"Infrared and Millimeter Waves", Vol. 3, ed. K.J. Button, Academic Press, p. 275 (1980)

9 M. Dobrowolska, W. Dobrowolski and A. Mycielski
Solid State Commun. 34, 441 (1980)

10 L. Esaki and R. Tsu
IBM J. Res. Rev. 14, 61 (1970)

11 M. von Ortenberg
Phys. Rev. Letters 49, 1041 (1982)

12 R. de L. Kronig and W.J. Penney
Proc. Royal. Soc. A130, 499 (1930)

13 W. Dobrowolski
private communication

Magnetically Stimulated Conductivity of Magnetic Semiconductor and Related Phenomena

K. Yamada, T. Koike, Saitama Univ., Urawa C., Saitama, Japan
H. Stubb, J. Heleskivi, J. Sinkkonen, Helsinki Univ. of Technology and Technical Research Center of Finland, SF02150 Espoo, Finland
W. Zinn, Inst. für Festkörperforschung der Kernforschungsanlage Jülich 5170 Jülich, Fed. Rep. of Germany

1. INTRODUCTION

In magnetic semiconductors the energies of localized states change with respect to the band edge in the paramagnetic region and in externally applied magnetic fields.[1]
Several theoretical models have been developed to account for the mechanism[2,3]. Also a lot of experimental results have been reported that show the changes of the ionization energies of localized states.[4,5,6]
The direct evidence of these phenomena has been reported[7] in the paper where not only thermally stimulated conductivity (TSC) but also magnetically stimulated conductivity (MSC) was found. In the case of MSC the ionization of trapped carriers is stimulated by a magnetic field controlled decrease of the ionization energy instead of an temperature increase as in the case of TSC.
In this paper the method of MSC is summarized and the experimental results of EuSe in paramagnetic temperature regions (T_N=4.6 K) are shown to reflect the magnetization dependence of transport phenomena in high magnetic fields up to 300 kOe.

2. EXPERIMENTAL DETAILS AND RESULTS

The EuSe single crystals were grown in sealed crucibles from a Eu-rich solution. As-grown crystals showed n-type conduction of the order of 1 to 10 $\Omega^{-1}cm^{-1}$ at room temperature and free carrier density of 10^{18} cm^{-3}. As-grown samples were studied as well as doped ones. In order to study the purity, an annealing procedure was carried out in an argon atmosphere at a temperature of 1600 C. After annealing the room temperature dark conductivity was about 10^{-9} $\Omega^{-1}cm^{-1}$. This high resistivity sample showed TSC and MSC mentioned above. However the samples which were hot-pressed before growing were found voidless. The volume ratio of the voids of two of samples differed by about a factor 1000.

The contacts to the samples were made with In using discharge method described elsewhere[8].
The TSC measurements were initiated by illuminating the sample with a He-Ne laser as intrinsic light (E_g=1.85 eV). After 30 min illumination the photocurrent fell to 10^{-13} A within several hours in the dark at a temperature of 10K. After this the sample was heated up with a constant speed of 0.5 K/min in different magnetic fields between 0 to 100 kOe.
In the MSC method the temperature was kept constant during the experiment. In contrast to TSC, the applied magnetic field was increased with a constant speed of 5 kOe/min from 0 to 90 kOe in MSC. Fig.1 shows the magnetically stimulated current buildup from 0 to several nA in which the applied magnetic field was stopped to increase after 10 min and held constant. In the 2nd run after several hours, the magnitude of the current was found to be much smaller than in the first one. The relaxation time constant of MSC was several hours at temperatures below 40K. From these results the analogy of MSC and TSC is established.
In Fig. 2 several profiles of TSC are shown. Curve a) shows a pho-

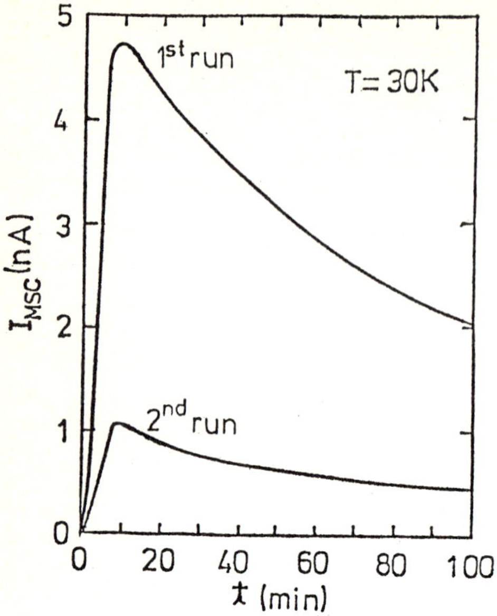

Fig. 1 Magnetically Stimulated current vs. time t under fixed temperature.

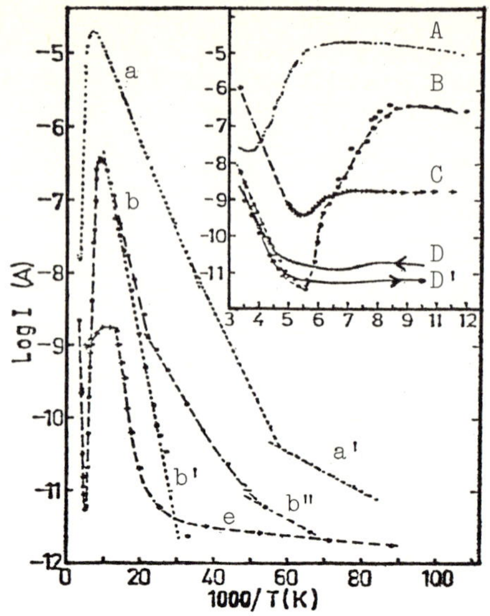

Fig. 2 Thermally stimulated currents and photocurrent vs. temperature.

tocurrent with strong illumination by a 15 mW He-Ne laser. Curve b-b' and b-b" are thermally stimulated currents in the magnetic field H=0 and H=100 kOe, respectively. Curve e shows also TSC of a voidless sample. An enlarged display at higher temperatures is shown in the insert of Fig. 2. Curve A and B correspond to a and b, respectively. Curve D-D' shows TSC of the hot-pressed sample in which no large scale voids were found.
It should be noted that the photocurrents and TSC have a similar form as indicated in the figures. Curves of type a were shown also in the papers[9,10] Penney and Kasuya[9] reported exactly the same characteristics.

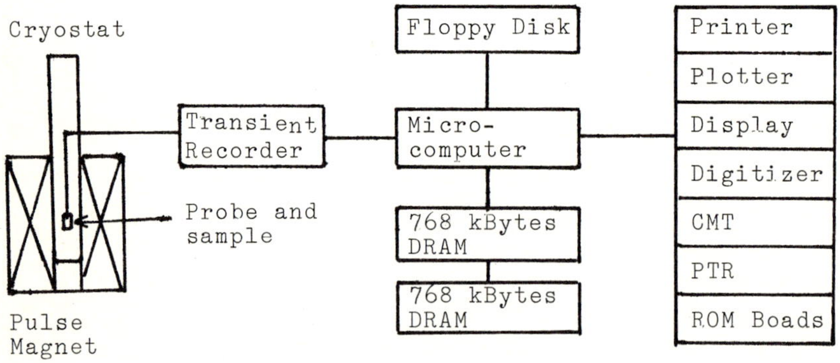

Fig. 3 Pulse Magnet and data Processing system.

Fig. 3 shows the experimental setup to measure the magnetization and the magnetoconductivity in pulsed magnetic fields up to 300 kOe. The number of data per shot reached 16 kwords and one set of data was derived from 3 pulses. The first one is the + direction of the magnetic field and the second its - direction. The third one has the purpose to measure the noise which is inevitably connected with the probes to the samples.
Fig. 4 shows the magnetization of EuSe at several temperatures. The saturation magnetization was normalized to 7 μ_B.
Fig. 5 shows the comparison of the B-H curves with Brillouin functions with different Neel temperatures T_N. From this figure the best fit can be deduced as T_N = 5 K.

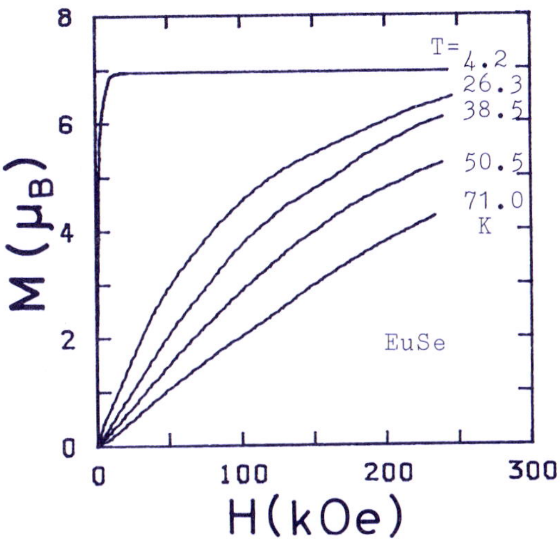

Fig. 4 Magnetization vs. applied magnetic field of EuSe at different temperatures.

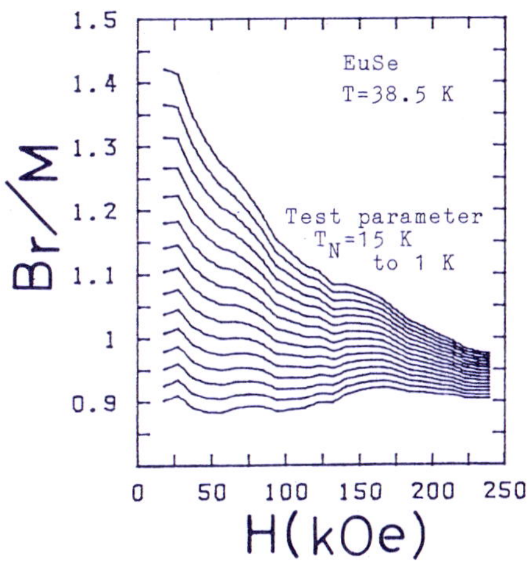

Fig. 5 Comparison of magetization with Brillouin function. The best fit: T_N≃5 K.

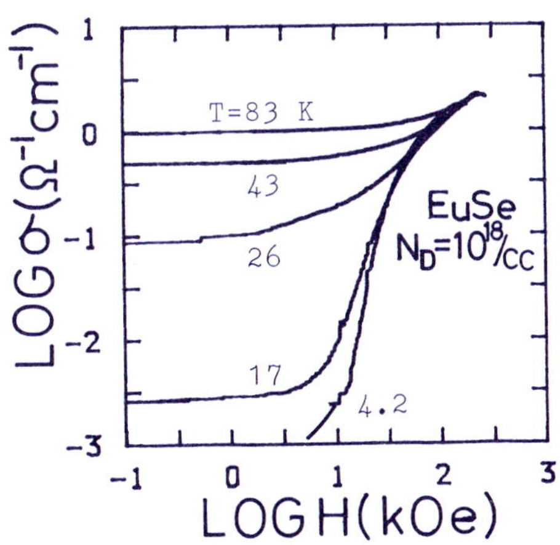

Fig. 6 Conductvity vs. magnetic field in various temperatures.

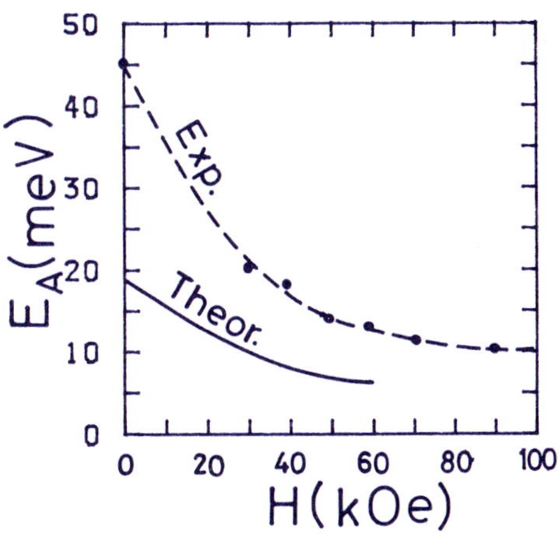

Fig. 7 Activation energy vs. magnetic field at T=40 K.

Fig. 6 shows the magnetic field dependence of conductivity at various temperatures. From Fig. 4 and Fig. 6 the magnetization dependence of the conductivity can be recognized.
Fig. 7 shows the activation energy of shallow donors as a function of applied magnetic fields up to 100 kOe and the theoretical values derived from the bound magnetic polaron model[3].

3. DISCUSSION

The fact that the temperature dependence of photoconductivity were scaled by TSC (as indicated in Fig. 2) reveals the physical mechanism. Apparently TSC is caused by carrier density variations. Therefore, the temperature dependence of photoconductivity of the sample is attributed to the carrier density changes. This conclusion is different from that by T. Penney and T. Kasuya[9] who attributed the photoconductivity to hopping mobility changes with ambient temperature.
These conclusions were derived from a comparison of the photocurrents in samples having large voids with those in voidless specimens. Large scale inhomogeneities cause photocurrents which persist up to days as observed with trigonal Se[11]. On the other hand, this long lifetime made it possible to identify the activation energy level of the traps at temperatures less than 50 K as shown in Fig. 7. It must be noted, however, tha the general shape of the activation energy of the shallow traps is in agreement with the bound magnetic polaron model.
The exact activation energy of donors is given by the following equation:

$$E_a = E_c^{(0)} - \frac{1}{2}\frac{MS}{M_0}\{\sum \langle\phi_c|J(r-R_n)|\phi_c\rangle - \sum_i \langle\phi_i\ |J(r-R_i)|\phi_i\rangle\ \}$$

where, $Ec^{(0)}$ and $Ed^{(0)}$ are the energy levels of the conduction band and the donors in H=0, M the magnetization. $M_0=-Ng\mu_B S$ the saturation magnetization at T=0, J() the function of the exchange interaction at the distance between electron and atom. If we substitute the experimentally obtained magnetization M_{exp} for M in the above equation, the 5d-4f exchange interaction is obtainable.

ACKNOLEDGEMENTS

We would like to thank Dr. G. Kido and Prof. N. Miura for helpful discussions, and Mr K. Shirata for technical asistance .

REFERENCES

1. P. Leroux-Hugon, New Developments in semiconductors, edited by P. R. Wallace et al (Noorhoff Int. Publ.g., Leyden, 1973).
2. Concalves da Silva, Solid State Commun., 17, 677 (1975).
3. P. Kuivalainen, J. Sinkkonen, K. Kaski and T. Stubb, Phys. Stat. Sol . (b)94, 181 (1979)
4. M. R. Oliver, J. O. Dimmock, A.L. McWorther and T. B. Reed, Phys. Rev. B5, 1078 (1972).
5. R. Faymonville, P. Grosse, G. Mützenich, J. Heleskivi and M. Mäenpää , J. Appl. Phys. 50, 1914 (1979).
6. J. Heleskivi and M. Mäenpää, Physca Scripta, 18, 441, (1978).
7. K. Yamada, J. Heleskivi, A. Salin, Solid State Commun., 37, 957, (1981)
8. K. Yamada, J. Heleskivi, H. Stubb, J. Phys. 41, c5-215 (1980).
9. T. Penney, T. Kasuya, J. Appl. Phys., 42, 1403 (1971).
10. P. Wachter, Solid State Commun., 8, 473 (1970)
11. K. Yamada, J. Heleskivi, M. Mäenpää, T. Stubb, Proc. Int. Conf. on Phys. of Se and Te, ed. by P. Grosse (Springer Verlag, Berlin 1979).

BINDING OF AN ELECTRON IN THE FIELD OF A FINITE ELECTRIC DIPOLE IN A MAGNETIC FIELD

Richard F. Wallis
Physics Department, University of California, Irvine
Irvine, California 92717, USA

Robert Herman
Center for Studies in Statistical Mechanics
University of Texas at Austin, Austin, Texas 78712, USA

The problem of the quantum-mechanical states of an electron in the field of a finite electric dipole is of great interest in several areas of physics and has been the subject of considerable study for more than thirty years. It has been found that, in the absence of a magnetic field, there is a minimum dipole moment required to produce binding of the electron. In view of the increase in the binding energy of a hydrogen atom produced by an external magnetic field, we have investigated the effect of a magnetic field on the binding of an electron by an electric dipole using the variational method. Our results indicate that, unlike the zero-field case, the presence of a magnetic field always leads to binding. This result is supported by analytic calculations carried out in the approximation of high magnetic fields and small dipole separations. The binding energy is found to be proportional to the first power of the magnetic field and the fourth power of the dipole separation.

1. Introduction

An elementary quantum mechanics problem which has received relatively little attention in the literature is that of a mobile charged particle such as an electron moving in the Coulomb field of a finite electric dipole, i.e., two charges of equal magnitude but opposite sign fixed in space a distance R apart. One example of this system is an electron moving in the field of a proton-antiproton pair. A second example, which is of particular interest in semiconductor physics, is provided by a conduction electron or hole moving in the field of a donor-acceptor pair. If we consider a compensated semiconductor, there is a certain probability that an ionized donor and an ionized acceptor will be in proximity to one another. Since they have charges of opposite sign, and if the magnitudes of their charges are equal, we have a finite electric dipole. We may ask what are the quantum mechanical energy levels of a conduction electron or hole moving in the field of such a dipole. Furthermore, since a dipole is electrically neutral, we may ask whether bound states exist regardless of the size of the dipole moment. In this paper we address these questions both in the absence and the presence of an external magnetic field.

2. No External Magnetic Field Present

Let us consider an electron of charge -e moving in the field of two fixed charged particles having charges +Ze and -Ze. The fixed charges are a distance R apart and form a finite electric dipole whose dipole moment has the magnitude ZeR. The configuration of the system is shown in Fig. 1. In the present section concerned with zero external magnetic field, the dipole axis will be taken parallel to the z-axis, so that the angle θ in Fig. 1 is zero.

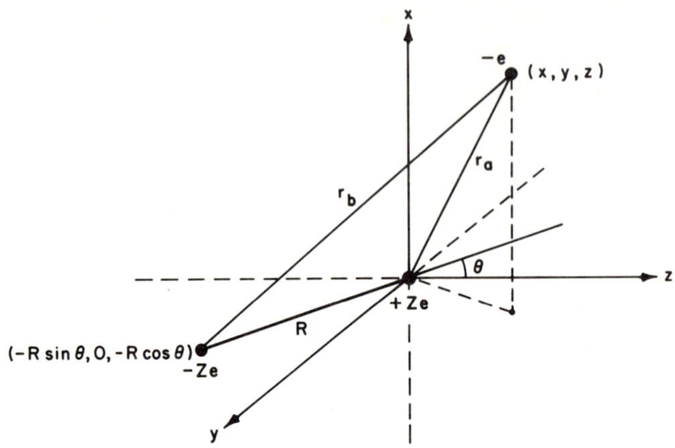

Fig. 1. Coordinate system for the electron in the field of a finite electric dipole.

For a semiconductor, we deal with a conduction electron which we assume is in a spherical parabolic energy band characterized by an effective mass m^*. The units of energy and length will be taken to be the effective Rydberg, $Ry^* = m^* e^4 Z^2 / 2\hbar^2 D^2$, and the effective Bohr radius, $a_B^* = \hbar^2 D / m^* e^2 Z$, respectively, where D is the static dielectric constant of the semiconductor. The Hamiltonian for the system can then be written as

$$H = -\nabla^2 - \frac{2}{r_a} + \frac{2}{r_b} \tag{1}$$

where r_a and r_b are the distances of the electron from the positive and negative fixed charges, respectively. This Hamiltonian can be separated in elliptic coordinates λ, μ, ϕ defined by

$$\lambda = \frac{r_a + r_b}{2}\,,\ \mu = \frac{r_a - r_b}{2}\,,\ \phi = \text{azimuthal angle}, \tag{2}$$

and the wave function ψ can be written as a product L(λ)M(μ)P(ϕ).

The question of the conditions under which the system described by the Hamiltonian H has bound states has a long history. Many years ago, Fermi and Teller [1] discussed the capture of negative mesotrons in matter and stated that an electric dipole has a minimum dipole moment for binding an electron. This minimum dipole moment is 0.6393 Zea_B which corresponds to a minimum dipole separation R of 0.6393 a_B. A similar result was obtained by Wightman [2] and much later by Levy-Leblond, [3] Brown and Roberts, [4] Coulson and Walmsley, [5] and others. Classically, however, there are bound states for all non-zero values of the dipole moment. [6]

The ordinary differential equations for $L(\lambda)$, $M(\mu)$, and $P(\phi)$ have been solved by Wallis, Herman, and Milnes [7] using expansions in series of complete orthonormal functions for a broad range of R values for several bound states. The results are shown in Fig. 2 where the binding energy is plotted against R. We see that the binding energy decreases very rapidly with decreasing R when R is comparable to the Bohr radius of the state in question. The minimum value of R for binding varies from one state to another; for the ground state it is 0.6393 a_B as mentioned above and is larger for excited states. [5] We should also mention that the Schrödinger equation has been solved using a variational method by Fox and Turner [8] who obtained a rather good approximate value for the minimum dipole moment to give binding.

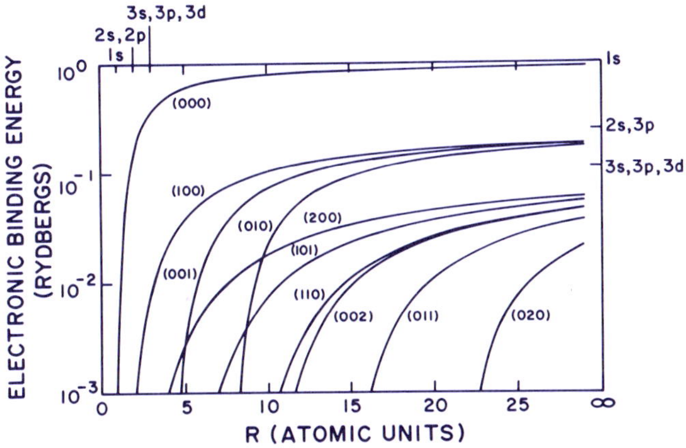

Fig. 2. Binding energy E_B versus dipole separation R for several electronic states.

3. External Magnetic Field Present: Numerical Results

It is well known that an external magnetic field increases the binding energy of a hydrogen atom. [9] One might anticipate that there would be a similar effect for an electron moving in the field of a finite electric dipole. Furthermore, one might ask what happens to the minimum dipole moment for binding. In this section we investigate these questions. Our treatment follows closely that of Herman and Wallis. [10]

We formulate the problem for a general orientation of a constant, uniform magnetic field relative to the dipole axis. Referring to Fig. 1, the magnetic field is in the z-direction, and the axis of the dipole makes an angle θ with respect to this direction. For general values of the dipole moment, we give specific results for the magnetic field parallel to and perpendicular to the dipole axis. For small values of the dipole moment, we give specific results for a general orientation of the magnetic field.

We find it convenient to use cylindrical coordinates ρ, ϕ, z where $\rho^2=x^2+y^2$. The Hamiltonian in the presence of the magnetic field B can be written as

$$H = H_o + H' \tag{3}$$

where

$$H_o = -\frac{\partial^2}{\partial x^2} - \frac{\partial^2}{\partial y^2} - i\gamma \frac{\partial}{\partial \phi} + \frac{1}{4}\gamma^2\rho^2, \tag{4}$$

$$H' = -\frac{\partial^2}{\partial z^2} - \frac{2}{r_a} + \frac{2}{r_b}, \tag{4b}$$

$\gamma = \hbar\omega_c/2Ry^*$, and ω_c is the cyclotron frequency eB/m^*c. We employ the variational method and take a trial function for the ground state of the form

$$\psi(\rho,\phi,z) = B_o\Phi_0(\sigma,\phi)f_0(z)+B_1[\cos\theta\Phi_0(\sigma,\phi)f_1(z)-\sin\theta\ \Phi_{-1}(\sigma,\phi)f_0(z)] \tag{5}$$

where $\Phi_m(\sigma,\phi)$ is an eigenfunction of H_o given by (for m = 0, -1)

$$\Phi_m(\sigma,\phi) = (-1)^m(\gamma/2\pi)^{1/2}e^{im\phi}\sigma^{|m|/2}e^{-\sigma/2} \tag{6}$$

with $\sigma = \gamma\rho^2/2$. The functions $f_n(z)$ are (for n = 0,1)

$$f_n(z) = (\varepsilon\sqrt{\gamma})^n(\gamma\varepsilon^2/2\pi)^{1/4}z^n e^{-\gamma\varepsilon^2z^2/4}. \tag{7}$$

The parameters B_o, B_1, and ε are variational parameters subject to the normalization condition $B_o^2 + B_1^2 = 1$. The form of the wave function given in Eq. (5) is particularly appropriate for high magnetic fields.

The expection value of the Hamiltonian H is evaluated with the wave function $\psi(\rho,\phi,z)$:

$$\langle H\rangle = \int \psi^* H\psi\ d\tau \quad . \tag{8}$$

Some of the integrals must be calculated numerically. For given values of the magnetic field B and dipole separation R, we minimize $\langle H\rangle$ with respect to the variational parameters. The binding energy E_B of the electron is then calculated using the equation

$$E_B = \gamma - \langle H\rangle_{min} \quad . \tag{9}$$

We now present and discuss the results for the ground electronic state.

For the case of the magnetic field parallel to the dipole axis, we plot the binding energy versus the dipole separation in Fig. 3 for several values of the magnetic field. As the value of R drops below unity, we see that the binding energy drops sharply for each magnetic field. It should be noted that the binding energy remains positive for values of R considerably smaller than the minimum value to give binding in the absence of a magnetic field. At a given value of R, the binding energy increases with increasing magnetic field just as in the case of the isolated hydrogen atom. This behavior is shown in detail in Fig. 4 where the binding energy is plotted against the magnetic field parameter γ for several values of R. The binding energy increases monotonically with increasing magnetic field. For the smallest values of R, the curves are in fact straight lines with unit slope in the log-log plot. This means that the relationship between the binding energy and the magnetic field parameter has the form

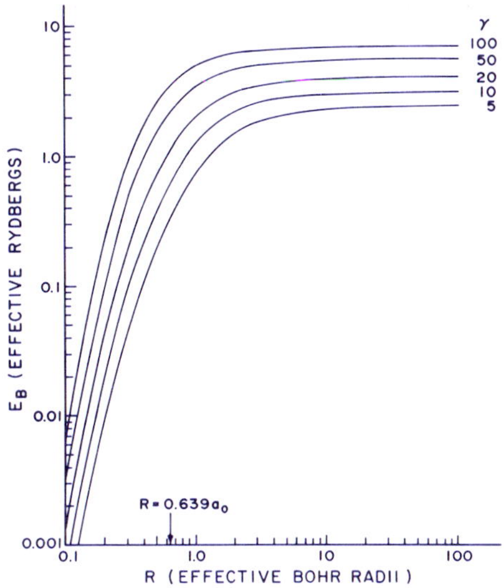

Fig. 3. Binding energy E_B versus dipole separation R for the case of the magnetic field parallel to the dipole axis for various values of the magnetic field parameter γ.

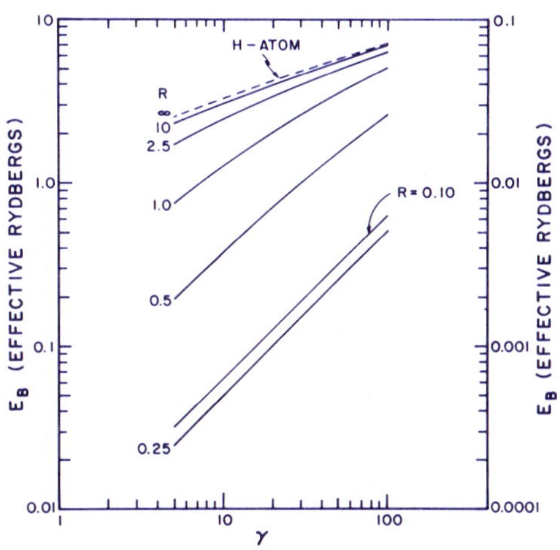

Fig. 4. Binding energy E_B versus magnetic field parameter γ for the case of a magnetic field parallel to the dipole axis for various values of the dipole separation R. The right-hand scale applies to the case R = 0.10.

$$\log E_B = \log \gamma + \log C \tag{10}$$

or

$$E_B = C\gamma \tag{11}$$

where C is independent of γ. Thus, for small R, the binding energy is simply proportional to the magnetic field. We shall return to this dependence later.

Precise values of the binding energy E_B and variational parameter ε are presented in Table I for a number of values of R and γ. The binding energy shows the trends illustrated in Figures 3 and 4. The variational parameter is nearly independent of magnetic field at small R, but decreases with increasing magnetic field at large R.

TABLE I. Binding Energy E_B (upper number) and variational parameter ε (lower number) for various values of the magnetic field γ and dipole separation R for the magnetic field parallel to the dipole axis. The numbers in parentheses are the powers of ten by which the entries are to be multiplied.

R \ γ	5	10	20	50
0.05	0.672367(-6) 0.195274(-2)	0.134493(-5) 0.195298(-2)	0.269043(-5) 0.195331(-2)	0.672890(-5) 0.195398(-2)
0.10	0.313183(-3) 0.297262(-1)	0.627847(-3) 0.297881(-1)	0.125986(-2) 0.298752(-1)	0.317013(-2) 0.300464(-1)
0.25	0.244243(-1) 0.175676	0.494515(-1) 0.178084	0.100275 0.180955	0.254419 0.184641
0.50	0.190659 0.393066	0.376390 0.393805	0.719152 0.385485	0.156506(1) 0.355736
1.0	0.748803 0.635373	0.128100(1) 0.581855	0.206349(1) 0.512931	0.354010(1) 0.409699
2.5	0.171120(1) 0.720008	0.244233(1) 0.603047	0.336023(1) 0.502148	0.490528(1) 0.391390
5	0.211775(1) 0.708122	0.285563(1) 0.595569	0.377406(1) 0.498239	0.531667(1) 0.389879
10	0.231963(1) 0.705753	0.305774(1) 0.594432	0.397597(1) 0.497707	0.551815(1) 0.389684
20	0.241992(1) 0.705412	0.315804(1) 0.594278	0.407622(1) 0.497638	0.561834(1) 0.389660
50	0.247996(1) 0.705363	0.321808(1) 0.594257	0.413626(1) 0.497628	0.567837(1) 0.389656
100	0.249997(1) 0.705360	0.323808(1) 0.594255	0.415626(1) 0.497628	0.569837(1) 0.389656
1000	0.251797(1) 0.705360	0.325608(1) 0.594256	0.417426(1) 0.497626	0.571637(1) 0.389657
∞ H atom	0.251997(1) 0.705360	0.325808(1) 0.594255	0.417626(1) 0.497627	0.571837(1) 0.389656

Let us now turn our attention to the case of the magnetic field perpendicular to the dipole axis. [10] The angle θ in Eq. (5) is now $\pi/2$. The resulting wave function is then used to calculate the expectation value of the Hamiltonian. The latter is minimized with respect to the variational parameters, and the binding energy is calculated using Eq. (9). The results for the binding energy as a function of dipole separation are plotted in Fig. 5 for several values of the magnetic field parameter. Qualitatively, the curves are very similar to those in Fig. 3 for the parallel case. However, for a given R and γ, the binding energy is larger in the perpendicular case than in the parallel case. This behavior is a consequence of the fact that magnetic field is more effective in pulling the electron charge distribution away from the negative fixed charge in the perpendicular case than in the parallel case.

The binding energy is plotted against the magnetic field parameter in Fig. 6 for several values of R for the perpendicular case. The curves are qualitatively similar to those in Fig. 4 for the parallel case. At the smallest R values, the binding energy is again proportional to γ.

The next problem to consider is that involving an arbitrary orientation of the magnetic field relative to the dipole axis. For general values of R, the variational calculations become very cumbersome and have not been attempted. However, at small R the Hamiltonian contribution H' can be expanded in power series in R to simplify the calculation. From Fig. 1 we see that the quantity r_b is given by

$$r_b = [(x + R\sin\theta)^2 + y^2 + (z + R\cos\theta)^2]^{1/2} \qquad (12)$$

where (x,y,z) are the coordinates of the electron. Substituting Eq. (12) into Eq. (4b) and expanding in power series in R yields to order R^2

$$H' \cong -\frac{\partial^2}{\partial z^2} - \frac{2R(x\sin\theta+z\cos\theta)}{(x^2+y^2+z^2)^{3/2}} - \frac{R^2}{(x^2+y^2+z^2)^{3/2}}\left[1 - \frac{3(x\sin\theta+z\cos\theta)^2}{x^2+y^2+z^2}\right]. \qquad (13)$$

The expectation value of the full Hamiltonian based on Eqs. (3), (4a), and (13) was evaluated using the wave function in Eq. (5), and the variational procedure was carried out. The binding energy was determined as a function of the angle between the magnetic field and the dipole axis for a variety of values of R and γ. The results are shown in Fig. 7. We see that the binding energy increases monotonically from $\theta = 0$ (parallel configuration) to $\theta = \pi/2$ (perpendicular configuration).

We now address the question of whether or not a minimum dipole moment for binding exists when a magnetic field is present. The results which we have presented in this section based on the variational method reveal no evidence of a minimum dipole moment. Nevertheless, it is difficult to prove in definitive fashion using the numerical procedures of the present study that a minimum dipole moment does not exist for $B \neq 0$. Analytic results would be very valuable in establishing the existence or absence of a minimum dipole moment. In the following section we develop an analytic approach.

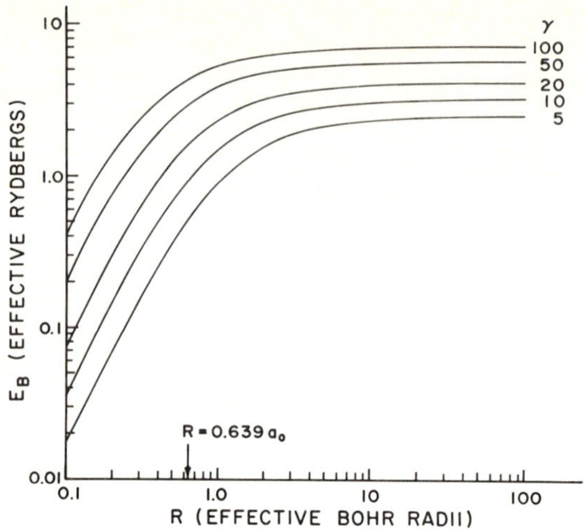

Fig. 5. Binding Energy E_B versus dipole separation R for the case of the magnetic field perpendicular to the dipole axis for various values of the magnetic field parameter γ.

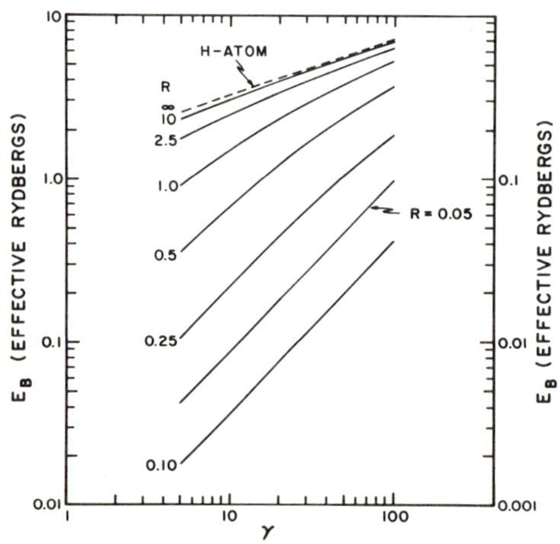

Fig. 6. Binding energy E_B versus magnetic field parameter γ for the case of a magnetic field perpendicular to the dipole axis for various values of the dipole separation R. The right-hand scale applies to the case R = 0.05.

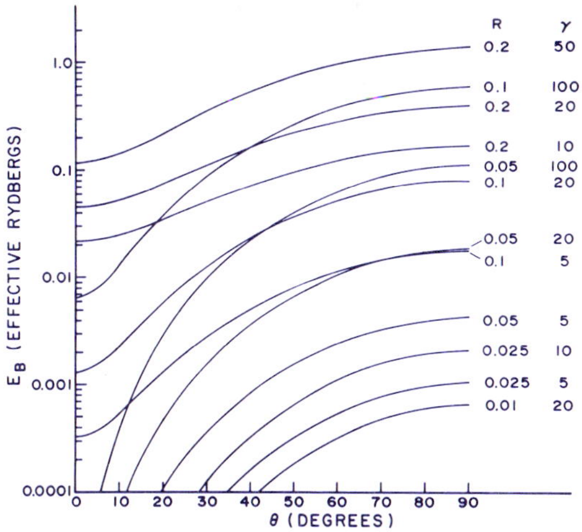

Fig. 7. Binding energy E_B versus angle between the magnetic field and the dipole axis for selected values of dipole separation R and magnetic field parameter γ.

4. External Magnetic Field Present: Analytic Results

So far we have used γ defined following Eq. (4b) as the measure of the magnetic field. However, for the dipole problem, one might argue that a better measure might be

$$\gamma' = \frac{\hbar\omega_c}{2E_B} \tag{14}$$

where E_B is the actual binding energy at the R value under consideration rather than that of an isolated hydrogen atom. For any given magnetic field, γ' would become large as R decreases and consequently E_B decreases. Under these circumstances we can follow the lead of Yafet, Keyes, and Adams [9] who have shown that the effective Coulomb potential in the presence of a high magnetic field is one-dimensional. Elliott and Loudon [11] have worked out a simple expression for this potential which for the dipole case (parallel configuration) becomes [12]

$$V(z) = \frac{2}{|z + \frac{R}{2}| + a} - \frac{2}{|z - \frac{R}{2}| + a} \tag{15}$$

where $a = \gamma^{-1/2}$. For R small, we can expand Eq. (15) in power series. Retaining only the lowest order nonvanishing terms, we obtain the one-dimensional Schrödinger equation in the form [12]

$$-\frac{d^2 f(z)}{dz^2} \mp \frac{2R}{(z \pm a)^2} f(z) = E\, f(z), \quad z \quad 0 \quad . \tag{16}$$

The solutions of Eq. (16) can be expressed in terms of Whittaker's functions, [13]

$$f(z) = \begin{cases} N_{o+}\, W_{0,m}(u) \quad , & z > 0 \\ N_{o-}\, W_{0,n}(u) \quad , & z < 0 \end{cases} \quad , \tag{17}$$

where $u = 2\alpha^{-1}(|z|+a)$, $\alpha^2 = -1/E$, $m = +(1-8R)^{1/2}/2$, $n = +(1+8R)^{1/2}/2$, and the quantities $N_{o\pm}$ are normalization constants. The requirement that f(z) and its first derivative be continuous at z = 0 leads to the eigenvalue equation

$$\frac{W'_{0,n}(\frac{2a}{\alpha})}{W_{0,n}(\frac{2a}{\alpha})} = - \frac{W'_{0,m}(\frac{2a}{\alpha})}{W_{0,m}(\frac{2a}{\alpha})} \quad . \tag{18}$$

Since we are interested in small R where the binding energy is small, we note from the definition $\alpha^2 = -1/E = 1/E_B$ that α will be large and hence the argument of the Whittaker functions will be small. We therefore expand the Whittaker functions and their derivatives in power series [14] and obtain the eigenvalue equation in the remarkably simple form [12]

$$E_B = 16\,\gamma\,R^4 \quad . \tag{19}$$

We observe that Eq. (19) has exactly the same form as Eq. (11) deduced from the variational calculations with the constant C given by $16R^4$. However, in quantitative terms, Eq. (19) predicts too large a value for E_B. This discrepancy is probably due to the fact that Eq. (19) is an infinite magnetic field limit, whereas the variational calculations are for finite fields.

5. Discussion

Both the numerical results based on the variational method and the analytic results in Eq. (19) suggest that there is no minimum dipole moment for binding in the presence of a high magnetic field. This is qualitatively different from the case of zero magnetic field where there is a minimum dipole moment. That a magnetic field can produce such a qualitative difference is not altogether surprising. Let us consider the binding of an electron to a neutral atom such as the hydrogen atom. In the absence of a magnetic field, the negative hydrogen atom H^- has only one bound state, [15] whereas in the presence of a magnetic field, it has an infinity of bound states. [16] This qualitative difference has been correlated by Avron, Herbst, and Simon [16] with the facts that a magnetic field converts the problem from a three-dimensional to a one-dimensional problem, and that a very shallow square well potential will bind a particle in one dimension but not in three. We feel that these considerations also apply to the dipole problem, and that there is a minimum dipole moment in the absence of a magnetic field but not in the presence of a magnetic field.

It is of interest to make specific application of our results to semiconductors. In the absence of a magnetic field, it is unlikely that a conduction electron or hole can be bound to a donor-acceptor pair on nearest neighbor sites in a typical semiconductor. Taking silicon as an example and using the density of states effective mass $m^* = (m_\ell m_t^2)^{1/3} = 0.328m$ and static dielectric constant D = 11.7, we find that the effective Bohr radius is ~ 20 A. The minimum dipole separation for binding would then be ~ 13 A which considerably exceeds the nearest neighbor distance of 2.35 A. Binding would occur only if the ions in the donor-acceptor pair are more than ~ 5.5 nearest neighbor distances apart. For germanium and InSb, the donor-acceptor pair separation would have to be even larger to give binding.

In the presence of an external magnetic field, binding can occur in principle even with the donor acceptor pair on nearest neighbor sites. However, the binding energy can be extremely small even for rather high magnetic fields. Taking silicon as an example, the nearest neighbor separation is roughly one-tenth of the effective Bohr radius. In a magnetic field corresponding to $\gamma = 5$ ($\sim 9 \times 10^6$ gauss for silicon) and parallel to the nearest neighbor bond, the binding energy of a conduction electron to a nearest neighbor donor-acceptor pair is only about 10^{-5} eV. The extent of the wave function parallel to the donor-acceptor pair axis is approximately 600 A or 250 times the nearest neighbor distance. The probability of finding a carrier bound under these conditions is exceedingly small even at liquid helium temperatures. At donor-acceptor pair separations closer to the critical value 0.6393 atomic units, however, the application of a magnetic field may tip the balance from no binding to binding depending on the temperature.

Acknowledgement

The work of one of the authors (R.F.W.) was supported by National Science Foundation Grant No. DMR-7809430.

References

1. E. Fermi and E. Teller, Phys. Rev. 72, 399 (1947).

2. A. S. Wightman, Phys. Rev. 77, 521 (1950).

3. J. M. Levy-Leblond, Phys. Rev. 153, 1 (1967).

4. W. B. Brown and R. E. Roberts, J. Chem. Phys. 46, 2006 (1967).

5. C. A. Coulson and M. Walmsley, Proc. Phys. Soc. London 91, 31 (1967).

6. J. E. Turner and K. Fox, J. Phys. A 1, 118 (1968); K. Fox, J. Phys. A 1, 124 (1968).

7. R. F. Wallis, R. Herman, and H. W. Milnes, J. Mol. Spectrosc. 4, 51 (1960).

8. K. Fox and J. E. Turner, J. Chem. Phys. 45, 1142 (1966).

9. Y. Yafet, R. W. Keyes, and E. N. Adams, J. Phys. Chem. Solids 1, 137 (1956).

10. R. Herman and R. F. Wallis, Phys. Rev. B23, 4902 (1981).

11. R. J. Elliott and R. Loudon, J. Phys. Chem. Solids 15, 196 (1960).

12. R. F. Wallis and R. Herman, Phys. Rev. B, B25, 7398 (1982).

13. E. T. Whittaker and G. N. Watson, A Course of Modern Analysis (Macmillan, New York, 1948), p. 337.

14. M. Abramowitz and I. A. Stegun, Handbook of Mathematical Functions (Dover, New York, 1965), p. 504.

15. R. N. Hill, Phys. Rev. Lett. 38, 643 (1977).

16. J. Avron, I. A. Herbst, and B. Simon, Phys. Rev. Lett. 39, 1068 (1977).

LOW-TEMPERATURE MAGNETOCAPACITANCE MEASUREMENTS ON n-TYPE SILICON: SPIN-DEPENDENT ELECTRICAL POLARIZABILITIES OF DONOR CLUSTERS

D. New and T. G. Castner
University of Rochester, Rochester, New York 14627

M. J. Naughton and J. S. Brooks
Boston University, Boston, Massachusetts 02215

Magnetocapacitance measurements on Si:P and Si:As samples have been made in a dilution refrigerator in the temperature range 60mK to 1K with magnetic fields to 19 Tesla. A break between low field B^2 behavior and B^n ($n \leq 1$) behavior shifts to lower fields as the temperature is lowered and also becomes increasingly sharp. The results support the concept of spin-dependent polarizabilities of donor clusters.

For many years the standard experimental approach for obtaining information on the shrinkage of donor impurity wave functions[1] in large magnetic fields has been the measurement of the magnetoresistance $(\rho(H)-\rho(0))/\rho(0)$ of doped semiconductors in the hopping regime. Many such studies have been made and Chroboczek[2] has reviewed this area. An alternative method, in principle, for obtaining the shrinkage of shallow impurity wave functions is afforded by magnetocapacitance studies. At sufficiently low temperatures the hopping transport contribution to the low frequency donor concentration-dependent dielectric response function $\varepsilon(N_D,\omega,T)$ is negligible[3] and capacitance measurements yield direct information on the contribution to the "dielectric constant" from the shallow donors in an n-type semiconductor. Application of a magnetic field permits the extraction of an effective field-dependent donor polarizability $\alpha_{eff}(N_D,T,B)$ which, roughly speaking, is proportional to the volume of the donor wave function. Essentially, the method determines, in the low field limit, the coefficient of a B^2E^2 term (E is electric field) in the free energy. From this standpoint the effect considered for a doped semiconductor is a magnetoelectric effect[4] of a random system of S=1/2 impurities coupled by pairwise antiferromagnetic exchange. This system shows no long range magnetic order at the lowest temperatures studied.[5]

The capacitor samples utilized evaporated Au electrodes (~1200Å thick), as in earlier dielectric anomaly studies,[3] and measurements were made with a 3-terminal bridge and dual-phase lockin amplifier. The effective capacitance $C_{eff} = C_bC_s/(2C_s+C_b)$ where C_b is the barrier capacitance of the two Schottky barriers and C_s is the bulk semiconductor capacitance. The magnetic field dependence of $C_{eff}(B)$ is given by

$$\frac{\partial C_{eff}}{\partial B} = \left(\frac{C_s}{2C_s+C_b}\right)^2 \frac{\partial C_b}{\partial B} + \left(\frac{C_b}{2C_s+C_b}\right)^2 \frac{\partial C_s}{\partial B} . \quad (1)$$

Since the Schottky barriers are depletion regions $(1/C_b)(\partial C_b/\partial B) < (1/C_s)(\partial C_s/\partial B)$. For the doping region studied $C_b >>> C_s$ and $\partial C_{eff}/\partial B \simeq \partial C_s/\partial B$, hence the Schottky barriers are of no consequence for the magnetocapacitance measurements. The earlier measurements[4] above 1K were on 1.42 cm. diameter samples while the dilution refrigerator measurements were made on the same samples, after recoring to a smaller diameter of order 1.00 cm. The field-dependent effective donor polarizability $\alpha_{eff}(N,T,B)$ is obtained via the Clausius-Mossotti relation, namely

$$\alpha_{eff} = \frac{3}{4\pi N_D}\left[\frac{\varepsilon(N_D,T,B)-\varepsilon_h(T,B)}{\varepsilon(N_D,T,B)/\varepsilon_h(T,B)+2}\right] = \frac{3}{4\pi N_D}\left[\frac{C(N_D,T,B)-C_h(T,B)}{C(N_D,T,B)/C_h(T,B)+2}\right] . \quad (2)$$

For the samples discussed below with $N_D > 2\times10^{17}$/cc. changes in the host value $\varepsilon_h(T,B)$ ($C_h(T,B)$) with T and B were negligible and were neglected. The best sensitivity was obtained at ν=10KHz where changes of order $\Delta C/C \simeq 3\times10^{-6}$ could be observed. All data reported below are for $\vec{E} \parallel \vec{B}$.

In Fig. 1 ℓnR $[R=\alpha_{eff}(B=0)/\alpha_{eff}(B)]$ is shown versus B for a Si:P ($N_D\sim1.0\times10^{18}$/cc) sample for 60 mK<T<4.2K. ℓnR exhibits a B^2 behavior, as expected, in the low field region, followed by a flattening to a much weaker field dependence at a characteristic value of B/T~5 Tesla/°K. At the lower temperatures (T<1K) the B-dependence above 10T. is approximately $B^{1/2}$ for this sample. In the high field region the temperature dependence is very much reduced compared to the low field region. At the maximum field shown in Fig. 1 $\alpha_{eff}(0)/\alpha_{eff}(14T.)\simeq1.07$. Fig. 2 shows ℓnR vs. B data for a more dilute Si:P sample ($N_D\simeq2.9\times10^{17}$/cc). The data is qualitatively similar to that in Fig. 1, but with certain quantitative differences. The values of ℓnR in the low field quadratic region are a factor of five smaller. Furthermore, the low temperature high field slope for B>12T is closer to 1 rather than 1/2 and the transition region is steeper. For the lowest temperature (T≃70mK) data there is no evidence of a quadratic B^2 region for B>0.2T. Data for Si:As samples exhibit features very similar to the Si:P data. To compare the magnitudes of $\ell nR)_P$ and $\ell nR)_{As}$ one should look at higher concentrations in the Si:As samples such that $\bar{n}$ (characterizing the Poisson distribution) is the same for both P- and As-doped samples.

A model, considering isolated donors and donor pairs only, has been developed[4] to explain the temperature dependence of ℓnR in the low field B^2 region. The donor pairs have spin-singlet (s) and spin-triplet (t) states split by an exchange splitting $J=J(R)/k$ (k = Boltzmann constant) where R is the donor pair spacing. The donor pair polarizabilities in zero field are $\alpha_s(0)=\alpha_s(0,R)$ and $\alpha_t(0)=\alpha_t(0,R)$. The effective polarizability $\alpha_{eff}(N,T,B)$ is given by

$$\alpha_{eff}=\frac{N_i}{N}\alpha_i(0)(1-\eta_i\gamma^2)+\frac{N_p}{N}\,\frac{\alpha_s(0)(1-\eta_s\gamma^2)+\alpha_t(0)(1-\eta_t\gamma^2)e^{-J/T}(e^{h/T}+1+e^{-h/T})}{1+e^{-J/T}(e^{h/T}+1+e^{-h/T})}\,,\qquad(3)$$

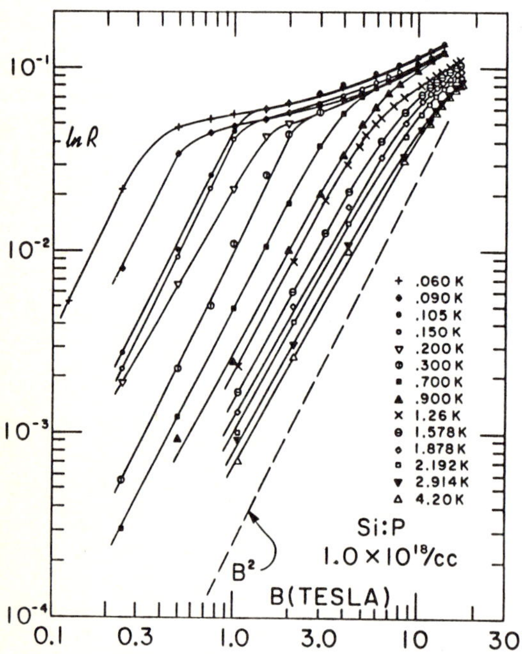

Fig.1 - ℓnR versus B for a $N_D\sim1.0\times10^{18}$/cc Si:P sample for 60mK<T<4.2K. At the highest fields and lowest temperature the slope approaches 1/2.

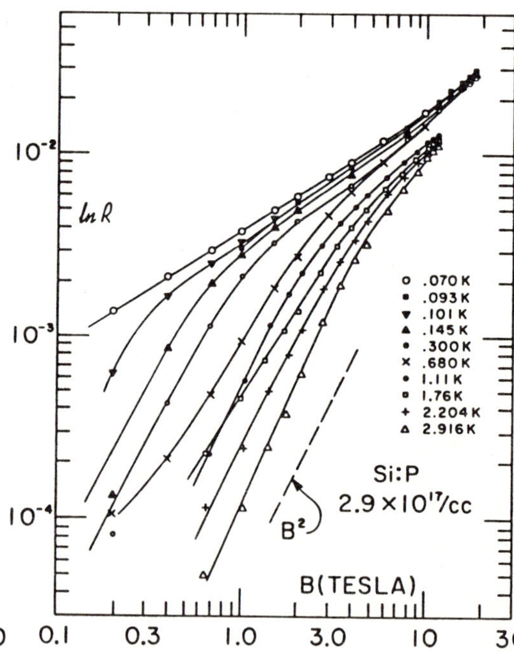

Fig.2 - ℓnR versus B for a $N_D\sim2.9\times10^{17}$/cc Si:P sample for 70mK<T<2.9K. At the highest fields and lowest temperature the slope approaches 1.

where $h=g\mu_B H/k$ is the Zeeman splitting divided by k, the η_j are the various diamagnetic reduction coefficients (tensor components), and γ^2 is the diamagnetic energy to isolated donor binding energy ratio ($\gamma^2 = \varepsilon_h a^{*3}B^2/m^*c^2$; a^* is an isotropic Bohr radius and m^* an isotropic effective mass). The isolated donors contribute an anisotropic temperature-independent term $\alpha_i(\gamma) = \alpha_i(0)(1-\eta_i\gamma^2+\lambda_i\gamma^4- \ldots)$ [the $\lambda_i\gamma^4$ and higher terms are negligible for $\gamma<0.12$ corresponding to B=19T. for Si] except for a negligible relativistic correction[6]. N_i/N and N_p/N are the fraction isolated donors and pairs respectively. The factor $e^{h/T}+1+e^{-h/T}$ results from the Zeeman effect of the pair triplet state. Keeping only terms of order γ^2 and neglecting small field-dependent changes in J(R) ℓnR is given by

$$\ell nR = \gamma^2\{\frac{N_i}{N}\,\alpha_i(0)\eta_i + \frac{N_p}{N}\left[\frac{\eta_s\alpha_s(0)+3\eta_t\alpha_t(0)e^{-J/T}}{1+3e^{-J/T}} + \frac{\Delta\alpha e^{-J/T}}{(1+3e^{-J/T})^2}\left(\frac{T_o}{T}\right)^2\right]\}\,, \qquad (4)$$

where $\Delta\alpha=\alpha_s(0)-\alpha_t(0)$ and $T_o = h/\gamma$. For Si T_o = 217K and the last term in Eq.(4), which exhibits a temperature dependence identical to the Schottky specific heat anomaly, dominates the pair contribution to ℓnR for T<4.2K and $\Delta\alpha\simeq 0.05\alpha_i(0)$. This term, resulting from triplet state Zeeman effect, and proportional to $\Delta\alpha$, is isotropic when the Zeeman effect is isotropic. The first three terms are anisotropic and produce a very weak temperature dependence unless η_s and η_t are very different. ℓnR has been averaged over a Poisson distribution of pair distances utilizing suitable expressions for J(R), $\alpha_s(0,R)$, and $\alpha_t(0,R)$, the latter from scaled calculated H_2 molecule results.[7-9] Previously calculated values[4] of $\langle\ell nR\rangle$ yield a temperature dependence in good agreement with the data (see Fig. 3) for T>0.3K, but below 0.3K the calculated curves drop rapidly. The data in Fig. 3 at B=1.0T is still rising at 60mK for both Si:P samples. The B=10T data shows a slow monatonic increase for the more dilute sample and a very broad maximum at about 0.1K for the more concentrated sample. The 1.0×10^{18}/cc sample shows a slight upturn at the highest temperature (4.2K) at 10T. which is believed to be the onset of a hopping magneto-transport contribution to ℓnR. This temperature dependence is somewhat different than the Schottky-type humps observed in the specific heat of Si:P in a magnetic field by Kobayashi et al.[10] The poor agreement between the present data and the pair model below 0.3K is not yet understood. However, a recent determination[11] of the characteristic volume V_c associated with the Poisson distribution from EPR spectra of donor clusters gives values of $\bar{n}$ 2.4 and 8.2 for the 2.9×10^{17}/cc and 1.0×10^{18}/cc Si:P samples respectively. This suggests larger clusters dominate the behavior of $\alpha_{eff}(N,T,B)$ for these samples. 3-donor clusters have a high energy S=3/2 state and two different S=1/2 states with different polarizabilities. The topological configurations of 3-donor clusters range from the improbable equilateral triangle ($J_{12}=J_{23}=J_{13}$, the two S=1/2 states are degenerate) to a high probability approximate elongated isoceles triangle to the three donors in a line (improbable). The topologically dependent energy splitting of the two S=1/2 states with different $\alpha_{1/2}(0)$ values introduces a new feature which may keep $\langle\ell nR\rangle$ from falling as rapidly at low temperatures. It appears necessary both to include larger clusters in the calculation of $\langle\ell nR\rangle$ and also to

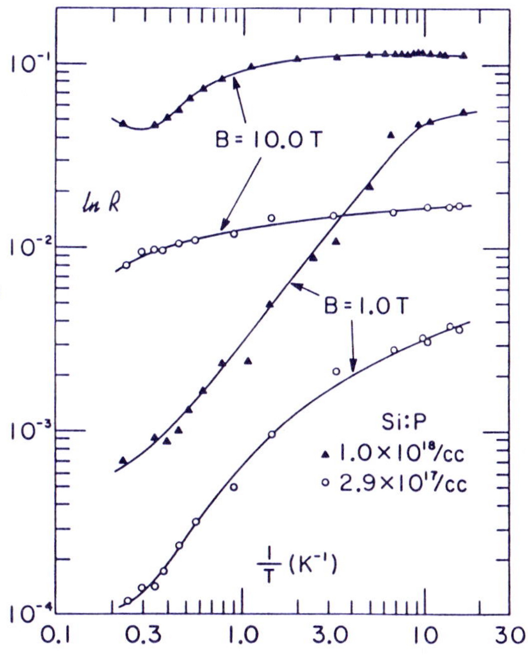

Fig.3 - ℓnR versus 1/T at B equal 1.0 and 10.0 Tesla for the 2.9×10^{17}/cc and 1.0×10^{18}/cc Si:P samples.

obtain data on $\alpha_{eff}(N,T,B)$ for samples with smaller values of $\bar{n}$ to obtain a more reliable comparison between theory[12] and experiment. For much more dilute samples, corrections due to field-induced changes in $\varepsilon_h(T,B)$ must be taken into account.

In sufficiently large B fields ($h/T >>> 1$ and $h >> J$ for the large majority of pairs) one can place most of the pairs in the lowest ($M_s = -1$) Zeeman state of the triplet. In this limit Eq.(3) leads to

$$\ell nR \simeq \gamma^2 \left[\frac{N_i}{N} \alpha_i(0)\eta_i + \frac{N_p}{N} \alpha_t(0)\eta_t \right] + \text{corrections} \ . \qquad (5)$$

This predicts a temperature-independent result still having a B^2 dependence for $\gamma < 0.16$, which is clearly inconsistent with the data which show a $B^{1/2}$ dependence in Fig. 1 and a B dependence in Fig. 2 for B>12T. Since $\bar{n} > 1$ for these samples and large clusters dominate the behavior, the strong concentration dependence of $\partial \ell nR / \partial B$ in the high field range suggests that clusters have a different diamagnetic behavior than isolated donors. The magnetic length $\lambda = (\hbar c/eB)^{1/2}$ is less than 100Å at 10T. Thus, for B>12T the magnetic length becomes less than the spacing of many pairs recalling that the radius of the characteristic volume V_c is 125Å. We are not aware of any theory of the diamagnetism of pairs or larger clusters for λ less than the pair spacing. For comparison the field dependence of the magnetoresistance in the hopping regime shows a B^2 dependence in the low field regime and a $B^{1/2}$ dependence in the high field regime. More work is required to understand the diamagnetic behavior of larger clusters, with particular emphasis on the high spin state having the most antisymmetric wave function.

An early objective of this work was to determine the diamagnetic susceptibility tensor components $\eta_{\parallel}$ and $\eta_{\perp}$ for isolated donors. Because of the dominant temperature-dependent contribution to $\alpha_{eff}(N,T,B)$ from the high spin states of clusters resulting from spin-dependent polarizabilities and the Zeeman effect of these spin states this objective has not been achieved. Accurate measurements on more dilute samples and high purity Si samples, to determine the host corrections, are required to allow the determination of the diamagnetic susceptibility tensor for isolated donors. We gratefully acknowledge the opportunity to use the facilities at the Massachusetts Institute of Technology National Magnet Laboratory. This work was supported in part by the National Science Foundation.

1. Yafet, Y., Keyes, R. W., and Adams, E. N., J. Phys. Chem. Solids 1 (1956) 137.
2. Chroboczek, J.A., Phil. Mag. 42 (1980) 933.
3. Castner, T. G., Lee, N. K., Cieloszyk, G. S., and Salinger, G. L., Phys. Rev. Lett. 34 (1975) 1627.
4. New, D., Lee, N.K., Tan, H.S., and Castner, T. G., Phys. Rev. Lett. 48 (1982) 1208.
5. Andres, K., Bhatt, R. N., Goalwin, P., Rice, T. M., and Walstedt, R. E., Phys. Rev. B24 (1981) 244.
6. Castner, T. G., Dexter, D. L., and Druger, S. D., Phys. Rev. A24 (1981) 2897.
7. Kolos, W. and Wolniewicz, L., J. Chem. Phys. 46 (1967) 1426.
8. Dupré, D. B. and McTague, J. P., J. Chem. Phys. 50 (1969) 2024.
9. Clarke-Hunt, K. L. and Buckingham, A. D., J. Chem. Phys. 72 (1980) 2832.
10. Kobayashi, N., Ikehata, S., Kobayashi, S., and Sasaki, W., Solid St. Commun. 32 (1979) 1147.
11. New, D. and Castner, T. G., Proc. 16 Int. Conf. on Phys. of Semiconductors, Montpellier, France (1982).
12. Lipari, N.O. and Dexter, D. L. Phys. Rev. B18 (1978) 1346.

PHASE TRANSITIONS IN HIGH MAGNETIC FIELDS

A.J. Arko[(a)], C.J. Beers, A.P.J. van Deursen, R.P.A.R. van Kleef, H.W. Myron, M.R. Parker[(b)], M. Pepper[(c)], D.A. Poole[(c)], Th.H.M. Rasing and P. Wyder
High Field Magnet and Physics Laboratory, Faculty of Science
University of Nijmegen
Toernooiveld, 6525 ED Nijmegen, The Netherlands

a) Material Science Division
Argonne National Laboratory
Argonne, Illinois, U.S.A.

b) Department of Pure and Applied Physics
University of Salford
Salford, England

c) Cavendish Laboratory
Cambridge University
Cambridge, England

The purpose of this paper is to summarize some of the research activities recently performed at the Laboratorium voor Hoge Magneetvelden at the University of Nijmegen. The scope here and unifying aspect is magnetically induced phase transitions. Here we summarize transitions in the settling velocity of paramagnetic aggregates, suppression of spin fluctuations in UAl_2, the phase diagram of a ferroelectric chiral smectic liquid crystal near the Lifshitz point and the transition from 3D to 2D conduction in a GaAs FET. In no way does this represent a complete review of transitions, but rather a summary of phase transitions observed at the magnet laboratory during the past year.

Magnetic Aggregation of Particles in High Fields

The stability of a colloidal suspension of particles is generally described by the Derjaguin-Landau-Verwey-Overbeek (DLVO) theory [1] where the two particle potential energy ϕ_T can be expressed by the sum of potential energy ϕ_R, due to the electrical double layers at the solid-liquid interface, and a London-van der Waals interaction term, ϕ_L. These interactions for particles of equal radii are given respectively by:

$$\Phi_R = 4\pi\varepsilon_r\varepsilon_0 a\psi_0^2 \frac{\exp[-\tau(r_a-2)]}{r_a} \tag{1}$$

and

$$\Phi_L = -\frac{A}{6}\left[\frac{2}{r_a^2-4} + \frac{2}{r_a^2} + \ln\left(\frac{r_a^2-4}{r_a^2}\right)\right] \tag{2}$$

where ε_r is the relative permittivity of the material of the particle, ε_0 the permittivity of free space, ψ_0 the double layer potential, $r_a = r/a$ the normalized interparticle distance (a particle size) and τ/a is the Debye-Hückel reciprocal double layer thickness. A potential energy barrier exists in a stable suspension which prevents the particles from approaching each other. The DLVO-theory can be extended [2] when an external magnetic field B_0 is turned on and the dipole-dipole interaction energy Φ_M is included. For a system of two particles, Φ_M is

$$\Phi_M = -\frac{32\pi^2 a^3 \chi^3 B_0^2}{9\mu_0 r_a^3}, \tag{3}$$

where μ_0 is the permeability of free space and χ is the net volume magnetic susceptibility of the particle. The total potential energy of interaction is given by the sum of Eqs. 1 to 3 and since Φ_M is negative, an applied magnetic field can reduce the potential energy to zero thereby allowing the particles to approach each other and form binary clusters.

In order to understand the dynamics of flocculation, a force balance equation can be constructed from the spatial derivatives of Φ_R, Φ_L and Φ_M [3]. Fixing a polar coordinate system at the center of the first particle and the direction of the applied magnetic field parallel to the x-axis, the dynamics of the radial and azimuthal components of the second are respectively given by:

$$\frac{dr_a}{dt} = - V_{Ma} \left(\frac{3\cos^2\theta - 1}{r_a^4} \right) + V_{La} \left[\frac{2}{r_a(r_a^2-4)} - \frac{r_a}{(r_a^2-4)^2} - \frac{1}{r_a^3} \right] + V_{Ra} \left(\frac{\tau r_a + 1}{r_a^2} \right) \exp[-\tau(r_a - 2)] \qquad (4)$$

$$r_a \frac{d\theta}{dt} = - V_{Ma} \frac{\sin 2\theta}{r_a^4} \qquad (5)$$

Here V_{Ma}, V_{La} and V_{Ra} are normalized characteristic velocity coefficients analogous to the magnetic velocity used in HGMS.

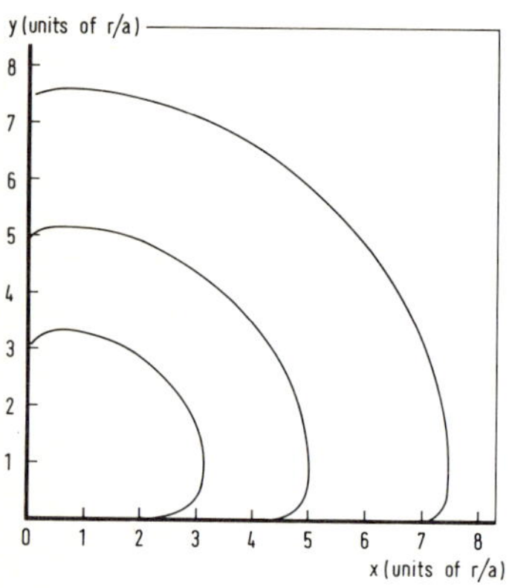

Fig. 1 Particle trajectories in a paramagnetic flocculation process with an external magnetic field B_0 along the x-axis. The initial angular position of the particles is at $\theta = 89^0$.

Using a forth order Runge-Kutta technique, particle trajectories have been calculated from Eqs. 5 and 6 and are shown in Fig. 1. A paramagnetic particle initially positioned at a point (r,θ) in the first quadrant, precesses clockwise about the "fixed" particle at the origin. If the particle is captured it moves along the x-axis towards the point (2a,0) forming a binary pair. For particles located at a distance greater than r_{ca}, the normalized capture radius, the particle is pushed away along the x-axis.

The capture radius r_{ca} can be estimated to a very good approximation, if θ is set equal to zero in Eqs. 4 and 5. Neglecting the London-van der Waals interaction and assuming τ is small, then the capture radius $r_{ca} = (2\ V_M/V_R)^{\frac{1}{2}}$. From the capture radius one can estimate the particle number density at which flocculation can be expected to be observed. This is of the order of

$$n_0 = \left(\frac{3}{8 \cdot 4\pi a^3} \right)(V_R/V_M)^{3/2}$$

Flocculation experiments were carried out on a number of commercially available mineral suspensions. The magnetic fields were produced by a water cooled Bitter magnet. Suspensions of paramagnetic materials were held in calibrated testtubes in the magnet. Field induced particle aggregation was detected by measuring the settling velocity V_S of the suspended particle at different values of the applied magnetic field. This was achieved by optically measuring the rate of fall of the meniscus between sedimenting particles and clear liquid using a cathetometer or videocamera.

The settling process can be described by a simple force balance equation between the dragforce and the combined influence of the gravitational and magnetic dipole forces:

$$\frac{4\pi}{3} a^3 \rho_{eff} g + \frac{4}{3} \pi a^3 \chi_{eff} \xi B_0^2/\mu_0 = 6\pi\eta a V_S \tag{6}$$

where ρ_{eff} is the effective density of the particle in the fluid, χ_{eff} is the net volume susceptibility, ξ is the fractional magnitude of the field gradient and η is the viscosity of the carrier fluid. When the magnetic field exceeds B_F two-particle clusters are formed which will align along the field direction, as shown in Fig. 1. With a vertical magnetic field this means that the Stoke's drag term on the RHS of Eq. 6 will remain relatively unaffected since the effective airodynamic diameter of the cluster will not be very different from that of the single particle. On the other hand, the volume dependent term on the LHS of Eq. 6 will double in size. We expect that the settling velocity plotted against B_0^2 will change discontinuously at $B_0 = B_F$ whereas its slope should approximately double.

Fig. 2.

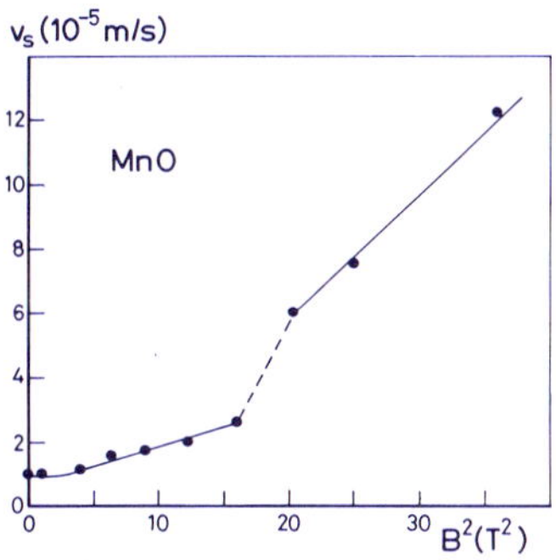

a) Settling velocity of a 30%-vol suspension of MnO in a vertical magnetic field. The discontinuity of V_S vs. B^2 indicates pair formation of particles. The volume susceptibility (MKSA units) is $\chi = 4.69 \times 10^{-3}$

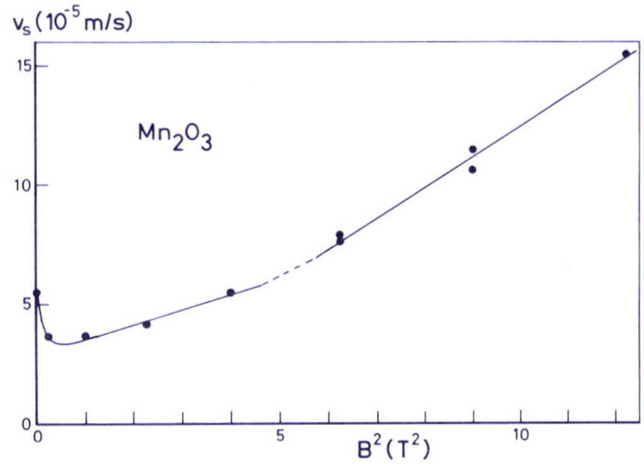

b) V_S in a 25%-vol suspension of Mn_2O_3. χ(MKSA) = 5.39 x 10^{-3}.

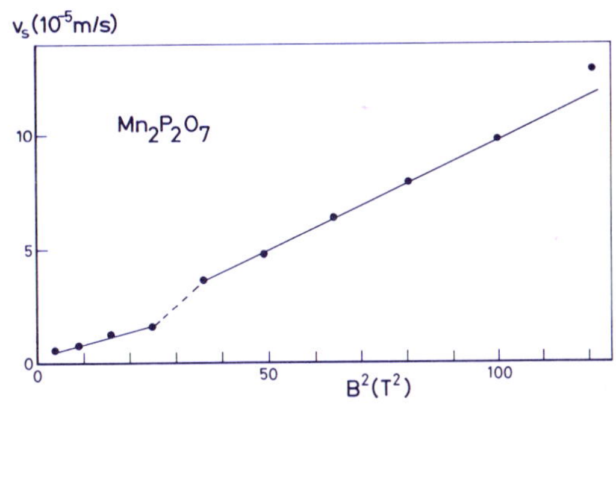

c) V_S in a 35%-vol suspension of $Mn_2P_2O_7$. χ(MKSA) = 4.72 x 10^{-3}.

Measurements of settling velocity as a function of field were carried out on MnO, Mn_2O_3 and $Mn_2P_2O_7$. Fig. 2 shows field induced binary flocculation for Mn_2O_3 and $Mn_2P_2O_7$. In the case of MnO the slope changes by more than a factor of two; this is not clearly understood. The minimum at low fields in Mn_2O_3 is thought to be caused by a combination of two effects: an anisotropic magnetic susceptibility and a nonspherical particle shape. Particles rotate that their strongest magnetic axis is directed parallel to the field, thus altering their effective diameter.

Suppression of Spin Fluctuations in UAl_2 in High Magnetic Fields

UAl_2 has been known for some time to be a spin-fluctuation (SF) system. Its temperature dependence of specific heat, resisitivity and magnetic susceptibility [4] fit neatly into various theories explaining these phenomena [5]. The characteristic SF temperature (T_{SF}) of UAl_2 is believed to be about 25 K. We have studied the transport properties of UAl_2 at low temperatures in high magnetic fields [6] where it has been argued that the application of high magnetic fields will reduce or freeze out paramagnon effects. The Zeeman splitting of the spin-up and spin-down electron states will prevent certain excitations with the reversal of spin direction to occur. The characteristic paramagnon excitations are of the order of $k_B T_{SF}$, with $g = 2$. The relationship between H_{SF} and T_{SF} (25 K) can be written as $g \mu_B H_{SF} = k_B T_{SF}$. H_{SF} is then calculated to be near 19 T.

Partial freezing out of SF in high magnetic fields has been previously reported for various compounds, for example in Pd [7] H_{SF} is believed to be 190 T! In contrast we have achieved complete suppression of SF in UAl_2.

The susceptibility of UAl_2 [6], derived from the magnetization at 1.4 and 4.2 K, decreases with increasing field reaching a field-independent value above 20 T. At 20 and 77 K the susceptibility is found to be independent of magnetic field. Values for the differential susceptibility, see Table I, for various field intervals as a function of temperature indicate that the high-field values at 4.2 and 1.4 K correspond to the low-field values of Brodsky and Trainor at 25 K.

Table I Molar susceptibilities of UAl_2. The units of $\frac{\partial M}{\partial H}$ are 10^{-9} m^3/mole.

T(K)	$\frac{\partial M^a}{\partial H}$ (1.5 T)	$\frac{\partial M}{\partial H}$ (2 - 5 T)	$\frac{\partial M}{\partial H}$ (20 - 35 T)
1.4	-	56.5	47.3
4.2	55.0	55.4	47.3
20	-	47.7	47.7
25	47.3	-	-
50	46.6	-	-
77	45.5	45.9	45.9

a) Ref. 4 - Brodsky and Trainor.

The temperature and field dependence of the magnetic susceptibility suggest that an equivalent contribution can be suppressed either by applying a magnetic field of about 20 T or increasing the temperature to T_{SF}.

The temperature dependent susceptibility has also been measured from 4.2 to 270 K at 1.2 and 8.0 T. The low temperature data corresponds to an expression $\chi = \chi_0 (1 - T^2/T_{SF}^2)$. The change of the specific heat in a magnetic field can be predicted from the susceptibility through the Maxwell relation,

$$\left(\frac{\partial S}{\partial \mu_0 H} \right)_{T,p} = \left(\frac{\partial M}{\partial T} \right)_{H,p}. \qquad (7)$$

On taking the temperature derivative and using $M = \chi H$, we find that

$$\left(\frac{\partial C_p}{\partial \mu_0 H}\right) = TH \left(\frac{\partial^2 \chi}{\partial T^2}\right)_{H,p} \tag{8}$$

Since the electronic contribution to the specific heat is linear in temperature and using the T^2 fit to the susceptibility, we find that Eq. 8 can be written as

$$\left(\frac{\partial \gamma}{\partial \mu_0 H}\right) = -2\ H\ (\chi_0/T_{SF}^2)_{H,p} \tag{9}$$

From Table I, χ_0 decreases with increasing field thus the specific-heat coefficient γ diminishes with a power law less than H^2. The specific heat experiments by Trainor et al. [4] are in very good agreement with the suppression of $\Delta\gamma/\gamma$ calculated here to be -1.25% at 4 T.

The magnetoresistance shows a bending over at fields of about 15 T, as shown in Fig. 3. The power law is $\frac{\Delta\rho}{\rho} = H^n$; for fields between 0 and 15 T, n = 1.45, while above 15 T, n = 1.3. Hertel, Appel and Fay [8] calculated the magnetic field dependence of UAl_2 where a 14% suppression of the SF contribution is predicted at 10 T. If we assume that the magnetoresistance due to paramagnons and other effects (band, e - e, e - p, etc.) are in series then the total resistivity may be written as $\Delta\rho = \Delta\rho_{SF} + \Delta\rho_{other}$. If we additionally assume that ρ_{SF} is maximum at zero field (contributing 30% at H = 0, from an extrapolation above 15 T) then the paramagnon contribution is suppressed by 75% at 10 T.

We find that the paramagnon effects at low temperature in UAl_2 are large and are frozen or suppressed in fields above H_{SF}. The characteristic field and temperature are related by $g\mu_B H_{SF} = k_B T_{SF}$, with g = 2. As a result of suppression of SF (above H_{SF}) it is quite plausible that the de Haas-van Alphen effect can only be seen when all paramagnon contributions to the effective mass are frozen out. In preliminary experiments, dHvA oscillations have been observed in UAl_2 only above 20 T [9]. In addition, these experiments suggest that field and temperature effects on the paramagnon contributions to the transport properties are closely related.

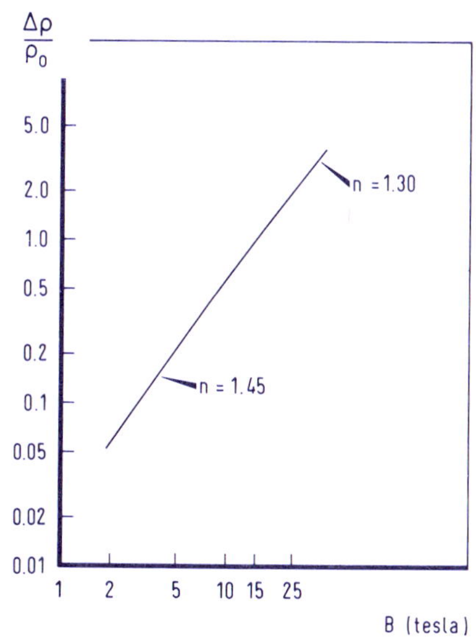

Fig. 3 Magnetoresistivey of UAl_2 at 4.2 K.

Magnetic Phase Diagram of a Ferroelectric Chiral Smectic Liquid Crystal near the Lifshitz Point

The Lifshitz point [10] is that point on a line of second-order phase transitions where an instability occurs in wave vector space. It represents a special case of a triple point between a disordered, homogeneously ordered ($\vec{k} = 0$), and modulated phase ($\vec{k} \neq 0$). It has been predicted [11] that a Lifshitz point exists in the T - H phase diagram of chiral smectic liquid crystals if the magnetic field is applied parallel to the smectic layers.

Here the phase diagram of the chiral ferroelectric smectic liquid crystal p-decyloxybenzilidene-p'-amino-2-methybutyl cinnamate (DOBAMBC) is presented [12] which suggests that a Lifshitz point exists between the disordered smectic-A, the helicoidally ordered smectic-$C^\star$, and the homogeneously ordered smectic-C phases. The observed T_c - H phase diagram in magnetic field up to 14.5 T is shown in Fig. 4. The A - $C^\star$ line was found to be of second order whereas the C - $C^\star$ transition line is of first order close to the λ line. The discontinuity in $\vec{k}$ becomes smaller as the Lifshitz point is approached along the $C^\star$ - C boundary. Above H = 8.5 T the C - $C^\star$ line changes its direction so that a reentrant smectic-$C^\star$ phase was found.

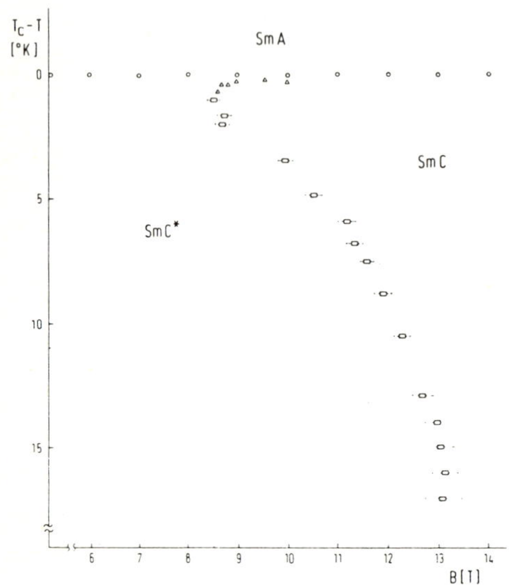

Fig. 4 Phase diagram of chiral DOBAMBC near the smectic A - $C^\star$ - C triple point. The magnetic field H is applied parallel to smectic layer.

The temperature and magnetic field dependence of the in-plane component of the dielectric constant ε and the pitch $2\pi/k$ of the smectic-$C^\star$ helix were measured in order to determine the phase boundaries. Simultaneous dielectric and optical measurements were performed using optically flat glass plates with transparent SnO_2 electrodes. The molecules were oriented parallel to the cell walls by slowly cooling ($1^oC/h$) the system through the isotropic-smectic A transition in a magnetic field of 10 T. The C - $C^\star$ phase boundary and the critical magnetic field H_c for unwinding the smectic-$C^\star$ helix was determined by measuring $\varepsilon = \varepsilon(H)$ at T = const. The sweep rate was 1.45 T/min. The A - $C^\star$ boundary was determined by measuring $\varepsilon = \varepsilon(T)$ at H = const.

The inverse pitch ($\vec{k}/\pi$) must tend to zero as the Lifshitz point is approached. If not, the $\vec{k}$ of the C - $C^\star$ transition is discontinuous at the A - $C^\star$ - C triple point. The magnetic field dependence of the wave vector of the $C^\star$ helix along the λ line has been given [2] as

$$k_\lambda^2(H) = k_0^2(1 - H^4/H_L^4) \quad , \quad H < H_L \tag{10}$$

while $k_\lambda = 0$ for $H > H_L$. Fig. 5 shows k measured as a function of H for constant T. The discontinuity at the $C^\star$ - C transition decreasing with increasing T as the Lifshitz point is approached. The isotherm (T_c - T = 0.2) can be reasonably fit by Eq. 10 with H_L = 10.2 T. However, for T_c - T = 0.1, H_L is larger than 14 T. The system behaves in the way predicted [2] for $T_c - T > 0.2$, however, for $T_c - T < 0.2$ the critical field for unwinding the helix strongly increases with increasing temperature.

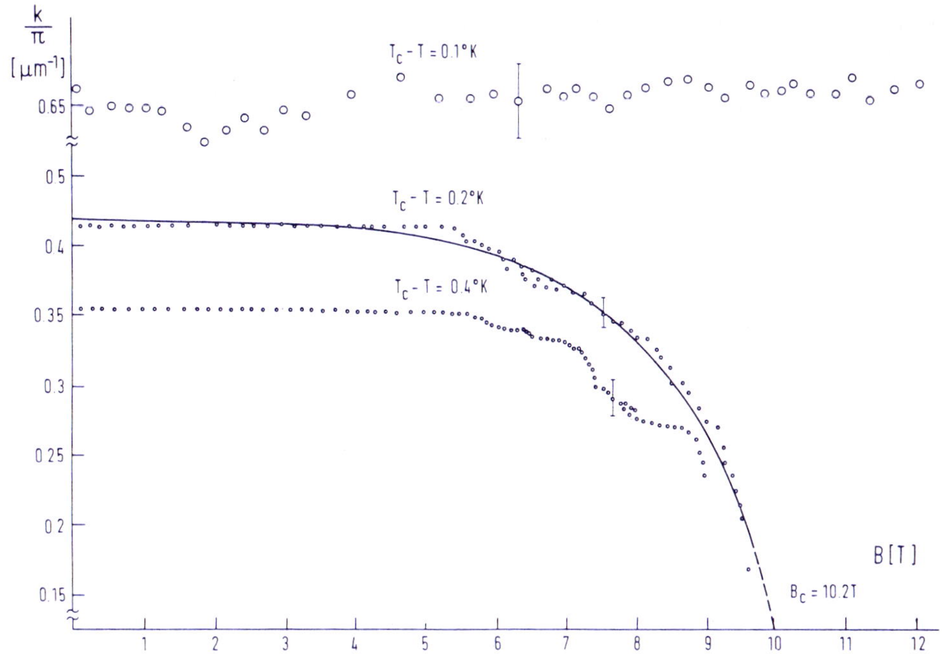

Fig. 5 Magnetic field dependence of the inverse pitch for various temperatures. The solid line through the T_c - T = 0.2 K is Eq. 10 with H_L = 10.2 T.

Transition from 3D to 2D Conduction in a GaAs-FET

It has been previously suggested [13] that it is possible to achieve a transition from a 3D to quasi 2D electron conduction in a GaAs field effect transistor (FET) by reducing the channel thickness with the gate bias V_g. In the hopping regime this occurs when the hopping distance is close to the channel thickness. In the metallic regime the electron motion perpendicular to the channel is confined to a narrow potential well, motion along the channel remains free electron like. The 3D density of states splits up into discrete sub-bands with only the ground state sub-band being occupied (2D limit) as t tends to zero. For large t, E_F should remain constant (3D limit) and fall for small t when electrons are removed from the channel (2D limit); finally E_F goes to zero at t = 0.

Four experimental measurements were made on each device [14] capacitance, resistance, Hall effect and Shubnikov-de Haas effect. The doping concentration N_D could then be independently confirmed from the capacitance measurements on one hand and the SdH and Hall effect on the other. SdH oscillations for various V_g's with constant substrate voltages were measured and from the periodicity of their oscillations, the Fermi energy is readily determined from

$$n_L = \frac{m^\star E_F}{\hbar e B} - \frac{1}{2} \qquad (11)$$

where n_L is the Landau level index. For the device with a doping concentration of $N_D = 2 \times 10^{17}$ cm^{-3}, the Fermi energy is plotted in Fig. 6. From the oscillatory part of the magnetoresistance

$$\cos\left(\frac{2\pi(E_F - E_n)}{\hbar\omega} - \alpha \right) \qquad (12)$$

each sub-band contributes a periodicity dependent on $\Delta E_n = E_F - E_n$. The theoretical oscillation periodicity obtained from Eq. 12 was used to calculate the sub-band energies in Fig. 6 [14]. As t is decreased towards zero it will eventually become less than the mean donor separaration $\bar{r}$ which is 174 Å. A transition from metallic behaviour was observed via the quenching of SdH oscillations for values of $t \simeq \bar{r}$. It can be seen, in Fig. 6, that there is a reasonable agreement between the experimental points and the theoretical predictions for E_F. The fall in E_F as the conducting electron gas in a GaAs FET is confined is confirmed resulting from a 3D to 2D transition.

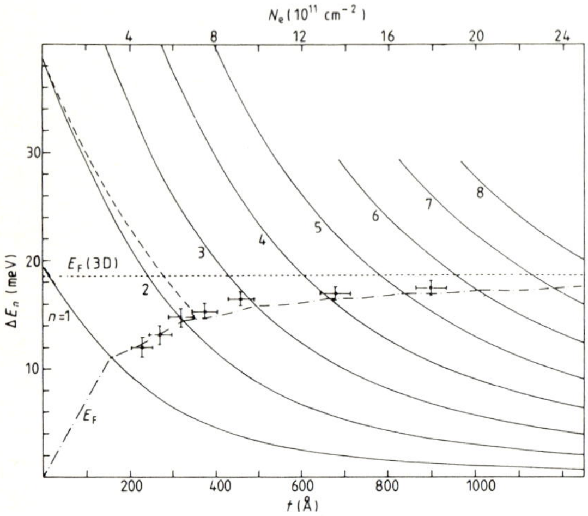

Fig. 6 Level quantization and E_F as a function of channel thickness (t) in a $N_D = 2 \times 10^{17}$ cm^{-3} device. Solid curves are WKB level separation (see ref. 14) ΔE_n and the chain curve to the corresponding E_F. The straight horizontal broken line is E_F (3D) for a uniform non-interacting gas. Measured E_F values are presented as full circles with error bars.

Here we have presented a variety of magnetically induced phase transitions measured at the High Field Magnet Laboratory of the University during this past year. The extent to which magnetic fields help us understand the properties of matter remains significant.

Acknowledgments

We would like to thank Klaas van Hulst, Henk Muileman and Jos Rook of the Nijmegen Magnet Laboratory for their valuable assistance and technical support of this research. Part of this work has been supported by the "Stichting voor Fundamenteel Onderzoek der Materie" (FOM) with financial support of the "Nederlandse Organisatie voor Zuiver Wetenschappelijk Onderzoek " (ZWO).

References

1. E.J.W. Verwey and J.Th.G. Overbeek, Theory of the Stability of Lyophobic Colloids, Elsevier, Amsterdam, 1948.
2. J. Svoboda, Int. J. of Mineral Processing 8, 377 (1981).
3. A more complete discussion can be found in M.R. Parker, R.P.A.R. van Kleef, H.W. Myron and P. Wyder, J. Magn. Magn. Mat. (to be published).

4. R.J. Trainor, M.B. Brodsky and H.V. Culbert, Phys. Rev. Lett. 34, 1019 (1975); A.J. Arko, M.B. Brodsky and W.J. Nellis, Phys. Rev. B 5, 4564 (1972); M.B. Brodsky and R.J. Trainor, Physica 91B, 271 (1977).
5. W.F. Brinkman and S. Engelsberg, Phys. Rev. 109, 417 (1968); A.B. Kaiser and S. Doniach, Int. J. Magn. 1, 11 (1970); M.T. Béal-Monod, S.K. Ma, and D.R. Fredkin, Phys. Rev. Lett. 20, 750 (1968).
6. For a more extensive report on this phenomena see: J.J.M. Franse, P.H. Frings, F.R. de Boer, A. Menovsky, C.J. Beers, A.P.J. van Deursen, H.W. Myron and A.J. Arko, Phys. Rev. Lett. 48, 1749 (1982).
7. I.Y. Hsiang, J.W. Reister, H. Weinstock, G. Crabtree, and J.J. Vuillemin, Phys. Rev. Lett. 47, 523 (1981).
8. P. Hertel, J. Appel, and D. Fay, Phys. Rev. B 22, 534 (1980).
9. A.J. Arko and J.E. Schirber, J. Phys. (Paris) Colloq. 40, C4 - 9 (1979).
10. R.M. Hornreich, M. Luban, and S. Shtrikman, Phys. Rev. Lett. 35, 1678 (1975).
11. A. Michelson, Phys. Rev. Lett. 39, 464 (1977).
12. For a more complete description of this work see: I. Muševič, B. Žekš, R. Blinc, Th. Rasing and P. Wyder, Phys. Rev. Lett. 48, 192 (1982).
13. M. Pepper, J. Phys. C 10, L173 (1977).
14. For a more complete description see: D.A. Poole, M. Pepper, K.-F. Berggren, G. Hill and H.W. Myron, J. Phys. C 15, L21 (1982).

HIGH MAGNETIC FIELD RESEARCH AT THE FRANCIS BITTER NATIONAL MAGNET LABORATORY

R.L. Aggarwal
Francis Bitter National Magnet Laboratory*
and
Department of Physics
Massachusetts Institute of Technology
Cambridge, Massachusetts 02139 USA

I. INTRODUCTION

A large number of water-cooled (Bitter), superconducting and combination ("hybrid") magnets are available to scientists and engineers from throughout the world for high field research at the Francis Bitter National Magnet Laboratory. Continuous fields ranging up to 30 T in a 3.3 cm bore and 25 T in a 5.4 cm bore are provided by the recently developed hybrid magnets. Field and homogeneity characteristics of water-cooled and hybrid magnets are summarized in Table I. In addition, there are four high-homogeneity superconducting magnets for nuclear magnetic resonance studies at 272, 304, 306 and 498 MHz. The highest frequency NMR spectrometer utilizes an 11.7 T niobium-titanium superconducting magnet operating at 2.3 K. This magnet has been operating continuously in the persistent mode since August 1979 with a cumulative decay of less than 1 part in 10^6.

High magnetic field research at the Francis Bitter National Magnet Laboratory covers a broad spectrum of basic and applied research programs. Most of the basic research is in the area of condensed matter physics, solid state chemistry and biophysics. In condensed matter physics, current emphasis is on superconductivity, which accounts for about half of all the research programs pursued by the Laboratory's research staff and the visiting scientists from within the United States and from a number of other countries.

In this paper we present a brief review of some of the recent high-field work in semiconductors at the Francis Bitter National Magnet Laboratory.

II. SEMICONDUCTOR RESEARCH

a. Raman Scattering from Donors in CdS

Feigenblatt, et al.[1] have measured Raman scattering from donors in CdS with magnetic fields up to 19 T, with the objective of studying electron-phonon interaction in this polar material. In this context, Raman spectroscopy has a potential advantage over the infrared spectroscopy, since strong infrared absorption due to polar phonons would prevent the observation of donor transitions via infrared absorption in the resonance region where the donor transition energies approach the phonon energies.

---

*Supported by the National Science Foundation through its Division of Materials Research.

Table I. Field and Homogeneity of Water-Cooled and Hybrid Magnets.

Category	Magnets	Bore (cm)	Radial Access (cm)	Field[a] (T)	Homogeneity Coefficients[b] c_1	c_2	c_3	c_4
Hybrid Systems	1-Z	3.3	—	29.0	0	-520	0	4
	2-Z	5.4	—	24.5	-29	7	1	-11
Small Bore	1-G 1-H	3.3	—	22.0	-5	-890	-1	-5
Standard	2-P, 2-T	5.4	—	19.5	-10	-360	-1	-1
	2-B,D,L, M,S,X	5.4	—	15.5	-11	-430	-1	-1
	2-A	5.4	—	10.0	-4	-430	0	-2
Homogeneous Field	2-Y	5.4	—	19.0	-10	-9	0	-14
	2-F	5.4	—	15.0	0	2	0	-4
	2-V	5.4	—	15.0	0	4	0	-9
	2-E	5.4	—	6.7	$\frac{B_{max}-B_{min}}{B_{avg}}$ = 5.2% over 16 cm length			
Radial Access	2-C	5.4	2.9	10.0	-11	1700	0	-55
	2-H	5.4	1.1	11.0	0	810	0	-52
	2-K	5.4	1.9	10.0	0	1600	0	-76
	2-R	5.4	5.4	8.0	0	2500	0	-35
Large Bore	4-D	10.5	—	15.0	-5	-420	0	1
	4-B	10.5	—	11.0	-7	-440	0	0
	4-E	10.5	—	10.0	-4	-59	0	-5
	6-B	14.7	—	12.5	0	-240	0	0
	10-A	24.8	—	7.7	-1	-180	0	0

[a]The numbers in this column represent average values. Variations of up to ±5% can be expected over a year's time, due primarily to changes in cooling water temperatures.

[b]$\Delta B/B_0 \approx c_1 z + c_2 z^2 + c_3 z^3 + c_4 z^4$; z measured in cm, $\Delta B/B_0$ measured in parts in 10^5.

Figure 1 shows the Raman spectra of n-type CdS at a temperature $T \simeq 2$ K. The $B = 0$ spectrum shows the $1s \rightarrow 2s$ donor transition labeled 2s, along with a TO-phonon of A_1/E_1 symmetry and a nonpolar phonon of E_2 symmetry. Application of the magnetic field splits this donor peak into four components corresponding to the transitions $1s \rightarrow 2p_-$, $1s \rightarrow 2p_0$, $1s \rightarrow 2s$ and $1s \rightarrow 2p_+$. In the absence of electron-phonon interaction, the Zeeman splitting between the $2p_+$ and $2p_-$ components should be equal to the

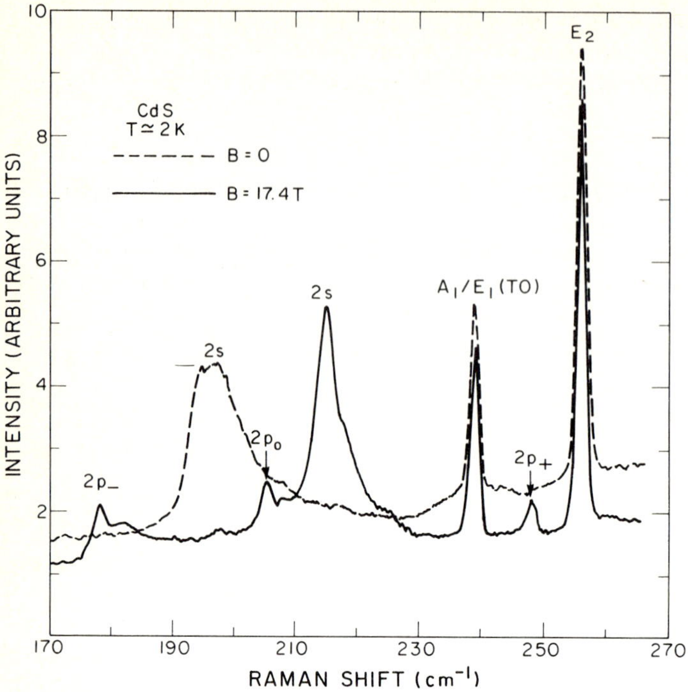

Figure 1. Raman spectra of n-type CdS with magnetic field $\vec{B} \parallel \vec{c}$, using 5017 Å argon-ion laser line. (After Feigenblatt, et al.[1]).

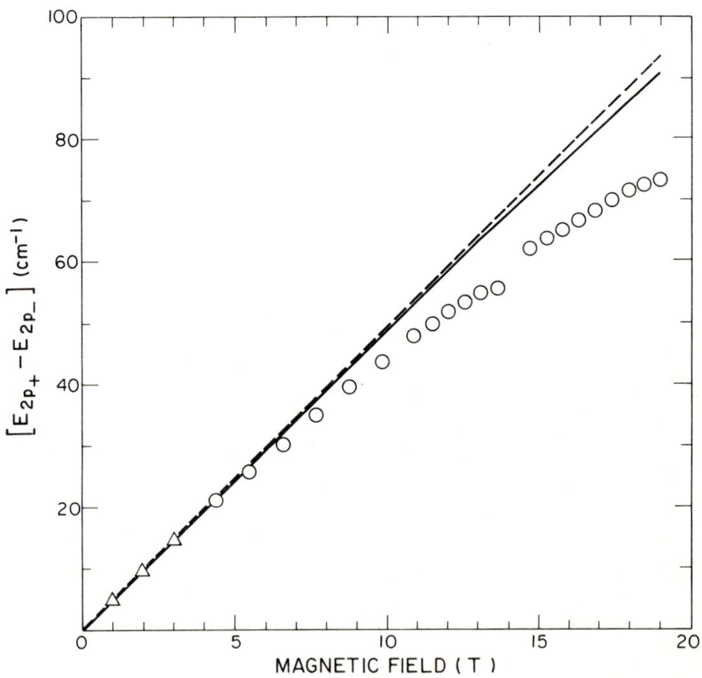

Figure 2. Zeeman splitting of the 2p± donor levels. Circles denote the data points. The dashed and solid lines represent theoretical curves for $\alpha = 0$, and $\alpha = 0.65$, respectively (After Feigenblatt, et Al.[1]).

cyclotron energy $\hbar\omega_c = \frac{\hbar eB}{m^*}$ where m^* is the electron effective mass. The observed splitting with magnetic field is shown in Fig. 2 where the dashed line indicates the cyclotron energy. The solid curve shows the theoretically expected behavior, assuming electron-LO phonon coupling constant $\alpha = 0.65$, LO-phonon energy $\hbar\omega_{LO} = 306\ cm^{-1}$, and donor Rydberg $R_y = 260\ cm^{-1}$. It is clear that the data shown by the circles in Fig. 2 shows a significant departure at high magnetic fields from the usually expected electron-LO phonon interaction. The data appears to show a pinning of the $1s \rightarrow 2p_+$ donor transition in the region of nonpolar E_2-phonon mode at 256 cm^{-1}, as can be seen in Fig. 3 which shows a plot of the energies for the transitions $1s \rightarrow 2p_-$ and $1s \rightarrow 2p_+$ as a function of the applied magnetic field.

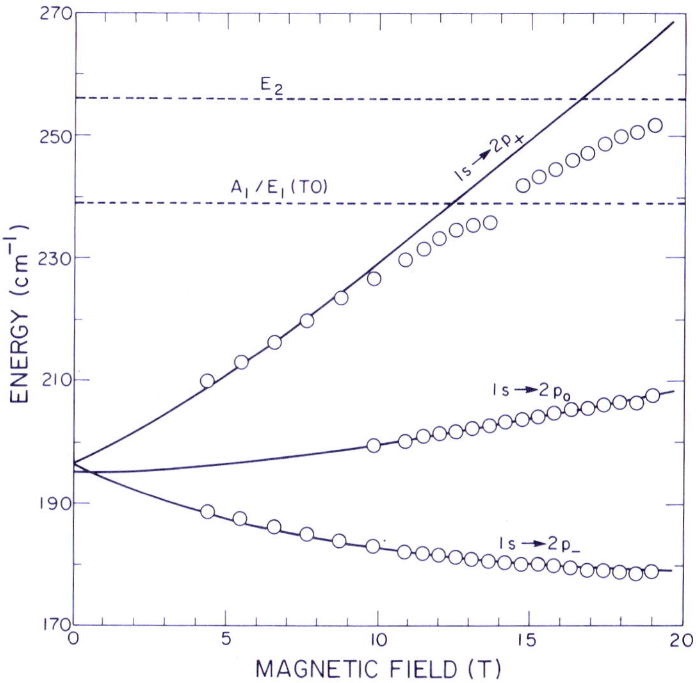

Figure 3. Variation of the $1s \rightarrow 2p$ donor transition energies with magnetic field. The solid lines indicate the calculated behavior for $\alpha = 0.65$, and $\hbar\omega_{LO} = 306\ cm^{-1}$. (After Feigenblatt, et al.[1]).

It is essential to extend the present measurements to higher fields in order to confirm the electron interaction with the E_2-phonon at 256 cm^{-1} through the observation of the upper branch. In addition, higher field measurements would enable us to tune the $1s \rightarrow 2p_+$ transition through the LO-phonon at 306 cm^{-1}.

Both the LO-phonon frequency (213 cm^{-1}) and electron effective mass (~0.12) for CdSe are smaller than those for CdS. Therefore, lower magnetic fields would be required to study the electron-LO phonon interaction in CdSe. Unfortunately, Raman transitions in CdSe were accompanied by considerable photoluminescence which made it difficult to follow the donor Raman transitions to fields required for tuning the $1s \rightarrow 2p_+$ transition through the LO-phonon.

b. Two-Photon Spectroscopy of A-exciton in CdS

Seiler et al.[2] have studied the Zeeman splitting and diamagnetic shifts of the n = 2 and n = 3 states of the free A-excitons in CdS, using two-photon spectroscopy. A schematic of the experimental setup is shown in Fig. 4. A Q-switched YAG laser is used to generate the two light beams, one tunable at $\hbar\omega_V \simeq 1.8$ eV from a dye laser and the other fixed at $\hbar\omega_{IR} \simeq 0.784$ eV, produced by stimulated Raman scattering in hydrogen gas.

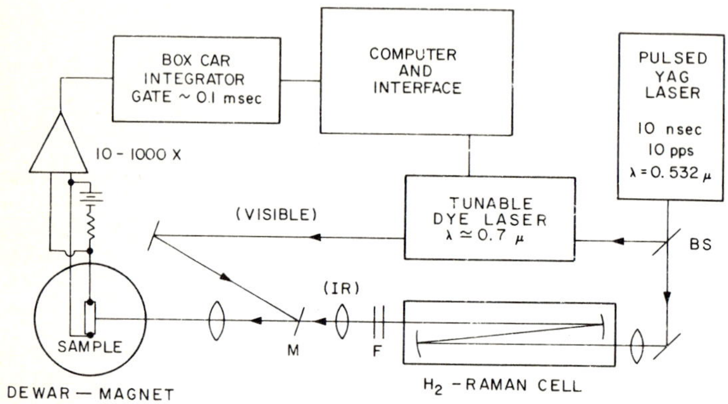

Figure 4. Schematic of the experimental arrangement for two-photon spectroscopy using photoconductivity detection. (After Seiler, et al.[2]).

High resolution spectra were obtained through the detection of two-photon photoconductivity. Transitions corresponding to $2P_0$, $2P_\pm$, $3P_0$, $3P_\pm$ were clearly resolved at B = 0. Zeeman splitting of the $2P_\pm$ and $3P_\pm$ states as a function of the applied magnetic field is shown in Fig. 5. As expected, the Zeeman splitting is found to increase linearly with magnetic field. Diamagnetic contribution to

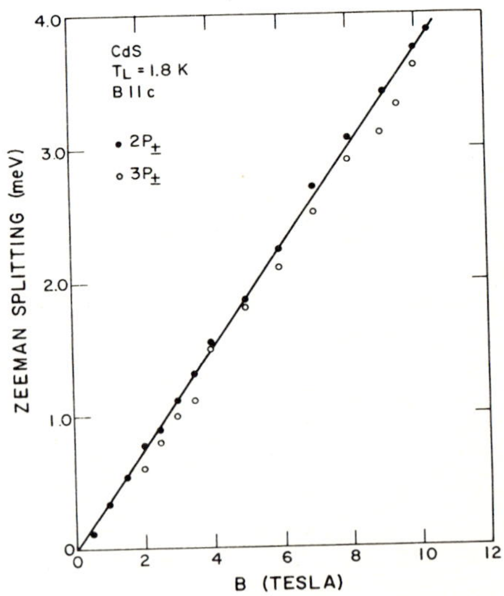

Figure 5. Zeeman splitting of the $2P_\pm$ and $3P_\pm$ A-exciton states in CdS. The solid line is a linear fit to the $2P_\pm$ data. (After Seiler, et al.[2]).

the energy shift is found by adding/subtracting one half of the Zeeman splitting from the P_-/P_+ state. The magnetic field dependence of the resulting diamagnetic energy shifts is shown in Fig. 6. As can be seen, $2P_\pm$ and $3P_\pm$ exciton states show considerable deviation from the quadratic behavior at high magnetic fields. This deviation has been explained by considering level mixing using Larsen's variational calculation.[3]

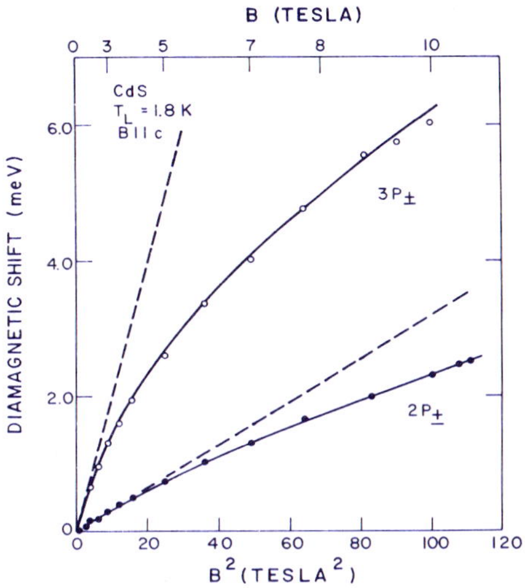

Figure 6. Diamagnetic shift of the 2P and 3P A-exciton states in CdS. The circles represent the data points. The solid curves represent theoretical values determined from a variational calculation which includes level interaction. The dashed curves are straight line fits to the data at low fields. (After Seiler, et al.[2]).

c. Four-Wave Spectroscopy of Arsenic Donors in Ge

Four-wave spectroscopy is one of the variants of coherent Raman scattering. Figure 7 shows a schematic of the experimental setup used by Jagannath et al.[4] for studying the magnetic field dependence of the valley-orbit splitting, $4\Delta(B)$, of phosphorous and arsenic donors in germanium. Two CO_2 lasers are used to produce ~100 ns pulses at frequencies ω_1 and ω_2. The two beams are combined using a beam splitter

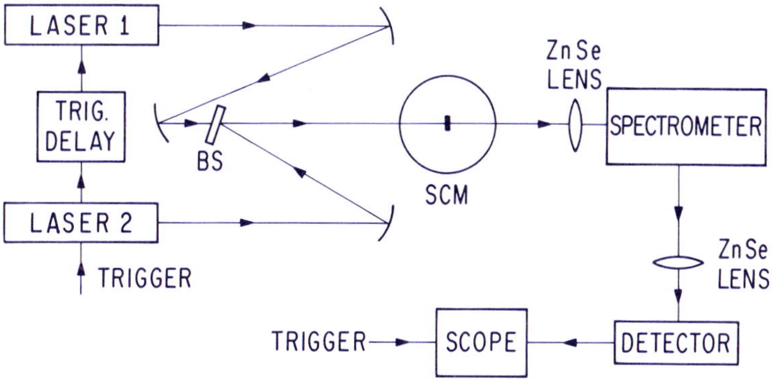

Figure 7. Schematic of the experimental setup for four-wave spectroscopy. (After Jagannath, et al.[4]).

and focused on the sample held at 1.8 K in a superconducting magnet. The four-wave output, at frequency $\omega_4 = 2\omega_1 - \omega_2$, is separated from the two incident beams using a Spex double grating monochromator. For a fixed difference, $\Delta\omega$, in the frequencies of the input laser beams, the power at ω_4, $P\omega_4$, is measured as a function of B. $P\omega_4$ exhibits a resonance at a value of B when $\Delta\omega = 4\Delta(B)$.

Figure 8 shows the four-wave power from arsenic-doped germanium as a function of the magnetic field applied along a [111] crystal axis, for $\Delta\omega = 34.83\ cm^{-1}$ and $36.09\ cm^{-1}$. Two resonance peaks can be clearly identified for each value of $\Delta\omega$. The two peaks are identified as originating from the spin-up and spin-down levels of ground state to the corresponding spin level of the upper state. This result

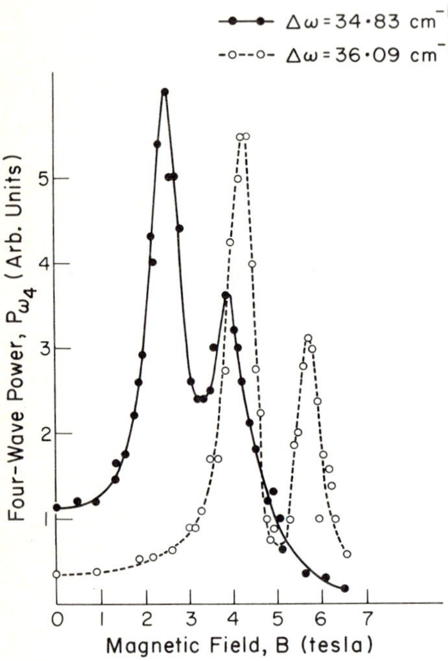

Figure 8. Four-wave power, $P\omega_4$, as a function of the magnetic field for arsenic-doped germanium at T = 1.8 K with magnetic field B ‖ [111]. (After Jagannath, et al.[4]).

indicates that considerable spin heating occurs during the 100 ns laser pulses; otherwise the spin-up peak would have been unresolvable at T = 1.8 K and B = 6 T for which spin-splitting/kT ≃ 3.5.

Figure 9 shows a plot of the valley-orbit splitting, $4\Delta(B)$, as a function of B^2 at which resonances occur for each of the two spin components. Solid squares, circles, and triangles represent data points for different runs on the same sample. The solid lines represent calculated values. As can be seen in Fig. 9, excellent agreement is obtained between measured and theoretical values of $4\Delta(B)$.

d. Spin-Flip Raman Scattering and Magnetization of $Cd_{1-x}Mn_xSe$

During the last few years, a number of novel effects have been reported in semimagnetic semiconductors[5] such as $Cd_{1-x}Mn_xSe$. These effects arise from a large exchange interaction between the spin of a free or a loosely bound electron and the spins of the magnetic ions. When a magnetic field $\vec{B}$ is applied, net alignment of the Mn^{++} spins $\langle \vec{S}_i \rangle$ leads to electron spin splitting ΔE which is considerably

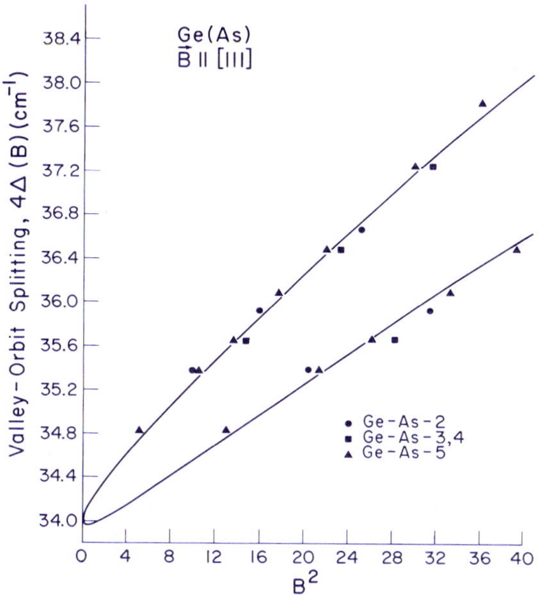

Figure 9. Valley-orbit splitting plotted as a function of B^2, where B is in tesla. The solid circles, triangles, and squares represent different runs made on the sample. The upper and lower lines are theoretical fits for the spin ↓ → spin ↓ and spin ↑ → spin ↑ transitions, respectively. (After Jagannath et al.[4]).

larger than the Zeeman splitting $g^*\mu_B B$ due to the direct action of $\vec{B}$ on the electron spin $\vec{s}$. Spin-flip Raman scattering (SFRS) provides a convenient probe for studying this exchange interaction. Nawrocki et al.[6] have reported giant effective g-values, $g_{eff} \sim 200$ compared to $g^* \simeq 0.5$ in some samples of $Cd_{1-x}Mn_xSe$. A finite spin-splitting was observed even at B = 0 and is attributed to the formation of a bound magnetic polaron (BMP).

Recently, Heiman et al.[7] have extended the SFRS measurements to higher magnetic fields and higher temperatures. Their results for ΔE vs. magnetic field for donor bound electrons are shown in Fig. 10. As can be seen in Fig. 10, ΔE increases linearly with B at moderate fields and saturates at high fields. This behavior can be understood in the framework of the mean-field approximation which yields

$$\Delta E = g^*\mu_B B + \alpha x\, N_0 \langle S_z \rangle$$

where αN_0 is the exchange energy and $\langle S_z \rangle$ is the absolute value of the average component of the Mn^{++} spins along $\vec{B}$. Since the magnetization M is proportional to $x\langle S_z \rangle$, αN_0 can be determined by combining the measurements of ΔE and M at high fields. Figure 11 shows a plot of M vs. magnetic field as obtained by Shapira et al.[8], who have deduced a value of 261 ± 13 meV for αN_0.

Variation of ΔE with temperature at low magnetic fields is found to be in good agreement with the BMP theory of Heiman et al.[7] As shown in Fig. 12, the spectral lineshape of the BMP at B = 0 is in excellent agreement with the predictions of BMP theory.

On account of the giant g-values for $Cd_{1-x}Mn_xSe$ with bandgap $E_g \sim 2$ eV in the visible region, this material is a good candidate for magnetically tunable radiation source. Heiman[9] has observed laser emission from stimulated SFRS by pumping near the bandgap. A threshold pump intensity of less than 1 MW/cm^2 was required.

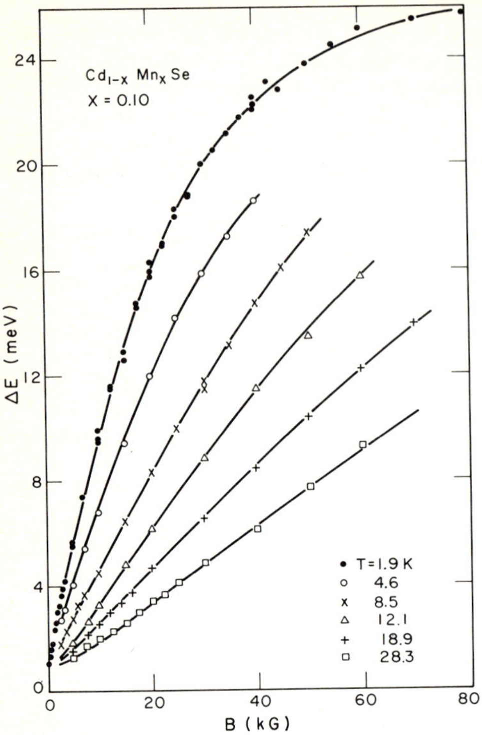

Figure 10. Spin-flip energy ΔE vs. magnetic field B in $Cd_{1-x}Mn_xSe$, x = 0.10 for various lattice temperatures. (After Heiman, et al.[7]).

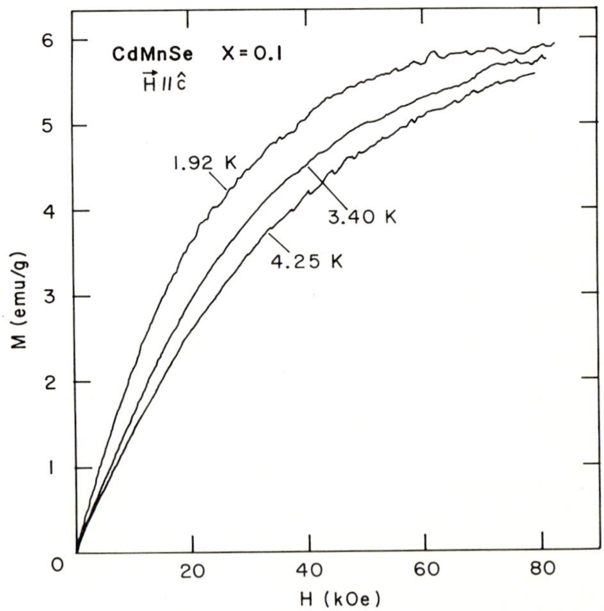

Figure 11. Magnetization (per gram) M vs. magnetic field for T = 1.92, 3.40 and 4.25 K. (After Shapira, et al.[8]).

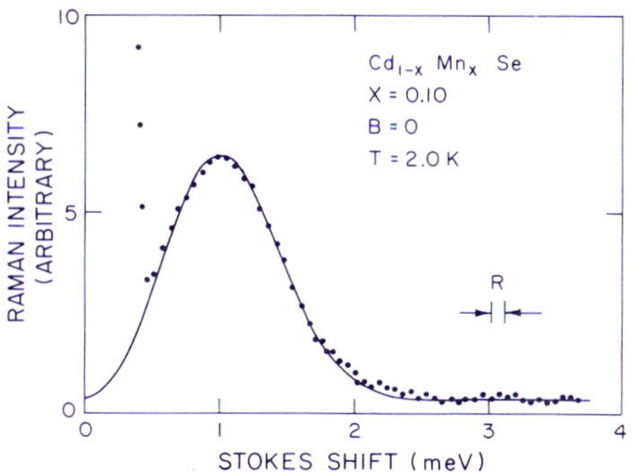

Figure 12. Spectral lineshape of bound magnetic polaron from spin-flip Raman scattering in $Cd_{1-x}Mn_xSe$, x = 0.10 at zero magnetic field. Solid curve is the theoretical fit to the data represented by the points. (After Heiman, et al.[7]).

e. Far Infrared Photoconductivity of n-Type GaAs

Afsar and Button[10] have studied the donors in GaAs via far infrared photoconductivity with a high resolution Fourier transform spectrometer. They have shown that the spectral lines due to transitions from the 1s ground state to the 2p state become considerably narrow in the presence of a large magnetic field. This is illustrated in Fig. 13 which shows the $1s \rightarrow 2p_-$ lines at B = 4 T and 19.33 T. In addition, the spacing between the peaks corresponding to different donor impurities increases with magnetic field. Both of these features facilitate the identification of the donors, using the tuning or "signature" curve for a given donor (see Fig. 14).

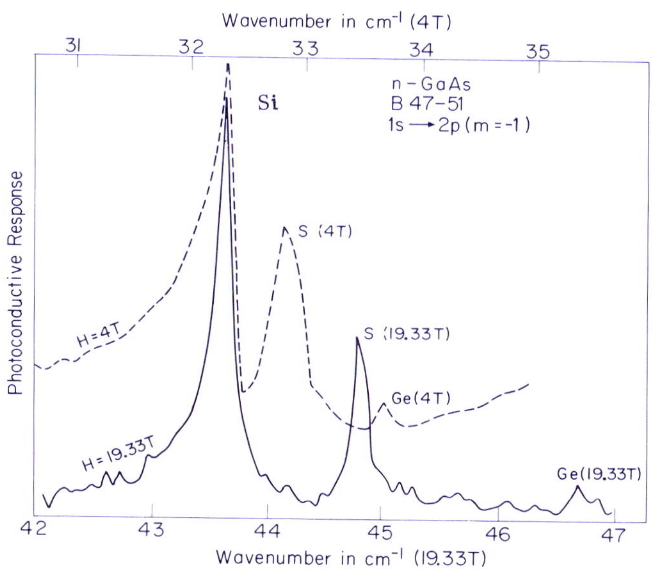

Figure 13. Far infrared photoconductivity spectra for the $1s \rightarrow 2p_-$ transition of Si, S, and Ge donors in GaAs at magnetic field B = 4 and 19.33 T. (After Afsar, et al.[10]).

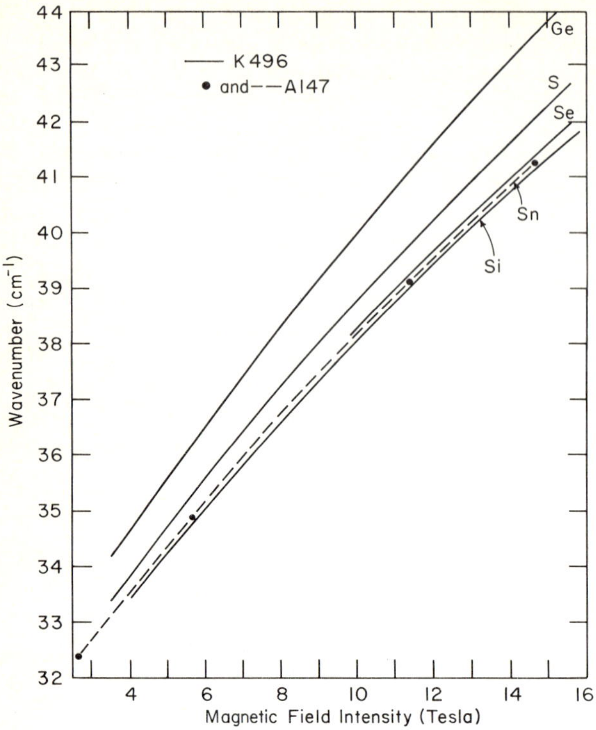

Figure 14. "Signature" curves for the energy of the 1s → 2p₋ transition for Si, Sn, Se, S and Ge donors in GaAs as a function of magnetic field. (After Afsar, et al.[10]).

f. Magneto-Transport of 2D Electrons

Tsui et al.[11] have recently measured the diagonal and the off-diagonal components ρ_{xx} and ρ_{xy} of the resistivity tensor for the 2D electrons in GaAs-AℓGaAs heterojunctions at high magnetic fields up to 22 T. Figure 15 shows plots of ρ_{xy}, ρ_{xx} as a function of B at four different temperatures $T < 5$ K. At integral values of the Landau level filling factor $\nu = nh/eB$, the data shows the characteristic features of the quantized Hall plateaus in ρ_{xy} and vanishing of ρ_{xx}. In the extreme quantum limit ($\nu < 1$), ρ_{xx} shows a dip at $\nu = 1/3$ and ρ_{xy} approaches a step of $3h/e^2$ at $\nu = 1/3$ as T decreases.

The appearance of the above features at $\nu = 1/3$ is a striking new result which may be evidence for a new electronic state involving a Wigner solid or a charge density wave state with triangular crystal symmetry.

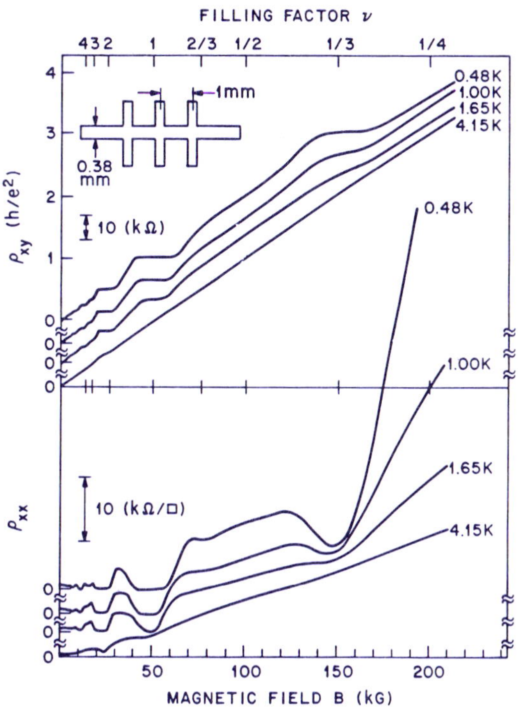

Figure 15. Magnetic field dependence of the diagonal and off-diagonal resistivity components ρ_{xx} and ρ_{xy} for 2D electrons in a GaAs-$A\ell_{0.3}Ga_{0.7}As$ sample with $n = 1.23 \times 10^{11}/cm^2$ and $\mu = 9 \times 10^4 cm^2/V$-sec. (After Tsui, et al.[11]).

Acknowledgements — We wish to thank R. Feigenblatt, D. Heiman, B. Lax, Y. Shapira and P.A. Wolff for communicating their results prior to publication.

References

[1] R. Feigenblatt, R.L. Aggarwal and B. Lax, to be published.
[2] D.G. Seiler, D. Heiman, R. Feigenblatt, R.L. Aggarwal and B. Lax, Phys. Rev. B25 7666 (1982).
[3] D.M. Larsen, J. Phys. Chem. Solids 29, 271 (1968).
[4] C. Jagannath, E.R. Youngdale, D.M. Larsen, R.L. Aggarwal and P.A. Wolff, Solid State Commun. 43, 267 (1982).
[5] For recent reviews see R.R. Galazka, Inst. Phys. Conf. Ser. 43, 133 (1979) and J.A. Gaj, J. Phys. Soc. Japan 49 Suppl. A, 797 (1980).
[6] M. Nawrocki, R. Planel, G. Fishman and R. Galazka, J. Phys. Soc. Japan 49 Suppl. A, 823 (1980).
[7] D. Heiman, P.A. Wolff and J. Warnock, to appear in Proceedings of the 16th International Conference on the Physics of Semiconductors, Montpellier, France, 1982.
[8] Y. Shapira, D. Heiman and S. Foner, to be published.
[9] D. Heiman, to appear in Appl. Phys. Letters, October 1982.
[10] M.N. Afsar, K.J. Button and G.L. McCoy, Int'l. Jour. IR and MM Waves 1, 145 (1980); 1, 513 (1980).
[11] D.C. Tsui, H.C. Stormer and A.C. Gossard, Phys. Rev. Letters 48, 1559 (1982).

HIGH MAGNETIC FIELDS AT THE CLARENDON LABORATORY, OXFORD

P.A. Hudson, H. Jones and H.M. Whitworth

The University of Oxford, The Clarendon Laboratory, Oxford. OX1 3PU. UK.

The Clarendon Laboratory in Oxford has been a centre for research in high magnetic fields for a number of years. The high-field facilities have been established around a 2MW dc motor-generator and today include a number of 10T solenoids which are energised by this machine as well as a hybrid resistive/superconductive magnet which is operated as an item of established laboratory equipment in which fields of 16T are routinely produced.

In parallel with the development of resistive magnets which we describe in this paper, we have also developed several large and versatile superconductive magnet systems for shared use as well as a number of smaller systems dedicated to particular fields of study. We thus also present details of several such systems, choosing those most relevant to the subject matter of this conference. Brief reference is also made to plans to improve the facilities by pulsed techniques.

INTRODUCTION

The Clarendon Laboratory is the only centre in the United Kingdom where magnetic fields greater than 15T are routinely produced. It has had a very long association with the production and use of high magnetic fields and improvements to the facilities reflect the changing requirements of research programmes. For example, the development of the high-field transverse access system was in response to the demand for a versatile experimental configuration to facilitate magneto-optical studies of defects in solids, magnetic compounds and semi-conducting and insulating materials. Likewise the present interest in ultra-high fields has grown from the desire to extend experimental work on the magneto-optical and magneto-transport properties of semiconductors to significantly higher intensities and in particular to investigate the properties of two dimensional systems such as GaAs-GaAlAs heterojunctions, (see the paper by R.J. Nicholas, this conference).

1. RESISTIVE MAGNETS AND THE HYBRID SYSTEM

Figure 1.

The 2MW system and its magnets have been described in detail elsewhere [1-3]. The system is based on a dc motor-generator which supplies power to four magnet stations, each of which is equipped with a 50mm bore water-cooled solenoid. At two stations the magnets are mounted on wheeled trolleys and are capable of being moved and repositioned up to ±1.5m horizontally and 0.5m vertically from their mean location. The remaining magnets are in fixed locations with their bores vertical, having free access from both above and below. The solenoids are constructed following the Tsai pattern [4] and are capable of producing maximum fields of 9T.

A fifth station is devoted to the Oxford Hybrid System (Figure 1). This is a co-axial arrangement of two solenoids. The outer is superconducting and is housed in an annular cryostat with a warm bore of 240 mm. This section produces a central field of 6.5T. The inner section is a 2MW water-cooled coil which gives 9.5T in a 50 mm bore, the combination thus producing a total field of 16.0T. One feature of this approach to generating high fields is that the rapid sweep rates associated with conventional technology can still be achieved over much of the field range.

Unlike the other water-cooled magnets of the 2MW system, the inner section of the hybrid combination is not built on the Tsai pattern but is of the "polyhelix" type. This consists of a nest of co-axially concentric single-layer helices which are individually bonded and insulated turn-to-turn with a suitable adhesive. The technique, which was pioneered at Oxford [5] possesses the high efficiency (Tesla per amp.) of the familiar Bitter arrangement whilst eliminating the need for massive clamping structures. The particular advantages of adopting the more complex polyhelix approach in hybrid systems have been discussed at length in an earlier paper [6].

2. DEVELOPMENT OF RESISTIVE MAGNETS

Figure 2.

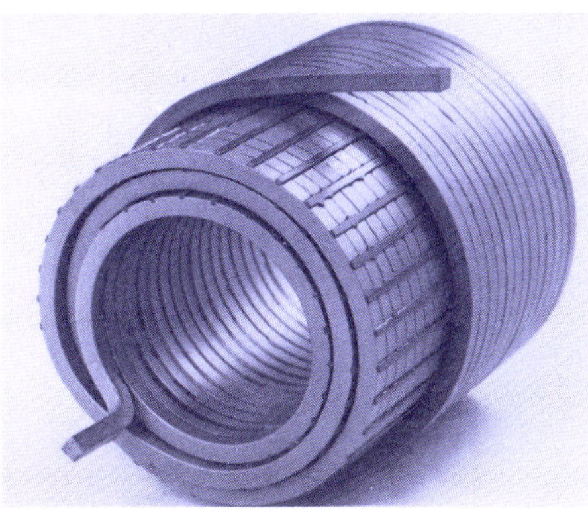

Figure 3.

Recently a burn-out of the inner four helices of the polyhelix section of the Hybrid Magnet caused us to look at new techniques for simplifying their construction. The most significant advance is our use of a dry-film adhesive, Redux 308A, manufactured in the U.K. by Ciba-Geigy. The film is a nylon-modified epoxy containing a small quantity of glass beads of 0.07 mm diameter which assist in defining the thickness of the glue-line. The adhesive has a high elongation and can be simply edgewound into the gaps between turns as shown in Figure 2. Curing is typically for 1 hour at 150C. The same adhesive is also used to bond spills of SRBF to the outside surfaces of the coils to form axial cooling channels. The new-style coils were fitted in November 1981 and have performed well.

Computations carried out by our colleagues at the Hochfeldmagnetlabor in Grenoble predict that the performance of the Oxford Magnet could be enchanced to ~ 20T by reducing its working bore to 30 mm. The trade-off is considered worthwhile and this development is being pursued.

An alternative to the polyhelix scheme is also being considered. As an extension of our work with Redux bonded single-layer coils, we are developing a continuous winding technique for multi-layer solenoids (Figure 3) in which the conductor changes section discretely two or three times within the space of the winding. The objective of this work is the production of a simple yet reasonably efficient replacement package for the resistive section of the Hybrid system, avoiding the complex inter-coil connecting arrangement used in the present magnet.

3. SUPERCONDUCTING MAGNETS

Since 1969 the High Field Support Group of the Clarendon Laboratory has been involved in producing and developing superconducting magnets. Initially the requirement was for relatively small systems dedicated to particular experiments. These allowed a number of users to become independent of the 2MW magnets which became more freely available for experiments for which they were uniquely suitable. Work has continued on the development of magnets for special purposes and to date about thirty such systems [1,7] have been produced. These have been of widely varying designs, tailored directly to the users' requirements. The most recent is described later.

For general high-field applications, the Clarendon has produced two portable superconducting magnet systems which are available as centrally administered research

COMPARATIVE DATA OF THE TWO PORTABLE SUPERCONDUCTING MAGNET SYSTEMS

	16T System	T/A System	
Magnet type	Solenoid, Nb_3Sn tape pancake wound.	Split solenoid, Nb-Ti/Nb_3Sn (filamentary)* duplex. *Demountable (under development)	
Maximum Induction at centroid, Bo_{max}	15.5T (4.2K) 16.5T (< 3.0K) (stress limited)	with Nb_3Sn section 9.4T(4.2K) ∿ 11.0T(2.0K)	without Nb_3Sn section 7.9T(4.2K) 9.3T(2.0K)
Working bore	25.4 mm	with Nb_3Sn section 25.4 mm	without Nb_3Sn section 50.0 mm
Width of axial split	N/A	10 mm with 4 x orthogonal rectangular section ports tapered to give maximum optical access in the azimuthal plane	
Homogeneity	∿ 0.13% over 10 mm diameter sphere volume	∿ 0.7% over 10 mm diameter sphere volume	
Ramp time	∿ 30 mins	∿ 20 mins	
Stability	∿ 1 in 10^4 at Bo_{max}	∿ 5 in 10^5 at Bo_{max}	
Magnet orientation	vertical	vertical or horizontal	
Top plate to field centre	967 mm	967 mm	
Cryostat neck access	150 mm	150 mm	
Optical access	1 x axial window group* *consists of 1x25mm 300K window 1x19mm 4K window	1 x axial window group* 4 x orthogonal transverse window groups* *consists of 1x25mm 300K window 1x19mm 77K window 1x19mm 4K window	
Window materials	Sapphire	Sapphire or BaF_2	
Optical access f numbers	Axial: f/7.6	Axial: f/7.6 Transverse: f/8 (azimuthal plane) f/10 (axial plane)	
300K access to magnet bore	A 19 mm working bore re-entrant finger set*, which can be fitted in the bottom axial window ports, and a straight through room temperature bore tube are available as accessories. *Used for n.m.r. field calibrations	As 16T system	
Vertical positioning of cryostat	The cryostat can be used at floor level or raised 640 mm in a high level portable platform to allow access to the bottom window	An integral feature of the system is a winch operated, variable height, single mast, portable platform which allows the transverse optical window level to be varied between 300 mm & 1300 mm above floor level	
Total operating time from 1 fill of liq. He reservoir	Typically > 10 h	Typically > 15 h	

TABLE 1

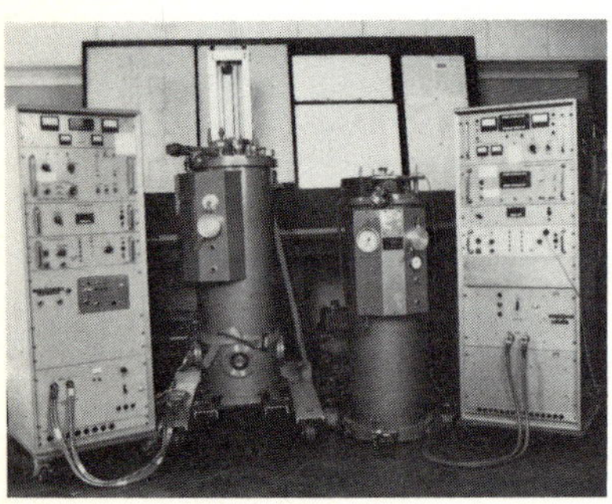

Figure 4.

facilities. The philosophy of rendering large expensive magnets available to a wide range of research projects by making both the dewar and the associated power supply and instrumentation fully mobile, is widely recognised as having been pioneered at the Clarendon [8] and over ten years has proved highly successful in satisfying the high field requirements - up to 16T - of a considerable number of research programmes. Frequently, the facilities are extended to visitors from other universities - both U.K. and overseas - government research establishments and industry. The design and mode of operation of both systems, which are shown together in Figure 4, have been described fully elsewhere [3,9]. Their main parameters, of interest to a user, are summarised in Table 1 and the special features of each are indicated. From an experimental point of view it should be noted that the two are fully compatible i.e. an insert suitable for one system can be used, without modification, in the other.

Both systems are used, on a pre-booked rota basis, by several research groups and their flexibility allows a diverse range of experiments to be covered although the current emphasis is on studies of the properties of semiconductors. Examples of work carried out using these systems can be found in Reference [10]. The 16T system is becoming increasingly important to the Clarendon's high-field superconducting materials test facility which has been established recently with S.E.R.C. support [11].

The latest in the series of small dedicated systems, referred to above, serves as a good illustrative example of a superconducting magnet and its associated equipment designed to perform a specific function. It is especially appropriate as this particular system is required for magneto optical studies of semiconductors. The system consists of two cryostats. One contains the superconducting magnet which is mounted horizontally and fitted with a room temperature bore of 33 mm diameter. The other cryostat is equipped with a proboscis which houses a cold finger on which is mounted the sample under investigation. This can be inserted into the magnet bore by moving the sample cryostat sideways on low friction rails. Thus the two cryostats are independent of each other allowing the sample to be studied at temperatures other than 4.2K and to be changed relatively easily. An additional benefit is that this arrangement results in a short optical path. The magnet power supply and all instrumentation is carried underneath the cryostats and the whole forms a self-contained, portable unit. As the system is to be used by a visually handicapped post-doctoral research worker, Dr. Stephenson Fung, a number of features have been incorporated which make the facility easier to use. For example, the digital meter which displays the magnet current can be hand-held for close viewing, rotary-switch knobs are shaped such that their position can be sensed by touch and each demountable connection is unique. The magnet produces 5T at an operating current of only 30 amps. Thus the power supply is air-cooled and the system can be used in virtually any location.

4. FUTURE DEVELOPMENTS

In addition to the further development of superconducting systems and of the Hybrid Magnet into a 20T facility, we are currently examining the feasibility of producing much higher fields using pulsed techniques. The area of greatest interest bridges the gap between the present limits of dc field generation and 50T and is coupled with pulse times $>$60 ms. This is relatively long compared with the inertial time constants of typical coils and poses a formidable problem in quasi-static field containment.

The work is at a very preliminary stage but as we consider it to be of benefit to the physics community to be kept aware of developments in magnet technology, we are taking the opportunity presented by this conference to announce our intentions to pursue the possibility of augmenting the dc systems in our laboratory by this approach. In collaboration with INSA Toulose, [12] we are proposing to test a design for a 20 mm bore, liquid nitrogen cooled, pulsed magnet in which an attempt to solve the mechanical problem is linked to modifications to the conductor from which the coil is wound. Pilot lengths of stainless-steel reinforced copper composite have been produced by the Chemistry Division of Harwell Laboratory (Figure 5) and initial tests on its tensile properties have proved encouraging. In the construction of the prototype coil we shall draw on the experience we have gained in the manufacture of resistive magnets and also on techniques for resin impregnation, for example, which have been developed in conjunction with our superconductive magnet programme.

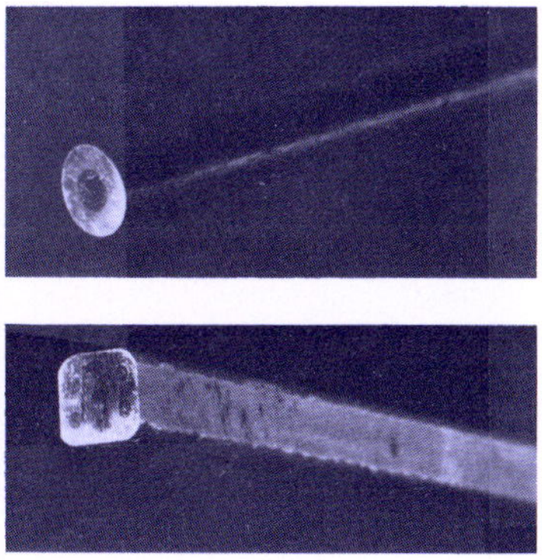

Figure 5.

ACKNOWLEDGEMENTS

The development of the facilities we have described has only been possible with the help of many people, too numerous to mention individually. Financial support for most of the projects has come principally from Departmental and University sources and also the SERC.

REFERENCES

[1] P.A. Hudson, Proc. 4th Int. Conf. on Magnet Technology, USAEC CONF-720908, 1972, p444
[2] P.A. Hudson and H. Jones, Proc ICEC8, Genoa, 1980, p 477.
[3] P.A. Hudson and H. Jones, IEEE Trans. Magnetics, Vol MAG-17, 1981, p 2242.
[4] J.M. Daniels, Proc. Phys.Soc.B, LX 111, 1950, p 1028.
[5] M.F. Wood, High Magnetic Fields, (MIT Press, Cambridge, Mass), 1962, p 387.
[6] P.A. Hudson, P.E. Hanley and P.O. Carden, Proc 5th Int. Conf. on Magnet Technology, 1975, p 414.
[7] P.A. Hudson, H. Jones and M.C.K. Wiltshire, Cryogenics, Vol. 17, 1977, p 355.
[8] High-Magnetic-Field Research and Facilities, US Nat.Acad.Sci.,1979.
[9] P.A. Hudson and H. Jones, Cryogenics, Vol. 16, 1976, p 593.
[10] See, for example
C.L. West, W. Hayes, J.F. Ryan and P.J. Dean, J. Phys. C, Vol. 13,1980, p 5631
D.J. Robbins, P.J. Dean, C.L. West and W. Hayes, Phil. Trans.R.Soc; Vol. A304, 1982, p 499.
[11] P.A. Hudson, F.C. Yin and H. Jones, IEEE Trans. Magnetics, Vol MAG-17, 1981, p 1649
[12] J.C. Portal, private communication.

SEMICONDUCTOR EXPERIMENTS IN MEGAGAUSS AND SUBMEGAGAUSS FIELDS

N. Miura, G. Kido and S. Chikazumi
Institute for Solid State Physics
University of Tokyo
Roppongi, Minato-ku, Tokyo, Japan

This paper reviews the recent progress of the semiconductor research using the high magnetic field facility at ISSP (Institute for Solid State Physics). A variety of experiments have been performed on semiconductors and semimetals under very high magnetic fields up to 280 T (2.8 Megagauss). Among others, we will discuss in the present paper, technical aspects of the pulsed high magnetic fields, far-infrared magneto-optical spectroscopy in Bi and graphite, the Faraday rotation in EuS and EuSe, and quantum transport phenomena in Si MOS-FET and GaAs-AlGaAs heterojunction interfaces.

1. Introduction.

During the last decade, a considerable progress has been made in techniques for generating very high magnetic fields in the megagauss (10^6 G or 100 T) range[1]. Although the duration time of the megagauss fields is relatively short (order of 10^{-6} s), data acquisition techniques in such short pulse fields have also advanced owing to the recent progress of electronic instruments. As a result, the available magnetic field range for solid state experiments has been greatly extended.

The megagauss magnetic fields are capable of having extremely large influence on electrons in semiconductors, so that it opens up various new possibilities for semiconductor physics as well as for other solid state problems. At ISSP of University of Tokyo, we have developed techniques for generating the megagauss fields by electromagnetic flux-compression[2,3]. Non-destructive submegagauss fields up to 45 T are also produced for precision experiments[3]. These megagauss and submegagauss pulse fields have been conveniently employed for various kinds of solid state research.

In this paper, we review the latest activities of the high magnetic field group of ISSP in semiconductor physics since the last High Field Conference at Hakone in 1980[4]. The topics to be discussed are : studies of electronic phase transitions in semimetals by far-infrared magneto-optical spectroscopy, the Faraday rotation in magnetic semiconductors, and studies of magneto-transport properties in two-dimensional electronic systems in Si MOS-FET and GaAs-AlGaAs heterojunction interfaces. We also describe the technical aspects of the field generation and solid state experiments in pulsed fields.

2. Experimental Techniques for Pulsed High Magnetic Fields.

2.1. Overview of Techniques for the Megagauss Field Generation.

As is well known, it is difficult to produce magnetic fields in the megagauss range by a conventional pulse magnet technique. The reason is mainly due to the large electromagnetic force called Maxwell stress exerted on the magnet itself. The Maxwell stress which is proportional to the square of the field reaches 40 k bar at 100 T (1MG), so that coils made of any material are usually destroyed before the produced field attains to such magnitude. Various means have been invented to produce the megagauss fields overcoming the problem of coil destruction. Fig. 1 schematically illustrates typical examples of devices for generating the megagauss fields. Magnetic flux-compression by the energy of either chemical explosives or electricity stored in a condenser bank has been widely employed for megagauss generation. Explosive driven flux-compression techniques (method (a) and (b)) utilize the energy of high explosives to squeeze a metal ring called a liner, or metal flat-plate, in which the magnetic flux is confined. If the velocity of the liner or the metal plate is high enough, the initially injected flux in them is almost conserved, so that a very high field can be generated when the crosssection of the circuit becomes sufficiently small. Fowler et al. first generated fields up to 1400 T by using a cylindrical liner[5]. Although the flux-compression can be carried out quite efficiently, this technique has inevitably a very destructive aspect, and is difficult to apply to fine solid state experiment. The bellows type flux-compression (method (b)), on the other hand, is more suitable for solid state experiments because the magnetic field is produced in a coil placed slightly apart from the explosive. The maximum field is limited to 100 or 200 T. The Los Alamos group succeeded in many magneto-optical experiments employing this technique[1].

The electromagnetic flux-compression is an alternative powerful means to produce the megagauss fields. It is more controllable than the explosive method, as it relies entirely on the electric energy stored in capacitor banks. The Θ-pinch technique (method (c)) was first tried by Cnare who reported the generation of a field up to 200 T[6]. The liner can be squeezed also by Z-pinch (method (d))[7]. The efficiency of the flux-compression is generally better in Θ-pinch than in Z-pinch.

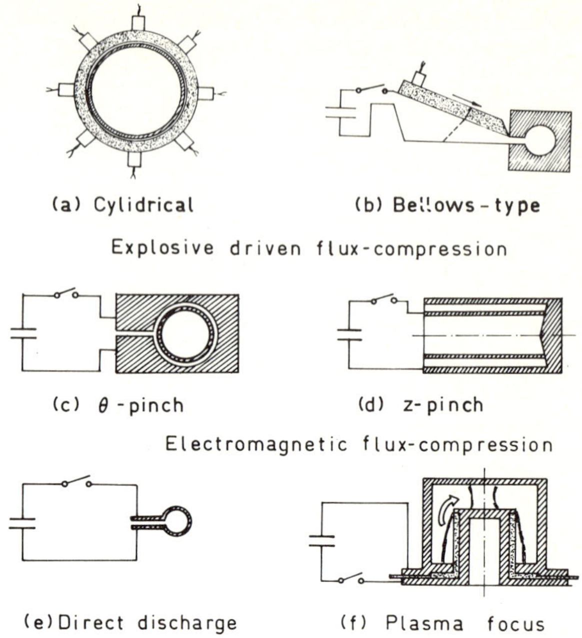

Fig. 1. Various devices to generate the megagauss fields.

At ISSP, we have succeeded in generating the fields up to 280 T by the Θ-pinch technique using a main capacitor bank with an energy of 285 kJ. Details of this technique have been described in previous papers[2,3]. In later sections, we present the latest experimental results on semiconductor physics obtained employing this system. At present, a new research project is being conducted at ISSP, to produce even higher fields up to 1000 T by installing a 5 MJ (40 kV) main capacitor bank and a 1.5 MJ (10 kV) sub-capacitor bank for seed field. These facilities have been almost completed including a new building.

Some other methods have been tried for the megagauss generation besides the flux-compression technique. Direct discharge of a fast pulse current into a one turn coil (method (e)) confines the fields by the inertia of the coil itself. This enables the compact design of the coil and capacitor bank system. The rise time of the current through the coil must be less than 2 μs for the field to attain the megagauss range before the coil is destroyed. Therefore we need a fast capacitor bank capable of supplying a few megaamperes of current. Herlach et al. utilized the high fields up to 140 T generated by this technique for the cyclotron resonance experiments[8].

Plasma focus (method (f)) is a new means for the flux-compression. At first, an azimuthally symmetric current sheath is created across the insulator between electrodes at the initial stage of the current discharge. Then the current sheath is accelerated towards the upper direction by the electromagnetic force, and collapses rapidly towards the central electrode axis. Finally, the current sheath pinches around the electrode axis coaxially. During this process, the dense plasma rakes up initially applied axial magnetic flux and generates a very high field on the electrode axis. Since the ambient gas possesses much smaller density than a metal liner, the flux-compression takes place in a fairly short time of the order of 10^{-7}s. Therefore the rise time of the field is generally short in comparison to the method (c) or (d). However, it has a great advantage that the sample may be saved undestroyed over several shots of experiments. Hirano et al. recently succeeded in observing the compression of the initially injected flux by the Faraday rotation in CdS[9]. Although the generated fields were limited to 14 T at the moment, further development of the technique seems promising.

Besides these destructive methods. Date and co-workers are trying to generate a field up to a megagauss in a non-destructive manner using a specially designed multilayered coil[10].

2.2. Generation of Non-destructive Submegagauss Fields.

Submegagauss fields up to about 60 T can be produced by a pulse magnet without destroying a coil. Considering the fact that the steady magnetic fields are now available only up to about 30 T even by means of a large hybrid magnet, non-destructive pulse fields above 30 T are very useful for various solid state experiments.

Conventional pulse magnets consisting of a wire-wound coil strengthened by a stainless steel frame can produce a field exceeding 30 T fairly easily, allowing their use for many shots. We construct strong pulse magnets winding Cu wire strengthened by Zr impurity. Epoxy-resin is impregnated under high pressure of about 50 kg/cm^2 in oil at 130 °C. Pressing the coil by a stainless steel cylinder, we can make long-life magnets with various sizes. The available bore of the magnets is in the range 13-22 mm. By supplying a pulse current from a capacitor bank of 200 kJ (10 kV/5 kV), pulsed magnetic fields up to 45 T are generated[3]. The rise time of the field can be varied from 1 ms to 10 ms by changing the capacitance of the bank or the inductance of the coil. As compared with the megagauss fields, it is far easier to carry out solid state experiments in these non-destructive pulse fields because of a longer rise time and larger available space. The non-destructive pulse fields are also applied to various semiconductor experiments complementary to the megagauss fields.

2.3. Techniques for Solid State Experiments in Pulsed Magnetic Fields.

The megagauss fields rise from a level of about 30 T to 100 T in several microseconds. Accordingly the induction of a voltage across any lead wires to the sample is enormous, so that the electrical measurements are extremely difficult in the megagauss fields. Magneto-optical measurements using an image-converter camera[11] are among the most suitable experiments for the megagauss fields because no wires are needed.

Laser spectroscopy is another useful means for the megagauss fields. Far-infrared radiation interacts with conduction electrons in samples, so that far-infrared magnetoreflection or magneto-transmission using various molecular gas lasers provides information concerning the electronic states and the transport properties. Extrinsic photoconductors such as Ga doped Ge and GaAs have very fast response time, and so they can be used as a far-infrared detector. Cyclotron resonance studies in various semi-conductors[12], and studies of magneto-plasma wave transmission as described in Section 3 were done using the technique mentioned above.

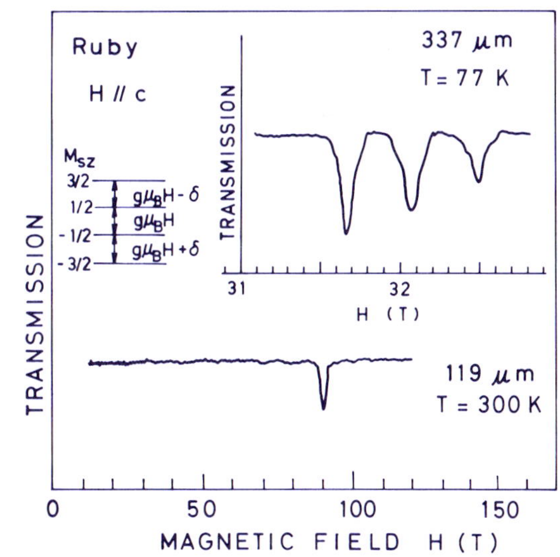

Fig. 2. Electron spin resonance (ESR) in ruby. The main figure shows the resonance for a wavelength of 118.6 μm at 300 K. The inset shows the resonance for 336.6 μm at 77 K with an enlarged horizontal scale.

Faraday rotation is also a convenient measuring technique in the megagauss fields. Near-infrared radiation at 1.15 μm wavelength from a He-Ne laser is employed for the study of Faraday rotation in magnetic semiconductors as Eu-chalcogenides and ferrimagnetic garnets, because these substances are transparent at this wavelength. The Ge-avalanche photo-diode or InGaAsP PIN diode are used as a detector. At this wavelength, optical fiber can be used as a light transmitting system in order to avoid the electromagnetic noise arising from the generation of the high fields.

The signals of the magnetic fields and light intensity are recorded by transient recorders with digital memories. The memorized data can be treated by a computer. The time resolution of the transient recorder is 10 ns which is fast enough for the present experiments. The temperature of the samples can be reduced to 4 K in the megagauss range by using a specially designed refrigerating system[13)].

The non-destructive submegagauss fields have a much longer rise time, so that not only magneto-optical measurements, but also transport measurements can be performed with high precision. For measuring the Shubnikov-de Haas effect in Si MOSFETs we used a long pulse field whose rise time was 10 ms, and the gate voltage was swept at the top of the field pulse within a time when the field is regarded as constant.

As an example of the experimental recordings of the far-infrared spectroscopy, the data of electron spin resonance (ESR) in ruby is shown in Fig. 2[14)]. The measurement was performed at a wavelength of 119 μm. A very sharp absorption peak is observed at 91 T. This peak is actually composed of three absorption peaks which are not resolved because of the small splitting as compared with the resonance energy. The inset shows the resonance spectra observed at a wavelength of 337 μm in the non-destructive fields. The splitting can be seen there. The central peak among the three observed at 32 T corresponds to the energy $g_{\parallel}\mu_B H$. The value of $g_{\parallel}$ in ruby is accurately known as 1.9840. The observed resonance peak positions 91 T and 32 T for 119 μm and 337 μm agree well with theoretically predicted values. This means that the g-factor is constant up to a megagauss. Consequently, it turned out that the ESR in ruby provides a good means for an accurate calibration of high magnetic fields.

3. Far-infrared Magneto-optical Spectroscopy in Semimetals.

3.1. Magnetoplasma Wave Transmission in Semimetals.

In typical semimetals such as Bi or graphite, the plasma frequency ω_p is of the order 10^{13} Hz corresponding to wavelengths of about 10 μm. Therefore, the far-infrared radiation which has frequency ω lower than ω_p cannot penetrate the crystal. When the magnetic field is applied along the direction of the incident radiation in such a way that the frequency ω_c of the cyclotron motion of electrons or holes is higher than ω, the radiation becomes allowed to propagate through the crystal as so called Alfvén wave. The wavelength of the Alfvén wave in the crystal is given by

$$\lambda_c = \frac{2\pi c}{\omega} \left[\varepsilon_\ell + \frac{4\pi c^2}{H^2} (n_e m_e^* + n_h m_h^*) \right]^{-1/2} , \qquad (1)$$

where ε_ℓ is the lattice dielectric constant, n_e and n_h are respectively the concentrations of electrons and holes, and m_e^* and m_h^* are the effective masses of electrons and holes. When the radiation passes through the crystal with a shape of parallel plate, Fabry-Pérot interference fringes can be observed. The interference occurs when twice of the sample thickness d becomes equal to the multiple of λ_c. Usually, the Alfvén wave in semimetals is observed in the microwave range[15,16)]. In high magnetic fields as in the present experiment, it can be observed in the far-infrared range.

The Fabry-Pérot interference condition of the Alfvén wave is derived from (1) as in the following,

$$\rho_{ij} = \frac{H^2}{4\pi\omega^2} \left\{ \left(\frac{N\pi}{d}\right)^2 - \left(\frac{\omega}{c}\right)^2 \varepsilon_{jj} \right\} , \qquad (2)$$

where ρ_{ij} is a component of the mass density tensor, when the magnetic field is applied in the i-th direction and the electric field of the radiation is in the j-th direction. ε_{jj} is a component of the dielectric tensor, and N is an integer. ρ_{ij} is represented as a sum of the quantities with a dimension of the effective mass multiplied by the carrier density. In Bi for H // binary axis, for example, one obtains[15)]

$$\rho_{xy} = N_A m_3 + (N_B + N_C)\,\frac{(m_1 + 3m_2)m_3 - 3m_4^2}{m_1 + 3m_2} + N_h M_3 \qquad (3a)$$

$$\rho_{xz} = N_A m_2 + (N_B + N_C)\,\frac{4m_1 m_2}{m_1 + 3m_2} + N_h M_1 \quad , \qquad (3b)$$

where x, y, z are the binary, bisectrix and trigonal axes, respectively, N_A, N_B, N_C and N_h are densities of electrons in the A, B and C pockets and holes in the hole band. m_1, m_2, m_3 and M_1 are the effective mass parameters of electrons and holes.

An example of the Fabry-Pérot interferograms in Bi at λ = 337 μm is shown in Fig. 3. It can be seen that the different interference patterns are observed in samples with different thicknesses. At the same time, the envelope of the interferogram is modulated with a longer period. This is because the imaginary part of the dielectric constant relates to the carrier scattering, so that the transmission reflects the magnetoresistance of the sample. Hence, we can observe the Shubnikov-de Haas oscillation in the envelope. Thus by the measurement of the Alfvén wave transmission, information can be obtained on the mass density, the lattice dielectric constant, the energy band parameters and so on.

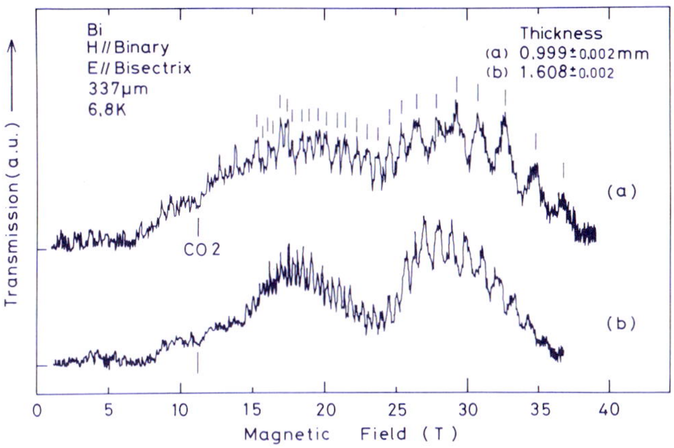

Fig. 3. An example of the Fabry-Pérot interferograms of the Alfvén wave observed in Bi. CO2 is the transmission minimum arising from the combined cyclotron resonance of holes.

3.2. Anomalous Increase of the Dielectric Constant in Bi.

By analysing the Fabry-Pérot interferograms of the Alfvén wave, the mass density in Bi was obtained for H // binary. Fig. 4 and Fig. 5 show the mass density for the electric field polarization of

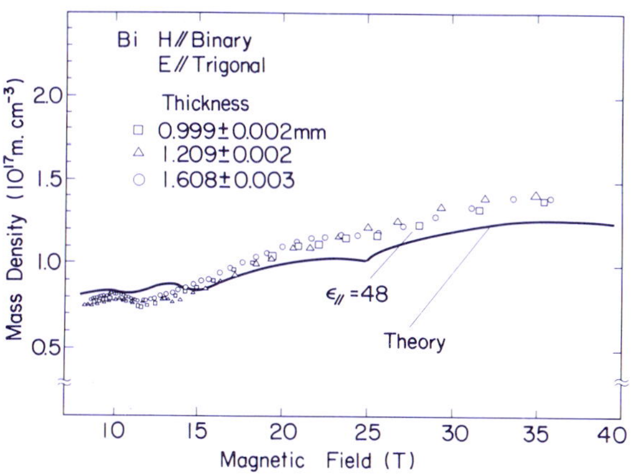

Fig. 4. Mass density ρ_{xz} in Bi as a function of magnetic field.

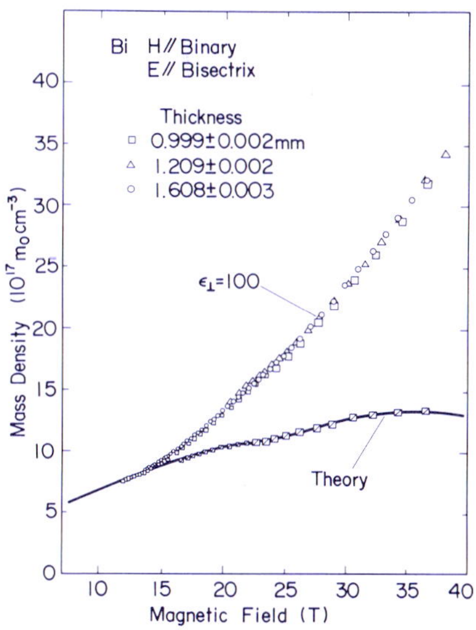

Fig. 5. Mass density ρ_{xy} in Bi as a function of magnetic field.

the radiation parallel to the trigonal and bisectrix axes, respectively. These figures also show the theoretically calculated mass density based on the band parameters obtained from the Shubnikov-de Haas effect[17,18]. For E // trigonal, the experimental values of ρ_{xz} agree well with the theoretical curve assuming $\varepsilon_{\parallel}=48$. However, for E // bisectrix, it was found that ρ_{xy} assuming a constant lattice dielectric constant $\varepsilon_{\perp}=100$ deviates significantly from the theoretical curve in high magnetic fields.

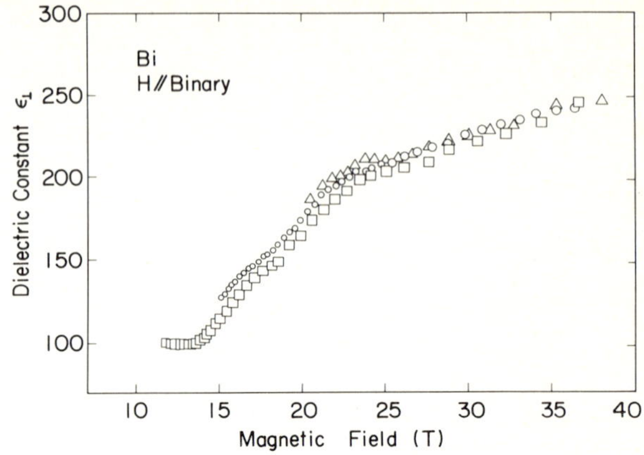

Fig. 6. Magnetic field dependence of the lattice dielectric constant $\varepsilon_{\perp}$ in Bi.

The reason of the discrepancy in ρ_{xy} is ascribed to the irrelevant assumption of a constant $\varepsilon_{\perp}$. If we allow that the lattice dielectric constant $\varepsilon_{\perp}$ increases with increasing magnetic field, a good agreement between theory and experiment can be obtained. Fig. 6 shows the magnetic field dependence of $\varepsilon_{\perp}$ thus obtained from the data of ρ_{xy}. We can see that $\varepsilon_{\perp}$ starts to increase with increasing field above 14 T. The increase of $\varepsilon_{\perp}$ can be explained if we take into account the contribution of the interband transition between the valence band and the conduction band at L-point. Since the band gap at L-point is small and field dependent, this contribution greatly affects the field dependence of $\varepsilon_{\perp}$. Using the interband matrix elements estimated by Golin[20], we can deduce that the interband contribution is only important for E // bisectrix[19].

3.3. Magnetic-Field-Induced Semimetal-Semiconductor Transition in Bi.

As the energy band overlap between the conduction band at L-point and the hole band at T-point is small in Bi, the magnetic-field-induced semimetal-semiconductor (SM-SC) transition[18] is expected to occur at high enough magnetic fields. For H // binary, the band gap between the two bands at L-point first increases, but above about 10 T, it starts to decrease with increasing magnetic field. In the framework of the Vecchi model[21], the SM-SC transition should occur near 40 T. Brandt et al. estimated the transition field to be about 150 T[22]. On the other hand, the analysis of the Shubnikov-de Haas effect predicts that the transition would occur at 85 - 100 T[17]. Near the transition, the zero gap state is realized and hence there is a possibility that the electronic phase transition such as excitonic phase transition[23] or the gas-liquid type phase transition[24] might take place.

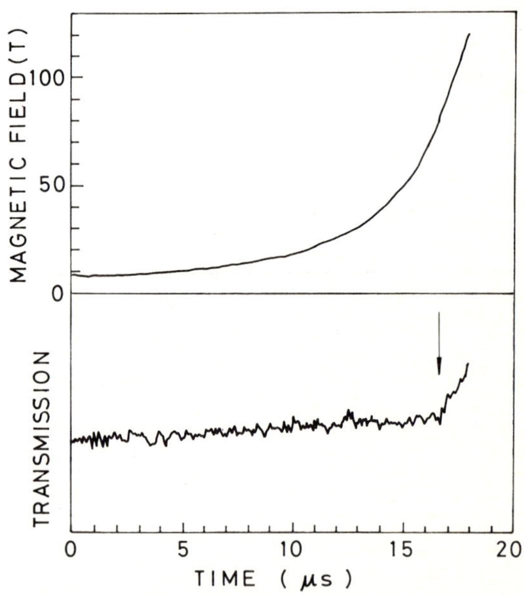

Fig. 7. Signals of magnetic field and the far-infrared transmission in Bi at 337 μm. T=4.5 K. H // binary. SM-SC transition takes place at the position indicated by an arrow.

Since the measurement of the dc magnetoresistance in the megagauss fields is difficult, the far-infrared transmission at 337 μm wavelength was measured. As is evident in Fig. 3, the far-infrared transmission reflects the dc magnetoresistance. Fig. 7

shows the observed SM-SC transition in the far-infrared magnetotransmission. It was found that the transition occurs at 88 T. This value of the transition field agrees well with the theoretical calculation based on the Vecchi model using newly determined parameters. The Landau levels and the Fermi level E_F are shown in Fig. 8. The pentagon point is marked on the line for the 0^- level of holes at 88 T, and it can be seen that the theoretical lines of lowest Landau levels of electrons and holes and E_F all meet together near this point. We have not seen a precursor of many body effects near the transition, but further investigation at even lower temperatures would be of interest.

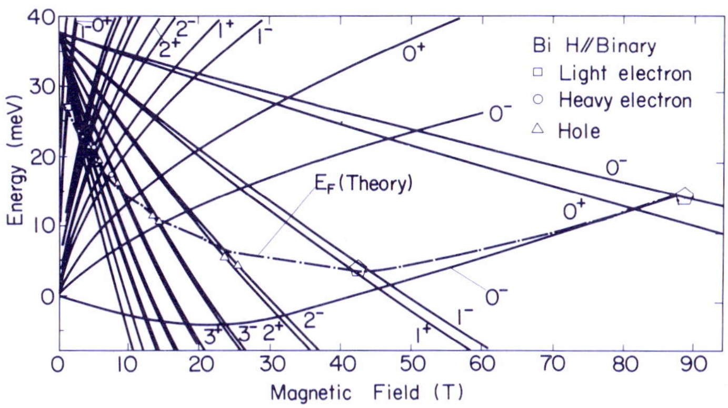

Fig. 8. Landau levels and Fermi level E_F as a function of magnetic field in Bi. B // binary. Experimental points for E_F were obtained in Shubnikov-de Haas effects.

3.4. Electronic Phase Transition in Graphite.

Recently a drastic anomaly was discovered by Tanuma et al.[25] in the magnetoresistance in graphite in high magnetic fields around 30 T at low temperature. The magnetic field where the anomaly occurs increases as the temperature is raised. Iye et al. extended the work to lower temperature down to 0.48 K[26]. Fig. 9 shows the magnetoresistance in single crystal graphite observed in a non-destructive pulse field. The Shubnikov-de Haas oscillation appeared at low field region below 10 T. The anomaly is seen as a kink at about 30 T as indicated by an arrow. The magnetoresistance sharply rises at this field. When we raise the temperature, the critical field of the anomaly increases. The inset of Fig. 9 shows the temperature dependence of the critical field. Yoshioka and Fukuyama explained this anomaly by a model of change density wave (CDW) in the electronic system which becomes nearly one-dimensional due to a strong magnetic field[27]. Their theory roughly reproduces the experimentally observed temperature dependence as indicated in Fig. 9. There is a slight discrepancy between the present experimental results in pulsed fields and the extrapolated curve of the data obtained at lower fields and temperatures by Iye et al.[26].

In order to investigate this anomaly by the magneto-optical means, we perform measurements of magnetoreflection and magnetotransmission in the far-infrared region since we can expect the anomaly should appear in the magneto-optical spectra as well. Fig. 10 shows the magnetotransmission spectra in highly oriented pyrolytic graphite (HOPG) at two different wavelengths 337 μm and 311 μm. The Fabry-Pérot interferogram of the Alfvén wave transmission was observed[28]. By analyzing the interferograms in three samples, we obtained the mass density in HOPG as shown in Fig. 11.

A previous measurement was analyzed assuming a field independent mass density[16]. However, the present data revealed that the mass density increases as the field is increased. This increase is quite large. Because both the effective mass and the carrier density increase with increasing field, we can expect that the mass density would increase with field. However, the calculated mass density based on the classical equation of motion of carriers cannot explain the remarkably large increase of the mass density. So far the measurements have been performed only at relatively high temperatures, so that the anomaly observed in the magnetoresistance has not appeared in the mass density curve. Experiments at even lower temperature are in progress.

Nakao predicted theoretically that graphite would become a zero gap semiconductor above 70 T[29]. On the basis of the CDW model, a more drastic anomaly due to CDW might appear at this critical field of 70 T. This is also being explored using the megagauss fields.

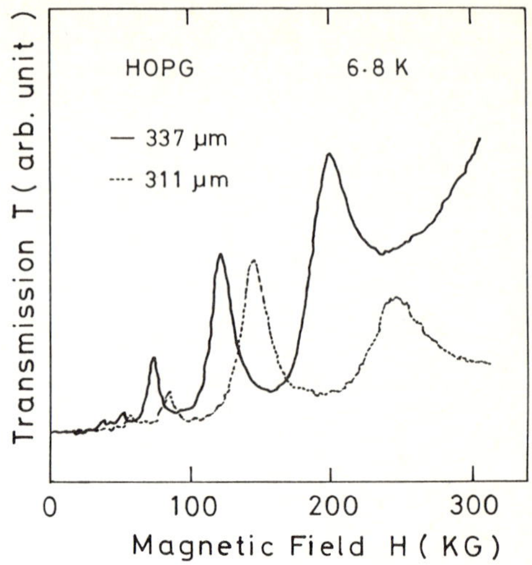

Fig. 10. Fabry-Pérot interferogram of the Alfvén wave transmission in highly oriented pyrolytic graphite (HOPG). H // c T = 6.8 K. λ = 337 µm and 311 µm. The thickness of the sample is 15.2 µm.

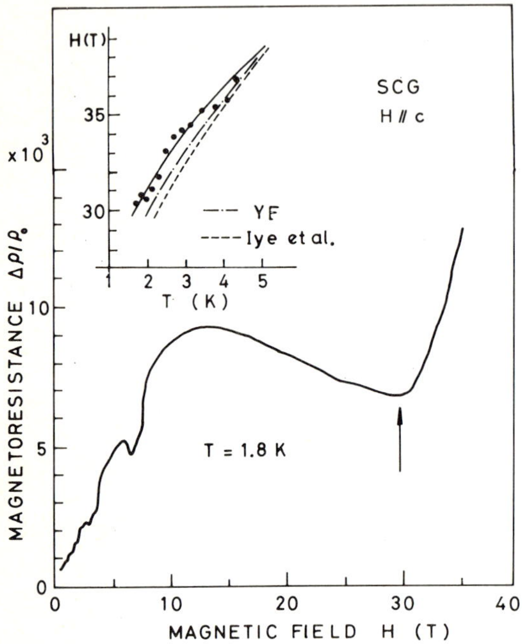

Fig. 9. Magnetoresistance in single crystal graphite (SCG). T = 1.8 K. H // c. The anomaly is observed at a point indicated by an arrow. The inset shows the temperature dependence of the critical field of the anomaly. A theoretical line by Yoshioka and Fukuyama (YF), and the extrapolation of the data of Iye et al. from the low field side are also shown.

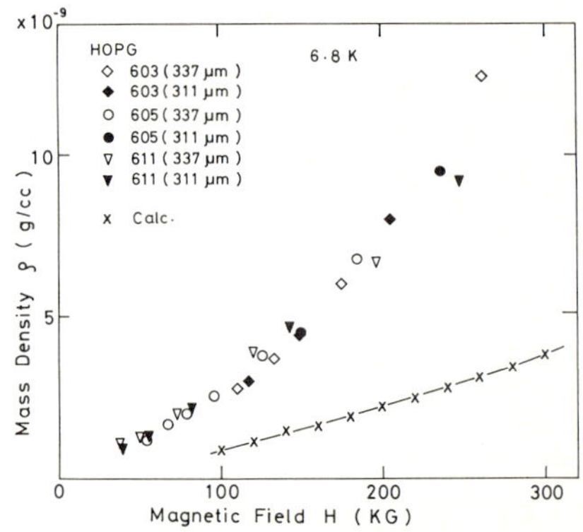

Fig. 11. Mass density in graphite as a function of magnetic field. The theoretical line was calculated from known band parameters.

4. Faraday Rotation in Magnetic Semiconductors.

Europium chalcogenides are known to exhibit large magneto-optical effects. However, previous measurements are limited only to low magnetic fields of the order of 1T. The Faraday rotation in Eu-chalcogenides is of interest in relation to their magnetism. We have measured the Faraday rotation in EuS and EuSe in the megagauss fields [30]. A He-Ne laser was employed as a radiation source at 1.15 µm wavelength.

Fig. 12 shows the signal of the Faraday rotation as a function of magnetic field for EuS at room temperature. The oscillation amplitude decreases with increasing magnetic field because of the the magnetoabsorption. The Verdet constant, the increase rate of the Faraday rotation, was found to decrease as the magnetic field was increased. When the rotation angle θ at room temperature was plotted as a function of field, it was found to be closely matched to the Brillouin function with J=7/2 and g=2 for EuS and EuSe. As the Curie temperature is much lower than 300 K in these materials, the magnetization is almost proportional to the Brillouin function.

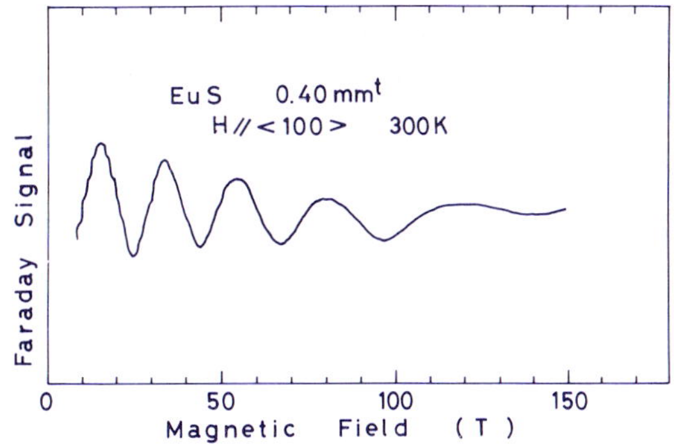

Fig. 12. Signal of Faraday rotation in EuS. λ = 1.15 μm. T = 300 K. H// <100>. The thickness of the sample is 0.40 mm.

At low magnetic field range below 30 T, the Faraday rotation is found to be almost proportional to the magnetic field, so that the field-independent Verdet constant can be defined. Fig. 13 shows a plot of the inverse of the Verdet constant against temperature for EuS. The extrapolation of the best fit line hits the temperature axis at T = 17 K very close to the Curie temperature T_c = 16.5 K. The result shown in Fig. 13 corresponds to the Curie-Weiss law.

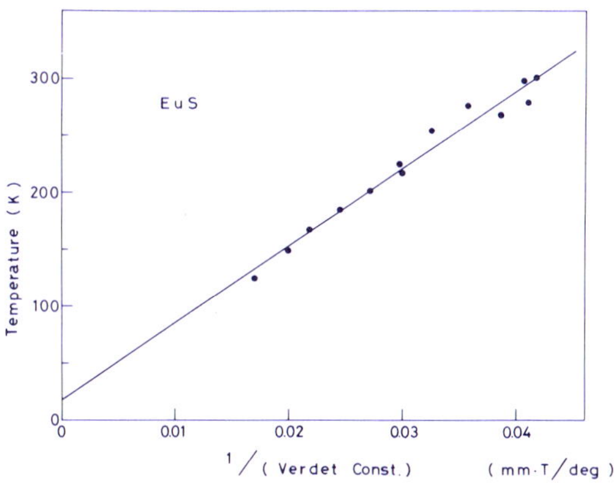

Fig. 13. Inverse of the Verdet constant measured below 30 T vs temperature for EuS. A straight line is drawn as a least square fit.

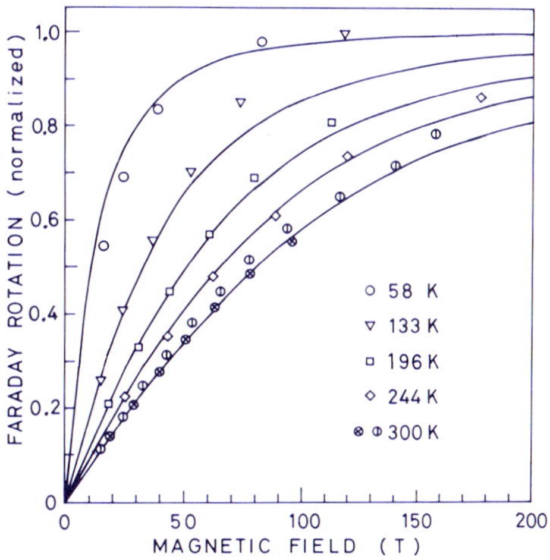

Fig. 14. Faraday rotation as a function of magnetic field at various temperatures. The solid lines show the theoretically calculated magnetization of EuS.

The results of the present experiments strongly suggest that the Faraday rotation is proportional to the magnetization of Eu^{2+} ions. In order to confirm this hypothesis more generally, the Faraday rotation in EuS is plotted as a function of magnetic field for many different temperatures in Fig. 14. The theoretically calculated magnetization is also shown by solid lines. By measuring the magnetization of the samples up to 32 T, these theoretical curves were justified. It can be seen that the Faraday rotation can be matched closely to the magnetization curves, except for some deviation at 133 K in high field region. The reason of this deviation is not clear at the moment.

5. Quantum Transport Phenomena in Quasi-Two-Dimensional Electron System.

5.1. Shubnikov-de Haas Effect in Si MOS-FETS.

The electronic states and the transport properties of Si MOS-FETS under high magnetic fields have collected much attention. In the Shubnikov-de Haas (S-dH) effect, many anomalous behaviors have been found associated with immobile electronic states in the lowest Landau level[31]. As the high magnetic fields increase not only the spacings between the Landau levels but also the density of states of each Landau level, some new features are expected to appear in the S-dH oscillations in high fields. We have performed S-dH experiments in Si MOS-FETS in pulsed high magnetic fields up to 37 T[32]. Fig. 15 shows a typical example of the observed S-dH oscillation at various field intensities. Here, the gate voltage V_g was rapidly varied at the top of the pulsed magnetic field. The n-channel MOS devices of Corbino-type fabricated on a (100) surface were measured. When the source-drain field E_{sd} and temperature are low, the height of the S-dH peak arising from the first of the four split N = 0 Landau level (0↑+) is very small. As is shown in Fig. 15, we found that the height of the (0↑+) peak, $(\sigma_{xx})_{peak}$, increases with increasing magnetic field, whereas the magnetic field dependence of the other peaks is relatively small.

Fig. 16 shows a plot of the $(\sigma_{xx})_{peak}$ as a function of magnetic field H. This also shows the field dependence of the theoretically calculated density of states n_t in a Landau level and the number of immobile electrons estimated from the gate voltage width of the non-conducting region before the conductivity occurs in the σ_{xx} vs V_g plot. We can see that the extrapolation of the $(\sigma_{xx})_{peak}$ vs H line crosses the field axis at the intercept of 4.2 T. The lines representing n_t and n_{im} cross also at nearly the same field. These findings can be interpreted in terms of the mobility edge model, because the increase of the state density more than the localized

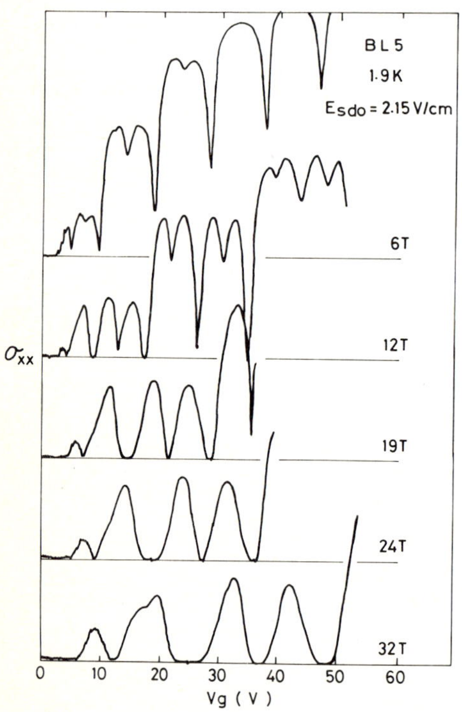

Fig. 15. Shubnikov-de Haas Oscillation in Si MOS-FET. T = 1.9 K. E_{sd} = 2.15 V/cm. t_{ox} = 3200 A.

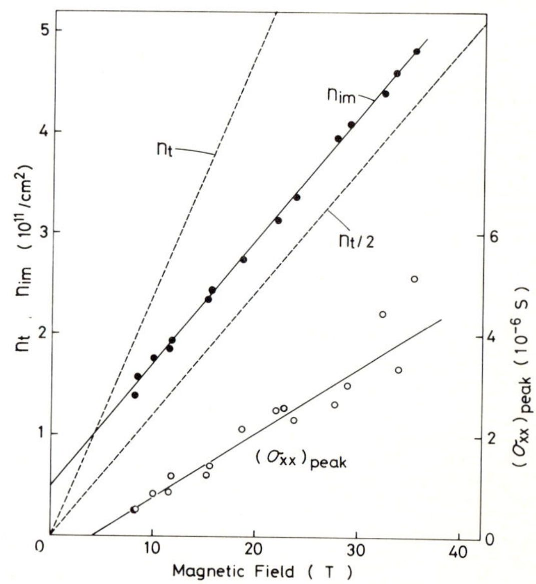

Fig. 16. Number of immobile electrons n_{im} and the total number of electrons in the lowest Landau level n_t as a function of magnetic field (the left ordinate scale), and the magnitude of $(\sigma_{xx})_{peak}$ for (0↑+) as a function of magnetic field (the right ordinate scale). T = 1.9 K. E_{sd} = 2.15 v/cm.

state density by field would cause the conductivity. However, an explanation may be also possible based on the pinned CDW model[33].

Recently we have measured the S-dH effect with a high frequency source-drain field in the range 100 kHz - 20 MHz. The real part of the ac conductivity was measured. A typical example is shown in Fig. 17. A remarkable feature is that as the frequency is increased, the width of the non-conducting region in the (0↑+) peak or in gaps between each peak is increased. Namely, as the frequency of the E_{sd} bias is increased, the S-dH oscillation shows a feature characteristic of the one observed at lower temperatures for the dc bias. Further investigations are being made to clarify this striking phenomenon.

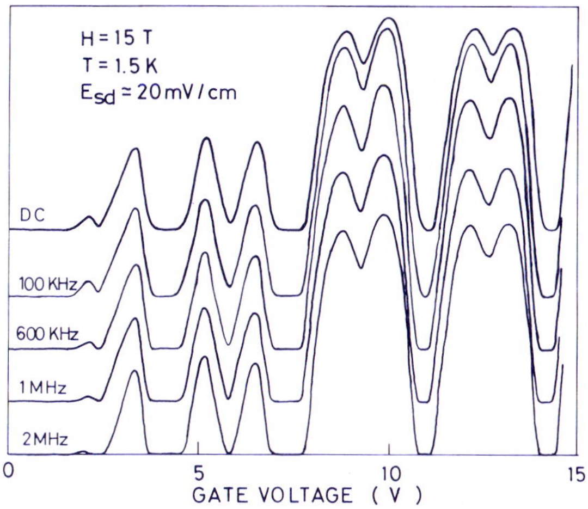

Fig. 17. Shubnikov-de Haas oscillation for dc and high frequency ac bias between source-drain electrodes. H = 15 T. T = 1.6 K.

5.2 Magnetophonon Resonance in GaAs-AlGaAs Heterojunction Interface.

The GaAs-AlGaAs heterojunction interface is another system in which we can investigate the behavior of the two-dimensional electron system. . We have observed the magnetophonon resonance in GaAs-$Al_{0.3}Ga_{0.7}As$ heterojunction interface under high magnetic fields up to 35 T[34]. The heterojunction consisting of p-type GaAs layer, non-doped $Al_{0.3}Ga_{0.7}As$ layer and a selectively Si-doped $Al_{0.3}Ga_{0.7}As$ layer was fabricated by Sakaki group of Univ. Tokyo. Fig. 18 shows the observed magnetophonon resonance oscillation in the interface at various angles θ between the magnetic field and the normal to the interface. The magnetophonon resonance in bulk GaAs is also shown for comparison. The oscillatry part of in the transverse magnetoresistance was extracted from the total magnetoresistance by subtracting the monotonic part proportional to the field. For θ = 0, the full resonance series to the N = 1 peak was observed.

As θ was increased, each peak was found to shift to higher field in accordance with the two-dimensional character of the electron system. The damping factor γ was estimated to be

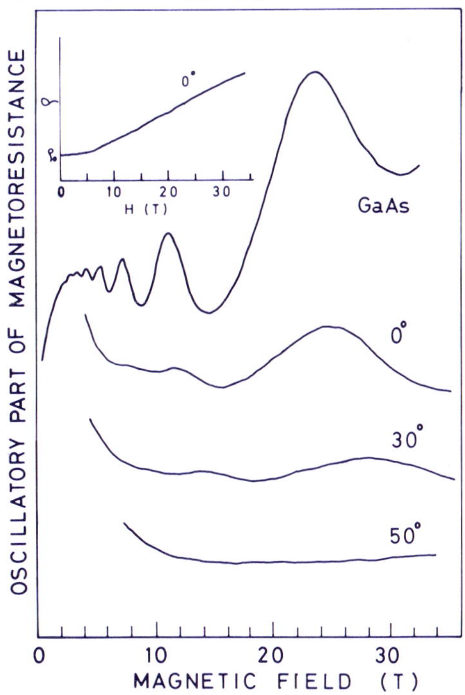

Fig. 18. Magnetophonon resonance in the transverse magnetoresistance for epitaxial GaAs and GaAs-$Al_{0.3}Ga_{0.7}As$ heterojunction. For the curves of heterojunction, the tilted angle of the magnetic field is shown. The inset shows the total magnetoresistance in the heterojunction. T = 180 K.

2.1. It is noted that this value is remarkably larger than $\gamma \simeq 1$ for 3-D electron system in epitaxial GaAs, although the difference in mobility between the two electron systems is not so large. The difference in γ-factor mentioned above is suggestive of a different carrier scattering process characteristic of the 2-D electrons being involved in case of heterojunction.

Taking into account the correction needed by the effect of the large damping factor, we obtained the magnetophonon mass of 2-D electrons $m^* = 0.0745\ m_e$ which is slightly smaller than in previous data by Tsui et al[35]. The above value is 4 % larger than the magnetophonon mass in bulk GaAs. The difference of 4 % was successfully explained by the non-parabolicity of the conduction band taking into account the 2-D character.

Acknowledgement

The authors are indebted to the members of their group, K. Hiruma, K. Nakamura, T. Itakura, M. Suekane, Y. Iwasa and T. Osada for every contribution to the works described in this paper.

References

1) C. M. Fowler: Science 180 (1973) 261.
2) N. Miura, G. Kido, M. Akihiro and S. Chikazumi: J. Magn. Magn. Magn. Mater. 11 (1979) 275.
3) N. Miura, G. Kido, H. Miyajima, K. Nakao and S. Chikazumi: in *Physics in High Magnetic Fields*, edited by S. Chikazumi and N. Miura (Springer-Verlag, 1981) pp. 64.
4) G. Kido, N. Miura, M. Akihiro, H. Katayama and S. Chikazumi: the same bood as 3) pp. 72.
5) C. M. Fowler, W. B. Garn and R. S. Caird: J. Appl. Phys. 31 (1960) 588.
6) E. C. Cnare: J. Appl. Phys. 37 (1966) 3812.
7) S. Alikhanov, V. G. Belan, A. I. Ivanchenko, V. N. Karasjuk and G. N. Kichigin: J. Phys. E. Sci. Inst. 1 (1968) 543.
8) F. Herlach, J. Davis, R. Schmidt and H. Spector: Phys. Rev. B 10 (1974) 682.
9) K. Hirano, M. Hattori, K. Shimoda, G. Kido, N. Miura and S. Chikazumi: J. Phys. Soc. Jpn. 51 (1982) 297.
10) M. Date, M. Motokawa, K. Okuda, H. Hori and T. Sakakibara: the same book as 3) pp. 44.
11) G. Kido, N. Miura, H. Katayama and S. Chikazumi: J. Phys. E. Sci. Instr. 14 (1981) 349.
12) N. Miura, G. Kido, M. Suekane and S. Chikazumi: To be presented at the Int. Conf. Phys. Semiconductors (Montpellier, 1982).
13) G. Kido, N. Miura, H. Kawauchi, I. Oguro and S. Chikazumi: J. Phys. E. Sci. Instr. 9 (1976) 587.
14) G. Kido and N. Miura: Appl. Phys. Lett. 41 (1982) (in press).
15) S. Takano and H. Kawamura: J. Phys. Soc. Jpn. 28 (1970) 348.
16) J. K. Furdyna and A. R. Krauss: J. Phys. Chem. Solids 32 Suppl. (1971) 127.
17) K. Hiruma, G. Kido and N. Miura: Solid State Commun. 31 (1979) 1019.
18) K. Hiruma, G. Kido, K. Kawauchi and N. Miura: Solid State Commun. 33 (1980) 257.
19) K. Hiruma and N. Miura: to be published.
20) S. Golin: Phys. Rev. 166 (1968) 643.
21) M. P. Vecchi, J. R. Pereira and M. S. Dresselhaus: Phys, Rev. B14 (1976) 298.
22) N. B. Brandt, E. A. Svistova and Yu. G. Kashirskii: JETP Lett. 9 (1969) 136.
23) N. B. Brandt and S. M. Chudinov: J. Low Temp. Phys. 8 (1972) 339.
24) S. Nakajima and D. Yoshioka: J. Phys. Soc. Jpn. 40 (1976) 328.
25) S. Tanuma, R. Inada, A. Furukawa, O. Takahashi, Y. Iye and Y. Onuki: same book as 3) pp. 316.
26) Y. Iye, P. Tedrow, G. Timp, M. Shayegan, M. S. Dresselhaus, G. Dresselhaus, A. Furukawa and S. Tanuma: Phys. Rev. B (Submitted)
27) D. Yoshioka and H. Fukuyama: J. Phys. Soc. Jpn. 50 (1981) 725.
28) K. Nakamura and N. Miura: Solid State Commun. 42 (1982) 119.
29) K. Nakao: J. Phys. Soc. Jpn. 40 (1976) 761.
30) M. Suekane, G. Kido, N. Miura and S. Chikazumi: To be presented at the Int. Conf. Magnetism, (Kyoto, 1982).
31) See for example; S. Kawaji and J. Wakabayashi: same book as 3) pp. 284.
32) N. Miura, Y. Iwasa, T. Itakura and G. Kido: J. Phys. Soc. Jpn. 51 (1982) 1228.
33) H. Fukuyama and P. A. Lee: Phys. Rev. B18 (1978) 6245.
34) G. Kido, N. Miura, H. Ohno and H. Sakaki: J. Phys. Soc. Jpn. 51 (1982) 2168.
35) D. C. Tsui, Th. Englert, A. Y. Cho and A. C. Gossard: Phys. Rev. Lett. 44 (1980) 341.

HIGH FIELD SUPERCONDUCTORS

H. Hillmann
VACUUMSCHMELZE GMBH
D-6450 Hanau, Germany

Introduction

The application of superconducting material to the construction of high field magnets has been established in a wide field since few years. The first superconducting magnet with cold worked niobium wire was constructed by G.B. Yntema 1955 [1]. The first high field superconductor consisting of Nb_3Sn-strip material carrying 10^5 A/cm² in a field of 8.8 T was manufactured 1961 [2]. A great progress in the further development was the invention of solid solution ductile superconductors such as niobium-zirconium and niobium-titanium in the following years [3-7]. In the early time of technical high field superconductors a real application was only established for laboratory scale magnets followed by large fully stabilized magnets for bubble chambers and different magnets applied in R and D centers. Since few years technical superconductors have found access to routine applications which are nuclear magnetic resonance in chemistry and even in medicine, superconducting machinery, magnetic separation and magnetic suspension of levitated trains. Another application being under development is hot plasma inclusion for nuclear fusion and energy storage. In comparison to the natural hard magnetic materials such as iron cobalt base or cobalt rare earth alloys the magnetic fields which can be reached by superconductor application are higher by an order of magnitude. Superconducting magnets can be operated loss free in a persistant mode condition,working comparable to a high field permanent magnet system.
The main parameters governing the material choice and the conductor design are:

1. Operating temperature and magnetic field to be generated. That is related to critical temperature T_c and upper critical induction $B_{c_2}(T)$ of the superconductor.

2. Magnet size, i.e. the magnitude of stored energy.
This mainly determines the degree of stabilization, i.e. the amount of highly conductive normal conducting material such as copper shunted continuously in parallel with the superconductor and the ne-

cessity of reinforcement of the superconductors by high strength material in large magnets.

3. Operation mode, i.e. dc or ac superimposed dc or pulsed field operation.

This mainly determines the number, the diameter of superconducting filaments and the application of additional highly resistive substrate material governing the losses during energizing the magnet. The magnet size in connection with the operation mode also influences whether a monolithic filament conductor or a fully transposed cable should be applied!

Additional requirements for the conductor materials have to be fulfilled by the fact that superconductors have to be stabilized by continuous shunting with highly conductive normal metal and current carrying parts must be fine filaments having diameters less than about 50 µm. Fabrication technology has to be a composite technology. Thus coprocessing with highly conductive substrate should be feasable. Conductor properties must be reproducible. Fabrication in large lengths should be reliable. Metallurgical processing during fabrication must be compatible with creating and optimizing flux pinners. Presently two kinds of conductors can fulfill these conditions

a) Solid solution based conductors such as NbTi.

b) Intermetallic A15 compound based solid diffusion conductors such as Nb_3Sn/tinbronze and V_3Ga/galliumbronze conductors.

Table 1 compares the main properties, advantages and disadvantages of solid solution - and A15 base superconductors.

Fig. 1 shows a typical cross section of a standard NbTi base filament superconductor containing 85 filaments embedded in low temperature highly conductive copper with a copper to superconductor area ratio of 1.4 carrying a current of about 130 A at 5 T.

Fig. 1: Cross section of a 85 filament conductor [8,9]

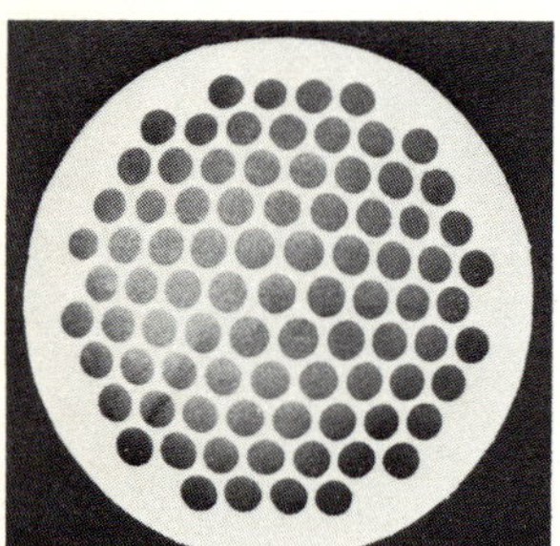

	Solid solution superconductors NbTi with 45-52 w/o Ti	A15-phase intermetallic compounds Nb_3Sn	V_3Ga
Electrical T_c	9 K	18,1 K	16,9 T
properties B_{c2}	14,5 T	24,5 T	20,5 T
Stabilization	Simple coprocessing with high purity copper or mixed substrate	complicated; coprocessing with high purity copper only possible in solid diffusion conductors by application of diffusion barriers	
Insulation	conventional different insulating processes (varnish, cotton, tape) are possible	complicated; in prereacted insulated conductors high thermal stability of the insulation is necessary; in after-reacted insulated conductors: much care must be taken in mechanical handling	
Advantages	economic fabrication by composite technique, high ductility and unsensitivity of physical properties to mechanical treatment; simple magnet technology	application to high fields or high temperatures	
Disadvantages	application limited to 8 T at 4.2 K	brittleness of the A15-layers, restricted mechanical handling, high filament numbers are necessary (filament diameter must be $\leq$5 µm) complicated magnet winding technique	

Table 1: Comparison of solid solution and A15 superconductors

Basic properties of technical high field superconductors

Beside the material dependent basic properties B_{c2} and T_c the critical current density $j_c(B,T)$ limits the maximum current which can be carried by a conductor of a certain diameter in a superconducting magnet.
The critical current density is a highly structure dependent property and is determined by the kind, the density, the size and the distribution of microscopic imperfections and precipitates [10-14]. They act as flux pinners and their geometrical size is in the range between the superconducting coherence length and the penetration depth i. e. 10^{-6} - 10^{-5} cm. Flux pinners prevent flux line migration by which the conductor goes normal when magnetic field and transport current are simultaneously applied to the conductor. Flux pinners are grain bounderies, vacancy agglomerates, dislocations and precipitates.
The formation and distribution of flux pinners is highly connected with the treatment of the conductor during fabrication. In the case of NbTi-base conductors a large area reduction of the conductor by cold work is necessary to create a high dislocation density arranged in so called subbands [10]. By additional thermal treatment normal conducting α-Ti-precipitates having diameters of the order of the subband width (10^{-6}-10^{-5} cm) can be created and their distribution can be changed during subsequent cold work.

Fig. 2 shows a trans electron micrograph of subbands in a 50 weight percent NbTi alloy and Fig. 3 shows a dark field micrograph of these subbands showing bright normal conducting α-Ti flux pinners [10].

Fig. 2. Subbands in Nb50 wt% Ti after heavy cold work.

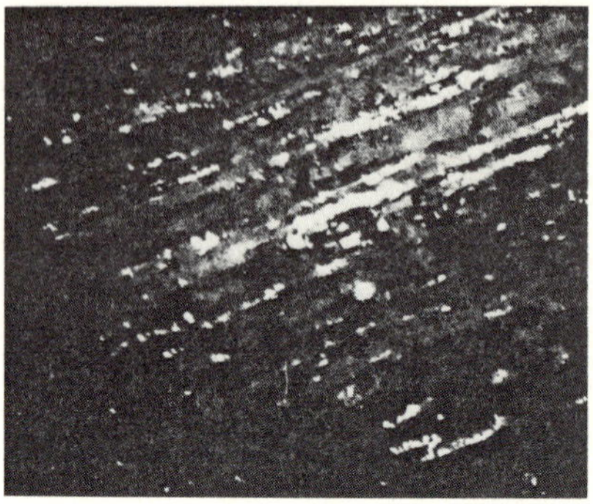

Fig. 3. Dark field electron micrograph of α-Ti precipitates.

To optimize the critical current density either a long time heat treatment at about 400 °C with subsequent cold work of about 95 % reduction in area or a repeated intermediate heat treatment with subsequent final heat treatment is necessary [10, 15].
In the first case current densities of $1.8 \cdot 10^5$ A/cm² and in the second case up to $3 \cdot 10^5$ A/cm² can be reached at 5 T.
By the necessity of intermediate heat treatment for reaching high current densities the actual conductor critical currents depend on the final conductor cross sectional area because large cross sections permit only a restricted number of intermediate heat treatments. On the other hand the density of α-Ti-precipitates increases with increasing Ti-concentration of the alloys. Fig. 4 shows the influence of the final diameter on the optimum current density of NbTi-conductors at 5 T and 8 T [9].
Because of increasing Ti-concentration, the current density at medium magnetic fields increases with increasing Ti-content. Unfortunately with increasing Ti-concentration in the range of 50 weight percent Ti the B_{c2} values and the current densities near B_{c2} are decreasing. Fig. 5 shows the influence of the Ti-content on the upper critical field B_{c2} and Fig. 6 shows typical current densities in dependence of the field of the two presently applied alloys NbTi 47 and NbTi 50 weight percent [16, 9].

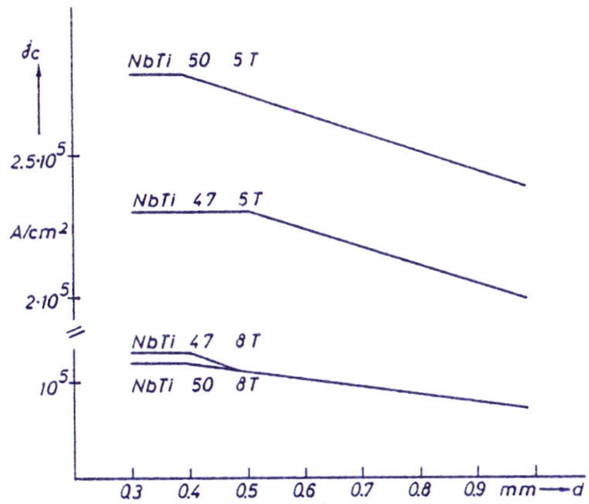

Fig. 4.

Typical optimized critical current densities of NbTi superconductors in dependence of the final conductor diameter.

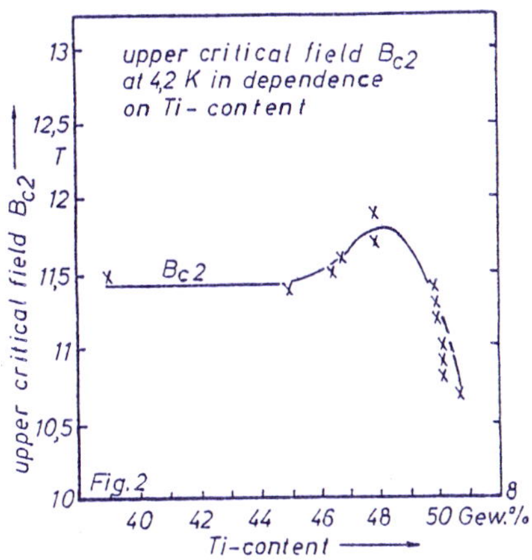

Fig. 5. Upper critical field B_{c2} at 4.2 K versus Ti concentration.

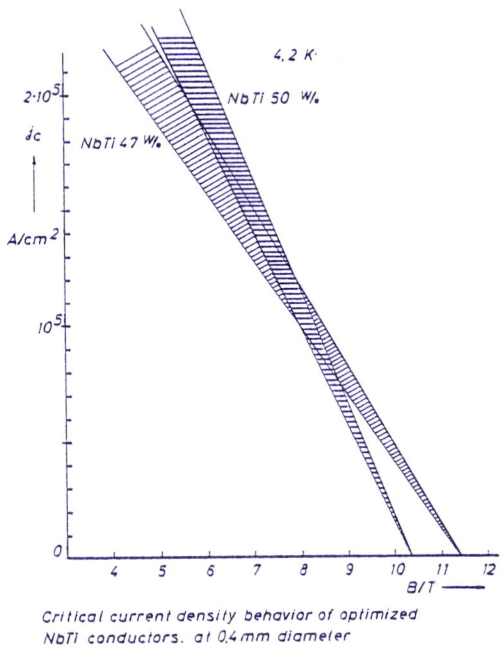

Fig. 6. Critical current density behavior of optimized NbTi conductors at 0.4 mm diameter.

Stabilization

For technical applications in magnets the reliable and safe performance is an important point. Superconductors carrying current in a magnetic field are in a non equilibrium condition. That means local disturbances, i.e. mechanical movements, friction, epoxy cracking cause hot spots releasing flux jumps and failure of the magnet. Therefore the superconducting state must be stabilized by continuously parallel shunting the superconductors with highly conductive material.
The simplest way to do so is to bond the superconductor continuously with low temperature highly conductive copper, being able to carry locally the whole current without warming up to the critical temperature of the superconductor by resistive heating. By this provision, the hot spot can again cool down without propagating increasing hot zones and carry again the superconducting current. Fig. 7 shows the simple criterion for the so called cryostatic stabilization after Stekly [17]. The left hand side of the equation contains the resistive heating and the right hand side the heat transfer to the helium bath. I is the current in A, ρ the substrate resistivity in Ωcm, q the substrate cross section in cm², h the heat transition coefficient to the He-bath in W/cm²K, T_C the critical temperature of the superconductor in K and S the cooled perimeter of the conductor in cm.

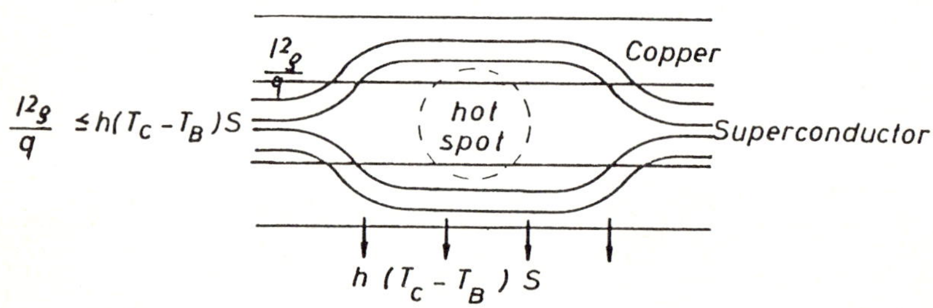

Fig. 7. Principle of cryostatic stabilization and Stekly criterion.

In practical cases cryostatic stabilization leads to very high copper to superconductor area ratios which can be reduced by using the criterion developed by Maddock, James and Norris regarding the heat conduction in the conductor and its temperature dependence to about one half of the copper cross section calculated with the Stekly criterion [18].
The cross section of a fully stabilized conductor having 990 filaments is shown in Fig. 8.

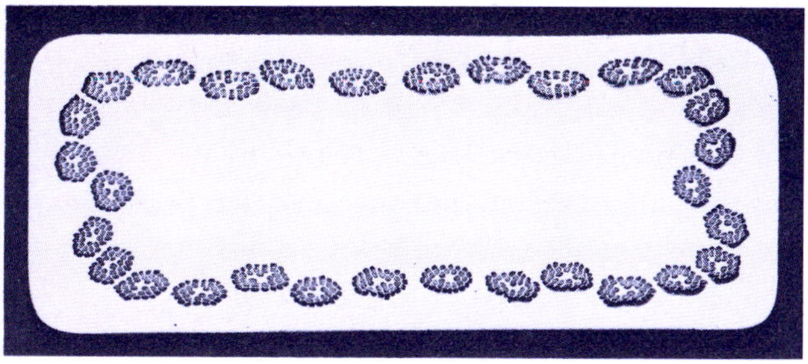

Fig. 8. Cross section of a rectangularly-shaped 5x2 mm^2 conductor with 30x33 filaments, α = 9 [27].

To overcome the necessity of applying large copper cross-sections to the superconductors further stability criteria have been developed. The adiabatic stability criterion answers the question how large is the adiabatically created temperature rise by flux penetration into the conductor in dependence on the filament diameter and which is the maximum filament diameter which does not lead to growing normal zones [19]. The dynamical stability criterion leads also to a maximum filament diameter by asking for an equilibrium between the rate of local flux penetration in the conductor and the rate of local heat diffusion [20]. Both criteria result in the requirement for making conductors with filament diameters of about 50 μm. In that cases only small copper to superconductor area ratios can be applied. Practical conductors for laboratory scale and medium size magnets have copper to superconductor area ratios of 1,2 - 4.
Fig. 9 shows the cross section of a 1500 filament adiabatically stabilized conductor having a copper to superconductor area ratio of 1.8. Fig. 10 shows a double extruded filament conductor with 163x163 filaments at 8 mm diameter.

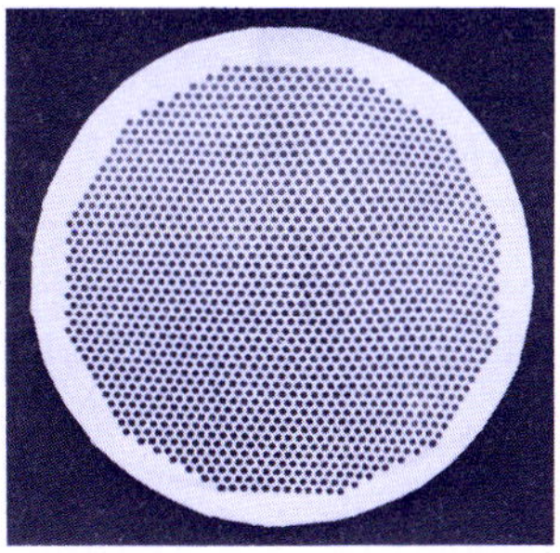

Fig. 9. Cross section of a 1500 filament conductor

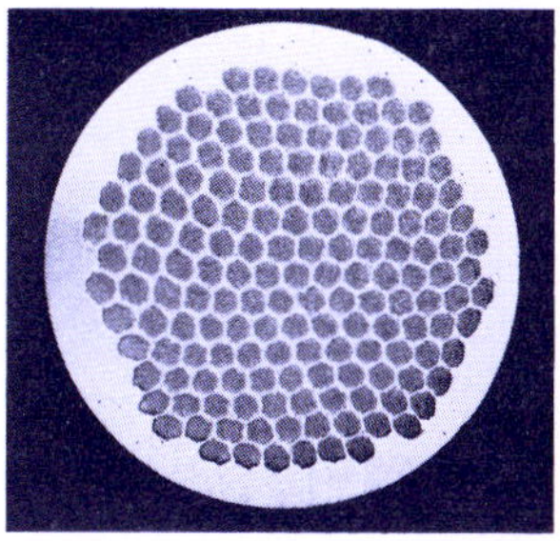

Fig. 10. Cross section of a double step extrusion 163x163 filament conductor

Twist

Energizing a magnet means applying a change in induction B on the conductor. Changing applied field causes transverse voltages and currents crossing the copper substrate with time constants of the order of one year. These coupling currents cause a filament conductor to behave like a conductor having only one single thick superconducting core.
To prevent coupling, filament conductors must be twisted. The twist pitch 4 l which reduces the coupling currents by a factor $(l/l_c)^2$ can be calculated using:

$$l_c^2 = \frac{10^4 \pi^2}{4} \quad \frac{\rho_N \Lambda^{\frac{1}{2}} j_c d}{\dot{B}} \quad \frac{w}{w+d} \qquad [21]$$

l_c is the length of the conductor where transverse currents are crossing the matrix in cm
$\dot{B}$ is the frequency of the changing field in T/s
ρ_N is the normal metal resistivity in Ω cm
Λ is the area fraction of superconductor
j_c is the superconductor current density in A/cm²
d is the filament diameter
w is the filament distance.

Taking $\dot{B}$ = 0.5 T/s; $j_c = 2 \cdot 10^5$ A/cm², $\rho = 2 \cdot 10^{-8} \Omega$ cm
$\Lambda = \frac{1}{2.3}$; d = 35 μm w = 10 μm and $(\frac{l}{l_c})^2 = 0.1$
l = results in 0.4 cm, that means the twist pitsch 4 l will be 1.6 cm.

In the case of short risetimes $\dot{B}$, resistive barriers between the copper surrounded filaments are applied to keep the twist pitch large enough for technical realization.

A15 Solid diffusion superconductors

Because of the brittleness of the intermetallic compound A15 phases it is impossible to apply cold work to them. Early A15 conductors have been tapes with A15 phase layers. The large extension of the layers causes instabilities in the changing field so that magnets wound of tape material have large energizing times and are unsuitable for rapidly energized or ac superimposed magnets.
The problem to make filamented A15-base conductors was solved by the so called solid state diffusion process [21]. It is based on the fact that copper is nearly unsoluble in Nb and V. In the case of Nb_3Sn base conductors a composite 13 w/o copper tin bronze and Nb is cold worked to the final diameter and subsequently heat treated at about 700 °C. By this long time high temperature heat treatment tin diffuses into

the niobium forming a A15 phase layer Nb_3Sn according to Fig. 11. By the fact that tin comes out of a copper rich solution only the Nb rich Nb_3Sn phase is formed and the formation of the tin rich phases $NbSn_2$ and Nb_6Sn_5 is prevented.

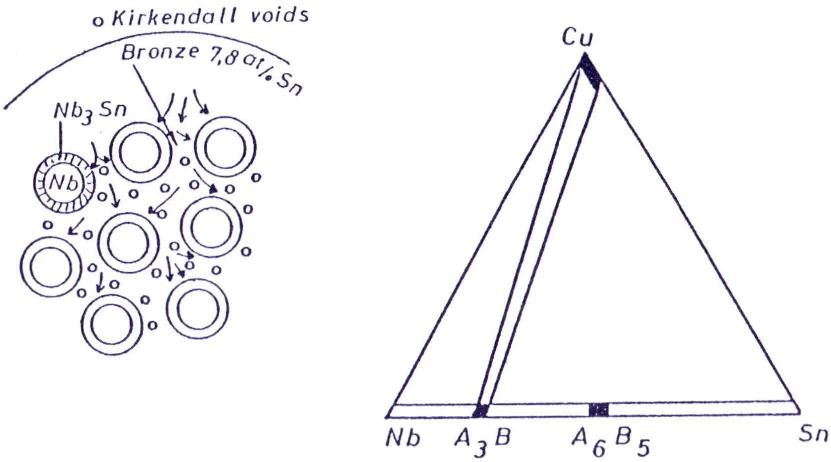

Fig. 11. Principle of solid state diffusion of Nb_3Sn and diffusion path in the ternary system Cu-Sn-Nb

The diffusion treatment governs mainly two processes - layer growth increasing the overall current density and recristallisation of the A15 layer decreasing the overall current density because grain bounderies are flux pinners. To reach optimum current densities layer thicknesses of about 1 μm are applied. This means filaments of 3-5 μm diameter must be produced. Therefore the filament numbers of technical conductors range from 4000 to 70 000 depending on the final conductor diameter. Fig. 12 shows the cross section of a 10 000 Nb_3Sn filament conductor.

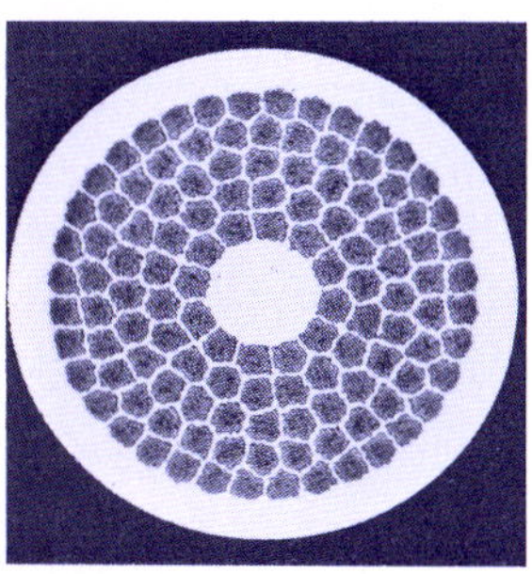

Fig. 12. Unstabilized 10 000 filament conductor 0.75 mm diameter.

Fig. 13 shows measured optimized current densities of unstabilized Nb_3Sn solid diffusion conductors.

Stabilization of Nb_3Sn-base filament conductors is provided by using Ta or Nb diffusion barriers preventing Sn-diffusion into the high purity copper which can be integrated into the conductor cross section in varions configurations. Fig. 14-16 shows differently stabilized Nb_3Sn filament conductors [22].

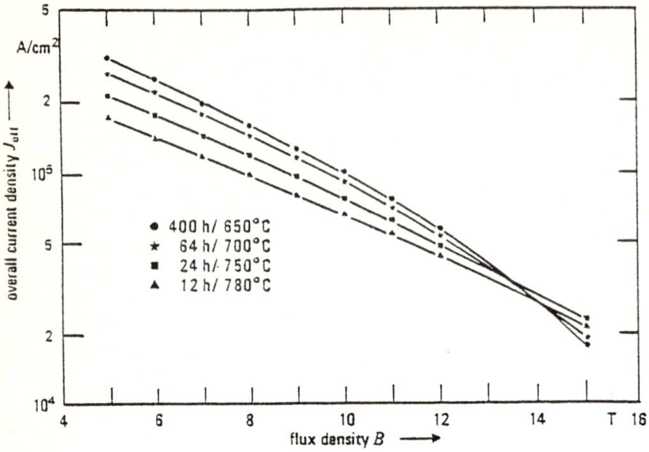

Fig. 13. Critical current density I_{eff}(B) of a 10 000 filament Nb_3Sn conductor for different diffusion temperatures.

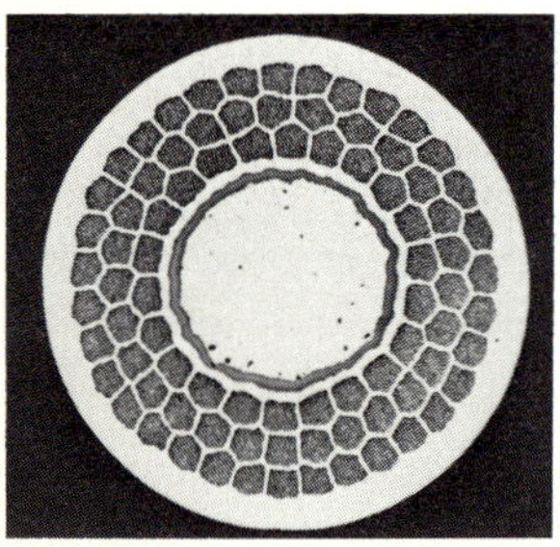

Fig. 14. Cross section of an internally stabilized 10.000-filament conductor. Wire diameter, 0.9 mm, critical current, 295 A at 10 T

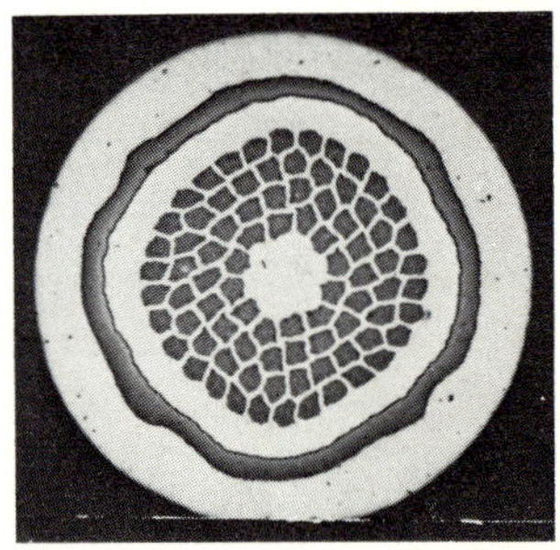

Fig. 15. Cross section of a peripherally stabilized 15.000-filament conductor. Wire diameter, 1.17 mm.

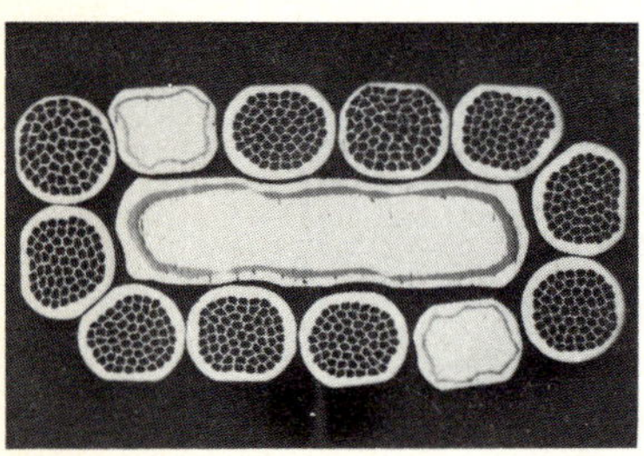

Fig. 16. Twelve strands cabled around a CuTa core: Wire diameter, 0.4 mm, 3.700 filaments per strand; total cross section: 1.1 mm x 2.25 mm. Critical current: 750 A at 10 T.

An important point for handling the material with brittle A15 layers and its performance in magnets exposed to electromagnetic Lorentz forces is the prestrained condition of the A15 layers at low temperatures. Prestrain arises by the different thermal contraction of bronze and Nb_3Sn or Nb. In the cooled condition the A15 layers are compressively strained in longitudinal direction. This prestrain reduces the electrical properties j_c and B_{c2} but on the other hand it provides a mechanical reservoir for tensile strain. Maximum critical current densities and critical fields appear in the strain free condition according to Fig. 17.

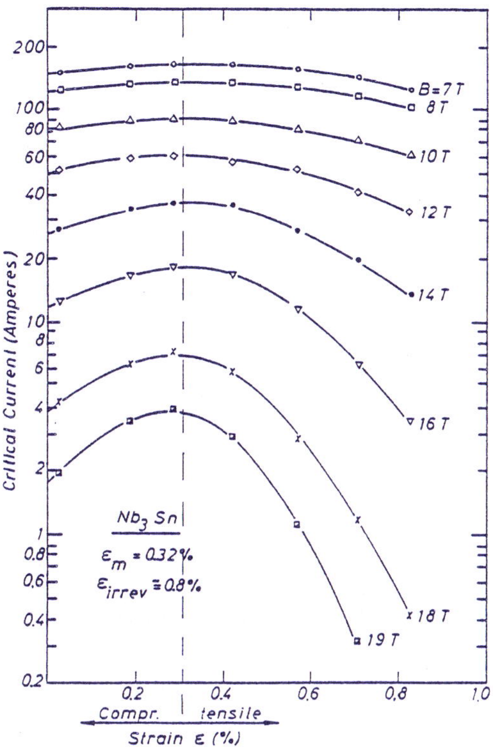

Fig. 17. Influence of compressive and tensile strain on the critical current for a multifilamentary Nb_3Sn conductor at different fields [23].

Insulation, magnet winding and magnet performance

For NbTi superconductors common insulation types like varnish, cotton and capton foils can be used. The choice of the insulation type depends on the available insulation layer thickness, the required break through voltage and the type of epoxy impregnation. In the case of cabled conductors either cotton or capton foil insulation are recommended. Another possibility is to use bare conductors, winding layer insulation tapes and insulation strips between the conductor sides. In the case of A15 conductors the insulation must resist the high temperature heat treatment. Therefore quartz or glassfiber insulation is necessary. Because of the brittleness of these braided insulations a dressing which is applied during the insulation process must be driven out either immediately before winding or after winding the magnet. During this operation carbon will be produced by decomposition of the dressing giving rise to some electrical conductivity across the windings and layers.

Main sources for failure of magnet winding by degradation are training caused by small wire movement, friction heating by mechanical bending when the magnets are energized and hot spots caused by impregnation cracking. Therefore magnet windings must be mechanically stable by using coil former material which resists high mechanical forces and which fits thermal expansion as far as possible. The same is valid for impregnation material. A common method to prevent loose windings after cooling down the magnet is the application of precompression of the winding by additional high strength surface windings or massive steel clamps.

To overcome the problem of cracking epoxy, fillers reducing the thermal expansion and glass fiber crack resistors can be used. In these cases wet winding techniques must be used because of the poor viscosity of the impregnation.

In the case of A15 base magnets a wet winding technique being also applicable to NbTi magnets was developed by K.J. Best [25,26] which overcomes the problem of carbon bridges and epoxy cracking. The windings are impregnated with a sodium silicate base material containing ceramic fillers having a comparable thermal expansion to the superconductors. This impregnation resists temperatures up to 1000 °C forming a cement like structure after heat treatment even at low temperatures in the case of NbTi magnets. By this fact also the final coil geometry is well fixed and cannot be influenced during the final high temperature heat treatment.

The maximum operating current of a magnet depends on the perfection and the uniformity of the magnet winding. It is limited by the maximum heat pulse which drives the magnet normal by the growth of a normal conducting spot inside the magnet winding. Fig. 18 shows the size of the energy pulse needed to quench a certain coil as a function of I/I_c calculated by N. M. Wilson [24]. It can be seen that with increasing magnetic field and operating current this energy becomes smaller and smaller decreasing drastically in the low field region near the conductor critical current density.

General points of view of magnet winding can be summarized as follows.

1. The choice and design of a conductor depends on the magnet design. Mechanical winding tensions are running up in the inner part during winding. Very long winding lengths can cause loose inner windings. Therefore keep the number of turns as small as possible.

2. Stabilization by larger copper to superconductor area ratio improves the reliability of reaching short sample currents in the magnet, but the more the conductor is stabilized the larger is the needed quantity of superconducting material. Therefore use low copper to superconduc-

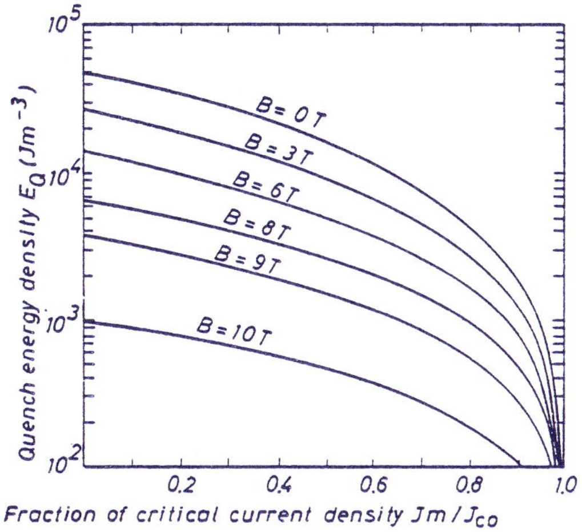

Fig. 18. The energy density of a distributed disturbance needed to quench a typical NbTi-magnet winding under various conditions of current density a hole field.

tor area ratio material for small magnets, make winding and impregnation carefully preventing mechanical conductor movement which leads to training and degradation.

3. Special tailor made conductor designs are more expensive than standard conductors. Therefore use standard material as far as possible.

4. To establish a good layer to layer insulation use additional layer insulation beside conductor insulation.

5. In the case of larger magnets rectangular shaped conductor cross sections result in larger packing densities and windings are better protected against mechanical movement.

6. Take note of differential thermal expansion of structural materials such as coil formers and impregnation material and superconductors.

7. Keep the design current of the magnet operation not near the short sample current. A good design value is 0.6 I_c.

References

1. Yntema, G.B.: Phys.Rev. 98, 1197 (1955)
2. Kunzler, J.E.: Rev. Med. Phys. 33, 1 (1961)
3. Aron, P.B., Hitchcock, H.C.: Superconductors edited by Tannenbaum,L.M. and Wright,W.K., New York, Interscience 119 (1962)
4. Wong, J., Superconductors 83 (1962)
5. Walker, M.S., Fraser, M.J., Superconductors 83 (1962)
6. Kneipp, G.D., Betterton, J.O., Easton, D.S., Scarborough, F.R.: High Magnetic Fields (1961)b, edited by Kolm,H.et al. New York, John Wiley 574
7. Treuting, R.Y., Wernik, J.H., Hsu, F.S.: High Magnetic Fields, 597 (1961)
8. Best, K.J., Hillmann, H., Hoeflich, H., Rudolph, J., Weber, H.: Forschungsbericht BMFT T81-009 (1981)
9. Hillmann, H.: Superconductor Materials Science, edited by Foner,S. and Schwartz,B., Plenum publishing corp. (1981)
10. Pfeiffer, I., Hillmann, H.: Acta Met. 16, 1429 (1968)
11. Hillmann, H., Pfeiffer, I.: Z.f.Metallkde. 58, 129 (1967)
12. Neal, D.F., Barber, A.C., Woolcock, A., Gidley, J.A.: Acta Met. 19, 143 (1971)

13. Larbalastier, D.C.: Superconducting Materials Science, edited by Foner, S. and Schwartz, B., Plenum publishing corp. (1981)
14. Willbrand, J., Schlump, W.: Z.f.Metallkde. 66, 714 (1975)
15. Hillmann,H., IEEE Trans. on Magn. Mag-17, 1614 (1981)
16. Hillmann,H., Best, K.J.: IEEE Trans. on Magn. Mag-13, 1568 (1977)
17. Stekly, Z.J.J.: Proc.of the First Intern. Cryogenic Eng. Conf., Kyoto, 112 (1967)
18. Maddock, B.J., James, G.B., Norris, W.T: Cryogenics 9, 261 (1969)
19. Hancox, R.: phys. Let. 16, 208 (1965)
20. Chester, P.F.: Rep. on Progr. on Phys. XXX, II, 561 (1967)
21. Tachikawa, K.; Proc. of the 3. Int. Cryog.Conf., Berlin 339 (1970)
22. Hillmann, H., Pfister, H., Springer, E., Wilhelm, M., Wohlleben,K.:
23. Ekin, J.W.: Superconducting Materials Science. edoted by Foner,S. and Schwartz, B., Plenum publ. Corp. (1981)
24. Wilson, N.M.: Superconducting Materials Science edited by Foner,S. and Schwartz, B., Plenum publ. Corp. (1981)
25. Best, K.J., Rothe, B., IEEE Trans. on Magn., MAG-17,1, 478 (1981)
26. Best, K.J., Hillmann, H.: IEEE Trans. on Magn., MAG-17,5, 2312 (1981)

THE GENERATION OF HIGH MAGNETIC FIELDS IN GRENOBLE - INSTALLATIONS AND MAGNETS

H.-J. Schneider-Muntau
J.C. Picoche*, P. Rub*, J.C. Vallier*

Max-Planck-Institut für Festkörperforschung, Hochfeld-Magnetlabor
166X, F-38042 Grenoble Cédex, France

* Service National des Champs Intenses, C.N.R.S.
166X, F-38042 Grenoble Cédex, France

We give a description of the High Field Magnet Laboratory of Grenoble. We present the 10 MW power supply, the 400 m^3/h cooling water circuit and the associated technical installations. We describe the water-cooled high field magnets (up to 25 T) and their rate of utilization. Technical details of the superconducting magnets (up to 14 T) and the hybrid magnet which is expected to give a continuous field of 30 T in a 5 cm diameter room temperature bore are presented.

1. INTRODUCTION

The High Field Magnet Laboratory of Grenoble is run as a joint venture by the Centre National de la Recherche Scientifique (C.N.R.S.) and the Max-Planck-Gesellschaft (MPG). It was formed originally by the 'Service National des Champs Intenses' (S.N.C.I.), a service laboratory of the C.N.R.S., which was founded in 1971 with the aim of creating a research institute for experiments on matter subjected to extreme conditions such as high magnetic fields, low temperatures and high pressures [1]. In 1972 the C.N.R.S. and the MPG signed a contract with the aim of an enlargement and the common exploitation of the existing installations, creating the Hochfeld-Magnetlabor as partner of the S.N.C.I. [2]. With the generous support of the Volkswagenstiftung, the available power installations were doubled to 10 MW. This measure made the High Field Magnet Laboratory competitive with the Francis Bitter National Magnet Laboratory, the leading laboratory in this field. Because of the international importance of the High Field Magnet Laboratory, the C.N.R.S. and the MPG also committed themselves to place the scientific equipment and the technical installations at the disposal of all scientists, i.e. the access is not limited to scientists from only these two organizations.

Today 60 French and German scientists, engineers and technicians are working continuously and in close cooperation at the S.N.C.I. and the Hochfeld-Magnetlabor and about 20 visiting scientists and many short-time guests from all over the world come to perform experiments in high magnetic fields.

2. THE TECHNICAL INSTALLATIONS

The High Field Magnet Laboratory is housed in a three story building with the magnet cells on the ground floor, work shops and laboratories on the second floor and offices, library, conference- and computer room on the top floor. The machine hall, where the power supplies and the pumps and the heat exchanger for the water cooling of the magnets are installed, adjoins the building. The electrical power and the cooling water are distributed in the basement and fed from there directly to the eight magnet sites on the ground floor, resulting in very short connections. Last year a new building was added which will house the hybrid magnet, a large superconducting magnet and the liquefier for these two magnets. Altogether a space of about 1200 m^2 is available for scientific experiments.

2.1 THE POWER SUPPLIES

The total installed power of 10 MW is generated by four identical power supplies each giving 7500 A at 335 V. They can be connected in parallel, therefore either two 5 MW magnets (up to 15 T) or one 10 MW magnet (up to 25 T) can be run at the same time. Each power supply is fed independently from a 15 kV high power line via a transformer which generates two three-phase systemswith a phase shift of 15 degrees by two different secondary windings, one in star, and one in delta (see Figure 1). The alternating current passes two bridge rectifiers with silicon controlled rectifiers (SCRs) and a balance inductance LO. The fundamental frequency of the current is therefore 600 Hz and is reduced by an LC-filter with a resonance frequency of 30 Hz. The harmonics, generated by the SCRs, are damped by inductances in front of the SCRs and by a special arrangement of the zig-zag windings and double shielding of the transformers. The voltage in this first part of the power supply is stabilized in a slow control loop (0-100 Hz). The voltage is monitored and commands the time of switching of the SCRs in such a way that voltage fluctuations are reduced to a value of $\Delta U/U \leqslant 1 \times 10^{-3}$. The additional amplifiers guarantee equal current in the two bridge rectifiers. To obtain a higher stability an active filter was added. The alternating-current component of the voltage is amplified and fed in with opposite phase via a transformer compensating any deviations up to $\pm 5 \times 10^{-6}$ in the frequency range from 50 Hz to 10 kHz. The shunt voltage (1 V at 7500 A) is used as current control; the difference between the shunt voltage and the stabilized variable reference voltage being amplified and fed into the two voltage regulation circuits.

The current rise time from 0 to maximum is approximately one minute. The magnet current can also be controlled and/or modulated by an external voltage (0-10 V).

We measured the long time stability of the magnetic field with a proton NMR system [3] and found $\pm 5 \times 10^{-6}$ over one day with an integration time of two seconds and $\pm 1 \times 10^{-5}$ with an integration time of 100 ms. The ripple is essentially independent of the field value, having an amplitude of 20 μT. In addition, there are also fluctuations of 200 μT in the frequency range from 0.1 to 1 Hz.

2.2. THE COOLING CIRCUIT

The magnets are cooled by deionized water of high resistivity in a closed loop circuit, which itself is cooled with river water via a heat exchanger. A demineralizer keeps the resistivity to about 5 MΩcm. The cooling circuit and also the flow rate and the pressure drop in the magnets (25 bar) are designed in such a way that the cooling power of one pump corresponds to the electrical power of one power supply. A 5 MW magnet uses two power supplies and consequently two pumps with 100 m^3/h each, a 10 MW magnet uses all four of them. A constant water pressure of 27 bar is therefore maintained in the cooling circuit and excellent cooling conditions are assured. Two of the four 10 MW magnets are equipped with additional valves allowing the inner coils to be used alone. Filters with 200 μm mesh apertures are installed in front of the magnets to avoid obstruction of the cooling holes of the Bitter stacks. A 100 μm filter is installed in front of the heat exchanger to eliminate particulate material which may originate from burnt-through stacks or broken parts. A 5 μm by-pass filter was added for the extraction of smaller particles. The shunts of the power supplies are cooled with well water because it changes its temperature only by a few degrees during the year, and therefore no thermostatic regulation of the water flow is required. The general lay-out of the cooling water circuit is shown in Figure 2.

2.3. THE COMPUTER CONTROL

Because of the high costs of a Bitter stack, we have devoted much care to the question of how to increase the longevity of our magnets and we have successfully introduced a few modifications [3].

The best measure to prevent the burning-through of a stack is of course to switch off the magnet as soon as any irregulaities of the resistance or the magnetic field

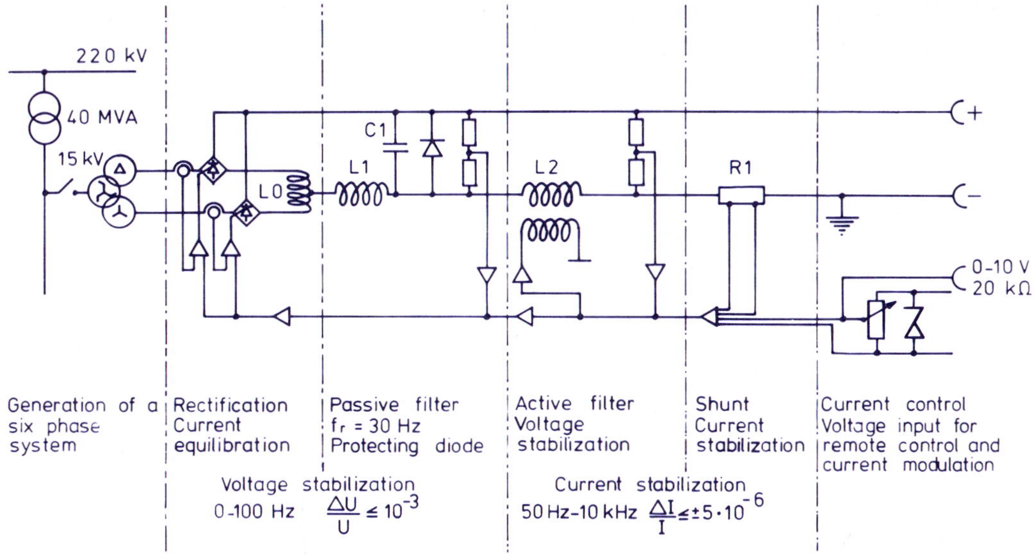

Figure 1: Basic circuit diagram of a 2.5 MW power supply

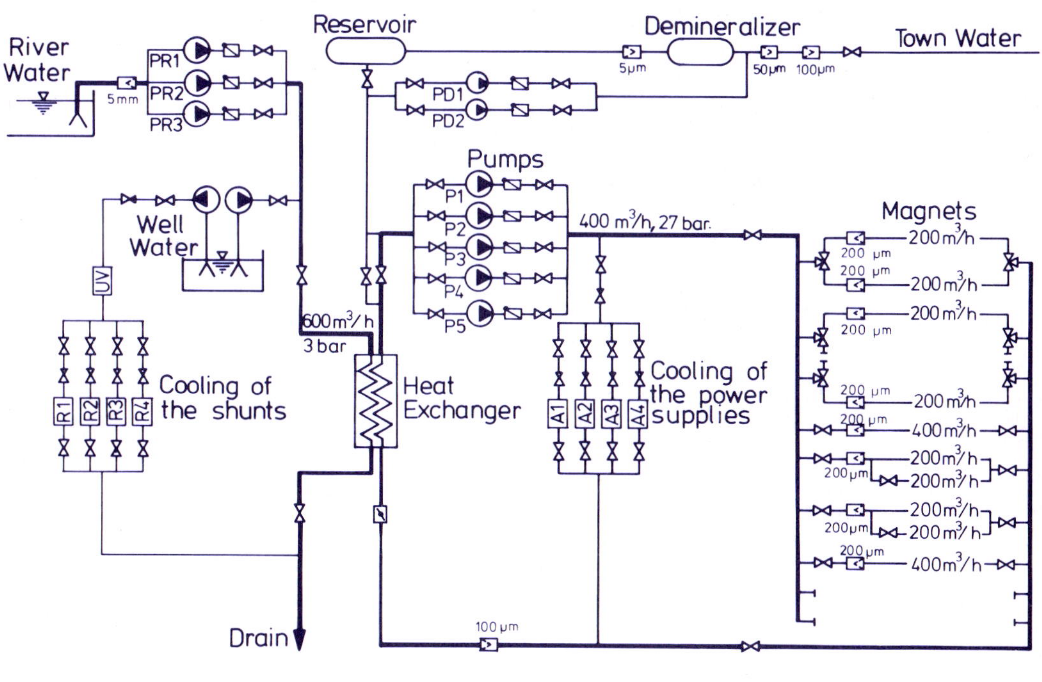

Figure 2: Cooling water flow diagram

is observed and to replace the damaged discs. The check of the resistance of the magnet before or after its daily run is of limited help only. Our experience has shown that a stack with a few short-circuited plates can still be used without any further risk if the power level is not too high. At maximum power, however, the short circuit propagates through the whole stack in less than a second. Consequently a fast and sensitive system is required to detect a short circuit between two discs, i.e. 1 % of the whole stack, in less than a second.

This is accomplished by a computer (PDP 11/20)which scans the voltage, the current and the inlet and outlet water temperature of each magnet coil three times a second, calculates the reduced resistance, and compares it with the calibration value at the corresponding power. Any deviation which exceeds a preset limit, releases an alarm.

3. RESISTIVE MAGNET

The magnet laboratory has at present eight magnet sites at its disposal which may be equipped with different coils or coil combinations, as shown on Fig. 3.

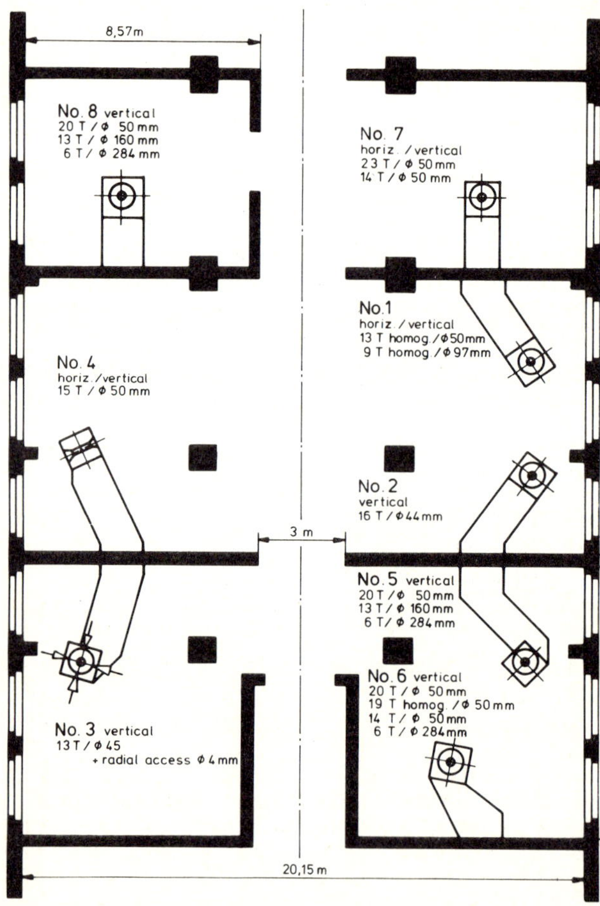

Figure 3: Magnet sites of the High Field Magnet Laboratory

Magnet site	Coil no.	Room temperature bore ∅ [mm]	Magnetic field [T] Max.	Nom.	Power at nominal field [MW]	Homogeneity in 1 cm SVD	Average life time [h]	Direction of magnetic field	Remarks
1	1a	50	14.1	13	3.6	$2\cdot10^{-5}$	800	vert./horiz.	high homogeneity coil
	1b	97	9.2	9	4.1	$1\cdot10^{-5}$	700	vert./horiz.	high homogeneity large bore coil
2	2a	44	16	15.5	4.0	$1.5\cdot10^{-3}$	1000	vert.	
3	3a	45/4	14	13	3.7	$2.5\cdot10^{-3}$	1000	vert.	split coil with 4 radial 4 mm ∅ bores
4	4a	50	15.8	15	3.9	$2\cdot10^{-3}$	700	vert./horiz.	
5	5a	50	20.9	20	8.3	$1.5\cdot10^{-3}$	500	vert.	
	5b	160	13.8	13	8.3	$1\cdot10^{-3}$	400	vert.	
	5c	284	6.5	6	4.1	$1\cdot10^{-3}$	1400	vert.	
6	6a	50	21	20	7.6	$1.5\cdot10^{-3}$	600	vert.	
	6b	50	19	19	8.1	$2\cdot10^{-5}$	200::	vert.	high homogeneity coil
	6c	50	14.6	14	3.6	$2\cdot10^{-3}$	600	vert.	
	6d	284	6.4	6	4.0	$1\cdot10^{-3}$	1400	vert.	
7	7a	50	25	23	8.4	$2\cdot10^{-3}$	400	vert./horiz.	polyhelix magnet
	7b	50	14.5	14	4.9	$1\cdot10^{-3}$	500::	vert./horiz.	
8	8a	50	20.9	20	8.3	$1.5\cdot10^{-3}$	500	vert.	
	8b	160	13.8	13	8.3	$1\cdot10^{-3}$	400	vert.	
	8c	284	6.5	6	4.1	$1\cdot10^{-3}$	1400	vert.	

:: new coil, the figure given represents the running time up to now

Table 1: The resistive magnets of the High Field Magnet Laboratory of Grenoble

The magnets (Table 1) consist of one or two coils, each using two power supplies with a combined rated power of 5 MW. The single coil magnets nos. 1 to 4 are therefore called 5 MW magnets, the two coil magnets nos. 5 to 8 10 MW magnets, although they use considerably less power than 5 or 10 MW respectively. Our magnets cover a wide range from 6 T in 284 mm to 25 T in 50 mm, both horizontal and vertical. We distinguish between maximum and nominal fields; the first being the design value up to which the magnets are run only for test purposes and on special requirements. In their daily operation the magnets are limited to their nominal field values as listed in Table 1. All coils but one are Bitter magnets, stacked in the original Bitter version which gives an equal force distribution around the circumference.

Magnet no. 7 is an entirely new magnet which was designed to achieve the maximum attainable field in a room temperature bore of 50 mm diameter with the given power of 10 MW.

In order to accomplish this task, we had to adopt the following new features:

- development of a new technology to build magnets, the so-called polyhelix magnets, to cope with the extremely high Lorentz forces in the inner part without losing too much in field
- optimization of the power distribution of the outer part in radial and axial directions
- optimization of the geometry and of the power distribution between the inner and outer parts.

The inner, highly stressed part consists of a series of concentric, single-layer coils, each of them being optimized for maximum field output at constant stress level [4]. The small annular spaces between the helices are used for cooling; for optimal cooling conditions required by the high power density of more than 2 MW per liter of volume, the water velocity is set to 20 m/sec.

As a consequence of the optimization of the power distribution between the inner polyhelix magnet and the outer Bitter magnet, the design of the outer magnet is not limited by stresses. Therefore the optimum current density distribution, the so-called Kelvin distribution, was applied for the Bitter magnet and approximated by two concentric sets consisting of a split coil and a central coil with higher current density. In this way an improvement of 10 % in field compared to a single Bitter magnet was achieved.

The magnet generated 25 T with 10 MW of power in a room temperature bore of 50 mm diameter. This is the highest continuous field ever achieved with resistive magnets. Predictions and calculations were confirmed by the performance of the magnet to within 1 %. The magnet has been in full operation for more than 400 hrs. of scientific experiments.

A common feature of all our magnets is their high filling factor. It results from the high water pressure (requiring only small cooling holes and giving relatively low copper temperatures), the silver plating of the discs which reduces contact surfaces, and the rather thin insulators. For this reason our magnets compare favourably with those used in other magnet laboratories.

As a basis for comparison we derive from the well-known Fabry relation a criterion of efficiency, which we define as:

$$G^{*} = B\ (d/W)^{1/2}$$

It relates the power W (MW), which is required to generate the obtained field, to the magnetic induction B (T) and to the diameter d (cm) of the room temperature bore in which the field is produced.

The efficiency factor G^{*} reflects all the decisions and the courage of the magnet designer in the choice of conductor and insulator materials, coil shape, current distribution and dimensions of the coils and in his design of the mechanical structure, coolant channels, the thickness of the central tube etc.. In Figure 4 the efficiencies of the best magnets of some magnet laboratories are given. It dramatically displays the

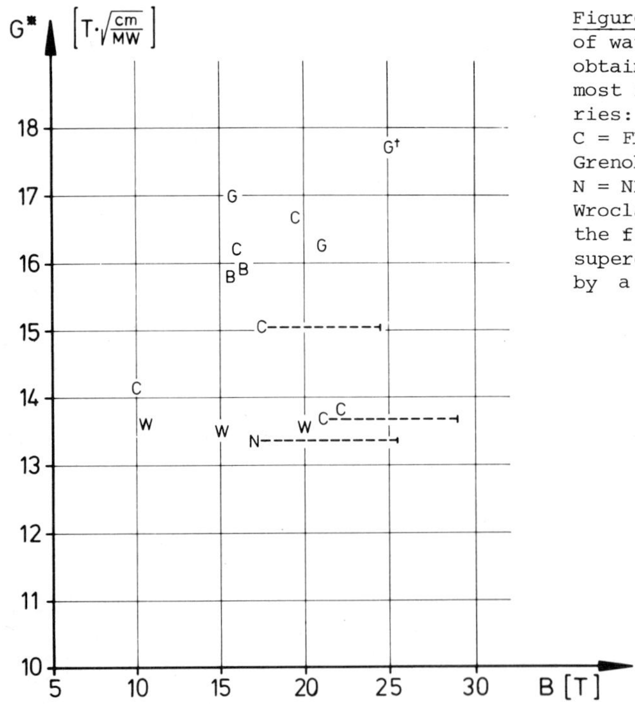

Figure 4: The efficiency factor of water-cooled magnets and the obtained magnetic fields of the most important magnet laboratories: B = HMFA Braunschweig, C = FBNML Cambridge, G = HFML Grenoble († = polyhelix magnet), N = NLHMF Nijmegen, W = IL HMF Wroclaw. For hybrid magnets the field contribution of the superconducting part is marked by a dotted line.

increased difficulties and reduced efficiencies encountered in building magnets for higher fields. Only the polyhelix magnet, with its new technology, shows an improved efficiency factor. Hybrid magnets gain by the additional booster field from the superconducting coil, as indicated by the dotted lines in the figure.

4. RATE OF UTILIZATION OF THE TECHNICAL INSTALLATIONS

The High Field Magnet Laboratory has been operating for ten years with its full power of 10 MW. Table 2 gives a list of the hours of use of our eight magnet sites during this period; it reflects the important demand for magnet time utilizing fields above 15 T (magnet sites nos. 5 to 7).

In 1981 the magnets consumed a total energy of 3.7 GWh. On the average 950 kWh were consumed per hour of magnet time. The average electrical running costs for one hour of magnet time, including cooling pumps and infrastructure, amounted to FFR 270.- (US $ 39.-).

Taking into account the time when the installations were shut-down for maintenance and repairs, a utilization rate of 70 % is calculated, i.e. the laboratory runs practically 17 hrs. a day for scientific operation.

Year	Magnet site							Total
	1	2	3	4	5	6	7	
1972	253	150						403
1973	127	781	111	108				1127
1974	282	1184	1166	1079	20	30		3071
1975	132	1489	1100	821	394	205		4141
1976	646	1215	1233	382	602	336		4414
1977	256	1613	1230	455	988	430		4972
1978	199	1379	1750	591	845	226	49	5039
1979	512	1498	1075	261	477	479	49	4345
1980	134	1322	1022	118	650	660	35	3951
1981	332	1110	765	324	650	488	245	3914

Table 2: Hours of use of the different magnet sites

4. SUPERCONDUCTING MAGNETS

The superconducting magnets of the laboratory are listed on Table 3.

The first three magnets were built in the laboratory. They were designed for magnetization measurements and offer, therefore, a high homogeneity along the axis.

The magnet no. 1 built in 1971 consists of two coils, the outer being wound from NbTi wire and the inner coil from Nb_3Sn tape. This magnet is still working and is able to produce 13.5 T at 4.2 K and 14.2 T at 2.1 K in saturated liquid helium.

The second magnet was constructed in 1974 with Nb_3Sn tape from IGC which also provided technical support. This magnet was particularly convenient to test new cryogenic technologies as described by Roubeau [5] and by Claudet et al. [6].

We have shown that running this second magnet at 1.8 K improves its performance by 1 T (this would correspond to 2.5 to 3 T for a magnet wound from NbTi) provided, of course, that the magnet is properly designed to withstand the increased forces [7]. The very important characteristic of this technology is that even at 1.8 K the helium bath is at atmospheric pressure, and consequently continuous operation is easily possible.

Magnet no. 3 produces 9.0 T at 4.2 K and 11.5 T at 2.0 K. This magnet forms a part of a dilution refrigerator for magnetization measurements at 10 mK.

The magnets nos. 4 and 5 are commercial magnets. The 10.3 T split coil has rather large opening angles of 12 to 20 degrees respectively and serves for optical measurements. The 7.5 T magnet has a 50 mm room temperature bore for preparatory experiments at lower fields.

The magnet no. 6 was built during the last three years using NbTi MF wire. The coil generates at 1.9 K a magnetic field of 10 T in a cold bore of 325 mm. The construction of this magnet is the result of a cooperation of many years between the CENG and the High Field Magnet Laboratory. The cryostat at 1.8 K and atmospheric pressure for the magnet was designed by Bon Mardion et al. [8]. This apparatus is an enlargement (3:1) of the cryostat built previously for the magnet no. 2.

Magnet	Clear bore at liquid helium temperature	Magnetic field (T) at 4.2 K	Magnetic field (T) at 2.0 K	Homogeneity	Stored energy (kJ)	Remarks
1	30	13.5	14.2	2×10^{-3} over ± 20 mm	350	Nb_3Sn tape and NbTi wire saturated liquid helium bath
2	40	11.9	13.0	2×10^{-3} over ± 20 mm	120	Nb_3Sn tape; working at 4.2 K or in unsaturated superfluid liquid helium
3	50	9.0	11.5	1×10^{-3} over ± 25 mm	80	NbTi MF for 10 mK dilution refrigerator
4	23	10.3	--	7×10^{-4} in 1 cm DSV	--	Nb_3Sn tape; split coil with Ø 15-43 mm conical radial access
5	79	7.5	--	10^{-5} in 1 cm DSV	41	Nb-Ti MF; horizontal room temperature bore of 50 mm Ø
6a	325	7.2	10.0	2×10^{-4} 1 cm DSV	4500	Nb-Ti MF; working at 4.2 K or in unsaturated superfluid helium
6b	254	--	11.0	2×10^{-4} 1 cm DSV	4600	Nb-Ti MF at 1.8 K as 6a
6c	120	--	14.0	1.7×10^{-4} 1 cm DSV	4500	Nb_3Sb tape at 1.8 K as 6a

Table 3: The superconducting magnets of the High Field Magnet Laboratory of Grenoble

The magnet no. 6 can be used with two inserts:
- a NbTi coil with a bore of 254 mm to generate a total field of 11 T and
- a Nb_3Sn magnet (built by IGC following our specifications) that can be used at 1.8 K to produce a total field of 14 T.

The combination of these three coils has not been tested up to now, but we expect to produce 15.5 T. An additional Nb_3Sn insert coil is planned for a total field of 16.5T.

A small V_3Ga magnet, wound from tape, was tested in the background field of 13 T of a resistive magnet. It generated 3.5 T at 1.8 K and confirmed our conviction that the definitively designed V_3Ga magnet will achieve the same field in a background field of 16.5 T [9].
We think, therefore, that it is possible to generate 20 T with commercially available materials, provided that the advantages of the 1.8 K technology are exploited.

5. THE GRENOBLE HYBRID MAGNET

In Table 4 we have listed all high field magnets above 20 T which are in service in the different magnet laboratories and can be used for scientific experiments. Fields above 25 T in 50 mm Ø can be generated economically only with hybrid magnets, i.e. a resistive magnet surrounded by a superconducting magnet which provides a booster field [10].
The C.N.R.S. and the MPG agreed recently to built a hybrid magnet in Grenoble. The design goal is 30 T in a room temperature bore of 50 mm. It is obvious from Table 4 that this magnet represents the biggest challenge, since it has to produce the highest field in the largest bore.

Field [T]	Contribution of supercond. coil	Diameter of room temper. bore [mm]	Power [MW]	Location
20	-	26	5.7	Wroclaw
21	-	50	8.4	Grenoble
22	-	33	8.6	Cambridge
24.5	7.0	54	7.3	Cambridge
25	-	50	10	Grenoble
25	6.9	28	5.6	Moscow
25.5	8.5	32	5.2	Nijmegen
29	7.9	33	7.9	Cambridge
In construction:				
30	∿ 10	50	∿ 10	Grenoble

Table 4: High field magnets above 20 T in service in the different magnet laboratories

To achieve this goal, we had to utilize two new technologies, which have never been used up to now for hybrid magnets but about which much experience is available at the High Field Magnet Laboratory in Grenoble:
- the polyhelix concept for the resistive part and
- the 1.8 K helium bath at atmospheric pressure for the superconducting part.

Both technologies have proven to give satisfactory results and have attained high reliability. The 25 T magnet is built using the polyhelix technique (no. 7 from Table 1) and the superconducting magnet, that generated 10 T in 325 mm diameter (no. 6 from Table 3), is run in unsaturated superfluid helium at 1.8 K. The design of the hybrid magnet is therefore an extrapolation of these two magnets and we are confident of achieving the goal of 30 T.

Figure 5 shows a model of the Grenoble hybrid magnet. The outer part consists of a big cryostat, which houses the two concentric superconducting coils. The inner coil has to have a large cooling surface and therefore will be wound in 22 separated layers; this way the winding exposes the maximum surface to the coolant. The NbTi wire has a cross section of 5.6 x 2.8 mm and consists of 10.000 filaments of a diameter of 22 µm. For the outer coil the cooling requirements are less demanding, so the pancake technique is applied. 60 double pancakes with an inner diameter of 742 mm and an outer diameter of 1087 mm are necessary to generate a field of 7.7 T. The total field is expected to be 11 T at 850 A. As the stored energy is 22 MJ, special measures to extract the energy in case of a quench are applied. The cryostat containing the two coils consists of a cold box at 1.8 K and atmospheric pressure. The coils sit on a fiberglass epoxy resin cylinder of about 1000 mm diameter and 8 mm thickness which is fixed to the bottom flange of the cryostat.

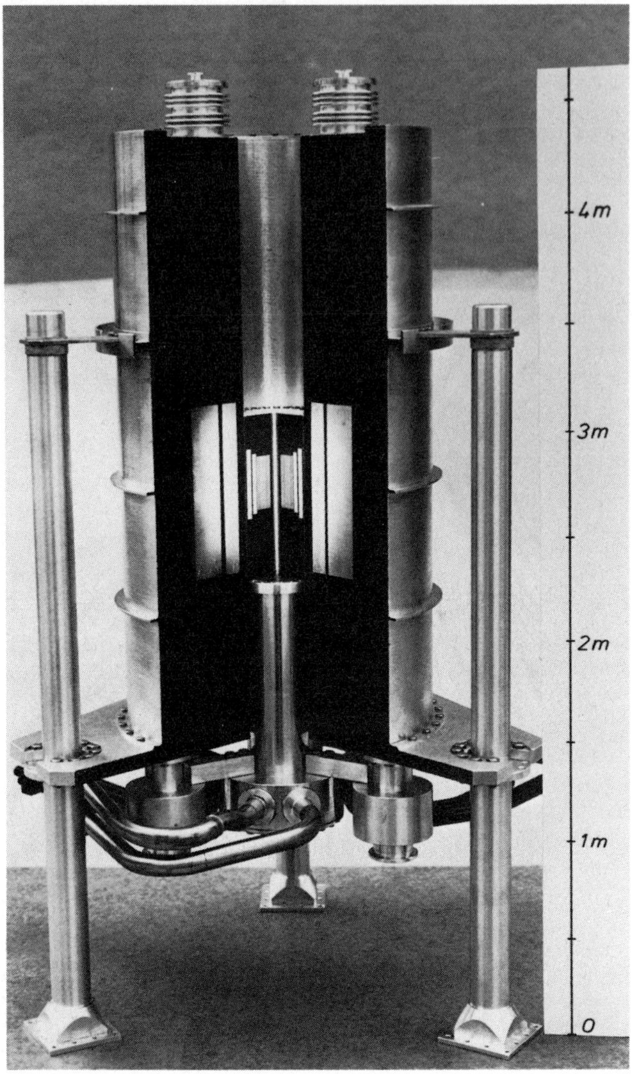

Figure 5: The Grenoble hybrid magnet (in construction)

Four large tie rods press the cold box to the bottom flange. They are calculated to withstand any vertical or radial forces, which may arise from a partial burn-out of the resistive magnet via the mutual magnetic coupling of the superconducting and the resistive system.

The inner, resistive part is also fixed to the bottom flange of the cryostat. It can be adjusted in both axial and radial directions in order to make the magnetic centers of the two systems coincide. For the case of axial misalignment, a force of 7.5 kN per mm of displacement was calculated. The resistive system consists of two groups of coils: two Bitter magnets, electrically in series, surround a set of twelve helices. A stress level of 320 N/mm^2 was chosen for all 14 coils. A field of close to 15 T will be generated by the helices with 5 MW. The two Bitter magnets are expected to give 5 T also with 5 MW. All connections to the resistive part are made from the bottom side to assure maximum accessibility to the magnet.

The hybrid magnet will have a total length of about 4.5 m and an outer diameter of 1.6 m; its weight will be almost 12 tons. Detailed technical data have been published in [11].

ACKNOWLEDGEMENT

The authors acknowledge the enthusiasm and commitment of their technical staff in designing and constructing the different magnets. The reliable and trouble-free running of the magnets and the technical installations with their high rate of utilization was only possible because of their untiring support.

REFERENCES

[1] G. Aubert et al., Proc. ICM-73, Vol. I(1), 1973, p. 246
[2] H.-J. Schneider-Muntau, Proc. Conf. Application of High Magnetic Fields in Semiconductor Physics, Würzburg, 1974, p. 120
[3] P. Rub, M. Lombardi and H.-J. Schneider-Muntau, Proc. 6th Conf. Magnet Technology, Bratislava, 1977, p. 406
[4] H.-J. Schneider-Muntau, I.E.E.E. Transactions on Magn., Vol. MAG-17, no. 5, 1981, p. 1755
[5] P. Roubeau, C.R. Acad. Sci. Paris, Vol. 273, 1971, no. 14, p. 581
[6] G. Claudet, A. Lacaze, P. Roubeau and J. Verdier, Proc. 5th International Cryogenic Engineering Conf. Kyoto, 1974, p. 265
[7] G. Bon Mardion, G. Claudet and J.C. Vallier, Proc. 6th International Cryogenic Engineering Conference, Grenoble, 1976, p. 159
[8] J.L. Augueres, R. Aymar, G. Bon Mardion, G. Claudet, G. Faure-Brac, J. Plancoulaine and L. Senet, Cryogenics, Sept. 1980, p. 529
[9] J. Marty and J.C. Vallier, I.E.E.E. Transactions on Magn., Vol. MAG-17, no. 5, 1981, p. 1635
[10] H.-J. Schneider-Muntau, Proc. Joint Intermag-MMM Conf., Montreal, 1982, I.E.E.E., to be published
[11] MPI-FKF, C.N.R.S., I.E.E.E. Transactions on Magn., Vol. MAG-17, no. 5, 1981, p. 1783

MAGNETS I HAVE KNOWN

N. Kurti
Department of Engineering Science,
University of Oxford

INTRODUCTION

I had reservations about accepting the flattering invitation to give a lecture on "High magnetic fields : in the past and now". In 1974 I gave a talk here in Grenoble entitled "Magnetic fields in retrospect and prospect"[1]. I could add very little to what I said then: Since my retirement in 1975 I had no direct connexion with high field work. I was therefore glad when Professor Landwehr accepted my suggestion that the talk should be entitled "High fields I have known" and agreed that it should be anecdotal and light hearted. However, even this title is somehow misleading since "high" is a relative term. 30 or 40 years ago an induction of 10 to 20 kilogauss in a volume of a few cm^3 could be termed as"high"- anyone calling 1 or 2 Tesla "high" today would be greeted with ironical laughter. Hence I have changed the title to "Magnets I have known".

Let me first briefly discuss a question which every physicist wanting to do experiments under certain extreme conditions should always bear in mind, namely to what extent can one make use of the conditions provided by nature. Here a few examples :

1. Experiments in zero gravity : In old days the only way you could do it was to put your apparatus into an evacuated tube and let it drop. But in order to do a "free fall" experiment of 10 sec. duration you need a 500 m long evacuated vertical tube. Today, thanks to satellites, we can make use of the fact that there are many places in the solar system where gravity is zero or negligible and many experiments are performed in this way.

2. Extreme temperature :
a) High temperatures: It is true that the temperature inside the sun is of the order of 10^8K but that does not help, and to do experiments at this sort of temperature one must rely on laboratory techniques. The surface of the sun has the still respectably high temperature of about 6000 K and work in the last 30 years, especially in France, led to the development of solar furnaces which use the energy emitted from the surface of the sun without however degrading the temperature. There are other methods for producing this sort of temperature but if one wants it in a clean, contamination-free environment, solar energy is perhaps the most economical way of doing it; this at least was the result of a calculation I made some 10 years ago. I therefore regard the Odeilho installation conceived and built by Professor Felix Trombe not as an expensive toy but as a sensible use of nature's provision of very high temperatures.

b) Turning to low temperatures we see immediately that nature is hopelessly inadequate. We know that interstellar space has a temperature of about 3°K, the aftermath of the "big bang". We on earth can produce in the laboratory 10^{-4} or 10^{-6}, even 10^{-7}°K, temperatures which do not exist in the universe, unless there are intelligent beings somewhere. It's a consequence of the 2nd law of thermodynamics that temperatures below that of the surroundings (3°K for the universe) must be produced artificially, i.e. by the intervention of intelligence.

3. Extreme pressures : Nature does not help us with high pressures and they must be produced in the laboratory. When it comes to very low pressures, i.e. very high vacua, one is tempted to think of interstellar space as a suitable environment, but it is possible to beat nature in the laboratory. By the adiabatic demagnetization of certain paramagnetic substances one can reach 10^{-2} K or even 10^{-3} K. If we have

a sealed capsule containing a suitable paramagnetic salt and surrounded by the same substance, and if we cool this assembly to 0.15 K, the pressure inside the capsule will correspond to the vapour pressure of helium which, at this temperature, is 10^{-18} mm Hg or about one atom per cm^3: roughly the density of interstellar gas in the galaxy. If we cool the assembly to 0.03 K the vapour pressure is so low that at equilibrium there would be 1 atom in the whole galaxy. All this shows that there are certain extreme conditions which do not occur in the universe but can readily be produced in the laboratory.

4. High electric fields : It was in the early 1930s that two German physicists, Brasch and Lange [2] thought that Rutherford's famous experiments on the artificial transmutation of elements by α-ray bombardment should be repeated by artificially accelerated particles, and they proposed to harness atmospheric electric potentials which, under thunderstorm conditions, can be very high. They set up their equipment on Monte Generoso in Italy but waited in vain for a thunderstorm; they hit a period of unusually fine weather. After spending a fortnight on the mountain-top they were longing for a bit of civilization and so went to nearby Lugano and while they were there the district experienced one of the greatest thunderstorms in living memory ! Discouraged, they gave up the project and soon afterwards Cockroft and Walton performed the first fully artificial transmutation of elements.

5. Elementary particle research : A lot of money is spent nowadays on particle accelerators, mainly to discover new elementary particles. It's well worth remembering that the first elementary particles, apart from the proton, neutrons and electrons, were discovered entirely thanks to cosmic rays, which are particles having energies much higher than we can produce at present in the laboratory. The question is, do we spend enough on cosmic ray research as compared with particle accelerators. The ratio at present is between 300 to 1 and 1000 to 1 in favour of accelerators. If one were to increase the funds allocated to cosmic rays by a factor of 2 or 5 or even 10, would the results make it worthwhile ? The stumbling block is the urge to do things in a hurry, because if you use cosmic rays you often have to wait for a long time before observing a sufficient number of events.

PRE - 1939

Let me now turn to the magnets I have known. The first one was a solenoid I designed myself when, in 1929, I started my thesis work [3] in the Laboratory of Physical Chemistry of the University of Berlin under the late Professor Simon. I was going to do some preliminary investigations about the magnetic cooling method which had been proposed a few years earlier by Giauque [4] and by Debye [5]. I needed a field of 3 to 4 kGauss but the laboratory possessed neither a magnet nor the money to by one. So Simon proposed that, since my apparatus would be immersed in liquid hydrogen, I should build a hydrogen-cooled copper solenoid. At the boiling point of hydrogen the resistivity of copper is 1/100th of that at room temperature so that the power needed to produce the field was modest. Moreover, since I needed the field for only a few seconds for each measurement, the additional amount of hydrogen evaporating was tolerable. I built the solenoid weighing about 2 kg and it was with great expectations that late one evening I pressed the switch which sent a current of 40A through a wire of 0.5 mm diameter. The result was spectacular - a deafening explosion, the apparatus disappeared, all windows were blown in or out, a wall caved in, and thus ended my pioneering experiment on liquid hydrogen-cooled coils. However, there is nothing wrong with the concept of liquid hydrogen-cooled coils. The mishap was caused by the lack of experience and poor soldering technique of a green research student who precooled the solenoid with liquid air instead of with liquid nitrogen, who did not wait for all the air to evaporate before immersing the coil into liquid hydrogen and who left one of the electrical joints "dry".

When,in 1933, Simon and I started work on the magnetic cooling method in the Clarendon Laboratory we had similar magnet difficulties and our first experiments were done with an old horse-shoe electromagnet better suited to be an archaeological exhibit than a piece of research equipment. So we made arrangements to do some of our work in

Bellevue near Paris with the famous electromagnet of the Académie des Sciences, conceived, designed and built by Professor A. Cotton [6]. It incorporates the well known principle of concentrating the field by conical pole-pieces. The difficulty of producing really high fields by this means is that the field increases logarithmically with the ratio of core diameter to pole-tip diameter (see Fig.1). For a tenfold increase in d_2/d_1 the field increases by 1.5 T.

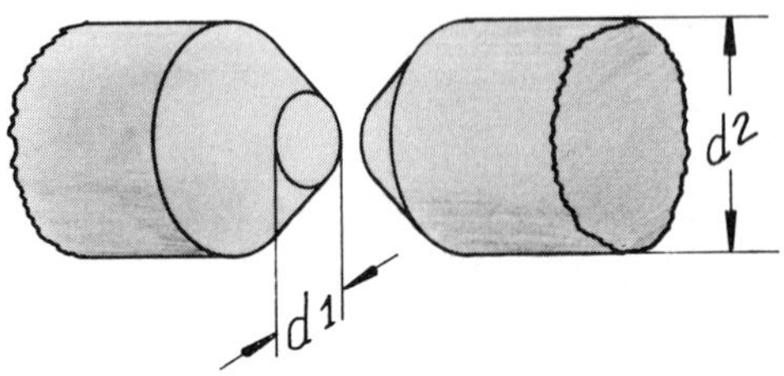

Fig.1 : Schematic drawing of conical iron-pole pieces of an electromagnet.

Fig.2 gives you an idea of the size of the magnet. It weighs about 100 tons and needs 100 kW to energize it by the water-cooled coils wound on the core which has a diameter d_2=1.5 m. Thus with a pole-tip diameter d_1=50mm - and similar polegap - it produces about 3 T.

Fig.2 :

View of the famous Bellevue electromagnet built by Prof. A. Cotton.

Thanks to the easy accessibility of the polegap, the apparatus could be rapidly installed and removed and several experiments could be carried out in parallel, with only brief intervals, measured in hours, between consecutive experiments. Figs.3 and 4 show the adiabatic demagnetization apparatus before and during an experiment [7]. Having several experimenters tending their apparatus in the space near the magnet had its risks. On one occasion, not noticing the red warning light and venturing too near the magnet with a steel liquid nitrogen container in my hand I pinned an unsuspecting colleague, who was adjusting his apparatus placed in the polegap, against the conical pole-piece and it took the better part of a minute to extricate him.

Fig.3 :

Early experimental equipment for studying adiabatic demagnetization before insert in the Bellevue magnet.

HIGH-POWERED AND HYBRID SOLENOIDS

In order to get 6 T with the same polegap dimensions of 50mm d_2 would have to be increased 100 fold and the whole magnet might weigh about 10^6 tons since the volume of the iron and the copper scales roughly as d_2^2. Clearly this is not the way to reach what we now call high fields, and high-powered, water-cooled solenoids are the answer. These were pioneered by Deslandres and Perot [8] who did their historic tests in the basement of the Paris department store "Au Bon Marché" which had a 350 kW D.C. generator-set for lighting. This technique was, as it were, rediscovered by Francis Bitter [9] who was led to it by considering the optimum position of the magnetising coils of an iron magnet. When high field work was started in the Clarendon Laboratory in 1948 we decided to adopt this technique. We bought an old 2 MW D.C. motor-generator from the city of Manchester. It was a machine built in 1919 and used for some 15 years to energize the arc street lamps of that city.

Our first high-powered solenoid cam from the Bellevue laboratory. It was constructed by B.Tsai [10] and with a few exceptions all our solenoids were of the Tsai design. In one important respect the Clarendon's "high field philosophy" differed from that of other laboratories in that we insisted on having mobile solenoids so that, instead of bringing the apparatus to the field we brought the field to the apparatus. The reason for this was that solenoids were often used for work at very low temperatures, and often the relative position of the magnet and the apparatus had to be changed. It is well known that vibrations generate heat through internal friction and this can produce prohibitively large effects at low temperatures, especially in the millikelvin and microkelvin range.

A notable advance was the design and introduction of the polyhelix solenoids at the Clarendon Laboratory by Martin Wood [11]. This technique was further developed by Carden [12], Hudson [13], Schneider-Muntau and Rub [14] and others [15] and has now replaced the "Bitter coil" for many applications. Much of this development took place in close cooperation between the Clarendon and the SNCI (Service National des Champs Intenses) in Grenoble, mainly under a collaborative NATO contract.

Another important development was the construction and installation in 1975 of a hybrid magnet system (combination of a superconductive and a resistive coil) in the Clarendon Laboratory which produces fields of up to 16 T and is used both by members of the Clarendon Laboratory and by visitors [16].

CNRS/MPG NOT CNRS/SRC

Let me finish by telling you about an episode of some historical interest which is not well known. In fact I only remembered it when, while preparing this talk, I consulted my old files. It was in the autumn of 1970, during a visit to Grenoble that I heard from Professor Pauthenet the disturbing news that, because funds were running out, two of the four 2.5 MW D.C. sets would have to be sold, unless $2x10^6$ FF could be found within the next few months. My first reaction was to exclaim that it dit not make sense to establish a "high field" laboratory with only 5 MW. So a few days later in the company of Professor Pauthenet, I visited Professor Chabbal who, at that time, was Directeur de Physique at the CNRS, and asked him what would be the reaction in France if Britain were to join the Grenoble high field project, e.g. by paying for the two 2.5 MW sets and making further regular annual payments. Chabbal's reply was "that's a good idea". I was then given a lot of technical and financial information about the Laboratoire des Champs Magnétiques Intenses (LCMI) as it was then called, which enabled me to address a report (dated 11 November 1970) to the Physics Committee of the (UK) Science Research Council (SRC) from which the following quotations are taken. It should be remembered that all the prices refer to 1970.

"If the UK were prepared to pay £ 150,000, the full 10 MW could be commissioned by the summer of 1971. This would have the following advantages :
a) a 20 T 70 mm I.D. solenoid could be built which would have a longer life, would be more reliable and would provide a larger experimental space than the 35 mm I.D. coil.
b) with an inner diameter of 35 mm the field could be increased to 26 T.

For the UK to participate formally in the LCMI the following seem to be the minimum conditions : a) a contribution of £ 150,000=2 MF i.e. the cost of the two 2.5 MW, towards the capital cost of the laboratory. The French would need a formal assurance by the end of March 1971, and the money would have to be paid before 31 December 1971. b) A contribution to the running costs of the water-cooled solenoids which, as stated earlier, amount to about £ 140,000 (=1.8 MF) p.a.. About 1/6th of this, say £ 25,000 p.a. has been suggested. One part of this UK contribution might be fixed, the rest depending on the frequency of use, the amount of electric power and liquid helium consumed and other similar expenditure.

In exchange the UK would have the right to use one of the high-powered solenoid-stations for about 2/3 of the total running time. Furthermore, the SRC would be represented on the scientific committee of the LCMI which, under the Chairmanship of Professor Chabbal, is in charge of the broad scientific direction of the laboratory, of allocation of funds and the phasing of further developments.

The annual expenditure for travelling and subsistence incurred by UK users can be estimated. If we assume full use of the facilities allocated to us (see previous paragraph) and occasional overlapping of two UK teams (each comprising on an average two members), the number of man-days per year may be about 500 : with a per diem allowance of £ 5 (LCMI will apparently have a few guestrooms available for visitors) the annual subsistence bill comes to about £ 2500. Add to this 30 round trips costing at an average £ 40 each (air: about £ 48; rail, with 2nd class couchette £ 26), making £ 1200. Hence total annual cost about £ 4,000 .

These considerations are brought to the notice of the Physics Committee so that they may decide whether the envisaged expenditure (£ 150,000 capital, £ 25,000 p.a. running cost) on modest participation in the Grenoble High Field Laboratory would be justified by the resulting benefits to UK solid state physicists."

The SRC decided against even discussing such a participation. Fortunately, the Max-Planck Institute of the Federal German Republic was interested in such a collaboration and, as a result, there is now in Grenoble a Franco-German High Field Laboratory. British physicists have, however, the consolation that they are welcome in Grenoble and in fact there are usually one or two of them working here. I think I can speak on behalf of my colleagues in Britain when I wish the Service National des Champs Intenses CNRS-Hochfeld-Magnetlabor, MPG continuing success and prosperity.

Fig. 4 :

Glass dewar for adiabatic demagnetization experiments after mounting in the 100 ton Bellevue magnet.

REFERENCES

[1] N.Kurti, Proc.Int.Conf.Physics in High Magnetic Fields, CNRS, Grenoble 1974, p.15

[2] A.Brasch and F.Lange, Z. Physik, 1931, 70, 10

[3] N.Kurti, Z.Physikal.Chem. 1933, 20, 305

[4] W.F.Giauque, J.Am.Chem.Soc. 1927, 49, 1864

[5] P.Debye, Ann.Phys.1926, 81, 1154

[6] A.Cotton, C.R.Acad.Sci., 1928, 187, 77

[7] N.Kurti, P.Laine, B.V.Rollin, F.E.Simon, C.R.Acad.Sci.1936, 202, 1421

[8] H.Deslandres et A.Perot, C.R.Acad.Sci.1914,159, 439

[9] F.Bitter, Rev.Sci.Instruments, 1936, 7, 479

[10] B.Tsai, Proc.Int.Conf.Fund.Part.and Low Temp.Phys. Cambridge 1947, Vol 2, 89

[11] M.F.Wood, Proc.Int.Conf.High Magnetic Fields, MIT, John Wiley, 1962, p.387

[12] P.O.Carden, J.of Physics, E. 1972, 5, 654-

[13] P.A.Hudson, Proc.Int.Conf.Physics in High Magnetic Fields, CNRS, Grenoble, 1974, p.167

[14] H.J.Schneider-Muntau and P.Rub, Proc.Int.Conf.Physics in High Magnetic Fields, CNRS, Grenoble, 1974, p.161

[15] P.A.Hudson, P.E.Hanley and P.O.Carden, Proc.5th Int.Conf.Magnet Technology, Frascati, 1975, p.414

[16] D.B.Montgomery and M.F.Wood, Proc.Int.Conf."Les Champs Magnétiques Intenses", Grenoble, Sept.1966, Publ.CNRS 1967, p.91,
P.A. Hudson and H.Jones, Cryogenics, 1976, 16, 593
For the most recent account of high field facilities at the Clarendon Laboratory see the paper by P.A.Hudson, H.Jones and H.M.Whitworth presented at this conference, p.500.

AUTHOR INDEX

Physics in High Magnetic Fields

Proceedings of the Oji International Seminar
Hakone, Japan, September 10–13, 1980
Editors: **S. Chikazumi, N. Miura**

1981. 257 figures. X, 358 pages
(Springer Series in Solid-State Sciences, Volume 24)
ISBN 3-540-10587-5

Contents: Recent Progress in the Steady High Field System. – Physics in Pulsed High Magnetic Fields. – Cyclotron Resonance and Laser Spectroscopy of Semiconductors. – Impurity States in High Magnetic Fields. – Magneto-Transport Phenomena. – Excitons and Magneto-Optics. – Electron-Hole Drops and Semimetals. – Narrow Gap Semiconductors. – Space Charge Layer and Superlattice. – Layered Materials and Intercalation. – Magnetism and Magnetic Semiconductors. – Photograph of the Participants of the Seminar. – List of Persons in the Photograph. – Index of Contributors.

The book contains the contributions of leading scientists to an international seminar on the application of high magnetic fields to the physics of semiconductors and magnetic materials. They describe recent developments of semiconductor physics, magnetism in high magnetic fields, as well as techniques for generating high magnetic fields in both continuous and pulsed forms. They also report on progress in steady field magnet technology and the ultra-high magnetic fields, and the cyclotron resonance, infrared spectroscopy, electron-hole drops, narrow gap semiconductors, magneto-transport phenomena, magneto-optics, surface inversion layers, superlattice layered materials, intercalated substances, magnetism, and magnetic semiconductors in high magnetic fields. The physics in pulsed high magnetic fields, including the megagauss range, is given its first indepth coverage in this book.

Springer-Verlag
Berlin
Heidelberg
New York
Tokyo

Lecture Notes in Physics

Vol. 144: Topics in Nuclear Physics I. A Comprehensive Review of Recent Developments. Edited by T.T.S. Kuo and S.S.M. Wong. XX, 567 pages. 1981.

Vol. 145: Topics in Nuclear Physics II. A Comprehensive Review of Recent Developments. Proceedings 1980/81. Edited by T. T. S. Kuo and S. S. M. Wong. VIII, 571-1.082 pages. 1981.

Vol. 146: B. J. West, On the Simpler Aspects of Nonlinear Fluctuating. Deep Gravity Waves. VI, 341 pages. 1981.

Vol. 147: J. Messer, Temperature Dependent Thomas-Fermi Theory. IX, 131 pages. 1981.

Vol. 148: Advances in Fluid Mechanics. Proceedings, 1980. Edited by E. Krause. VII, 361 pages. 1981.

Vol. 149: Disordered Systems and Localization. Proceedings, 1981. Edited by C. Castellani, C. Castro, and L. Peliti. XII, 308 pages. 1981.

Vol. 150: N. Straumann, Allgemeine Relativitätstheorie und relativistische Astrophysik. VII, 418 Seiten. 1981.

Vol. 151: Integrable Quantum Field Theory. Proceedings, 1981. Edited by J. Hietarinta and C. Montonen. V, 251 pages. 1982.

Vol. 152: Physics of Narrow Gap Semiconductors. Proceedings, 1981. Edited by E. Gornik, H. Heinrich and L. Palmetshofer. XIII, 485 pages. 1982.

Vol. 153: Mathematical Problems in Theoretical Physics. Proceedings, 1981. Edited by R. Schrader, R. Seiler, and D.A. Uhlenbrock. XII, 429 pages. 1982.

Vol. 154: Macroscopic Properties of Disordered Media. Proceedings, 1981. Edited by R. Burridge, S. Childress, and G. Papanicolaou. VII, 307 pages. 1982.

Vol. 155: Quantum Optics. Proceedings, 1981. Edited by C.A. Engelbrecht. VIII, 329 pages. 1982.

Vol. 156: Resonances in Heavy Ion Reactions. Proceedings, 1981. Edited by K.A. Eberhard. XII, 448 pages. 1982.

Vol. 157: P. Niyogi, Integral Equation Method in Transonic Flow. XI, 189 pages. 1982.

Vol. 158: Dynamics of Nuclear Fission and Related Collective Phenomena. Proceedings, 1981. Edited by P. David, T. Mayer-Kuckuk, and A. van der Woude. X, 462 pages. 1982.

Vol. 159: E. Seiler, Gauge Theories as a Problem of Constructive Quantum Field Theory and Statistical Mechanics. V, 192 pages. 1982.

Vol. 160: Unified Theories of Elementary Particles. Critical Assessment and Prospects. Proceedings, 1981. Edited by P. Breitenlohner and H.P. Dürr. VI, 217 pages. 1982.

Vol. 161: Interacting Bosons in Nuclei. Proceedings, 1981. Edited by J.S. Dehesa, J.M.G. Gomez, and J. Ros. V, 209 pages. 1982.

Vol. 162: Relativistic Action at a Distance: Classical and Quantum Aspects. Proceedings, 1981. Edited by J. Llosa. X, 263 pages. 1982.

Vol. 163: J. S. Darrozes, C. Francois, Mécanique des Fluides Incompressibles. XIX, 459 pages. 1982.

Vol. 164: Stability of Thermodynamic Systems. Proceedings, 1981. Edited by J. Casas-Vázquez and G. Lebon. VII, 321 pages. 1982.

Vol. 165: N. Mukunda, H. van Dam, L.C. Biedenharn, Relativistic Models of Extended Hadrons Obeying a Mass-Spin Trajectory Constraint. Edited by A. Böhm and J.D. Dollard. VI, 163 pages. 1982.

Vol. 166: Computer Simulation of Solids. Edited by C.R.A. Catlow and W.C. Mackrodt. XII, 320 pages. 1982.

Vol. 167: G. Fieck, Symmetry of Polycentric Systems. VI, 137 pages, 1982.

Vol. 168: Heavy-Ion Collisions. Proceedings, 1982. Edited by G. Madurga and M. Lozano. VI, 429 pages. 1982.

Vol. 169: K. Sundermeyer, Constrained Dynamics. IV, 318 pages. 1982.

Vol. 170: Eighth International Conference on Numerical Methods in Fluid Dynamics. Proceedings, 1982. Edited by E. Krause. X, 569 pages. 1982.

Vol. 171: Time-Dependent Hartree-Fock and Beyond. Proceedings, 1982. Edited by K. Goeke and P.-G. Reinhard. VIII, 426 pages. 1982.

Vol. 172: Ionic Liquids, Molten Salts and Polyelectrolytes. Proceedings, 1982. Edited by K.-H. Bennemann, F. Brouers, and D. Quitmann. VII, 253 pages. 1982.

Vol. 173: Stochastic Processes in Quantum Theory and Statistical Physics. Proceedings, 1981. Edited by S. Albeverio, Ph. Combe, and M. Sirugue-Collin. VIII, 337 pages. 1982.

Vol. 174: A. Kadić, D.G.B. Edelen, A Gauge Theory of Dislocations and Disclinations. VII, 290 pages. 1983.

Vol. 175: Defect Complexes in Semiconductor Structures. Proceedings, 1982. Edited by J. Giber, F. Beleznay, J. C. Szép, and J. László. VI, 308 pages. 1983.

Vol. 176: Gauge Theory and Gravitation. Proceedings, 1982. Edited by K. Kikkawa, N. Nakanishi, and H. Nariai. X, 316 pages. 1983.

Vol. 177: Application of High Magnetic Fields in Semiconductor Physics. Proceedings, 1982. Edited by G. Landwehr. XII, 552 pages. 1983.